Atomic and Molecular Beams

Dear MyCopy Customer,

This Springer book is a monochrome print version of the eBook to which your library gives you access via SpringerLink. It is available to you at a subsidized price since your library subscribes to at least one Springer eBook subject collection.

Please note that MyCopy books are only offered to library patrons with access to at least one Springer eBook subject collection. MyCopy books are strictly for individual use only.

You may cite this book by referencing the bibliographic data and/or the DOI (Digital Object Identifier) found in the front matter. This book is an exact but monochrome copy of the print version of the eBook on SpringerLink.

Springer
*Berlin
Heidelberg
New York
Barcelona
Hong Kong
London
Milan
Paris
Singapore
Tokyo*

Physics and Astronomy ONLINE LIBRARY

http://www.springer.de/phys/

Roger Campargue (Ed.)

Atomic and Molecular Beams

The State of the Art 2000

With 436 Figures
Including 11 Color Figures

 Springer

Professor Roger Campargue
Laboratoire d'Aerothermique, CNRS
4ter Route des Gardes
92140 Meudon, France
email: RCampargue@aol.com

Cover Picture:

(*frontcover, top*) Femtosecond dynamics of the charge-transfer reaction of Benzene (Bz) with Iodine (I-I), a century-old reaction now clarified thanks to Femtochemistry. Shown are the snapshots at different reaction times: 0 fs, 200 fs, and 1500 fs, from the transition state* prepared at the initial femtosecond pulse, to the secondary Bz–I dissociation, as described in the article. Bz . . . I–I→[Bz$^+$ . . . (I-I)$^-$]* → Bz · I + I→Bz + I + I. (From: A.H. Zewail with courtesy of D. Zhong, see also Article IV.1)

(*frontcover, bottom left*) Image of a Bose-Einstein condensate of Rubidium 87 atoms, containing 1 million atoms, released from a magnetic trap. The false color code reveals the atomic density from white (highest density) to dark (lowest density). The elliptical structure at the center (200 × 100 micrometers) corresponds to what is properly referred to as the 'condensate' and the surrounding circular feature represents the non-condensed atoms, at a temperature of the order of 0.3 microkelvin. (From: C. Cohen-Tannoudji with courtesy of P. Desbiolles, D. Guéry-Odelin, J. Soeding, and J. Dalibard, see also Article I.1)

(*frontcover, bottom right*) Photograph of a corona discharge supersonic free-jet used to generate a beam of metastable nitrogen molecules for the growth of nitride semiconductors. A 1 inch long conical graphite skimmer extracts the isentropic core of the expansion used in supersonic beam epitaxy. (From: R.B. Doak with courtesy of D.C. Jordan, see also Article VII.8)

(*back cover*) Schematic impression of a molecular beam (green) traversing a hexapole quantum state-selector. The state-selected parent molecules are dissociated by a laser pulse (blue) and the neutral fragments are quantum state-selectively ionized by a second laser pulse (red). The three-dimensional recoil distribution of photofragments (CD$_3$ in this case) are detected by a position sensitive detector located at the end of a time-of-flight tube. (From: M.H.M. Janssen with courtesy of A. van den Brom and M. Janssen, see also Article III.1)

DOI 10.1007/978-3-642-56800-8

Library of Congress Cataloging-in-Publication Data. Atomic and Molecular beams: The state of the art 2000/ Roger Campargue (ed.) p. cm. Includes bibliographical references.
1. Atomic beams. 2. Molecular beams. I. Campargue, Roger, 1932– QC173.4.A85 A87 2000 539'.6–dc21
00–046411

This work is subject to copyright. All rights are reserved, whether the whole or part of the material is concerned, specifically the rights of translation, reprinting, reuse of illustrations, recitation, broadcasting, reproduction on microfilm or in any other way, and storage in data banks. Duplication of this publication or parts thereof is permitted only under the provisions of the German Copyright Law of September 9, 1965, in its current version, and permission for use must always be obtained from Springer-Verlag. Violations are liable for prosecution under the German Copyright Law.

Springer-Verlag Berlin Heidelberg New York
a member of BertelsmannSpringer Science+Business Media GmbH

© Springer-Verlag Berlin Heidelberg 2001
MyCopy version of the original edition 2001

The use of general descriptive names, registered names, trademarks, etc. in this publication does not imply, even in the absence of a specific statement, that such names are exempt from the relevant protective laws and regulations and therefore free for general use.

Typesetting: Data conversion by LE-TEX Jelonek, Schmidt & Vöckler GbR, Leipzig
Cover design: Erich Kirchner, Heidelberg

SPIN: 10759805   55/3141/ba   5 4 3 2 1 0
www.springer.com/mycopy

Preface

This book is published about 90 years after the first atomic beam experiment (Dunoyer, 1911). This effusive beam technique has been used for over half a century in many historical beam experiments to establish the basic principles of Modern Physics. Such accomplishments have been celebrated by Nobel prizes in Physics awarded to eleven laureates whose names are mentioned below with an asterisk* (two stars** in reactive scattering are used for Nobel laureates in Chemistry!). Among these achievements are the verification of the Maxwell-Boltzmann velocity distribution of gaseous atoms (O. Stern*, 1911), the discovery of the space quantization (O. Stern* and W. Gerlach, 1922) and its important applications in nuclear magnetic resonance (I. Rabi*, 1938). The NMR technique was extended to liquids and solids (F. Bloch* and E. M. Purcell*, 1945) and later remarkably developed to determine structures of proteins and observe even fleeting thoughts in human brains by magnetic resonance imaging. Other atomic beam experiments of the highest interest also exploited space quantization, for example in the resolution of the splitting in the fine structure of the first electronically excited H atom (W. Lamb* and R. Retherford, 1947). This led to a major theoretical advance, the development of modern quantum electrodynamics. Likewise the magnetic moment of the electron was measured revealing a 0.1 % discrepancy, small but highly significant, from Dirac's quantum theory (P. Kusch*). Also, optical pumping was discovered using a sodium beam (A. Kastler*, 1950), although this application is not restricted to molecular beams. Finally the combination of an ammonia beam (NH_3) and an electric field, instead of a magnetic field, provided space quantization and state selection which have led to the invention of the maser and hence the laser (Ch. Townes* and also N. Basov* and A. Prokhorov*, in the 1950s). The historical beam experiments listed above are discussed extensively (except for optical pumping and lasers) in a classic book "Molecular Beams," published in 1955 by N.F. Ramsey* (1989) and briefly described in this volume in the introductory article by Dudley Herschbach** (1986).

Chemistry in crossed molecular beams, under single collision conditions, was not shown to be feasible until after the second half of the century began (Bull & Moon, 1954; Taylor and Datz, 1955). The early era dealt with alkali atom reactions, by virtue of the ease of detection by means of surface ionization and the high yield of many of these reactions, shown already in the 1920s by Michael Polanyi. Happily, during the past 30 years, beam experiments have gone far beyond the pioneering alkali age, probing a host of collision processes involving exchanges of momentum, energy, charge, or atoms. Most of these advances resulted from marked improvements in detection capability (dating from 1968 and largely due to Y.T. Lee**) as well as the development of versatile supersonic beams based on theoretical predictions by A. Kantrowitz and J. Grey (1951). When compared with effusive beams, this powerful technique allows great enhancements in density, intensity, velocity (or de Broglie wavelength) resolution, and the kinetic energy. By heating or cooling the nozzle, as well as having the additional option of seeding the gas of interest and even clustering it during the expansion, the sources can be operated as monochomators and over a wide energy range, of about 0.01 to 50 eV, or higher for large clusters. The use of light atoms, such as helium, at low kinetic energy, results in de Broglie wavelengths comparable to interatomic distances (of order one Angstrom unit) and hence an exceptional tool to probe solid surfaces.

The very low temperature achieved in the free jet is extremely useful in high resolution molecular spectroscopy which can be performed down to the kelvin range at the maximum source pressure allowed by the largest diffusion pumps (J.B. Fenn, 1963, 1996). The millikelvin range is accessible, at least for the translational temperature, when higher pressure is acceptable by exploiting the free jet shock wave structure to shield the expansion in a "zone of silence" unaffected by the background gas (R. Campargue, 1964, 1984). Such a method has been used beautifully in the pioneering works in molecular jet spectroscopy (R.E. Smalley**, D.H. Levy, and L. Wharton, 1975-1980) and later in work using the pulsed jet technique (W.R. Gentry, 1978). Finally, it should be pointed out that the great advances during the last decades, and mainly during the recent years, have been due to the progressive coupling of atomic and molecular beams with lasers (J.C. Polanyi**, 1986) and other light sources such as the synchrotron radiation and the free electron laser. Thus, spectacular developments have been obtained after initial studies on the physics of the free-jet expansion involving the translational and internal energy relaxation processes. The most important research works have been devoted to molecular jet and beam spectroscopy, photodissociation, single-collision reactions, van der Waals molecules, clusters and nanoparticles, as well as gas-surface interactions. The impressive accomplishments on chemistry-reaction dynamics have been rewarded by the 1986 Nobel Prize in Chemistry awarded to D.R. Herschbach**, Y.T. Lee**, and J.C. Polanyi**.

The predecessor book "Atomic and Molecular Beam Methods" (1988 and 1992), edited by Giacinto Scoles, was presented by the Editor as "more concerned with describing how to carry out beam experiments than describing the Physics and Chemistry, underlying them". The present volume will be complementary to this excellent predecessor from two points of view: (i) the improvements and novel techniques obtained since 1988 for using atomic and molecular beams and, increasingly, their combination with lasers, (ii) the state of the art in 2000 largely described in up-to-date review articles, as well as in specialized articles on original works, all dealing with beam research and applications in Physics and Chemistry, with an extensive coupling with lasers.

The present book was born as a project early last year after the scientific organization, by the Editor, of the 21st International Symposium on Rarefied Gas Dynamics, held in Marseille, France, July 26–31, 1998 (RGD-21). It should be pointed out that the supersonic beam techniques have been developed largely in the RGD community alongside the work of the aerodynamicists. Also their cooperation with physicists and chemists, during about 30 years, has been useful to get the balance between experiment and theory in the RGD symposia. Nevertheless, during the last decade, the work in RGD have more and more been concerned with only theoretical aspects of flows of rarefied gases: kinetic theory, Boltzmann equation, numerical solutions, and Monte Carlo methods. In order to include many more experimental works in the RGD-21 symposium program, the Editor of this book organized a third session with the scope and style of the Molecular Beam Symposia issued from the RGD series a few decades earlier. This has been possible thanks to 34 individual invitations accepted spontaneously by experts in the field, who together with resulting additional participants, allowed us to include 62 papers in the Molecular Beam session of the RGD-21 symposium.

This volume arose initially as only an alternative to publish separately in about 300 pages (instead of 1000 in the present book) this body of predominantly experimental work alongside the theoretical treatments on rarefied gas flows and aerodynamics. A much better venue has been found later by undertaking an ambitious project consisting of collecting in a topical volume the outstanding works in Atomic and Molecular Beams, reflecting the state of the art 2000. This tremendous project was encouraged by the manifest accomplishments in the field, rewarded by 11 Nobel prizes during the last 13 years, and also by the kind support of six Nobel laureates, as well as many other important scientists who joined successively the project in the course of 1999 and 2000.

Finally, the current state of the art in Atomic and Molecular Beams, is described in the volume by scientists at the highest level and worldwide groups in the field:

- four Nobel laureates and other important contributors invited individually,
- invited speakers of the Molecular Beam session at the RGD-21 Symposium, having acted as the initial and essential "cristallites" to attract and gather together all the contributors,
- invited speakers of five Gordon type Conferences held in 1999:
  - Physics School on Atom Optics Applications, Les Houches, France, May 23–28, 1999
  - 18 th International Symposium on Molecular Beams, Ameland, The Netherland, May 30–June 4, 1999
  - COMET 16, Perugia, Italy, June 25–27, 1999
  - 31st EGAS (European Group for Atomic Spectroscopy), Marseille, France, July 6–9, 1999.
  - Dynamics of Molecular Collision Conference, Lake Harmony, USA, July 18–23, 1999

The volume contains 68 invited contributions including a few surveys. The papers extend to work throughout the 20th Century and include up-to-date reviews corresponding to the seven sections of the book, and a large variety of specialized chapters on the advances on Atomic and Molecular Beams largely coupled with lasers. Despite being invited, the articles have been carefully reviewed at least by two or three experts and then largely revised by the contributors. The great interest of the book is due essentially to the diversity of modern and fundamental subjects in Physics and Chemistry, as well as their actual and potential applications, the high scientific level of the contributions, and the great effort of the authors to improve the manuscripts, especially for being accessible to non specialists and address a wide audience. The 68 invited contributors and their coauthors total 231 authors originating from 17 countries. The articles and the seven Parts of the book are summarized, reviewed, highlighted and even enlarged in the Foreword including, as in the book, the following successive sections:

Introductory Perspectives by D.R. Herschbach (Nobel Prize in Chemistry, 1986) Atomic and Molecular Beams in Chemical Physics: A Continuing Odyssey
Part I: Laser Cooling and Manipulation of Atoms and Molecules – Atomic and Molecular Optics and Applications Foreword by J. Dalibard and C. Cohen-Tannoudji (Nobel Prize in Physics, 1997)

Part II: Translational and Internal Energy Relaxation in Supersonic Free Jets, Including Possible Alignment and Condensation Phenomena
Part III: Photodissociation Dynamics and Electronic Spectroscopy in Molecular Beams: From Simple Molecules to Clusters and Ions
Part IV: Dynamics of Elementary Molecular Collisions and Femtochemistry. Theory, Real-Time Probing, and Imaging of Crossed-Beam Reactions Femtochemistry by A. H. Zewail (Nobel Prize in Chemistry, 1999)
Part V: Clusters and Nanoparticles: Diffraction, Size Selection, Polarizability, and Fragmentation by Photodissociation, Electron Impact, and Atom Collision
Part VI: Spectroscopy and Reaction Dynamics of Molecules Isolated, Cooled (or Conditioned) by Techniques of Molecule, Cluster, Droplets, or Liquid Beams
Part VII: Interactions of Molecular Beams and Cluster Beams with Surfaces, and Applications
Looking Ahead: J.C. Polanyi (Nobel Prize in Chemistry, 1986) Macro, Micro and Nanobeams.

We would like to express our deep appreciation and gratitude to Claude Cohen-Tannoudji*, Dudley Herschbach**, John Polanyi** and Ahmed Zewail,** for their huge encouragement and stimulation of the project and finally for enhancing, in 136 pages, the scientific level and the interest of the volume. Our appreciation and gratitude are extended to Will Castleman, Helmut Haberland, Joshua Jortner, Yuan Lee**, Rick Smalley**, Dick Zare and other leading scientists, not able to join us despite of attempts of most of them, but generous in their support for our project. Obviously, our thanks are also largely due to all the other invited contributors and their co-authors for reflecting thoroughly and so beautifully the various perspectives of Atomic and Molecular Beams. Additional acknowledgements are addressed to 41 authors, and, especially, 36 non contributors (together listed below) all having reviewed carefully and improved widely the manuscripts.

We are indepted also to the Organizations and Laboratories having supported the preparation and finally the realization of our ambitious project, as well as the sponsors for allowing one free copy of the volume to be offered to the authors for each contribution. Finally, our sincere acknowledgements are given to Springer-Verlag and especially to the persons mentioned by names below for their flexible and supportive arrangements during the preparation and publication of this book.

In conclusion, it is a happy outcome, for the contributors as well as the editor, to finally heft a lovely volume that provides a rather comprehensive (and comprehensible!) topical survey of the state of the current art in Atomic and Molecular Beams. This research field has been commemorated by Nobel Prizes awarded to 21 Laureates during the past century, including eleven during the last thirteen years. We are proud and honored that most of these Laureates are still active in our community and several have contributed to our volume. Finally, in view of the recent outburst that yielded seven new Laureates in just the past four years, it seems likely that other Nobel Prizes will be awarded to some of the outstanding contributors to this volume.

Roger Campargue, Editor   September 15, 2000

Acknowledgements

68 Invited Contributors and their Co-Authors totalling 231 Authors originating from 17 countries

77 Reviewers originating from 13 countries:

Aguillon, F.(fr), Allison, W. (uk), Aquilanti, V. (it), Ashfold, M. (uk), Auweter-Kurtz, M. (de), Barat, M. (fr), Baudon, J. (fr), Behmenburg, K. (de), Berden, G. (nl), Beswick, A. (fr), Bowker, M. (uk), Bréchignac, Ph.(fr), Broyer, M. (fr), Buck, U. (de), Cahuzac, Ph. (fr), Campargue, R. (fr), Casavecchia, P. (it), Cattolica, R.J. (us), Chandler D.W. (us), Dagdigian, P. (us), de Vries, M.S. (il), DeMartino, A. (fr), Demtroeder, W. (de), Dimicoli, I. (fr), Dudeck, M. (fr), Duncan, M.A. (us), Ernst, J. (fr), Faubel, M. (de), Fayeton, J. (fr), Gallagher, Th. (us), Gaveau, M.A. (fr), Girard, B. (fr), Guet, C. (fr), Haberland, H. (de), Hagena, O. (de), Hancock, G. (uk), Hodgson, A. (uk), Hotop, H. (de), Huisken, F. (de), Kawasaki, M. (jp), Kleyn, A.W. (nl), Knuth, E.L. (us), Kolodney, E. (il), Kummel, A. C. (us), Kusunoki, I. (jp), Labastie, P. (fr), Lebéhot, A. (fr), Liu, K. (tw), Meijer, G. (nl), Mestdagh, J.M. (fr), Miller, R. (us), Moeller, Th. (de), Mons, M. (fr), Montero, S. (es), Mueller-Dethlefs, K. (uk), Parker, D.H. (nl), Pillet P. (fr), Piuzzi, F. (fr), Rebrov A.K. (ru), Schinke, R. (de) Schmiedmayer, J. (at), Schram, D.C. (nl), Scoles, G. (us), Seideman, T. (ca), Shermann, J.P. (fr), Shizgal, B. (us), Smith, Y. (uk), Soep B. (fr), Stace, T. (uk), Stolte, S. (nl), Stwalley, W.C. (us), Suits, A. (us), Takahashima, K. (jp), Toennies, J.P. (de), Vigué, J. (fr), Visticot, J.P. (fr), Weiner, J. (fr),

Supporters: Organization and Laboratories:

- Laboratoire d'Aérothermique, CNRS, Meudon, France.
- Université de Provence, IUSTI, Marseille, France.

Sponsors:

- Centre National de la Recherche Scientifique (CNRS),
  Direction des Relations Internationales, Paris.
- Laser Manufacturers: Coherent, Continuum, Quantel, Spectra-Physics
- Laser and Optical Components: Technoscience, Orsay, France.
- Mechanics: Physiméca, Villiers Le Bacle, France.

Springer-Verlag: Prof. W. Beiglböck, Dr. C. Caron, Mrs. B. Reichel-Mayer, Mr. C.-D. Bachem

Foreword

Beginning with an introductory survey "Atomic and Molecular Beams: A Continuing Odyssey" by Dudley Herschbach and concluding with a study of future prospects "Looking Ahead: Macro, Micro and Nanobeams" by John Polanyi, this topical book presents the current state of the art of atomic and molecular beams. This Foreword is intended to complement the Preface by summarizing, reviewing, highlighting and even extending the seven parts of this volume. A grateful acknowledgement is due to those who contributed to the summaries and helped the editor to write this Foreword.

Drawn from among the authors and referees of the articles in this volume, these experts wrote two of the summaries in their entirety and provided corrections, completions, and improvements to the remainder. We are particularly indebted to the following persons: J. Baudon, U. Buck, Ph. Cahuzac, B. Girard, G. Hancock, D. Herschbach, A. Hodgson, A.W. Kleyn, K. Liu, R. Miller, P. Pillet, J.C. Polanyi, T. Seideman, S. Stolte, J.P. Schermann, B. Shizgal, and J.P. Toennies.

Introductory Perspectives: Atomic and Molecular Beams in Chemical Physics: a Continuing Odyssey by Dudley Herschbach

In his fascinating Introductory Perspectives, addressed to nonspecialists, Dudley Herschbach (who shared the 1986 Nobel Prize in Chemistry with John Polanyi and Yuan Lee, "for the development of a new field of research in chemistry-reaction dynamics") provides what he terms "appetizers for a smorgasbord." After sketching the historical evolution of research employing atomic and molecular beams in Physics and Chemistry, he serves up morsels representative of the seven parts of this book. His "menu" includes the elucidation of molecular dynamics in interstellar clouds, in meteor trails, in clusters generated in supersonic expansions, and in surface-catalyzed reactions. Other appetizers exemplify techniques for enhancing resolution and control of the intimate mechanics of collisions. Among these are means to arrest molecular rotation; to slow translation markedly, thereby endowing molecules with pronounced wave properties; and to select by laser light a specific reaction pathway, thus fulfilling a modern alchemical quest. As well as emphasizing thematic motivations and methods, these enticing vignettes also convey much about the adventurous spirit of a continuing odyssey.

Part I: Laser Cooling and Manipulation of Atoms and Molecules – Atomic and Molecular Optics, and Applications

Part I is devoted to the present state of the art of laser manipulation and cooling of gas phase matter. The interaction of atoms with coherent light has been studied since the invention of the laser. Nevertheless, apart from extensive study of the (undesired) Doppler effect, little attention was devoted to the effect of light on centre-of-mass motion of the atom other than the pioneering work on "lumino-réfrigération" (A. Kastler, 1950). The use of the radiation pressure of light to slow down and cool atoms was introduced only in the seventies and paved the way to the development of atomic traps. During the last fifteen years, a wide variety of devices combining cooling and trapping, such as the magneto-optical trap (MOT), have been developed, with applications ranging from new time standards (atomic clock) to the realization of Bose-Einstein condensation. At very low temperature, provided that the density is sufficient, the quantum character of the statistics becomes dominant. For identical bosonic atoms this may result in an accumulation of almost all atoms in the ground state of the trap, leading to a macroscopic wave function – a Bose-Einstein condensate.

The remarkable progress in atom cooling technology going beyond all the standard limits, down almost to the nanokelvin range, was acknowledged by the 1997 Nobel Prize in Physics awarded to S. Chu, C. Cohen-Tannoudji and W.D. Phillips.

The laser cooling technique for atoms cannot be extended directly to molecules, but laser-assisted cold atom collisions can form ground state molecules through photoassociation. In this process, one photon is absorbed to form resonantly excited dimers. For well-chosen excitation configurations, photoassociated excited molecules spontaneously decay to the ground state. This method opens access to high-resolution photoassociative spectroscopy and provides information about the long-range parts of internuclear potentials that is not available from conventional spectroscopy. The photoassociation technique extends the field of ultracold diluted matter to molecules in the microkelvin range, with promising applications in optics and interferometry, as well as in the development of an ultracold photochemistry. Bose-Einstein condensation of a molecular gas is at present being actively pursued.

The field of atom optics has similarly reached maturity, providing new tools for fundamental research, as well as for applied purposes. Polarization or

Stern-Gerlach interferometers are based upon the atomic spin evolution in a magnetic field. The very high sensitivity of such devices in a spin-echo configuration is evidenced by the study of inelastic and quasi-elastic ^{3}He scattering by clean and adsorbate-covered gold surfaces. The versatility of Stern-Gerlach interferometers is demonstrated by the variety of interference effects that have been produced by special magnetic configurations. Atomic mirrors are routinely produced and serve as a key optical element in a number of experiments. For example, at very low energy, atoms are prevented from sticking to the surface by the phenomenon of quantum reflection. An intense atomic micro-probe has been produced by focusing helium atom beams by a concave mirror (a crystal surface deformed by a static electric field). Finally, atom lithography is now going beyond the resolution of conventional lithographic techniques. Sub-micrometric structures can be deposited on a substrate, or engraved on a resist, using a phase mask, i.e., a standing optical wave with a resolution of a few tens of nm. Interference of atomic and molecular wave packets has been observed using time delayed ultra-short laser pulses.

The well-established techniques of atom optics cannot be extended to molecular systems. Nonetheless, recent work illustrated the possibility to spatially manipulate molecules by use of the intensity property of laser light. The newly introduced field of molecular optics is based on the interaction of an intense, nonresonant laser field with the molecular polarizability tensor. Since the polarizability of molecules (by contrast to that of atoms) is anisotropic, a polarized laser field (where both spatial and orientational gradients are available) can serve to align molecules along given space-fixed axes, as well as focus their center-of-mass motion in space.

Part II: Transnational and Internal Energy Relaxation in Supersonic Free Jets, Including Possible Alignment and Condensation Phenomena

The supersonic expansion of neutral atoms, or molecules, is a very useful technique to investigate relaxation phenomena with possible ultracooling, special alignments of molecules, and interesting nonequilibrium effects on the internal degrees of freedom, including population inversions as exploited in a gas dynamics laser. Furthermore, this technique is uniquely suited to create and study "exotic" species such as van der Waals molecules, clusters, and nanoparticles. The physics of supersonic free jet expansion is extended to plasmas (first state of the matter and major part of the Universe) ranging from the astrophysical sizes (supernovae, solar flares) to intermediate lengths as used in nuclear fusion (Tokamak) and plasma chemistry (DC, RF, and microwaves torches), and to very small scales such as in cathode spots and micro-sized laser spots for laser ablation. Plasma expansions, for temperatures not exceeding about 20 000 K (low degree of ionization) appears to be describable (at least for neutral species) in terms of local density, velocity, and translational temperature, analogous to a neutral gas expansion. Electronic relaxation, as deduced from fluorescence measurements, leads to excitation temperatures close to the calculated electron temperatures but much higher than the measured and calculated translational temperatures. The applications of these plasma flows include such topics as plasma chemistry, deposition and surface modification, also the simulation of space experiments in low earth orbit or re-entry, and the generation of beams of H (proton sources), O (atmosphere and space interests), and metastable nitrogen molecules for the growth of nitride semiconductors.

The relaxation of O^1D, as produced in O atom sources of atmosphere interest, is studied experimentally in argon and theoretically in neon using a solution of the Boltzmann equation, as well as a Monte Carlo simulation. In the last few decades, the cooling and the natural alignment of rotational angular momentum, in seeded supersonic jets, has been investigated both experimentally and from a quantum mechanical view. The internal energy distributions in molecules, such as K_2, HBr, and OH, seeded in argon or nitrogen free jets, deviate from rotational and vibrational equilibrium Boltzmann distributions at the translational temperature. As shown using the techniques of laser induced fluorescence (LIF) and resonance-enhanced multiphoton ionization (REMPI), the overpopulations increase with both the rarefaction in the expanding flow and the rotational quantum numbers. Nitrogen is more efficient than argon as a relaxant of rotational states, but this may be influenced by possible Ar condensation enthalpy. Complete diagnostics and mapping of CO_2 and N_2 free jets are obtained by the Raman technique which provides the absolute local densities and local vibrational and rotational temperatures. These data, together with the momentum and energy conservation, make it possible to deduce the corresponding translational temperatures, flow velocities, and condensation energies. More information on the aggregation is monitored through Rayleigh scattering. Finally, the average size, composition, and structure of mixed Ar-N_2 clusters, as well as the amount of uncondensed monomers, can be deduced by coupling beam scattering in gases and at surfaces, with high energy electron diffraction.

Part III: Photodissociation Dynamics and Electronic Spectroscopy in Molecular Beams: From Simple Atoms to Clusters and Ions

Studies of photoexcitation of species in molecular beams continuously reveal new and exciting phenomena, and the results reported in Part III are no exception. Much progress results from new or refined techniques, and in this section two themes are dominant.

The first area is the probing of the photodissociation dynamics of small molecules, particularly with photoionization detection (REMPI). The technique provides a highly selective and sensitive method for extracting both the scalar and vector properties of a photodissociation event. Scalar properties are those of fragment energies, available through the spectroscopic signatures of quantum states and of the kinetic energies from time-of-flight measurements. Laser light also has inherent vector properties – for example it has a propagation vector, simply the direction of the laser beam, and is normally polarized so that there exists an electric vector which is perpendicular to the propagation direction. These properties can be used to unravel the correlations between parent transition moments, fragment velocities and angular momenta - for example, does the departing fragment spin away from the parent like a frisbee or does it take off like a helicopter? Velocity map imaging, state selection of reagent molecules, and femtosecond excitation feature as recently refined techniques to explore in ever increasing detail the properties of the dissociating molecule.

The second area concerns the study of the dissociation dynamics of van der Waals complexes as revealed by the spectral signature of fragmentation paths. Such molecules can most easily be prepared when their component species are co-expanded, from a high to a low pressure environment, to form a jet and then a beam, with the expansion providing exactly the cooled conditions required for the complexes to survive. Careful reagent preparation and precise spectroscopic probing, inevitably with lasers, are the keys to successful interpretation of the behaviour of electronically excited complexes. The electronic spectra of van der Waals complexes, involving metal atoms, can be used to infer the non-bonding interactions of both the ground and excited states, of relevance for an understanding of the spectra of atom-doped solid matrices. Molecular beams can yield internally cold species, and a striking example is in the formation of cold cations of polycyclic aromatic hydrocarbons (PAHs) for spectroscopic investigation. This example illustrates an increasingly important feature of molecular beam technology, that the results can have pronounced importance outside the field of laboratory spectroscopy. PAHs have been suggested as the carriers of the long known Diffuse Interstellar Bands, and only recently has it been possible to prepare such species for laboratory characterisation.

Dissociation of simple molecules such as O_3, N_2O and NO_2, has important atmospheric consequences. For ozone, the observation of both vibrationally mediated and spin forbidden photodissociation processes through molecular beam time-of-flight methods, has helped to lead to a quantitative reassessment of the oxidizing capacity of the troposphere, and for N_2O a detailed knowledge of its photolysis is required to understand its global budget and hence its participation in ozone loss processes. These examples of the application of molecular beam and laser techniques show the relevance of such studies to problems in applied photochemistry.

Part IV: Dynamics of Elementary Collisions and Femtochemistry: Theory, Real-Time Probing, and Imaging of Crossed-Beam Reactions

Chemistry is a subject of bond breaking and bond making. In addition to its fundamental importance, gas phase chemical dynamics also has a wide range of applications in many practical disciplines such as combustion, atmospheric chemistry, astrophysics, plasmas, and chemical vapor deposition. Recent years have witnessed a remarkable progress in our deeper understanding of the microscopic details of chemical reactivity. Novel experimental techniques make it nowadays feasible to actually watch how the chemical transformation occurs (seeing is believing) and to carry out fully quantum state resolved studies for a wide range of chemical systems far superseding the former limitations of the alkali age. The combination of molecular beams and lasers has played a pivotal role in this advancement. The crossed-beam technique allows experimentalists to measure the product, speed, and angular distributions under well defined single-collision conditions. The implementation of lasers in the beam experiment provides the additional full capability of quantum state, polarization, and real-time resolution in interrogating and/or manipulating a bimolecular scattering event. The experimental progress has been accompanied by an equally impressive development in theory. In fact, the hallmark of recent advances in the field of chemical dynamics is the synergy of experiment and theory.

Part IV starts with a broad review by A. H. Zewail, the 1999 Nobel laureate in Chemistry. His revolutionary femtochemistry techniques make it possible to follow continuously the motions of the atoms in molecules, from reagents to products, by taking femtosecond real-time snapshots of the chemical transformation which typically occurs within a few pico-seconds. This feature article is then followed by a theoretical perspective on femtochemistry and by an alternative optical-collision approach in which the colliding pair is optically interrogated directly while they are still in close encounter.

The remaining eight articles focus on the recent advances and applications of crossed-beam techniques in exploring various collisional processes with unprecedented fine details. The method of counterpropagating pulsed beam scattering, combined with high resolution ion time-of-flight (TOF) analysis and state-specific resonance-enhanced multiphoton ionization (REMPI) detection, is reviewed and illustrated by the accessible state resolved integral and differential cross sections and even single collision induced alignment. Then a powerful multiplexed ion imaging detection scheme is presented in which the state-specific product, after being ionized, is projected onto a position sensitive detector so that its entire angular distribution can be revealed simultaneously. The stereodynamic effects are explored in the rotational energy transfer of NO+Ar using an oriented molecular beam. Thus, it is found that, for the formation of highly rotationally excited products, a collision at the O-end is more effective than a collision at the N-end, in agreement with theoretical prediction. Also the Doppler-selected TOF method is introduced, in which the conventional Doppler and high resolution ion TOF techniques are blended in an innovative manner, so that the product three dimensional velocity distribution, in the center-of-mass coordinates, can be mapped out directly. Its application to the study of a prototypical insertion reaction is beautifully illustrated. Another chapter highlights recent applications of an ultra-high resolution translational energy spectroscopy, the H-atom Rydberg tagging TOF technique achieving fully state-resolved angular distributions. Also the reaction kinetics are studied in uniform supersonic flows at very low temperatures and the results obtained for a variety of molecular processes are highly relevant to astrophysical chemistry. The last two articles present ion/neutral crossed-beam collisions. In particular, the dependence of the reaction cross section on the initial collision energy is studied for ion-molecule reaction at low energies. In a combined experimental and theoretical study of an ion-atom collisional system, the measured and calculated cross sections and its energy dependence are in good agreement.

Finally, these eleven chapters together provide a quick, yet balanced glimpse of state-of-the-art crossed beam techniques, and offer exciting new perspectives on the fundamental nature of chemical transformation, as we march into the New Millennium.

Part V: Clusters and Nanoparticles: Diffraction, Size Selection, Polarizability, and Fragmentation by Photodissociation, Electron Impact, and Atom Collision

This section deals with the physics of clusters, a new state of matter between isolated atoms and molecules and the condensed phase of bulk matter, as generated in jet and beam experiments. These aggregates differ considerably both from their original constituents and from bulk matter in several ways. Their physical and chemical properties are strongly dependent on the number of atoms in the cluster, and even for one specific size there may exists a considerable number of different isomeric structures. In spite of their finite number of particles, clusters can undergo phase transitions and exhibit collective excitations. In contrast to bulk matter, a large fraction of their atoms or molecules are on the surface. All these properties make them very attractive for investigations which provide a new microscopic understanding of macroscopic phenomena such as solvation, chemical reactions, and catalysis.

In some cases, cluster research even resulted in the discovery of new materials. The most prominent example is the discovery of a new modification of carbon, the fullerenes C_{60} and C_{70}, in a molecular beam experiment in which laser vaporized graphite was seeded and expanded in helium. The apparatus was not expressly designed for the study of carbon but for the production of silicon and germanium clusters. This accomplishment has been acknowledged by the 1996 Nobel prize in Chemistry awarded to H.W. Kroto, R.F. Curl and R.E. Smalley. Aside from these strongly bound fullerenes, clusters are also observed for weakly bound van der Waals and hydrogen bonded, as well as for ionic and metallic systems. The latter ones are of special interest, since such bond does not exist for small molecules so that the transition to metallic behavior is an important research goal.

The articles in Part V deal with a variety of systems ranging from small clusters, where fundamental quantum aspects are predominant, to nanoscale particles with properties relevant for applications. An important topic is the production of neutral, size-selected clusters. Here the most advanced methods are discussed which include examples of small rare gas clusters (especially He) produced by diffraction from a free-standing transmission grating, or which are generated by deflection from an atomic beam with additional velocity analysis. The results of measured and calculated dipole polarizabilities of alkali metal clusters reveal, already for small cluster sizes, the transition to metallic behavior based on non-local effects during the interaction. This also clarifies the optical response of these systems.

Dynamical processes are studied in the fragmentation of argon clusters by electron impact, the photodissociation of water clusters, and in a detailed series of collisions of metal cluster ions with atomic partners, using state-of-the-art techniques. The large fragmentation during the ionization process is essentially caused by the energy release when the system undergoes the transition from the weakly interacting neutral to the strongly bound ionic configuration. The photoabsorption and the subsequent dissociation of water clusters is strongly dependent on the formation process. In collision-induced dissociation processes with rare gas atoms, the different mechanisms for impulsive and electronic energy transfer are elucidated as a function of the cluster size. In charge transfer experiments of metal, metal oxide, and metal hydroxide ions with metal atoms, the onset of the non-metal to metal, or other bonding behavior, is observed. Another important problem, which is addressed in several contributions, is how the energy is distributed after the clusters collided with electrons. Applications range from electron attachment to molecular clusters, via the radiative cooling of fullerenes, to the assessment of internal energy distributions for solvated metal ions and the fragmentation of these systems by evaporation.

The small molecular species allow us, aside from solving problems of fundamental interest, to study also questions relevant to solvation and atmospheric chemistry, whereas larger covalent species are investigated with the aim of designing nanostructured materials with novel electronic properties which can be exploited for various applications, particularly in optoelectronics. As an example, nanosized silicon crystallites are generated by combining a laser-driven flow reactor containing gas phase precursor molecules (SiH_4 in He) with a supersonic expansion to form a beam of pure silicon clusters with velocities strongly dependent on their size.

Part VI: Spectroscopy and Reaction Dynamics of Molecules Isolated, Cooled (or Conditioned), by Techniques of Molecule, Cluster, Droplet, or Liquid Beams

Considerable progress has been made in recent years in the development of experimental methods for studying large clusters, nanodroplets, and liquid beams. These media are not only interesting in their own right, but are also providing novel spectroscopic matrices capable of isolating, ultracooling (or conditioning at the gas-liquid beam interface) molecules, mediating chemical reactions, and forming new molecular aggregates. Free jet expansions can be used to make clusters of quite large size (10^3 to 10^5 atoms or molecules) for the rare gases and hydrogen. These large clusters generally cool by evaporation and, with the exception of helium, are thought to be effectively solid.

At low pressures, helium remains liquid to absolute zero and the corresponding nanodroplets are the only ones that are known to be liquid under molecular beam conditions. These large clusters and droplets can be doped with other atoms and molecules by passing them through a "pick-up cell", namely a chamber containing a low pressure of molecules of interest. For a typical pick-up cell 10 cm in length, one microtorr pressure is needed to capture a single molecule. While the alkali atoms have been shown to reside on the surface of liquid helium droplets, molecules are immersed in the helium and thus reside on the inside. Owing to the weak interactions between an immersed molecule and the helium, the environment is both homogeneous and non-dissipative (superfluid) so that high resolution spectroscopy is often possible. The temperature of the clusters and droplets depend directly upon the binding energy of the matrix atoms or molecules, resulting in 4 K for hydrogen clusters and 0.38 K for ^{4}He. Even lower temperatures (0.15 K) are possible using ^{3}He, owing to the higher zero point energy of this system, which reduces the energy required to evaporate an atom. In contrast with conventional matrix spectroscopy, liquid helium gives much smaller solvent frequency shifts, again owing to the weak interactions, and the molecules undergo free rotation. Also the technique is useful in high resolution spectroscopy on collimated beams of cold molecules spontaneously desorbed from droplet surfaces, or on molecules laser desorbed from liquid beams.

In the case of supercooled hydrogen droplets, the coagulation of CO would argue also in favor of a matrix of metastable superfluid liquid. Thus, the technique of using finite-size, isolated and ultracold droplets appears as an exciting new general approach to study both properties of the liquid solvent and interactions between the molecule and solvent, as well as providing a suitable growth medium to form and investigate novel cluster structures. Liquid beams offer another interesting approach for studying the liquid-gas interface and areas such as the dynamics of solute molecules on solution surfaces. Fragile biomolecules, such as DNA bases and their pairs, or peptides, can be seeded in pulsed supersonic free jets, without using excessive temperatures. They are just now being investigated for the first time by high resolution spectroscopy, using REMPI detection combined with laser desorption obtained by heating the substrate rather than the molecules of interest.

Finally, the clusters constitute novel nanoreactors for studying solvation effects on chemical reactions at the microscopic level, as shown on neon clusters with dramatic increases of reaction rates, as compared to those in the gas phase.

As with all good ideas, this is a divergent field that is finding applications in many different areas.

Part VII: Interactions of Molecular Beams and Cluster Beams with Surfaces, and Applications

Molecular beams offer a powerful technique for studying surfaces and surface phenomena, providing a flexible tool which is entirely specific to the surface. Since their adoption to investigate surfaces, beam techniques have helped to transform our understanding of surface and phonon structure, energy transfer, molecular dissociation and chemical reactions, giving detailed information against which new models can be tested. The combination of molecular beam scattering with state resolved techniques (such as LIF and REMPI) has opened up gas-surface dynamics to detailed scrutiny and is providing a unique understanding of how small molecules and clusters scatter and dissociate at surfaces.

Foreword XXIII

The applications described here range from simple reflection of noble gas atoms to the impact of large and complex species on surfaces. The opening chapter describes theoretically the unique environment and chemistry which occur during high energy cluster-surface collisions. In this case the surface collision reflects an incoming cluster back on itself, creating a super hot, dense gas in which each collision nevertheless remains essentially isolated. The high energy and gas density created in the reflecting cluster promote multi-centre reactions, which do not occur directly under thermal conditions, while preserving new products as the cluster disintegrates.

All the other contributions focus on modern research into gas-surface interactions. These begin with a chapter describing the reflection and adsorption of a simple molecule at an inert surface, using beam and state selective detection to probe reaction dynamics and reveal how small molecules dissociate and react at metal surfaces. Molecular beam scattering is used to investigate the energy exchange as a molecule approaches a dissociation barrier, providing information on the shape of the potential energy surface and the role of short lived molecular states during scattering. Also the reverse process is described : the recombinative desorption of two adsorbed atoms to form a diatomic molecule. Nitrogen bond formation and product desorption are investigated using state selective techniques to explore the dynamics of surface recombination, using detailed balance arguments to link desorption to dissociation and predict how molecular motion influences the dissociation probability.

Subsequent chapters describe experiments in which more complex species interact with a surface under a variety of conditions. For the scattering of low energy van der Waals clusters of mixed rare gases, from a graphite surface, attention is focussed on the enrichment of the surviving species. Electrical charging of clusters is described for the more cohesive water clusters, scattering from a number of solid targets. A fullerene may be considered as a very strongly bound cluster but to regain a similar energy transfer to that of a lightly bound cluster the interaction energy of a fullerene must be much higher, and charge exchange and ionization can easily occur. Electron emission from clusters colliding with graphite reflects electronic transitions during scattering and the neutralization dynamics. This is analysed by combining experimental and theoretical approaches.

Next, the growth of surface films is performed by supersonic free-jet epitaxy of wide bandgap semiconductors of particular relevance to III-N deposition (nitrogen), and also by carbonization of Si surfaces during reaction with alkenes, as shown by beautiful chemical map analysis using X-ray Photoelectron Spectroscopy (XPS) and Scanning Electron Microscope (SEM).

Looking Ahead: Macro, Micro and Nanobeams by John Polanyi

In the final chapter entitled "Looking Ahead", John Polanyi (who shared the 1986 Nobel Prize in Chemistry with Dudley Herschbach and Yuan Lee, "for the development of a new field of research in chemistry-reaction dynamics") considers some tempting approaches to molecular-beam chemistry that he refers to as "Macro, Micro and Nanobeams." He begins with the "micro" in which the pulsed beam sources would be located fractions of a millimetre apart, so as to maximise the density in the crossing region. This could be achieved by pulsed laser-induced desorption of the beams from tilted crystal faces. Unprecedentedly low reactive cross-sections should be measurable.

Microbeams lead naturally to nanobeams in which the interacting photofragments encounter one another after travelling a distance measured in angstroms. This approach, termed Surface-Aligned Photochemistry, SAP, has been realized in a number of laboratories.

Particular interest attaches to the cases in which the photorecoiling molecular fragment is aimed in a downward-directed beam. In Localised Atomic Scattering, LAS, such a beam (in contrast to the generalised beam + surface scattering described in Part VII) is aimed with a restricted impact parameter at preferred atomic sites on the surface, scattering therefore at angles far from specular.

Still more interesting is the analogous case in which the atomically-localised collision with the surface results in chemical reaction at the surface, this being Localised Atomic Reaction, LAR. The site of the parent molecule relative to the site of the new bond at the surface will have much to tell us about the molecular dynamics of the surface reaction, since it will allow us to map the molecular motions onto the substrate.

Finally, moving back from nano to macro, mention is made of beam chemistry on the largest scale that occurs when "pulsed" dense clouds of interstellar gas encounter one another, leading, one may presume, to characteristic forward, backward or sideways scattering (see Dudley Herschbach's opening article) depending on the molecular dynamics of the all-important interstellar reactive events.

Not surprisingly phenomena on the grandest-scale depend upon the finest molecular details. If this were not the case, why would we study molecular beam physics and chemistry?

Contents

Introductory Perspectives

**Atomic and Molecular Beams in Chemical Physics:
A Continuing Odyssey**
D. Herschbach... 3

Part I   Laser Cooling and Manipulation of Atoms and Molecules – Atomic and Molecular Optics and Applications

1. Foreword: Laser Cooling and Trapping of Neutral Atoms
J. Dalibard, C. Cohen-Tannoudji 43

2. Optics and Interferometry with Atoms and Molecules
J. Schmiedmayer... 63

3. Some New Effects in Atom Stern-Gerlach Interferometry
R. Mathevet, K. Brodsky, F. Perales, M. Boustimi, B. Viaris de Lesegno,
J. Reinhardt, J. Robert, J. Baudon 81

**4. Atom Optics and Atom Polarization Interferometry
Using Pulsed Magnetic Fields**
O. Gorceix, E. Maréchal, S. Guibal, R. Long, J.-L. Bossennec, R. Barbé,
J.-C. Keller 95

5. Making Molecules at MicroKelvin
W.C. Stwalley.. 105

6. Molecular Photoassociation and Ultracold Molecules
P. Pillet, F. Masnou-Seeuws, A. Crubellier 113

7. Molecular Optics in Intense Fields: From Lenses to Mirrors
T. Seideman 133

**8. Ultrashort Wavepacket Dynamics and Interferences
in Alkali Atoms**
B. Girard, M.A. Bouchene, V. Blanchet, C. Nicole, S. Zamith 145

9. Atomic Beam Spin Echo.
Principle and Surface Science Application
M. DeKieviet, D. Dubbers, S. Hafner, F. Lang 161

10. Sufficient Conditions for Quantum Reflection
with Real Gas–Surface Interaction Potentials
R.B. Doak, A.V.G. Chizmeshya 175

11. Focusing Helium Atom Beams Using Single Crystal Surfaces
B. Holst, W. Allison 183

12. Atom Lithography with Cesium Atomic Beams
F. Lison, D. Haubrich, D. Meschede 195

Part II Translational and Internal Energy Relaxation in Supersonic Free Jets, Including Possible Alignment and Condensation Phenomena

1. The Physics of Plasma Expansion
D.C. Schram, S. Mazouffre, R. Engeln, M.C.M. van de Sanden 209

2. Laser Sustained Plasma Free Jet and Energetic Atom Beam
of Pure Argon or Oxygen Seeded Argon Mixture
A. Lebéhot, J. Kurzyna, V. Lago, M. Dudeck, R. Campargue 237

3. Doppler Profiles of the Distribution of $O(^1D)$ Relaxing in Ne
B.D. Shizgal, K. Hitsuda and Y. Matsumi 253

4. Natural Alignment and Cooling
in Seeded Supersonic Free Jets:
Experiments and a Quantum Mechanical View
V. Aquilanti, D. Ascenzi, D. Cappelletti, M. de Castro, F. Pirani 263

5. Nonequilibrium Distributions of Rotational Energies of K_2
Seeded in a Free-Jet of Argon
H. Hulsman 273

6. Rotational and Vibrational Relaxation of Hydrides
in Free Jets: HBr and OH
A.E. Belikov, M.M. Ahern, M.A. Smith 283

7. Raman Studies of Free Jet Expansion
(Diagnostics and Mapping)
S. Montero, B. Maté, G. Tejeda, J.M. Fernández, A. Ramos 295

8. Mixed Clusters Produced in Argon-Nitrogen Coexpansions
as Evidenced by Two Experimental methods
E. Fort, A. De Martino, F.Pradère, M. Châtelet, H. Vach, G. Torchet,
M.-F. de Feraudy, Y. Loreaux 307

Part III Photodissociation Dynamics and Electronic Spectroscopy in Molecular Beams: From Simple Molecules to Clusters and Ions

1. Imaging of State-to-State Photodynamics of Nitrous Oxide
in the 205 nm Region of the Stratospheric Solar Window
M.H.M. Janssen, J.M. Teule, D.W. Neyer, D.W. Chandler,
G.C. Groenenboom 317

2. The Photodissociation Dynamics of Tropospheric Ozone
G. Hancock, R.D. Johnson, J.C. Pinot de Moira, G.A.D. Ritchie,
P.L. Tyley 331

3. Photodissociation of NO_2 near 225 nm
by Velocity Map Imaging
M. Ahmed, D.S. Peterka, A.G. Suits 343

4. The $(Ba \cdots FCH_3)^*$ Photofragmentation Channels:
Dynamics of the Laser Induced Intracluster
$(Ba \cdots FCH_3)^* \to BaF^* + CH_3$ and $Ba^* + FCH_3$ Reaction
S. Skowronek, A. González Ureña 353

5. Electronic Spectroscopy and Excited State Dynamics
of Aluminum Atom-Molecule Complexes
P.J. Dagdigian, Xin Yang, I. Gerasimov, Jie Lei 367

6. Electronic Spectra of Cold Polycyclic Aromatic Hydrocarbons
(PAH) Cations in a Molecular Beam
Ph. Bréchignac, Th. Pino, N. Boudin 379

7. Structure and Dynamics of van der Waals Complexes
by High Resolution Spectroscopy
G. Pietraperzia, M. Becucci, I. López-Tocón, Ph. Bréchignac 393

8. Femtosecond Pump-Probe Experiments
with a High Repetition Rate Molecular Beam
W. Roeterdink, A.M. Rijs, G. Bazalgette, P. Wasylczyk, A. Wiskerke,
S. Stolte, M. Drabbels, M.H.M. Janssen 405

Part IV Dynamics of Elementary Molecular Collisions and Femtochemistry: Theory, Real-Time Probing, and Imaging of Crossed-Beam Reactions

1. Femtochemistry: Recent Progress in Studies of Dynamics and Control of Reactions and Their Transition States
A.H. Zewail 415

2. Femtosecond Spectroscopy of Collisional Induced Recombination: A Simple Theoretical Approach
V. Engel 477

3. Direct Observation of Collisions by Laser Excitation of the Collision Pair
J. Grosser, O. Hoffmann, F. Rebentrost 485

4. Counterpropagating Pulsed Molecular Beam Scattering
H. Meyer 497

5. Ion Imaging Studies of Chemical Dynamics
D.W. Chandler, J.R. Barker, A.J.R. Heck, M.H.M. Janssen, K.T. Lorenz, D.W. Neyer, W. Roeterdink, S. Stolte, L.M. Yoder 519

6. Orientation of Parity-Selected NO and Its Steric Asymmetry in Rotational Energy Transfer Collisions
M.J.L. de Lange, S. Lambrechts, J.J. van Leuken, M.M.J.E. Drabbels, J. Bulthuis, J.G. Snijders, S. Stolte 529

7. Collisional Energy Dependence of Insertion Dynamics: State-Resolved Angular Distributions for $S(^1D) + D_2 \to SD + D$
S.-H. Lee, K. Liu 543

8. High Resolution Translational Spectroscopic Studies of Elementary Chemical Processes
X. Liu, J.J. Lin, D.W. Hwang, X.F. Yang, S. Harich, X. Yang 555

9. Reaction Kinetics in Uniform Supersonic Flows at Very Low Temperatures
B.R. Rowe, A. Canosa, C. Rebrion-Rowe 579

10. Ion-Neutral Collisions in Beam Experiments. Production of $ArN^{n+}(n = 1, 2)$ in the Reaction of Ar^{n+} with N_2
P. Tosi, D. Bassi 591

11. Crossed Beams and Theoretical Study of the $(NaRb)^+$ Collisional System
A. Aguilar, J. de Andrés, T. Romero, M. Albertí, J.M. Lucas, J.M. Bocanegra, J. Sogas, F.X. Gadea 599

Part V Clusters and Nanoparticles: Diffraction, Size Selection, Polarizability, and Fragmentation by Photodissociation, Electron Impact, and Atom Collision

1. Diffraction of Cluster Beams from Nanoscale Transmission Gratings
L.W. Bruch, W. Schöllkopf, J.P. Toennies 615

2. Electron Impact Fragmentation of Size Selected Ar_n ($n = 4$ to 9) Clusters
P. Lohbrandt, R. Galonska, H.-J. Kim, M. Schmidt, C. Lauenstein, U. Buck 623

3. Static Dipole Polarizability of Free Alkali Clusters
Ph. Dugourd, E. Benichou, R. Antoine, D. Rayane, A.R. Allouche, M. Aubert-Frecon, M. Broyer, C. Ristori, F. Chandezon, B. A. Huber, C. Guet 637

4. Photodissociation of Water Clusters
K. Imura, M. Veneziani, T. Kasai, R. Naaman 647

5. Collision Induced Fragmentation of Molecules and Small Na_n^+ Clusters: Competition Between Impulsive and Electronic Mechanisms
J.A. Fayeton, M. Barat, Y.J. Picard 657

6. Charge Exchange in Atom–Cluster Collisions
C. Bréchignac, Ph. Cahuzac, B. Concina, J. Leygnier, I. Tignères 667

7. Electron Attachment to Oxygen and Nitric Oxide Clusters
G. Senn, P. Scheier, T.D. Märk 683

8. The Radiative Cooling of C_{60} and C_{60}^+ in a Beam
A.A. Vostrikov, D.Yu. Dubov, A.A. Agarkov, S.V. Drozdov, V.A. Galichin 693

9. Internal Energy Distributions in Cluster Ions
J.M. Lisy 701

10. Molecular Beams of Silicon Clusters and Nanoparticles Produced by Laser Pyrolysis of Gas Phase Reactants
M. Ehbrecht, H. Hofmeister, B. Kohn, F. Huisken 709

Part VI Spectroscopy and Reaction Dynamics of Molecules Isolated, Cooled (or Conditioned), by Techniques of Molecule, Cluster, Droplets, or Liquid Beams

1. Spectroscopy in, on, and off a Beam
of Superfluid Helium Nanodroplets
*J.P. Higgins, J. Reho, F. Stienkemeier, W.E. Ernst,
K.K. Lehmann, G. Scoles* 723

2. Spectroscopy of Single Molecules and Clusters
Inside Superfluid Helium Droplets
*E. Lugovoj, J.P. Toennies, S. Grebenev, N. Pörtner,
A.F. Vilesov, B. Sartakov* 755

3. The Spectroscopy of Molecules and Unique Clusters
in Superfluid Helium Droplets
K. Nauta, R.E. Miller 775

4. Electronic Structure and Dynamics of Solute Molecules
on Solution Surfaces by Use
of Liquid Beam Multiphoton Ionization Mass Spectrometry
F. Mafuné, N. Horimoto, T. Kondow 793

5. Shedding Light on Heavy Molecules, One by One
M.S. de Vries 805

6. Molecular Beam Studies of DNA Bases
C. Desfrançois, J.P. Schermann 815

7. Reaction Between Barium and N_2O on Large Neon Clusters
M.A. Gaveau, M. Briant, V. Vallet, J.M. Mestdagh, J.P. Visticot 827

8. Capture and Coagulation of CO Molecules to Small Clusters
in Large Supercooled H_2 Droplets
E.L. Knuth, S. Schaper, J.P. Toennies 839

Part VII Interactions of Molecular Beams and Cluster Beams with Surfaces, and Applications

1. Essentials of Cluster Impact Chemistry
T. Raz, R.D. Levine 849

2. Probing the Dynamics of Chemisorption
Through Scattering and Sticking
A.E. Kleyn 873

3. Product State Measurements
of Nitrogen Formation at Surfaces
M.J. Murphy, P. Samson, J.F. Skelly, A. Hodgson 887

4. Enrichment of Binary van der Waals Clusters
Surviving Surface Collision
E. Fort, A. De Martino, F. Pradère, M. Châtelet, H. Vach 901

5. Collision Dynamics of Water Clusters on a Solid Surface:
Molecular Dynamics and Molecular Beam Studies
A.A. Vostrikov, I.V. Kazakova, S.V. Drozdov, D.Yu. Dubov,
A.M. Zadorozhny 909

6. Hyperthermal Fullerene-Surface Collisions:
Energy Transfer and Charge Exchange
A. Bekkerman, B. Tsipinyuk, E. Kolodney 917

7. Electron Emission Induced by Cluster-Surface Collisions:
a Fingerprint of the Neutralization Dynamics?
M.E. Garcia, O. Speer, B. Wrenger, K.H. Meiwes-Broer............. 933

8. Supersonic Beam Epitaxy of Wide Bandgap Semiconductors
V.M. Torres, D.C. Jordan, I.S.T. Tsong, R.B. Doak 945

9. Chemical Maps and SEM Images of the Reaction Products
on Si Surfaces Irradiated with Cold and Hot C_2H_4 Beams
I. Kusunoki... 959

Looking Ahead

Macro, Micro and Nanobeams
J.C. Polanyi 973

Subject Index 989

Author Index... 995

Introductory Perspectives

Atomic and Molecular Beams in Chemical Physics: A Continuing Odyssey

Dudley Herschbach

Department of Chemistry and Chemical Biology
Harvard University, Cambridge, MA 02138 U.S.A.

1 Introduction

This introductory review presents vignettes intended to convey to the nonspecialist something of the style and scope of research with atomic and molecular beams. Necessarily, the episodes described are idiosyncratic and impressionistic, merely appetizers for the profuse smorgasbord of papers offered in this volume. Only a few pertinent references and kindred reviews are cited, as the smorgasbord provides ample guidance to to a burgeoning literature. In a typical format, replete with terse descriptions, the appetizer menu reads:

Historical Perspectives
Evolutionary role in quantum physics, magnetic resonance, lasers, and chemical reaction dynamics

Let There Be Light in the Night Sky
Chemical implications of meteor trails and nightglow

Paradoxical Proliferation of Organic Molecules in the Heavens
Mutual distaste of helium ions and hydrogen molecules gives rise to complex organic molecules in interstellar clouds

Molecular Clusters, Superstrong or Superfluid
Discovery of buckyballs, leading to nanotubes; reactions and spectra within solvent clusters

Liberating Catalysts from Thermodynamic Constraints
Supersonic expansions through nozzles fashioned from catalytic material reveal reaction intermediates suppressed under equilibrium conditions

Pursuit of Forbidden Fruit
How to undo averaging over random impact parameters

Bringing Molecules to Attention
Conversion of molecular pinwheels into pendula by interaction with static or laser fields

Slowing Molecules to Make Nanomatter-Waves
Current efforts to slow and spatially confine translational motion, prerequisite to obtaining ultracold molecules with deBroglie wavelengths longer than their size or separation

Laser control of reaction pathways
The coherence of laser light enables selecting pathways by exploiting quantum interference

Like any appetizer menu, this is not at all nutritionally balanced; rather it samples a few of the chef's favorite delectations, largely chosen from previous repasts [1–6]. Some topics reach back decades, others are recent or ongoing, selected to emphasize thematic motivations or methods inviting further exploration or application.

2 Historical Perspectives

The first molecular beam experiments were carried out 90 years ago, immediately after the invention of the high speed vacuum pump had made it possible to form directed "rays" of neutral atoms or molecules traveling in vessels maintained at sufficiently low pressures to prevent disruption of the beams by collisions with background gas. That key feature destined the method to become an extraordinarily fruitful research technique [7]. Figure 1 indicates the historical progression through mid-century. Systematic research began with Otto Stern, who developed many aspects of beam techniques. He had retired and was living in Berkeley when I arrived there in 1959, and I had the opportunity to hear from him memorable stories about his work [2].

In his first experiment, done in 1919, Stern undertook to test the Maxwell–Boltzmann velocity distribution by analyzing the speed of a silver atom beam with a rotating device. His next experiment, done with Walther Gerlach in 1921–22, begat a host of others of great consequence. In the ancestral experiment, a beam of silver atoms sent through an inhomogeneous magnetic field was found to split in two. This discovery, shocking to physicists of the day, first revealed the existence of space-quantization It showed that the angular momentum associated with electrons within atoms (and consequent atomic magnetic moments) can point only in certain discrete directions.

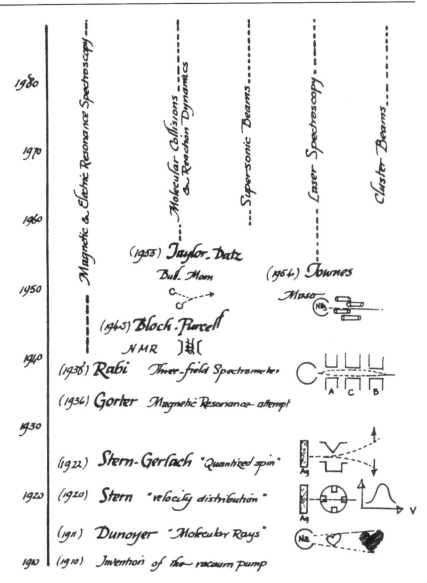

Figure 1: Evolution of molecular beam and kindred methods. From [2].

(That the doublet splitting was due to angular momentum from electron spin rather than orbital motion was not actually recognized until several years later [8].) Space quantization provided one of the most compelling items of evidence that a new mechanics was required to describe the atomic world.

2.1 Descendants of Space Quantization

The prototypes for nuclear magnetic resonance (NMR), radioastronomy, and the laser all derived from space quantization. NMR spectroscopy, developed by Isidor Rabi in the late 1930s, exploits the orientation of nuclear spins in an external field. Radiofrequency radiation tuned to the slight difference in energy of spins oriented "up" or "down" induces a change or "flip" in the spin orientation. It was not obvious that the minute nucleus, with dimensions 10^5-fold smaller than the atomic electron distribution, would interact appreciably with the electrons. If not, the different spin orientations would remain equally probable even when subjected to an external magnetic field. There would then be no net absorption of resonant radio waves, since spin flips induced by the radiation are equally likely up or down. Thus, in bulk matter, net absorption can only occur if interactions permit the populations of the spin orientation states to be unequal.

Rabi devised an elegant way to use atomic beams to escape this constraint. In his apparatus, he introduced two inhomogeneous magnets like Stern's, but with their fields in opposite directions. A beam of atoms traversing the first field (denoted A) is split into its space quantized components, but on passing through the second field (denoted B) these are recombined. Between these magnets, which act like diverging and converging lenses, Rabi introduced a third magnet (the C-field). This was homogeneous, so has no lens action, but serves to define the "up" and "down" quantization directions. Since atoms in the beam experience virtually no collisions, the populations of the component spin orientation states remain equal. Radio waves sent into the C-field region will at resonance flip equal numbers of spins up and down. But now any spin changing its orientation after the A-field will not be refocused in the B-field. That enabled Rabi to detect which radiofrequencies produced resonances. He thus created a versatile new spectroscopic method with extremely high resolving power. His work provided a wealth of information about nuclear structure and led to many further developments, including atomic clocks of fabulous accuracy.

The scope of NMR has expanded vastly since 1945, when resonances were first detected in liquids and solids, by Edward Purcell and by Felix Bloch and their coworkers. The key discovery was that small differences in the population of the space quantized nuclear spin orientation states did after all arise from interactions in the bulk media, enabling detection of spectroscopic transitions. Now NMR spectra are routinely used to determine detailed structural features of proteins containing thousands of atoms. Another marvelous offspring is magnetic resonance imaging, which provides far higher resolution than x-rays. It is applicable to soft tissues such as the brain, even enabling researchers to literally catch sight of glimmering thoughts.

Two other atomic beam experiments in the 1940s, also direct descendents of Stern and Rabi in exploiting space quantization, led to a profound theoretical advance. By means of magnetic resonance spectroscopy, Willis Lamb and Robert Retherford resolved a miniscule splitting in the fine structure of the first electronically excited state of the hydrogen atom. This was highly significant, because Dirac's relativistic quantum theory could not account for such a splitting. Likewise, Polykarp Kusch and coworkers were able to measure the magnetic moment of the electron to unprecedented accuracy, and found a slight discrepancy ($\sim 0.1\%$) with Dirac's theory. These discoveries spurred the development of quantum electrodynamics, in which charged particles interact by interchange of virtual photons. Modern QED, now regarded as the most exact theory in physics, also greatly influenced high-energy physics, particularly the field theory of nuclear forces, quantum chromodynamics.

Atomic beam magnetic resonance also launched an extremely fruitful experimental technique, termed optical pumping, that stemmed from work undertaken in the 1950s by Alfred Kastler and colleagues. In this, atoms or molecules subjected to a magnetic field are illuminated by polarized light. Thereby electronic states are excited with unequal populations among the space-quantized components that correspond to different orientations in the magnetic field. The excited states revert quickly to the ground state by emitting fluorescence whose polarization reveals the preferred orientation. It also provides sensitive detection of transitions induced among the space-quantized components by radiofrequency or microwave radiation. By virtue of coupling effects, nuclear spins as well as atomic magnetic moments are oriented. The initial experiments used atomic beams, to avoid anticipated relaxation by collisions with vessel walls or ambient gas. It was soon found, however, that relaxation could be suppressed by means such as wall coatings. Thus, like NMR, optical pumping is not at all restricted to molecular beams and has found many applications. From a genealogical perspective, the key aspect is the use of the angular momentum of light (rather than Stern–Gerlach magnets) to control and analyze the space-quantized states.

Radioastronomy, which has hugely extended what we can see of the heavens, likewise grew from examining consequences of space quantization. The prototype experiment, conducted by Edward Purcell in 1951, detected radiofrequency emission from the spin flip of the hydrogen atom nucleus, subject to the magnetic field generated by the orbital motion of the atom's lone electron. The ability to map the distribution of hydrogen, by far the most abundant element in the universe, soon revealed unsuspected aspects of the structure of galaxies. Kindred techniques employing radiofrequency and microwave spectroscopy have revealed a remarkable variety of molecules in interstellar clouds, products of previously unsuspected galaxies of chemical reactions.

Molecular beam techniques and space quantization also enabled the ancestral experiments that led to lasers. The prototype, conceived in the 1950s by Charles Townes and also by Nicolay Basov and Aleksandr Prokhorov, employed a beam of ammonia molecules subjected to an electric field which acted as a state-selection device analogous to Stern's magnet. The field selected an energetically unfavorable molecular state, which when illuminated by microwaves reverted to the energetically favorable state. The excess energy was emitted as radiation, coherent with and amplifying that stimulating the emission. (The acronym laser means light amplified by stimulated emission of radiation.) Thus was born molecular amplifiers and oscillators and other wonders of quantum electronics [9], now ubiquitous both in science laboratories and the wider world, including music shops and grocery stores.

2.2 Chemistry with Molecular Beams

As Stern was a physical chemist, he would be pleased that his beam techniques, augmented both by magnetic and electric resonance spectroscopy and especially by lasers, have also evolved powerful tools for the study of molecular structure and reactivity. Under ordinary conditions, in the gas phase or solution, myriad collisions wash out all information about the intimate molecular dynamics of reactive encounters. By the 1930s it was recognized that this could be avoided by crossing two reactant beams in a vacuum and detecting in free flight the products, unbattered by any collisions subsequent to the one that gave them birth. However, it was not until the mid-1950s that such experiments were shown to be feasible.

In a crossed-beam study of the K + HBr → KBr + H reaction, Taylor and Datz (1955) provided a detection scheme that launched a fledgling field. Surface ionization on a tungsten filament had long been used as a sensitive and specific means to detect alkali species, but was about equally responsive to K and KBr. Taylor and Datz found that a platinum alloy filament was much more effective for K than for KBr. From the difference in the signals read on the two detector filaments, they were able to distinguish the small amount of reactively scattered KBr from the large background of elastically scattered K atoms. Small as it was, the little difference bump was a joyful sight. In the 1920s, Michael Polanyi had shown that many alkali reactions proceeded at rates corresponding to "reaction at every collision". The differential surface ionization detector enabled crossed-beam studies of many of these reactions to be made with relatively simple apparatus.

In another notable experiment, Bull and Moon (1954) bombarded a stream of Cs vapor with a pulsed accelerated beam of CCl_4 produced by swatting with a paddle attached to a high-speed rotor. Signal pulses due to scattered Cs and CsCl were detected by surface ionization on a tungsten filament. Although there was no direct means to distinguish between Cs and CsCl, the observed signal pulses appeared to come primarily from reactively scattered CsCl, on the basis of time-of-flight analysis and blank runs with the CCl$_4$ replaced by mercury vapor. Unjustly, this work was long discounted or ignored (although later shown to be quantitatively correct; [10]). This was due to the misconception that elastic scattering would always predominate. The high-speed rotor technique also intimidated other prospective experimenters. Later the rotor technique was revived, to good effect [11].

The 1960s became the "early alkali age" of molecular reaction dynamics. As techniques were developed, the range of properties accessible in single-collision experiments grew to include the disposal of energy among translation, rotation, and vibration of the product molecules; angles of product emission; angular momentum and its orientation in space, and variation of reaction yield and other attributes with impact energy, closeness of collision, rotational orientation or vibrational excitation of the target molecule [12]. The experimental results stimulated and responded to a vigorous outburst of theoretical developments, particularly computer simulations. Dozens of alkali atom reactions were studied, and unexpected variety was found in the dynamics for reactions with different target molecules. In some reactions, the reagents execute a folk dance, lingering together long enough to rotate completely around several times before the products whirl away. Other reactive encounters are very brisk. In some, the incident alkali atom swoops in and almost surreptitiously plucks a halogen atom from the target molecule, which continues on its way as if unaware of the theft. Still other brisk reactions are brutal encounters involving strong repulsive forces. In these, the products rebound backwards at high velocities. Since alkali reactions typically involve transfer of the alkali valence electron into the lowest available orbital of the target molecule, these systems were particularly amenable to relating the dynamical properties to electronic structure [1, 4].

Happily, in 1968 reaction dynamics leapt beyond the alkali age. This became possible with the advent of an extremely sensitive mass spectrometric detector, designed chiefly by Yuan Lee, then a postdoctoral fellow in my lab. (He had intended to try his hand at theory, but was too shy to demur when I asked him to undertake the new apparatus!) How to discriminate against interference from background was the key problem. Most molecules in crossed beams do not react where the two diffuse beams intersect. Unless those that "miss" can be quenched at the apparatus walls (easy with alkali beams), many bounce about and form the same product in the background, obscuring the signal from reactions that occur in single collisions at the intersection zone of the crossed beams. Most crucial is background in the inner sanctum of the detector, where a small fraction (less than 10^{-4}) of the incident product molecules are converted by electron bombardment into the ions actually detected. These difficulties were overcome by a carefully designed pumping scheme, nested about the electron bombarder so that unionized molecules passed through without creating background in the sanctum. As so often in demanding experiments, reducing noise proved more important than enhancing the signal itself.

This mass spectrometric detector, today still a mainstay of the field, enabled study of a wide range of reactions. A striking aspect soon emerged: many dynamical features and electronic aspects of other reactions could be readily understood in terms of what had already been learned from alkali atom reactions. Figure 2 shows a favorite example. The reaction of $H + Cl_2$ to form HCl served as the prototype in the development by John Polanyi of

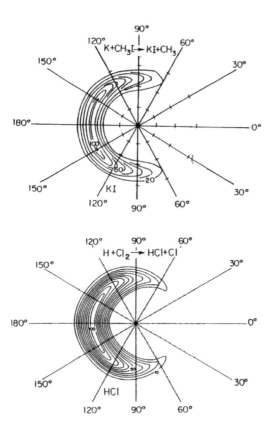

Figure 2: Contour maps displaying distributions in angle and velocity of KI product from $K + CH_3I$ reaction (*top*) and HCl product from $H + Cl_2$ reaction (*bottom*). Angular scale indicates direction (same for 0°, opposite for 180°) in which the product molecules emerge relative to the incident reactant atom beam (K or H). Tick marks along radial lines indicate velocity intervals of 200 meters/sec. Contours show intensity relative to peak (100%). From [1].

his fruitful infrared chemiluminescence method. With it, he had mapped out the energy disposal in vibration and rotation of the HCl product; our beam experiments provided the distributions of relative translational energy and scattering angle. Quite unexpectedly, we found that a contour map of the product distribution in angle and velocity is virtually identical, except for scale, to the map for KI from the K + CH₃I reaction, the first alkali reaction we had studied, years before. The 19th-century notation still used to write down chemical reactions gives no hint of such kinships.

Both these reactions are dominated by repulsive energy release, and for both it is very similar in magnitude and form to that found in photodissociation of the reactant molecule. For K + CH₃I, that was readily accounted for because the electron transfered from the alkali enters an antibonding sigma orbital with a node between the carbon and iodine atoms, the same orbital excited in photodissociation. For H + Cl₂, the resemblance to photodissociation is less obvious, since the high ionization energy of the H atom prohibits electron transfer. However, an explanation emerged from the "frontier orbital" concept of Fukui; this orbital, the highest occupied in the H-Cl-Cl reaction complex, does indeed have a node about midway between the chlorine atoms, producing strong repulsion just as in photodissociation. For many other reactions, the nodal structure of the frontier orbital was likewise found to account for major features of the dynamics observed in beam experiments. This established useful generalizations in terms of familiar chemical notions such as electronegativity [2, 4].

We now take leave from historical episodes in reaction dynamics, but must emphasize that in the past 30 years the scope and sophistication of the field has grown immensely. This has resulted chiefly from coupling lasers with molecular beams, exemplified particularly in leading work by Kent Wilson, Richard Zare, and Ahmed Zewail.

2.3 The Versatile Supersonic Beam

The canonical physics literature on molecular beams, from Stern onwards, stressed that the pressure within the source chamber should be kept low enough so that molecules, as they emerge from the exit orifice, do not collide with each other. In this realm of effusive or molecular flow, the emergent beam provides a true random sample of the gas within the source, undistorted by collisions. Chemists, in desperate need of intensity for studies of reactions in crossed beams, violated the canonical ideal by using much higher source pressures. Collisions within the orifice then produce hydrodynamic, supersonic flow. The properties of gas flow in this realm had been avidly explored from the mid-1950s on, especially by chemical engineers, and since the 1970s most molecular beam experiments employ supersonic beams [13]. This versatile technique offers marked advantages.

When a gas expands isentropically into a vacuum through a pinhole nozzle, the pressure and temperature both drop abruptly. The nozzle imposes collisional communication that brings the gas molecules to nearly the same direction and velocity. It also efficiently relaxes thermal excitation of molecular rotation and (less so) vibration. Thus, not only is the intensity of a supersonic beam far higher than that from an effusive source, but the spreads in velocity and rotational states are markedly narrowed. The effective temperature for relative motion of molecules within such a beam is typically only a few °K. Temperatures much lower, in the millikelvin range, can readily be attained either by using a pulsed nozzle [14] to permit operation at higher source pressures; or by elegantly exploiting a shock wave structure formed by interaction of a continuous supersonic flow with ambient gas [15]. The beam translational kinetic energy also can be varied over a wide range. This is done by seeding the molecules of interest in a large excess of carrier gas that is heavier or lighter; the expansion then accelerates or decelerates the beam species to the exit velocity of the light gas. Likewise, by using a pulsed laser to ablate the sample into a carrier gas, beams of nonvolatile materials can be obtained.

At the low internal temperatures attained in supersonic beams only a few of the lowest energy states are populated. This greatly facilitates spectroscopy of polyatomic molecules [16]. Also, within the markedly nonequilibrium environment of a supersonic expansion, chemical interactions are liberated from thermodynamic constraints. In effect, in the free energy, $\Delta H - T\Delta S$, the entropy term ΔS is suppressed by the low internal temperature. Even a weakly favorable enthalpy term ΔH then can suffice to produce large yields of molecular clusters, solvation complexes, or other species that could not be synthesized under equilibrium conditions.

3 Let There Be Light in the Night Sky

Everyone has seen meteors flash across the night sky, leaving a glowing trail stretching far behind. But very few viewers are aware of chemical implications, still being unraveled. Most of the light in the meteor's wake comes from the yellow D-line emitted by electronically excited sodium atoms. As this is a fully allowed transition, the lifetime of the excited atoms is only of the order 10^{-8} sec. Thus, if the excitation resulted simply from ablating sodium from the meteorite, there would be no trail behind but only a yellow glow at the meteorite's nose. The bright trail must come from a chemiluminescent reaction between ambient atmospheric species and the sodium left in the wake of the meteor. The elucidation of this process over the past 60 years is an unfinished saga. It exemplifies well how incisive molecular beam experiments can enhance results contributed by other techniques and theory.

In 1939, Chapman proposed a two-step reaction sequence to explain the sodium D-line nightglow in the mesosphere and in meteor trails. He suggested that the ablated Na reacts with O_3 to form NaO, which then reacts with O to reform Na atoms, some of them electronically excited. The net effect is thus $O_3 + O \to 2O_2$, with sodium acting as a catalyst. In fact, the catalytic cycle that creates the stratospheric "ozone hole", identified by Rowland and Molina and notorious since the 1970s, is a variant of the Chapman mechanism with a chlorine atom instead of sodium. In the mesophere (at ~ 90 km altitude) the concentrations of ozone and oxygen atoms are of course far lower than in the stratosphere (at ~ 30 km). However, the flux of meteorities maintains a layer of mesospheric sodium (~ 3000 /cm^3 at its peak). It is readily observable from the earth's surface by the lidar technique, which sends a laser beam to excite the distant sodium and observes the resulting fluorescence. This method, familiar to laboratory scientists as "laser-induced fluorescence", has provided a means to correct ground-based astronomical observations for the effects of atmospheric fluctuations by using the mesospheric sodium fluorescence to monitor those fluctuations. In its natural state, unperturbed by passing meteors or laser beams, the chemiluminesence produced by the sodium layer appears as a weak but ubiquitous nightglow.

By 1986 models of mesospheric chemistry derived from much atmospheric and laboratory data appeared to confirm the Chapman mechanism, but to account for the nightglow intensity required that the NaO + O step give a large yield of excited sodium. Soon thereafter, however, a flow-tube experiment designed to test this found no detectable sodium emission. This contradiction prompted a molecular beam experiment, done in my laboratory. The aim was to find out whether the Na + O_3 reaction produced a substantial yield of NaO in a low-lying electronically excited state ($A^2\Sigma^+$). From theory, we suspected that state, in the subsequent reaction with O atoms, might be the actual source of excited Na atoms. Such a possibility was not tested in the flow-tube experiment; it had been performed at atmospheric pressure, so the reactant NaO could be assumed to be collisionally relaxed to the ground electronic state ($X^2\Pi$). Since the excited state is paramagnetic whereas the ground state is diamagnetic, we used a Stern–Gerlach deflecting magnet to analyze the NaO and found that the excited state is indeed the predominant and possibly the sole product [17].

Figure 3 gives a theoretical perspective on the NaO + O reaction [18]. There are 18 energetically accessible potential surfaces linking NaO($X^2\Pi, A^2\Sigma^+$) +O(^3P) with Na($^2S, ^2P$) + $O_2(X^3\Sigma_g^-, a^1\Delta_g, b^1\Sigma_g^+)$. It would be a daunting task to compute all these surfaces, but key aspects emerge from symmetry analysis and qualitative considerations. Like alkali halides, NaO states are expected to be strongly ionic, so we assumed that reaction could occur only via the 6 surfaces that involve electron transfer and which have doublet spin configurations (rather than quartet states with three spins parallel).

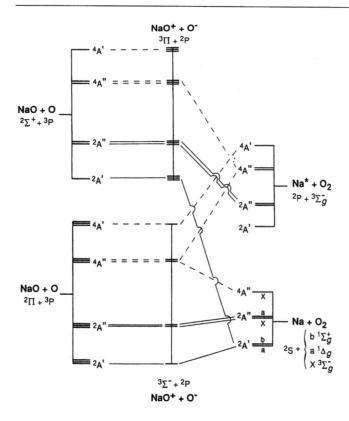

Figure 3: Correlation of electronic states of reactants, kindred ion-pairs and products for NaO + O reaction. The four distinct types of states react separately, because the electronic symmetry species and total spin do not change during reaction. For states of symmetry species A' the wavefunction does not change sign on reflection in the plane of the three atoms; for species A'' it does change sign. Full lines show correlations of states with total electronic spin of 1/2 (doublets), dashed lines those for spin 3/2 (quartets). Potential surfaces arising from the quartet states are likely to be too repulsive to permit reaction. From [18].

The resulting correlation diagram indicates that in reacting with oxygen, the excited NaO($A^2\Sigma^+$) state should produce a large yield of excited sodium atoms, whereas the ground NaO($X^2\Pi$) state should produce none. Experimental evidence for this has now been obtained in a fast-flow study, and a rich spectrum of transitions to the $A^2\Sigma^+$ state has been observed [19]. Knowledge of the $A^2\Sigma^+ \leftarrow X^2\Pi$ spectrum will enable a much more incisive characterization of the role of NaO in a dozen or more other atmospheric processes.

The underlying electronic dynamics in the Na + O_3 reaction also poses a puzzle: why is the excited $A^2\Sigma^+$ state of NaO formed predominantly? That seems very strange in view of the ionic bonding. For the ground $X^2\Pi$ state, the hole in the O^- valence shell is in a $p\pi$ oxygen orbital transverse to the internuclear axis, so two electrons reside in the $p\sigma$ orbital aimed at the Na^+ cation. For the excited $A^2\Sigma^+$ state, however, the O^- hole is instead in the $p\sigma$ orbital, so only one electron is directed toward the cation. There is an appealing heuristic interpretation of the electronic state specificity. If an S atom attacks a molecule along any symmetry element, electron transfer to form an S cation cannot occur unless the parent molecule and its anion have the same symmetry. Electron transfer in a coplanar attack thus is forbidden for O_3 (A_1 parent, B_1 anion, so $A' \to A''$) because the transferred electron enters a $2b_1$ orbital that is asymmetric to the plane of the molecule. This transfer thus will occur most readily when the sodium atom attacks perpendicular to the molecular plane. For such broadside attack the electron affinity is maximal; hence, the radial distance at which the electron transfer occurs is also maximal. This fosters reaction with the $2b_1$ orbital of the anion aimed toward the incoming cation. If the population of this orbital goes into the $p\sigma$ orbital of O^- as the product Na^+O^- molecule is formed, the $p\sigma$ would be singly occupied ($A^2\Sigma^+$ state). A molecular beam experiment to probe the orientation dependence of electron transfer to a planar molecule may now be feasible. This requires aligning the *plane* of the molecule in the laboratory frame. As described below (Sect. 8), it might be done either by combined action of static and laser fields [20] or by using an elliptically polarized laser field [21].

4 Paradoxical Poliferation of Organic Molecules in the Heavens

Over the past 30 years, radioastronomy has revealed a rich variety of molecular species in the interstellar medium of our galaxy and even others. Well over 100 molecules have now been identified in the interstellar gas or in cicumstellar shells [22]. These include H_2, OH, H_2O, NH_3 and a few other small inorganic species, but most are organic molecules, many with sizable carbon chains involving double or triple bonds. In most cases, identification of these molecules depends on comparing spectra obtained in molecular beam spectroscopic experiments with those from radiotelescopes. To appreciate how surprising is the proliferation of organic molecules in the heavens, we need to review some aspects of the interstellar environment.

The cosmic abundance of the elements is drastically nonuniform. Hydrogen comprises over 92%, helium over 6%; next come oxygen at 0.07%, carbon at 0.04%, and nitrogen at 0.009%. The average density within our galaxy is only about one H atom per cubic centimeter. Yet the density is about a hundredfold higher in what are called diffuse interstellar clouds and up to a millionfold higher in dark clouds. In diffuse clouds, a molecule collides with another (H_2 or He) only once every two months or so. The observed molecular abundances depart enormously from estimates derived by assuming chemical equilibrium. For instance, next to H_2, carbon monoxide is the most abundant interstellar molecule (although typically down by a factor of 10^{-4} or more). But thermodynamic calculations predict that under typical dark cloud conditions (20 °K, density of $H_2 \sim 10^5$ cm^{-3}), at chemical equilibrium there would be less than one CO molecule in the volume (10^{84} cm^3) of the observable universe! Likewise, the prevalence of organic molecules containing many carbon atoms and relatively little hydrogen is inexplicable by thermodynamics.

4.1 Synthesis of interstellar molecules

This situation led Klemperer [23] to propose a nonequilibrium kinetic scheme for the synthesis of interstellar molecules, to show how "chemistry can, in the absence of biological direction, achieve complexity and specificity". The scheme invokes sequences of exoergic, bimolecular ion-molecule reactions. Extensive laboratory ion-beam experiments have shown that these processes are typically quite facile and uninhibited by activation energy barriers, unlike most gas-phase chemical reactions not involving ions. Uninhibited reactions in two-body collisions are the only plausible candidates for gas-phase chemistry at the low density and temperature of an interstellar cloud. The clouds also contain dust particles, of unknown composition. Formation of hydrogen molecules from atoms is probably catalyzed on the surface of dust particles, but the host of other molecules seem more likely to be produced by nonequilibrium gas-phase kinetics.

The dark clouds where most interstellar molecules have been seen are immense, typically comprised of hydrogen and helium with a million times the mass of our Sun. In our galaxy such clouds loom as huge dark blotches obscuring regions of the Milky Way. Ionization by the pervasive flux of 100 MeV cosmic rays seeds the clouds with a little H_2^+ and He^+ (about one ion per 500 cm^3), from which sprout many reaction sequences. The H_2^+ rapidly reacts with H_2 to form H_3^+ which, as known from laboratory studies, itself readily transfers a proton to many other molecular species. Most of the H_3^+ is converted to HCO^+, a very stable species. This prediction was a triumph for Klemperer's model. Soon thereafter interstellar emission from a species dubbed Xogen, which had not yet been seen on earth, was shown to come from the HCO^+ ion. It has proved to be the most abundant molecular ion in dark clouds and has even been observed in several distant galaxies. Much else offers support for the kinetic model. For instance, proton transfer from H_3^+ to nonpolar molecules such as N_2 and CO_2 converts them to polar species HN_2^+ and $HOCO^+$ which are capable of emitting rotational spectra.

Again, laboratory observation of these spectra confirmed the detection of interstellar emissions from these species.

Most striking are offspring of the He$^+$ ions, which exemplify how chemical kinetics can produce paradoxical results. The abstraction by He$^+$ of a hydrogen atom from H$_2$, the most abundant molecule in interstellar clouds, would be very exoergic. Yet, for reasons described below, that reaction does not occur. Instead, He$^+$ reacts with CO, the second most abundant molecule, to form C$^+$ and O. The ionization of helium is almost quantitatively transferred to C$^+$, enhancing its concentration a thousandfold (by the He/CO abundance ratio). In turn, the C$^+$ ion reacts only feebly with H$_2$ (via radiative association), but reacts avidly with methane, CH$_4$, and acetylene, C$_2$H$_2$, to launch sequences that build up many organic compounds, including chains punctuated with double and triple bonds. The paradoxical irony is that the mutual distaste of the simplest inorganic species, He$^+$ and H$_2$, gives rise to the proliferation of complex organic molecules in the cold interstellar clouds.

4.2 Electronic structure and reaction specificity

The three-electron system involving only helium and two hydrogen atoms offers a prototypical example for interpretation of chemical dynamics in terms of electronic structure [24]. As shown in ion-beam scattering experiments, the reaction He + H$_2^+$ → HeH$^+$ + H is endoergic by 0.8 eV but occurs readily if at least that amount of energy is supplied, either as relative kinetic energy of the collision partners or as vibrational excitation of H$_2^+$. In contrast, the reaction He$^+$ + H$_2$ → HeH$^+$ + H is exoergic by 8.3 eV, but appears not to occur at all; the less exoergic pathway to form He + H$^+$ + H has been observed, but its reaction rate is four orders of magnitude smaller than for comparable exoergic ion-molecule reactions.

Figure 4 provides an explanation, due to Mahan [24], for the drastic difference in reactivity of He + H$_2^+$ and He$^+$ + H$_2$. Plotted are diatomic potential energy curves for the reactants and products; these represent cuts through the triatomic potential energy surfaces in the asymptotic entrance and exit channels. Consider first the lowest lying trio of separated atoms, He+H$^+$+H. Since both the reactants He+H$_2^+$ and the products HeH$^+$+H correlate adiabatically to He + H$^+$ + H, the reaction can be expected to proceed on a single triatomic potential energy surface.

However, for the upper trio of atoms, He$^+$ + H + H, this does not hold. The ground-state H$_2$ diatomic potential curve, $^1\Sigma_g^+$, which arises from bringing together two H atoms with antiparallel spins, represents a cut in the asymptotic reactant region through the potential surface for He$^+$ + H$_2$ collisions. The corresponding cut in the product region, generated by bringing together He$^+$ + H, yields an excited singlet state that is totally repulsive, according to electronic structure calculations. Likewise, the accompanying excited triplet

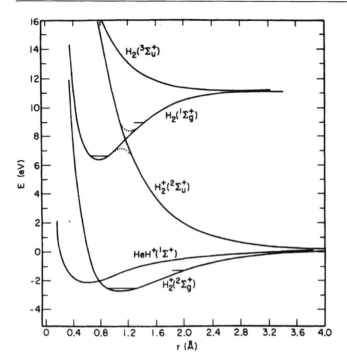

Figure 4: Potential energy curves for the diatoms in the asymptotic reactant and product regions of the $(\mathrm{He} - \mathrm{H}_2)^+$ system. Since the energies of He and He$^+$ are included, here the ground $^1\Sigma_g^+$ state of H_2 lies above the states of H_2^+. Note the crossing of the curves for $\mathrm{H}_2(^1\Sigma_g^+)$ and $\mathrm{H}_2^+(^2\Sigma_u^+)$ which occurs in the reactant region of He$^+$ + H_2, but which becomes an avoided intersection (indicated by dashes) when all three atoms are close to each other. From [24].

state with parallel spins is at best only very weakly bound. Accordingly, colliding He$^+$ + H$_2$ is very unlikely to form a stable HeH$^+$ molecule.
As seen in Figure 4, in the asymptotic reactant region, the ground-state H$_2$($^1\Sigma_g^+$) potential curve crosses at about 1.1 Å that for H$_2^+$($^2\Sigma_u^+$), a strongly repulsive state. When He$^+$ draws nigh to H$_2$, however, the resulting interaction induces these states to mix, as both then acquire the same symmetry (A' under the C_s point group). The crossing thus evolves into an avoided intersection (indicated schematically by dashes). If electron transfer occurs, the adiabatically formed products initially are He + H$_2^+$($^2\Sigma_u^+$), which dissociate directly to He + H$^+$ + H.
Exemplary of chemical physics, the profusion of organic molecules tumbling in the heavens is linked to devilish details that govern an electron hopping between helium and hydrogen.

5 Molecular Clusters, Superstrong or Superfluid

The properties and interactions of molecules are often much influenced by the company they keep. Molecular clusters, generated by supersonic expansion of gas into a vacuum apparatus, have now become a favorite medium for study of reactions and spectra. Such clusters, comprised of from two to up to a billion molecules, offer means to interpolate between gaseous and condensed phase or solution media [25]. We visit two exemplary episodes in this cornucopian field, made possible by the supersonic nozzle.

5.1 Balls and Tubes of Carbon

The discovery of carbon-60 and kindred fullerene molecules ranks among the most important achievements of molecular beam science [26]. It also affirms the value of fostering eclectic collaborations and the playful pursuit of curious observations. The crucial ingredient was a technique, devised by Richard Smalley, to generate clusters from solid samples. This uses a laser to vaporize material, enabling it to be entrained in a supersonic gas flow. In the early 1980s, several laboratories had adopted this technique, chiefly to study clusters of metals or semiconductor materials, of interest for catalysis or microelectronics. Among many curious results were features of a mass spectrum of carbon clusters from laser-vaporized graphite, published by a group at the Exxon laboratory as part of an Edisonian survey [27]. For C_{40} and larger clusters, only those with an even number of carbon atoms appeared, in a broad distribution extending above C_{100}. Especially prominent in the mass spectrum was the C_{60} peak, about twice as tall as its neighbors.

What is now justly regarded as the discovery of C_{60} did not come until nearly a year later. Smalley's group at Rice University was visited by Harry Kroto from Sussex, who had long pursued work on carbon-containing interstellar molecules. Kroto wanted to examine carbon clusters and their reactions with other molecules, in hopes of identifying candidates for unassigned interstellar spectra. Smalley was reluctant to interrupt other work, particularly since vaporizing carbon would make the apparatus very dirty. Fortunately, hospitality and willing graduate students prevailed. On repeating the Exxon work, the Rice group found that by varying conditions, the C_{60} peak became far more prominent. That led them to play with models and propose as an explanation the celebrated soccer-ball structure, dubbed Buckminsterfullerene. It contains 12 pentagonal and 20 hexagonal carbon rings, with all sixty atoms symmetrically equivalent and linked to three neighbors by two single and one double bond. Soon other fullerene cage molecules were recognized, differing from C_{60} by addition or subtraction of hexagonal rings, in accord with a theorem proved in the 18th century by Euler.

These elegant structures, postulated to account for cluster mass spectra, remained unconfirmed for five years. Then, in 1990, it was not chemists but astrophysicists who found a way to extract C_{60} in quantity from soot produced in an electric arc discharge. As well as enabling structural proofs, that opened up to synthetic chemistry and materials science a vast new domain of molecular structures, built with a form of carbon that has sixty valences rather than just four. It is striking, however, that despite the great stability of C_{60} and its self-assembly after laser ablation or arc discharge of graphite, as yet all efforts to synthesize C_{60} by conventional chemical means have failed. Such means, which operate under thermodynamic equilibrium conditions, evidently cannot access facile reaction pathways.

Among the families akin to fullerene molecules are carbon nanotubes, first discovered by Sumio Iijima at NEC Fundamental Research Laboratories in Tsukuba, Japan. Particularly intriguing is a single-walled nanotube (designated 10,10) of the same diameter as C_{60} (7.1Å). In principle, its chicken-wire pattern of hexagons and pentagons can be extended indefinitely; in practice, nanotubes of this kind have now been made which contain millions of carbon atoms in a single molecule. The electrical conductivity of this hollow carbon tube is comparable to copper, and it forms fibers 100 times stronger than steel but with only one-sixth the weight. Already carbon nanotubes have provided much enhanced performance as probe tips in atomic force microscopy. Chemically modifying the nanotube tips has even been shown to create the capability of chemical and biological discrimination at the molecular level, in effect directly reading molecular braille [28]. A host of other applications is in prospect.

5.2 Reactions and Spectra in Clusters

Much current work examines the effect of solvation on reaction dynamics or spectra by depositing solute reactants or "guests" on a cluster of solvent or "host" molecules, bound by van der Waals forces or by hydrogen bonds [29]. Often photoinduced reactions, particularly involving electron or proton transfer, are studied in this way, or processes involving ion-molecule reactions within clusters [25], including "cage" effects due to the solvent. Among many variants is work colliding high-velocity clusters with metal or crystal surfaces. Such collisions can induce even guest species that are ordinarily inhibited by a high activation barrier to react [30]. Here we consider a quite different special realm, employing spectra of a guest molecule to study clusters that are finite quantum fluids.

In the prototype experiment, shown in Figure 5, a supersonic expansion generates large He or Ar clusters, each with 10^3 to 10^5 atoms. These pick up in flight one or more small guest molecules while passing through a gas cell, without suffering appreciable attenuation or deflection. The cluster beam is probed downstream by a laser, coaxial or transverse to the flight path,

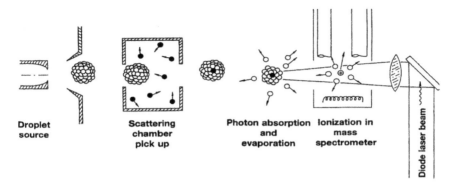

Figure 5: Schematic diagram (omitting vacuum pumps!) of molecular beam apparatus for depletion spectroscopy of molecules embedded in helium clusters. The large clusters or droplets of He, formed in a nozzle, pick up a guest molecule while passing thorough a scattering chamber. En route to the mass spectometric detector, the cluster is irradiated by a tunable, coaxial laser. When the laser is in resonance with the guest molecule, the absorbed energy induces evaporation from the cluster and a depletion in the mass spectrometer signal. From [32].

and spectroscopic transitions of the guest molecule are detected by laser-induced fluorescence or by beam depletion. This pickup technique, originally developed by Giacinto Scoles [31], is well suited to the study of unstable or highly reactive chemical species. Indeed, these may be synthesized *in situ* by using more than one pickup cell, or introducing into the cell species generated in a discharge or pyrolytic decomposition. Instead, a stable guest molecule can serve to probe the environment within its solvent cluster. In this way, the group of Peter Toennies at Göttingen [32,33] has recently obtained striking results for superfluid helium clusters.

These ^{4}He clusters are produced by a strong supersonic expansion (*e.g.*, He at 5 bar pressure and 6.6 °K behind a 5 mm nozzle), which drops the internal temperature to about 0.4 °K, well below the transition temperature for superfluidity ($T_\lambda = 2.12$ °K). When a hot guest molecule comes aboard (from the pickup cell at 10^{-5} mbar and ~ 300 °K), the cluster rapidly evaporates away a few hundred He atoms, thereby cooling itself and the guest to the original internal temperature (in about 10^{-6} sec). In the process, the guest molecule also migrates from the surface to the center of the cluster (as deduced from mass spectroscopic experiments and predicted by theory). For a variety of guest species, among them the linear triatomic molecule OCS, the Göttingen group found spectra with well resolved rotational structure. This indicates the guest molecules are rotating freely within the clusters, although with an effective moment of inertia larger (by a factor of 2.7 for OCS) than that for an isolated, gas-phase molecule.

In normal liquids, rotational structure in spectroscopic transitions is destroyed by diffusional and librational processes; free rotation is seen only for light, weakly interacting molecules such as H_2 or CH_4. Free rotation within the ^{4}He clusters thus can plausibly be attributed to superfluidity, but it might instead result from the exceptionally cool and feeble guest-host interactions, further blurred by the large zero-point oscillations of the helium atoms.

As a diagnostic test for the role of superfluidity, the experiments were repeated using ^{3}He clusters. These have lower density, so are more weakly interacting and somewhat colder (about 0.15 °K), although far above the superfluid range ($T_\lambda = 3 \times 10^{-3}$°K). Indeed, in ^{3}He clusters the OCS spectrum shows no rotational structure, but rather has the broad, featureless form typical for heavy molecules in normal liquids. A further elegant test was obtained in a series of runs made with increasing amounts of ^{4}He added to the OCS vapor in the pickup cell. Because of their high diffusivity and lower zero-point energy, the ^{4}He atoms entering a ^{3}He cluster gather around the guest molecule. The average number of such friendly ^{4}He atoms that are picked up was estimated from a Poisson distribution. When this number reached about 60, the rotational structure in the OCS spectrum had again grown in, just as sharp as for pure ^{4}He clusters.

Thus, in the pure, nonsuperfluid ^{3}He clusters the guest molecule does not rotate freely, but it does so in pure, superfluid ^{4}He clusters or when surrounded by about 60 atoms of ^{4}He, enough to form about two shells around the OCS molecule. Rough estimates suggest the increase in effective moment of inertia may be due chiefly to dragging along the vestigial normal fluid component of these shells. The Göttingen experiments offer strong evidence that a sharp guest rotational spectrum is diagnostic for superfluidity and it can occur even in ^{4}He clusters with as few as 60 atoms.

6 Liberating Catalysts from Thermodynamic Constraints

A predilection for supersonic molecular beams recently led my laboratory to an excursion into surface chemistry. This was motivated by the notion that simply fashioning the nozzle from a catalytic material might markedly change the consequences of catalytic processes. According to the traditional textbook definition, a catalyst speeds (or retards) the attainment of chemical equilibrium. Under supersonic conditions, however, the gas flow is far from equilibrium and the contact time with the nozzle surface can be quite short ($\sim$ 1–10 msec). The catalytic process then can deliver a host of product species that would not appear in appreciable yield under equilibrium conditions. Detection of such products is also facilitated, as shortly downstream

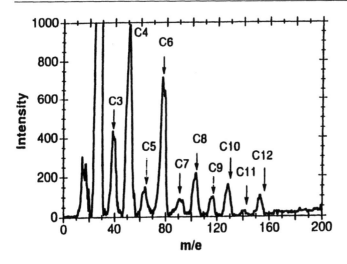

Figure 6: Low resolution mass spectrum of products from Ni catalyzed reaction of ethane (1060 °C, 85 Torr) expanded from supersonic nozzle. Product peaks for $C_n H_m$ are labeled by n, the number of carbon atoms. Arrows indicate masses for $m = n$ (located at $n \times 13$ mass units); thus $m = n$ up to $n = 8$ but the most probable m is lower by 1 to 3 units for higher n. From [34].

from the nozzle the products travel serenely in a collisionless molecular beam.

In exploratory experiments [34], we indeed found that a dramatically different regime becomes readily accessible by expanding a reactive gas through a pinhole nozzle made of a catalytic metal. Thus, on flowing ethane gas at 80 torr through a nickel nozzle heated to 1000 °C, about 40% is converted to higher hydrocarbons. Mass spectra, as illustrated in Figure 6, show these hydrocarbons contain from three to 12 or more carbon atoms and a range of hydrogen atoms, mostly equal or near the number of carbon atoms. The six-carbon product appears to be largely or solely benzene, formed in about 15% yield. In contrast, at thermodynamic equilibrium ethane at 1000 °C would decompose entirely, to form graphite on the metal surface and hydrogen in the gas phase.

The strong gas flow, in addition to imposing short contact time, fosters desorption into the gas-phase, thereby further restricting the interaction of reactive intermediates or products with the surface. Much or most of the association reactions producing the remarkably facile formation of higher hydrocarbons must occur in the supersonic gas; a predominance of even-carbon species suggests a kinship with the mechanism for formation of fullerenes. Replacing ethane by ethylene or by C_3 or C_4 alkanes or alkenes enhanced the conversion to higher hydrocarbons, with up to about 45% yield of benzene. Disappointingly, with methane no appreciable net conversion was observed, although runs with CD_4 mixed with C_2 reactants gave extensive scrambling of deuterium atoms among the products.

This catalytic nozzle technique has now been improved by Romm and Somerjai, with striking results [35]. In similar experiments with metal nozzles, they found that preconditioning the catalytic surface with a hydrogen flow enhanced its effectiveness. They obtained nearly 100% conversion of ethane at 1000 °C. The major products included methane, acetylene, and ethylene as well as benzene and higher hydrocarbons; as the ethane source pressure was increased (from 150 to 670 torr), the benzene yield increased (from 15 to 25%) whereas that of the C_2 products declined and that of methane remained unchanged (at 5%). With methane as reactant, they found no detectable conversion using metal nozzles. Remarkably, however, they found a quartz nozzle just as effective as the metal nozzles for ethane and acetylene conversion. Moreover, with the quartz nozzle operated at 1150 °C, they obtained for methane 20% conversion, to yield about 4% each of acetylene, C_3H_3, C_4H_n and benzene. This opens a promising avenue in the quest for practical means to attain methane activation.

7 Pursuit of Forbidden Fruit

As Otto Stern liked to emphasize, the great appeal of the molecular beam method is its conceptual simplicity and directness. However, no matter how much care is devoted to defining or analyzing directions, velocities, and internal states of the collision partners, the dynamical resolution of beam scattering experiments is limited by random aspects of the initial conditions. Even if the reactant molecules are oriented (by means discussed in Sect. 8), there remains the "dartboard" distribution of impact parameters in the collision. Dynamical information obscured by averaging over the random orientations of impact parameters was long thought to be irretrievable. Yet in principle and in practice such "forbidden fruit" can be harvested [36].

Figure 7 indicates how the averaging over impact parameter can be undone [3, 4]. The impact parameter **b** denotes the closest distance of approach of the collision partners, in a hypothetical encounter with the forces switched off. Both the magnitude of **b** and its azimuthal orientation about the initial relative velocity vector **k** are random, but here we are concerned just about the azimuthal orientation. The magnitude of **b** is in effect observable, since it enters directly into many collision properties. For instance, a head-on collision has $b = 0$ and the particles scatter at wide angles; a grazing collision has b large compared with the range of the forces, so the particles scatter at small angles. However, the random azimuthal distribution of impact parameters makes the final relative velocity vector **k'** azimuthally

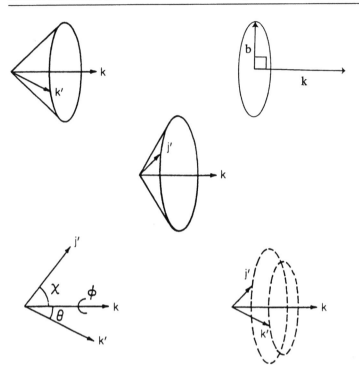

Figure 7: Three-vector correlation among initial and final relative velocity vectors, denoted by **k** and **k'**, and product rotational angular momentum vector **j'**. Upper pair of diagrams indicate the azimuthal symmetry about **k** of the **k'** and **j'** vectors inherent when these are observed separately, as in the two-vector correlations (**k, k'**) and (**k, j'**). Lower pair of diagrams indicates how the three-vector correlation (**k, k', j'**) can give information about the dihedral angle ϕ, in effect undoing the azimuthal averaging about the initial relative velocity. From [3].

symmetric about the initial relative velocity velocity vector **k**, although the forces acting in the collision do not usually have such symmetry (unless both partners are spherical). In the same way, darts thrown at a board exhibit azimuthal symmetry even when the thrower has astigmatism. Likewise, the distribution of any other vector property of the collision such as the final rotational angular momentum **j'** of a product molecule must be azimuthally symmetric about **k**. The redeeming strategy is to measure two product vectors such as **k'** and **j'** simultaneously and thereby to determine the *triple-vector correlation* among **k**, **k'**, and **j'**.

When a subset is selected of **k'** vectors with particular **j'** (or vice versa), this subset in general will not have azimuthal symmetry about **k**. Accordingly, the dihedral angle ϕ between the **k**, **k**′ and **k**, **j**′ planes need not be uniformly distributed. Evaluating the distribution of ϕ recoups information about the reaction dynamics that otherwise would be lost by the azimuthal average over impact parameters.

As a chemical analogy, consider a helium atom in its ground state. The distribution of each electron, viewed separately, is spherically symmetric. But if the two electrons are observed simultaneously, their positions are strongly correlated as a consequence of their repulsive interaction. If not known beforehand, the presence of this repulsion would be revealed by the simultaneous observation, although not by viewing the electrons singly.

Although it is convenient to use the product rotational angular momentum **j**′ as the "extra" quantity in the triple-vector correlation, this poses an instructive question. According to quantum mechanics, only the magnitude and one projection of an angular momentum vector can be specified. Since we envision a measurement that specifies both the magnitude of **j**′ and the polar angle χ between **j**′ and **k**, the azimuthal angle about **k** should be unobservable. Is not the dihedral angle ϕ therefore unobservable after all? That is so, for any particular measurement. But this can be circumvented by measuring the angular momentum distribution using several different choices for the axis of quantization. The data can then be combined to obtain moments of the ϕ distribution as well as those of χ and the scattering angle θ between the **k** and **k**′ velocity vectors. The classical version of the three-vector correlation hence can be resurrected, including its ϕ dependence [3]. In this sense, quantum mechanics in effect allows a dihedral angle such as ϕ to be observed even when the uncertainty principle does not permit this in any particular measurement.

This undoing of impact parameter averaging in molecular collisions has some heuristic resemblance to the celebrated "phase problem" encountered in x-ray crystallography. Both are resolved by introducing additional observables that provide new reference points, and by combining data from variant experiments.

An experimental demonstration [36] of the utility of the triple-vector correlation was made in 1974, showing that **j**′ was not azimuthally symmetric about **k** but rather strongly aligned (preferred $\phi \sim 90°$) perpendicular to **k** × **k**′. Our method harked back to Otto Stern, determining the alignment of **j**′ by means of the interaction of an inhomogeneous electric deflecting field (analogous to the Stern–Gerlach magnet) with the dipole moment of the product molecule. Theoretical aspects of vector correlations were also examined by my group, including models for impulsive [36, 37] and statistical [38, 39] dynamics, trajectory calculations illustrating the use of the ϕ-distribution as a diagnostic property [40], and analysis of how moments of the triple-vector correlation could be extracted from laser induced fluorescence experiments [41]. Despite evangelical efforts [1–4, 42], renewed here, vector correlation analysis has not yet been widely employed in chemical reaction dynamics. Recently, however, a laser technique has shown **j'** to be perpendicular to **k** × **k'** for HCl from the Cl + CH$_4$ reaction [43] and further theoretical aspects have been explored [44].

8 Bringing Molecules to Attention

In typical collisional or spectroscopic experiments, molecules are not only randomly oriented but rotate freely, imposing isotropic averaging of interactions. Hopes of avoiding this entropic curse have motivated many efforts to develop means to restrict molecular rotation, reviewed elsewhere [45]. Here we consider two widely applicable molecular beam methods; one produces alignment of rotational angular momentum, the other orientation or alignment of a molecular axis. Both methods exploit the drastic rotational cooling attainable in supersonic expansions. Note that the anisotropy created by *alignment* behaves like a double-headed arrow (↔), and that created by *orientation* behaves like a single-headed arrow (→).

Alignment of rotational angular momentum by macroscopic transport of nonspherical molecules had been established by the 1960s, although in bulk media the observable effects are small. In supersonic expansions, such alignment via collisions seemed likely to be appreciable. But before 1989, it appeared to be negligibly small, except for dimer molecules formed as a small fraction of an alkali atom beam. This led to suspicion that collisional alignment depended on velocity slip between beam components (appreciable for dimers, intrinsically seeded in the alkali beam). In experiments provoked by this suspicion, we indeed found quite large alignment occurs for iodine molecules seeded in a supersonic beam of a light carrier gas [45]. As in bulk transport, the rotational angular momentum is typically aligned perpendicular to the beam flow velocity. Under suitable conditions, however, we found alignment parallel to the flow could be obtained, indicating a substantial role for anisotropic rotational cooling when the temperature becomes very low. Subsequent work, especially in the laboratories of Vincento Aquilanti, David Nesbitt, and Alex Wodtke, likewise found marked alignment for other molecules in seeded beams; this now appears as the rule rather than the exception. The ability to obtain rotational alignment is useful, for instance in contrasting the interaction with a surface of molecules rotating parallel or perpendicular to the surface.

Of more chemical interest is the ability to align or orient a molecular axis (rather than just the rotational momentum), as required for studies of the stereodynamics of gas phase collisions. To attain this for typical pinwheeling molecules, the tumbling rotational motion must be quenched, in order that the molecular axis remain confined to a restricted range of directions. A generally applicable means to arrest molecular tumbling did not emerge until a decade ago. In the meanwhile, the only method for producing beams of oriented molecules, developed in the late 1960s, exploited the fact that symmetric top molecules (for which two of the three moments of inertia are equal) in certain rotational states precess rather than tumble. As a result, the dipole moment does not average out but rather has a constant projection on a space-fixed axis, giving rise to a first-order Stark effect. State-selection by an inhomogeneous electric focusing field thus suffices to pick out molecules with substantial intrinsic orientation of the figure axis. This is an excellent method; it has enabled incisive studies of "head vs. tail" reaction probabilities in collisions with both gas molecules and surfaces. However, besides its limitation to symmetric tops, the method requires an elaborate apparatus.

A different method, much wider in chemical scope and far simpler to implement, became feasible by exploiting the low rotational temperatures attainable in supersonic beams. This method, introduced in 1990, uses a strong homogeneous electric or magnetic field to create oriented or aligned states of polar or paramagnetic molecules [46, 47]. In the presence of the field, the eigenstates become coherent linear superpositions or hybrids of the field-free rotational states. These hybrids coincide with the familiar second-order Stark or Zeeman states when the dipole and/or the moment of inertia is small or the field is weak; then the molecule continues to tumble like a pinwheel. When the interaction is sufficiently strong, however, the hybrids become librational; then the molecule swings to and fro about the field direction like a pendulum. Such pendular states can be produced for linear or asymmetric rotors as well as symmetric tops. The magnetic version produces alignment rather than orientation, but is applicable to many molecules not accessible to the electric version; this includes paramagnetic nonpolar molecules and molecular ions (which would just crash into an electrode if subjected to an electric field). Either version requires that the interaction of the molecular dipole with the external field exceeds the kinetic energy of tumbling, hence the key role of drastic rotational cooling by a supersonic expansion.

The experimental simplification is major because a focusing field (typically a meter long and expensive to fabricate) is not needed. Instead, the molecular beam is merely sent between the plates of a small condenser (usually about $1\,\mathrm{cm}^2$ in area and a few mm apart) or between the pole pieces of a compact magnet. The uniform field which creates the hybrid eigenstates need only extend over the small region in which the beam actually interacts with its target.

A kindred variety of pendular states can be obtained by utilizing the induced dipole moment created by nonresonant interaction of intense laser radiation with the molecular polarizability [48]. This can produce alignment whether the molecule is polar, paramagnetic, or neither, as long as the polarizability is anisotropic. That is generally the case; *e.g.* for a linear molecule the polarizability is typically about twice as large along the axis as transverse

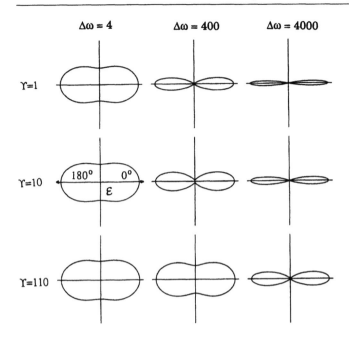

Figure 8: Polar plots of ensemble-averaged angular distributions produced by laser-induced alignment via interaction with the polarizability of a linear molecule. The distributions are governed by two dimensionless parameters: $\Delta\omega$ is proportional to the anisotropy of the polarizability and the laser field strength, Υ to the gas temperature; both are inversely proportional to the molecular rotational constant. From [48].

to it. Although the electric field of the laser rapidly switches direction, since the interaction with the induced dipole is governed by the square of the field strength, the direction of the aligning force experienced by the molecule remains the same. With a pulsed laser, the field strength can readily be made high enough to attain strong alignment. Figure 8 illustrates theoretical estimates [48]; comparably sharp alignment has been obtained experimentally [49].

The polarizability anisotropy does not distinguish between the two ends of a molecule, whether or not they differ. Therefore even for a heteropolar molecule such as ICl the induced dipole is the same in both directions and the interaction with the laser creates a symmetric double-well potential. The energy levels associated with the librational motion of an aligned molecule thus are split by tunneling between the two potential wells. The tunneling rate can be varied over many orders of magnitude by scanning the laser intensity over a modest range [50]. In a molecular beam experiment, it should be feasible to observe directly the rate of interconversion of "orientational isomers" by such angular tunneling; the process would be akin to the interconversion of enantiomeric molecules. In many cases, the tunneling splitting will be very small for the lowest pair of pendular energy levels. If the molecule is polar, introducing in addition a static electric field, congruent with the laser field, connects the nearly degenerate tunneling doublets, which have opposite parity. Even a weak static field can thus produce a strong pseudo-first-order Stark effect. Almost any polar molecule, linear or asymmetric, can thereby be made to act like a symmetric top for the duration of the laser pulse. This method can be exploited in schemes to control and manipulate molecular trajectories [51].

Of special interest is the possibility of aligning or orienting molecules with respect to more than one lab-fixed axis. This could be done by a static field orienting a permanent dipole moment while a linearly polarized laser field aligns an induced dipole perpendicular to the permanent moment [20]. A more elegant method, recently demonstrated, uses an intense, elliptically polarized laser field to simultaneously align all three principle axes of a molecule [21], thereby exploiting fully the anisotropy of the polarizability tensor.

Pendular states make accessible many stereodynamical properties. Studies of steric effects in inelastic collisions or chemical reactions are a chief application [46]. The ability to turn the molecular orientation on or off enables modulation of angular distributions and other collision properties, thereby revealing anisotropic interactions not otherwise observable. In photodissociation of oriented molecules [52] pendular hybridization renders the laboratory photofragment distributions much more informative. For all applications, the spectroscopy of pendular states has an important role, as the field dependence of suitable transitions reveals the extent of molecular orientation or alignment. Conversely, the anisotropic distribution of the molecular axis imposes a corresponding anisotropy in the spatial distribution of transition moments, so alters the polarization of transmitted light. This has led to a powerful form of polarization spectroscopy with dramatic resolution and sensitivity [53]; moreover, because pendular hybridization is most effective for low J states, the spectra obtained for a room temperature gas become quite sparse, simplifed to an extent otherwise attainable only by cooling to a few degrees kelvin. Other features arising from the hybrid character of pendular states also prove valuable in spectroscopy, including the ability to tune transitions over a wide frequency range and to access states forbidden by the field-free selection rules.

9 Slowing Molecules to Make Nanomatter Waves

The advent of powerful methods for cooling, trapping, and manipulating neutral atoms with laser light has led to dramatic achievements [54]. With atomic vapors cooled to the microkelvin range, Bose–Einstein condensation has been attained [55] as well as hugely nonlinear optical effects [56], and an atom laser, atom interferometry, and atom lithography, all exploiting coherent matter waves [57]. To pursue such phenomena with molecules is an appealing prospect, as molecules offer a vast range of properties not available with atoms. In collisions involving very long deBroglie wavelengths, energy transfer processes and chemical reaction dynamics with "nanomatterwaves" will have distinctive features, including prominent tunneling and resonances [5].

However, for molecules it is much more difficult to reach this hyperquantum domain than for atoms. The forces available to manipulate the trajectories or spatially confine neutral particles are weak, and inversely proportional to their translational kinetic energy. A key requisite for trapping thus a means to lower markedly the kinetic energy. Typically, the initial confinement requires the kinetic energy be below 1 °K; if trapped in sufficient quantity, the particles may then be cooled to the microkelvin range by distilling away the warmer components [58]. Slowing alkali atoms is readily accomplished by repeated absorption and stimulated emission to transfer momentum in a laser-induced "optical molasses". Such a method fails for molecules because they have myriad vibrational and rotational levels, so even a dilute optical syrup is not attainable.

In the past two years, three elegant means of slowing molecules have achieved trapping [59]. One method creates alkali dimer molecules within an alkali atom trap by photoassociation [60]. The dimers are obtained at microkelvin temperatures, but the yield is very small. Another technique employs collisional relaxation by ^{3}He buffer-gas, maintained by a dilution refrigerator at about 0.3 °K, to load a magnetic trap [61]. The third method decelerates polar molecules by means of multiple stages of time-varying electric field gradients, bringing them down to kinetic energies well below 0.3 K and confining them in an electrostatic trap [62]. Although for the latter two methods the trapping times attained thus far are brief ($\sim$ 0.2 sec) and conditions otherwise not amenable to cooling to the microkelvin range, the results justify optimism. Here we note a few other promising approaches to manipulating, slowing, and trapping molecules, some not yet demonstrated experimentally.

In addition to its utility for molecular alignment, the polarizability interaction with an intense, directional laser field provides a lensing effect acting on the translational motion of molecules [63]. The interaction with the field produces molecular states, "high-field seekers", whose energy levels decrease as the field increases. This generates a force that moves the molecule towards the spatial region of highest laser intensity. Thereby focusing occurs, as demonstrated by sending a molecular beam through the intensity gradient near the focus of a laser beam [64]. The extent of focusing depends on the ratio between the molecular translational kinetic energy and the maximum attractive field-induced potential. In the strong interaction limit, where that ratio becomes much less than unity, the molecules can become trapped in the laser field [48].

Friedrich has recently proposed a versatile method to slow and trap molecules by means of the nonresonant polarizability interaction [65]. His method envisions use of a laser beam steered by a scanning device with variable angular speed and a pulsed supersonic beam which precools the molecules internally and narrows their velocity spread. High-field seeking molecules would be scooped from the supersonic jet by the sweeping laser beam, decelerated by gradually reducing the angular velocity of the laser, and deposited in the resonator of a build-up cavity or other type of trap.

Another versatile and relatively simple method, demonstated in preliminary experiments in my laboratory, relies on mechanical means to produce intense beams of slow molecules [66]. This involves mounting a supersonic nozzle near the tip of a high-speed rotor, with peripheral velocity high enough to cancel the flow velocity of the beam. The high-speed rotor also functions as a gas centrifuge, thereby enhancing the supersonic character of the gas flow and further reducing the temperature within the beam. The current apparatus is far from optimal, but for example has reduced the kinetic energy of an O_2 beam (seeded in Xe) in the lab frame to below $10\,°K$ (from $170\,°K$ for the stationary seeded beam). Improvements are in prospect, such as stronger pumping to enable use of higher source pressures and to avoid attenuation of the slow molecules by background gas. Already the technique is capable of enhancing many experiments that call for slow molecules. The apparatus required is compact (rotor of length 10 cm, driven within vacuum by a small motor), and inexpensive, as it does not call for cryogenics or lasers. Indeed, Otto Stern or Philip Moon might have operated something similar, had slow molecules been wanted in their day.

An electrostatic storage ring for polar molecules has been proposed [67], modeled on a neutron storage ring. This would employ an inhomogeneous hexapolar toroidal field, within which molecules in "low-field seeking" states would be confined and follow orbits determined by their rotational state and translational velocity. Design calculations limited to practical parameters indicate that storage lifetimes of the order of 10^3–10^4 s can be expected. Since the molecular trajectories must bend to stay in the ring, only molecules with low translational kinetic energy can be stored, but such devices are well suited to harvest molecules slowed by either the laser scoop or rotating nozzle source.

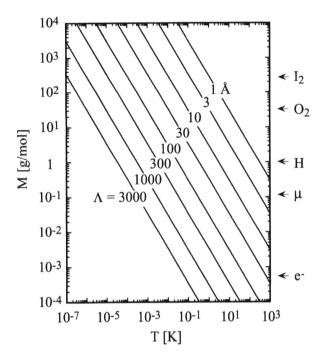

Figure 9: Nomogram for thermal deBroglie wavelength Λ (in Å units) as a function of molecular mass (g/mol) and temperature (K), with $\Lambda = 17.4/(MT)^{1/2}$. From [5].

Figure 9 gives a nomogram displaying the deBroglie wavelength Λ (in Å units) corresponding to the average thermal momentum for molecules of mass M (in g/mol) at thermal equilibrium at temperature T (in °K). Thus, at room temperature a hydrogen atom has a thermal deBroglie wavelength of about one Angstrom unit (0.1 nanometers in current textbook parlance). But at 10 millikelvin, an iodine molecule has $\Lambda = 10$ Å and at 1 microkelvin, $\Lambda = 1000$ Å. The latter is about four times longer than the wavelength of a room temperature electron. Thus, at such low temperatures, collisions of even fairly heavy molecules exhibit very marked quantum effects.

At first blush, ultracold temperatures would seem to preclude any appreciable chemical reaction. Even a modest activation energy would greatly exceed the available thermal collision energy. However, when the deBroglie wavelength for relative motion becomes large, a reactive encounter becomes something like the coupling of amoebas. The familiar concept of a potential energy surface still pertains, but it is no longer appropriate to think of particle trajectories traversing it; rather what matters is how the surface slows or speeds the progress of surging waves. If as usual the potential surface has a substantial activation barrier, in the hyperquantum regime tunneling becomes the dominant reaction pathway. When Λ is much larger than the thickness of the barrier, the rate constant is governed by a threshold law derived by Wigner [68]. Accordingly, for an exothermic bimolecular reaction, below some low temperature, reaction occurs entirely by tunneling, and the rate does not change with a further drop in temperature. If the barrier thickness for the reactants in their ground internal states is denoted by D and the reduced mass for the collision partners by m, the criterion $\Lambda \gg D$ requires $T \ll 300/mD^2$. A generous estimate is $D \sim 10$ Å. Hence, to attain the Wigner limit the temperature needs to be well below one degree Kelvin for a hydrogen atom reaction and about a hundredfold lower for heavier reactants like chlorine molecules.

At ordinary temperatures, tunnel effects are often appreciable for hydrogen atom reactions. That is even more so for reactions of muonium; indeed, for the Mu + F_2 reaction nice evidence has been found for a transition towards the Wigner threshold tunneling in the gas phase [69]. The low-temperature limiting behavior expected when tunneling is dominant has also been seen for several reaction processes in the solid phase [70]. Accordingly, at least for exothermic bimolecular reactions with activation barriers, such results indicate the general nature of the tunneling behavior that can be anticipated in an ultracold molecular trap. For the study of tunneling processes, the trap environment offers a great advantage, however, since the imprisoned molecules can keep colliding with each other until all have undergone the tunneling reaction.

10   Laser control of reaction pathways

In "real world" chemistry, reaction pathways and yields have to be cajoled or conjured, by adjusting macroscopic conditions (temperature, concentrations, pH) or catalysts. Chemical physicists have sought genuine control of molecular pathways [71]. For bimolecular reactions, this has been done by varying the collision energy, orientation, or vibrational excitation of reactant molecules, or selecting particular alignments of excited electronic orbitals of atoms. For unimolecular processes, which we consider briefly, control has been achieved by utilizing the coherence of laser light [72]. This enables the outcome to be governed by quantum interference arising from the phase difference between alternate routes or by the temporal shape and spectral content of ultrashort light pulses.

The method employing phase control, first proposed by Brumer and Shapiro [73], offers a molecular analogue of Young's two-slit experiment. An upper state of a molecule is simultaneously excited by two lasers, of frequencies ω_n and ω_m, absorbing n photons of one color and m of the other, with $n\omega_m = m\omega_n$. This prepares a superposition of continuum eigenstates,

$\Psi_n + \Psi_m$, that correlate asymptotically with different product channels. The coefficients of the components of this superposition state are determined by the relative phases and amplitudes of the two lasers. Since the wavelengths differ markedly, the light waves do not interfere, but the wave functions produced by them strongly interfere. The cross term in $\langle \Psi_n + \Psi_m \rangle^2$ thus governs the product distribution and the outcome can be controlled by varying the relative phases and amplitudes of the lasers. This scheme has received several experimental demonstrations in which product yields from two competing channels exhibit large modulations, 180° out of phase.

Another method, operating in the time domain, was introduced by Tannor and Rice [74]. This employs a sequence of ultrashort light pulses to create a wave packet of molecular eigenstates. The frequencies, amplitudes and phases of the pulse sequence are tailored to favor a particular outcome as the wave packet evolves in time. By means of optimal control theory, the pulse shape and spectral content can be adjusted to maximize the yield of a desired product [75, 76]. Moreover, with iterative feedback, the experimenter can be taught by the molecule how best to tailor the light pulses [77]. In 1997 such molecular instruction, aided by a computer-controlled pulse shaper was first reported [78], and other such studies have rapidly followed [79]. This work fulfills a quest ardently pursued by the late Kent Wilson [80], to conduct "conversations" with molecules and thereby learn how to negotiate a desired chemical choreography of atomic motions. Ancient alchemists sought in vain a philosopher's stone; modern chemical physicists have now found its equivalent in a laser beam.

Acknowledgments

I am grateful for the opportunity, in company with extremely able students and colleagues, to pursue research with molecular beams for more than 40 years, captivated by the conceptual "simplicity and directness" extolled by Otto Stern. Our work has been supported mainly by the National Science Foundation. It is a special pleasure to applaud the contributions presented in this volume as well as to thank Roger Campargue for his invitation to prepare this appetizer menu and his patience and advice that had much to do with shaping and seasoning it.

References

[1] D. R. Herschbach, Faraday Disc. Chem. Soc. **55**, 233 (1973).

[2] D. R. Herschbach, Angew. Chem. Int'l. Ed. Engl. **26**, 1221 (1987).

[3] D. R. Herschbach, Faraday Disc. Chem. Soc. **84**, 465 (1987).

[4] D. R. Herschbach, in *The Chemical Bond*, edited by A. Zewail (Academic Press, New York, 1992), p. 175.

[5] D. R. Herschbach, in *Chemical Research–2000 and Beyond: Challenges and Visions*, edited by P. Barkan, (ACS Books/Oxford Univ. Press, New York, Oxford, 1998) p.113.

[6] D. Herschbach, Rev. Mod. Phys. **71**, S411 (1999).

[7] N. F. Ramsey, *Molecular Beams* (Oxford University Press, New York, Oxford, 1956). See also the Nobel e-Museum on the web at www.nobel.se/.

[8] B. Friedrich and D. Herschbach, Daedalus **127**, 165 (1998).

[9] C. H. Townes, *How the Laser Happened*, (Oxford University Press, New York, Oxford, 1999).

[10] S. J. Riley, P. E. Siska, and D. R. Herschbach, Chem. Soc. Faraday Disc. **67**, 27 (1979).

[11] P. B. Moon, C. T. Rettner, and J. P. Simons, Faraday Disc. **77**, 630 (1977).

[12] R. D. Levine and R. B. Bernstein, *Molecular Reaction Dynamics*, (2nd Ed., Oxford Univ. Press, New York, Oxford, 1987).

[13] J. B. Fenn, Ann. Rev. Phys. Chem. **47**, 1 (1996).

[14] W. R. Gentry and C. F. Giese, Rev. Sci. Instrum. **49**, 595 (1978).

[15] R. Campargue, J. Phys. Chem. **88**, 4466 (1984).

[16] R. E. Smalley, L. Wharton, and D. H. Levy, J. Chem. Phys. **63**, 4977 (1975); D. H. Levy, Annu. Rev. Phys. Chem. **31**, 197 (1980).

[17] X. Shi, D. R. Herschbach, D. R. Worsnop, and C. E. Kolb, J. Phys. Chem. **97**, 2113 (1993).

[18] D. R. Herschbach, C. E. Kolb, D. R. Worsnop, and X. Shi, Nature **356**, 414 (1992).

[19] S. Joo, D. R. Worsnop, C. E. Kolb, S. K. Kim, and D. R. Herschbach, J. Phys. Chem. **103**, 3193 (1999).

[20] B. Friedrich and D. R. Herschbach, Phys. Chem. Chem. Phys. **2**, 419 (2000).

[21] J. J. Larsen, K. Hald, N. Bjerre, H. Stapelfeldt, and T. Seideman, Phys. Rev. Lett. **85**, 2470 (2000).

[22] P. Thaddeus, M. C. McCarthy, M. J. Travers, C. A. Gottlieb, and W. Chen, Faraday Discuss. **109**, 1 (1998).

[23] W. Klemperer, Ann. Rev. Phys. Chem. **46**, 1 (1995); Proceedings of the Royal Institution, London, 209 (1997).

[24] B. H. Mahan, Accts. Chem. Res. **8**, 55 (1975).

[25] A. W. Castleman and K. H. Bowen, J. Phys. Chem. **100**, 12911 (1996).

[26] M. S. Dresselhaus, G. Dresselhaus, and P. C. Eklund, *Science of Fullerenes and Carbon Nanotubes* (Academic Press, New York, 1996).

[27] E. A. Rohlfing, D. M. Cox, and A. Kaldor, J. Chem. Phys. **81**, 3322 (1984).

[28] S. S. Wong, E. Joselevich, A. T. Woolley, C. L. Cheung, and C. M. Lieber, Nature **394**, 52 (1998).

[29] J. M. Mestdagh, M. A. Gaveau, C. Gee, O. Sublemontier, and J. P. Visticot, Int'l Rev. Phys. Chem. **16**, 215 (1997).

[30] T. Raz and R. D. Levine., J. Phys. Chem. **99**, 7495 (1995).

[31] K. K. Lehmann and G. Scoles, Science **279**, 2065 (1998).

[32] M. Hartmann, R. E. Miller, J. P. Toennies, and A. F. Vilesov, Science **272**, 1631 (1996).

[33] S. Grebenev, J. P. Toennies, and A. F. Vilesov, Science **279**, 2083 (1998).

[34] L. Shebaro, S. R. Bhalotra, and D. R. Herschbach, J. Phys. Chem. **101**, 6775 (1997).

[35] L. Romm and G. A. Somorjai, Catalysis Lett. **64**, 85 (2000).

[36] D. S. Y. Hsu, G. M. McClelland, and D. R. Herschbach, J. Chem. Phys. **61**, 4927 (1974).

[37] G. M. McClelland and D. R. Herschbach, J. Phys. Chem. **91**, 5509 (1987).

[38] D. A. Case and D. R. Herschbach, Mol. Phys. **30**, 1537 (1975).

[39] J. D. Barnwell, J. G. Loeser, and D. R. Herschbach, J. Phys. Chem. **87**, 2781 (1983).

[40] S. K. Kim and D. R. Herschbach, Faraday Disc. Chem. Soc. **84**, 159 (1988).

[41] D. A. Case, G. M. McClelland, and D. R. Herschbach, Mol. Phys. **35**, 541 (1978).

[42] R. B. Bernstein, D. R. Herschbach, and R. D. Levine, J. Phys. Chem. **91**, 5365 (1987).

[43] A. J. Orr-Ewing, W. R. Simpson, T. P. Rakitzis, S. A. Kandel, and R. N. Zare, J. Chem. Phys. **106**, 5961 (1997); **107**, 9382, 9392 (1997).

[44] M. P. de Miranda, D. C. Clary, J. F. Castillo, and D. E. Manolopoulos, J. Chem. Phys. **108**, 3142 (1998).

[45] B. Friedrich, D. P. Pullman, and D. R. Herschbach, J. Phys. Chem. **95**, 8118 (1991).

[46] H. J. Loesch, Ann. Revs. Phys. Chem. **46**, 555 (1995).

[47] B. Friedrich and D. Herschbach, Comments At. Mol. Phys. **32**, 47 (1995); Int'l. Revs. Phys. Chem. **15**, 325 (1996).

[48] B. Friedrich and D. Herschbach, Phys. Rev. Lett. **74**, 4623 (1995); J. Phys. Chem. **99**, 15686 (1995).

[49] J. J. Larsen, H. Sakai, C. P. Safvan, I. Wendt-Larsen, and H. Stapelfeldt, J. Chem. Phys. **111**, 7774 (1999).

[50] B. Friedrich and D. Herschbach, Z. Phys. D **36**, 221 (1996).

[51] B. Friedrich and D. Herschbach, J. Phys. Chem. A **103**, 10280 (1999).

[52] D. T. Moore, L. Oudejans, and R. E. Miller, J. Chem. Phys. **110**, 197 (1999).

[53] A. Slenczka, J. Phys. Chem. A **101**, 7657 (1997); Phys. Rev. Lett. **80**, 2566 (1998); Chem. Eur. J. **5**, 1006, 1136 (1999).

[54] S. Chu, Rev. Mod. Phys. **70**, 685 (1998); C. N. Cohen-Tannoudji, ibid. **70**, 707, (1998); W. D. Phillips, ibid. **70**, 721 (1998).

[55] D. G. Fried, T. C. Killian, L. Willmann, D. Candhuis, S. C. Moss, D. Kleppner, and T. J. Greytak, Phys. Rev. Lett. **81**, 3811 (1998).

[56] L. Hau, S. E. Harris, Z. Dutton, and C. H. Behroozi, Nature **397**, 594 (1999).

[57] C. E. Wieman, D. E. Pritchard, and D. Wineland, Rev. Mod. Phys. **71**, S263 (1999).

[58] H. J. Metcalf and P. van der Straten, *Laser Cooling and Trapping* (Springer, New York, 1999).

[59] J. M. Doyle and B. Friedrich, Nature **401**, 749 (1999).

[60] R. Wynar, R. S. Freeland, D. J. Han, C. Tyu, and D. J. Heinzen, Science **287**, 1016 (2000).

[61] J. D. Weinstein, R. deCarvalho, T. Guillet, B. Friedrich, and J. M. Doyle, Nature **395**, 148 (1998).

[62] H. L. Bethlem, G. Berden, F. M. H. Crompvoets, R. T. Jongma, A. J. A. van Roij, and G. Meijer, Nature **406**, 491 (2000).

[63] T. Seideman, J. Chem. Phys. **106**, 2881 (1996); **107**, 10420 (1997).

[64] H. Stapelfeldt, H. Sakai, E. Constant, and P. B. Corkum, Phys. Rev. Lett. **79**, 2787 (1997).

[65] B. Friedrich, Phys. Rev. A **61**, 025403 (2000).

[66] M. Gupta and D. Herschbach, J. Phys. Chem. A **103**, 10670 (1999).

[67] D. P. Katz, J. Chem. Phys. **107**, 8491 (1997).

[68] T. Takayanagi, N. Masaki, K. Nakamura, M. Okamoto, S. Sata, and G. C. Schatz, J. Chem. Phys. **86**, 6133 (1987); **90**, 1641 (1989).

[69] S. Baer, D. Fleming, D. Arseneau, M. Senba, and A. Gonzalez, in *Isotope Effects in Gas-Phase Chemistry*, edited by J. A. Kaye (ACS Symposium Series No. 502, American Chemical Society, Washington, D. C., 1992), p. 111.

[70] V. A. Benderskii, D. E. Makarov, C. A. Wight, *Chemical Dynamics at Low Temperatures* (John Wiley & Sons, New York, 1994).

[71] R. N. Zare, Science **279**, 1875 (1998).

[72] R. J. Gordon and S. A. Rice, Ann. Rev. Phys. Chem. **48**, 601 (1997).

[73] P. Brumer and M. Shapiro, Chem. Phys. Lett. **126**, 541 (1986); J. Chem. Soc. Faraday Trans. **93**, 1263 (1997).

[74] D. J. Tannor and S. A. Rice, J. Chem. Phys. **83**, 5013 (1985).

[75] R. Kosloff, S. A. Rice, P. Gaspard, S. Tersigni, and D. J. Tannor, Chem. Phys. **139**, 201 (1989).

[76] A. P. Peirce, M. A. Dahleh, and H. Rabitz, Phys. Rev. A **37**, 4950 (1988); **42**, 1065 (1990).

[77] R. S. Judson and H. Rabitz, Phys. Rev. Lett. **68**, 1500 (1992).

[78] C. J. Bardeen, V. V. Yakovlev, K. R. Wilson, S. D. Carpenter, P. M. Weber, and W. S. Warren, Chem. Phys. Lett. **289**, 151 (1997).

[79] A. Assion, T. Baumert, M. Bergt, T. Brixner, B. Kiefer, V. Seyfried, M. Strehle, and G. Gerber, Science **282**, 919 (1998).

[80] D. Herschbach, Nature **405**, 902 (2000).

Part I

Laser Cooling and Manipulation of Atoms and Molecules-Atomic and Molecular Optics and Applications

Foreword: Laser Cooling and Trapping of Neutral Atoms

J. Dalibard and C. Cohen-Tannoudji,
Laboratoire Kastler Brossel*and Collège de France,
24 rue Lhomond, 75005 Paris, France

1 Introduction

Electromagnetic interactions can be used to act on atoms, to manipulate them, to control their various degrees of freedom. With the development of laser sources, this research field has considerably expanded during the last few years. Methods have been developed to trap atoms and to cool them to very low temperatures. The purpose of this paper is to review the physical processes which are at the basis of this research field, and to present some of its most remarkable applications [1].

Two types of degrees of freedom can be considered for an atom: the internal degrees of freedom, such as the electronic configuration or the spin polarization, in the center of mass system; the external degrees of freedom, which are essentially the position and the momentum of the center of mass. The manipulation of internal degrees of freedom goes back to optical pumping [2], which uses resonant exchanges of angular momentum between atoms and circularly polarized light for polarizing the spins of these atoms. These experiments predate the use of lasers in atomic physics. The manipulation of external degrees of freedom uses the concept of radiative forces resulting from the exchanges of linear momentum between atoms and light. Radiative forces exerted by the light coming from the sun were already invoked by J. Kepler to explain the tails of the comets. Although they are very small when one uses ordinary light sources, these forces were also investigated experimentally in the beginning of this century by P. Lebedev, E. F. Nichols and G. F. Hull, R. Frisch [3].

It turns out that there is a strong interplay between the dynamics of internal and external degrees of freedom. This is at the origin of efficient laser cooling mechanisms, such as *Sisyphus cooling* or *Velocity Selective Coherent Population Trapping*, which were discovered at the end of the 80's (for a historical survey of these developments, see for example [4]). These mecha-

---

*Unité de Recherche de l'Ecole Normale Supérieure et de l'Université Pierre et Marie Curie, associée au CNRS.

nisms have allowed laser cooling to overcome important fundamental limits, such as the *Doppler limit* and the *single photon recoil limit*, and to reach the microKelvin, and even the nanoKelvin range.

This paper is organized as follows. In section 2 we briefly review the basic processes which enter into play when an atom modelled as a two-level system interacts with a monochromatic laser beam. Then, in section 3, we introduce the idea of Doppler cooling, and of magneto-optical and dipole trapping. The notions of sub-Doppler and sub-recoil cooling are presented respectively in sections 4 and 5. In section 6 we briefly outline the complementary technique of evaporative cooling and we present its most remarkable achievement, namely the realization of Bose-Einstein condensates with dilute atomic gases. Finally in section 7, we present another very promising application of laser cooled atoms, leading to a spectacular improvement of the performances of atomic clocks.

2 Interaction of a Two-Level Atom with a Quasi-Resonant Light Beam

To classify the basic physical processes which are used for manipulating atoms by light, it is useful to distinguish two large categories of effects : dissipative (or absorptive) effects on the one hand, reactive (or dispersive) effects on the other hand. This partition is relevant for both internal and external degrees of freedom.

Consider first a light beam with frequency ω_L propagating through a medium consisting of atoms with resonance frequency ω_A. The index of refraction describing this propagation has an imaginary part and a real part which are associated with two types of physical processes. The incident photons can be absorbed, more precisely scattered in all directions. The corresponding attenuation of the light beam is maximum at resonance. It is described by the imaginary part of the index of refraction which varies with $\omega_L - \omega_A$ as a Lorentz absorption curve. We will call such an effect a dissipative (or absorptive) effect. The speed of propagation of light is also modified. The corresponding dispersion is described by the real part n of the index of refraction whose difference from 1, $n - 1$, varies with $\omega_L - \omega_A$ as a Lorentz dispersion curve. We will call such an effect a reactive (or dispersive) effect. Dissipative effects and reactive effects also appear for the atoms, as a result of their interaction with photons. Consider for simplicity an atom sitting in point **r**, which can be modelled by a two-level system involving a ground state g and an excited state e, with a radiative lifetime Γ^{-1} (see figure 1).

For relatively low laser intensities, these dissipative and reactive effects can be understood as a broadening and a shift of the atomic ground state. The broadening Γ' is the rate at which photons are scattered from the incident beam by the atom. The shift $\hbar\Delta'$ is the energy displacement of the ground

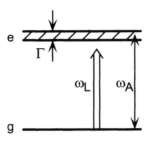

Figure 1: Modelling of an atomic transition in terms of a two-level system.

level, as a result of virtual absorptions and stimulated emissions of photons by the atom within the light beam mode; it is called light-shift or AC Stark shift. The expressions for Γ' and Δ' can be derived in various ways, using either the optical Bloch equation formalism or the dressed atom approach [5]. We simply give here the results, as functions of the local intensity $I(\mathbf{r})$ of the light beam, the saturation intensity I_s, and the detuning $\Delta = \omega_L - \omega_A$:

$$\Gamma'(\mathbf{r}) = \Gamma \frac{I(\mathbf{r})}{2I_s} \frac{1}{1 + 4\Delta^2/\Gamma^2} \qquad \Delta'(\mathbf{r}) = \Delta \frac{I(\mathbf{r})}{2I_s} \frac{1}{1 + 4\Delta^2/\Gamma^2} \qquad (1)$$

These expressions are valid as long as Γ' and Δ' are respectively small compared with Γ and Δ. The saturation intensity for a typical atomic resonance transition is quite low, in regard of the intensities achievable with usual laser sources. For instance, for the resonance line of sodium atoms, one gets $I_s = 6$ mW cm^{-2}.

Both Γ' and Δ' are proportional to the light intensity. They vary with the detuning as Lorentz absorption and dispersion curves, respectively, which justifies the denominations absorptive and dispersive used for these two types of effects. For large detunings ($|\Delta| \gg \Gamma$), Γ' varies as $1/\Delta^2$ and becomes negligible compared to Δ' which varies as $1/\Delta$. On the other hand, for small detunings, ($|\Delta| \ll \Gamma$), Γ' is much larger than Δ'. In the high intensity limit, the expressions (1) are not valid anymore, and the physical understanding of the atom-laser interaction is more subtle [5]. Let us simply indicate that in this case the atomic scattering rate saturates to the value $\Gamma/2$, indicating that the average population of the atomic excited state reaches $1/2$.

3 Radiative Forces on a Two-Level Atom

3.1 The two types of radiative forces

There are two types of radiative forces, associated respectively with dissipative and reactive effects.

Dissipative forces, also called radiation pressure forces or scattering forces, are associated with the transfer of linear momentum from the incident light beam to the atom in resonant scattering processes. They are proportional to the scattering rate Γ'. Consider for example an atom in a laser plane wave with wave vector $\mathbf{k}$. Because photons are scattered with equal probabilities in two opposite directions, the mean momentum transferred to the atom in an absorption-spontaneous emission cycle is equal to the momentum $\hbar \mathbf{k}$ of the absorbed photon. The mean rate of momentum transfer, *i.e.* the mean force, is thus equal to $\hbar \mathbf{k} \Gamma'$. Since Γ' saturates to $\Gamma/2$ at high intensity, the radiation pressure force saturates to $\hbar \mathbf{k} \Gamma/2$. The corresponding acceleration (or deceleration) which can be communicated to an atom with mass M, is equal to $a_{\max} = \hbar k \Gamma / 2M = v_R / 2\tau$, where $v_R = \hbar k/M$ is the recoil velocity of the atom absorbing or emitting a single photon, and $\tau = 1/\Gamma$ is the radiative lifetime of the excited state. For sodium atoms, $v_R = 3 \times 10^{-2} \mathrm{m\,s^{-1}}$ and $\tau = 1.6 \times 10^{-9}$s, so that $a_{\max}$ can reach values as large as $10^6 \mathrm{m\,s^{-2}}$, *i.e.* $10^5 g$ where g is the acceleration due to gravity. With such a force, one can stop a thermal atomic beam in a distance of the order of one meter, provided that one compensates for the Doppler shift of the decelerating atom, by using for example a spatially varying Zeeman shift [6] or a chirped laser frequency [7].

Dispersive forces, also called dipole forces or gradient forces, can be interpreted in terms of position dependent light shifts $\hbar \Delta'(\mathbf{r})$ due to a spatially varying light intensity [8]. Consider for example a laser beam well detuned from resonance, so that one can neglect Γ' (no scattering process). The atom thus remains in the ground state and the light shift $\hbar \Delta'(\mathbf{r})$ of this state plays the role of a potential energy, giving rise to a force which is equal and opposite to its gradient : $\mathbf{F} = -\nabla[\hbar \Delta'(\mathbf{r})]$. Such a force can also be interpreted as resulting from a redistribution of photons between the various plane waves forming the laser wave in absorption-stimulated emission cycles. If the detuning is not large enough to allow Γ' to be neglected, spontaneous transitions occur between dressed states having opposite gradients, so that the instantaneous force oscillates back and forth between two opposite values in a random way. Such a dressed atom picture provides a simple interpretation of the mean value and of the fluctuations of dipole forces [9].

3.2 Doppler cooling and magneto-optical trapping

The principle of Doppler cooling has been suggested by T.W. Hänsch and A.L. Schawlow [10] for neutral atoms, and by D. Wineland and H. Dehmelt [11] for trapped ions. In the proposal [10] this cooling results from a Doppler induced imbalance between two opposite radiation pressure forces. The two counterpropagating laser waves have the same (weak) intensity and the same frequency. They are slightly detuned to the red of the atomic

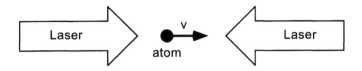

Figure 2: Doppler cooling in 1D, resulting from the imbalance between the radiation pressure forces of two counterpropagating laser waves. The laser detuning is negative ($\omega_L < \omega_A$).

frequency ($\omega_L < \omega_A$) (see fig. 2). For an atom at rest, the two radiation pressure forces exactly balance each other and the net force is equal to zero. For a moving atom, the apparent frequencies of the two laser waves are Doppler shifted. The counterpropagating wave gets closer to resonance and exerts a stronger radiation pressure force than the copropagating wave which gets farther from resonance. The net force is thus opposite to the atomic velocity v and can be written for small v as $F = -\alpha v$ where α is a friction coefficient. By using three pairs of counterpropagating laser waves along three orthogonal directions, one can damp the atomic velocity in a very short time, on the order of a few microseconds, achieving what is called an *optical molasses* [12].

The Doppler friction responsible for the cooling is necessarily accompanied by fluctuations due to the fluorescence photons which are spontaneously emitted in random directions and at random times. These photons communicate to the atom a random recoil momentum $\hbar k$, responsible for a momentum diffusion described by a diffusion coefficient D. As in usual Brownian motion, competition between friction and diffusion usually leads to a steady-state, with an equilibrium temperature proportional to D/α. The theory of Doppler cooling [13, 14, 15] predicts that the equilibrium temperature obtained with such a scheme is always larger than a certain limit T_D, called the Doppler limit, and given by $k_B T_D = \hbar\Gamma/2$ where Γ is the natural width of the excited state and k_B the Boltzmann constant. This limit, which is reached for $\Delta = \omega_L - \omega_A = -\Gamma/2$, is, for alkali atoms, on the order of 100 μK. In fact, when the measurements became precise enough, it appeared that the temperature in optical molasses was much lower than expected [16]. This indicates that other laser cooling mechanisms, more powerful than Doppler cooling, are operating. We will come back to this point in the next section.

The imbalance between two opposite radiation pressure forces can be also made position dependent though a spatially dependent Zeeman shift produced by a magnetic field gradient. In a one-dimensional configuration, initially proposed by one of us (J.D.), the two counterpropagating waves, which are detuned to the red ($\omega_L < \omega_A$) and which have opposite circular polarizations are in resonance with the atom at different places. This

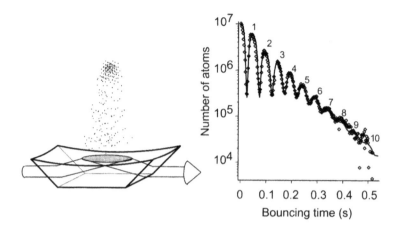

Figure 3: Gravitational cavity for neutral atoms (from [26]). (a) Trampoline for atoms: atoms released from a magneto-optical trap bounce off a concave mirror formed by a blue detuned evanescent wave ($\omega_L > \omega_A$), propagating at the surface of a glass prism. (b) Number of atoms at the apex of the average trajectory, as a function of the time after the magneto-optical trap has been switched off.

results in a restoring force towards the point where the magnetic field vanishes. Furthermore the non zero value of the detuning provides a Doppler cooling. Such a scheme can be extended to three dimensions and leads to a robust, large and deep trap called *magneto-optical trap* (MOT) [17]. It combines trapping and cooling, it has a large velocity capture range and it can be used for trapping atoms in a small cell filled with a low pressure vapour [18].

3.3 Dipole traps and atomic mirrors

When the detuning is negative ($\omega_L < \omega_A$), light shifts are negative. If the laser beam is focussed, the focal zone where the intensity is maximum appears as a minimum of potential energy, forming a potential well where sufficiently cold atoms can be trapped [19, 20, 21, 22].

If the detuning is positive, light shifts are positive and can thus be used to produce potential barriers. For example an evanescent blue detuned wave at the surface of a piece of glass can prevent slow atoms impinging on the glass surface from touching the wall, making them bounce off a "carpet of light" [23]. This is the principle of mirrors for atoms. Plane atomic mirrors [24, 25] have been realized as well as concave mirrors [26] (see fig. 3).

4 Sub-Doppler Cooling

In the previous section, we discussed separately the manipulation of internal and external degrees of freedom, and we have described physical mechanisms involving only one type of physical effect, either dispersive or dissipative. In fact, there exist cooling mechanisms resulting from an interplay between spin and external degrees of freedom, and between dispersive and dissipative effects. We discuss in this section one of them, the so-called "Sisyphus cooling" or "polarization-gradient cooling" mechanism [27, 28], which leads to temperatures much lower than Doppler cooling. One can understand in this way the sub-Doppler temperatures observed in optical molasses [16].

4.1 The Sisyphus mechanism

Most atoms, in particular alkali atoms, have a Zeeman structure in the ground state. Since the detuning used in laser cooling experiments is not too large compared to Γ, both differential light shifts and optical pumping transitions exist for the various Zeeman sublevels of the ground state. Furthermore, the laser polarization varies in general in space so that light shifts and optical pumping rates are position-dependent. We show now, with a simple one-dimensional example, how the combination of these various effects can lead to a very efficient cooling mechanism.

Consider the laser configuration of Fig. 4, consisting of two counterpropagating plane waves along the z-axis, with orthogonal linear polarizations and with the same frequency and the same intensity. Because the phase shift between the two waves varies linearly with z, the polarization of the total field changes from σ^+ to σ^- and vice versa every $\lambda/4$. In between, it is elliptical or linear.

Consider now the simple case where the atomic ground state has an angular momentum $J_g = 1/2$. The two Zeeman sublevels $M_g = \pm 1/2$ undergo different light shifts, depending on the laser polarization, so that the Zeeman degeneracy in zero magnetic field is removed. This gives the energy diagram of Fig. 4 showing spatial modulations of the Zeeman splitting between the two sublevels with a period $\lambda/2$.

If the detuning Δ is not too large compared to Γ, there are also real absorptions of photons by the atom followed by spontaneous emission, which give rise to optical pumping transfers between the two sublevels, whose direction depends on the polarization: $M_g = -1/2 \longrightarrow M_g = +1/2$ for a σ^+ polarization, $M_g = +1/2 \longrightarrow M_g = -1/2$ for a σ^- polarization. Here also, the spatial modulation of the laser polarization results in a spatial modulation of the optical pumping rates with a period $\lambda/2$.

The two spatial modulations of light shifts and optical pumping rates are of course correlated because they are due to the same cause, the spatial modulation of the light polarization. These correlations clearly appear in

Fig. 4. With the proper sign of the detuning, optical pumping always transfers atoms from the higher Zeeman sublevel to the lower one. Suppose now that the atom is moving to the right, starting from the bottom of a valley, for example in the state $M_g = +1/2$ at a place where the polarization is σ^+. Because of the finite value of the optical pumping time, there is a time lag between internal and external variables and the atom can climb up the potential hill before absorbing a photon and reach the top of the hill where it has the maximum probability to be optically pumped in the other sublevel, i.e. in the bottom of a valley, and so on.

Like Sisyphus in the Greek mythology, who was always rolling a stone up the slope, the atom is running up potential hills more frequently than down. When it climbs a potential hill, its kinetic energy is transformed into potential energy. Dissipation then occurs by light, since the spontaneously emitted photon has an energy higher than the absorbed laser photon. After each Sisyphus cycle, the total energy E of the atom decreases by an amount of the order of U_0, where U_0 is the depth of the optical potential wells of Fig. 4. When E becomes smaller than U_0, the atom remains trapped in the potential wells.

4.2 The limit of Sisyphus cooling

The previous discussion shows that Sisyphus cooling leads to temperatures T_{Sis} such that $k_B T_{\text{Sis}} \simeq U_0$. According to Eq. (4), the light shift U_0 is proportional to I/Δ. Such a dependence of T_{Sis} on the laser intensity and on the detuning has been checked experimentally [29].

At low intensity, the light shift is much smaller than $\hbar\Gamma$. This explains why Sisyphus cooling leads to temperatures much lower than those achievable with Doppler cooling. One cannot however decrease indefinitely the laser intensity. The previous discussion ignores the recoil due to the spontaneously emitted photons which increase the kinetic energy of the atom by an amount on the order of E_R, where

$$E_R = \hbar^2 k^2 / 2M \qquad (2)$$

is the recoil energy of an atom absorbing or emitting a single photon. When U_0 becomes on the order or smaller than E_R, the cooling due to Sisyphus cooling becomes weaker than the heating due to the recoil, and Sisyphus cooling no longer works. This shows that the lowest temperatures which can be achieved with such a scheme are on the order of a few E_R/k_B, which is on the order of a few microKelvins for rubidium or cesium atoms. This result is confirmed by a full quantum theory of Sisyphus cooling [30] and is in good agreement with experimental results.

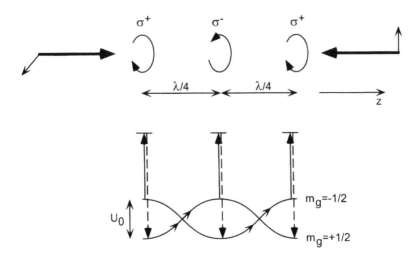

Figure 4: One dimensional Sisyphus cooling. The laser configuration is formed by two counterpropagating waves along the z axis with orthogonal linear polarizations. The polarization of the resulting field is spatially modulated with a period $\lambda/2$. For an atom with two ground Zeeman sublevels $M_g = \pm 1/2$, the spatial modulation of the laser polarization results in correlated spatial modulations of the light shifts of these two sublevels and of the optical pumping rates between them. Because of these correlations, a moving atom runs up potential hills more frequently than down.

4.3 Optical lattices

For the optimal conditions of Sisyphus cooling, atoms become so cold that they get trapped in the quantum vibrational levels of a potential well. More precisely, one must consider energy bands in this perodic structure [31]. Experimental observation of such a quantization of atomic motion in an optical potential was first achieved in one dimension [32, 33]. Atoms are trapped in a spatial periodic array of potential wells, called a *1D–optical lattice*, with an antiferromagnetic order, since two adjacent potential wells correspond to opposite spin polarizations. 2D and 3D optical lattices have been realized subsequently (see the review papers [34, 35]).

5 Sub-Recoil Cooling

5.1 How to circumvent the one photon recoil limit

In most laser cooling schemes, fluorescence cycles never cease. Since the random recoil $\hbar k$ communicated to the atom by the spontaneously emitted

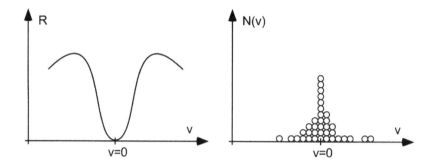

Figure 5: Sub recoil cooling. The random walk in velocity space is characterized by a jump rate vanishing in $v = 0$. As a result, atoms which fall in a small interval around $v = 0$ remain trapped there for a long time.

photons cannot be controlled, it seems impossible to reduce the atomic momentum spread δp below a value corresponding to the photon momentum $\hbar k$. The condition $\delta p = \hbar k$ defines the "single photon recoil limit", the recoil temperature being set as $k_B T_R/2 = E_R$. The value of T_R ranges from a few hundred nanoKelvin for alkalis to a few microKelvin for metastable helium. It is in fact possible to circumvent this limit and to reach temperatures T lower than T_R, a regime called *subrecoil laser cooling*. The basic idea is to create a situation where the photon absorption rate Γ', which is also the jump rate R of the atomic random walk in velocity space, depends on the atomic velocity $v = p/M$ and vanishes for $v = 0$ (Fig. 5). Consider then an atom with $v = 0$. For such an atom, the absorption of light is quenched. Consequently, there is no spontaneous reemission and no associated random recoil. One protects in this way ultracold atoms (with $v \simeq 0$) from the "bad" effects of the light. On the other hand, atoms with $v \neq 0$ can absorb and reemit light. In such absorption-spontaneous emission cycles, their velocities change in a random way and the corresponding random walk in $v-$ space can transfer atoms from the $v \neq 0$ absorbing states into the $v \simeq 0$ dark states where they remain trapped and accumulate.

Up to now, two subrecoil cooling schemes have been proposed and demonstrated. In the first one, called *Velocity Selective Coherent Population Trapping* (VSCPT), the vanishing of $R(v)$ for $v = 0$ is achieved by using destructive quantum interference between different absorption amplitudes [36]. The second one, called Raman cooling, uses appropriate sequences of stimulated Raman and optical pumping pulses for tailoring the appropriate shape of $R(v)$ [37].

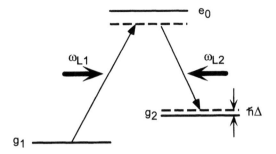

Figure 6: Coherent population trapping. A three level atom is driven by two laser fields When the detuning Δ of the stimulated Raman process is zero, the atom is optically pumped into a linear superposition of g_1 and g_2, which no longer absorbs light.

5.2 Brief survey of VSCPT

We first recall the principle of the quenching of absorption by *coherent population trapping* [38]. Consider the 3-level system of Fig. 6, with two ground state sublevels g_1 and g_2 and one excited sublevel e_0, driven by two laser fields with frequencies ω_{L1} and ω_{L2}, exciting the transitions $g_1 \leftrightarrow e_0$ and $g_2 \leftrightarrow e_0$, respectively. Let $\hbar\Delta$ be the detuning from resonance for the stimulated Raman process consisting of the absorption of one photon ω_{L1} and the stimulated emission of one ω_{L2} photon, the atom going from g_1 to g_2. One observes that the fluorescence rate R vanishes for $\Delta = 0$. In this case indeed, atoms are optically pumped into a linear superposition of g_1 and g_2 which is not coupled to e_0 because of a destructive interference between the two absorption amplitudes $g_1 \to e_0$ and $g_2 \to e_0$.

The basic idea of VSCPT is to use the Doppler effect for making the detuning Δ of the stimulated Raman process of Fig. 6 proportional to the atomic velocity v. The quenching of absorption by coherent population trapping is thus made velocity dependent and one achieves the situation of Fig. 5. This is obtained by taking the two laser waves ω_{L1} and ω_{L2} counterpropagating along the z-axis and by choosing their frequencies in such a way that $\Delta = 0$ for an atom at rest. Then, for an atom moving with a velocity v along the z-axis, the opposite Doppler shifts of the two laser waves result in a Raman detuning $\Delta = (k_1 + k_2)v$ proportional to v.

A more quantitative analysis of the cooling process [39] shows that the dark state, for which $R = 0$, is a linear superposition of two states which differ not only by the internal state (g_1 or g_2) but also by the momentum along the z-axis:

$$|\psi_D\rangle = c_1 |g_1, -\hbar k_1\rangle + c_2 |g_2, +\hbar k_2\rangle \qquad (3)$$

This is due to the fact that g_1 and g_2 must be associated with different

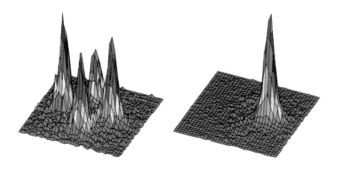

Figure 7: Velocity selective coherent population trapping (VSCPT) in two dimensions with metastable helium atoms (from [44]). The velocity distribution on the left has been obtained using four laser beams in a horizontal plane, giving rise to four peaks located at $(v_x = \pm v_R, v_y = 0)$ and $(v_x = 0, v_y = \pm v_R)$. When three of the four VSCPT beams are adiabatically switched off (right figure), the whole atomic population is transferred into a single wave packet. For a better visibility the vertical scale of the left figure has been expanded by a factor 4 with respect to the right one.

momenta, $-\hbar k_1$ and $+\hbar k_2$, in order to be coupled to the same excited state $|e_0, p = 0\rangle$ by absorption of photons with different momenta $+\hbar k_1$ and $-\hbar k_2$. Furthermore, when $\Delta = 0$, the state (5) is a stationary state of the total atom + laser photons system. As a result of the cooling by VSCPT, the atomic momentum distribution thus exhibits two sharp peaks, centered at $-\hbar k_1$ and $+\hbar k_2$, with a width δp which tends to zero when the interaction time θ tends to infinity. Experimentally, temperatures as low as $T_R/800$ have been observed [40].

VSCPT has been extended to two [41] and three [42] dimensions. For a $J_g = 1 \leftrightarrow J_e = 1$ transition, it has been shown [43] that there is a dark state which is described by the same vector field as the laser field. More precisely, if the laser field is formed by a linear superposition of N plane waves with wave vectors $\mathbf{k}_i$ ($i = 1, 2, ...N$) having the same modulus k, one finds that atoms are cooled in a coherent superposition of N wave packets with mean momenta $\hbar \mathbf{k}_i$ and with a momentum spread δp which becomes smaller and smaller as the interaction time θ increases. Furthermore, because of the isomorphism between the de Broglie dark state and the laser field, one can adiabatically change the laser configuration and transfer the whole atomic population into a single wave packet chosen at will (Fig. 7).

5.3 Subrecoil laser cooling and Lévy statistics

Quantum Monte Carlo simulations using the delay function [45, 46] have provided new physical insight into subrecoil laser cooling [47]. They show that the random walk of the atom in velocity space is anomalous and dominated by a few rare events whose duration is a significant fraction of the total interaction time. More precisely one can proof that the distribution $P(\tau)$ of the trapping times τ in a small trapping zone near $v = 0$ is a broad distribution which falls as a power-law in the wings. These wings decrease so slowly that the average value $\langle \tau \rangle$ of τ (or the variance) can diverge. In such cases, the central limit theorem (CLT) can obviously no longer be used for studying the distribution of the total trapping time after N entries in the trapping zone separated by N exits.

It is possible to extend the CLT to broad distributions with power-law wings [48]. We have applied the corresponding statistics, called *Lévy statistics*, to subrecoil cooling and shown that one can obtain in this way a better understanding of the physical processes as well as quantitative analytical predictions for the asymptotic properties of the cooled atoms in the limit when the interaction time θ tends to infinity [47]. For example, one predicts in this way that the temperature decreases as $1/\theta$ when $\theta \to \infty$, and that the wings of the momentum distribution decrease as $1/p^2$, which shows that the shape of the momentum distribution is closer to a Lorentzian than a Gaussian. This is in agreement with experimental observations [40].

One important feature revealed by this theoretical analysis is the non ergodicity of the cooling process. Regardless of the interaction time θ, there are always atomic evolution times (trapping times in the small zone of Fig. 5 around $v = 0$) which can be longer than θ. Experimental evidence for non ergodic effects has recently been obtained [49]. Another advantage of such a new approach is that it allows the parameters of the cooling lasers to be optimized for given experimental conditions. For example, by using different shapes for the laser pulses used in one-dimensional subrecoil Raman cooling, it has been possible to reach for Cesium atoms temperatures as low as 3 nK [50].

6 Evaporative Cooling and Bose–Einstein Condensation

A spectacular use of cold atom technology has been the demonstration in 1995 of Bose–Einstein condensation (BEC) in a dilute atomic gas. The first gaseous Bose–Einstein condensate has been obtained with rubidium atoms (^{87}Rb) [51]; subsequently BEC has also been demonstrated with sodium [52], lithium [53] and hydrogen [54]. A detailed discussion of this very lively field of research is outside the scope of the present paper, and we refer the interested reader to the reference [55]. In the following we simply outline the basic features of BEC for atomic gases.

BEC has been initially predicted by Einstein in 1925, who was considering an ideal gas of indistinguishable material particles. If the density n of the

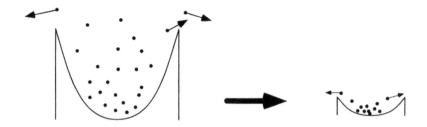

Figure 8: Principle of forced evaporative cooling. Atoms are confined in a truncated potential well (in practice a magnetic trap), whose depth is a few times larger than the thermal energy $k_B T$. The fastest atoms are ejected, while the remaining atoms thermalize to a lower temperature. The truncation value is decreased accordingly, to maintain a constant evaporation rate.

gas is larger than the critical value

$$n_c \simeq \frac{0,166}{\hbar^3}(mk_B T)^{3/2} \qquad (4)$$

one expects that a macroscopic fraction of atoms condense in the quantum ground state of the box confining the gas. For a long period, the only experimental example of a Bose-Einstein condensate has been liquid helium, below the superfluid transition temperature. However this dense system is far from Einstein's ideal gas model. For instance due to the interactions within the liquid, the condensed fraction never exceeds 10% of the atoms, while it should reach 100% at zero temperature for an ideal gas. Therefore numerous efforts have been made to achieve BEC with dilute systems, closer to the initial model.

Laser cooled and trapped atoms were *a priori* good candidates for observing the condensation phenomenon. Unfortunately, it has not been possible to increase the atomic density in these samples to the value given in Eq. (4). Up to now, light-assisted inelastic collisions between laser cooled atoms have limited the density to a small fraction only of n_c. As a consequence the realization of BEC with Li, Na and Rb uses laser cooling and trapping only as first step. The atoms are then transferred into a magnetic trap, formed around a magnetic field minimum, which confines the atoms whose magnetic moment is anti-parallel with the local magnetic field. To increase further the density and decrease the temperature, one then proceeds with forced evaporative cooling.

Evaporative cooling consists in using a truncated confining magnetic potential, so that the fastest atoms are ejected from the trap (see fig. 8). Due to elastic collisions, the remaining atoms reach a lower temperature. In practice the truncation of the potential is chosen 5 to 6 times larger than

(a)　　　　　　　　　　　　(b)

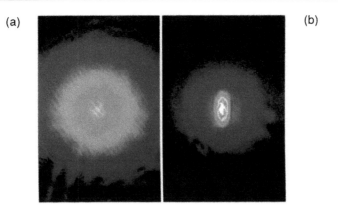

Figure 9: Velocity distributions of a ^{87}Rb gas. Both pictures represent the absorption of a probe laser beam by the cloud after a 31 ms free fall. (a) Temperature above condensation: isotropic velocity distribution. (b) Temperature below condensation: the anisotropic central feature corresponds to a macroscopically occupied ground state (figure obtained at ENS by P. Desbiolles, D. Guéry-Odelin and J. Söding).

the instantaneous thermal energy $k_B T$, and one lowers this truncation continuously as the remaining atoms get colder. Typically a reduction of the temperature by a factor 1000, and an increase of the density by a factor 30 is obtained through the evaporation of 99.9% of the atoms. One starts with $\sim 10^9$ atoms at a temperature of ~ 1 mK, and ends up at the condensation point with 10^6 atoms at 1 μK (see Fig. 9).

Since the first realization of an atomic BEC, there has been numerous experimental and theoretical studies of these systems: phase coherence properties, superfluidity, excitation spectra, non linear atom optics (see [55] for a review). In direct relation with the topic of this book, pulsed diffraction limited and coherent atomic beams have been generated, using these condensates as a source [56]. These beams are often referred to as "atom lasers", to emphasize the analogy between them and the coherent light beams issued from a laser device. There is no doubt that these coherent atomic sources will play a crucial role in future precision experiments dealing with atom optics and interferometry, metrology, or nanolithography.

7 Cold atom clocks

Cesium atoms cooled by Sisyphus cooling have an effective temperature on the order of 1 μK, corresponding to a r.m.s. velocity of 1 cm s^{-1}. This allows them to spend a longer time T in an observation zone where a microwave field induces resonant transitions between the two hyperfine levels g_1 and g_2 of the ground state. Increasing T decreases the width $\Delta \nu \sim 1/T$ of the microwave resonance line whose frequency is used to define the unit of time. The stability of atomic clocks can thus be considerably improved by using ultracold atoms [57, 58].

In usual atomic clocks, atoms from a thermal cesium beam cross two microwave cavities fed by the same oscillator. The average velocity of the atoms is several hundred m s^{-1}, the distance between the two cavities is on the order of 1 m. The microwave resonance between g_1 and g_2 is monitored and is used to lock the frequency of the oscillator to the center of the atomic line. The narrower the resonance line, the more stable the atomic clock. In fact, the microwave resonance line exhibits Ramsey interference fringes whose width $\Delta \nu$ is determined by the time of flight T of the atoms from one cavity to another. For the longest devices, T, which can be considered as the observation time, can reach 10 ms, leading to values of $\Delta \nu \sim 1/T$ on the order of 100 Hz.

Much narrower Ramsey fringes, with sub-Hertz linewidths can be obtained in the so-called Zacharias *atomic fountain* [59]. Atoms are captured in a magneto-optical trap and laser cooled before being launched upwards by a laser pulse through a microwave cavity. Because of gravity they are decelerated, they return and fall back, passing a second time through the cavity. Atoms therefore experience two coherent microwave pulses, when they pass through the cavity, the first time on their way up, the second time on their way down. The time interval between the two pulses can now be on the order of 1 sec, *i.e.* about two order of magnitudes longer than with usual clocks. Atomic fountains have been realized for sodium [60] and cesium [61]. A short-term relative frequency stability of $4 \times 10^{-14} \tau^{-1/2}$, where τ is the integration time, has been recently measured for a one meter high Cesium fountain [62]. This stability reaches now the fundamental quantum noise induced by the measurement process: it varies as $N^{-1/2}$, where N is the number of detected atoms. The long term stability of 6×10^{-16} is most likely limited by the Hydrogen maser which is used as a reference source. The real fountain stability, which will be more precisely determined by beating the signals of two fountain clocks, is expected to reach $\Delta \nu / \nu \sim 10^{-16}$ for a one day integration time. In addition to the stability, another very important property of a frequency standard is its accuracy. Because of the very low velocities in a fountain device, many systematic shifts are strongly reduced and can be evaluated with great precision. With an accuracy of 2×10^{-15}, the BNM-LPTF fountain is presently the most accurate primary standard [63]. A factor 10 improvement in this accuracy is expected in the near future. In addition cold atom clocks deisgned for a reduced gravity environment are currently being built and tested, in order to increase the observation time beyond one second [64]. These clocks should operate in space in relatively near future.

References

[1] A more detailed presentation can be found with the text of the three Nobel lectures of 1997: S. Chu, Rev. Mod. Phys. **70**, 685 (1998), C. Cohen-Tannoudji, Rev. Mod. Phys. **70**, 707 (1998), and W.D. Phillips, Rev. Mod. Phys. **70**, 721 (1998).

[2] A. Kastler, J. Phys. Rad. **11**, 255 (1950).

[3] For a historical survey of this research field, we refer the reader to the following review papers: A. Ashkin, Science, **210**, 1081 (1980); V.S. Letokhov and V.G. Minogin, Phys.Reports, **73**, 1 (1981); S. Stenholm, Rev.Mod.Phys. **58**, 699 (1986); C.S. Adams and E. Riis, Prog. Quant. Electr. **21**, 1 (1997).

[4] C. Cohen-Tannoudji and W. Phillips, Physics Today **43**, No 10, 35 (1990).

[5] C. Cohen-Tannoudji, J. Dupont-Roc and G. Grynberg, *Atom-photon interactions – Basic processes and applications*, (Wiley, New York, 1992).

[6] J.V. Prodan, W.D. Phillips and H. Metcalf, Phys. Rev. Lett. **49**, 1149 (1982).

[7] W. Ertmer, R. Blatt, J.L. Hall and M. Zhu, Phys. Rev. Lett. **54**, 996 (1985).

[8] C. Cohen-Tannoudji, *Atomic motion in laser light*, in *Fundamental systems in quantum optics*, J. Dalibard, J.-M. Raimond, and J. Zinn-Justin eds, Les Houches session LIII (1990), (North-Holland, Amsterdam 1992) p.1.

[9] J. Dalibard and C. Cohen-Tannoudji, J.Opt.Soc.Am. **B2**, 1707 (1985).

[10] T.W. Hänsch and A.L. Schawlow, Opt. Commun. **13**, 68 (1975).

[11] D. Wineland and H. Dehmelt, Bull. Am. Phys. Soc. **20**, 637 (1975).

[12] S. Chu, L. Hollberg, J.E. Bjorkholm, A. Cable and A. Ashkin, Phys. Rev. Lett. **55**, 48 (1985).

[13] J.P. Gordon and A. Ashkin, Phys. Rev. **A21**, 1606 (1980).

[14] V.S. Letokhov, V.G. Minogin and B.D. Pavlik, Zh. Eksp. Teor. Fiz. **72**, 1328 (1977) [Sov.Phys.JETP, **45**, 698 (1977)].

[15] D.J. Wineland and W. Itano, Phys.Rev. **A20**, 1521 (1979).

[16] P.D. Lett, R.N. Watts, C.I. Westbrook, W. Phillips, P.L. Gould and H.J. Metcalf, Phys. Rev. Lett. **61**, 169 (1988).

[17] E.L. Raab, M. Prentiss, A. Cable, S. Chu and D.E. Pritchard, Phys.Rev.Lett. **59**, 2631 (1987).

[18] C. Monroe, W. Swann, H. Robinson and C.E. Wieman, Phys.Rev.Lett. **65**, 1571 (1990).

[19] V.S. Letokhov, Pis'ma. Eksp. Teor. Fiz. **7**, 348 (1968) [JETP Lett., **7**, 272 (1968)].

[20] S. Chu, J.E. Bjorkholm, A. Ashkin and A. Cable, Phys.Rev.Lett. **57**, 314 (1986).

[21] C.S. Adams, H.J. Lee, N. Davidson, M. Kasevich and S. Chu, Phys. Rev. Lett. **74**, 3577 (1995).

[22] A. Kuhn, H. Perrin, W. Hänsel and C. Salomon, OSA TOPS on Ultracold Atoms and BEC, 1996, Vol. 7, p.58, Keith Burnett (ed.), (Optical Society of America, 1997).

[23] R.J. Cook and R.K. Hill, Opt. Commun. **43**, 258 (1982).

[24] V.I. Balykin, V.S. Letokhov, Yu. B. Ovchinnikov and A.I. Sidorov, Phys. Rev. Lett. **60**, 2137 (1988).

[25] M.A. Kasevich, D.S. Weiss and S. Chu, Opt. Lett. **15**, 607 (1990).

[26] C.G. Aminoff, A.M. Steane, P. Bouyer, P. Desbiolles, J. Dalibard and C. Cohen-Tannoudji, Phys. Rev. Lett. **71**, 3083 (1993).

[27] J. Dalibard and C. Cohen-Tannoudji, J. Opt. Soc. Am. **B6**, 2023 (1989).

[28] P.J. Ungar, D.S. Weiss, E. Riis and S. Chu, JOSA **B6**, 2058 (1989).

[29] C. Salomon, J. Dalibard, W. Phillips, A. Clairon and S. Guellati, Europhys. Lett. **12**, 683 (1990).

[30] Y. Castin and K. Mølmer, Phys. Rev. Lett. **74**, 3772 (1995).

[31] Y. Castin and J. Dalibard, Europhys.Lett. **14**, 761 (1991).

[32] P. Verkerk, B. Lounis, C. Salomon, C. Cohen-Tannoudji, J.-Y. Courtois and G. Grynberg, Phys. Rev. Lett. **68**, 3861 (1992).

[33] P.S. Jessen, C. Gerz, P.D. Lett, W.D. Phillips, S.L. Rolston, R.J.C. Spreeuw and C.I. Westbrook, Phys. Rev. Lett. **69**, 49 (1992).

[34] G. Grynberg and C. Triché, in *Proceedings of the International School of Physics Enrico Fermi*, Course CXXXI, A. Aspect, W. Barletta and R. Bonifacio (Eds), p.243, IOS Press, Amsterdam (1996); A. Hemmerich, M. Weidemüller and T.W. Hänsch, same Proceedings, p.503.

[35] P.S. Jessen and I.H. Deutsch, in *Advances in Atomic, Molecular and Optical Physics*, **37**, 95 (1996), ed. by B. Bederson and H. Walther.

[36] A. Aspect, E. Arimondo, R. Kaiser, N. Vansteenkiste, and C. Cohen-Tannoudji, Phys. Rev. Lett. **61**, 826 (1988).

[37] M. Kasevich and S. Chu, Phys. Rev. Lett. **69**, 1741 (1992).

[38] G. Alzetta, A. Gozzini, L. Moi, G. Orriols, Il Nuovo Cimento **36B**, 5 (1976); Arimondo E., Orriols G., Lett. Nuovo Cimento **17**, 333 (1976).

[39] A. Aspect, E. Arimondo, R. Kaiser, N. Vansteenkiste, and C. Cohen-Tannoudji, J. Opt. Soc. Am. **B6**, 2112 (1989).

[40] B. Saubamea, T.W. Hijmans, S. Kulin, E. Rasel, E. Peik, M. Leduc and C. Cohen-Tannoudji, Phys. Rev. Lett. **79**, 3146 (1997).

[41] J. Lawall, F. Bardou, B. Saubamea, K. Shimizu, M. Leduc, A. Aspect and C. Cohen-Tannoudji, Phys. Rev. Lett. **73**, 1915 (1994).

[42] J. Lawall, S. Kulin, B. Saubamea, N. Bigelow, M. Leduc and C. Cohen-Tannoudji, Phys. Rev. Lett. **75**, 4194 (1995).

[43] M.A. Ol'shanii and V.G. Minogin, Opt.Commun. **89**, 393 (1992).

[44] S. Kulin, B. Saubamea, E. Peik, J. Lawall, T.W. Hijmans, M. Leduc and C. Cohen-Tannoudji, Phys. Rev. Lett. **78**, 4185 (1997).

[45] C. Cohen-Tannoudji and J. Dalibard, Europhys. Lett. **1**, 441 (1986).

[46] P. Zoller, M. Marte and D.F. Walls, Phys. Rev. **A35** 198 (1987).

[47] F. Bardou, J.-P Bouchaud, O. Emile, A. Aspect and C. Cohen-Tannoudji, Phys. Rev. Lett. **72** 203 (1994).

[48] J.P. Bouchaud and A. Georges, Phys. Rep. **195**, 127 (1990).

[49] B. Saubamea, M. Leduc, and C. Cohen-Tannoudji, Phys. Rev. Lett. **83**, 3796 (1999).

[50] J. Reichel, F. Bardou, M. Ben Dahan, E. Peik, S. Rand, C. Salomon and C. Cohen-Tannoudji, Phys. Rev. Lett. **75**, 4575 (1995).

[51] M. H. Anderson, J. Ensher, M. Matthews, C. Wieman, and E. Cornell, Science **269**, 198 (1995).

[52] K. B. Davis, M.O. Mewes, N. Van Druten, D. Durfee, D. Kurn, and W. Ketterle, Phys. Rev. Lett. **75**, 3969 (1995).

[53] C. C. Bradley, C. A. Sackett, and R. G. Hulet, Phys. Rev. Lett. **78**, 985 (1997); see also C. C. Bradley, C. A. Sackett, J. J. Tollett, and R. G. Hulet, Phys. Rev. Lett. **75**, 1687 (1995).

[54] D. Fried, T. Killian, L. Willmann, D. Landhuis, S. Moss, D. Kleppner, and T. Greytak, Phys. Rev. Lett. **81**, 3811 (1998).

[55] M. Inguscio, S. Stringari, and C.E. Wieman, *Proceedings of the International School of Physics Enrico Fermi*, Course CXL, *Bose–Einstein Condensation in Atomic Gases* (IOS Press, Amsterdam, 1999).

[56] K. Helmerson, D. Hutchinson, K. Burnett and W. D. Phillips, *Atom lasers*, Physics World, August 1999, p. 31; Y. Castin, R. Dum and A. Sinatra, *Bose condensates make quantum leaps and bounds*, Physics World, August 1999, p. 37.

[57] K. Gibble and S. Chu, Metrologia, **29**, 201 (1992).

[58] S.N. Lea, A. Clairon, C. Salomon, P. Laurent, B. Lounis, J. Reichel, A. Nadir, and G. Santarelli, Physica Scripta **T51**, 78 (1994).

[59] J. Zacharias, Phys. Rev. **94**, 751 (1954). See also : N. Ramsey, *Molecular Beams*, Oxford University Press, Oxford, 1956.

[60] M. Kasevich, E. Riis, S. Chu and R. de Voe, Phys. Rev. Lett. **63**, 612 (1989).

[61] A. Clairon, C. Salomon, S. Guellati and W.D. Phillips, Europhys. Lett. **16**, 165 (1991).

[62] G. Santarelli, Ph. Laurent, P. Lemonde, A. Clairon, A. G. Mann, S. Chang, A. N. Luiten, and C. Salomon, Phys. Rev. Lett. **82**, 4619 (1999).

[63] E. Simon, P. Laurent, C. Mandache and A. Clairon, Proceedings of EFTF 1997, Neuchatel, Switzerland.

[64] Ph. Laurent, P. Lemonde, E. Simon, G. Santarelli, A. Clairon, N. Dimarcq, P. Petit, C. Audoin, and C. Salomon, Eur. Phys. J. D **3**, 201 (1998).

Optics and Interferometry with Atoms and Molecules

Jörg Schmiedmayer

Institut für Experimentalphysik, Universität Innsbruck
Physikalisches Institut, Universität Heidelberg

1 Introduction

The development of wave optics for light brought many new insights into our understanding of physics, driven by fundamental experiments like the ones by Young, Fizeau, Michelson-Morley and others. Quantum mechanics, and especially the de Broglie's postulate relating the momentum p of a particle to the wave vector k of an matter wave: $k = 2\pi/\lambda = p/\hbar$, suggested that wave optical experiments should be also possible with massive particles (see table 1), and over the last 40 years electron and neutron interferometers have demonstrated many fundamental aspects of quantum mechanics [1].
Even though first optical experiments with atoms were performed already very early (diffraction of atoms at a LiF surface was demonstrated by I. Estermann and O. Stern in 1930 [6] and 1969 J. Leavitt and F. Bills observed Fresnel diffraction of atoms on a slit [7]), only recent years brought a fast development of atom and molecular optics, driven by two technologies:

- Using *nanofabrication* one can build the small structures needed for diffractive optical elements for atoms and molecules. Furthermore one can microfabricate traps and guides to manipulate neutral atoms at the mesoscopic scale, which will allow to integrate atom optical devices on an *Atom Chip*.

	mass	energy	velocity	wave length
neutron (300K)	1 amu	25 meV	2200 m/s	2.2 Å
electron	1/1823 amu	100 eV	6000 km/s	1.2 Å
Na (jet beam)	23 amu	110 meV	1000 m/s	0.17 Å
Cs (laser cooled)	133 amu	10 peV	1 cm/s	3000 Å

Table 1: Examples for parameters used in matter waves experiments.

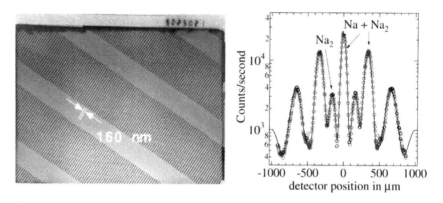

Figure 1: Diffraction of Na atoms and Na_2 molecules (right) from a nanofabricated transmission grating (left).

- The *laser* enables us to manipulate atoms by enhancing the interactions with the electro magnetic field by tuning close to resonance. This allows us to use light to cool, trap and prepare atomic samples for the experiments and in addition one can design atom optical elements using light field potentials.

Using these technologies a variety of optical elements for atoms/molecules were realized, and as early as 1973 an atom interferometer was patented [8]. Shortly afterwards several papers discussed the close similarity between multiple pulse laser spectroscopy and atom interferometers [9]. By now more than half a dozen experiments have demonstrated various types of atom and molecular interferometers [2], and recently atom optical elements were demonstrated at the mesoscopic scale [10].

2 Diffracting Atoms and Molecules

One of the basic optical elements important for many experiments with matter waves is a beam splitter. Because of the close analogy between the time independent Schrödinger equation and the Helmholz equation describing the electro magnetic field, diffraction of matter waves will be the same as for light. Diffraction from stationary objects can then be described by the Kirchhoff diffraction integrals. To study atom/molecule diffraction there are two approaches.

- Nanofabricated free standing structures can be used as amplitude (transmission) gratings [11] to diffract atoms [12, 13].

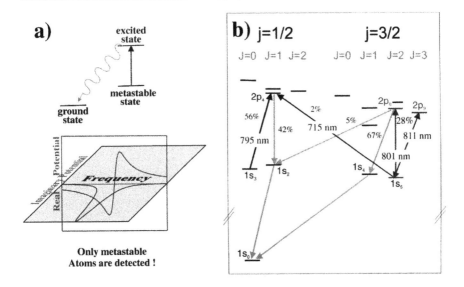

Figure 2: **a)** The complex interaction potential between a light field and a two-level atom with an additional decay channel of the excited state to a third not detected state. **b)** Level scheme of metastable Ar used for the experiments. We use the $1s_5$ metastable state and there the *closed* transition ($1s_5 \to 2p_9$) at 811 nm and the *open* transition ($1s_5 \to 2p_8$) at 801 nm.

- One can use interaction between near resonant light and the atom to design periodic structures with an (in general complex) index of refraction [14].

The advantage of using a nanofabricated free standing transmission grating is, that the diffraction depends only on the de Broglie wavelength λ_{dB} of the particle and is independent of the species or its internal state. We used this in our experiments studying diffraction of atoms [12] and molecules [13] (Fig. 1). All constituents in a supersonic molecular beam have the same velocity, therefore their de Broglie wavelength and consequently the diffraction angle depends only on the mass of the particle. One can use diffraction to separate out the different species in the beam, effectively building a mass spectrometer for neutral particles. We used this technique to separate atomic and molecular sodium in a molecular beam. This technique was also applied by the group of D. Thoenneis to verify the existence of the He_2 dimer [15]
Similarly one can study diffraction of atoms by using a standing light wave to build a periodic refractive index structure for atoms (Fig. 3) [16, 17, 18, 19] . The interaction between a light field and a two level atom, with an additional decay channel to a third non interacting state (shown for metastable *Ar* in

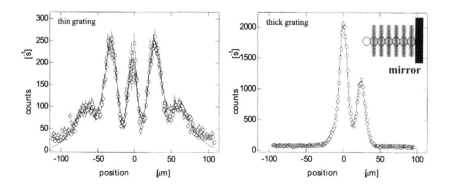

Figure 3: Diffraction of atoms from a standing light wave. *(left)* Raman-Nath diffraction from a thin light grating, *(right)* Bragg scattering from a thick *light crystal*

Fig. 2), can be described by the complex optical potential [14]:

$$U(x,y) = \frac{1}{\hbar} \frac{(d\, E(x,y))^2}{\Delta + i\gamma/2}.$$

Here $E(x,y)$ is the electric field connected to the light, d the dipole matrix element of the transition, Δ the detuning between the light and the atomic transition, and γ is the loss rate from the excited level to the non interacting state. The potential has a Kramers-Kronig like dependence on the light frequency as typical for a resonant processes.

Since in matter wave optics a potential U is equivalent to a complex index of refraction $n \simeq 1 - \frac{U}{2E_{kin}}$, any light field structure can, be regarded as a complex refractive index structure acting as a hologram for atomic de Broglie waves. Realizing a refractive index for atomic matter waves by light is a beautiful example of the possibility to exchange light and matter in wave optical experiments. Diffraction from such a light grating can be viewed in two ways:

- As diffraction from a periodic structure, the diffraction angle ϑ_{diff} is given by the ratio between wave vector of the atom $\vec{k}_{atom}$ and the grating vector $\vec{G}$ ($\vartheta_{diff} = \frac{\vec{G}}{\vec{k}_{atom}}$).

- As absorption and stimulated emission of photons in the standing light wave resulting in momentum transfer in multiples of $\Delta \vec{k}_{atom} = \vec{k}_1 - \vec{k}_2$, where $\vec{k}_1$ and $\vec{k}_1$ are the wave vectors of the two light fields creating the standing light wave. Using the relation for the grating vector of

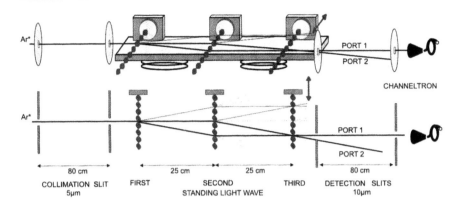

Figure 4: Mach-Zehnder atom interferometer built from three standing light waves. The first standing light wave acts as a beam splitter, the second redirects the beams back to be superposed by the third standing light wave. The interference pattern can be observed in either of the two output ports.

the standing light wave $\vec{G} = \vec{k}_1 - \vec{k}_2$ one easily sees that the deflection angle is the same as the diffraction angle ϑ_{diff} discussed above.

Depending on its thickness, the standing light wave acts either as a thin diffraction grating exhibiting many diffraction orders [16], or as a thick *"light crystal"* showing the characteristics of Bragg diffraction (Fig. 3) [17, 18, 19] One other application of the interaction potential with light is to build thin amplitude structures with *on*-resonant light [20]: Atoms crossing the standing light wave near the antinodes of the field will scatter many photons and will consequently be pumped to the undetected state (=absorbed). Atoms near the nodes of the light field will have a small probability to scatter light. For sufficiently high light intensities amplitude structures with open slits $< \lambda/10$ can easily be obtained. These optical mask for neutral atoms with nanometer size, resolution, and accuracy are a very useful tool to create and probe small atomic matter wave patterns [20].

3 Atom and Molecular Interferometry

Using three diffraction gratings one can build a Mach-Zehnder interferometer for atoms and molecules. Since atomic beams have a very short coherence (typically 1 Å) the three grating Mach-Zehnder geometry, which is *white light*, is essential. In such an interferometer the interference pattern is independent of the incoming wavelength or direction.

An interesting example is an atom interferometer based on diffraction at three standing light waves [22, 18]. In such an atom interferometer the roles

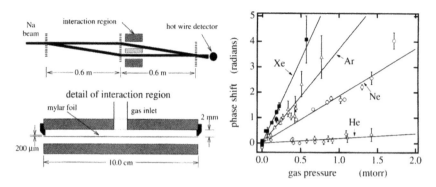

Figure 5: Measuring the refractive index of a gas for Na matter waves. *(left)* Schematics of the interferometer setup, including a detailed drawing of the gas scattering cell. *(right)* Experimental phase shifts for Na matter waves when moving through a 10 cm long noble gas targets.

of atoms and light are switched. The atoms, behaving as waves, are split and recombined using standing light fields representing diffraction gratings. At MIT we used three nanofabricated gratings to construct an interferometer for atoms and molecules [23, 13, 21]. In this interferometer the two interfering portions of each particle's wavefunction are spatially separated and physically isolated from each other by a stretched metal foil, $10\,cm$ long and $10\,\mu m$ thick. This foil, placed between the two interferometer arms allows us to perform measurements on single states with spectroscopic precision [21].

We have performed several experiments probing interactions that shift the energy of an atomic/molecular state by observing the resulting phase shift in the interferometer [21]. For example we measured the electric polarizability of the ground state of sodium [24], determined the coherence length of our beam, and studied Zeeman coherences with a magnetic field applied differentially to the two sides of the interferometer [25].

By inserting a gas cell in one arm of the interferometer we observed both the attenuation and the phase shift of Na matter waves (λ from $0.1 \rightarrow 0.26$ Å, velocity from $700 m/s \rightarrow 1800 m/s$) when transmitted through sample of monoatomic (He, Ne, Ar, Kr, Xe) and molecular (N_2, CO_2, NH_3, H_2O) gases . In this way we could for the first time measure the complex index of refraction of an atomic matter wave in a gas and determined with high accuracy the ratio of the real and imaginary parts of the forward scattering amplitude [26] (see Fig. 5) and estimate the absolute magnitude of the real part of the forward scattering amplitude. We also measured the index of refraction for molecular Na_2 matter waves [13]. These experiments give new insights in atomic and molecular scattering.

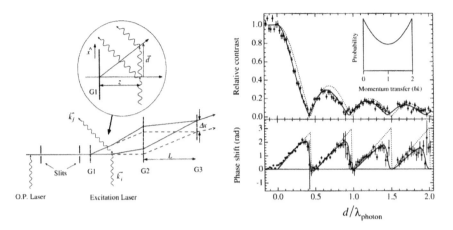

Figure 6: Experiment demonstrating the loss of coherence by photon scattering. *(left)* Schematics of the setup with a detail of the photon scattering geometry. *(right)* Experimental data demonstrating the loss of contrast in the atomic interference pattern depending in the path separation d at the point of photon scattering.

Furthermore, we have used scattering of single photons from an atom to investigate the effect of a quantum measurement on a coherently split, and spatially separated, atomic de Broglie wave [27]. The separation of the atomic wave function was thereby determined by distance of the photon scattering laser from the first grating of the interferometer (Fig. 6 left). We observed the decrease of the contrast of the atomic interference pattern (decrease of single particle coherence) with the increasing separation of the atomic de Broglie waves (Fig. 6 right) and the "revivals" in the interference contrast due to the overlap of a maximum in the light wave emitted from one path with the other path thereby diminishing the *welcher Weg information* carried by the emitted photon. We also observed the phase shift imprinted on the atomic wave function by the photon scattering. In addition we could demonstrate that the lost single particle coherence can be recovered when one reduces the possible final states of the scattered photon [27]. This is equivalent to reducing the resolving power of the path detection by the scattered photon.

Atom interferometer are especially sensitive to inertial effects. This sensitivity arises because the freely propagating particles form fringes in an inertial frame. The fringes appear shifted if the interferometer moves in respect to its inertial frame. For constant area of the interferometer, the observable phase shift increases with the mass of the interfering particle. Recently atom interferometers were tested as inertial sensors (for example rotation sensors, accelerometers, or gravity gradiometers) in pilot experiments [28, 29, 30], reaching sensitivities comparable to or better than the best commercially available sensors.

4 Model Systems: Atoms in Light Crystals

Since any light field can be regarded as a complex refractive index structure, like a hologram, for atomic de Broglie waves, one can imagine building many model potentials for atomic de Broglie waves using light. We used this to investigate the coherent motion of atomic de Broglie waves in *periodic* (space and/or time) structures made of light, which we call *'light crystals'* [19]. By changing the intensity and the frequency of the laser light one can change the interaction between the light crystal and the atom at will, so that one may consider

- a very weak, elastic interactions such as those in dynamical diffraction,

- a very strong interactions as in channeling

- and interactions which are dominated by dissipative processes using on-resonant light.

Starting from the similar and well developed fields of dynamical diffraction in neutron, electron and x-ray physics, we investigated the atomic wavefields inside the light crystal in detail [19] and demonstrated some basic effects like anomalous transmission (Fig. 7) [31] or a violation of Friedels law [32]. In the former *on*-resonant light allows us to create purely imaginary (absorptive) periodic potentials. One experimentally observes that the total number of atoms transmitted through the *on*-resonant standing light wave increases if the angle of incidence is the Bragg angle (see Fig. 7a). This observation is similar to *anomalous transmission* [33] discovered for X-rays by Borrmann in 1941, and is caused by a *gray* state, which has reduced coupling to the light field when propagating through the crystal. This *gray* state is identical with the weakly coupling Bloch state known from the theory of dynamical diffraction and solid state physics.

Even more interesting is the possibility to switch the light crystals faster than the typical time scales of motion in the periodic potential and investigate *time dependent* matter wave optics. As an example we built a frequency shifter for atomic matter waves and observed beating between matter waves of different energy (frequency) as shown in Fig. 8 [34]. In these experiments we measure the intensity of the diffracted atoms as a function of the mirror angle (rocking curves). The top graph of the series shows the rocking curve for Bragg scattering off a static (unmodulated) light crystal. Only one peak at the static Bragg angle is observed.

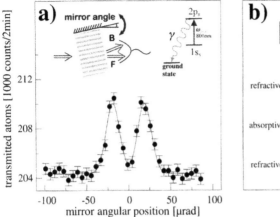
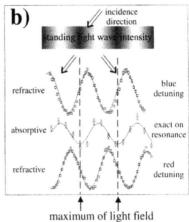

Figure 7: **a)** Total intensity of the metastable Ar^* beam after transmission through a standing light wave tuned exactly *on*-resonance to an open transition (see insert) as a function of incidence angle. The transmission increases anomalous for incidence at the Bragg angle. **b)** Standing atomic wave fields in the light crystal. The anomalous transmitted wave field (middle trace) has its maxima at the nodes of the standing light field. For light tuned far *off*-resonance, the standing atomic wave is shifted by $\frac{\pi}{2}$ to the left for blue detuning (positive potential) and to the right for red detuning (negative potential).

The next curves show the same experiment, but with intensity modulation frequencies in the range from $25kHz$ to $250kHz$. Two pronounced side peaks appear, indicating two additional Bragg resonances. They are located symmetrically around the central Bragg angle, and their angular separation from the central peak increases linearly with the modulation frequency.

For an explanation of the observed Bragg scattering at a time-dependent potential we first examine the usual case of static Bragg diffraction. In the time independent case Bragg diffraction conserves the kinetic energy of the atoms $|\vec{k}_B| = |\vec{k}_F|$ (elastic scattering) but changes the direction of the atomic momentum by one reciprocal grating vector $\vec{k}_B = \vec{k}_F + \vec{G}$. First order diffraction can occur only at a specific incidence angle, θ_B, which fulfills the Bragg condition: $\sin(\theta_B) = k_L/k_A$. The modulated light field exhibits, besides the carrier frequency ω_c, also sidebands with frequencies $\omega_c \pm \omega_m$ These sidebands in the light field form *moving* light crystals with a velocity of $v = \pm \frac{c\omega_m}{\omega_c}$. Atoms can now be Bragg diffracted from these moving crystals [34]. The incident angle for satisfying the Bragg condition is now defined in the moving frame and will be different in the laboratory

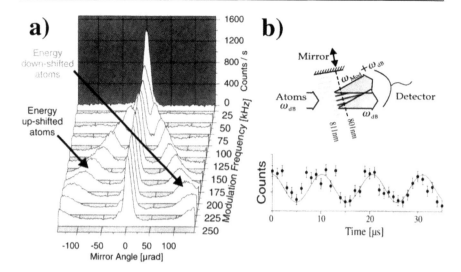

Figure 8: **a)** Diffracted atoms as a function of their incidence angle at the amplitude modulated standing light wave, at different modulation frequencies. The central peak corresponds to static Bragg diffraction; the two symmetric side peaks to atoms with a de Broglie wave frequency shifted by ± the modulation frequency. **b)** A thin stationary absorptive grating ($801 nm$ focused standing light wave) probes the traveling atomic interference pattern behind the modulated ($100 kHz$) $811 nm$ light crystal. The total atomic transmission is detected as a function of time showing the interference between matter waves of different frequencies.

frame of the experiment, and the energy of the Bragg diffracted beam will be shifted by $\pm\hbar\omega_m$. This frequency shift is clearly seen in the beating experiment shown in Fig. 8b.

5 Miniaturizing Atom Optics: Atom Chips

In electronics and light optics, miniaturization of components and integration into networks has lead to new very powerful tools and devices, for example in quantum electronics [35] or integrated optics [36]. It is essential for the success of such designs, that the size of the structures is, at least in one dimension, comparable to the wavelength of the guided wave. Similarly we anticipate that atom optics [4], if brought to the microscopic scale, will give us a powerful tool to combine many atom optical elements into integrated quantum matter wave circuits.

Such microscopic scale atom optics can be realized by bringing cold atoms (de Broglie wavelength λ_{dB} of 100 nm or larger) [3, 5] close to nanostructures [37, 38, 39] which can easily be designed and built using standard nanofabrication techniques as used in the semiconductor industry. Such microscopic potentials for neutral atoms can be created using:

(i) The electric interaction between a neutral, polarizable atom and a charged nanostructure, where the potential is $V_{el} = -\frac{1}{2}\alpha_{el}E^2$.

(ii) The magnetic interaction between the atomic magnetic moment μ and a magnetic field B, described by the potential $V_{mag} = -\vec{\mu} \cdot \vec{B}$.

These potentials can be combined with traditional atom optical elements like atom mirrors and evanescent light fields. A variety of novel atom optical elements like quantum wells, quantum wires and quantum dots for trapping and guiding of neutral atoms can then be constructed at the microscopic scale on an *Atom Chip*.

Having the atoms trapped or guided and, well localized near the surface, will allow integration of atom optics and light optics with, for example light cavities and waveguides fabricated on the surface.

To study the basic principles of constructing *Atom Chips* we first preformed experiments studying micro manipulation of atoms with small free standing current carrying structures starting in the early 90's where we demonstrated the guiding of thermal Na atoms along a 1 m long wire around a beam stop [40].

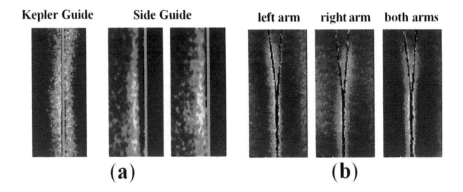

Figure 9: Wire guides and traps for atoms: (a) Kepler guide and side guide for different wire currents. (b) Beamsplitter for guided atoms.

Recently we studied guiding and trapping using current carrying wires with laser cooled Li atoms from a MOT. We performed experiments studying a Kepler guide with atoms orbiting around the wire, (Fig. 9a) and with the side guide, with atoms guided in a potential minimum along the side of the wire, and studied extensively the scaling laws [41, 42]. Furthermore, we could demonstrate a simple beamsplitter for guided atoms by combining two guides in the form of a Y. By controlling the current through the arms of the Y-shaped structure we can send cold atoms along either arms of the Y or into both arms (Fig. 9b) [42], and studied simple wire traps using a Z-shaped wire. In the latter case one easily obtains trap parameters which exceed those needed for BEC with only a few Watts of power consumption [43].

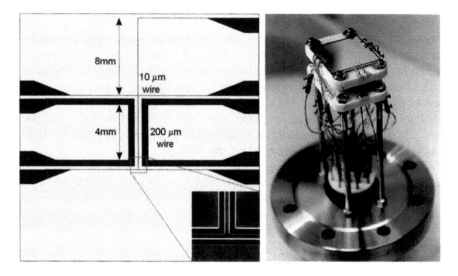

Figure 10: (a) A schematic of the chip surface design. For simplicity, only wires used in the experiment are shown. The wide wires are $200\mu m$ wide while the thin wires are $10\mu m$ wide. The insert shows an electron microscope image of the surface and its $10\mu m$ wide etchings defining the wires. (b) The mounted chip before it is introduced into the vacuum chamber.

Recently we demonstrated that standard nanofabrication technology can be used to miniaturize our wire experiments and to build atom optical elements on a chip which allow trapping and guiding of atoms at a mesoscopic scale interesting for quantum information proposals [45]. The *Atom Chip* used in the experiments consists of a $2.5\mu m$ thick gold layer deposited onto a GaAs substrate. This gold layer is then patterned using standard nanofabrication techniques [10, 47]. This leaves the chip as a gold mirror (with $10\mu m$ etchings) which can be used to reflect the laser beams for the MOT during cooling and collecting of atoms. A schematic of the wires on such an *Atom Chip* is shown in Fig. 10a and include a series of magnetic traps to transfer atoms into smaller and smaller potentials: the large U-shaped wires are $200\mu m$ wide and provide a quadrupole potential if combined with a homogeneous bias field [10, 43, 44], while the thin wires are $10\mu m$ wide.

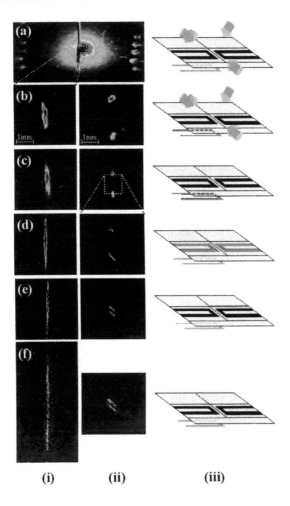

Figure 11: Experiments with an Atom Chip: column (i) shows the view from the top, column (ii) the front view and (iii) a schematic of the wire configuration. Current carrying wires are highlighted in red. The front view shows two images: the upper is the actual atom cloud and the lower is the reflection on the gold surface of the chip. The distance between both images is an indication of the distance of the atoms from the chip surface. Rows (a)-(f) show the various steps of the experiments. (a)-(d) show the step wise process of loading atoms onto the chip while (e) and (f) show atoms in a microscopic trap and propagating in a guide. The pictures of the magnetically trapped atomic cloud are obtained by fluorescence imaging using a short laser pulse (typically $0.5\,ms$)

The atoms are loaded onto the *Atom Chip* using a standard procedure [10, 44]: Typically 10^8 cold ^7Li atoms are accumulated in a "reflection MOT" [46]. Transferring atoms from the MOT to the mesoscopic potentials is carried out in the following steps (Fig. 11) [10]: Atoms are first transferred into a MOT generated by the quadrupole field of a U-wire located underneath the chip. Then, the laser light is switched off, leaving the atoms confined only by the magnetic quadrupole field of the U-wire. Atoms are then further compressed by increasing the bias field, and transferred into a magnetic trap generated by the two $200\mu m$ wires on the chip, compressed again and transferred into the $10\mu m$ guide. Each compression is easily achieved by decreasing the current generating the larger trap to 0 and switching on the current generating the smaller trap over a time of 10 ms. Typically we transfer 10^6 into the microscopic guide with a typical transverse trap frequency of up to 100 kHz and a typical ground state size down to $100nm$, surpassing the parameters required for recent quantum information proposals [45].

6 Conclusion

In the last years optics and interferometry with atoms and molecules became a mature field with many techniques and tools available for fundamental and applied experiments, especially in atomic and molecular physics, precision experiments and building model systems.

Integrating many elements to control atoms onto a single device, an *Atom Chip* will make atomic physics experiment much more robust and simple. This will allow much more complicated tasks in atom manipulation to be performed in a way similar to what integration of electronic elements allowed in the development of new powerful devices. This is of special interest since it has been suggested that the high degree of control achieved over neutral atoms and their weak coupling to the environment (long decoherence time) will allow the realization of Quantum Information Processing (QIP).

The experiments described in this overview were performed at MIT in collaboration with D. Pritchard and the atom diffraction experiments at Innsbruck in collaboration with A. Zeilinger. The *Atom Chips* were fabricated at the Institut für Festkörperelektronik, Technische Universität Wien, Austria, and the Sub-micron center, Weizmann Inst. of Science, Israel. We thank E.Gornik, C. Unterrainer and I. Bar-Joseph of these institutions for their assistance. The work at MIT was supported by the Army Research Office (DAAL03-89-K-0082 and ASSERT 29970-PH-AAS), the Office of Naval Research (N00014-89-J-1207), NSF (9222768-PHY), and the Joint Services

Electronics Program (DAAL03-89-C-0001). The experiments at Innsbruck were supported by the Austrian Science Foundation (S065-04 PHY, S065-05 PHY, SFB F15-07), the Jubiläums Fonds der Österreichischen Nationalbank, project 6400 and by of the European Union (TMRX-CT96-0002 and IST-1999-11055 ACQUIRE).

References

[1] for an overview see: *Matter Wave Interferometry* Ed.: G. Badurek, H. Rauch and A. Zeilinger, Physica **B151**, (1988) and references therein.

[2] for an overview see: *Atom Interferometry* Ed.: P.Berman, Academic Press (1997) and references therein; *Catching the Atom Wave*, Science **268**, 1129 (1995); *Measuring the Refractive Index in Atom Optics*, Physics World June 1995, p25; *Atom Interferometers Prove Their Worth in Atomic Measurements*, Physics Today, July 1995, p17.

[3] A good overview of laser cooling is given in: *Laser Manipulation of Atoms and Ions*, edited by E. Arimondo, W.D. Phillips, F. Strumia (North Holland, 1992); S. Chu, Rev. Mod. Phys. **70**, 685 (1998); C. Cohen-Tannoudji, Rev. Mod. Phys. **70**, 707 (1998); W.D. Phillips, Rev. Mod. Phys. **70**, 721 (1998).

[4] For an overview see: C.S. Adams, M. Sigel, J. Mlynek, Phys. Rep. **240**, 143 (1994); reference [2]; and references therein.

[5] M.H. Anderson, J. R. Ensher, M. R. Matthews, C. E. Wieman, E. A. Cornell, Science **269**, 198 (1995); K.B. Davis, M.O. Mewes, M.R. Andrews, N.J. van Druten, D.S. Durfee, D.M. Kurn, and W. Ketterle, Phys. Rev. Lett. **75**, 3969 (1995); M.-O. Mewes, M. R. Andrews, N. J. van Druten, D. M. Kurn, D. S. Durfee, C. G. Townsend, and W. Ketterle, Phys. Rev. Lett. **77**, 988 (1996); C.C. Bradley, C.A. Sacket, and R.G. Hulet, Phys. Rev. Lett. **78**, 985 (1997); see also C.C. Bradley, C. A. Sackett, J. J. Tollett, and R. G. Hulet, Phys. Rev. Lett. **75**, 1678 (1995). For a list of recent references see the BEC homepage http://amo.phy.gasou.edu/bec.html

[6] I. Estermann, O. Stern, Z.Physik **61**, 95 (1930).

[7] J. Leavitt F. Bills, Am.J.Phys. **37**, 905, (1969).

[8] S. Altschuler, L.M. Franz, (1973). US Patent Number 3,761,721

[9] V.P. Chebotayev, B.Y. Dubetsky, A.P. Kasantsev, V.P. Yakovlev, J. Opt.Soc.Am. **B2**, 1791 (1985); Ch. Bordé, Phys.Lett. **A140**, 10 (1989).

[10] R. Folman, P. Krüger, D. Cassettari, B. Hessmo, T. Maier, J. Schmiedmayer, quant-ph/9912106, Phys. Rev. Lett. in print, D. Cassettari, A. Chenet, R. Folman, A. Haase, B. Hessmo, P. Krger, T. Maier, S. Schneider, T. Calarco, J. Schmiedmayer, Appl. Phys. B in print, For experiments at larger scale see: [40, 41, 42, 43, 44, 48]

[11] D.W. Keith, R.J. Soave and M.J. Rooks, J. Vac. Sci. Technol. **B 9**, 2846 (1991). M. Rooks, R.C. Tiberio, M.S. Chapman, T.D. Hammond, E.T. Smith, A. Lenef, R.A. Rubenstein, D.E. Pritchard, S. Adams, J. Vac. Sci. Technol. B **13**, 2745 (1995).

[12] D.W. Keith, M.L. Shattenburg, H.T. Smith, D.E. Pritchard, Phys. Rev. Lett. **61**, 1580 (1988).

[13] M. Chapman, C.R. Ekstrom, T.D. Hammond, R.A. Rubenstein, J. Schmiedmayer, S. Wehinger, D.E. Pritchard, Phys. Rev. Lett. **74**, 4783 (1995)

[14] D.O. Chudesnikov and V.P. Yakovlev, Laser Physics **1**, 110 (1991).

[15] W. Schöllkopf, J. P. Toennies, Science **266**, 1345 (1994).

[16] P.E. Moskowitz, P.L. Gould, S.R. Atlas, D.E. Pritchard, Phys. Rev. Lett. **51**, 370 (1983); P.L. Gould, G.A. Ruff, D.E. Pritchard, Phys. Rev. Lett. **56**, 827 (1983).

[17] P.J. Martin, B.G. Oldaker, A.H. Miklich, D.E. Pritchard, Phys. Rev. Lett. **60**, 515 (1988).

[18] H. Batelaan, St. Bernet, M. Oberthaler, E. Rasel, J. Schmiedmayer, A. Zeilinger, in *Atom Interferometry* Ed.: P.Berman, Academic Press (1997) p.85-119.

[19] M. Oberthaler, R. Abfalterer, St. Bernet, C. Keller, J. Schmiedmayer, A. Zeilinger, Phys. Rev. A 60, 456 (1999).

[20] R. Abfalterer, C. Keller, St. Bernet, M. Oberthaler, J. Schmiedmayer, A. Zeilinger, Phys. Rev. A **56** R4365 (1997); C. Keller, R. Abfalterer, St. Bernet, M. Oberthaler, J. Schmiedmayer, A. Zeilinger, J. Vac. Sci. Tech. **16**, 3850 (1998). K.S. Johnson, J.H. Thywissen, N.H. Dekker, K.K. Berggren, A.P. Chu, R. Younkin, and M. Prentiss, Science **280**, 1583 (1998).

[21] J. Schmiedmayer, M. Chapman, C.R. Ekstrom, T. Hammond, A. Lenef, R. Rubenstein, E. Smith, D.E. Pritchard, in *Atom Interferometry* Ed.: P.Berman, Academic Press (1997) p.1-84.

[22] E. Rasel, M. Oberthaler, H. Batelaan, J. Schmiedmayer, A. Zeilinger, Phys. Rev. Lett. **75**, 2633 (1995)

[23] D. Keith, C.R. Ekstrom, Q.A. Turchette, and D E. Pritchard, Phys. Rev. Lett. **66**, 2693 (1991)

[24] C. Ekstrom, J. Schmiedmayer, M. Chapman, T.Hammond, D.E. Pritchard, Phys. Rev. **A51**, 3883 (1995)

[25] J. Schmiedmayer, C.R. Ekstrom, M. Chapman, T.Hammond, D.E. Pritchard, J.Phys. II **4**, 2029 (1994)

[26] J. Schmiedmayer, M. Chapman, C.R. Ekstrom, T.Hammond, S. Wehinger, D.E. Pritchard, Phys. Rev. Lett. **74**, 1043 (1995)

[27] M. Chapman , T.D. Hammond, A. Lenef, J. Schmiedmayer, R.A. Rubenstein, E. Smith, D.E. Pritchard, Phys. Rev. Lett. **75**, 3783 (1995)

[28] rotation: F. Riehle, Th. Kisters, A. Witte, J. Helmcke, and Ch. J. Bord'e, Phys. Rev. Lett. **67** 177 (1991); A.Lenef, M. Chapman, T.Hammond, R. Rubenstein, E. Smith and D.E. Pritchard, Phys. Rev. Lett. **78** 760 (1997); T. L. Gustavson, P. Bouyer, M. A. Kasevich, Phys. Rev. Lett. **78** 2046 (1997);

[29] acceleration: M. Kasevich and S. Chu, Phys. Rev. Lett. **67** 181 (1991); M. J. Snadden, J.M. McGuirk, P. Bouyer, K. G. Haritos and M. Kasevich, Phys. Rev. Lett.**81** 971 (1998);

[30] classical Moire deflectometer: M. Oberthaler, St. Bernet, E.Rasel, J. Schmiedmayer and A.Zeilinger, Phys. Rev. **A 54**, 3165 (1996).

[31] M. Oberthaler, St. Bernet, R. Abfalterer, J. Schmiedmayer, A. Zeilinger, Phys. Rev. Lett. **77**, 4980 (1996)

[32] C. Keller R. Abfalterer, St. Bernet, M. Oberthaler, J. Schmiedmayer, A. Zeilinger, Phys. Rev. Lett. **79**, 3327 (1997)

[33] G. Borrmann Zeitschr.Phys **42**, 157 (1942); anomalous transmission was also observed for neutrons: S. Sh. Shilshtein, V.J. Marichkin, M. Kalanov, V. A. Somenkov, and L. A. Sysoev, Zh. ETF Pis. Red. **12**, 80 (1970); and electrons: A. Mazel and R. Ayroles, J. Microscopie **7**, 793 (1968).

[34] St. Bernet, M. Oberthaler, R. Abfalterer, J. Schmiedmayer, A. Zeilinger, Phys. Rev. Lett. **77**, 5160 (1996); St. Bernet, R. Abfalterer, C. Keller, J. Schmiedmayer, A. Zeilinger, JOSA-B **15**, 2817 (1998); St. Bernet, R. Abfalterer, C. Keller, J. Schmiedmayer, A. Zeilinger, Proc. of Royal Society (London) **455**, 1509 (1999).

[35] See for example: *Quantum Coherence in Mesoscopic Systems*, edited by B. Kramer, NATO ASI Series B: Physics Vol. 254, Plenum (1991).

[36] See for example: *Fundamentals of Photonics*, B.E.A. Saleh, M.C. Teich, J. Wiley & Sons (1991).

[37] J. Schmiedmayer, Eur. Phys. J. D **4**, 57 (1998).

[38] E A Hinds, I G Hughes, J. Phys. D: Appl. Phys. **32** 119 (1999)

[39] J.D. Weinstein, K. Libbrecht, Phys. Rev. A. **52**, 4004 (1995); M. Drndic, K.S. Johnson, J.H. Thywissen, M. Prentiss, and R.M. Westervelt, Appl. Phys. Lett. **72**, 2906 (1998); J.H. Thywissen M. Olshanii, G. Zabow, M. Drndic, K.S. Johnson, R.M. Westervelt, M. Prentiss, Eur. Phys. J. D **7**, 361 (1999).

[40] J. Schmiedmayer in *XVIII International Conference on Quantum Electronics: Technical Digest*, edited by G. Magerl (Technische Universität Wien, Vienna 1992), Series 1992, Vol. 9, 284 (1992); Appl. Phys. B **60**, 169 (1995); Phys. Rev. A **52**, R13 (1995).

[41] J. Denschlag, D. Cassettari, J. Schmiedmayer, quant-ph/9809076, Phys. Rev. Lett. **82**, 2014 (1999).

[42] J. Denschlag, D. Cassettari, A. Chenet, S.Schneider, J. Schmiedmayer, Appl. Phys. B 69 (1999) p. 291.

[43] A. Haase, D. Cassettari, B. Hessmo, J. Schmiedmayer, submitted to Phys. Rev. A (1999).

[44] J. Reichel, W. Haensel, T.W. Haensch, Phys. Rev. Lett. **83**, 3398 (1999).

[45] T. Calarco, D. Jaksch, E.A. Hinds, J.Schmiedmayer, J.I. Cirac, P. Zoller, Phys. Rev. A **61**, 022304 (2000).

[46] K.I. Lee, J.A. Kim, H.R. Noh, W. Jhe, Opt. Lett. **21**, 1177 (1996).

[47] T. Maier *et al.*, in preparation.

[48] J. Fortagh, A. Grossmann, and C. Zimmermann, Phys. Rev. Lett. **81**, 5310 (1998); D. Müller, D.Z. Anderson, R.J. Grow, P.D.D. Schwindt, E.A. Cornell, Phys. Rev. Lett **83**, 5194 (1999); N. H. Dekker, C. S. Lee, V. Lorent, J. H. Thywissen, S. P. Smith, M. Drndic R. M. Westervelt and M. Prentiss, Phys. Rev. Lett **84**, 1124 (2000).

Some New Effects in Atom Stern-Gerlach Interferometry

R. Mathevet, K. Brodsky, F. Perales, M. Boustimi,
B. Viaris de Lesegno, J. Reinhardt, J. Robert, J. Baudon

Laboratoire de Physique des Lasers
Université Paris-Nord, Villetaneuse, France

1 Introduction

Stern-Gerlach interferometry is based on the interaction of the magnetic moment of the atom (or molecule) with external magnetic fields. It owes its name to the use of magnetic field gradients to macroscopically split the incident beam into separated partial beams, as in the original Stern and Gerlach experiment. It is a polarization interferometry, which has much in common in its principle with the birefringent crystal-plate optical interferometry : the incident light wave is first linearly polarized at 45° with respect to the neutral axes of the crystal. Then at the entrance side of the plate, the state is a superposition with equal amplitudes of the polarization basis set vectors inside the crystal. Because of the multivalued refraction index, each polarization component accumulates a different phase shift inside the crystal. At the output side, the total state is a combination of two orthogonal polarization states. To be able to observe the interference pattern, one needs to project it on a single state ; this is achieved by the second polarizer, also called the analyzer.

Two different atomic versions are based on this scheme in our laboratory, one (I) using a beam of metastable hydrogen atoms ($2s_{1/2}$), the other one (II) a beam of ground-state potassium atoms ($4s_{1/2}$). In both experiments, the atomic beam travels through a magnetic field $\vec{B}(\vec{r},t)$ created by specific configurations of wires or coils. The interaction of the atomic magnetic moment $\vec{\mu}$ give rise to a magnetic potential energy $V(\vec{r},t) = \vec{\mu}.\vec{B}(\vec{r},t)$. As the incident kinetic energy is very high compared to it ($E_{kin}/|V| \sim 10^7$), the semi-classical limit applies: one can see the potential V as a refractive medium whose index n differs from 1 (refraction index out of the potential) by an amount of the order of V/E_{kin} :

$$n(\vec{r},t) = \sqrt{1 - \frac{V(\vec{r},t)}{E_{kin}}} \sim 1 - \frac{V(\vec{r},t)}{2E_{kin}}.$$

In (I), the preparation and analysis filters rely on the Lamb-Retherford method (Fig. 1): the beam passes through a 600 G transverse magnetic field, where the $2s_{1/2}$, $F = 0$ and $F = 1$, $m_F = -1$ metastable sublevels cross the radiative $2p_{1/2}$ levels. The motional electric field is then strong enough to yield a very efficient dipolar coupling, resulting in a quick decay of these two 2s sublevels (as the typical time of flight through the device, ~ 50 µs, is much longer than the 1.6 ns lifetime of the $2p_{1/2}$ levels). The emerging beam is thus partially polarized in the $F = 1$ manifold as it contains two incoherent populations corresponding to the sublevels $m_F = 0$ and $m_F = +1$. Actually, except for a continuous background, this is equivalent to a complete polarization in the sublevel $m_F = -1$. It is the very simplest method available for hydrogen atoms. Atoms are analysed the same way at the other end of the apparatus.

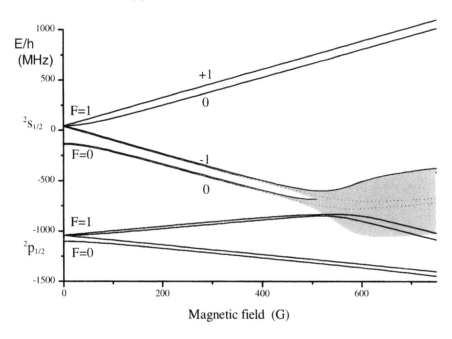

Figure 1: Zeeman diagram of hydrogen illustrating the Lamb-Retherford partial polarisation method. In a 600 G magnetic field, the $2s_{1/2}$, $F = 0$, $m_F = 0$ and $F = 1$, $m_F = -1$ are quenched leaving a partially polarised beam in the $2s_{1/2}$, $F = 1$, $m_F = 0, +1$ sublevels.

To make interferences, one has to prepare, before and after the interaction region a coherent superposition of the three polarization components ($F = 1$, $m_F = 0, \pm 1$), for each incoherent incoming population. In order to do so, atoms undergo a non-adiabatic evolution in a region where the magnetic field magnitude becomes very low while its direction rotates quickly by 90°. In the sudden transition limit ($\omega_{\text{Larmor}} \ll \omega_{\text{rotation}}$), one obtains the resulting state by projecting the initial state onto the new basis set vectors, defined by the new magnetic field direction. So (see Fig. 2), atoms are successively, polarised (P), superposed (BS_1), sent into the multirefringent potential V (IR), superposed again (BS_2) and analysed (A). Detection (D) is realised by quenching the atoms in a static electric field : the Lyman-α emitted photons are then collected by a channel electron multiplier. It should be noted that, when operating continuously, the counting rate is about 100 s^{-1}(1 s^{-1} in the time-of-flight mode, using a chopped source (S)).

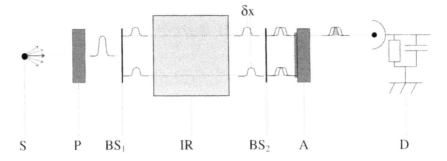

Figure 2: Global scheme of the interferometer: S (source), P and A (polarizer and analyser), BS (beam splitters where the coherent superposition is realised), IR (interaction region with $V(\vec{r},t)$) and D (detector).

In (II), the polarization and analysis are realized by Stern-Gerlach magnets, which create strong magnetic field gradients of $\sim 10^4$ G cm^{-1}, able to transversally split the two spin components $\pm 1/2$ (J=1/2). By setting appropriate collimating slits, one allows only one of the two components to pass through. The superposition of the two magnetic components inside the potential V is obtained by sending the atoms inside RF-field regions, where transitions between the $F = 2, m_F = -1$ (J=1/2 manifold) and the $F = 2$, $m_F = -2$ (J=1/2 manifold) sublevels are induced. The interaction time is chosen (via the length of the RF zone) such that atoms see a $\pi/2$ pulse and end in a linear superposition of the two components with the same amplitude (maximum contrast). Finally, atoms are ionized on a hot wire and ions are collected by an electron multiplier. In this device, the counting rate in a continuous operating mode is about 1000 s^{-1}. Note also that here the filtering process is velocity selective, as the magnetic gradients act on the external motion according to the incident kinetic energy. If required, the velocity selection can be enhanced by reducing the width of the collimating slit. One can then achieve up to 10^{-2} velocity dispersion relative to the mean velocity. In comparison, in the metastable hydrogen experiment, the velocity distribution relative width is unity as is usual for thermal effusive beams.

In this paper, we present the most recent experiments that we have carried out in our lab using for the most part apparatus (I). In section 2 we use time-and-space dependent magnetic fields moving at atomic velocities ("comoving" fields). Then, in section 3, we present a double-interferometer configuration as an approach to the quantum Zeno effect. Section 4 is devoted to quadrupolar magnetic fields that produce interference patterns in the transverse plane with a potential for novel applications. In the end, section 5, we present our new metastable Argon interferometer showing the versatility of the Stern-Gerlach approch to atom inteferometry.

2 Comoving fields with a beam of metastable hydrogen atoms [1]

In the hydrogen atom interferometer, we have used a potential V longitudinally moving with the atoms :

$$V(x,t) = V(x, x - ut),$$

where x denotes the beam incident direction, and u the propagation velocity of the potential. Actually, the optical counterpart of such a device is an electrooptic modulator. The important feature of "comoving" potentials, as we called them, is that a matching condition is required for the atomic velocity : the "resonant" velocity class $v = u$ plays obviously a particular role, and yields characteristic interference effects.

A comoving magnetic field can be created by a set of eight helices H_n^p, $n = 1, ..., 4$, $p = \pm 1$, aligned on x axis, of radius R and of pitch λ, parametrized as follows :

$$\begin{cases} x \\ y = R\cos(2\pi x/\lambda - n\pi/2) \\ z = pR\sin(2\pi x/\lambda - n\pi/2). \end{cases}$$

The helices are supplied with the alternating currents

$$I_n^p = I_0 \cos(2\pi \nu t - n\pi/2).$$

The Biot-Savart law then yields the following expression for the magnetic field on x axis (as was expected by symmetry considerations) :

$$\vec{B}(x,t) = \frac{2\mu_0 I_0}{\pi R} f(\lambda/R) \cos\frac{2\pi}{\lambda}(x - ut)\, \vec{e_y},$$

I.3 Some New Effects in Atom Stern-Gerlach Interferometry

where f is a geometrical factor such that $f(0) = 0$ (opposite solenoids), $f(\infty) = 1$ (parallel wires). Moreover, the current distribution is such that the first order term in the expansion of the field in powers of the distance to x axis is zero, i. e. the field is stationary around x axis. On the above expression of $\vec{B}$, one sees that the propagation velocity of the field is $u = \lambda \nu$. As λ and ν are experimental parameters, the propagation velocity is easily tuned just by varying one or the other (rather the current frequency in practice) ; for example, for $\lambda = 1$ cm, one can scan velocities up to 100 km/s by varying ν up to 10 MHz, so that any velocity in the velocity distribution can be reached.

The phase shift induced in such a comoving potential, starting at $x = d$ and of length L, can then be calculated as :

$$\Phi(x,t) = -\frac{1}{\hbar v} \int_d^{d+L} dx' \, V\left(x', \, t - \frac{x-x'}{v}\right),$$

where $V(x,t) = -m_F \mu_B B(x,t)$ for sublevel m_F (the Landé factor is found to be one). Now, the parameter t can be expressed in terms of the atomic motion, assumed to be classical (high-energy limit), in so far as the field chronology is completely determined by the atom history, that is, the field is synchronized with the atom emission time t_0 and applied with a delay τ :

$$t = t_0 + \frac{x}{v} + \tau.$$

The phase difference between two consecutive sublevels therefore reads :

$$\Delta\Phi = \frac{\mu_B B_0}{\hbar v} \int_d^{d+L} dx' \, \cos\frac{2\pi}{\lambda}\left[x'\left(1 - \frac{u}{v}\right) - u\left(t_0 + \tau\right)\right],$$

where B_0 stands for the pre-factor $2\mu_0 I_0 f(\lambda/R)/\pi R$. With all factors reckoned up, and taking into account that in the experiments τ is chosen such that the resonant velocity class sees the potential at the time it arrives at distance d, i. e. $\tau = d/u$, one finds :

$$\Delta\Phi = \frac{2\mu_B B_0}{\hbar \nu^*} \sin 2\pi\nu^* \frac{L}{2v} \cos 2\pi\left[\nu^* \frac{L}{2v} - \nu\left(t_0 + \frac{d}{v}\right)\right],$$

where :

$$\nu^* = \nu\left(\frac{v}{u} - 1\right).$$

ν^* can be interpreted as a "Doppler-shifted" field frequency. The dependency of the phase difference on ν^* clearly indicates a resonant behavior and a velocity matching condition. Let us re-write the expression of the phase difference in terms of the number N of field periods ; we have $L = N\lambda$ and therefore :

$$\Delta\Phi = \frac{\mu_B B_0 L}{\hbar v} \frac{\sin \pi N^*}{\pi N^*} \cos 2\pi\left[\frac{N^*}{2} - N\frac{u}{L}\left(t_0 + \frac{d}{v}\right)\right],$$

where we have defined :

$$N^\star = N\left(1 - \frac{u}{v}\right).$$

On this expression, it is clear that, when the number N of field periods is important, $\Delta\Phi$ is almost zero for all atomic velocities except for $v = u$, for which the atom experiences a fixed field. This means that the device is velocity selective and acts as an interference filter whose finesse is given by the number of field periods.

This velocity selective effect has been experimentally demonstrated, as well as the enhancement of the velocity sensitivity of the device when the number of comoving field periods in the interaction zone is increased. The interference signal versus the atom time-of-flight through the device is observed. The magnetic field magnitude is chosen such that $\Delta\Phi = (2p+1)\pi$, p integer, for the resonant velocity class. Figure 3 (a) shows the signal obtained for $N = 4$, and (b) for $N = 18.5$. As expected, the velocity selectivity is really enhanced. However, the ultimate value of 5% for the velocity finesse $\Delta v/v$ is not reached due to a broadening effect at higher field magnitude ($p > 1$). Unfortunately, as our counting rate is as low as 1 s^{-1} in these pulsed experiments, it is quite difficult to think of a smaller field magnitude than the one that has been used (50 mG, $p = 1$), as the transmitted peak width would become smaller concerning then too few atoms. It should be stressed that the device can actually work *continuously*, the time-of-flight operating mode being used here only to demonstrate as clearly as possible the desired effect. This is a real advantage over traditional velocity selection techniques using chopped sources and selective detection.

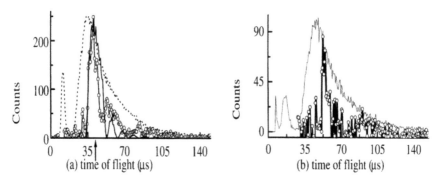

Figure 3: (a) Velocity selection for $N = 4$ (unpolarized background substracted). Dotted line represents the velocity distribution at the source point; One can distinguish two contributions: a high velocity (short time-of-flight) gaussian shaped one and a thermal low velocity. These two components correspond to different H_2 dissociation channels (b) for $N = 18.5$, which shows the sensitivity enhancement.

At this stage, one can think of supplying the helices with currents at different frequencies, in order to select several velocity classes. When N is large enough, choosing an appropriate frequency spectrum, one could obtain any given dispersion relation $\Delta\Phi(v)$. This property, called "genericity", could be used, for example, to create a medium whose dispersion relation balances the dispersion of matter wave packets in vacuum.

3 A double interferometer with a beam of metastable hydrogen atoms

When a static potential V is used, the calculation of the phase difference simply gives :

$$\Delta\Phi = \frac{1}{\hbar v}\int^{x} dx'\ \mu_B B(x'),$$

which turns into

$$\Delta\Phi = \frac{\mu_B B_0}{\hbar}\frac{L}{v}$$

if B is assumed to be constant over the interaction-zone length ($B(x) = B_0$ for $d \leq x \leq d+L$, and zero outside). The interference signal involves phaseshifts $\Delta\Phi$ and $2\Delta\Phi$, and, as a function of the time of flight of the atom through the interferometer (T), it has the following expression, which takes into account perfect polarization and analysis (i. e. the $F = 1$, $m_F = -1$ sublevel is completely eliminated) and perfect mixing in the splitting zone (the magnetic field direction makes a $\pi/2$ rotation) :

$$S(T) = \rho(T)\left(\frac{11}{16} + \frac{1}{4}\cos\frac{\mu_B B_0 L}{\hbar\mathcal{L}}T + \frac{1}{16}\cos 2\frac{\mu_B B_0 L}{\hbar\mathcal{L}}T\right),$$

where $\mathcal{L}$ denotes the total length of the interferometer, and $\rho(T)$ is the incident time-of-flight distribution. Around the most probable interaction time t_0, $\rho(T)$ is found to be well approximated by a Maxwellian function :

$$\rho(T) \sim \alpha\left(\frac{t_0}{T}\right)^5 \exp\left[-2.5\left(\frac{t_0}{T}\right)^2\right],$$

where α is a normalization constant.
Now, we can consider $S(T)$ as the time-of-flight distribution characterizing a new source, exhibiting specific coherence properties since it comes from an interferometer. This source can be used for a second interferometer. Let us denote $\rho_0(T)$ the initial time distribution, and $\rho_1(T)$ the transformed one after the atoms have passed through the interferometer. Similarly, we can deduce the time-of-flight distribution $\rho_2(T)$ after the atoms have passed through a second interferometer. We call $B_{1,2}$ and $L_{1,2}$ the magnetic field

magnitude and the length of the interaction zone in the first or second interferometer respectively, and we define

$$\omega_{1,2} = \frac{\mu_B B_{1,2} L_{1,2}}{\hbar \mathcal{L}}.$$

Then we have :

$$\rho_2(\mathcal{T}) = \rho_1 \left(\frac{11}{16} + \frac{1}{4} \cos \omega_2 \mathcal{T} + \frac{1}{16} \cos 2\omega_2 \mathcal{T} \right)$$

$$= \rho_0 \left(\frac{11}{16} + \frac{1}{4} \cos \omega_1 \mathcal{T} + \frac{1}{16} \cos 2\omega_1 \mathcal{T} \right) \left(\frac{11}{16} + \frac{1}{4} \cos \omega_2 \mathcal{T} + \frac{1}{16} \cos 2\omega_2 \mathcal{T} \right).$$

Note that ρ_1 is mainly ρ_0 modulated by a sinusoidal function at frequency ω_1 (the $2\omega_1$ factor is four times weaker). In the same way, the overmodulation of ρ_2 is mainly at frequency ω_2, and should be of maximum amplitude when $\omega_1 = \pm \omega_2$.

This has been verified experimentally by measuring time-of-flight distributions in a double-interferometer configuration, B_1 being fixed, and B_2 being varied around $-B_1$. The results are shown on Fig. 4 : B_1 is created by a current I_1 of 1 A, supplying a coil identical to that used for B_2, which is supplied with a current I_2 varying from -1 A to -1.28 A. The modulation is of maximum amplitude when $I_2/I_1 = -1.14$, which means that then $|\omega_2| = |\omega_1|$. For smaller or greater values of ω_2, the amplitude of the modulation decreases, and the lateral peaks slightly shift (overmodulation at a frequency ω_2 slightly shifted from ω_1).

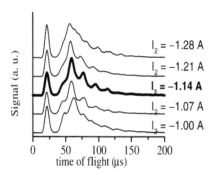 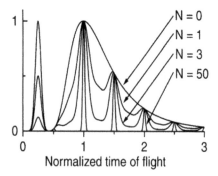

Figure 4: Signal for different values of B_2, B_1 being fixed.

Figure 5: Simulation of the signal for N successive interferometers.

One could carry on the process using a series of N interferometers, which would give an outgoing time-of-flight distribution :

$$\rho_N(\mathcal{T}) = \rho_0(\mathcal{T}) \prod_{i=1}^{N} \left(\frac{11}{16} + \frac{1}{4} \cos \omega_i \mathcal{T} + \frac{1}{16} \cos 2\omega_i \mathcal{T} \right).$$

If all frequencies ω_i are set equal, let say to ω, which means that we have a series of N identical interferometers (in the sense they all induce the same phase shift), the device can then allow us to study the so-called quantum Zeno effect, arising when a dynamical system is kept for a time under continuous observation. Usually, the continuous observation process is considered as the limiting case of a series of quasi-instantaneous measurement processes, as the delay between each measurement tends to zero. In our apparatus (in the double-interferometer configuration), each phase shifter is followed by a spin-splitting stage (in the non-adiabatic evolution region), then by an analysis filter. This means that one of the three outgoing magnetic components - which are in a *coherent* superposition - is selected and eliminated, which is completely equivalent to the detection of this spin component. As the number of interferometers is increased for a fixed overall length, the individual length of each interferometer scales as:

$$\frac{L_i}{\mathcal{L}} \sim \frac{1}{N}.$$

We remember that the definition of ω_i is :

$$\omega_i = \frac{\mu_B B_i L_i}{\hbar \mathcal{L}}.$$

so we can finally write, as long as all ω_i are equal :

$$\omega_i = \frac{1}{N}\tilde{\omega},$$

where $\tilde{\omega}$ is a certain finite constant. Under these conditions, we have :

$$\rho_N = \rho_0 \left(\frac{11}{16} + \frac{1}{4}\cos\frac{1}{N}\tilde{\omega}\mathcal{T} + \frac{1}{16}\cos\frac{2}{N}\tilde{\omega}\mathcal{T} \right)^N \stackrel{N\to\infty}{\longrightarrow} \rho_0.$$

This result meets the standard formulation of the quantum Zeno effect, as the paradoxical statement that when a particle in some initial quantum state is continuously observed to see whether it transits to another probable state (in the sense that the probability that the particle, not being observed, be in that state after a time t differs from 0), it is never be found in that state ! From the point of view of the quantum measurement theory, which states that to measure is to reduce the wave function, this also means that when the particle is observed frequently enough, it is necessarily reduced into the initial quantum state. As raised by Nakazato *et al.* [4], the problem remains to know whether a continuous observation (in the sense of the limit defined above) is physically relevant. It can be first argued that experimental uncertainties are unavoidable, and prevent, for example, the ratio $L_i/\mathcal{L}$ from being as small as desired, as would require the limit $N \to \infty$. On the other hand, as claimed by Misra and Sudarshan [5], there is no principle inherent to quantum theory that forbids the duration of a single measurement, or the delay between successive measurements, from being arbitrarily small, nor does it state the existence of an elementary interval of time. However, it seems that there is a deeper *a priori* limitation to continuous observation coming from the quantum mechanical uncertainty principle [2, 3, 4]. Then, the $N \to \infty$ is no longer physically significant as N admits a finite maximum value. Anyway, paradoxical or not it may seem, it should nevertheless be of interest for our purpose to come to higher value of the number of successive interferometers, and see what the tendency is concerning the outgoing time-of-flight distribution (Fig. 5 shows a simulation of the expected time-of-flight spectra for different values of N).

4 Interference patterns in the transverse plane

When a magnetic field depends on transverse coordinates $\vec{\rho}$ (with respect to the axis x of the beam), the resulting phase shift $\Delta\Phi$ as well as the intensity also depend on these coordinates. In a recent experiment with H* metastable atoms we have used a quadrupolar static field produced by 4 rectilinear bars parallel to x axis, supplied by alternate direct curents $\pm i_A$. The field magnitude is $B_A(\vec{\rho}) = \frac{2\mu_0}{\pi a^2} i_A \rho$ where a is the distance from the bars to the x axis. In so far as the atomic spin adiabatically follows the field direction – which is verified for almost all trajectories [6] – the resulting phase shift for a given velocity v is

$$\Delta\Phi = \frac{2\mu_0 \, g \, \mu_B \, i_A \, L}{\hbar \, \pi \, a^2 \, v} \rho_M$$

ρ_M being the value of ρ at the middle point of the field region $(d, d+L)$. The interference signal expected at a given point of the transverse plane, once averaged over the velocity distribution, has the form :

$$i(\rho) = \frac{1}{16} \left[11 + 4 < \cos \Delta\Phi > + < \cos 2\Delta\Phi > \right]$$

where $<\ >$ is the average value. The interference pattern is the "phase portrait" of the magnetic configuration : it consists of a central bright spot ($\Delta\Phi \equiv 0$) surrounded by less and less contrasted annular fringes. The spot diameter is proportional to i_A^{-1}. When an homogenous field is added to B_A the interference pattern is translated as a whole in the transverse plane. The experiment has entirely confirmed these predictions. The transverse beam profile has been explored through a moving hole ($0.1 \times 0.3 \, \text{mm}^2$), with $i_A = 1.5 \, \text{A}$ i.e. a field gradient of 47 mG/mm. It exhibits a central fringe the radius of which at half maximum is 0.36 mm. It should be noticed that (i) this size is much smaller than that of the collimating diaphragms (radius 2 mm); (ii) the rather tiny value of the gradient results from a compromize, namely to obtain a spot significally narrow but also wide enough to be explored with a hole able to provide not a too small signal (neverthless the acquisition time is about 25000 s). By adding an homogenous field of about 9.5 mG, the interference pattern has been displaced by half an interfringe perpendiculary to the explored vertical line, which gives a dark fringe on the axis.

5 Metastable Argon interferometer

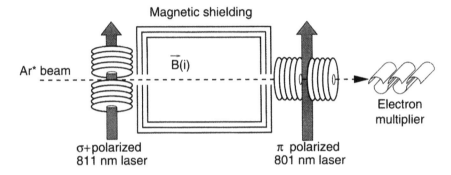

Figure 6: Experimental setup for rare gas metastable atoms Stern-Gerlach interferometry.

In principle the Stern-Gerlach interferometry works with any atom with spin. Therefore the previous experiment can be done as well with rare gas metastable atoms, such as Ar* ($3p^5\,4s$, $^3P_2$), with the great advantage of a flux 10^4 times larger and a mean velocity 20 times lower ($\Delta\Phi$ is multiplied by 20 for a given field). Here the Lamb-Retherford polarization technique can no longer be used. The polarization is achieved by a σ^+ optical pumping of the closed $^3P_2\,4s(3/2)_2 - 4p(5/2)_3$ transition ($\lambda = 811.5$ nm) in a transverse magnetic field, whereas the analysis uses the open $^3P_2 - 4p(5/2)_2$ transition ($\lambda = 801$ nm) with a π-polarization in a longitudinal magnetic field. Only atoms emerging in the $m_j = 0$ Zeeman state are allowed to pass through and be detected by an electron multiplier. The "phase-object", i.e. the region where the depahsing field is applied and its magnetic shielding, is basically the same as in the H* experiment (see fig. 6). The experiment has been successfully tested by scanning a simple homogenous transverse magnetic field. An interference pattern is shown in fig. 7. The contrast is good ($\simeq 1/3$) and the number of visible fringes ($\simeq 20$) reflects the narrow velocity distribution of the Ar* beam ($\delta v/v \simeq 10\%$). This new experimental setup is aimed to produce, by use of the method discribed in the previous section, a narrow Ar* spot (less than 100 nm in diameter) driven by an additional homogenous field to engrave lines of arbitrary shape on a resist. Contrary to other classical methods of atomic micro-lithography, the resolution of this application is not limited by an optical wavelength but by atomic diffraction. It is in the range of the de Broglie wavelength (a fraction of an Ångström for thermal beams). It does not need a drastic collimation but rather low transverse velocities (< 1 m/s), which could be obtained by a 2D laser cooling [7].

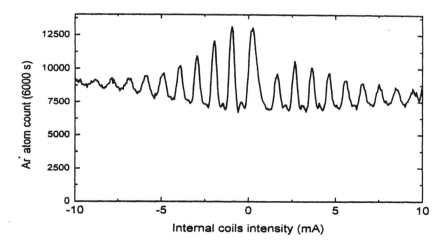

Figure 7: Interference pattern obtained with Ar* by scanning a simple homogenous transverse magnetic field.

6 Conclusion

Stern Gerlach atom interferometers are very simple devices as regards the techniques involved. Up to now they have been used in a wide variety of studies, most of them regarding fundamental aspects of Quantum Mechanics, such as Aharonov-Anandan topological phases for a spin 1 in a conical magnetic field, delayed choice interferences,... and the experiments presented here. It should be worthwhile to undertake some improved experiments, such as setting three or four successive interferometers in order to better approach the limiting Zeno behaviour, or use co-moving fields in a double interferometer configuration. Another series of experiments about non dispersive effects (cf. the so-called "scalar Bohm-Aharonov effect") have been carried out using a beam of ground-state potassium atoms (apparatus II) and temporally pulsed magnetic fields. The use of transversally inhomogeneous static fields such as a quadrupolar field results into a modulation of the transverse intensity profile of an atom beam, e.g. a narrow spot surrounded by less and less contrasted annular fringes. This pattern is at the same time a phase portrait of the magnetic configuration and a manifestation of the angular coherence of the atom beam, as it is collimated in the experimental set up. Once transposed to rare gas metastable atoms, e.g. $Ar^*(3p_54s, {}^3P_2)$, this technique can be used either in surface microscopy or in atom sub-micro lithography. The satisfactory contrast already observed in this new experiment can be further improved by a 2D transverse laser cooling providing a highly parallel and wide incident beam, and by the use of a multiple-interferometer configuration. The possibility of moving the spot at will in the transverse plane by adding an homogeneous field is a great advantage since it allows us to engrave a line of an arbitrary shape. Any atom with a spin is convenient, therefore this method can also be used for the deposition of metals (alkalis, chromium) on a substrate. All these experiments emphasize the great versatility of Stern-Gerlach atomic interferometry in the study of various effects in atom beam physics as well as in applications. Falling cold atoms initially trapped, e.g. in a magneto-optical trap (MOT), or even a Bose-Einstein condensate, can be used as well in such interferometers. This has been recently demonstrated on Cesium atoms released from a MOT [8]. With a source of such cold atoms, accurate acceleration and gravity measurements could be performed, using a tuned accelerated (i.e. chirped) co-moving field. Up to now Stern-Gerlach interferometers have been essentially used as filters (in position, velocity, angle, etc.). A very promising prospect is to use them to increase the atomic density in the phase space, by means of appropriate inhomogeneous and time-dependent fields.

Acknowledgments

M. Boustimi thanks the *Association Louis de Broglie d'Aide à la Recherche* for providing him with a Nicolas Claude Fabri de Peiresc grant.

References

[1] R. Mathevet, K. Brodsky, B. J. Lawson-Daku, Ch. Miniatura, J. Robert, and J. Baudon, Phys. Rev. A. **56**, 2954 (1997).

[2] A. DeGasperis, L. Fonda, and G. C. Ghirardi, Nuovo Cimento A **21**, 471 (1974).

[3] L. Fonda, G. C. Ghirardi, and A. Rimini, Rep. Prog. Phys. **41**, 587 (1978).

[4] H. Nagazato, M. Namiki, S. Pascazio, and H. Rauch, Phys. Lett. A **199**, 27 (1995).

[5] B. Misra and E. C. G. Sudarshan, J. Math. Phys. **18**, 756 (1977).

[6] M. Boustimi, V. Bocvarski, B. Viaris de Lesegno, K. Brodsky, F. Perales, J. Baudon, and J. Robert, Phys. Rev. A. **61**, 033602 (2000).

[7] E. Rasel, F. Pereira dos Santos, F. Pavone, F. Perales, C. S. Unnikrishnan, and M. Leduc, Eur. Phys. J. D **7**, 311 (1999).

[8] E. Maréchal, R. Long, J.-L Bossenec, R. Barbé, J.-C Keller and O. Gorceix, Phys. Rev. **A60**, 3197 (1999)

Atom Optics and Atom Polarization Interferometry Using Pulsed Magnetic Fields

O. Gorceix, E. Maréchal, S. Guibal, R. Long,
J.-L. Bossennec, R. Barbé and J.-C. Keller

Laboratoire de Physique des Lasers, Institut Galilée,
Université Paris-Nord, 93 430 - Villetaneuse, France

1 Introduction

Optical elements such as mirrors, lenses and splitters have been demonstrated recently for atomic matter waves. Atom interferometry and optics have opened exciting opportunities both for studying fundamental issues of quantum theory and for the realization of ultra-high sensitivity measurements [1]. The two fields of atom optics and atom interferometry are interrelated since breakthroughs in either of them have immediate repercussions on the other one. Both have benefited from the development of laser cooling and trapping techniques during the last fifteen years [2]. Laser manipulation of atoms provides a means to tailor the spatial and velocity distributions of atomic samples for use in subsequent experiments. Furthermore, the realization of atom condensates [3] promises the emergence of a new branch of physics where the techniques developed in atom optics will find new direct applications.

In this work we report how pulsed magnetic fields can be employed to realize most of the standard optical components needed for applications (lenses, polarizers, wave plates, etc.). Compared to alternative techniques that make use of the radiative forces exerted on atoms by laser beams, the present one is much cheaper and features much greater transverse aperture capabilities. Compared to other magnetic atom optical experiments involving permanent magnets [4], the present technique is more versatile. Indeed, changing the element characteristics only requires the control of a voltage supply without any mechanical intervention on the setup.

The basic ingredients to implement magnetic atom geometrical optics (MAGO) and atom polarization interferometry (API) are the Stern–Gerlach force and the Larmor precession, respectively. The force experienced by a paramagnetic atom in inhomogeneous magnetic fields is used to change the sample external degree of freedom distributions. The dephasing effect of magnetic fields is used to coherently manipulate the atom Zeeman coherences. The demonstrated methods are general since they are suitable for any paramagnetic species. Their validity is established for cold cesium atoms released from a standard laser cooling setup.

2 Experimental setup: Longitudinal Stern–Gerlach effect

Various experiments have been performed with a falling cold Cesium atom cloud. Figure 1 gives a scheme of the experimental setup. A standard magneto-optical trap (MOT) is repeatedly loaded with atoms inside a high-vacuum cell [2]. For each cycle, about one hundred million atoms in the $F = 4$ hyperfine level of the 6s ground state are released from this trap at a temperature of about 5 μK after a molasses phase (see [5] for details). The molasses beam cut-off time is chosen as the cycle time origin and the trap center as the spatial origin. After its release, the atom sample travels ballistically under gravity. Its properties are monitored by standard time-of-flight (TOF) methods i.e. by recording the absorption induced by the falling atoms on an horizontal counterpropagating pair of laser probe beams. Two TOF spectra are registered separately at two distinct altitudes beneath the source (160 mm for probe 1 and 950 mm for probe 2) to get the cloud initial and modified properties (rms vertical extension and rms longitudinal velocity dispersion). The rms initial cloud radius value is about 3 mm and the cloud temperature is about 5 μK. The corresponding rms velocity spread is 18 mm/s.

As a first example of a pulsed magnetic atom optical element, we analyze the sample magnetic distribution through Stern–Gerlach longitudinal separation. A 90-turn coil having its axis along the vertical atom mean trajectory, is wrapped around the glass neck of the upper cell. The atoms cross the plane of the coil located at $z_C = 95$ mm below the trap center at the mean time $t_C = 140$ ms. In order to apply the steering magnetic field B_{SG}, the coil current I is pulsed during the $[t_1, t_2]$ time interval. Here, t_1 is greater than t_C in order to keep constant the sign of the field gradient vertical component. Under the chosen experimental conditions, the ground state hyperfine splitting is much greater than the Zeeman energy shift; thus, the effective magnetic moment remains very close to its weak-field value $\mu_{eff} = -m_F \mu_B/4$. The definition of a quantization axis and the adiabatic following condition are ensured by an approximately uniform 0.04 mT vertical field $\mathbf{B}_q$ that is switched on 10 ms after the atoms are launched from the trap. The pulsed force acting on the atoms, is given by $\mathbf{F}_{SG} = \mu_{eff} \nabla |\mathbf{B}_{SG}|$ since both the Helmholtz coil induction and gradient overwhelm that of $\mathbf{B}_q$ in the relevant region. Moreover, the field modulus gradient is mainly longitudinal thanks to the geometrical parameters

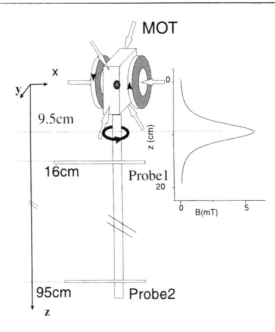

Figure 1: Schematic overview of the experimental setup. The inset shows the z-dependence of the $|\mathbf{B}_{SG}|$ magnetic field modulus for a 5 A current in the coil.

of the setup. Switching times being about 1 ms are much shorter than the $\Delta t = t_2 - t_1$ pulse duration. As demonstrated in Fig. 2 for a 5 A current pulsed during the $[t_1 = 140\,\text{ms}, t_2 = 190\,\text{ms}]$ time interval, we get a pulsed and longitudinal version of the historical Stern–Gerlach experiment [6]. Since the mean velocity at probe 2 altitude is about 4 m/s, the magnetic state selection is demonstrated with spatial separations in the 20 cm range. This demonstrates how inhomogeneous pulsed magnetic fields can be employed to analyze the atom sample spin polarization (see [5] for complementary studies).

3 Magnetic Atom Geometrical Optics (MAGO)

If the intensity I of the current supplied during the pulse is raised, strong spatial beam compressions show up in the TOF splitted spectra. This outcome is related to the second and higher-order spatial derivatives of the applied magnetic potential. The peak temporal widths are related in a non-trivial way to the spatial and velocity spreads of the atom cloud at

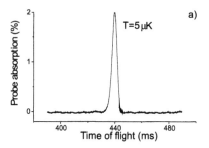 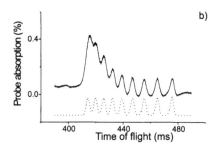

Figure 2: TOF spectra: a) Reference spectrum registered without any magnetic force applied; b) LSG resolved spectrum for a $I = 5$ A current applied during the $[140\,\text{ms}, 190\,\text{ms}]$ time interval after the cloud launching (full line); calculated TOF spectrum (vertically shifted dotted line). Experimental spectra are averaged for 10 repetitions.

the magnetic interaction cut-off time t_2. The choice of the experimental t_1, $\Delta t = t_2 - t_1$ and I parameters can be made in order to yield either a large peak separation, or a strong TOF peak compression, or a compression of the velocity distribution in the moving frame. An essential feature of this method is thus to provide ensembles of atoms in a well-defined quantum state with either a narrow longitudinal extension or a very narrow velocity spread along the vertical direction.

We have performed a thorough analysis of this cheap and versatile method for reshaping cold atom clouds [7] using the previously described experimental setup. In this experiment, the atoms are optically pumped in the $F = 4$, $MF = 4$ sublevel of the ground state before switching-on the magnetic interaction. Once again, the rms velocity and diameter along the vertical direction are measured by time-of-flight (TOF) spectra registered at two distinct altitudes (Fig. 1). Lensing effects are induced by the almost parabolic magnetic pulsed potential applied when the cloud passes in the coil vicinity. In Fig. 3, the cloud diameters at the two probe altitudes are plotted as a function of the coil current intensity for a given pulse time interval. At the curve minima, longitudinal imaging of the source is clearly demonstrated. This imaging is associated with a reduction of the arrival time dispersion and thus with spatial compression. On another hand, the occurrence of a crossing of the two curves implies that the extension of the cloud can be frozen by a proper choice of the experimental parameters [7]. The effective longitudinal temperature at that experimental point is $200\,\text{nK}$. Usual imaging and focusing capabilities (in the transverse plane) have not yet been experimentally investigated for lack of a suitable imaging system. Nevertheless, the properties of the magnetic field (deriving from Maxwell's equations) imply that longitudinal focusing comes with transverse defocusing and vice versa. As shown by numerical simulation, the reported method allows for the realization of an atomic imaging device.

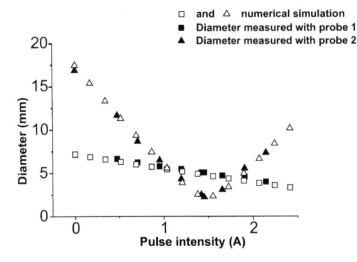

Figure 3: Cloud diameter at probe 1 (square) and probe 2 (triangle) altitudes. The first curve crossing occurs under the experimental conditions for which the thermal expansion of the cloud is frozen by the magnetic interaction.

This work shows that parabolic or quasi-parabolic pulsed magnetic fields can be used to focus an atom cloud either longitudinally or transversely. This is of interest for many experiments where a large number of atoms and a high density are needed. Applications in the field of integrated atom optics and in atom lithography can be foreseen. Strong velocity compressions have also been demonstrated that might be valuable to e.g. collision experimental studies.

4 Atom Polarization Interferometry (API)

We report here the experimental realization of an atom polarization interferometer (API). The demonstrated interferences are the atomic analogs of the patterns obtained with polarized light crossing crystal plates. It is also an extension of the Stern–Gerlach Atom Interferometry previously developped in our laboratory [8] using a fast metastable hydrogen atomic beam. In the present work, the previously described falling cold Cesium atom cloud is used. Additional coils with horizontal axis are activated to apply an almost-homogeneous horizontal magnetic field in coincidence with the arrival of the free-falling atoms at their symetry plane. The following operations are performed. First, atoms are optically pumped in the $F = 4$, $MF = 4$ sublevel by means of a circularly polarized resonant laser beam.

This first step parallels the polarization of light. After this step, atoms are polarized along their vertical trajectory while the definition of a quantization axis is ensured by a static weak vertical field $\mathbf{B_q}$. Second, a pulsed transverse homogeneous magnetic field $\mathbf{B_0}(\approx 10\ \mu\text{T})$ is applied during time τ (typically a few ten μs). Switching-on and -off times ($\approx 1\ \mu$s) are such that adiabatic following is forbidden. This step parallels the crossing of a crystal plate by light. The last step then follows: state-selective detection is realized by means of the mechanical longitudinal Stern–Gerlach effect implemented as reported above. The TOF spectra of Fig. 4 demonstrate the realization of atomic wave-plates.

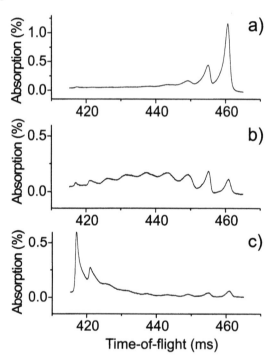

Figure 4: Time of flight spectra registered one meter below the trap. State selection is obtained via Stern–Gerlach effect. The homogeneous magnetic pulse duration τ is chosen such that the pulse induces a relative phase shift of 0 in a), $\pi/2$ in b) and π in c) between the MF $= 0$ and MF $= +1$ state. In the photonic analogy, b) stands for a quarter-wave plate, c) for a half-wave. a) is the reference spectrum.

Another way to study this interferometric method is to selectively measure the population of the MF $= 4$ (by proper time gating in the data acquisition) as a function of the dephasing pulse duration. As shown in Fig. 5, this yields a large number of high-contrast sharply-peaked nine-beam interference fringes. This experimental outcome is a manifestation of the so-called Scalar Bohm–Aharonov effect [8, 9]. In more common words, this outcome reveals Larmor precession as an implicit quantum interference effect: a coherent superposition of the nine sublevels is built through Majorana transitions at the switching-on of the current. Each of these nine components dephases under the influence of the applied field. These relative phases manifest themselves after the subsequent steps of the procedure (second coherent mixing and analysis).

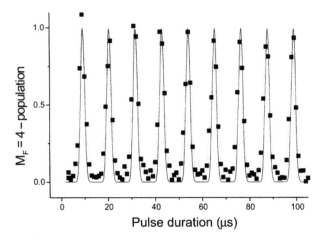

Figure 5: The modulation of the MF = 4 peak height reveals atom polarization interferences. Each point (full square) corresponds to one shot. The full line is the theoretical pattern under optimum conditions.

The demonstrated waveplates have also been combined in the time domain to realize atomic spin-echo experiments. Namely, a $\pi/2$-pulse is followed by an evolution of duration T under the influence of the ambiant slightly inhomogeneous vertical guiding magnetic field $\mathbf{B}_q$. During this phase, the dispersion of the dephasing across the cloud spatial extension results in a vanishing contrast of the fringes if a second $\pi/2$-pulse followed by state-selection is applied. Now, if a π-pulse is applied after the stage of duration T, the atoms are forced to undergo a rephasing evolution. By applying after time T' a $\pi/2$-pulse followed by state-selection, and by plotting the population of the MF = 4-state as a function of T', one gets a spin-echo signal demonstrating the revival of the spin coherence on a macroscopic space-time scale (Fig. 6). The atom ensemble longitudinal and transverse coherences can be analyzed by slight modification of the above procedure. Related experiments are in progress.

The reported results also calls for further developments of the API technique. For example, addition of inhomogeneous magnetic fields opens the

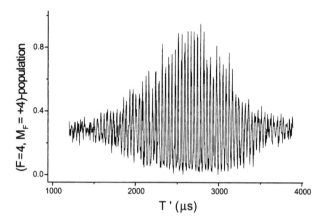

Figure 6: Spin-echo profile obtained by plotting the MF = 4-state population as a function of T' using the $\pi/2 - T - \pi - T' - \pi/2$ sequence.

way for the realization of several high-precision inertial sensors and the implementation of new methods to measure atomic ensemble spatial correlation functions.

5 Conclusion

We have presented a variety of experimental outcomes demonstrating the enormous potential of pulsed magnetic atom optics both for applied physics and for experimental tests of fundamental principles of physics. We have reported on the use of pulsed magnetic fields to realize various atom optical components. Large aperture lenses, analyzers, coherent mixers, wave plates and beam splitters for atoms have been implemented. Strong velocity or spatial compressions have been demonstrated. This cheap and versatile technique has been applied to operate a spin-polarization longitudinal atom interferometer. This latter device yields a large number of high-contrast sharply-peaked fringes well-suited for precise measurements and quantum physics experiments.

Acknowledgements

The authors thank Jacques Baudon, Christian Bordé and Christian Chardonnet (LPL, Villetaneuse) for fruitful discussions. Laboratoire de Physique des Lasers is Unité Mixte 7538 du CNRS. This work was supported by the Conseil Régional Ile-de-France (contract LUMINA #E-946-95198) and the DGA/DSP (contract STTC n 97.060).

References

[1] P. Berman Ed. Atom Interferometry, Academic Press, San-Diego (1997) and Arimondo, E. and Bachor, H., Eds. Special issue on Atom Optics, Quantum Semiclass. Opt. **8**, 495 (1996)

[2] C.S. Adams and E. Riis, Prog. Quant. Elec. **21**, 1 (1997), and references therein

[3] M.H. Anderson, J.R. Enscher, M.R. Mathews and C.E. Wienman, Science **269**, 198 (1995); K.B. Davis, M.O. Mewes, M.R. Andrews, M.J. van Druten, D.S. Durfee, D.M. Kurn and W. Ketterle, Phys. Rev. Lett. **75**, 3969 (1995) and Bradley, C. et al.: ibid. p. 1687

[4] T.M. Roach, H. Abele, M.G. Boshier, H.L. Grossmann, K.P. Zetie and E.A. Hinds, Phys Rev. Lett. **75**, 629 (1995); A.I. Sidorov, R.J. Mc Lean, W.J. Rowland, D.C. Lau, J.E. Murphy, M. Walkiewicz, G.I. Opat and P. Hannaford, Quantum Semiclass. Opt. **8**, 713 (1996); W.G. Kaenders, F. Lison, I. Mller, A. Richter, R. Wynands and D. Meschede, Phys. Rev. A **54**, 5067 (1996)

[5] E. Maréchal, S. Guibal, J.-L. Bossennec, M.-P. Gorza, R. Barbé, J.-C. Keller and O. Gorceix, Eur. Phys. J. D **2**, 195 (1998)

[6] W. Gerlach and O. Stern, Zeit. für Phys. **8**, 110 (1922); **9**, 349 (1922)

[7] E. Maréchal, S. Guibal, J.-L. Bossennec, R. Barbé, J.-C. Keller and O. Gorceix, Phys Rev. A **59**, 4636 (1999)

[8] S. Nic Chormaic, D. Miniatura, O. Gorxeix, B. Viaris, J. Robert, S. Féron, V. Lorent, J. Reinhardt, J. Baudon and K. Rubin, Phys. Rev. Lett. **72**, 1 (1994)

[9] E. Maréchal, R. Long, J.-L. Bossennec, R. Barbé, J.-C. Keller and O. Gorceix, Phys Rev. A **60**, 3197 (1999)

Making Molecules at MicroKelvin

William C. Stwalley

Department of Physics University of Connecticut Storrs, CT U.S.A

1 Introduction

The production of molecules with translational kinetic energies small compared to kT at 1 mK ("ultracold") has only just been observed in 1998. Here proposals and experiments based on photoassociation (free $\to$ bound absorption) of ultracold atoms [1] (recently reviewed in references [2, 3, 4]) which produce translationally ultracold molecules will be briefly surveyed. The use of ultracold photoassociation (Sect. 2) to form ultracold molecules (Sect. 3) is a major long term goal at the University of Connecticut. Such ultracold molecules could be used for molecule trapping, for the molecular analog of atom optics, for study of highly quantum-mechanical, resonance-dominated ultracold collisions, for fundamental nucleation studies, for formation of molecular Bose–Einstein condensates (BECs) and for production of molecule lasers (i.e. coherent beams of state-selected molecules). A separate and much more extensive review of ultracold molecule formation has recently appeared [5].

2 Photoassociation of Ultracold Atoms

Recent advances in atom cooling and trapping have provided samples of isotopically-selected atomic gases at densities of $> 10^{11}/\text{cm}^3$ and ultracold temperatures (without using cryogenics) below 1 mK. Here the thermal kinetic energy $[kT/h = 21\,\text{MHz at 1 mK}]$ is comparable to or smaller than many terms in the Hamiltonian (Table 1), including, for example, the natural energy linewidth of excited states! For example, for two K atoms, the long range potential (J = 0) is well approximated by an inverse power sum $(-C_6R^{-6} - C_8R^{-8} - C_{10}R^{-10})$ [6]. For the effective potentials

$$U_J(R) = V(R) + \frac{\hbar^2[J(J+1) - \Omega^2]}{2\mu R^2},$$

at a collision energy of $\sim 3kT$ (1 mK) in our trap, only J = 0 (s-wave) and J = 1 (p-wave) collisions directly reach short distances. J = 2 (d-wave) collisions are reflected at ~ 80 Å unless tunneling occurs (which has significant probability because of the long DeBroglie wavelength) and J = 3 (f-wave) collisions are reflected at ~ 120 Å. Thus our photoassociation spectra are dominated by lower s, p and d-wave free states, giving very simple rotational spectra corresponding to J = 0, 1 and 2 only in the initial free state of the colliding atomic pairs.

This provides new (and relatively simple and inexpensive) opportunities to study high resolution long range molecular spectroscopy by free → bound photoassociation of ultracold atoms as first pointed out in [1]. This is because, compared to kT, the vibrational splittings are large ($kT \ll \Delta G_{v'+1/2}$) and even the rotational splittings are large ($kT \ll B_{v'}$). Moreover, the 4p fine structure splitting (57.7 cm^{-1}) is large and even the ^{39}K ground state hyperfine splitting (462 MHz) is large. Only the excited state hyperfine splittings, natural linewidths, Zeeman splittings and AC Stark shifts and widths are less than $10kT$ in our experiments.

Table 1: Ultracold collisions ($T \lesssim 1$ mK) take place in a regime where the kinetic energy of the colliding atoms is comparable to or smaller than the following quantities

Electronic Splittings	AC/DC Stark Splittings
Vibrational Splittings	Electronic and Nuclear Zeeman Splittings
Rotational Splittings	Isotope Shifts
Fine Structure Splittings	Centrifugal Barriers
Hyperfine Structure Splittings	Natural Linewidths

For the purposes of this summary, it is sufficient to understand the operation of a so-called Cell Dark Spot Magneto-Optical Trap (CDSMOT) [6]. Operationally our CDSMOT is a robust sample of isotopically selected ultracold atoms with a density of 10^{11} K/cm^3 and $T = 300\,\mu$K. It is composed by three major parts: 1) an ultrahigh vacuum chamber, 2) three pairs of circularly polarized, red-detuned, counterpropagating laser beams, and 3) a pair of anti-Helmholtz coils. Our potassium cell MOT was built in a ten arm stainless steel UHV chamber with a background pressure of 5×10^{-10} Torr and a room temperature potassium vapor pressure of $\sim 10^{-8}$ Torr. The three pairs of orthogonal laser beams are provided by a single mode tunable ring Ti:Sapphire laser. The ring laser frequency is locked to the ^{39}K $4^2S_{1/2}(F'' = 2) \to 4^2P_{3/2}(F' = 3)$ transition using saturated absorption for long-term frequency stability ($\Delta \nu < 1$ MHz). Further details may be found in [4, 6].

The photoassociation is induced by a second Ti:Sapphire laser with typical output power of ~ 500 mW. This intense laser beam is focused to a diameter of 0.5 mm at the trap region. The absolute laser frequency is calibrated by uranium atomic lines as well as the potassium resonance lines. The detection schemes are discussed below.

The ultracold photoassociative process is exemplified by the reaction

$$K + K + h\nu_1 \to K_2^*(v', J') \qquad (1)$$

where the small magnitude of kT allows for excitation of a single low J' rovibrational level in a specific electronically excited state, just as in laser-induced fluorescence of K_2. The singly excited $K_2^*(v', J')$ molecules (here assumed to be near the 4s + 4p asymptotes) then decay radiatively in bound $\to$ bound

$$K_2^*(v', J') \to K_2(v', J') + h\nu_2 \qquad (2)$$

or bound $\to$ free emission

$$K_2^*(v', J') \to K + K + h\nu_3, \qquad (3)$$

(where $K_2(v', J')$ is either the ground $X^1\Sigma_g^+$ state or the lowest triplet state $(a^3\Sigma_u^+)$).

Since significant atomic fluorescence ($4p_{3/2} \to 4s_{1/2}$) is excited by the very near resonance trap laser, the atom density is readily monitored. Process 1 yields a decrease in atomic density, dependent on the fate of the excited molecules. If process 2 occurred exclusively, trap loss would occur with maximum efficiency (two atoms lost per photoassociative photon absorbed) assuming the excited molecule cannot emit bound-bound photons with a wavelength in the narrow bandpass of the filter ($4p_{3/2} \to 4s_{1/2}$) (which is very likely). However, process 2 is a relatively minor process for levels near dissociation and process 3 dominates as is well known from earlier laser-induced fluorescence studies starting from ground state molecules. If process 3 occurs nearly exclusively, the question becomes the distribution of final kinetic energies in the bound $\to$ free emission. In particular, if the identical kinetic energies of the two separating atoms are greater than the trap depth (typically 1 K for a MOT such as ours) the atoms will escape and "trap loss" will be detected by diminished atomic fluorescence. If no such "hot" atoms (KE $\geq$ 1 K) are formed by bound-free emission, there will be no trap loss and no photoassociation detection by trap loss. Examples of levels showing negligible trap loss in a MOT are the zero-point levels of the pure long range states (0_g^- and 1_u) at 28 and 39 Å, respectively, and are discussed in detail in [6, 7].

Alternately, the singly excited $K_2^*(v', J')$ molecules can be further excited [8]

$$K_2^*(v', J') + h\nu_4 \to K_2^{**}(v, J) \qquad (4)$$

or single- or multi-photon ionized to form molecular

$$K_2^*(v', J') + nh\nu_5 \to K_2^+(v^+, N^+) + e^- \qquad (5)$$

or atomic ions

$$K_2^*(v', J') + nh\nu_6 \to K + K^+ + e^- . \qquad (6)$$

The singly excited $K_2^*(v', J')$ molecules in some cases nonradiatively decay (predissociate) to fragments [9]

$$K_2^*(v', J') \to K^*(4p_{1/2}) + K(4s_{1/2}) . \qquad (7)$$

Finally, the doubly excited molecules $K_2^{**}(v, J)$ can decay radiatively [as in (2) and (3)], or nonradiatively by predissociation as in (7) or by autoionization [8]

$$K_2^{**}(v, J) \to K_2^+(v^+, N^+) + e^- \qquad (8)$$

or by ion pair formation

$$K_2^{**}(v, J) \to K^+ + K^- . \qquad (9)$$

The $K_2(v', J')$ and $K_2^{**}(v, J)$ can also be single- or multi-photon ionized as in (5). In addition, singly- and doubly-excited atomic fragments from predissociation of K_2^* and K_2^{**} undergo radiative decay such as

$$K^*(4p_{1/2}) \to K(4s_{1/2}) + h\nu_7 . \qquad (10)$$

The doubly-excited fragments (e.g. K^{**} (5d)) undergo associative ionization as well

$$K^{**}(5d) + K(4s_{1/2}) \to K_2^+(v^+, N^+) + e^- , \qquad (11)$$

and ion pair formation (e.g. K^{**} (6d))

$$K^{**}(6d) + K(4s_{1/2}) \to K^+ + K^- . \qquad (12)$$

Collisional energy transfer is a final possibility, e.g.

$$K^{**}(6s) + K(4s_{1/2}) \to K^{**}(4d) + K(4s_{1/2}) . \qquad (13)$$

All these processes are significantly constrained by the well-known energetics [4] of the K atom (e.g. IP = 35004.18 cm^{-1}) and K_2 molecule (e.g. $D_0^0(K_2) = 4404.583 \pm 0.072$ cm^{-1}, IP$(K_2) = E_x(v^+ = 0, N^+ = 0) - E_x(v'' = 0, J'' = 0) = 32775.5 \pm 0.15$ cm^{-1} and $D_0^0(K_2^+) = 6633.26 \pm 0.16$ cm^{-1}).

The above processes suggest a wide variety of detection schemes for ultracold photoassociation, four of which have been implemented as summarized in Table 2. We measure the trap loss rate (1. in Table 2) by monitoring the $4p_{3/2} \rightarrow 4s_{1/2}$ atomic fluorescence of trapped K atoms using a photomultiplier-filter system. An additional CW ring laser provides the second photon for the two color optical-optical double-resonance photoassociative spectroscopy (2. in Table 2) and the fragmentation atomic resonance-enhanced multiphoton ionization spectroscopy (3. in Table 2). Molecular and atomic ions generated in techniques 2. and 3. are collected by a channeltron particle multiplier. Alternatively, the translationally ultracold ground state molecules formed by photoassociation followed by bound-bound emission (4. in Table 2) are detected with pulsed laser resonance-enhanced multiphoton ionization and similarly detected.

Table 2: Ultracold Photoassociative Spectroscopy Detection Techniques Used in Studies of $^{39}K_2$

1. Trap Loss (Decrease of Atomic Fluorescence) [6]
$K + K + h\nu \rightarrow K_2^* \rightarrow K + K + h\nu'$ "hot" loss
 "cold" no loss
$ \rightarrow K_2 + h\nu''$ loss

2. Direct Molecular Ionization [6, 8]
$K + K + h\nu \rightarrow K_2^* \xrightarrow{(1 \text{ or } 2)h\nu'} K_2^+ + e^-$ or $K^+ + K + e^-$

3. Fragmentation Spectroscopy [9]
$K + K + h\nu \rightarrow K_2^* \rightarrow K^*(^2P_{1/2}) + K$
$K^*(^2P_{1/2}) + 2h\nu' \rightarrow K^+ + e^-$ REMPI (via 5d)

4. Ground State Molecule Detection [10]
$K + K + h\nu \rightarrow K_2^* \rightarrow K_2(v') + h\nu'$
$K_2(v') + h\nu'' + h\nu''' \rightarrow K_2^+ + e^-$ REMPI (via B $^1\Pi_u$)

3 Formation of Ultracold Molecules

Ultracold photoassociation is the leading technique for producing ultracold molecules (reviewed in [5]) and has recently been used to produce translationally ultracold K_2 [10, 11] and Cs_2 [12, 13, 14, 15] molecules.
The first results [12], on Cs_2, produced unspecified levels clearly in the metastable $a^3\Sigma_u^+$ state, which were detected by resonance-enhanced multiphoton ionization. Subsequent experiments [13] have partly assigned the spectra and also provided evidence that some $X^1\Sigma_g^+$ ground state Cs_2 molecules are also being formed.
Our first results on K_2 [10] clearly indicate that we are producing $v = 36$ of the $X^1\Sigma_g^+$ ground state of $^{39}K_2$ via photoassociation to a specific level (tentatively $v = 191$) of $A^1\Sigma_u^+ \sim O_u^+(4p_{1/2})$ state. Level $v = 36$ is then resonance-enhanced two-photon-ionized through various levels of the $B^1\Pi_u$

state (e.g. $v = 26$). The photoassociative spectrum detected by trap loss for the $O_u^+(4p_{1/2})$ state agrees with that detected by molecule formation.
Other results [14, 15], where the trap laser photoassociation and/or three body recombination produces the molecules, report an unspecified distribution of vibrational levels (probably near dissociation) in the $a^3\Sigma_u^+$ and possibly the $X^1\Sigma_g^+$ state of Cs_2. These results [15] include clear evidence of trapping a few molecules in an optical trap.
Recently we have focussed on finding more efficient and selective methods for production of translationally ultracold molecules [11] and on employing methods giving narrower distributions involving only low vibrational levels of the ground $X^1\Sigma_g^+$ state for which an internal cooling scheme is available [16]. Interesting suggestions include the two-color excitation scheme of [17] and the triplet state formation scheme of [18]. Our recent results [11], based on the two-color excitation scheme [17], produce $\sim 10^6$ molecules per second ($\sim 10\%$ in certain individual vibrational levels such as $v'' = 0$ and 25). This is far more than our earlier one-color results [10] of $\sim 10^3$ molecules/second.
Also of significance are the stimulated Raman proposals [19, 20] which could directly produce state-selected molecules (perhaps even $v = 0$, $J = 0$ [19] and perhaps even a "molecule laser" from an atomic BEC [20]).

Acknowledgments

This work was carried out in collaboration with Professors Phillip Gould and Edward Eyler, Drs. He Wang, John Bahns and Jason Ensher, and Xiaotian Wang, Jing Li and Anguel Nikolov, and was supported in part by the National Science Foundation.

References

[1] H.R. Thorsheim, J. Weiner, P.S. Julienne: Phys. Rev. Letters **58**, 2420 (1987)

[2] P.D. Lett, P.S. Julienne, W.D. Phillips: Ann. Rev. Phys. Chem. **46**, 423 (1995)

[3] J. Weiner, V.S. Bagnato, S.C. Zilio, P.S. Julienne: Rev. Mod. Phys. **71**, 1 (1999)

[4] W.C. Stwalley, H. Wang: J. Mol. Spectrosc. **195**, 194 (1999)

[5] J.T. Bahns, P.L. Gould, W.C. Stwalley: Adv. At. Mol. Opt. Phys. **42**, 177 (1999)

[6] H. Wang, P.L. Gould, W.C. Stwalley: J. Chem. Phys. **106**, 7899 (1997)

[7] X. Wang, H. Wang, P.L. Gould, W.C. Stwalley: Phys. Rev. A **57**, 4600 (1998)

[8] H. Wang, X.T. Wang, P.L. Gould, W.C. Stwalley: Phys. Rev. Lett. **78**, 4173 (1997)

[9] H. Wang, P.L. Gould, W.C. Stwalley: Phys. Rev. Lett. **80**, 476 (1998)

[10] A.N. Nikolov, E.E. Eyler, X. Wang, H. Wang, J. Li, W.C. Stwalley, P.L. Gould: Phys. Rev. Lett. **82**, 702 (1999)

[11] A.N. Nikolov, J.R. Ensher, E.E. Eyler, H. Wang, W.C. Stwalley, P.L. Gould: Phys. Rev. Lett. **84**, 246 (2000)

[12] A. Fioretti, D. Comparat, A. Crubellier, O. Dulieu, F. Masnou-Seeuws, P. Pillet: Phys. Rev. Lett. **80**, 4402 (1998)

[13] A. Fioretti, D. Comparat, C. Drag, C. Amiot, O. Dulieu, F. Masnou-Seeuws, P. Pillet: Eur. Phys. J. D **5**, 389 (1999)

[14] T. Takekoshi, B.M. Patterson, R.J. Knize: Phys. Rev. A **51**, R5 (1999)

[15] T. Takekoshi, B.M. Paterson, R.J. Knize: Phys. Rev. Lett. **81**, 5105 (1998)

[16] J.T. Bahns, W.C. Stwalley, P.L. Gould: J. Chem. Phys., **104**, 9689 (1996)

[17] Y.B. Band, P.S. Julienne: Phys. Rev. A **51**, R4317 (1995)

[18] R. Côté, A. Dalgarno: Chem. Phys. Letters, **279**, 50 (1997)

[19] A. Vardi, D. Abrashkevich, E. Frishman, M. Shapiro: J. Chem. Phys. **107**, 6166 (1997)

[20] P.S. Julienne, K. Burnett, Y.B. Band, W.C. Stwalley: Phys. Rev. A **58**, R797 (1998)

Molecular Photoassociation and Ultracold Molecules

Pierre Pillet, Françoise Masnou-Seeuws, Anne Crubellier
Laboratoire Aimé Cotton,* CNRS II, Bât. 505,
Campus d'Orsay, 91405 Orsay cedex, France

1 Introduction

During the last fifteen years, the laser manipulation of neutral atoms has experienced impressive development. The experimental techniques of laser cooling of atoms in the mK-μK range and below, as well as the trapping of neutral atomic samples, based on radiation pressure, are now well established. However their extension to molecules seems to be very difficult because of the lack of a two-level optical pumping scheme for recycling population [1]. During the last years, molecules have hardly been touched by the developments in laser cooling. One can quote the deflection of a molecular beam [2] or the demonstration by Djeu and Whitney [3] of laser cooling by spontaneous anti-Stokes scattering, introduced long ago by Kastler as "luminorefrigeration" [4]. The latter method presents however poor efficiency and poor control. An alternative interesting specific scheme for obtaining of a cold molecular sample is to start from cold and dense atomic samples and to form cold molecules by molecular photoassociation (PA) of two cold atoms [5].

In the PA process, a pair of free cold atoms absorb resonantly one photon and produce an excited molecule in a well-defined ro-vibrational level. In a dilute medium PA is a collisional process, essentially efficient at long-range interatomic distance (R), the probability to find two atoms at a distance R, varying as R^2. The molecular PA of cold atoms is therefore a technique particularly well adapted to the study of long-range molecular states. For homonuclear dimers of alkali atoms, the molecular potential curves, below the first $s + p$ dissociation limit, present a R^{-3} asymptotic behavior. It is thus possible to populate very efficiently by PA vibrational levels presenting very large elongation, up to few hundred atomic units ($1a_0 = 0.529177$ Å).

---
*Laboratoire Aimé Cotton is associated with Université Paris-Sud

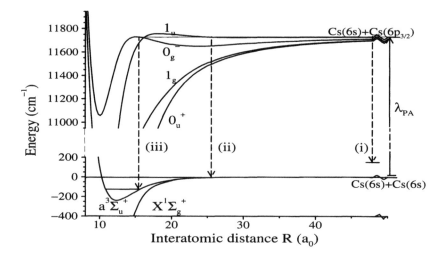

Figure 1: Cs$_2$ relevant potential curves. The right vertical arrow represents the absorption by a pair of atoms. The dashed arrow (i) represents the long-range spontaneous emission to continuum states, with dissociation of the molecule; the dashed arrows (ii) and (iii) represent the spontaneous emission to bound states, with formation of cold molecules.

We notice that this R-dependence is R^{-3}, for the homonuclear systems, which corresponds to the electrostatic dipole-dipole long-range interaction, but that it is R^{-6} for the heteronuclear systems which present a van der Waals asymptotic behavior, as for the ground-state alkali molecular potentials. PA is so expected to be less efficient for heteronuclear systems. PA has already been demonstrated for alkali atoms from Li to Cs [7, 9, 6, 8, 10], and more recently for hydrogen atom [11] and for metastable helium atom [12]. Experiments with alkaline earth atoms are now in progress. Preliminary results for heteronuclear alkali systems have also been reported [13]. The principle of the PA experiments is quite similar for all alkali atoms. We will present here the case of the cesium atom, which is studied at Laboratoire Aimé Cotton and which is an especially good candidate for the formation of cold molecules. The cesium dimer presents four attractive long-range Hund's case (c) states below the $6s_{1/2} + 6p_{3/2}$ limit (see figure 1): 1_g, 0_u^+, 0_g^-, and 1_u (an electric dipole transition from the ground state to the attractive 2_u states is forbidden), which can be populated by photoassociation. While the states 1_g and 0_u^+ correspond to deep molecular potential wells with a "hard wall" very repulsive at short interatomic distances, we notice that the 0_g^- and 1_u states present a double-well shape for the molecular potential curves, so that the ro-vibrational levels inside the outer long-range well can be populated by PA. The origin of this long-range well is due to an avoided crossing with the curve of the same symmetry correlated to the $6s_{1/2}+6p_{1/2}$ dissociation limit. In the case of the cesium, due to the large value of the spin-orbit splitting, these pseudo-crossings are however located at shorter distance, if we compare with the other alkali atoms. We can finally notice the smooth hump separating the inner and outer wells, due to the competition between the leading R^{-3} term and the higher order terms in the multipole expansion [14]. We will see that those properties are particularly important to understand the efficient formation of cold molecules with the cesium atom.

In a cold atomic sample, the obtained excited photoassociated molecules are translationally cold (neglecting the recoil energy, the molecular sample has the same temperature as the atomic one). Subsequent deexcitation of these photoassociated molecules appears thus as an obvious way to form cold ground state molecules. However, as PA corresponds to a long-range excitation, the bound-bound spontaneous decay probability is generally very small [15]. Spontaneous emission leads back mostly to dissociation of the transient cold excited molecules into two free atoms, with on average, much more relative kinetic energy than initially. This is the case for the 0_u^+ and 1_g states of cesium dimer. The pair of restored atoms escapes outside the atomic trap, and the analysis of the trap losses is a common way for the detection of the PA process. In contrast, in the case of the 0_g^- and 1_u states of the cesium dimer, the excited molecule oscillates in an external well between long-range ($> 40a_0$) and intermediate distances (~ 15 or $25a_0$). Spontaneous emission can occur at intermediate distance in a bound-bound transition, and cold stable ground state or metastable lowest triplet state molecules can be formed. Spontaneous emission at long-range distance is of course also present and leads always to dissociation. The formation of translationally cold molecules is therefore due here to a peculiarity of the external potential wells which offers at the same time an efficient photoassociation rate and a reasonable branching ratio of spontaneous emission towards the lowest two states.

The paper is organized as follows. Section 2 is devoted to the description of the experiments. The PA spectra obtained are then described and analyzed in section 3. Three aspects are discussed: the long-range molecular spectroscopy, an unexpected tunneling effect and the application of PA to the determination of nodal structure of the initial radial wave-function of the two colliding atoms. In section 4, we report the evidence for the formation of cold molecules in the ground state and the lowest triplet state. In particular, the measurements of the temperature of the cold molecules and of the formation rate are described. Finally, in section 5 we conclude and we analyze the perspectives in the research field of cold molecules.

2 Experimental Setup

The PA experiments with cold alkali atoms use the same tools, used to provide cold atomic samples: magneto-optical trap (MOT), dark SPOT (dark spontaneous force optical trap) , or dipole trap such as FORT (far off resonance trap) [9]. The experimental procedure consists in illuminating the cold atoms by a PA cw laser beam. The analysis of trap losses due to the PA process has been used for all the alkali atoms and allows one to obtain the PA spectra. Other detection techniques have also been used such as associative ionization through excitation of autoionizing states [7], direct photoionization [16], or predissociation of levels formed by PA [17]. In the case of formation of cold molecules in the ground state and the lowest triplet state, their direct detection by photoionization is also the signature of PA [10]. The figure 2 shows a schema of the experimental set-up used at Laboratoire Aimé Cotton. The cold atom source is provided by the use of either a Cs vapor loaded MOT or a dark SPOT. Using a dark SPOT instead of a MOT implies a different initial state of the pair of atoms in the PA collisional process : the atoms are in the $6s_{1/2}$, $f = 3$ hyperfine level, instead of being in $f = 4$. The trapped, cold Cs atoms are illuminated with a cw laser to produce the photoassociative transitions; one has:

$$2 \text{ Cs}\,(6s, f = 4) + h\nu_1 \to \text{Cs}_2\,(\Omega_{u,g}(6s + 6p_{3/2}; v, J)), \tag{1}$$

for the MOT experiment, and

$$2 \text{ Cs}\,(6s, f = 3) + h\nu_1 \to \text{Cs}_2\,(\Omega_{u,g}(6s + 6p_{3/2}; v, J)), \tag{2}$$

for the dark SPOT experiment. More details about the experimental set-up can be found in references[14, 18].

The dimension (FWHM) of the bright cold MOT sample ranges between 400-600 μm, and the number of atoms in the trap ranges between $1-5 \times 10^7$, leading to a peak density of the order of 10^{11} atoms/cm^3. Under these experimental conditions, the temperature of the cold atomic sample is $T \simeq 200$ μK. The MOT is shifted into a dark SPOT configuration by modifying the repumping laser arrangement to transfer most of the atoms (>90%) in the "dark" state $f = 3$ in the center of the trap. In the MOT and dark SPOT devices, atoms can be further cooled below the Doppler limit down to $T \leq 30$ μK [19]. PA is achieved by continuously illuminating the cold Cs atoms with the beam of a Ti:Sapphire laser (Coherent 899 ring laser) pumped by an Argon ion laser. The maximum available power in the experiment zone is 600 mW, focussed on a spot with ~ 300 μm diameter, leading to an intensity in the MOT zone ranging 100-500 W cm^{-2}. The frequency scale is calibrated using a Fabry Perot interferometer and the absorption lines of iodine. The maximum absolute uncertainty is estimated to be ± 150 MHz, mainly due to the uncertainties in the positions of the iodine lines. The local uncertainty is about ± 10 MHz.

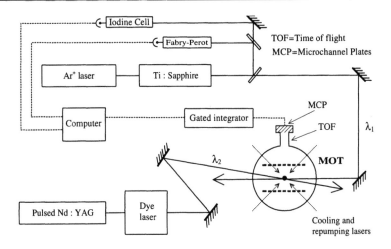

Figure 2: Scheme of the experimental setup. The optical and laser setup for the MOT is not shown, as well as the two vertical trapping laser beams.

Two kinds of detection are used for observing the photoassociation process and for recording vibrational progressions. First, one can observe with a photodiode the fluorescence yield from the trap, which allows one to analyze the trap losses. Second, in the case of formation of cold molecules in the ground state or the lowest triplet state, we can use photoionization of the translationally cold Cs_2 molecules into Cs_2^+ photo-ions, which are detected through a time of flight mass spectrometer. Ionization is made using a pulsed dye laser (pulse duration: 7 ns; pulse energy: 1 mJ) pumped by the second harmonic of a Nd-YAG laser, running at 10 Hz repetition rate. The dye laser is tuned at the wavelength $\lambda_2 \sim 716$ nm. The photoionization process is a REMPI process (Resonance Enhanced Multi-Photon Ionization), via the vibrational levels of an electronic molecular state correlated to the $6s_{1/2} + 5d_{3/2,5/2}$ dissociation limit [10, 20]. At the trap position, a high-voltage field is applied by means of a pair of grids. The produced ions are expelled out of the interaction region in a 6 cm field-free zone which constitutes a time-of-flight mass spectrometer discriminating Cs_2^+ ions from Cs^+ ones. The ions are detected by a pair of micro-channel plates and the Cs_2^+ ion signal is recorded with a gated integrator.

3 Photoassociative Spectroscopy

3.1 Photoassociation of cold atoms: a new tool for molecular spectroscopy

The PA process with cold atoms corresponds to a transition from a pair of atoms to a molecule. An interesting property is the resonant character of this process, which has consequences for molecular spectroscopy. In particular, accurate determinations for the long-range part of molecular potential curves have been made [5, 21, 22]. The resolution of the PA spectroscopy is limited by the width of the statistical distribution of the relative colliding kinetic energy, of the order of $k_B T$ (k_B is the Boltzmann constant and T the atomic sample temperature), which corresponds to about 2 MHz at T≈100 µK. Taking into account the natural width of the resonance line and the spectral width of the PA laser, we should be able to reach a resolution better than 10 MHz. In the previous articles [10, 14, 18, 23], we have reported the photoassociative spectroscopy with cold cesium atoms. A summary of our data is shown in figure 3 with typical spectra obtained by using a MOT atomic sample. The fluorescence and the Cs_2^+ ion spectra, recorded as a function of the PA laser frequency, are quite different. Clearly first we observe resonance lines up to a PA laser detuning of 80 cm^{-1} in the case of the Cs_2^+ ion spectrum, and only of 40 cm^{-1} for the trap-loss one, demonstrating the high sensitivity of the ion detection. Second, the density of resonance lines in the trap-loss spectrum is much more important. In section 4, we will come back to the reasons for these differences, which are deeply linked with the formation of cold molecules. In this section, to show the richness of photoassociative spectroscopy, we will focus on the spectroscopic analysis of the Cs_2^+ ion spectrum, which exhibits the vibrational progressions of the 0_g^- and 1_u states.

3.2 At the frontier of atomic and molecular physics: Cs_2 1_u pure long-range state

The large structures in the Cs_2^+ ion spectrum in the range (3 − 7 cm^{-1}) are assigned to the 1_u state. Figure 4 shows the well resolved structure of the vibrational level $v = 1$ of the 1_u state, obtained by using a dark SPOT device. The spectrum does not exhibit any simple structure, neither hyperfine nor rotational. To interpret these complex structures, systematic asymptotic calculations including both hyperfine structure and molecular rotation, for all the electronic states involved in 1_u photoassociation, have been performed [23]. A precise calculation of the hyperfine structure of all adiabatic asymptotic curves for the excited $(6s + 6p)$ asymptotes (without rotation) are calculated, by using the following procedure. The matrix elements of the relevant terms of the multipole expansion (R^{-3}, R^{-6}, R^{-8}),

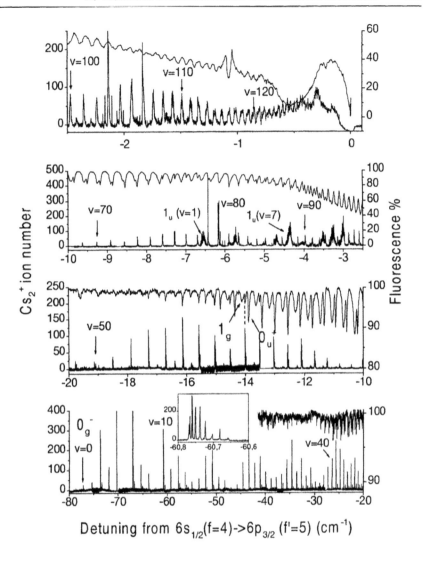

Figure 3: Cs_2^+ ion signal (lower curve) and trap fluorescence yield (upper curve) versus detuning of PA laser. In the inset, details of the rotational progression of the 0_g^- $v = 10$ level are shown. The dashed line indicates the correspondance of a vibrational level of the 0_g^- state on both spectra. The origin of the energy scale is fixed at $6s_{1/2}$, $f = 4 \rightarrow 6p_{3/2}$, $f\prime = 5$ atomic transition. For detunings smaller than 0.1 cm^{-1}, the MOT is destroyed by PA laser. Notice the different scales for the axis.

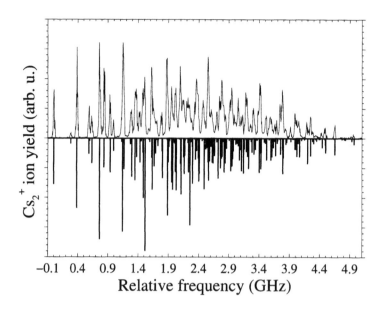

Figure 4: Structure (upper) of the $v = 1$ level of the 1_u vibrational progression, compared (lower) in mirror with the calculated intensities.

of the fine and hyperfine interaction are determined. No exchange energy is introduced in the s+p calculations because it is still small at the position of the inner wall of the 1_u potentials (about 25 a_0). As rotation and hyperfine structure are quite entangled, the two effects have to be treated simultaneously and details are found in references [23, 17]. The complete calculation of the intensities in which the only adjustable parameters are the initial populations of the different partial waves (from s to g) agrees well with experiment (see figure 4). The 1_u state can really be labelled as a pure long-range state, corresponding to a pair of atoms, the cohesion of which is given by the electrostatic long-range multipole interaction [25]. Such a molecular state is really at the frontier of atomic physics.

3.3 An unexpected tunneling effect in an heavy molecule

The recorded Cs_2^+ ion spectrum exhibits 133 well resolved structures assigned as the vibrational progression of the 0_g^- state, starting at $v = 0$. The rotational structure, shown for $v = 10$ in the inset of the figure 3, is observed up to $J = 8$ for most of the vibrational levels below $v = 74$. The energies of the spectral lines of the 0_g^- vibrational progression have been analyzed

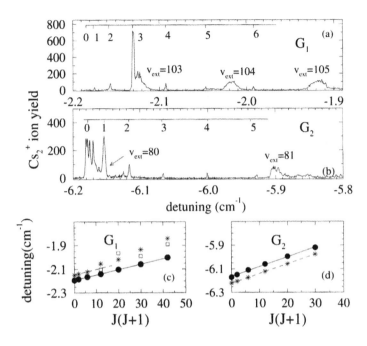

Figure 5: Details of the photoassociation spectrum in the region of (a) the G_1 and (b) G_2 structures, with their J labelling. The rotational structure of the "regular" levels $v_{ext} = 103, 104, 105$ is unresolved. (c) and (d) show the variation, as a function of $J(J+1)$, of the binding energies for the rotational levels identified in the G_1 and G_2 structures. The slope of the curves yields the rotational constants. Black circles: experiment. Stars and open squares: computed energies.

with a Rydberg-Klein-Rees (RKR) and near dissociation expansion (NDE) approach, yielding, for the outer well, an effective potential curve with a 77.94 ± 0.01 cm^{-1} depth and an equilibrium distance $R_e = 23.36 \pm 0.10$ a_0 [14]. This approach provides a good knowledge for the inner and outer turning points of the classical vibrational motion up to $v = 74$ and for the vibrational wave-functions. The PA ion spectrum of figure 3 manifests also unexpected structures in the 0_g^- vibrational progression around the resonances corresponding to the vibrational levels $v = 103 - 104$ and $v = 80 - 81$. We present in figure 5 details of these unassigned structures labelled G_1 and G_2, which are significantly more intense and display larger rotational structure than the neighboring "regular" lines (presenting unresolved rotational structures). The structure G_1 displays rotational lines from $J_1 = 0$ to 6 (Fig.5a). The $J_1=3$ line, at $\delta_1 = -2.14$ cm^{-1}, on top of the $v = 103$ line, is about 7 times more intense. A rotational constant $B_v^1 = 137 \pm 4$ MHz has been fitted to the variation of the energies according to $E_v^1 = B_v^1 \times J(J+1)$ (Fig.5c). The structure G_2 exhibits a more intense rotational component at $\delta_2 = -6.15$ cm^{-1} on top of the $v = 80$ line (Fig.5b). The identification of the $J_2 = 0$ component is not easy as the G_2 structure is strongly embedded in the ill-resolved rotational structure of the $v = 80$ level. A rotational constant $B_v^2 = 243 \pm 8$ MHz can be extracted (Fig. 5d), once the position for the $J_2 = 0$ line is chosen as shown in Fig. 5c.

From the definition of the rotational constant of a vibrational level v: $B_v = \langle \hbar^2/2\mu R^2 \rangle_v$ (μ is the reduced mass), the mean value of the internuclear distance can be estimated according to $\bar{R} = \hbar/(2\mu B_v)^{-1/2}$. For G_1 and G_2 structures respectively, this yields $\bar{R}_1 \approx 14 a_0$ and $\bar{R}_2 \approx 12 a_0$, that are typical internuclear distances in the region of the inner well of the 0_g^- curve. As in the photoassociation process the vibrational levels of the excited state are populated at large internuclear distances [21], it seems justified to assign the structures G_1 and G_2 to ro-vibrational levels of the inner well populated via tunneling through the barrier. The complete treatment [24] necessitates that one considers both tunneling of vibrational wavefunctions through the barrier, and electronic channel mixing due to the dynamical radial coupling in the region of the inner well: the $0_g^-(6s+6p)$ and $0_g^-(6s+5d)$ (not shown in the figure) potential curves display an avoided crossing at $R \approx 10 a_0$.

3.4 Nodal structure of the radial wave-function of two colliding atoms

In the Cs$_2^+$ ion spectrum, one notices also a modulation of the line intensities. The experimental intensities of the spectral lines depend on several parameters [21], but the modulation is linked to the variations of the Franck-Condon factors of the transition between the initial state and the final ro-vibrational level of the 0_g^- state. As discussed in [10, 21, 26] for vibrational levels with large v, a good approximation is to consider the PA resonance lines as being proportional to $|\Psi(R_P(v))|^2$. R_P is the classical outer turning point of the excited ro-vibrational wave-function and $\Psi(R)$ is the radial wave-function of the pair of the colliding free atoms. One understands in this description the existence of minima in the spectrum, reflecting the nodes of the radial wave-function $\Psi(R)$, which can be built from the 0_g^- spectrum. Considering the case of a s-wave-function, the asymptotic behavior of $\Psi(R)$ is given by:

$$\sin\left(\sqrt{2\mu E/\hbar^2}\,(R - b(E))\right) \tag{3}$$

where E is the kinetic energy associated to the relative motion of the two atoms. The collision parameter, a, the so-called scattering length is defined as the limit of $b(E)$ at zero energy ($E = 0$). The modulation in PA spectrum can thus be used for the determination of scattering length parameters [27,

28, 29], which is a key parameter to determine the stability of a Bose-Einstein condensate.

The figure 6 shows, in the asymptotic region of the ground state potential (in C_6/R^6), a dimensionless graph showing the positions of the nodes versus the scattering length values. The knowledge of the molecular parameter, C_6, allows one to determine the scattering length by considering any zero PA signal in the spectrum. Indeed, calculations with a more realistic asymptotic potential (including the C_6, C_8, C_{10} and exchange terms) have been performed.

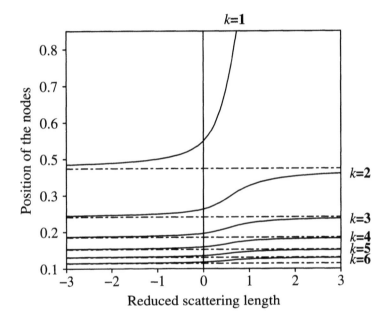

Figure 6: Dimensionless positions ($x_k = R_k/\alpha$) of the last nodes, k, versus the dimensionless scattering length ($l = a/\alpha$). $\alpha = (2\mu C_6/\hbar^2)^{1/4}$ is a scaling factor (of the order of 200 a_0 for cesium). x_k are the solutions of the differential equation $y'' + y/x^6 = 0$ with asymptotic behavior $(x - l)$.

To determine the scattering length of the molecular triplet state of the cesium dimer, we have recorded the Cs_2^+ ion spectrum by using doubly polarized atoms prepared in the Zeeman sublevel $f = 4$, $m_f = 4$. The colliding atoms in such a state are only correlated with the molecular Cs_2 lowest triplet state. The interpretation of the photoassociation minima is thus a one-channel problem and this allows one to determine the triplet scattering length of the Cs atom. Unfortunately, the value of the C_6 parameter, which is the leading term of the multipole expansion, is not known very accurately.

The published theoretical values, ranging in 6300-6900a.u. [30], provide different predictions of the triplet scattering length for the different nodes. The figure 7 shows the dependence of the scattering length versus the C_6 parameter for the different nodes. The compatibility between the different minima leads us to establish a triplet scattering length value, $-530a_0$, with limiting value $-825a_0$ and $-370a_0$, but also to fix the value of the molecular C_6 parameter $6510\pm70 a.u.$. We emphasize that this approach is completely self-sufficient and demonstrates that PA spectroscopy is a wonderful tool for accurate determination of collisional parameters.

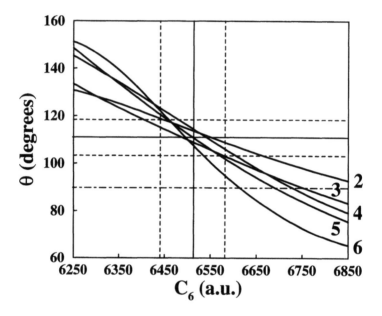

Figure 7: Scattering length characteristic angle, $\theta = \tan^{-1}(a/\alpha)$, versus C_6 parameter (α is the scaling factor defined in figure 6). The different curves correspond to the θ values obtained for different (second to sixth) intensity minima. For θ values superior to 90 degrees (dot-dashed line), the triplet scattering length a is negative.

4 Ultracold Molecules

4.1 Translationally cold molecules

As already mentioned, the trap loss spectrum is very different from the Cs_2^+ ion spectrum (see figure 3). In the trap-loss spectrum, about 80 lines for each vibrational progression of the 1_g, 0_u^+ and 0_g^- states are well resolved in the range 2-40 cm^{-1} of the figure 3. Only the vibrational progressions of the 0_g^- and 1_u states are present in the Cs_2^+ ion spectrum. The ion-detection is sensitive enough up to a detuning range of 80 cm^{-1} for the PA laser. To understand the difference between both spectra, we have made the following experimental temporal sequence. The PA laser beam is applied during a duration of 15 ms and the ionizing laser pulse (7 ns width) is delayed compared to the switch on of the PA laser. The Cs_2^+ ion signal decreases with a characteristic time of the order of 10 ms. This time is five orders of magnitude larger than the radiative lifetime of any singly excited molecular state with electric-dipole allowed transition to the ground or lowest triplet state. Indeed, it is of the order of the time during which molecules can move significantly out of the trap because of gravity. This result clearly indicates that ions are not produced by direct photoionization of PA excited molecules, but by photoionization of the stable ground or metastable lowest triplet state molecules. This is the proof for the formation of cold ground-state and lowest-triplet-sate Cs_2 molecules, which are produced for the 1_u and 0_g^- PA, respectively, but not for the 0_u^+ and 1_g PA.

4.2 Cold molecule temperature in microkelvin range

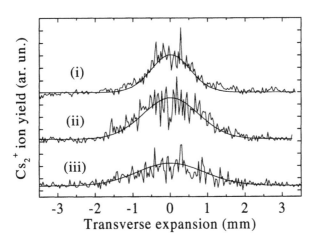

Figure 8: Temperature measurement through ballistic expansion: recordings (i-iii) correspond to the spatial analysis of the Cs_2^+ ion signal at the MOT position, and at 1.9 mm (ii) and 3.8 mm (iii) below the MOT position.

These cold ground-state Cs$_2$ molecules are not trapped by the MOT and can be detected below the trap zone [10, 18]. The spatial analysis of the ballistic expansion of the falling molecular cloud yields a measure of the temperature of the molecular cloud. The cold Cs$_2$ molecules in the ground or lowest triplet state, are photoionized to produce Cs$_2^+$ ions, using here the dye laser focused to a spot of 300 μm diameter. Figure 8 depicts the spatial analysis of the falling molecular cloud, for different heights precisely determined through time-of-flight measurements. Atoms are here further sub-Doppler cooled down to $T \leq 30\mu K$ by detuning the trap laser and simultaneously reducing the beam's intensity [19]. A model, taking into account the formation and the fall of the cold molecules, allows one to derive the molecular temperature from the data [23]. A temperature as low as $20^{+15}_{-5}\mu K$ has been determined. The atomic temperature has been measured similarly by photoionizing the cold Cs atoms to produce Cs$^+$ ions. The measured atomic temperature lies in the range 20-30 μK, with no significant difference between molecular and atomic temperatures.

4.3 Formation process for cold molecules: an optimized Franck-Condon pumping scheme

The efficiency of the mechanism of the formation of cold molecules comes from the existence of a Condon point at intermediate distance. Two different PA processes are schematically shown in figure 1. In both cases photoassociation happens at long-range distance. If spontaneous emission occurs at a short enough interatomic distance (cases ii or iii), cold ground or lowest triplet state molecules can be formed, while spontaneous emission at a long-range distance (case i) leads always to dissociation of the excited molecules. In the case of the 1_g or 0_u^+ states, the vibration of the excited molecules mostly keeps the two atoms at too large interatomic distance to get the formation of cold molecules after spontaneous decay. In the cases 0_g^- or 1_u, the molecule oscillates between long-range and intermediate interatomic distances and the formation of stable and metastable cold molecules is possible. The formation of translationally cold molecules in the latter cases is due to the particular shape of the external potential wells which offers, because of large Franck-Condon factors, at the same time an efficient free-bound photoassociation rate and a reasonable branching ratio of bound-bound spontaneous emission towards the ground or lowest triplet state. Cold molecules are expected to be obtained either in the singlet ground state or the lowest triplet state, if we consider the case of either 1_u or 0_g^- PA excitation, respectively. We will discuss this point in a later paragraph.

The previous scheme can be generalized to other systems presenting such an outer well for an excited state. For alkali dimers, only Rb$_2$ presents a similar well for the $0_g^-(5s + 5p_{3/2})$ state. For the other alkali dimers, because of the smaller fine structure splitting, the outer wells are too far out to possess a Condon point at intermediate distance. The formation of translationally cold K_2 molecules in their singlet ground state through PA has also been demonstrated [33]. The relatively low formation rate for cold molecules is here not due to a molecular configuration corresponding to a long-range outer well, but to a large enough Franck-Condon factor for a given bound-bound transition. More complex schemes of formation of cold molecules can be proposed, with photoassociation of highly excited molecular states [31, 32]. Recently an efficient production of ground-sate potassium molecules has been obtained by a two-step photoassociation [34].

4.4 Rate of cold molecule formation

From the measured number of Cs_2^+ ions, N_{ion}, detected at a detuning corresponding to a given ro-vibrational level, it is possible to estimate the corresponding number of cold molecules N_{cold} produced in the trap; one can write:

$$N_{ion} = \eta_{ion} N_{cold},$$

where η_{ion} includes the microchannel plate efficiency, which is about 35%, the ion recollection rate, which is about 80%, and the efficiency of the ionization process, including both the excitation to the intermediate step (levels of the $(2)^3\Pi_g(6s+5d)$) excited molecular states are thought to serve as intermediate step) and the ionization itself. The efficiency of the latter process is estimated around 10%. We obtain for the η_{ion} parameter a value ~ 0.03. Typically up to 500 ions per shot are detected, which correspond to 15000 cold molecules in the trap zone. The characteristic duration of the stay of cold molecules in the trap zone is 10 ms. One can thus infer a rate of cold molecule formation of about one million per second. Similar rates are reached with the formation of cold K_2 molecules [34].

4.5 Rotational and vibrational motions

PA of cold atoms for formation of stable cold molecules opens a very promising field of investigation. Nevertheless several difficulties must still be resolved . The molecules formed are indeed translationally cold. It is also possible to prepare them cold rotationally. For the 0_g^- spectrum, the observed rotational quantum number J is related to the relative orbital angular momentum l through the formula $\vec{J} = \vec{L} + \vec{S} + \vec{l}$, where $L = 1$ is the electronic angular momentum for the two atoms, and $S = 1$ is their spin, with the selection rule $J + l$ even [27]. The observation of rotational lines is thus a direct measure of the number of partial waves participating in the PA process. Rotational states with J as high as 8 are clearly visible, implying a maximum relative angular momentum of the colliding atoms of

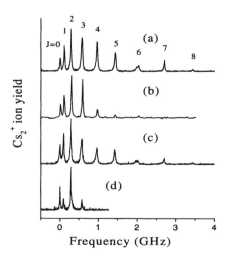

Figure 9: Rotationally resolved spectra of the 0_g^- ($v = 4$) vibrational level. (a) Trap and PA lasers always on ($T = 200$ μK); (b) same conditions as in (a) but trap laser off 10 ms before laser PA pulse. (c) sub-Doppler cooling of the atomic sample ($T \sim 30$ μK); (d) same condition as (c) but trap laser off as in (b).

$l = 6$. As the heights of the centrifugal barriers for the p, d, and f waves are 36, 190, 530 μK, and that for the i wave ($l = 6$) is large as 3.5 mK, we observe higher values of l than should be the case with the 200 μK atomic temperature. The large number of observed rotational levels is due to a conjugate effect of the cooling laser and the PA laser [19]. The presence of high partial waves is imputed to the attractive force experienced by the quasimolecule, excited at long range by the trap laser [35]. In fact if the cooling is switched off during the PA phase, at a temperature, T $\leq$ 30 μK, only the s-wave has to be considered in the experiment with the excitation only in $J = 0$ and $J = 2$ rotational levels (a very small contribution of the p-wave is also observed), at T$\simeq$ 200 μK s-, p- and d-waves are present and the excitation of rotational states up to $J = 4$ is possible (see figure 9). In any case, if we consider the excitation of $J = 1$ rotational levels we can obtain translationally cold molecules in $J = 0$ and $J = 2$ rotational levels. To obtain vibrationally cold ground-state molecules is an interesting challenge. Formed after spontaneous emission of the photoassociated molecules, the cold molecular sample is in a statistical mixture of different vibrational states. In the case of 0_g^- PA excitation, we obtain cold molecules in the lowest triplet state in vibrational levels with binding energies half the depth of the $a^3\Sigma_u^+$ molecular potential well. In the case of 1_u PA excitation, we would expect cold molecules in the singlet ground state $X^1\Sigma_g^+$, but as the relevant ground-state vibrational levels are located in energy in the vicinity of the dissociation limit, the gerade/ungerade symmetry is broken due to hyperfine coupling, and the molecular ground-states correspond to a mixture or singlet and triplet characters. Stimulated Raman photoassociation, where the emission on a given bound-bound transition is stimulated, is probably a good way to obtain all the cold molecules in a well-defined ro-vibrational level. Recent experiments, still in progress, performed in Laboratoire Aimé Cotton have demonstrated that this process is possible. Further Raman transition will allow to bring them to the lowest energy level, $v = 0$, $J = 0$ of the molecular ground state, if it is not possible to obtain them directly. Stimulated Raman molecular photoassociation of cold atoms should thus allow one to prepare cold molecules, which are simultaneously cold translationally, vibrationally and rotationally!

4.6 Cold molecule trapping

In order to develop applications of cold molecules, it will often be necessary to store them. The trapping of translationally cold Cs_2 molecules has already been performed with the use of a CO_2 laser [36]. In this experiment, the trapped cold molecules are not formed through photoassociation, but correspond to the cold molecules initially present in the Cs vapor-cell magneto-optical trap. Further experiments should confirm the possibility of photoassociative formation of cold molecules in such a dipolar trap. At Laboratoire Aimé Cotton, we have already demonstrated the formation of cold Cs_2 molecules through photoassociation in a dipolar trap, performed with a cw Nd:YAG laser. However, the formed ground-state molecules are not trapped in such a dipolar trap.

Magnetic traps can also be settled for this purpose. Magnetic trapping of ground state Calcium monohydride molecules has been demonstrated [37]. The technique used to obtain cold molecules is totally different from photoassociation: CaH molecules are produced via laser ablation of solid CaH_2. Loading of these molecules into a 2K magnetic trap is accomplished trough the use of a cryogenically cooled ^{3}He buffer gas. 10^8 molecules have been confined at a temperature of 400 mK. Magnetic trapping can also be used in the case of ground-state Cs_2 molecules in the lowest triplet state.

5 Conclusion

By starting from colder and denser atomic samples, colder and denser molecular clouds will certainly be obtained through PA, in the near future. Even if the difficulties are still significant, one can reasonably hope to produce a sample of one million molecules, which will be cold in all the degrees of freedom (translation, vibration and rotation), with temperatures in the 10-100 μK range. Other approaches can also be fruitful. We have already cited the case of the CaH molecules [37]; we can also for instance mention the development of time-varying electric fields to change the longitudinal velocity of polar molecules in a beam [38].

PA of cold atoms is first a molecular spectroscopic method which readily permits the study of long-range molecules, where the cohesion of the molecular system is given only by the electrostatic multipole interaction between both atoms. The use of polarized atoms opens new avenues in the PA experiments, for the determination of scattering lengths or Feshbach resonances, which will be very helpful in the development of Bose-Einstein condensation of atoms. The possibility for PA to lead to cold molecule formation is, in our experiments, due to the fact that the long-range 0_g^- and $1_u(6s_{1/2} + 6p_{3/2})$ potentials present a Condon point at intermediate distance. This feature is responsible for the existence of a rather efficient channel in spontaneous emission for the creation of ground state molecules. Other schemes can certainly be found, extending the field of cold molecules to other species. Future developments and applications in the cold molecule field will be similar to those performed in the cold atom research field: molecule optics, molecule interferometry, nonlinear optics, high precision spectroscopy, metrology, nanolithography... The Bose-Einstein condensation of molecules will be investigated. Raman PA with cold atoms of a Bose-Einstein condensate opens the way to laser molecules [39].

C. Amiot, D. Comparat, C. Drag, O. Dulieu, A. Fioretti, V. Kokoouline, B. Laburthe Tolra, and B. T'Jampens, M. Vatasescu participated in the reported works. The authors thanks them for many stimulating discussions.

References

[1] J.T. Bahns, W.C. Stwalley, P.L. Gould, J. Chem. Phys. **104**, 9689 (1996).

[2] A. Hermann, S. Leutwyler, L. Wöste, E. Schumacher, Chem. Phys. Lett. **62**, 444 (1979).

[3] N. Djeu, W.T. Whitney, Phys. Rev. Lett. **46**, 236 (1981).

[4] A. Kastler, J. Phys. Radium **11**, 255 (1950).

[5] H.R. Thorsheim, J. Weiner, P.S. Julienne, Phys. Rev. Lett. **58**, 2420 (1987).

[6] E.R.I. Abraham, N.W.M. Ritchie, W.I. McAlexander, R.G.Hulet, J. Chem. Phys. **103**, 7773 (1995);

[7] P.D. Lett, K. Helmerson, W.D. Phillips, L.P. Ratliff, S.L. Rolston, M.E. Wagshul, Phys. Rev. Lett. **71**, 2200 (1993).

[8] H. Wang, P.L. Gould, W.C. Stwalley, Phys. Rev. A **53**, R1216 (1996).

[9] J.D. Miller, R.A. Cline, D.J. Heinzen, Phys. Rev. Lett. **71**, 2204 (1993).

[10] A. Fioretti, D. Comparat, A. Crubellier, O. Dulieu, F. Masnou-Seeuws, P. Pillet, Phys. Rev. Lett. **80**, 4402 (1998).

[11] A.P. Mosk, M.W. Reynolds, T.W. Hijmans, J.T.M. Walraven, Phys. Rev. Lett. **82**, 307 (1999).

[12] N. Herschbach, P.J.J. Tol, W. Vassen, W. Hogervorst, G. Woestenenk, J.W. Thomsen, P. van der Straten, A. Niehaus, Phys. Rev. Lett. **84**, 1874 (2000).

[13] J.P. Shaffer, W. Chalupczak, N.P. Bigelow, Phys. Rev. Lett. **82**, 1124 (1999).

[14] A. Fioretti, D. Comparat, C. Drag, C. Amiot, O. Dulieu, F. Masnou-Seeuws, P. Pillet, Eur. Phys. J. D**5**, 389 (1999).

[15] R. Coté, A. Dalgarno, Chem. Rev. Lett. **279**, 50 (1997).

[16] D. Leonhardt, J. Weiner, Phys. Rev. A**52**, R4332 (1995).

[17] Xiaotian Wang, He Wang, P.L. Gould, W.C. Stwalley, E. Tiesinga, P. Julienne, Phys. Rev. A**57**, 4600 (1998).

[18] D. Comparat, C. Drag, A. Fioretti, O. Dulieu, P. Pillet, J. Mol. Spectrosc. **195**, 229 (1999).

[19] A. Fioretti, D. Comparat, C. Drag, T.F. Gallagher, P.Pillet, Phys. Rev. Lett. **82**, 1839 (1999).

[20] A. Fioretti, D. Comparat, C. Drag, A. Crubellier, O. Dulieu, F. Masnou-Seeuws, C. Amiot, P. Pillet, Resonance Ionization Spectroscopy, edited by J.C. Vickerman, I. Lyon, N.P. Lockyer, and J.E. Parks, The American Institute of Physics CP**454**, 147 (1998).

[21] P. Pillet, A. Crubellier, A. Bleton, O. Dulieu, P. Nosbaum, I. Mourachko, F. Masnou-Seeuws, J. Phys. B**30**, 2801 (1997).

[22] W.C. Stwalley, He Wang, J. Mol. Spectrosc. **195**, 154 (1999).

[23] D. Comparat, C. Drag, B. Laburthe Tolra, A. Fioretti, P. Pillet, A. Crubellier, O. Dulieu, F. Masnou-Seeuws, Eur. Phys. J. D (2000).

[24] M. Vatasescu, O. Dulieu, C. Amiot, D. Comparat, C. Drag, V. Kokoouline, F. Masnou-Seeuws, P. Pillet, Phys. Rev. A**61**, 044701 (2000).

[25] W.C. Stwalley, Y.H. Uang, G. Pichler, Phys. Rev. Lett. **41**, 1164 (1978).

[26] P.S. Julienne, J. Res. Natl. Inst. Stand. Technol. **101**, 487 (1996).

[27] E. Tiesinga, C.J. Williams, P.S. Julienne, K.M. Jones, P.D. Lett, W. D. Phillips, J. Res. Natl. Inst. Stand. Technol. **101**, 505 (1996).

[28] E.R.I. Abraham, W.I. Alexander, J.M. Gerton, R.G. Hulet, Phys Rev. A **55**, R3299 (1997).

[29] H.M.J.M. Boesten, C.C. Tsai, J. R. Gardner, D.J. Heinzen, F.A. van Abeelen, B.J. Verhaar, Phys Rev. Lett. **77**, 5194 (1996).

[30] M. Marinescu, H. R. Sadeghpour, A. Dalgarno, Phys. Rev. A **49**, 982 (1994); S. H. Patil, K. T. Tang, Chem. Phys. Lett. **301**, 64 (1999); S. H. Patil, K. T. Tang, J. Chem. Phys. **106**, 2298 (1997); A. Derevianko, W.R. Johnson, WM.S. Safronova, J.F. Babb, Phys. Rev. Lett. **82**, 3589 (1999); F. Maeder, W. Kutzelnigg, Chem. Phys. **42**, 95 (1979).

[31] Y.B. Band, P.S. Julienne, Phys. Rev. A **51**, R4317 (1995).

[32] J.L. Bohn, P.S. Julienne, Phys. Rev. A **54**, R4637 (1996).

[33] A.N. Nikolov, E.E. Eyler, X.T. Wang, J. Li, H. Wang, W.C. Stwalley, P.L. Gould, Phys. Rev. Lett. **82**, 703 (1999).

[34] A.N. Nikolov, J.R. Ensher, E.E. Eyler, H. Wang, W.C. Stwalley, P.L. Gould, Phys. Rev. Lett. **84**, 246 (2000).

[35] S.D. Gensemer, P.L. Gould, Phys. Rev. Lett. **80**, 936 (1998).

[36] T. Takekoshi, B.M. Patterson, R.J. Knize, Phys. Rev. Lett. **81**, 5105 (1998).

[37] J.D. Weinstein, R. de Carvalho, A. Martin, B. Friedrich, J.M. Doyle, Nature **395**, 148 (1998).

[38] H.L. Bethlem, G. Berden, G. Meijer, Phys. Rev. Lett. **83**, 1558 (1999).

[39] R. Wynar, R.S. Freeland, D.J. Han, C. Ryu, D.J. Heinzen, Science **287**, 1016 (2000).

Molecular Optics in Intense Fields: From Lenses to Mirrors

Tamar Seideman
Steacie Institute for Molecular Science
National Research Council of Canada
Ottawa, Canada

1 Introduction

In his introduction to Scoles' book "Atomic and Molecular Beam Methods" [1], Fenn points out that atomic beams, which preceded molecular beams, were developed in, and for several decades confined to, physics departments. Today, however, notes Fenn, molecular beams have considerably more applications than their atomic precedents and much of their more recent development took place in chemistry departments [1]. Several papers in the present volume illustrate his point.

At present other tools, traditionally confined to physics departments and to atomic research, are finding a variety of applications in chemical reaction dynamics research. Two such tools, relevant to the present contribution, are intense laser fields and the gradient force of light. The latter forms the basis of atom optics, the application of lasers to spatially manipulate atomic motions [2]. Having been developed for nearly three decades atom optics has, more recently, been extended to advanced fields such as atom holography, atom lithography and atom interferometry [2]. Exciting applications are discussed elsewhere in this volume. Unfortunately, the techniques of atomic manipulation with light cannot be extended to spatially manipulate molecular motions, due to the complex level structure and weak transition dipole elements of molecules. Success in the atomic domain nevertheless suggests that an analogous field of molecular optics could open a rich variety of new opportunities.

Recently it was proposed [3] that general molecules could be focussed and guided using a non-resonant, intense laser field. Spatial manipulation in that case is based on the nonlinear interaction of the intense field with the quasi-static polarizability tensor, rather than on near-resonance interactions as in the atomic case [2]. The spatial intensity profile of the strong field produces an effective well for the center of mass motion which accelerates the molecular trajectories toward the high intensity region and brings them to a focus in a predetermined point in space. The method [3] is applicable to all molecules (and atoms), since all molecules are polarizable to some extend, and robust, since the deep well translates into weak sensitivity to velocity distribution and other aberrations [3]. Generalization to other molecular optics devices, forming the basis for nanoscale lithography and molecular separation techniques, is given in [4]. Molecular focussing was very recently demonstrated also experimentally [5].

An important advantage of atom optics which the molecular optics scheme [3, 4, 5] does not share, is the availability of both attractive and repulsive optical elements, obtained by red- or blue-detuning the laser frequency from resonance [2]. The molecular optics scheme of [3, 4, 5], by contrast, is based on attractive interactions alone. One of the purposes of Ref. [6] is to extend molecular optics to include repulsive optical elements and illustrate some of their potential applications. A second goal is to develop a general framework for describing the interaction of molecules with a nonperturbative intensity gradient, reducing in one limit to the well-known Hamiltonian of atom optics [2] and in another to the molecular optics Hamiltonian of [3].

Related to intense-light molecular optics is the problem of molecular alignment in intense laser fields [7, 8, 9]. From the formal view point both intense field alignment and photomanipulation of the center-of-mass motion rely on inhomogeneous field effects. Furthermore, both reduce in one limit to fully quantal cycles between two electronic states and in another to the nearly classical interaction of an electric field with a many level system. From a practical view point, the possibility of simultaneously aligning molecules and focussing their center-of-mass motion is intriguing, with potential applications in stereodynamics [10, 11, 12, 13], gas-surface research, surface catalysis and material processing [3]. It is clear, however, that to become useful, simultaneous focussing and alignment should be realized under field-free conditions. A means of simultaneously focussing and aligning molecules in a field-free region of space is proposed in [14], where a hybrid quantum-classical method for simulating such an experiment is developed.

We begin this contribution by briefly introducing intense-light alignment from a different view-point than that adopted previously [8], and illustrating the possibility of preserving and enhancing alignment subsequent to turn-off of the laser pulse [9]. Section 3 draws the formal analogy of alignment to atom optics in one limit and to molecular optics in another limit. In Sec. 4 we extend molecular optics to include repulsive optical elements [6] and the final section summarizes our conclusions. We do not review published theory in the present paper but concentrate on the qualitative physics and refer the reader to the literature for formal and technical details.

2 Intense-Field Alignment

Aligning or orienting molecules has been one of the important goals of reaction dynamics and the search for efficient alignment techniques has been ongoing for several decades [10, 11, 12, 13]. The method of aligning molecules using a strong DC field [12], in particular, has proved successful in a vast range of applications in recent years. It is, however, limited to polar and relatively heavy molecules and requires that study and applications of the aligned system be carried out in the presence of the strong field. Both the success and the limitations of the DC technique [12] suggest the application of an intense laser field to align molecules.

The qualitative physics underlying intense-laser alignment is readily understood. Consider first the case of a near-resonance, linearly polarized field [8]. In the weak field limit, the electric dipole selection rules allow the total angular momentum to change by at most one unit upon single photon excitation. At nonperturbative intensities, Rabi-type cycling (zone-to-zone, rather than level-to-level transitions) between the ground and excited electronic states allows sequential exchange of several units of angular momentum between the field and the system, producing a coherent rotational wavepacket in both electronic states. The degree of rotational excitation is limited either by the pulse duration (τ) or by the detuning from resonance $\Delta^{JJ'} \equiv \epsilon_{\xi'}^{J'} - \epsilon_{\xi}^{J} \simeq 2JB_e$, where ϵ_{ξ}^{J} are rotational energies in the $\xi = 0, 1$ electronic state, the laser frequency is tuned to the electronic origin and B_e is the rotational constant. In the former case $J \lesssim \tau \Omega_R^{J,J'}$ and in the latter $\Omega_R^{J,J'} \gtrsim \Delta^{J,J'}$, $\Omega_R^{J,J'}$ being the Rabi coupling [8]. In a rotationally cooled molecular beam environment, low initial angular momentum ensures the excitation of a broad rotational wavepacket of highly anisotropic orbitals. Such coherent superposition states can be analytically shown [8] to align along the polarization vector.

Figure 1a shows, for instance, the excited state component of a Li_2 wavepacket as a function of the angle between the field and molecular axes, $\Theta = \cos^{-1}(\hat{\epsilon} \cdot \hat{R})$ and time during the laser pulse. As the field turns on, higher rotations are populated and the probability density becomes better aligned. Figure 1b shows that the alignment survives and is significantly *enhanced* after turn-off of the field. It undergoes periodic dephasing and rephasing, remaining on the average well-aligned. The enhancement effect is explained by means of an analytical model in [9] and was found numerically also for other systems, including both parallel [9] and perpendicular [15] transitions.

At nonresonant frequencies [7], tuned below all electronic transitions of the molecule, $\omega_l \ll \omega_e$ the field-matter interaction is readily transformed to an effective Hamiltonian,

$$H_{\text{ind}} = -\frac{1}{4}\sum_{\rho\rho'}\varepsilon_\rho^*\alpha_{\rho\rho'}\varepsilon_{\rho'} \qquad (1)$$

through adiabatic elimination of all vibronic excited states. In Eq. (1) the radiation field has been written as $\epsilon = \text{Re}[\varepsilon e^{i\omega_l t}]$, $\alpha_{\rho\rho'}$ are components of the polarizability tensor and ρ are space-fixed coordinates. The factor half in Eq. (1) as compared to the familiar Stark effect Hamiltonian in a DC field is convenient to understand as arising from cycle averaging of the rapid oscillations of the field, proportional to $\cos^2\omega_l t$, but can be derived within a rigorous formulation [6]. Thus, a rotationally broad wavepacket is populated in this case through two-photon cycles, obeying the $\Delta J = 0, \pm 2$ selection rule.

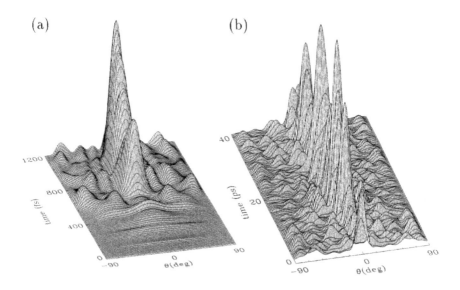

Figure 1: (a) A rotational wavepacket of Li_2 molecules vs the polar Euler angle Θ and time during the laser pulse. The intensity is 10^{11}Wcm^{-2} and the pulse duration is 200fs. As higher rotational levels are excited the wavepacket becomes better aligned. (b) The rotational wavepacket after turn-off of the laser pulse. The alignment survives and is significantly enhanced under field free conditions.

In the low frequency limit laser alignment is formally equivalent to the technique of alignment in an electric DC field [12], the practical advantage of the AC scheme being the possibilities of reaching higher field strengths and of preserving alignment after turn-off of the laser pulse.

The line alignment of Refs. [7, 8, 9, 15] extends to two- and three-dimensional alignment. Ongoing work exploits circularly and elliptically polarized fields to make use of all components of the molecular polarizability tensor, illustrating the possibilities of converting, using laser light, the free rotation about two or all three space-fixed angles into small amplitude libration.

3 Molecular Optics in Intense Light

The discussion concluding the previous section suggests the intriguing possibility of manipulating also the center-of-mass motion of molecules, by exploiting the *spatial* gradient of the field-matter interaction, as well as its orientational gradient. The former gradient arises from the inhomogeneous intensity of a laser beam and is familiar from the fields of optical atom trapping [2] and optical tweezers [16]. For a cylindrical Gaussian optical focus, for instance, $\varepsilon \propto \exp(-r^2/2w_0^2)$,

$$F_r = -\frac{d}{dr} H_{\text{ind}} \propto -r/w_0^2 \qquad (2)$$

where r measures distance from the laser axis and w_0 is the laser spot size. The latter (orientational) gradient arises from the anisotropy of the molecular polarizability. For a linear or a symmetric top molecule subject to linearly polarized radiation, for instance, the induced dipole interaction in Eqs. (1) and (2) takes the form

$$H_{\text{ind}} = -\frac{1}{4}\varepsilon^2(\alpha_\perp + \Delta\alpha \cos^2 \Theta), \qquad \Delta\alpha = \alpha_\parallel - \alpha_\perp, \qquad (3)$$

where $\alpha_\parallel$ and $\alpha_\perp$ are the parallel and perpendicular components of the polarizability tensor. The angular part of Eq. (3) is responsible for alignment while the radial part allows for manipulation of the center-of-mass motion.

In the limit of strong interaction, the depth of the laser-induced potential well for the angular motion becomes large as compared to the rotational energy and the free rotation transforms into nearly harmonic vibrations about the polarization axis, $\cos^2 \Theta \sim 1 - \Theta^2$. (This applies to positive anisotropy, $\Delta\alpha > 0$ as, for instance, in linear molecules. In the case of negative anisotropy the limiting motion corresponds to vibrations about the plane perpendicular to the polarization axis.) Similarly, in the limit where the depth of the field-induced well for the center-of-mass motion is large as compared to the translational energy, the system is trapped in the laser beam, undergoing close-to-harmonic vibrational motion about the laser focus, $\exp(-r^2/w_0^2) \sim 1 - r^2/w_0^2$. Although the analogy between field-induced confinement of the angular and center-of-mass modes is revealing, and in fact carries further, it is important to note two qualitative differences between the rotational and translational motions. First, the field-gradient inducing the latter motion is smaller by a factor or order ω_0^{-2} than that inducing the former, see Eq. (2). Hence much longer (4–6 orders of magnitude) pulses are required in order to affect the radial motion than are required to affect the angular motion, in practice few to few hundreds of nanoseconds. Second, while the potential wells for the two modes are comparable, it is considerably more difficult to cool the translational mode of molecules than it is to cool their rotations [17]. Since the present scheme produces orders of magnitude deeper potential wells than those produced in atom traps [2], its application to molecular trapping is possible at much higher translational energies, circumventing the notorious task of molecular cooling [17]. The price paid for the depth of the intense field trap, however, is its limitation to pulsed fields and its consequent short lifetime. In practice the pulse duration would be generally upper bounded by the onset of nonresonance ionization, rather than by the availability of appropriate laser sources.

Below we will concentrate on field-induced motion of translationally hot molecules, such that the center-of-mass kinetic energy is larger than the laser-induced potential well, translational cooling is not needed and the pulsed nature of the radiation is of advantage. The problem of optical trapping in intense laser fields is addressed in [3, 7, 14].

The large time-scale disparity between the center-of-mass and internal modes of the molecule, together with the long pulses required to focus molecules, suggest adiabatic separability. We first solve (quantum mechanically) for the internal modes at fixed values of the field amplitude and next study the evolution of the center-of-mass subject to the parametrically ε-dependent adiabatic potential. The latter motion is associated with vanishingly small deBroglie wavelength in all cases considered here and is confined to a single vibronic state. In the present work it is studied within the classical approximation. Quantum mechanical effects are considered in [14, 18]. The related problem of molecular interferometry is discussed in [19].

Figure 2a shows several classical trajectories in the $\{y, z\}$ plane, evolving subject to the adiabatic potential shown in Figs. 2b. We consider a typical molecular beam velocity of 10^3ms^{-1} and a laser intensity of $5 \times 10^{12} \text{Wcm}^{-2}$, resulting, for the case of I_2 molecules, in an energy ratio $R = E_{\text{kin}}/V_w \sim 12$ between the kinetic energy of the molecules and the laser-induced well depth. Figure 2a shows that the molecular density reaches a maximum at a point $\{y = 0, z = -f\}$, referred to below as the focus of the "molecular lens". A constant-z cut through the molecular beam at $z = -f$ is given in Fig. 2c, showing a sharp peak at $y = 0$. The image-size and image-distance of the molecular lens, defined as the width and the distance from the laser beam of the plane where the molecular beam attains its minimum width, are marked W and D, respectively, in Fig. 2a. All three parameters are fully determined by the energy ratio R, allowing characterization of the molecular lens in terms of a single, system independent parameter. The energy ratio

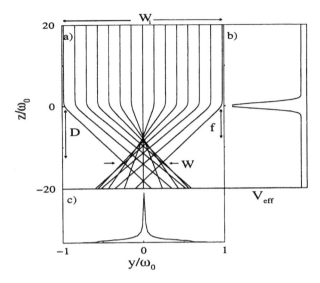

Figure 2: (a) Classical trajectories and (b) adiabatic potential for $J_i = 0$ I_2 molecules. The peak intensity is $I_m = 5 \times 10^{12} \text{Wcm}^{-2}$ and the initial beam velocity is 10^3 m/sec. The depth of the field-induced effective well is 1278 K. f marks the point of highest molecular density (the focus of the molecular lens), D marks the distance of the minimum molecular beam-size from the beam-axis (the image distance) and W is the corresponding size (the image-size of the lens). (c) A constant z-cut of the molecular density at $z = -f$.

can thus be regarded as an analog of the refractive index, which determines the properties of an optical lens. In the strong interaction limit, $R < 1$ the system is trapped in the laser beam. As $R \to 0$ only the harmonic part of the interaction is probed and the motion follows classical Lissajous trajectories. In the opposite limit, $R \to \infty$, the motion is essentially unaffected by the potential well and the trajectories focus at infinity, as expected from the laws of geometric optics.

For practical applications, such as nanolithography [3, 20], one wishes to decrease the demagnification ratio, that is, the ratio W/W_i between the image size and the aperture size (Fig. 2), while retaining the flexibility of the image distance. This can be done by replacing the spherical focus by an elliptical one. Figure 3 shows the image size vs. R for different aspect ratios of the elliptical optical focus. As the transverse focal size ω_{0y} increases at a constant beam aperture W_i, the demagnification ratio decreases. Spherical aberration, resulting from the anharmonic part of the intensity distribution and marked in Fig. 2c, is fully eliminated in Fig. 3 as the system samples a nearly harmonic potential in the direction transverse to its motion. The effects of chromatic aberrations, resulting from velocity distribution, transverse velocity and/or finite rotational temperature and neglected in Fig. 3, are discussed in [4].

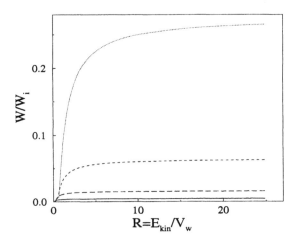

Figure 3: The image-size of the molecular lens W defined in Fig. 2a vs the energy ratio R for different values of the aspect ratio $\eta = \omega_{0y}/\omega_{0z}$: (······) $\eta=1$; (- - - - - -) $\eta =2$; (— — —) $\eta = 4$; (————) $\eta = 7.5$.

Molecular focussing extends to a general field of molecular optics. The same type of attractive interaction, Eq. (3), is used in [4] to collimate diverging molecular beams and to steer molecules in a predetermined direction. With different relative orientations of the laser and molecular beams the latter can be dispersed and separated into mass, velocity or rotational components [4]. Laser beam splitting techniques can serve to combine two or more of the above effects and to produce a composite molecular lens of improved demagnification ratio. The possibility of producing a short-pulse molecular beam, of duration determined by a laser pulse, suggests further potential applications.

4 Reflecting Molecular Beams with Light

Comparison of the molecular optics scheme outlined in the previous section with the well-established field of atom optics points out, as discussed in the introduction, a major limitation of the former as compared with the latter, namely the lack of repulsive molecular optics elements. This issue is addressed in the present section.

At low frequencies, $\omega_l \ll \omega_e$ the polarizability approaches its DC limit, which is always positive, producing a purely attractive interaction, e.g.

Eq. (3). Consider, on the other hand, the opposite limit, namely, the limit of large laser frequency with respect to molecular frequencies. This limit is not practical for low electronic states but is typically realized with high Rydberg states, where the electronic spacing scales as n^{-3}. For $n \sim 100$ even micrometer sources, such as CO_2 ($\lambda_l = 10.6\mu m$) or Nd:YAG ($\lambda_l = 1.06\mu m$) lasers are well into the high frequency ($\omega_l \gg \omega_e$) limit. Figure 4 shows

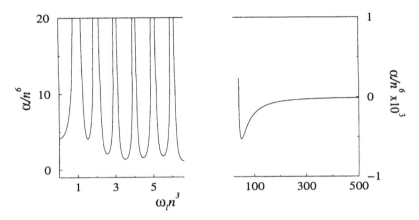

Figure 4: Schematic illustration of the dynamic polarizability for angular momentum $l = 5$ vs the field frequency, measured in units of the level spacing. α is estimated within a semiclassical approximation.

the average dynamic polarizability vs the laser frequency, measured in units of the electronic level spacing. The calculation [6] employed approximate dipole elements, based on a semiclassical approximation, and hence Fig. 4 should be considered *qualitative* only. In the low frequency limit the polarizability converges to its DC limit which is always positive. This is the regime considered in the previous sections. As the laser frequency is scanned through the discrete spectrum α undergoes a series of resonances and at high $\omega_l n^3$ it scales as $-\omega_l^{-2}$ [6]. In this limit the interaction converges to the ponderomotive energy of a free electron. Thus, high Rydberg states provide a negative, tunable polarizability which can be rather large in absolute magnitude, suggesting the possibility of manipulating molecules at lower intensities than those discussed previously for ground states [3, 4, 5].

An important issue to be addressed in connection with focussing of Rydberg states is the risk of ionizing the molecules. In the low frequency regime, $\omega_l n^3 \ll 1$, ionization takes place by tunneling and its probability scales as n^{10}. Since the polarizability scales as n^7 [6], ionization will become more dominant the higher n. For $\omega_l n^3 \gg 1$, by contrast, ionization can only take place by photon absorption and scales as n^{-3}. The latter scaling law follows from the n^3 scaling of the Kepler orbit which allows the electron to spend only a proportionally small fraction of the period in the vicinity of the core, where absorption can occur.

As one of the potential applications of the negative polarizability, we show in Fig. 5 a "negative lens", $I \propto y^2/\omega_{0y}^2 \exp(-y^2/\omega_{0y}^2 - z^2/\omega_{0z}^2)$. The classical trajectories are superimposed on the laser intensity contours. The trajectories are repelled from the high intensity centers and are funneled through the low-intensity "gate", focussing downstream. Other intensity distributions that could act as a negative lens can be readily envisioned. We note here that the intensity profile of Fig. 5 and other nonGaussian distributions can be readily produced with current technology [21].

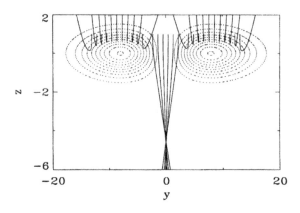

Figure 5: Focussing in a "negative lens". The dashed contours show the intensity distribution and the solid curves show the molecular trajectories. See the text for details. Reproduced with permission from Ref. [6]

One of the potential advantages of the negative lens of Fig. 5 over the positive lenses of the previous section is that a smaller demagnification ratio may be achievable, circumventing the need to skim the molecular beam. In Fig. 5 selective repelling of parts of the beam produces a demagnification ratio of $W/W_i \sim 0.5\%$ and a focus essentially free of spherical aberrations while relaxing the constraint of Fig. 3 on the relative dimensions of the molecular beam aperture and the transverse focal spot size. We assumed, in Fig. 5, an interaction of $-\alpha I/4 = 3 \times 10^{-4}$ a.u., *roughly* estimated (Fig. 4) to correspond to an intensity of $2 \times 10^8 \mathrm{Wcm}^{-2}$ at a Nd:YAG wavelength ($\lambda = 1.06 \mu$m). More accurate determination of the intensity requirement will have to await detailed modeling of the effective polarizability for a given system and frequency. The order of magnitude estimation of Fig. 4 suggests that focussing in a negative lens may be achievable at lower intensities than those predicted [3] for positive lenses. An experiment aiming to realize the negative polarizability molecular optics scheme suggested here was recently initiated [22].

5 Conclusions

This study discussed the possibility of manipulating external molecular modes with moderately intense light. Considering first control of the rotation angles, we described the qualitative physics underlying strong field alignment, noted its near- and far-off-resonance limits and discussed the phenomenon of enhanced alignment after the pulse turn off, a general effect whose origin lies in the peculiar revival structure of rotational wavepackets [9]. Proceeding to discuss control of the center-of-mass translation, we noted the formal analogy of strong field alignment and strong field molecular focussing and briefly outlined the physics of quasistatic ($\omega_l \ll \omega_e$) focussing and its extension to molecular optics in intense laser fields. Finally, we proposed a complementary molecular optics scheme which operates at the opposite frequency limit, $\omega_l \gg \omega_e$ and introduces the new possibility of reflecting molecular beams with intense light.

The present work suggests several avenues for future research. The possibility of simultaneously aligning molecules and focussing their center-of-mass motion may find applications in stereodynamical studies and in the study and manipulation of gas-surface processes. Perhaps most interesting, the possibility of depositing orientationally ordered nanostructures by focussing a molecular beam onto a substrate, together with the dependence of magnetic and electric properties on the degree of order, suggests applications in the production of new materials, possessing predesigned electric or magnetic properties.

The negative polarizability scheme of Sec. 4 may offer further opportunities. These include molecular waveguiding along the low intensity axis of a TEM_{01}^* (doughnut) mode laser, where long time confinement could be possible while ionization is minimized, trapping in a low intensity region where again ionization could be avoided or minimized, combining optical elements to optical devices and, by analogy to atom optics [23], studying problems of interest in scattering theory.

Acknowledgments I am grateful to P.B. Corkum and R.J. Gordon for many enjoyable conversations. The Department of Energy is acknowledged for support.

References

[1] J.B. Fenn in *Atomic and Molecular Beam Methods*, G. Scoles Ed., Vol.1, (Oxford Univ. Press., 1988).

[2] A collection of earlier works on atom cooling and trapping is given in J.Opt.Soc.Am. **2** (1985), and **6**, 1989; The related problem of atom

interferometry is reviewed in *Atom Interferometry* P.R. Berman Ed., (Academic Press, London, 1997).

[3] (a) T. Seideman, Phys.Rev. A **56**, R17 (1997).
(b) T. Seideman, J.Chem.Phys., **106**, 2881 (1997).

[4] T. Seideman, J.Chem.Phys. **107**, 10420 (1997).

[5] H. Stapelfeldt *et al*, Phys.Rev.Lett. **79**, 2787 (1997).

[6] T. Seideman, J.Chem.Phys. **111**, 44397 (1999).

[7] B. Friedrich and D.R. Herschbach, Phys.Rev.Lett. **74**, 4623 (1995).

[8] T. Seideman, J.Chem.Phys. **103**, 7887 (1995).

[9] T. Seideman, Phys.Rev.Lett. **83**, 4971 (1999).

[10] R.N. Zare, J.Chem.Phys. **69**, 5199 (1978); K. Bergman in [1], p. 293.

[11] V.A. Cho and R.B. Bernstein, J.Phys.Chem. **95**, 8129 (1991); S. Stolte in [1], p. 631; P.R. Brooks *et al*, Chem.Phys.Lett. **66**, 144 (1979).

[12] H.J. Loesch and J. Remscheid, J.Chem.Phys., **93**, 4779 (1990); B. Friedrich and D.R. Herschbach, Z.Phys. D **18**, 153 (1991); M. Wu, R.J. Bemish and R.E. Miller, J.Chem.Phys. **101**, 9447 (1994).

[13] See, for instance, V. Aquilanti *et al*, Phys.Rev.Lett. **74** 2929 (1995).

[14] Z.-C. Yan and T. Seideman, J.Chem.Phys. **111**, 4113 (1999).

[15] S.C. Althorpe and T. Seideman, J.Chem.Phys. **110**, 147 (1999).

[16] A. Ashkin, Phys.Rev.Lett. **24**, 156 (1970).

[17] Atoms of simple level structure can be cooled to the micro- and even nano-Kelvin regime [2]. Cooling molecules to subKelvin temperatures is considerably more difficult; see, for instance, J.D. Weinstein *et al*, Nature, **395**, 148 (1998).

[18] T. Seideman and V. Kharchenko, J.Chem.Phys. **108**, 6272 (1998).

[19] M.S. Chapman *et al*, Phys.Rev.Lett. **74**, 4783 (1995).

[20] See, for instance, J.J. McClelland *et al*, Science **262**, 877 (1993).

[21] P.A. Naik, S.R. Kuoobhare and P.D. Gupta, manuscript in preparation; E. Constant, private communication; A. Schroeder, private communication.

[22] R.J. Gordon, private communications.

[23] See, for instance, A. Landragin *et al*, Phys.Rev.Lett. **77**, 1464 (1996).

Ultrashort Wavepacket Dynamics and Interferences in Alkali Atoms

B. Girard, M. A. Bouchene, V. Blanchet, C. Nicole, S. Zamith
Lab. de Collisions Agrégats Réactivité (CNRS UMR 5589),
IRSAMC, Université Paul Sabatier, Toulouse, France

1 Introduction

Wavepacket dynamics in quantum systems has been the subject of numerous studies over the past decade [1–3]. It is generally studied in a pump-probe approach using two time-delayed ultrashort laser pulses. This powerful investigative method led to the observation of many phenomena that were not accessible using standard spectroscopic measurements. The first laser pulse creates a wavepacket in an electronic excited state, which evolves freely before the second laser projects the wavepacket to a final state after a variable delay. This final state is subsequently detected with time independent methods. To reveal the dynamics of the system, it is necessary that the probe step probability varies significantly with the time evolution of the wavepacket.

Wavepacket dynamics can also be attained using wavepacket interferences, as is the case of Ramsey fringes experiments [4–16]. This technique relies on building interferences between two time-delayed quantum paths leading to the same final state. The time delay is used to control the phase of the interference and hence the excitation probability of the system. Furthermore, this technique can provide access to wavepacket dynamics even in situations where the pump-probe scheme cannot be applied. Examples of such applications range from Rydberg states dynamics [6–8] to semiconductor quantum wells [17–19].

Wavepacket interferences have been performed in the weak field regime where the excitation paths towards the various excited states are independent [4–14] or in the strong field regime [15,16]. In the weak field regime, the linearity of the interaction leads to several equivalent interpretations [9,10,13,20] : quantum beats, wavepacket interferences or Fourier transform spectroscopy. In such a field regime, the experimental data can be understood simply from the pulse spectrum which is the only relevant quantity. This explains for instance why equivalent results have been achieved with incoherent light [9,10]. As a result, it is possible to obtain a direct measurement of the absorption lines of the system(s).

Although wavepacket interferences provide several advantages, it is impossible to determine from the data without any ambiguity whether a wavepacket is created or not. A wavepacket can be created only if the transitions belong to the same quantum system and share the same initial state. In the strong field regime, the non-linearity of the interaction results in a mixing of the transition frequencies and the population is modulated at the frequency difference [15,16]. This mixing exists only in case of a wavepacket and provides therefore an evidence of its creation [14]. Wavepacket interferences in the saturated regime provide therefore the same information on the system as the pump-probe technique.

We choose in this chapter to illustrate the features of these various techniques to study wavepacket dynamics in a simple system : the fine structure 4p doublet state in atomic potassium. We present the bright state - dark state framework which is particularly suitable to describe the dynamics of the wavepacket. This reveals in particular the spin precession which is induced in this system. Finally, we discuss the advantages and drawbacks of each scheme.

2 Experimental set-up

The experimental set-up combines an ultrashort laser oscillator with an atomic beam and a quadrupole mass spectrometer to detect the ions resulting from this interaction is depicted on Fig. 1 [12]. A Ti: sapphire oscillator produces pulses of 100 fs, 13 nJ at 769 nm. The spectral width is 150 cm^{-1}, much larger than the fine structure splitting (57.7 cm^{-1}) of the potassium 4p state studied in the following sections. For the two-color pump-probe experiments (sect. 4), the pulses are frequency doubled. The fundamental frequency is separated from the second harmonic and time delayed. For the one color interferometry experiments (sect. 5), each pulse is split into two with a Michelson type interferometer. The time delay between the pulses can be varied with steps down to 0.2 fs. The intensity ratio is 1.8. The pulses are focused in the interaction chamber with a 1 m and a 200 mm focal length lens for weak (3.10^8 W/cm^2) and intermediate intensities (9.10^9 W/cm^2) respectively.

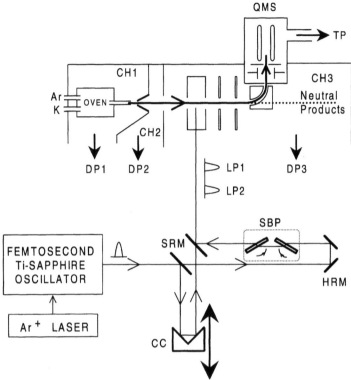

Figure 1: Typical experimental set-up used the one color experiments. For the two-color experiments, a frequency doubling crystal, followed by a dichroic mirror replaces the semireflecting mirrors. CH1, CH2, CH3 : chambers 1, 2 and 3. DP1, DP2, DP3 : Diffusion pumps. TP : Turbo pump. QMS : Quadrupole Mass-Spectrometer. LP1, LP2 : Laser pulses 1 and 2. SRM : Semireflecting mirrors. HRM : High reflectivity mirrors. CC : cube corner. SBP : scanning Brewster plates.

3 Excitation and evolution of the wavepacket : Bright state - Dark states formalism

3.1 General framework

Consider a quantum system consisting of a single initial state $|g\rangle$ and two excited states $|k\rangle$ of energies ω_k ($k = a, b$). This system interacts with a laser pulse $E^{(1)}(t) = E_0(t)\cos\omega_L t$ of duration τ_L. The spectral width of this pulse ($\Delta\omega_L \simeq \tau_L^{-1}$) is large enough to populate both excited states ($|\omega_{ba}| \leq \Delta\omega_L$ with $\omega_{ba} = \omega_b - \omega_a$), see Fig. 2. The excited state wavepacket at time $t \gg \tau_L$ is given by

$$|\psi(t)\rangle = \sum_k \tilde{E}_1(\omega_k) \mu_{kg} e^{-i\omega_k t} |k\rangle \qquad (1)$$

where μ_{kg} is the dipole moment matrix element for the $|g\rangle \to |k\rangle$ transition and $\tilde{E}_1(\omega_k)$ is the pump pulse electric field spectrum at frequency ω_k. The phase factors in Eq. 1 govern the evolution of the excited state wavepacket. Any detection of this evolution requires a transition which should result in a mixing of these phase factors. This requirement is met in two cases : transition from this coherent superposition to a common final state in the pump-probe case and transition towards or from the initial state in the wavepacket interferences scheme. This scheme is valid for any number of excited states and corresponds to the usual quantum beats description [21]. Wavepacket dynamics is easily illustrated in the case of Rydberg or vibra-

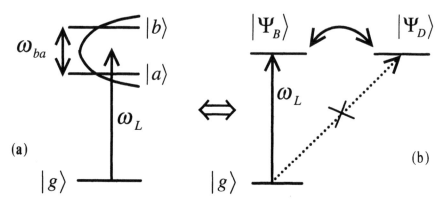

Figure 2: a) Principle of the experiment. A ground state $|g\rangle$ is excited towards a set of two excited states $|k\rangle$ (energy ω_k) by a two pulse sequence. b) Description in the bright state - dark state framework. The free evolution is driven by the Hamiltonian ω_{ab}.

tional wavepackets by projecting the wavepacket onto a basis set associated with the radial or internuclear coordinate respectively. These two examples are in fact particular illustrations of the bright state - dark states formalism developed in another context and which can be efficiently used to illustrate wavepacket dyanmics. The concept of bright and dark states has been introduced to describe states which are respectively coupled and uncoupled to a radiation field. These states can be eigenstates of the system or linear combinations of eigenstates [22]. This description has been used with success in many subtle applications such as in dark resonances or in subrecoil laser cooling [23].

Due to the broad spectrum of ultrashort laser pulses, several states can be

I.8 Ultrashort Wavepacket Dynamics and Interferences in Alkali Atoms

simultaneously reached. In the limit of ultrashort pulses, the excitation step can be separated from the free evolution of the excited state. Moreover, this implies that $\tilde{E}^{(p)}(\omega_a) = \tilde{E}^{(p)}(\omega_b)$. The bright state is therefore the linear combination of the excited states prepared at the initial time. The dark states are any linear combinations of excited states orthogonal to the bright state. No coupling between the ground state and the dark states is possible via an electric dipole transition and in general, the Hamiltonian of the system is not diagonal in the bright state - dark states basis set. The first pulse creates a wavepacket in the bright state. As a result of the coupling between the bright state and the dark states, the wave packet evolves freely oscillating periodically in and out of the bright state.

In a two-level system, the bright and dark states are

$$|\psi_B\rangle = \cos\alpha\,|a\rangle + \sin\alpha\,|b\rangle \tag{2}$$
$$|\psi_D\rangle = -\sin\alpha\,|a\rangle + \cos\alpha\,|b\rangle \tag{3}$$

respectively, with $\tan\alpha = \mu_{bg}/\mu_{ag}$. The evolution of the wavepacket is given, for $t \gg \tau_L$, by

$$|\psi(t)\rangle = e^{-i\frac{H}{\hbar}t}|\psi_B\rangle$$
$$= e^{i\omega_a t}\left[(\cos^2\alpha + \sin^2\alpha\,e^{-i\omega_{ba}t})|\psi_B\rangle + \sin\alpha\,\cos\alpha\,(-1 + e^{-i\omega_{ba}t})|\psi_D\rangle\right] \tag{4}$$

Depending on α, this wavepacket experiences either a total (for $\alpha = \pi/4$), or a partial oscillation between the bright and dark states. Eventually no oscillation at all appears if only one of the two states can be excited ($\alpha = 0$ or $\alpha = \pi$).

These oscillations are detected by a second ultrashort laser pulse. Two detection schemes are examined in details in the next sections : excitation towards a final state in the pump-probe scheme (sect. 4), creation of a second, identical, wavepacket which interferes with the first one in the wavepacket interferences scheme (sect. 5). We first analyse (subsect. 3.2) the physical content of the oscillating wavepacket in the case of fine structure doublet states.

3.2 Application to a fine structure doublet state

The eigenstates of the 4p state of the potassium atom are described in the coupled basis set $|J, M_J\rangle$, $|a\rangle = |3/2, 1/2\rangle$ and $|b\rangle = |1/2, 1/2\rangle$. Both of these states are accessible from the 4s ground state, $|g\rangle = |1/2, 1/2\rangle$ for a linear polarization (parallel to the quantization axis) since $\Delta M_J = 0$. An evaluation of the 3-j factors gives $\tan\alpha = \mu_{bg}/\mu_{ag} = -1/\sqrt{2}$. Here, for the sake of simplicity, we have chosen to restrict our discussion to the case $M_J = 1/2$, although the state with the opposite value $M_J = -1/2$ exhibits equivalent behaviour.

If the spin-orbit interaction is the only relevant relativistic effect, so that the L-S coupling is strictly valid, then the bright state corresponds exactly to an eigenstate of the uncoupled basis set $|L, M_L; S, M_S\rangle$. In our case, for a linear polarization, from the selection rules ($\Delta M_L = 0$ and $\Delta M_S = 0$) we get $|\psi_B\rangle = |1, 0; 1/2, 1/2\rangle$. The dark state is orthogonal to the bright space, in the subspace defined by the same value of M_J : $|\psi_D\rangle = |1, 1; 1/2, -1/2\rangle$. These expressions can also be deduced from Eq. 3. The free evolution of the wavepacket is therefore an oscillation between two states with opposite spin. It corresponds thus exactly to the classical precession of the spin and orbital angular momenta around the total angular momentum, as illustrated on Fig. 3.

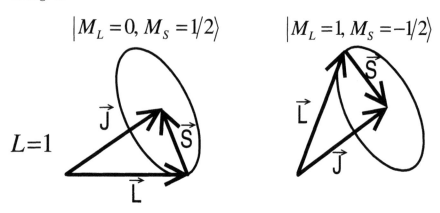

Figure 3: Drawing of the spin and orbital angular momentum precession. Left : bright state ($|L, M_L, S, M_S\rangle = |1, 0, 1/2, 1/2\rangle$). Right : dark state ($|L, M_L, S, M_S\rangle = |1, 1, 1/2, -1/2\rangle$).

4 Detection of the wavepacket in the pump-probe scheme

4.1 Principle

A second ultrashort laser pulse $E^{(2)}(t)$ (of the same duration τ_L), time-delayed by τ with respect to the first one, is used as a probe pulse to photoionize the atom. The wavelength is chosen outside the spectral range of the first transition to avoid any interference effect such as those described in sect. 5. The ion signal is given by

$$P = P_B + (P_D - P_B)\sin^2 2\alpha \left(1 - \cos\omega_{ba}\tau\right)/2 \qquad (5)$$

where P_B and P_D are the ionization probabilities from the bright and dark states respectively. Thus oscillations of the final state population can be observed only for $P_B \neq P_D$. This condition is for instance easily fulfilled for Rydberg electron radial wavepackets since the ionization probability is higher near the core, while the detection of vibrational wave packets requires an ionization probability depending on the internuclear distance [24,25].

In the fine structure doublet, the bright and dark states differ by spin-orientation but also by the orientation of the orbital angular momentum which is relevant for electric dipole transitions. Obviously, the transition amplitude depends directly on the relative orientation of the excited state angular momentum with respect to the electric field of the probe pulse. The orbital angular momentum of the bright state is always parallel to the polarisation of the pump pulse. With parallel probe and pump polarizations, the bright state has the highest ionization probability [26]. Conversely, since the orbital angular momentum of the dark state is perpendicular to the pump pulse electric field, its ionization probability is the largest in case of perpendicular polarizations. We therefore expect to observe oscillations out of phase by π between the two polarization configurations.

These qualitative results are confirmed by a quantitative evaluation of all the transition amplitudes towards the final ion states $|E_k; L^+, M_L^+; S^+, M_S^+\rangle$. The $L^+ = 0$ or 2 partial waves are open photoionization channels. However, the $L^+ = 0$ channel is in practice only open for the $M_L^+ = 0$ sublevel. It can therefore be accessed from the bright state only in case of parallel polarizations or from the dark state only with perpendicular polarizations. The opening or closing of the $L^+ = 0$ channel explains most of the contrast described above.

4.2 Results

The pump pulse at $\lambda = 769$ nm excites the 4p doublet states of potassium atoms. A time-delayed probe pulse (at $\lambda = 384.5$ nm) ionizes the atom. Fig. 4 displays the corresponding ion signal as a function of the time delay for parallel and perpendicular polarizations.

The signal for negative time delay is almost negligible since the probe pulse wavelength is not resonant with any transition from the ground state. The only contributions arise from (1+2) Resonance Enhanced Multiphoton Ionization (REMPI) from the pump pulse or from direct 2-photon ionization from the probe pulse. Another contribution is due to pump-probe sequences separated by approx. 13 ns.

A sharp increase of the ion signal takes place when the two laser pulses overlap in time. The (4s-4p) transition is excited by the pump pulse and the 4p state is immediately ionized by the probe pulse.

This process is the most efficient for a positive time delay of the order of the pulse duration [27]. For longer delays, the spin-orbit precession associated to the bright state - dark state oscillation takes place, with a period of 580 fs.

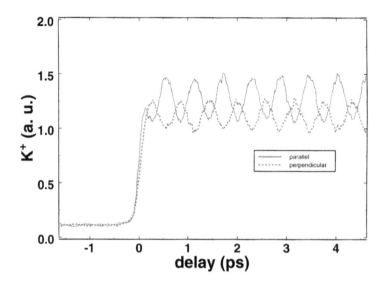

Figure 4: Measured ion signal as a function of the time delay between the pump and probe pulses, for parallel and perpendicular relative polarizations. The oscillations are out of phase by π.

As expected, the oscillations for the two polarization configurations are out of phase by π.

The contrast of the oscillation depends directly on the anisotropy of the transitions. For instance, for a two-photon transition probe step [11], the ionization probability of the bright and dark states are almost equal ($P_B \simeq P_D$). This means that a pump-probe scheme cannot be applied in such cases. We discuss in Sect. 5 how this limitation can be bypassed by using wavepacket interferences.

5 Detection with wavepacket interferences

Wavepacket interferences can be used as an alternative to the pump-probe scheme in order to observe the dynamics of the system. They also allow to control the amplitude of the excited state wavepacket and were therefore called temporal coherent control [11,13]. Two identical wavepackets are created in the same excited states with a sequence of two identical pulses.

By varying the time delay τ between the pulses, it is possible to adjust the state of evolution of the first wavepacket when the second is created. Moreover, it is also possible to control their respective absolute phase and thus the nature (constructive or destructive) of the interferences. As a constraint the delay must be controlled with an accuracy much better than the optical period (about 2.5 fs for near infrared pulses). We consider in the next subsections two cases corresponding to the weak field and to the saturated regimes.

5.1 The weak field regime

We first analyze the simpler situation in which the interaction can be described in the perturbative regime. This means that the ground state is not depleted after interaction with the laser sequence. The two wavepackets can therefore be added coherently. The total laser field is expressed as $E_T(t) = E^{(1)}(t) + E^{(2)}(t)$. The second laser pulse, $E^{(2)}(t) = \beta\, E^{(1)}(t-\tau)$, is generated by splitting the initial laser output. After the two pulse sequence, the wave function of the system is given by

$$|\psi(t)\rangle = |g\rangle + |\psi_{exc}(t)\rangle \tag{6}$$

$$|\psi_{exc}(t)\rangle = \sum_{p=1,2} \left|\psi^{(p)}(t)\right\rangle = \sum_{p=1,2}\left(\sum_k a_k^{(p)} e^{-i\omega_k t}|k\rangle\right) \tag{7}$$

where the amplitudes are

$$a_k^{(p)} = i\,\mu_{kg}\,\tilde{E}^{(p)}(\omega_k) \tag{8}$$

$$a_k^{(2)} = \beta a_k^{(1)} e^{i\omega_k \tau} \tag{9}$$

The excited state population $P_{exc} = \langle\psi_{exc}(t)|\psi_{exc}(t)\rangle$ contains contributions from each wavepacket $\langle\psi^{(p)}(t)|\psi^{(p)}(t)\rangle$ and an interference term:

$$\langle\psi^{(1)}|\psi^{(2)}\rangle = \beta e^{i\omega_m \tau}\left(e^{-i\Delta\omega\tau}\cos^2\alpha + e^{i\Delta\omega\tau}\sin^2\alpha\right)\langle\psi^{(1)}|\psi^{(1)}\rangle \tag{10}$$

where $\omega_m = (\omega_a + \omega_b)/2$ and $\Delta\omega = (\omega_{ba})/2$. Therefore,

$$P_{exc} = \frac{1}{4}\left(\theta^2 + \theta'^2 + 2\theta\theta'(\cos^2\alpha\cos\omega_a\tau + \sin^2\alpha\cos\omega_b\tau)\right) \tag{11}$$

with the generalized Rabi angle $\theta = \frac{\sqrt{\mu_{ag}^2 + \mu_{bg}^2}}{\hbar}\int_{-\infty}^{+\infty} E_0(t)\,dt$ and $\theta' = \beta\theta$.

As can be expected, this population depends only on the time delay τ but not on the evolution time t. The modulus of the interference term is linked to the overlap between the two wavepackets. It varies at the frequency ω_{ba} of the bright state - dark state oscillation. it corresponds to the contrast of the interferences. Its maximal value of 1 is reached for $\omega_{ba}\tau = 2p\pi$, when the first wavepacket is back into the bright state at the creation of the second wavepacket. Conversely, the lowest contrast, equal to $(cos^2\alpha - sin^2\alpha)$, is reached when the first wave packet is in the dark state, for $\omega_{ba}\tau = (2p+1)\pi$.

The phase of the interference term (Eq. 10), $e^{i\omega_m\tau}$, governs the constructive or destructive nature of the interferences. It oscillates at the average frequency of the transition ω_m, which is close to the laser frequency ω_L for a one-photon transition [13], or twice the laser frequency for a two-photon transition [11,13]. This rapid oscillation requires a control of the time delay on a scale much smaller than the optical period of the laser, in order either to control these interferences or simply to record the wavepacket dynamics (on a time scale ω_{ab}^{-1}).

The experiment has been performed on the same transition (4s-4p in potassium) as in the pump-probe case. The excited state population is detected through a two-photon ionization from one of the two exciting pulses, simultaneously to excitation. This step has the same efficiency for the bright and dark states. Fig 5 displays the experimental results. The envelope of the interferences reflects the wavepacket oscillation. The fast oscillations due to the interferences are displayed with more details in the insets.

Thanks to the linearity of the interaction, the excited wave function can be written as resulting from the Fourier transform of the total electric field $\tilde{E}_T(\omega)$. This means that knowledge of this quantity is sufficient to predict the results of the interaction. The rapid oscillations observed result only from the fringes in the two-pulse spectrum which scroll as a function of the time-delay [13]. This explains why similar interferograms were observed with incoherent light [9,10], or in Fourier transform spectroscopy.

The interferences can also be understood by looking at each excited state independently from the others. The amplitude is given by

$$b_k(t) = a_k^{(1)}(1 + e^{i\omega_k\tau})e^{-i\omega_k t} \qquad (12)$$

If the sample contains several independent quantum systems, with transitions at frequencies ω_k, $\omega_{k'}$, then the signal corresponds to a superposition of interferograms whose beats behave exactly as the wavepacket interferences given by Eq. 10, even though no wavepacket is created. The same result is obtained if transitions arise from different initial states of the same system. This is for instance the case with different hyperfine structure states [28], or for different rovibrational states [12].

Wavepacket interferences can be produced with a sequence of two identical ultrashort pulses. They can reveal the dynamics of the system as well as provide an efficient and sensitive way to control coherently the amplitude of the wavepacket. However, it is possible to analyse the data in terms of wave packets only if the pulses are coherent and if all the excited transitions share

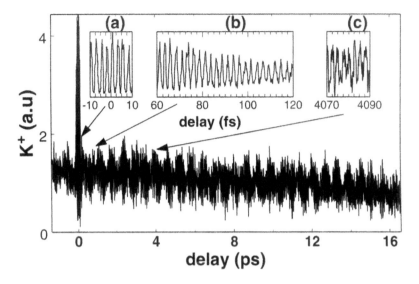

Figure 5: Wavepacket interferences in the weak field regime. The envelope of the beats (enlarged in the insets) corresponds to the wavepacket oscillation (period of 580 fs). The period of the beats (2.6 fs) corresponds to the frequency of the transition. For short time delays (inset a)), when the wavepackets overlap, the oscillations are due to optical interferences.

the same initial state. Changing the interaction regime towards saturation offers an efficient way to check that these two conditions are fulfilled. It provides also an alternative method for observing the wavepacket dynamics as detailed in the next section.

5.2 The saturation regime

At higher intensities, many new phenomena can be created and observed. In a simple two-level system, Rabi oscillations take place which can leave it in any coherent superposition of the two states. The final state depends only on the interaction time and the field strength, or more precisely on the Rabi angle.

With a multilevel system, the free evolution of the excited state starts as soon as some population has been created by the pulse, and it takes place simultaneously with the Rabi oscillation driven by the field. This behaviour results in many new phenomena, depending on the relative strengths and time scales of these two dynamics. Non Franck-Condon transitions were for instance predicted [29]. Likewise, schemes based on this principle have been proposed to control tunneling (enhancing or blocking it) [30]. We consider here only the onset of saturation which means that the number of Rabi oscillations, $\theta/2\pi$, is of the order of a few units.

For overlapping laser pulses, $\tau \leq \tau_L$, the total electric field is modulated as a result of optical interferences so that the excitation probability oscillates at the laser frequency ω_L as a function of the time delay τ, exactly as in the weak field regime [13]. For $\tau \geq \tau_L$, the two interactions are completely disconnected. A simple analytical treatment, obtained with realistic assumptions, provides a direct understanding of the various physical processes observed. Here the free evolution dynamics takes place on a time scale $(2\pi/\omega_{ba})$ much longer than the pulse duration (τ_L). Therefore the whole process can be described sequentially as the creation of a wavepacket by the first pulse in the bright state (two levels Rabi oscillation), a free evolution of the wavepacket between the bright and dark states followed by a second Rabi oscillation between the ground state and the bright state. This description is valid only if the pulse duration is small compared to the characteristic time of free evolution ($\tau_L \ll 1/\omega_{ba}$) and if the field-driven interaction dominates the free evolution during the interaction, $\theta \gg \omega_{ba}\tau_L$. After interaction with the two pulses, the population in the excited states is calculated analytically by solving the time-dependent Schrödinger equation in the frame of the rotating wave approximation. It can be separated in a fast and a slow oscillating component, $P_{exc} = P_{fast} + P_{slow}$ where

$$P_{fast} = \frac{1}{2} \sin\theta \sin\theta' \left(\cos^2\alpha \cos\omega_a\tau + \sin^2\alpha \cos\omega_b\tau \right) \quad (13)$$

$$P_{slow} = 1 - (\cos\theta/2 \cos\theta'/2)^2 - (\sin\theta/2 \sin\theta'/2)^2 \left| \cos^2\alpha + \sin^2\alpha\, e^{-i\omega_{ba}\tau} \right|^2 \quad (14)$$

In the weak field regime, $\theta, \theta' \ll 1$, the contribution P_{fast} dominates and is equivalent to the expression obtained in the weak field regime (Eq. 11).

As the field intensities rise, saturation takes place and for $\theta \geq \pi/2$ the contribution P_{fast} starts to drop. For half a Rabi cycle ($\theta \simeq \pi$) this contribution vanishes because the first pulse has completely transferred the population of the ground state to the excited states. The second pulse can no longer create a second wavepacket and hence there is no interference. The highest contrast in interference pattern is obtained for two "$\pi/2$" pulses. The first pulse leaves the system with equivalent populations in the ground and the excited states. The second pulse transfers totally the population towards the excited state or down to the ground state depending on the relative phase (time delay) of the two pulses.

The contribution P_{slow} contains an oscillating term at the relative frequency ω_{ba}. This term is a result of the coupling between the two excitation paths. It arises at intermediate field intensities when depopulation of the ground state becomes significant. Such a coupling term would not exist with two independent two-level systems (as with different isotopes, hyperfine structure states [28], or rotational levels [12]). Therefore its observation is a clear and direct signature of the existence of a wavepacket and of the underlying level structure and is a necessary complement to a weak intensity experiment. There are two different ways to understand this result. First, in terms of frequency mixing due to the nonlinearity of the interaction [15] and second, by analyzing the wavepacket motion. In the latter case, if the first wavepacket is back into the bright state, it can be pumped down to the ground state by the second pulse. The resulting excited state population drops. Thus, this term reflects directly the wavepacket motion between the bright state and the dark states. Its maximum magnitude is obtained for $\theta = \theta' = \pi$ when each pulse induces a total transfer.

The experimental results are presented in Fig. 6 for the intermediate ($\theta_{\max} \approx 1.4\pi$) field regime together with the weak field regime ($\theta_{\max} \approx 0.2\pi$) as a reference. The quantum interferences have disappeared as expected from Eq. 13 for $\theta \approx \pi$. The optical interferences can still be observed when the laser pulses overlap (for $\tau \simeq 0$), because they are independent of the interaction process. For longer delay times, there remains a slow modulation of the ion signal with a period of 577 fs. This modulation at ω_{ba} results from oscillation of the wavepacket between the bright state and the dark state. When it is in the bright state the second pulse can pump it down to the ground state by stimulated emission. This stimulated emission competes with ionization and reduces the ion signal. Therefore, minima in the ion signal correspond to the maxima of the envelope of the quantum interferences in the weak field regime (see the dashed line). For two independent two-level systems, this slow modulation would not be observed. As a result, we have a direct evidence of the three-level structure of the system and of the wavepacket creation and dynamics.

6 Conclusion

We have shown that the combination of weak field and intermediate field regime experiments provides a unique tool for preparing a wavepacket and monitoring its dynamics. The pump-probe scheme is a simple and robust method well adapted for these studies. However, in some systems, this basic investigation scheme is no longer time-dependent. wavepacket interferences not only offer the possibility of controlling the wavepacket after its creation but also provide insights into the wavepacket dynamics. However, some ambiguities regarding the nature of the wavepacket remain in this scheme. They can be removed when saturation takes place. Then a clear distinction between interferences arising from a mixture of quantum systems and a coherent wavepacket excited in a single quantum system is obtained.

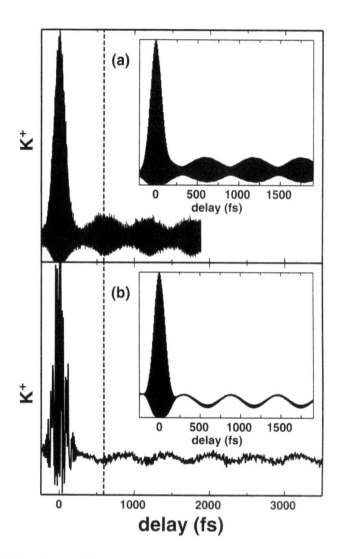

Figure 6: Experimental ion signal in the intermediate field regime (b) compared to the weak field regime (a) as a reference. The insets show the result of a numerical simulation [14]. The minimum of the oscillation in the saturation case (b) corresponds to the maximum beat envelope of the weak field regime (a), as underlined by the dotted vertical line.

The saturation regime can be viewed as a particular case of the pump-probe scheme. The probe step is in this case the stimulated emission from the excited state towards the ground step. This transition is always time-dependent since it depends only on the part of the wavepacket projected on the bright state. The depletion of the excited state population results in a decrease of the ionization probability when the wavepacket is in the bright state.

Whatever the investigation scheme chosen, the observed signal has a behavior which is well described in the bright state - dark state formalism. This formalism is generally applicable to cases with more than two excited states. In such cases, the definition of the bright state is $\sum_k \mu_{kg} |k\rangle$ and the dark states can be chosen arbitrarily in the sub-space orthogonal to the bright state. This scheme has been applied to prepare and observe a wavepacket in the potassium 4p doublet state. This dynamics corresponds to the precession of the spin and angular momentum around the total angular momentum.

This work has been financially supported by the C.N.R.S., the M.E.N.E.S.R and the Région Midi-Pyrénées, France.

References

[1] J. A. Yeazell and C. R. Stroud Jr, Phys. Rev. Lett. **60**, 1494 (1988).

[2] A. H. Zewail, *Femtochemistry: ultrafast dynamics of the chemical bond*, (World Scientific Publishing, Singapore, 1994).

[3] J. Manz and L. Wöste eds, *Femtosecond Chemistry*, (VCH, Weinheim, New York, 1995).

[4] J. T. Fourkas, W. L. Wilson, G. Wäckerle, A. E. Frost and M. D. Fayer, J. Opt. Soc. Am. B **6**, 1905 (1989).

[5] N. F. Scherer, A. J. Ruggiero, M. Du and G. R. Fleming, J. Chem. Phys. **93**, 856 (1990).

[6] L. D. Noordam, D. I. Duncan and T. F. Gallagher, Phys. Rev. A **45**, 4734 (1992).

[7] B. Broers, J. F. Christian, J. H. Hoogenaard, W. J. van der Zande, H. B. van Linden van den Heuvell and L. D. Noordam, Phys. Rev. Lett. **71**, 344 (1993).

[8] L. Marmet, H. Held, G. Raithel, J. A. Yeazell and H. Walther, Phys. Rev. Lett. **72**, 3779 (1994).

[9] R. R. Jones, D. W. Schumacher, T. F. Gallagher and P. H. Bucksbaum, J. Phys. B **28**, L405 (1995).

[10] L. C. Snoek, S. G. Clement, F. J. M. Harren and W. J. van der Zande, Chem. Phys. Lett. **258**, 460 (1996).

[11] V. Blanchet, C. Nicole, M. A. Bouchene and B. Girard, Phys. Rev. Lett. **78**, 2716 (1997).

[12] V. Blanchet, M. A. Bouchene and B. Girard, J. Chem. Phys. **108**, 4862 (1998).

[13] M. A. Bouchene, V. Blanchet, C. Nicole, N. Melikechi, B. Girard, H. Ruppe, S. Rutz, E. Schreiber and L. Wöste, Eur. Phys. J. D **2**, 131 (1998).

[14] C. Nicole, M. A. Bouchene, S. Zamith, N. Melikechi and B. Girard, Phys. Rev. A **60**, R1755 (1999).

[15] R. R. Jones, C. S. Raman, D. W. Schumacher and P. H. Bucksbaum, Phys. Rev. Lett. **71**, 2575 (1993).

[16] R. R. Jones, Phys. Rev. Lett. **75**, 1491 (1995).

[17] J. Y. Bigot, M.-A. Mycek, S. Weiss, R. G. Ulbrich and D. S. Chemla, Phys. Rev. Lett. **70**, 3307 (1993).

[18] A. P. Heberle, J. J. Baumberg and K. Köhler, Phys. Rev. Lett. **75**, 2598 (1995).

[19] X. Marie, P. Le Jeune, T. Amand, M. Brousseau, J. Barrau, M. Paillard and R. Planel, Phys. Rev. Lett. **78**, 3222 (1997).

[20] J. F. Christian, L. C. Snoek, S. G. Clement and W. J. Van der Zande, Phys. Rev. A **53**, 1894 (1996).

[21] S. Haroche, Quantum beats and time-resolved fluorescence spectroscopy, in *High-resolution laser spectroscopy*, edited by K. Shimoda (Springer-Verlag, Berlin, 1976), Vol. 13, pp. 256.

[22] T. W. Ducas, M. G. Littman and M. L. Zimmerman, Phys. Rev. Lett. **35**, 1752 (1975).

[23] A. Aspect, E. Arimondo, R. Kaiser, N. Vansteenkiste and C. Cohen-Tannoudji, Phys. Rev. Lett. **61**, 826 (1988).

[24] G. Grégoire, M. Mons, I. Dimicoli, F. Piuzzi, E. Charron, C. Dedonder-Lardeux, C. Jouvet, S. Martrenchard, D. Solgadi and A. Suzor-Weiner, Eur. Phys. J. D **1**, 187 (1998).

[25] C. Nicole, M. A. Bouchene, C. Meier, S. Magnier, E. Schreiber and B. Girard, J. Chem. Phys. **111**, 7857 (1999).

[26] E. Sokell, S. Zamith, M. A. Bouchene and B. Girard, J. Phys. B **33**, 2005 (2000).

[27] B. De Beauvoir, V. Blanchet, M. A. Bouchene and B. Girard, (in preparation).

[28] M. Bellini, A. Bartoli and T. W. Hansch, Opt. Lett. **22**, 540 (1997).

[29] B. Garraway and K.-A. Suominen, Phys. Rev. Lett. **80**, 932 (1998).

[30] N. Tsukada, Y. Nomura and T. Isu, Phys. Rev. A **59**, 2852 (1999).

Atomic Beam Spin Echo.
Principle and Surface Science Application

M. DeKieviet, D. Dubbers*, S. Hafner, F. Lang

Physikalisches Institut Universität Heidelberg,
Heidelberg, Germany

* Institut Laue Langevin, Grenoble, France

I Introduction

We describe a novel technique, Atomic Beam Spin Echo (ABSE), which aims at measuring extremely small changes of the kinetic energy in a flux of atoms (or molecules) carrying a magnetic moment. In the last few decades, atomic and molecular beam techniques have contributed invaluable information to many areas of science [1]. Over the years, especially with the implementation of supersonic expansions, the quality of these beams has improved steadily, both in intensity and monochromaticity. Thusfar, the Time of Flight (TOF) method is the best established technique for studying dynamical process in, or with molecular beams. The energy resolution hereby is generally limited by the velocity spread in the beam. In addition, the need for a beam chopper and the fact that with TOF detection is time critical seriously limit the available intensity, so that at present, the TOF method seems to have reached a practicable resolution of some tenths of meV. In order to improve on this, we combine the in-beam Spin Echo (SE) method, originally developed for neutron scattering by Mezei [2], with helium atom scattering.

In this paper, we first introduce the principle of ABSE for the case of a spin-$\frac{1}{2}$ particle, using a semi-classical picture. Subsequently, we develop a quantum mechanical description, leading to an almost intuitive interpretation of the method for scattering experiments. In the second part of the paper, we illustrate the performance of such an ABSE spectrometer operating in the area of surface science. We show ^{3}He scattering data on both the structure and the dynamics of a clean, and an adsorbate ($C_{24}H_{12}$) covered Au(111) surface. We will discuss the inelastic and the quasi-elastic contributions separately and demonstrate how they can be retrieved from spin echo data.

II The Principle

II.1 Semi-Classical Picture

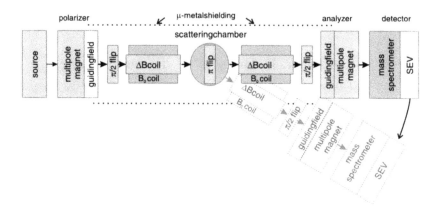

Figure 1: Schematics of an Atomic Beam Spin Echo spectrometer.

In Fig. 1, we show an overview of the components involved in an **ABSE** experiment. The basic idea behind ABSE is to use the Larmor precession of the spin of each particle exiting the atomic or molecular beam source (left in the Fig. 1) as an individual, internal clock. By polarizing the atomic beam, say in direction x perpendicular to the beam axis z, all clocks get synchronized. Depending on the species used, this can be achieved in a multipole magnet through the Stern-Gerlach effect (e.g. ^{3}He [4], ^{1}H, or ^{2}D [5]), via optical pumping (e.g. ^{7}Li [6]), or in an RF-cavity (as recently demonstrated on neutrons [7]). In the guiding field, the global polarization is conserved, untill the clocks start running, which is when the beam enters the first Spin Echo coil (B_0 in Fig. 1). In here, the magnetic field is orthogonal, say in z-direction, to the original polarization vector, so that the change in relative orientation between the spins exiting the guiding field and entering the Spin Echo coil corresponds to a $\frac{\pi}{2}$-flip. The total precession angle accumulated by each particle individually traversing the coil with velocity v_1 is proportional to the Larmor frequency ω_L, the strength of the SE field $B(z)$ and the time spent in it:

$$\varphi_1 = \int \omega_L dt = \frac{\gamma \int B(z) dz}{v_1} = \frac{\gamma \bar{B} l}{v_1} = \frac{\gamma \bar{B} l m}{h} \lambda_1, \qquad (1)$$

Herein, γ is the gyromagnetic ratio (e.g. $\gamma = -2\pi \cdot 32.433$ MHz/T for ^{3}He, $\gamma = +2\pi \cdot 14.160$ GHz/T for ^{1}H), λ_1 represents the de Broglie wavelength, and $\bar{B}$ is the magnetic field averaged over its length l (m is the mass of

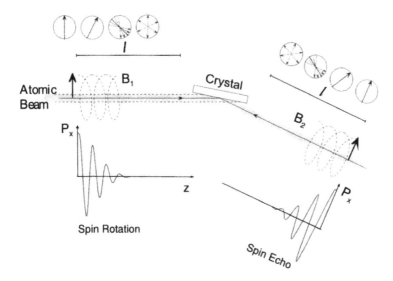

Figure 2: Semi-classical picture of the ABSE experiment.

the atomic beam species and h Planck's constant). The beam-averaged polarisation at any position along the beam axis is the overall magnetic moment generated by all individual spins together. In general, the atomic beam is polychromatic (characterized by its wavelength distribution $\frac{dn}{d\lambda}$) and different particles need differently long times for the same flight path, so that the beam-averaged polarisation decays with l:

$$P = \int \frac{dn}{d\lambda} \cos\left(\varphi_1(\lambda)\right) d\lambda = \int \frac{dn}{d\lambda} \cos\left(\frac{\bar{B}lm}{h}\right) d\lambda. \qquad (2)$$

In Fig. 2, we try to visualize this in a semi-classical way. Considering the left-hand side of Fig. 2 for now, an atom (starting in the middle left) traverses the first SE coil (of length l), while its spin (the vector) describes a spiral path (represented by the dashed line). On top, four snapshots of the beam-averaged polarization are plotted, each displaying a distribution of spin vectors as viewed along the beam axis. The further along in the SE coil, the more the internal clocks of the particles dephase. In the bottom panel of Fig. 2, the resulting polarization along the original x-direction is plotted as a function of the position along the flight path. Since it decays with the magnetic field integral $\bar{B}l$ (Eq. 2), one may as well plot the polarisation as a function of magnetic field strength for a fixed SE coil length. As an example of such a so-called spin rotation curve, we shown an actual measurement in the left-hand side of Fig. 3. This curve was obtained using a beam of ^{3}He atoms (see below), varying the strength of the SE field by detuning the

ΔB-coil, depicted in Fig. 1, and leaving $B_0 = 0$. According to Eq. 2, the envelope of this interference pattern is but the cosine-Fourier transform of the wavelength distribution $\frac{dn}{d\lambda}$. That means, the more monochromatic the atomic beam, the larger the number of fringes in Fig. 3.

After exiting the first SE coil (see Fig. 3), the atomic beam interacts with whatever system whose dynamics is subject of the investigation (in Figs. 2 and 3 this process is represented by scattering from a crystal sample), thereby changing its de Broglie wavelength from λ_1 into λ_2. Subsequently, the beam enters a second SE coil, which is equally strong as, but anti-parallel to the first. Note that the directional change by 180° between the magnetic fields in SE coil 1 and 2 corresponds to a π-flip, as indicated in Fig. 1. Trespassing the second SE field, the spins precess in reverse direction, so that upon exit each spin has accumulated a total precession angle of:

$$\varphi = \varphi_1 + \varphi_2 = \frac{\gamma \bar{B} l m}{h}(\lambda_1 - \lambda_2) \qquad (3)$$

Given an energy transfer $\hbar\omega$ between the initial and final state before and after scattering, respectively, energy conservation requires that $E_i = E_f + \hbar\omega$. Assuming that this transfer is small ($\hbar\omega \ll E_i$) and applying the dispersion relation for a free particle one obtains for the leading term in the expansion of φ around $\omega = 0$:

$$\varphi \simeq \frac{\gamma m^2 \lambda^3 \bar{B} l}{2\pi \hbar^2}\omega \equiv \omega t_{se}, \qquad (4)$$

This is the approximation for so-called quasi-elastic scattering and herein, λ is the mean wavelength of the beam.

Detuning the two SE coils from their inital value $B_1 = -B_2 = \bar{B}$ (using the ΔB coils indicated in Fig. 1) one obtains a so-called spin echo curve, shown on the right side of figure 3. This SE curve was measured (again using ^{3}He atoms) in the absence of a scattering center and merely demonstrates the kind of resolution ABSE may achieve. In this case, running a current of 10 A through each SE coil, the spins precess almost 5000 full revolutions forward and the same number back, to within better than a few degrees. As a consequence, energy changes as low as some 10^{-9}eV can and have been measured [4].

According to Eq. 4 the overall precession angle φ is proportional to the energy transferred upon scattering. The proportionality constant, basically depending on the mean wavelength of the beam and the strength of the SE fields, is called spin echo time t_{se}. From the full quantum mechanical description of the spin echo process [8, 9, 5] which will be sketched below, t_{se} turns out to be the correlation time of the dynamics at or within the scattering target. It can be measured with ABSE to as long as 10^{-8}s, which, through the uncertainty relation, corresponds to the energy resolution above.

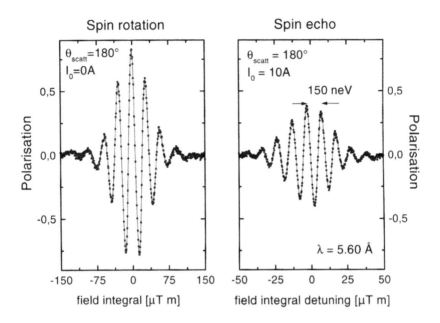

Figure 3: Left side: spinrotation signal. Right side: spin echo signal for about 4500 turns in each magnetic field region

In a spin echo experiment, the net polarisation of the beam is determined for different spin echo times t_{se}. The net polarisation of the scattered beam consists of the spin contribution of each atom, weighed with the (2-dimensional) scattering function $S(\vec{K}, \omega)$, which gives the probability for a scattering process with the (parallel) momentum transfer $\vec{K}$ and the energy transfer ω:

$$P = \langle \cos \varphi \rangle = \int S(\vec{K}, \omega) \cdot \cos(\omega t_{se}) d\omega = I(\vec{K}, t_{se}). \quad (5)$$

The intermediate scattering function $I(\vec{K}, t_{se})$ is the Fourier transform into the time domain of the scattering function $S(\vec{K}, \omega)$, or equivalently the Fourier transform into reciprocal space of the correlation function $G(\vec{R}, t_{se})$: $S(\vec{K}, \omega) \xleftrightarrow{FT} I(\vec{K}, t) \xleftrightarrow{FT} G(\vec{R}, t)$. $G(\vec{R}, t_{se})$, and therewith $I(\vec{K}, t_{se})$, describes the dynamics of the system and is composed of an auto- and a pair-correlation part. Due to the convolution theorem, these contributions can be factorized for the intermediate scattering function in the Vineyard approximation [10]: $I(\vec{K}, t_{se}) = I(\vec{K}, 0) I_a(\vec{K}, t_{se})$. Since diffusion is determined by the auto-correlation part $I_a(\vec{K}, t_{se})$, its dynamics can easily be extracted from the data through a simple division by the beam polarisation at $t_{se} = 0$.

II.2 ABSE: Atom Interferometry

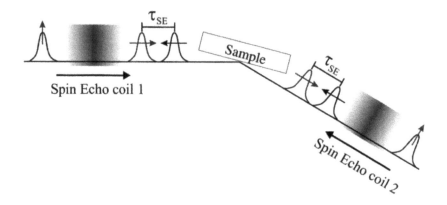

Figure 4: QM Picture of the ABSE experiment.

In the quantum mechanical picture of ABSE (see Fig. 4), the rather formal definition of SE time t_{SE} in Eq. 4 becomes more transparent. In analogy to the Faraday effect for photons, Larmor precession is here considered to originate from magnetic birefringence [8, 9]. Starting after the polarizer Fig. 1 the polychromatic, continuous flux of particles can be described by a wave packet, linearly polarized in x. Entering the first longitudinal SE coil, the quantization axis of the magnetic eigenstates becomes along the beam axis z. As a consequence, the wave packet splits into a coherent superposition of two new eigenstates (assuming an $S = \frac{1}{2}$-particle), namely the left- and right-handed, circularly polarized eigenfunctions ψ^+ and ψ^-, having the energies E^+ and E^-. Within the magnetic field, the temporal evolution of these two states follows $\psi^\pm(t) \propto exp[iE^\pm t]$, so that after the first SE coil ψ^+ and ψ^- have acquired a difference in phase $\approx exp[iE_0 t_{SE}]$. It appears as if the partial waves ψ^+ and ψ^- have wandered apart a time t_{SE} (see Fig. 4). In the second SE coil, the role of the eigenfunctions ψ^+ and ψ^- is inverted (because the magnetic field is inverted), so that the phase difference between ψ^+ and ψ^- reduces back to zero again. Since ψ^+ and ψ^- originate from the same wave packet, they coherently interfere at the analyzer and, if their overlap is perfect, the original linear polarization is fully recovered (i.e. $P = 1$). If, however, a process occurs between the first and the second SE coil, such that the scattering potential changes within the SE time (the time difference between the arrival of ψ^+ and that of ψ^- at the sample), then both packages scatter from a different potential to $\tilde{\psi}^+$ and $\tilde{\psi}^-$. In how far these $\tilde{\psi}^+$ and $\tilde{\psi}^-$ can be brought back to overlap by the second SE coil is reflected in their interference pattern at the analyzer, and hence in the amount of regained linear polarization. In this picture, the SE time t_{SE} thus represents the time interval over which correlations in the scattering potential are sampled. The SE method directly measures the correlation function $I(\vec{K}, t_{SE})$ in the time domain (in which the scattering geometry determines through $\vec{q}$ the length scale under investigation). ABSE herewith is a true atom interferometry experiment, with, as we will show below, practical use..

III Application of ABSE in Surface Science

III.1 ^{3}He Spin Echo from Coronene on Gold(111)

With the goal in mind to study slow motion in and on surfaces, we chose to develop the ABSE method for a helium beam. In the discipline of surface science, He scattering takes in a unique position among all other probes. With thermal beam energies of 10 - 100 meV, corresponding to de Broglie wavelengths of some Å, a beam of He atoms is inert (no electronic excitations or reactions) and exclusively surface sensitive (He atoms reflect some Angstrom above the surface and therefore exclusively probe the outermost layer of the sample. In addition, their de Broglie wavelength matches the atomic distances at the surface. Whereas He diffraction reveals structure, inelastic He scattering interrogates the dynamics at or on a surface. For the kind of processes that we will be investigating (diffusion of large molecules, phase transitions, etc.), the energy resolution of standard TOF methods is not sufficient (≥ 0.1meV). ^{3}Helium Spin Echo (^{3}He-SE) increases this resolution by at least four orders of magnitude. In addition, the facts that no chopper is needed, the velocity spread may be polychromatic, and the Stern-Gerlach polarizer focusses the beam there is a huge advantage in intensity as well. A detailed description of the experimental setup can be found in [11]. Here we only point out that, in order to be able to polarize the beam with an elctromagnetic quadrupole, the ^{3}He atoms expand from a source at 1.2, or 4.2 K. As a consequence, both the de Broglie wavelength of the beam (≈ 6 Å), and its coherence length (upto 10μm [12]) are relatively large, which is helpful when studying structures having extended periodic structures. On the other hand, the ^{3}He atoms need to keep a relatively low energy, so that their spin direction can be analyzed after scattering. This requirement practically limits the region of interest for the ^{3}He-SE spectrometer to small energies (≈ 1 meV) and small momentum transfers (≈ 0.25 Å^{-1}). The experiments reported on here were performed on the surface of single crystal Au(111). This commercially obtained sample was prepared by argon ion sputtering (300-600 eV) and annealing (850-900 K), such that the LEED pattern showed clear Bragg peaks belonging to the reconstruction. In subsequent elastic ^{3}He scattering, diffraction peaks appear up to the third order in the $[1\bar{1}0]$ direction, as shown in figure 4. From this data we extract the dimensions of the reconstructed unit cell to be $(p \times \sqrt{3})$, with $p = (21.5 \pm 0.5)$ expressed in the bulk gold nearest neighbor distance $a = 2.885$ Å (i.e. a periodicity of ≈ 62 Å).

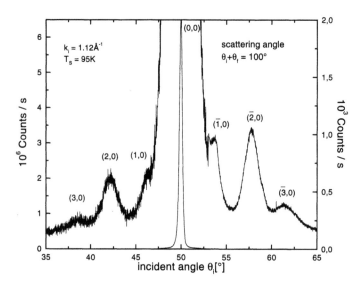

Figure 5: ^{3}He diffraction trace along the $[1\bar{1}0]$ azimuthal direction for the $(p \times \sqrt{3})$ reconstruction of Au(111). The wavelength of the beam is $\lambda = 6.5$Å. Note that the actual diffraction trace has been amplified 1000x with respect to the specular peak at $\theta_i = 50°$.

The adsorbate was evaporated onto the surface through freezing deposition by subliming coronene ($C_{24}H_{12}$) at 720K, while the gold substrate was kept at $T_s = 100$ K. After careful annealing and desorption at 400 K, the adsorbate at monolayer coverage was observed to form a (4×4) commensurable superstructure with respect to the unreconstructed gold substrate. A real space model of this superstructure is shown in figure 6.

For the purpose of studying diffusion, the adsorbate was diluted to sub-monolayer coverages ($\theta \ll 1$) by controlled evaporation. The 2-D dynamics of coronene on Au(111) was subsequently investigated using the spin echo method. Spin echo curves were taken at several temperatures (T_s from 95K to 437K) for different incident angles (θ_i, ranging from 51.5° to 60°). The total scattering angle was fixed at $\theta_s = 100°$. The intermediate scattering function at each temperature and incident angle was obtained from a series of spin echo curves at different values of t_{se}, i.e. for different applied longitudinal magnetic fields. A representative set of data showing the composition of $I(\vec{K}, t_{se})$ from the spin echo curves is presented in Fig. 7

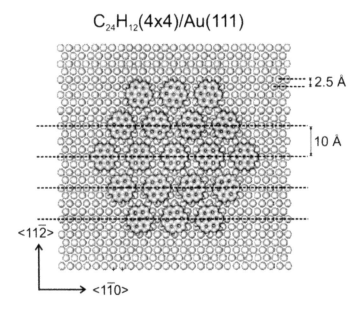

Figure 6: Coronene overlayer on Au(111) as determined by LEED. The light gray balls represent the gold atoms on the surface, whereas the dark balls are the carbon atoms of coronene.

for $T_s = 294$K and $\theta_i = 53.5°$. Note, that in accordance with the above discussion, $I(\vec{K}, t_{se})$ has been normalized to its value at $t_{se} = 0$

III.2 Discussion

The measured intermediate scattering function, an example of which is shown in the lower panel of Fig. 7, generally displays two features: An oscillatory modulation superimposed on a rapid decay with t_{se}. The exponential decay is believed to result from quasielastic scattering of ^{3}He on diffusing coronene, while the oscillation originates from inelastic processes between the helium and the gold.

Quasielastic scattering

Quasielastic scattering refers to processes in which the energy exchange with each ^{3}He atom is very small ($\omega \ll E_i$), and for the total beam averages out to zero. Within this framework, we will discuss two types of adsorbate motion, random, continuous diffusion, and non-continuous, jump diffusion. Random, continuous diffusion is likely to take place for large adsorbates at low coverage and high temperatures on very flat surfaces. The dynamics of

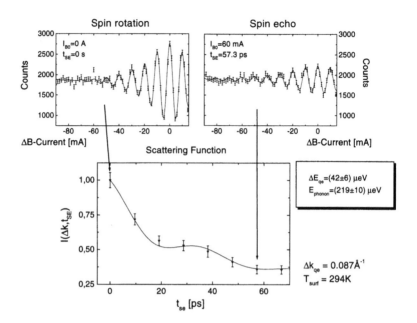

Figure 7: Composition of the intermediate scattering function at an incident angle of 53.5° and a surface temperature of 294 K.

random diffusion is described by Fick's second law:

$$\frac{\partial G_a(\vec{R}, t_{se})}{\partial t_{se}} = -D \cdot \Delta_{\vec{R}} G_a(\vec{R}, t_{se}), \tag{6}$$

which is solved for the autocorrelation function $G_a(\vec{R}, t_{se})$ and Fourier transformed to give:

$$I(\vec{K}, t_{se}) = I(\vec{K}, 0) \cdot \exp(-K^2 D t_{se}). \tag{7}$$

Herein, $\Delta_{\vec{R}}$ represents the Laplacian and D is the 2-D random diffusion coefficient. This model implies that the exponential decay increases quadratically with the parallel momentum transfer $K \equiv |\vec{K}|$.

If the adsorbate motion in 2-D is not free, as in the case of a corrugated surface with strong adsorption sites, this description is not adequate. Diffusion then becomes a quantum mechanical process, in which the adsorbate jumps in discrete steps from site to site. The corresponding rate equation is given by

$$\frac{\partial G_a(\vec{R}, t_{se})}{\partial t_{se}} = \sum_j \frac{1}{\tau_j} \left[G_a(\vec{R} + \vec{j}, t_{se}) - G_a(\vec{R}, t_{se}) \right], \tag{8}$$

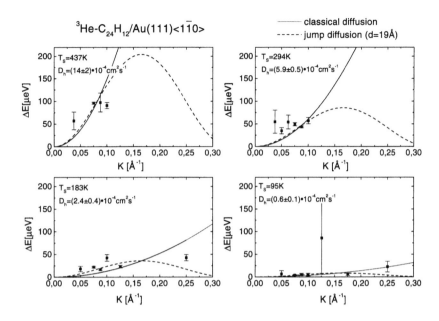

Figure 8: Dependence on the momentum transfer. Random diffusion and jump diffusion are compared to the data.

where the index j runs over all possible jump directions and lenghts, and τ_j is the jump trial frequency. The solution of this equation also leads to an exponential decay of the intermediate scattering function with t_{se}:

$$I(\vec{K}, t_{se}) = I(\vec{K}, 0) \cdot \exp(-\Gamma t_{se}). \qquad (9)$$

For jump diffusion, however, the decay constant Γ is an oscillatory function of $\vec{K}$:

$$\Gamma = 4 \sum_{j=1}^{\infty} \frac{1}{\tau_j} \sin^2\left(\frac{\vec{K} \cdot \vec{j}}{2}\right). \qquad (10)$$

For small momentum exchange ($K \to 0$) and a single jump length d, one can define an effective diffusion (hopping) constant D_h, in analogy to the classical case, $D_h = d^2/\tau_d$.

Although at first sight coronene on gold seems to be a perfect candidate for random 2-D diffusion, our data suggest that $C_{24}H_{12}$ moves along the $[1\bar{1}0]$-direction across the surface in discrete steps of about 19 Å(Fig. 8). This is quite surprising, since the coronene molecule has a diameter of more than 10 Å and the corrugation of gold is relatively small (that of the reconstruction is less than 0.2 Å). In order to determine the activation barrier to this diffusion, we semilogarithmically plot the obtained jump diffusion coefficients versus

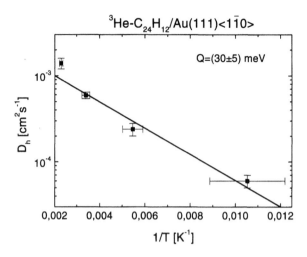

Figure 9: Arrhenius plot for the system $C_{24}H_{12}/Au(111)$ in the $[1\bar{1}0]$ direction.

inverse temperature. From the linear fit to the data in Fig. 9 we infer that when the motion is governed by Arrhenius' law $D_h = D_0 \exp(-Q/kT)$, the coronene molecules experience an activation barrier of $Q = (30 \pm 5)$ meV. This appears to be a reasonable value for the lateral variance in an estimated total binding energy of ≥ 500 meV for a coronene molecule on gold.

Inelastic scattering

In order to determine the inelastic contribution to the intermediate scattering function (see for example Fig. 7), we start from a scattering function $S(\vec{K}, \omega)$, having a Lorentzian resonance of width Γ centered around $\omega = \omega_0$ (in addition to the one located at $\omega = 0$ of the same width, responsible for the quasielastic scattering). Now, the total precession angle described in (4) is expanded around $(\omega = \omega_0)$, leading to

$$\varphi = 2t_{se}\omega_i(\beta - 1) - \beta^3 t_{se}(\omega - \omega_0), \quad \text{in which} \quad \beta = \frac{1}{1 + \frac{\omega_0}{\omega_i}} \quad (11)$$

Inserting this in equation (5), one obtains for the inelastic part of the intermediate scattering function

$$I_{in}(\vec{K}, \omega) = \cos((2t_{se}\omega_i(\beta - 1)) \cdot \exp(-\Gamma \cdot t_{se})). \quad (12)$$

Note, that it is the cosine term in this expression, which gives rise to the oscillatory behavior observed in the data (Fig. 7). From the period of the

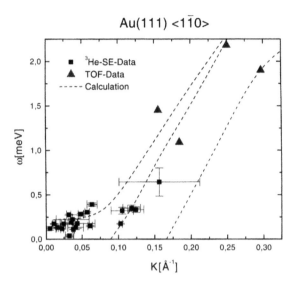

Figure 10: Dispersion curve for reconstructed Au(111). Squares: Spin Echo data; triangles: TOF-data [14], dashed line: calculation [13]

oscillations we obtain the inelastic energy transfer $\hbar\omega_0$. Relating this energy to the momentum transfer following from the kinematic conditions in each scattering geometry (i.e θ_i and θ_s), we obtain points on the gold surface dispersion curve, plotted in Fig. 10. In addition, the calculated dispersion curve [13] for the reconstructed Au(111) surface is plotted together with the lowest-lying experimental points from TOF measurements [14]. Our data seem to confirm the theoretical prediction that, due to the reconstruction, at least one of the two lowest-lying phonon branches becomes zero at finite momentum transfer $\vec{K} \neq 0$. As can be seen from Fig. 10, the ^{3}HeSE method is particularly sensitive to the very low energy range of the surface dispersion curve. This is because, firstly, the initial beam energy is so low (≈ 1 meV), and in addition, the energy resolution of the ^{3}He-SE spectrometer is relatively high. So far, little is known about the dispersion in solid surfaces at this low energy, but with Atomic Beam Spin Echo we now have experimental access to unobserved phenomena in this interesting region.

References

[1] G. Scoles, *Atomic and Molecular Beams Methods*, Volume **I** + **II**, Oxford Press (1988, 1992).

[2] F. Mezei, Z. Phys. **255**, 146-160 (1972).

[3] F. Mezei, *Neutron Spin Echo*, Lecture Notes in Physics **128**, Springer Verlag (1980).

[4] M. DeKieviet, D. Dubbers, C. Schmidt, D. Scholz, U. Spinola, Phys. Rev. Lett. **75(10)**, 1919 (1995).

[5] M. DeKieviet, S. Hafner, A. Reiner, *Hydrogen and Deuterium Atomic Beam Spin Echo*, Phys. Rev. A (submitted).

[6] M. Zielonkowski, J. Steiger, U. Schünemann, M. DeKieviet, R. Grimm, Phys. Rev. A**58**, 3993 (1998).

[7] H. Abele, A. Boucher, P. Geltenbort, M. Klein, U. Schmidt, C. Stellmach, Nucl. Instr. Methods **A440**, 760 (2000).

[8] R. Gähler, R. Golub, K. Habicht, J. Felber, Physica B **229**, 1 (1996).

[9] R. Gähler, J. Felber, F. Mezei, R. Golub, Phys. Rev. A **58**, 280 (1998).

[10] G.H. Vineyard, Phys. Rev. **110**, 999 (1958).

[11] DeKieviet, M., D. Dubbers and Ch. Schmidt, *A ^3He Spin Echo Spectrometer for Surface Dynamics Studies: Design and Performance*, Phys. Rev. A (submitted).

[12] DeKieviet, M., D. Dubbers, M.Klein, Ch. Schmidt, M. Skrzipczyk, Surf. Sci. **377-379**, 1112 (1997).

[13] Jayanthi C.S., H. Bilz, W. Kress, G. Benedek, Phys. Rev. Let., **59(7)**, 795-798, (1987).

[14] Harten, U., J. P. Toennies, Ch. Wöll, G. Zhang, Phys. Rev. Lett. **55**, 2308 (1985).

Sufficient Conditions for Quantum Reflection with Real Gas-Surface Interaction Potentials

R.B. Doak[†], A.V.G. Chizmeshya[‡]
[†]Department of Physics and Astronomy
[‡]Materials Research Group
Arizona State University, Tempe, AZ, USA

1. Introduction

In complete contradiction to intuition based on classical mechanics, an atom striking a solid surface at extremely low velocity does not stick. Understanding of this non-sticking emerges only through quantum theory, which predicts the sticking probability $S(k)$ of the atom to vanish linearly with decreasing incident wave vector k for sufficiently small k [1]. Accordingly this non-sticking of ultra-slow atoms has come to be known as "quantum reflection." Considerable theoretical effort [1]-[5] has been devoted to the phenomenon over the last two decades. Measurements of the sticking probability of H-atoms scattered from a surface of liquid helium at extremely low k have been reported [6], the only accredited empirical observation of quantum reflection to date. By criteria developed below those measurements did indeed broach the quantum reflection regime. Even more satisfying would be an explicit verification of the phenomenon as might be provided by molecular beam scattering, for example by monitoring elastic helium atom scattering from a single crystal solid surface while microscopically roughening the surface through ion bombardment. This would destroy coherent elastic scattering at normal k but should have no effect in the quantum reflection regime. There may be practical uses of quantum reflection as well: If energy exchange between the surface and the atom is truly absent, ultra-cold atoms can be stored (and even evaporatively cooled) by placing them in a simple container of any temperature under UHV conditions. The sole constraint is that the atoms be slow enough to quantum reflect from the wall material.

With such thoughts in mind and recognizing that it is now possible to generate very slow atoms by laser deceleration [7] or even by simple mechanical deceleration [8], it becomes crucial to know exactly how small k must be in order for quantum reflection to ensue. The literature yields predictions for specific system such as H:W [2] and of H:He(liq) [5] but offers no general specification of quantum reflection for arbitrary choice of real gas atom and real surface material. Most of the theory is limited to simple model systems and it is often not clear how the model parameters relate to real gas-surface potential properties. Communication with leading theorists in the field produced friendly and useful responses but not the real numbers needed for experimental design.

Accordingly we embarked on Numerov integrations of the Schrödinger equation, using static state-of-the-art LDA gas-surface interaction potentials [9] to inject real gas-surface parameters into the problem. With numerical calculations for real potentials in hand, it became clear that quantum reflection for real potentials - as befits a basic, fundamental phenomenon - can be understood to lowest order on an elementary level of both physics and mathematics. The phenomenon ultimately arises through disparities at small k in the magnitude and functional dependence of the potential energy and total energy terms in the Schrödinger equation. This yields distinct portions of the wavefunction in the region near the surface and far from the surface which we denote respectively as $\Psi_<(x)$ and $\Psi_>(x)$. The separation of the scattering wavefunction into separate inner and outer portions we term "spatial scission." Normalization fixes the amplitude far from the surface and the mathematical problem reduces to smoothly joining $\Psi_<(x)$ with $\Psi_>(x)$ at the scission boundary $x_{<>}(k)$. The boundary matching forces the amplitude of the near-surface wavefunction to vanish linearly with k in the limit $k \to 0$ and leads to quantum reflection. All of this is very reminiscent of the simple 1-D hard-wall/attractive square well problem [2] but here matching asymptotically across an extended continuous boundary region rather than abruptly at a discontinuity in the potential.

To establish sufficiency for quantum reflection we observe that under specific conditions $\Psi_<(x)$ becomes linear in x at large x whereas $\Psi_>(x)$ becomes linear in x at small x. The latter is easily understood: $\Psi_>(x)$ is a sine curve of argument $(kx + \delta)$ and will be linear in x with slope proportional to k provided the phase angle δ is sufficiently small and $x \ll \lambda = 2\pi/k$, the deBroglie wavelength of the free particle. The large x linearity of $\Psi_<(x)$ is only marginally more complicated to demonstrate. The Schrödinger equation may be solved in terms of Bessel functions for any potential having the power law form $V(x) = -C_s/z^s$ [10], [2]. This is, of course, the asymptotic large x form of any realistic potential, with $s = 3$ or 4 for for an unretarded or retarded gas-surface interaction potential, respectively. The Bessel function solution becomes linear in x for $x \gg B_s$, a characteristic length parameter which depends on s and C_s. Provided the scission boundary $x_{<>}(k)$ lies sufficiently far from the surface that the potential has attained its asymptotic power law form, it then follows that both $\Psi_<(x)$ and $\Psi_>(x)$ will be the linear at the scission boundary $x_{<>}(k)$ provided $B_s \ll x_{<>}(k) \ll \lambda(k)$. Smoothly connecting the the inner and outer portions of the wavefunction forces the k-proportional slope of $\Psi_>(x)$ onto $\Psi_<(x)$. Since $\Psi_<(x)$ is k-independent in the limit of small k, this boundary condition requires the amplitude of $\Psi_<(x)$ to vanish linearly with k, yielding quantum reflection. The above conditions therefore constitute sufficiency conditions for quantum reflection for any realistic interaction potential and require knowledge only of its limiting asymptotic power law form, specifically s and C_s.

Chung and George [2] discuss this within the "dynamical Levinson theorem" which restricts the scattering phase δ to a multiple of π. The real restriction, however, ought not to be on the scattering phase but rather on λ relative to yet another characteristic length scale of the problem, the outermost stationary zero of the wavefunction in the limit $k \to 0$. This we denote b, the "binding length" in direct analogy to a, the "scattering length" of effective range theory. The binding length is the largest length parameter of the potential. To guarantee that the ratio of outer to inner amplitude becomes large in the limit $k \to 0$, it is necessary that $\lambda \gg b$ which immediately yields the small δ of the dynamical Levinson theorem.

Introducing b also provides insight into situations where λ is comparable to or smaller than b. Most interesting is the case $\lambda \ll b$, in which instance the conditions laid out above lead not to a phase angle of $\delta = n\pi$ and quantum reflection but rather to $\delta = (2n+1)\pi/2$ and "anti-quantum reflection." The binding length b becomes large when the system supports a very weakly bound state. By heteroepitaxially depositing an overlayer of one material onto a substrate of another, it should be possible to alter the bound state spectrum of a surface and thereby to "tune" one or more bound states successively to zero energy [11]. Anti-quantum reflection at the corresponding overlayer thicknesses should produce abrupt and easily observable rises in sticking probability.

2. Mathematics of Spatial Scission

Proceeding now to the mathematical underpinnings we note, on very general grounds of length scale mismatch, that the scattering problem is expected to separate in the directions normal to and parallel to the surface provided λ is much greater than the surface lattice constant and surface corrugation. We require that the $\hbar^2 k^2/2M$ total energy term of the Schrödinger equation be much smaller in magnitude than the $V(x)$ potential energy term near the well minimum or, defining $U(x) = 2MV(x)/\hbar^2$, we require $k^2 \ll |U(x)|$ in this near-surface region. Since $V(x)$ decays as C_s/x^s far from the surface, it follows that $k^2 \gg |U(x)|$ at sufficiently large x. The inner region (wherein $U(x)$ dominates over k^2) and the outer region (wherein k^2 dominates over $U(x)$) are separated by the scission boundary $x_{<>}(k)$ defined by $|U(x_{<>}(k))| = k^2$.

In the direction normal to the surface the Schrödinger equation in the outer region is then

$$\frac{d^2\Psi_>(x)}{dx^2} + k^2\Psi_>(x) = 0 \quad , \tag{1}$$

which yields immediately the solution $\Psi_>(x) = A_> sin(kx+\delta)$. In the inner region the Schrödinger equation takes the form

$$\frac{d^2\Psi_<(x)}{dx^2} - U(x)\Psi_<(x) = 0 \quad . \tag{2}$$

The requirement that $k^2 \ll |U(x)|$ in the well region guarantees, for any real gas-surface interaction potential, that the potential will have attained its large x asymptotic form at $x_{<>}(k)$. In the the vicinity of the scission boundary (2) then becomes

$$\frac{d^2\Psi_<(x)}{dx^2} + \frac{2MC_s}{\hbar^2 x^s}\Psi_<(x) = 0 \quad , \tag{3}$$

which may be solved immediately in terms of Bessel functions,

$$\Psi(x) = \sqrt{\frac{x}{B_s}}\, \mathcal{C}_\nu(z) \quad , \tag{4}$$

where $\mathcal{C}_\nu(z)$ with $z = -(x/B_s)^{(2-s)/2}$ and $\nu = 1/(2-s)$ denotes either the Bessel functions $J_\nu(z)$ or $Y_\nu(z)$, the Hankel functions $H_\nu^{(1)}(z)$ or $H_\nu^{(2)}(z)$, or any linear combination thereof. The linear demarcation boundary B_s is

$$B_s = \left(\frac{8MC_s}{(2-s)^2\hbar^2}\right)^{\frac{1}{(s-2)}} \quad . \tag{5}$$

The Bessel functions $\mathcal{C}_\nu(z)$ of (4) will in general have negative argument and negative fractional index [10] but can be expressed in terms of positive arguments and positive integer index via the usual transformations [12], [10] to reveal that $\Psi_<(x) \sim x$ in this limit of $k \to 0$ for $x \gg B_s$ and for $s > 2$. Any short-range ($s > 2$) potential possessing the correct $1/x^s$ asymptotic form must therefore become linear in x for $x \gg B_s$ in the limit of small k. This is demonstrated in Fig. 1 with a numerical calculation of $\Psi(x)$ for a realistic unretarded He:Cu(111) potential with k set equal to zero (heavy solid line curve). Also shown in that figure is the outer plane wave solution (light dashed line). The amplitude $A_>$ of $\Psi_>(x)$ has been normalized to unity in the figure and the amplitude $A_<$ of $\Psi_<(x)$ adjusted to join the inner and outer curves smoothly in the vicinity of $x_{<>}(k)$. The spatial scission, linear regions of the two curves, and boundary matching at $x_{<>}(k)$ can all be clearly seen.

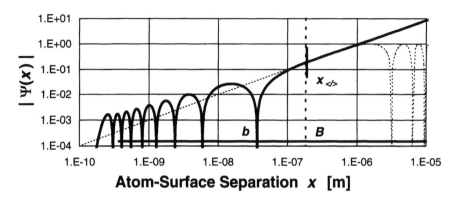

Figure 1: Log-log plot of wavefunction amplitude for He scattering from unretarded Cu(111) potential, illustrating boundary matching of $\Psi_<(x)$ inner and $\Psi_>(x)$ outer portions of scattering wavefunction. Light dashed curve is $\Psi_>(x)$ (a sine wave), calculated setting $U(x) = 0$ with $k = 10^6 \, m^{-1}$. Scission boundary $x_{<>}(k)$ for this k is marked with the heavy vertical bar. Heavy line curve is small k limiting form of $\Psi_<(x)$, calculated setting $k = 0$ and scaling in amplitude to smoothly join the outer solution near $x_{<>}(k)$. Bar at bottom marks $1/x^3$ region of potential. The eight zeros in $\Psi_<(x)$ curve reveal eight bound states of He in this potential. $\Psi_<(x)$ becomes linear in x beyond $x = B_s$ (vertical dashed line) in accordance with (3) and (4) for this $1/x^3$ potential. Since $x_{<>}(k) \approx B_s$ at this k, this is the k_{QR} for the onset of quantum reflection.

Of particular interest in (4) is the exponent $(2 - s)/2$ within the argument of the Bessel function. As has been discussed by Chung and George [2], this exponent is negative for $s > 2$ and positive for $s < 2$. Hence for increasing atom-surface separation x, the argument z of the Bessel function decreases in the former case and increases in the latter. From this simple fact alone the $1/x^2$ potential emerges as the mathematical demarcation between long-ranged ($s < 2$) and short-ranged ($s > 2$) potentials. It is also clear that the long-ranged potentials will not have the linear x-dependence at large x illustrated in Fig. 1. Rather, since the Bessel argument z increases with x to sample all of the oscillations of the Bessel function, those solutions will remain oscillatory for all x. Numerical calculations for a $1/x$ potential verify this. The extended linear boundary region of Fig. 1, which is in fact the region of WKB breakdown, is crucial for quantum reflection and hence it is no surprise that quantum reflection does not exist for the $s < 2$ long-ranged potentials. It is of interest that effective range theory fails for $s = 3$ but that quantum reflection does not.

In simplistic terms, quantum reflection is the "dog" $\Psi_>(x)$ wagging the "tail"

$\Psi_<(x)$. Overall normalization of the wavefunction is dominated by box normalization in the outer region, fixing $A_>$. To match $\Psi_<(x)$ and $\Psi_>(x)$ at the scission boundary, it is then the inner amplitude $A_<$ which must adjust. With decreasing k, the scission boundary $x_{<>}(k)$ moves outwards. Once it exceeds B_s, the inner wavefunction $\Psi_<(x)$ is linear at the boundary and above. $\Psi_>(x)$ is linear at the boundary and below provided λ is the largest length scale of the problem whereupon, for small δ,

$$\Psi_>(x) = A_> sin(kx + \delta) \approx (kA_>) x + A_> \delta \ . \qquad (6)$$

At still smaller k the amplitude of the wavefunction in the inner region drops linearly with decreasing k as illustrated in Fig. 2 with Numerov calculations for unretarded He:Cu(111). Quantum reflection thus occurs for $x_{<>}(k) \gg B_s$, establishing a critical k_{QR}. Defining $x_{<>}(k_{QR}) = B_s$ and using the large x asymptotic form of the potential to calculate $x_{<>}(k)$, it follows that

$$k_{QR} = [2MC_s/\hbar^2 B_s^s]^{1/2} \ , \qquad (7)$$

with B_s defined as in (5) and recalling the constraint that λ be the largest characteristic length scale of the problem (specifically, larger than b). This provides a sufficient condition for quantum reflection. The only parameters which enter are the mass of the atom, the power s and dispersion coefficient C_s of the asymptotic (large x) potential, and physical constants. Values of k_{QR} are given in Table I.

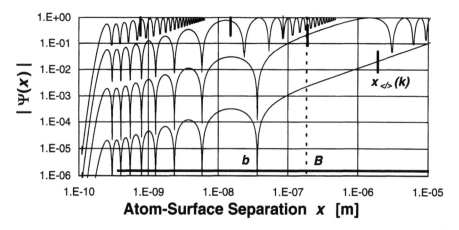

Figure 2: Log-log plot of scattering wavefunctions demonstrating decrease in near-surface amplitude with decreasing k for, top to bottom, $k = 10^{10}, 10^8, 10^6$, and 10^4 m^{-1}. Numerov calculations again for an uretarded He:Cu(111) potential. Short vertical bar on each curve marks scission boundary $x_{<>}(k)$ at that k. Dashed vertical line marks separation B_s.

Surface	Gas	C_s (meV-Å^3)	B_s (Å)	k_{QR} (Å^{-1})
Ag(111)	^{4}He	249	1940	2.6x10^{-4}
Au(111)	^{4}He	274	2130	2.3x10^{-4}
Cu(111)	^{4}He	235	1830	2.7x10^{-4}
Cs	^{4}He	58	450	1.1x10^{-3}
Cu(110)	Ne	488	19,000	2.6x10^{-5}
Cu(111)	Ar	1621	126,000	3.9x10^{-6}
Cu(111)	Xe	3080	502,000	9.9x10^{-7}
^{4}He(*liq*)	H	31	60	8.4x10^{-3}

Table 1: Quantum reflection parameters for representative atom–surface scattering systems, using $s = 3$ (unretarded) interaction potentials.

3. Sticking Coefficient

Finally, it remains to relate the above behavior to the sticking coefficient $S(k)$ in the quantum reflection regime. Within the DWBA [1]

$$S(k) \propto |<\Psi_n|V'|\Psi(k)>|^2/k \qquad (8)$$

namely sticking is proportional to the square modulus of the matrix element for a transition into a surface bound state, normalized with respect to the incident flux. V' is the perturbative potential mediating the transition. Ψ_n is independent of k and we consider k-independent $V'(x)$.

From the above discussion, $\Psi(k)$ is seen to be linearly proportional to k everywhere in the $x < x_{<>}(k)$ inner region and out into the $\sin kx \approx kx$ regime of the $x > x_{<>}(k)$ outer region. This comprises the entire region of overlap between the scattering and bound state wavefunctions other than in the presence of a very weakly bound state, and that possibility is specifically excluded by the restriction that $B_s \ll x_{<>}(k)$. Hence it follows immediately that $S(k) \propto |k|^2/k = k$, the hallmark of quantum reflection.

Acknowledgement

Support for this research in part by the U.S. National Science Foundation under grant PHY-9223053 is gratefully acknowledged.

References

[1] D.P. Clougherty and W. Kohn, Phys. Rev. B **46 II** 4921 (1992).

[2] S.G. Chung and T.F. George, Surf. Sci. **194** 347 (1988).

[3] J. Böheim, W. Brenig, and Stutzki, Z. Physik **B 48** 43 (1982).

[4] W. Brenig, Z. Physik **B 36** 227 (1980).

[5] C. Carraro andW.M. Cole, Z. Physik **B 98** 319 (1995) and references therein.

[6] I.A. Yu, J.M. Doyle, J.C. Sandberg, C.L. Cesar, D. Kleppner, and T.J. Greytak, Phys. Rev. Lett. **71** 1589 (1993).

[7] J. Lawall, J.S. Kulin, B. Saubamea, N. Bigelow, M. Leduc, and C. Cohen-Tannoudji, *Laser Spectroscopy. 12th Int. Conf.* (World Scientific, Singapore, 1995) and other papers in those proceedings.

[8] R.B. Doak, K. Kevern, A. Chizmeshya, R. David, and G. Comsa, Proc. Int. Soc. Opt. Eng. **2995** 146 (1997).

[9] A. Chizmeshya and E. Zaremba, Surf.Sci. **220** 443 (1989); *ibid.*, **268** 432 (1992).

[10] R.B. Doak and A.V.G. Chizmeshya, submitted, Europhysics Letters, 1999.

[11] R.B. Doak and A.V.G. Chizmeshya A, Symposium on Surface Science 1998, Park City UT, 1998 and to be published.

[12] G.N. Watson *A Treatise on the Theory of Bessel Functions* (University Press, Cambridge, 1962).

Focusing Helium Atom Beams Using Single Crystal Surfaces

B. Holst[1,2] and W. Allison[1]
[1] Cavendish Laboratory, University of Cambridge,
CB3 0HE, UK
[2] Max Planck Institut für Strömungsforschung
Bunsenstrasse 10, 37073, Göttingen, Germany

1 Introduction

The manipulation of molecular beams has a history dating back to the days of Stern and Gerlach [1]. In the last few years, new techniques have become available so that the field of atom-optics has now emerged, with several potential areas of application. Classical methods of manipulation, using inhomogeneous fields, as well as recent approaches, using laser manipulation, require atoms with accessible electronic states and/or a permanent moment; that is, atoms that are generally regarded as reactive. Inert atoms, such as helium, demand an alternative approach. The motivation for manipulating beams of helium atoms lies in their use as a tool in the study of surfaces. Inertness is the property that confers value as a surface probe since, at thermal energies, scattering from a surface occurs without reaction or surface damage. This short review discusses recent progress in focussing thermal beams of helium atoms, where the principal aim is to create a micro-focused atom probe.

For the focusing of a beam of neutral, ground state atoms, the only means of focusing are those of classical optics: lenses and mirrors. Fresnel zone plates offer the possibility of focusing in transmission mode and the effect was first demonstrated in 1991 [2] using metastable helium. These first results have been improved recently by several orders of magnitude [3]. However, current fabrication technology limits the performance of zone plates. Atom mirrors offer a number of advantages compared to zone plates. The intensity in the focused beam can be much greater since larger apertures are possible and the mirror offers true "white light" focusing without chromatic aberration. A mirror makes use of specular scattering, which is predominantly elastic for light atoms, such as helium. Thus the coherence of the beam is preserved. When choosing a surface for atom optical applications, the surface geometry must be considered on both a macroscopic length scale, where classical mechanics and optics are applicable, and on an atomic length scale, where quantum and diffraction effects dominate. On a macroscopic length scale the goal is to bend the crystal into the shape that provides the best possible approximation to the so called Cartesian surface [4]. The Cartesian surface focuses all rays from one given point onto another given point. It is always a rotational ellipsoid, with the foci as object and image points. On an atomic length scale the aim is to control diffraction effects and hence optimise the specular reflectivity.

The mirror thickness is also important since the mirror profile is obtained by applying external stress to the system. Thin plates are easier to deform into the desired shape and, for a given deformation, the strain energy density is smaller. For this reason we have concentrated our efforts on very thin crystals, typically less than 50 μm thick. In this regime the crystals can be deformed using external electrostatic fields (typically less than 10^7 Vm^{-1}), which are both easy to implement in vacuum and which offer a measure of external fine tuning.

In Section 2 we present the experimental arrangement used for our focusing experiments, including a discussion of the different surfaces which we have investigated. Section 3 moves on to present the first results obtained using a focusing mirror. We have used the focused microprobe to map the ionisation region in an ionisation detector and we demonstrate that He-focusing provides a new, direct method for investigating supersonic beam expansions. In the final section we discuss future developments of the mirrors, showing how the mirror profiles can be improved by changing the boundary conditions in the electrostatic deformation.

2 Experimental Arrangement

Figure 1 shows a schematic diagram, in cross section, of the sample mount used for electrostatic deformation. The sample, A, is mounted on a precision ground sapphire insulator spacer, B, with a central circular hole. The sapphire spacer separates the sample from a lower electrode, C. The central part of the sample is free-standing and deforms in an approximately parabolic shape under electrostatic pressure when a potential difference, E, is applied between the sample and lower electrode (the sample is kept at ground). The deflection and hence the radius of curvature of the mirror (sample) can thus be changed simply by adjusting the voltage. A metal ring, D, acts as an electrical contact and a clamp for the top of the sample. The first issue to address is the choice of mirror materials. The two primary aims are to maximise specular scattering, by using an atomically flat surface, and to minimise inelastic scattering, by using a heavy-atom substrate and/or one with a high Debye temperature. On an atomic scale, most metal

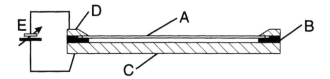

Figure 1: Schematic diagram (not to scale) showing the geometry used to apply uniform electrostatic pressure to the mirror. The mirror (A) forms the upper electrode of a parallel-plate capacitor. For clarity the mirror thickness is greatly enlarged compared to the space between the electrodes and the overall diameter. The lower electrode (C) is a metal plate and (B) acts as an insulating spacer. Electrode (D) is both an electrical connection to the mirror and a mechanical clamp. The electrostatic pressure arises from the field created by the applied potential difference (E).

surfaces have a low corrugation and act as almost perfect specular reflectors for helium atoms [5]. In contrast, semiconductors have significant corrugation. The specular reflectivity is reduced as intensity is fed into a large number of diffraction peaks. Surface reconstruction is also more common in semiconductors and is often associated with significant densities of point defects [6]. However, semiconductors offer excellent bulk crystalline properties and generally have high Debye temperatures. Thus, it is not clear, *a priori*, whether a semiconductor or a metal system offers the best prospect for creating an atom mirror.

Our initial search for a suitable mirror material concentrated on two prototype systems, Au(001) and Si(001). The samples were prepared using deposition and etching techniques respectively [7]. All samples had a thickness between 0.2 μm and 0.4 μm. The macroscopic behaviour of these samples was investigated using optical phase step interferometry, which produced contour maps of the films during deformation. These contour maps can be seen in Fig. 2. The full contour maps were used in Monte Carlo simulations to investigate the focusing properties of the samples for atom-focusing applications [7]. Neither system was perfect. The Au films (Fig. 2a,b) show significant deviation from the expected parabolic profile and the optical quality was limited by imperfect contact between the sample and the holder. Deformations in the silicon system (Fig. 2c,d) were more parabolic and the optical quality of the surfaces was better. The mechanical properties of both systems were dominated by internal stress [7].

The superior optical quality of the silicon samples led us to consider free standing wafers of Si(111) [8], which are available in diameters up to 50 mm and less than 100 μm thick. Si(111) was chosen in preference to Si(001) as it is possible to produce surfaces of high quality and low defect density by hydrogen passivation. Thus, our focusing mirror was created from a passivated

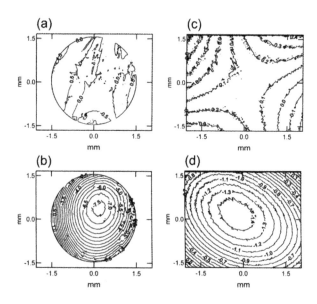

Figure 2: a) Contour map for an Au(001) sample, thickness 2000 Å, 5 mm in diameter (central 3 mm diameter area shown) with no applied field. b) Same sample with an applied field of 1×10^6 Vm^{-1}. c) Contour map for a Si(001) sample, thickness 3000 Å, 10 mm in diameter (central 3.5 mm × 3 mm shown) with no applied field. d) Same sample with an applied field of 1×10^6 Vm^{-1}. Contours in the images are derived from phase-step interferometry measurements. The contour separation is 0.5 μm for the Au(001) sample and 0.1 μm for the Si(001) sample.

Si(111)-(1 × 1)H wafer, with a thickness of 50 μm. Preparation of the mirror surface consisted of cycles of oxide stripping and hydrogen passivation, which were performed *ex situ* [9]. The sample was transferred to the vacuum in an inert atmosphere and the quality of the surface remained unchanged over a period of several months at a pressure of about 10^{-6} mbar. Hydrogen passivation offers clear advantages over most uncoated surfaces, which would become contaminated within a very short time at such a pressure.

A further advantage of Si(111)-(1 × 1)H is that its reflectivity for helium is high. We have determined the specular reflectivity from the known He/Si(111)-(1 × 1)H potential [10]. Exact quantum calculations of the atom dynamics for this potential enable all the diffracted intensities to be determined for a wide range of incident energies and angles [11]. At normal incidence, an energy of 20 meV gave the highest specular reflectivity, with a value in the region of 0.6. The specular reflectivity also depends strongly on

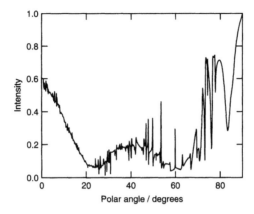

Figure 3: Calculated variation of specular He-intensity from Si(111)-(1×1)H with incident polar angle at a fixed beam energy of 20 meV and 0° azimuth.

the angle of incidence. The variation with incident polar angle at an energy of 20 meV is given in Fig. 3. Sharp selective adsorption resonance features are visible in the data together with a slowly varying component, which is determined by the surface corrugation. The results show that the intensity is high both at near-normal incidence and at near grazing incidence, with lower reflectivity at intermediate angles. A reflectivity of 30-40% is possible for a realistic experimental setup; that is, one slightly away from normal incidence. For the experimental data presented below, it was not possible to work at the optimum scattering geometry and beam energy and the reflectivity was correspondingly lower.

The sample holder used for the focusing experiments using Si(111)-(1 × 1)H was similar to that in Fig. 1. A diagram of the mirror in relation to the atom source and detector is shown in Fig. 4. The mirror is arranged to scatter atoms through approximately 90° to a fixed mass spectrometer detector. The source to mirror and mirror to detector distances are 700 mm and 900 mm respectively. The helium beam source was produced by expansion through a 10 µm diameter nozzle operated at a stagnation pressure of 150 bar and a temperature of 310 K, giving a helium beam wavelength of 0.52 Å and an energy of about 70 meV. The beam extended an area of about 3.2 mm² on the mirror. The spatial resolution of the ion source was increased to the desired level by placing a pinhole of diameter 100 µm, between the mirror and the detector. The focussed beam profile was obtained by scanning the beam across the pinhole.

The intensity profile of the beam can be seen in Fig. 5 for various values of the electrostatic deflection-field. The difference between the initial flat surface, Fig. 5a, and the field giving maximum intensity, Fig. 5c, is evident. The beam comes to focus first in the beam scattering plane and then in the perpendicular plane. These positions correspond approximately with Fig. 5b and Fig. 5d, respectively. The disk of least confusion, lies in between, Fig. 5c. The focusing in the geometry used here is necessarily astigmatic because the mirror is parabolic. The minimum spot diameter obtained was 210 ± 50 μm, where the uncertainty arises from the effect of the finite size of the pinhole. The solid angle subtended by the beam at the detector is reduced by a factor of about 100. The spot size is limited by the geometry of the system (the object and image distances) and the size of the object (the surface of last scattering in the supersonic expansion, see [12] and next section).

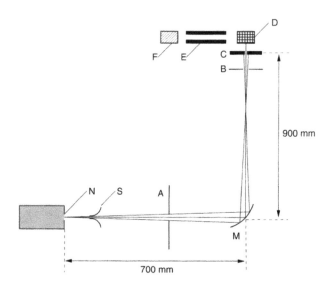

Figure 4: Diagram showing the arrangement of the source, mirror and detector in the Cambridge apparatus. Supersonic expansion through the nozzle (N) produces the He-beam. The skimmer (S) and a limiting aperture (A) separate the expansion chamber from the main vacuum chamber which houses the mirror (M). The beam passes through a slit aperture (B) and pinhole (C) to reach a differentially pumped detector chamber. The detector consists of an electron impact ioniser (D), a quadrupole mass spectrometer (E) and a single electron multiplier tube (F).

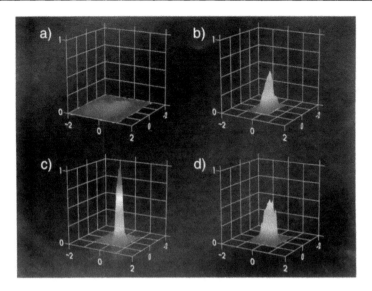

Figure 5: Beam cross-sections for various mirror curvatures (R). The horizontal axes span a plane perpendicular to the beam direction of travel (scale in mm). The vertical axis shows normalized intensity. a) Unfocused beam. b) First focus in scattering plane (R=1.2 m). c) Disk of least confusion (R=0.8 m). d) Second focus perpendicular to scattering plane (R=0.5 m).

3 Applications

In this section we discuss two applications of the atom-focusing mirror. The first relies on imaging, in the detector plane, trajectories from the gas flow at the beam source. By measuring the intensity distribution in the detector plane we have been able to obtain a direct picture of the properties of the original supersonic expansion. In the second application we demonstrate how the micro-focussed atom beam can be used to explore the spatial variation of efficiency in the atom detector.

The initial dynamics of a supersonic expansion are complex, but eventually a molecular flow regime is reached in which the atoms travel in straight lines without further collisions [13]. Reflection at the mirror changes the direction of the atoms without altering their energy or speed. The situation is therefore analogous to classical optics where the light rays travel in straight lines between optical elements. The last scattering region of the beam, where the transition to molecular flow occurs, determines the image in the detector plane. The focusing mirror serves as a non-interfering tool for measuring beam properties. When the molecular flow regime is reached, the individual trajectories can be projected backwards onto a plane perpendicular to the beam. In this virtual-source plane a spatial distribution function, the virtual source, is obtained [13]. In the present experiment, for a nozzle diameter of only 10 μm, the virtual source plane is essentially the same as the nozzle plane.

We have performed a series of calculations where rays were traced from the virtual source, through collimating apertures over the mirror and on to the detector. The profile of the mirror was taken to be a paraboloid with a curvature given by the known applied field and known mechanical properties of the mirror [12, 14]. Following Beijerinck et al. [13] the virtual source was modelled with a Gaussian distribution function. The measured beam profiles were thus fitted with a single variable parameter, the effective radius of the virtual source.

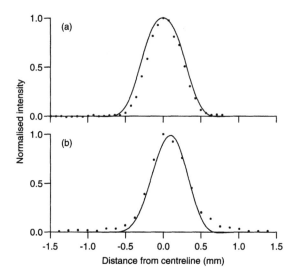

Figure 6: The focused helium beam profile with a mirror radius of curvature $R_C = 0.78$ m. The points show experimental measurements of the distribution in: (a) the scattering plane and (b) the perpendicular plane. The solid lines show the corresponding simulation results, giving a fit to experiment with a virtual source radius of 0.2 ± 0.05 mm.

The beam profile measurements are compared with a best fit for the disk of least confusion in Fig. 6 (the fitted simulations are shown as a solid curve). The agreement is good. The best fit was obtained with a source radius of 0.2 ± 0.05 mm. The width of the broad virtual source is predicted by Beijerinck to be: $R = 0.08$ mm [13]. Beijerinck's model has hitherto only been tested with source pressures up to 8 bar. Thus our measurements, carried out at a source pressure of 150 bar, show that the model breaks

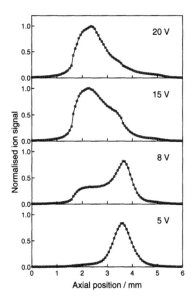

Figure 7: The normalised ion signal as a function of the micro-probe position along the axis of the ion source. Each curve corresponds to a different nominal ion energy, as indicated. The same normalisation factor has been used for all curves.

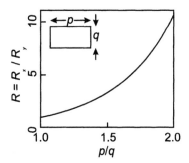

Figure 8: Calculated ratio of the principal radii of curvature, $R = R_x/R_y$ for a mirror with a rectangular boundary. The boundary shape is defined by the ratio $p/q = 1.5$ as shown in the inset. Calculations apply in the small deflection limit

down in the higher pressure regime.

We now turn to the properties of the atom detector and illustrate the first application of a scanned atom micro-probe. The work gives new insights into the operation and efficiency of electron impact ionisers, which form the basis of many atom detectors. The basic principle of ionisation using energetic electrons is well understood. However, most practical instruments display a complex dependence on operating conditions that is poorly understood [15],[16]. A micro-focused atom beam offers the possibility of exploring the spatial variation of ionisation efficiency within an ioniser by scanning the atom-probe across the active region of the source.

The experimental arrangement is the same as in Fig. 4 and the focused atom beam is scanned across the active volume of the ion source by tilting the mirror through small angles [17]. Fig. 7 shows the results of scans along the axis of the ion source (i.e. in the plane of the diagram in Fig. 4). Results for various ion energies are shown. The most obvious feature of the data is the trend for the peak to move to the right, away from the extraction aperture, when the ion energy is reduced. The effect can be explained in terms of the potential energy at the point of ionisation. A combination of electron space charge and penetration of the extraction field into the ionisation region reduces the energy of ions created closest to the extraction electrode. Those with insufficient kinetic energy are unable to enter the mass filter and the sensitivity is reduced. The data demonstrates that ionisation takes place in a confined region and, more importantly that the sensitivity is a complex function of position and varies markedly with operating conditions. The results have been interpreted in detail [17] and related to earlier work of ioniser performance [18],[19].

4 Future prospects

Further developments of micro-probe atom beams will require a smaller spot size at the focus. To achieve this it is necessary to have a better approximation to the ideal Cartesian Surface. For non-normal incidence, an ideal mirror surface has a radius of curvature in the scattering plane that is larger than that out of the plane. The mirror described above has the same curvature in every direction. It is necessarily astigmatic and hence displays large aberrations. The problem of reducing aberrations, and hence the size of the focused spot, is thus one of producing a more complex mirror profile. In particular, the first requirement is to generate a curvature in the plane of scattering that differs from the curvature out of the plane. One advantage of electrostatic actuation is that it is possible, in principle, to remove aberrations by suitable design of the mirror mount and by using an appropriate actuating field. The principles of the method are similar to those used in systems of adaptive optics [20].

The simplest approach to obtain the desired mirror shape is to choose boundary conditions, at the periphery of the mirror, that reflect the desired symmetry of the final surface. A simple rectangular boundary is one example. The properties of a range of possible boundary conditions have been calculated using accurate simulations of the mirror displacement [4]. The ratio of the principal radii of curvature is shown as a function of the boundary conditions in Fig. 8. Here, a rectangular boundary, with long side of length p and short side q, has been used and the results are presented for different values of p/q. The data shows that it is possible to vary the ratio of the radii of curvature by a factor of more than ten, simply by tuning the relative dimensions of the rectangular boundary. Thus the results in Fig. 8 demonstrate that the main source of astigmatism in atom mirrors can be eliminated with relative ease.

In summary, we have presented a range of work that demonstrates the possibility and the potential of atom mirrors. The results show that elastic scattering of thermal energy atoms from a suitable mirror surface can be used to give a small, intense micro-probe of atoms. A helium micro-probe has already yielded insights into the operation of atom detectors [17], and into the gas dynamics of beam source [12]. We have also described work that will allow the mirror geometry to be optimised, in terms of the choices of incident energy and incident angle. The remaining challenge is to minimise optical aberrations arising from the macroscopic surface profile. Preliminary results suggest that these problems can be managed by a suitable choice of the mirror boundary and the application of an inhomogeneous electrostatic field [4], [21].

Acknowledgements

We would like to thank D. A. MacLaren and A. P. Graham for valuable discussions and careful reading of the manuscript.

References

[1] N. F. Ramsey, Molecular Beams (Clarendon Press, Oxford, 1956).

[2] O. Carnal, M. Siegel, H. Takuma and J. Mlynek, Phys. Rev. Lett. **67**, 3231 (1991).

[3] R. B. Doak, R. E. Grisenti, S. Rehbein, G. Schmahl, J. P. Toennies and C. Wöll, Phys. Rev. Lett. **83**, 4229 (1999).

[4] R. J. Wilson, B. Holst and W. Allison, Rev. Sci. Instrum. **70**, 2960 (1999). D. A. MacLaren, B. Holst and W. Allison, Rev. Sci. Instrum. (submitted).

[5] B. Poelsema and G. Comsa, Springer Tracts in Modern Physics **115** (1989).

[6] D. M. Rohlfing, J. Ellis, B. J. Hinch, W. Allison and R. F. Willis, Vacuum **38**, 347 (1988).

[7] B. Holst, J. M. Huntley, R. Balsod and W. Allison, J. Phys. D: Appl. Phys. **32**, 2666 (1999).

[8] B. Holst and W. Allison, Nature **390**, 244 (1997).

[9] G. S. Higashi, R. S. Becker, Y. J. Chabal and A. J. Becker, Appl. Phys. Lett. **58**, 1656 (1991).

[10] J. R. Buckland and W. Allison, J. Chem. Phys. **112**, 970 (2000).

[11] J. R. Buckland, B. Holst and W. Allison, Chem. Phys. Lett. **303**, 107 (1999).

[12] B. Holst, Ph.D. Thesis (University of Cambridge, 1997).

[13] H. C. W. Beijerinck and N. F. Verster, Physica **C 111**, 327 (1981).

[14] B. Holst, J. R. Buckland and W. Allison, Chem. Phys. (submitted).

[15] L. Lieszkovszky, A. R. Filippelli and C. R. Tilford, J. Vac. Sci. Technol. A **8**, 3838 (1990).

[16] M. C. Cowen, W. Allison and J. H. Batey, Meas. Sci. Technol. **4**, 72 (1993).

[17] B. Holst, J. R. Buckland and W. Allison, Vacuum **53**, 207 (1999).

[18] K. Kuhnke, K. Kern, R. David and G. Comsa, Rev. Sci. Inst. **65**, 3458 (1994).

[19] M. C. Cowen and W. Allison, J. Vac. Sci. Technol. A **12**, 228 (1994).

[20] G. Vdovin, Opt. Eng. **34**, 3249 (1995).

[21] D. A. MacLaren, B. Holst and W. Allison, Rev. Sci. Instrum. (submitted).

Atom Lithography with Cesium Atomic Beams

F. Lison, D. Haubrich, D. Meschede
Institute for Applied Physics
University of Bonn
Bonn, Germany

1 Introduction

Optical lithography is the dominant method of manufacturing lateral micro- and nanostructures in nearly all areas of technology, but it is predicted to be limited to feature sizes of about 100 nm due to diffraction. At this scale the miniaturization is not yet impaired by quantum limits of the substrates and materials used for construction, i.e. transistors will still be governed by the same physical laws as the currently available components. Therefore nanofabrication of known devices may continue beyond this border without a conceptual change of important components involved, provided suitable lithographic processes with sub 100 nm resolution are available. According to the "roadmap" published by the semiconductor industry [1] it is expected that the technological 100 nm barrier will be reached by 2005. The applicability of sub 100 nm methods for nanostructure fabrication will not only be determined by technological and physical reasons, however, but more importantly by economical factors.

Charged particle beams (electron beams, ion beams) have been extensively investigated as an alternative method. Beams with very short wavelengths of order 1 nm or below can be produced and controlled, and it has already been demonstrated that their application gives access to lateral structures below 10 nm.

In recent years neutral atom beams have emerged as yet another species of particle beams with a natural potential for nanostructure fabrication. At velocities of several 10 m/s atomic beams have de Broglie wavelengths of less than one nanometer. Hence diffraction does not prevent focusing down to nanometer spot sizes. In contrast to charged particle beams the influence of interparticle forces which limits for instance the speed of writing and the ultimate resolution in electron beam lithography is strongly suppressed. Experimental results with neutral atomic beams to date have been obtained with laboratory equipment but may be scaled up to larger instruments offering a highly parallel approach to nanostructure fabrication.

The reason for the recent emergence of atomic beams for this purpose is the maturing of laser cooling methods [2] which now allow to prepare neutral atomic beams with narrow velocity spread. A fast evolution of atom optical elements based on light forces was furthermore initiated during the last decade [3].

2 Atom Lithography

In the following we will give a brief overview of what is called "atom lithography", i.e. the methods used to control the lateral spatial structure of atomic beams and to transfer this structure to a suitable substrate.

The first demonstration of atom lithographic processes was achieved by Timp et al. [4] by depositing narrow lines of sodium atoms on a substrate. Unfortunately sodium atoms adsorbed on a surface are chemically unstable when extracted from the vacuum, but the application of chromium [5, 6] and aluminum [7] atomic beams removed this deficiency. In all cases periodic one- and two-dimensional structures with linewidths well below 100 nm were written in a parallel process by focusing atoms with standing wave light fields onto a substrate. The crucial point in these deposition experiments was the transverse laser cooling to prepare a high flux and well collimated atomic beam. This necessitated the construction of suitable, intense sources of blue or UV laser light.

The field of atom lithography can be extended to atomic species that are better suited for the application of laser cooling but do not form permanent structures. A collaboration of Harvard University and NIST has developed a resist for metastable argon atoms using self assembled monolayers (SAM) of alkanthiols ($CH_3(CH_2)_nSH$) on gold [8]. In the meantime it has been shown that the same type of resist can be used for metastable helium and neon atoms [9, 10], and also for cesium atoms [11, 12]. Moreover, in a more recent result it was shown that the internal state structure of atoms can be exploited during the deposition process to generate nanostructures [13].

2.1 Standing wave focusing

An inhomogenous alternating electric field induces an oscillating dipole moment in a neutral atom. This phenomenon is called AC–Stark effect. A strong spatial modulation of the electric field and the associated potential energy of the atom is for instance created by an optical standing wave. In this intensity grating a mechanical dipole force is exerted on the atomic motion. If the frequency of the electric field is near an atomic resonance the force on the atoms is resonantly enhanced and low laser intensities of a few mW/cm^2 are sufficient to manipulate atomic trajectories.

The interaction between the atoms and the light field can be described by

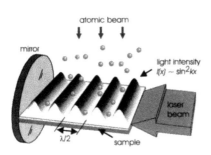
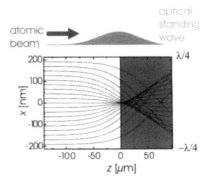

Figure 1: Standing wave focusing of an atomic beam. The sample is mounted close to the mirror creating the standing wave. Optimum conditions are achieved when the transverse profile of the light field for the optical standing wave is cut in half by the sample, i.e. the light force lenses are operated in the thick lens limit

Figure 2: Calculated atomic trajectories in one period ($\lambda/2$) of a transverse standing wave light field. On *top* the Gaussian beam profile of the light field is shown. The atomic beam is assumed to be monoenergetic and perfectly collimated. The parameters for the laser beam match our experimental situation. The *vertical line* at $z = 0$ is where the substrate surface is placed

the potential [14]

$$U = \frac{\hbar\Delta}{2} \ln(1 + S) \tag{1}$$

where the saturation parameter S is given by

$$S = \frac{I}{I_s} \frac{\Gamma^2}{\Gamma^2 + 4\Delta^2} . \tag{2}$$

Here $\Gamma = 1/\tau$ is the spontaneous decay rate of the excited state and Δ is the laser detuning with respect to the atomic resonance. In the case of $S \ll 1$ and $\Delta \gg \Gamma$ spontaneous emission can be neglected and the potential (1) is approximated by

$$U \approx \frac{\hbar\Gamma^2}{8I_s} \cdot \frac{I}{\Delta} . \tag{3}$$

The principle arrangement for the application of a standing wave for focusing of an atomic beam is shown in Fig. 1. In a one-dimensional standing wave light field with wavelength $\lambda = 2\pi/k$ the intensity pattern is given by $I = I_0 \sin^2(k \cdot z)$. In our experiment the cesium D_2 line with $\lambda_{Cs} = 852.1$ nm and $I_s = 1.1$ mW/cm^2 was used. If the laser frequency is larger (smaller)

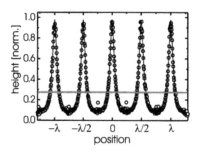
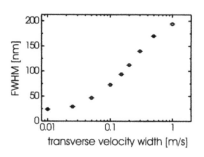

Figure 3: Atomic flux density distribution in the focal plane (on the substrate) according to trajectory calculations. The *horizontal line* corresponds to the homogeneous density distribution with equal total flux. The simulation assumes a beam with a thermal Maxwell–Boltzmann velocity distribution and a transverse velocity width comparable with our experimental situation

Figure 4: FWHM of the atomic flux density distribution in the focal plane as a function of the initial transverse velocity width of the atomic beam. The longitudinal velocity distribution is included in the calculation

than the frequency of the atomic resonance line the force on the atoms is directed towards the intensity nodes (antinodes). A Taylor expansion of the potential around each of its minima shows a harmonic behavior for small excursions from the node (antinode). Therefore the standing wave light field acts as an array of cylindrical lenses for a traversing atomic beam (Fig. 1 and Fig. 2). The properties of the light lenses can be chosen such that the focus lies within the light field (thick lens limit). The simulation in Fig. 2 clearly shows that atoms on trajectories entering far from the lens axis deviate from a perfect focusing behaviour [15]. We have therefore numerically estimated the atomic flux density distribution in the focal plane which shows that narrow line widths of 100 nm and below can be achieved inspite of the spherical aberrations (Fig. 3). The most important contribution to the focal spot size is due to the initial transverse velocity width. In Fig. 4 we show the results of a numerical analysis of the FWHM of the spot size as a function of the initial transverse velocity for the parameters of our cesium atomic beam. The numerical analysis furthermore shows that the influence of the longitudinal velocity distribution (chromatic aberration) is small in comparison with the above described contributions. This is an important result since it demonstrates that it is not necessary to have an atomic beam with a small longitudinal velocity spread – an experimental configuration that requires significantly less effort.

2.2 Experimental

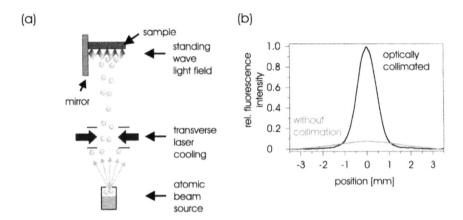

Figure 5: **a** Sketch of the experimental setup showing atomic beam source (*bottom*), two dimensional optical collimation stage (*middle*), standing wave light field, and sample (*top*). **b** Beam profiles of the atomic beam with (*dark line*) and without (*light line*) transverse optical collimation. The profiles were taken after 1 m of free flight

Most experimental setups look quite similar to Fig. 5. In our case, an effusive cesium atomic beam source (Knudsen cell) is precollimated by apertures. The transverse velocity width is further reduced by a transverse laser cooling stage. It consists of two near resonant standing wave light fields intersecting at right angle. The polarization in each arm corresponds to the so called lin⊥lin configuration which is known to produce very low atomic velocities [16]. A 100 cm ballistic flight section for diagnosis of transverse velocities did not show any noticable divergence of the beam with 1 mm diameter after the collimation stage. From this result an upper limit substantially below the so called Doppler limit for cesium atoms ($v_{\text{Dopp}} = 8.6$ cm/s) can be derived, showing that polarization gradient cooling is effective in our apparatus.

In summary, just before entering the lithography zone the conditioned cesium beam has a mean longitudinal velocity of 300 m/s, transverse velocity width below 8 cm/s, and a total flux density of $2 \cdot 10^{12}$ atoms/s.

This beam traverses the optical standing wave light field which is aligned such that its profile is cut in half by the substrate. The detuning and power of the optical standing wave light field are chosen such that atoms with a velocity corresponding to the maximum of the Maxwell–Boltzmann distribution get focused onto the substrate.

3 Direct deposition

Atom lithography enables the direct control of the spatial intensity of an atomic beam at nanometer scales during deposition on a substrate. If the atoms remain where they hit the surface, this structure is directly transferred to the substrate, in contrast to other methods which involve exposure of a resist and multiple chemical processing steps. Although our subject, atom lithography with cesium atomic beams, is also achieved by a resist and hence resembles those methods more closely, direct deposition holds one of the attractive promises of atom lithography, i.e. fabrication of nanostructures with a single or very few processing steps. Therefore we briefly summarize the results achieved with this method.

As already mentioned, the first demonstration of atom lithography with a chromium atomic beam [5] made it possible to extract samples from the vacuum without immediate disintegration and constituted the first success of this method. The extension of this process to other elements seems feasible, provided suitable laser sources for the generation of light forces are available. Although the application of intense cw laser radiation in the blue and ultraviolet region down to 300 nm still poses a major technological challenge, a significant number of elements of technological interest could be accessible with this methods, including for instance the III. group of the periodic table (of which aluminum has already been demonstrated [7]), the magnetic 3d elements Co, Ni, Fe, or noble metals Ag and Au.

Direct deposition methods have been used to demonstrate that not only linear periodic arrays of lines can be written but also two dimensional structures with cubic [17] or hexagonal [18] symmetry, depending on the superposition of light fields. In principle it is possible to generate even more complex patterns through the addition of several standing wave light fields. We have previously argued that the intensity gradient of a simple standing wave light field is the origin of the nanoscale light forces resulting in structured deposition of an atomic beam. Instead one may take advantage of a polarization standing wave having constant intensity but a sinusoidal variation of the polarization state, as is the case for two counterpropagating light waves with equal intensities but orthogonal linear polarizations. In such a light field polarization gradient forces exist due to the multilevel nature of an atom which may also be exploited for atom lithography. This has been demonstrated in an experiment with chromium atomic beams, which showed a periodicity of $\lambda/8$ [19]. Note that the internal structure of atoms adds a new degree of freedom to particle beam methods. This freedom has also been used in a recent experiment by Johnson et al. [13].

4 Resist techniques

4.1 Sample preparation

The lithographic process consists of sample preparation, exposure to cesium atoms and etching (Fig. 6). Sample preparation was carried out with an established procedure used for metastable atomic beam printing [9], which is simple enough to be adopted even by laboratories with limited chemical equipment. A polished silicon wafer was first coated by evaporation with a

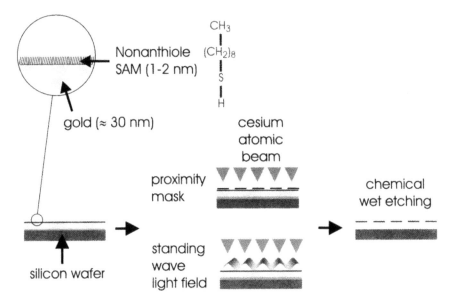

Figure 6: The lithographic process consists of sample preparation, exposure to cesium atoms and etching in a wet gold etching solution (development)

1.5 nm chromium layer for improved sticking properties followed by a 30 nm gold layer. Immediately after the evaporation, the samples were immersed in a 1 mM solution of nonanethiole ($CH_3(CH_2)_nSH$) in pure ethanole, inducing the formation of a self assembling monolayer on the gold surface within 24 hours [20]. The monolayer has a thickness of about 1 nm and consists of $5 \cdot 10^{14}$ molecules/cm^2 [21]. It protects the underlying gold surface during a wet gold etching process due to the hydrophobic character of the methyl end groups.

The prepared samples were mounted on a holder and covered by a proximity mask made from a nickel mesh with 12.5 μm period and 8 μm square openings. The assembly was then inserted into a vacuum chamber where the samples were exposed to an effusive cesium atomic beam for typically 5 to 15 minutes. After exposure the samples were immersed in a gold etching solution for 12 minutes and thereafter investigated by means of an atomic force microscope.

We have observed an optimum response with maximum contrast at doses of 6 to 10 cesium atoms per nonanethiole molecule, in agreement with [12]. While many methods of structuring at microscopic scales are loosely called *lithography* the method described here is indeed close to the original method invented by A. Senefelder about 200 years ago which relied on the modification of the wetting properties of a stone surface.

4.2 Results

We have used two types of masks for fabricating periodic gold structures on silicon substrates: A nickel mesh which was used as an amplitude mask in the proximity configuration, and the standing wave light field which is equivalent to a phase mask.

The first series of experiments was used to establish the lithographic process [11]. Using the processing steps described above the structure of a nickel mesh of period 12.5 μm is clearly transferred into the gold surface. The grainy structure of the remaining gold film is due to the growth process of this thin film and not to the etching procedure. In order to analyze

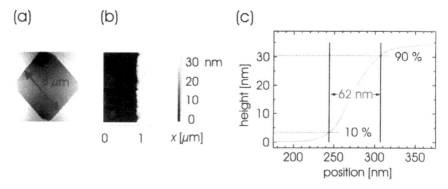

Figure 7: Atomic force microscope images of structures produced with a proximity mask. In both images bright structures represent areas of gold, silicon is dark. **a** Overview of a single 8 μm square opening. **b** Magnification. **c** Line profile of the AFM tip across a Si Au border averaged over a distance of 2 μm

the spatial resolution achievable by this method we have investigated the structure of the edges between the exposed and the unexposed areas in detail. While we find edges of 40–50 nm width for single line scans of the AFM tip, the averaged edge resolution along 2 μm borderline is on the order of 60 nm (Fig. 7c), which is in agreement with the deviations of our mask from a perfectly straight mesh.

In a second series of experiments we have used a standing wave light field, or phase mask, for atom lithography at the cesium resonance wavelength of $\lambda = 852$ nm. We have found that this method is well suited to generate a periodic array of lines with widths below 100 nm and extending over the whole cross section of the atomic beam (Fig. 8) [22]. Only 5 to 10 monolayers of cesium atoms are required for efficient exposure of the thiol resist, resulting in short exposure times on the order of minutes which may however be significantly shortened by more intense atomic beam sources.

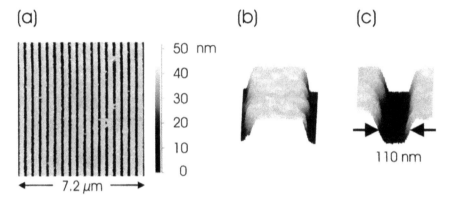

Figure 8: Atomic force microscope images of the etched structures. In all images bright structures represent areas of gold, silicon is dark. **a** 7.2×7.2 μm^2 area. **b** Magnification of a bridge. **c** Magnification of a ditch

At this point the resolution of this method is severely limited by the processing method involving a 30 nm thick gold layer which due to isotropic etching limits the edge resolution at this scale. It is conceivable to develop more suitable atom–resist combinations [23], and ultimate resolutions in the 10 nm domain may be possible.

5 Conclusion and Outlook

Over the past years a number of suitable combinations of atoms and surfaces have been found and several advantages of atom lithography have been demonstrated: it can be highly parallel; it can be used to modify suitable surfaces at nanometer scales in a traditional print and lift–off technique; it is capable of directly writing nanostructures in a single step process; the position of dopants may be controlled during the deposition and this may be used to generate fully three dimensional nanostructures which are necessary for instance to create photonic band gap materials at optical wavelengths [24].

Lithographic experiments with technologically more interesting elements e.g. indium and gallium are currently in preparation. The main difficulty in these experiments is to set up the light sources necessary for laser cooling. In order to write complex patterns we are currently investigating imaging schemes using magnetic atom optical components like mirrors and lenses. Magnetic components may offer viable alternatives to their equivalents constructed from light fields: they are simple to construct from maintenance free permanent magnetic materials and they have large apertures which are difficult to achieve with light fields. In an earlier experiment [25] we have already demonstrated that a magnetic lens can be used to image a transparent mask using a cesium atomic beam slowed with the frequency chirp method [26]. However, the resolution in these experiments was mainly limited by the quality of the atomic beam used. We now have set up a high flux slow cesium atomic beam prepared by the Zeeman slowing technique [27, 28]. This beam will be used to explore whether atomic beams can be imaged with a resolution rivaling established methods and whether atom beam projection lithography is possible at the nanometer scale.

Acknowledgements

We wish to thank M. Kreis, H.-J. Adams, and S. Nowak for their invaluable contributions to the experiments. The BMBF financially supported this work under contract 13N6636/4.

References

[1] Semiconductor Industry Roadmap,
   http://notes.sematech.org/97pelec.htm

[2] W.D.Phillips, in Laser Cooling and Trapping of Neutral Atoms, *Laser Manipulations of Atoms and Ions*, edited by E.Arimondo, W.D.Phillips, and F.Strumia, (North-Holland, Amsterdam, 1992).

[3] C.S.Adams, M.Sigel, and J.Mlynek, Phys. Rep. **240**, 143 (1994).

[4] G.Timp, R.E.Behringer, D.M.Tennant, J.E.Cunningham, M.Prentiss, and K.K.Berggren, Phys. Rev. Lett. **69**, 1636 (1992).

[5] J.J.McClelland, R.E.Scholten, E.C.Palm, and R.J.Celotta, Science **262**, 877 (1993).

[6] U.Drodowsky, J.Stuhler, B.Brezger, T.Schulze, M.Drewsen, T.Pfau, and J.Mlynek, Microelectron. Eng. **35**, 285 (1997).

[7] R.W.McGowan, D.M.Giltner, and S.A.Lee, Opt. Lett. **20**, 2535 (1995).

[8] K.K.Berggren, A.Bard, J.L.Wilbur, J.D.Gillespy, A.G.Helg, J.J.McClelland, S.L.Rolston, W.D.Phillips, M.Prentiss, and G.M.Whitesides, Science **269**, 1255 (1995).

[9] S.Nowak, T.Pfau, and J.Mlynek, Appl. Phys. B **63**, 203 (1996).

[10] S.J.Rehse, A.D.Glueck, S.A.Lee, A.B.Goulakov, C.S.Menoni, D.C.Ralph, K.S.Johnson, and M.Prentiss, Appl. Phys. Lett. **71**, 1427 (1997).

[11] M.Kreis, F.Lison, D.Haubrich, S.Nowak, T.Pfau, and D.Meschede Appl. Phys. B **63**, 649 (1996).

[12] K.K.Berggren, R.Younkin, E.Cheung, M.Prentiss, A.Black, G.M.Whitesides, D.C.Ralph, C.T.Black, and M.Tinkham, Adv. Mat. **9**, 52 (1997).

[13] K.S.Johnson, J.H.Thywissen, N.H.Dekker, K.K.Berggren, A.P.Chu, R.Younkin, and M.Prentiss, Science **280**, 1583 (1998).

[14] J.Dalibard and C.Cohen-Tannoudji, J. Opt. Soc. Am. B **2**, 1707 (1985).

[15] J.J.McClelland, and M.R.Scheinfein, J. Opt. Soc. Am. B **8**, 1974 (1991).

[16] J.Dalibard and C.Cohen-Tannoudji, J. Opt. Soc. Am. B **6**, 2023 (1989).

[17] R.Gupta, J.J.McClelland, Z.J.Jabbour, and R.J.Celotta, Appl. Phys. Lett. **67**, 1378 (1995).

[18] U.Drodofsky, J.Stuhler, Th.Schulze, M.Drewsen, B.Brezger, T.Pfau, and J.Mlynek, Appl. Phys. B **65**, 755 (1997).

[19] R.Gupta, J.J.McClelland, R.J.Celotta, and P.Marte, Phys. Rev. Lett. **76**, 4689 (1996).

[20] C.D.Bain, E.B.Troughton, Y.-T.Tao, J.Evall, G.M.Whitesides, and R.G. Nuzzo, J. Am. Chem. Soc. **111**, 321 (1989).

[21] L.Strong and G.M.Whitesides, Langmuir **4**, 546 (1988).

[22] F.Lison, H.-J.Adams, D.Haubrich, M.Kreis, S.Nowak, and D.Meschede Appl. Phys. B **65**, 419 (1997).

[23] R.Younkin, K.K.Berggren, K.S.Johnson, D.C.Ralph, M.Prentiss, and G.M.Whitesides, Appl. Phys. Lett. **71**, 1261 (1997).

[24] J.D.Joannopoulos, R.D.Meade, and J.N.Winn, in Photonic Crystals, (Princeton University Press, Princeton, 1995).

[25] W.G.Kaenders, F.Lison, A.Richter, R.Wynands, and D.Meschede, Nature **375**, 214 (1995). W.G.Kaenders, F.Lison, I.Müller, A.Richter, R.Wynands, and D.Meschede, Phys. Rev. **A54**, 5067 (1996).

[26] W.Ertmer, R.Blatt, J.L.Hall, and M.Zhu, Phys. Rev. Lett. **54**, 996 (1985).

[27] W.D.Phillips and H.Metcalf, Phys. Rev. Lett. **48**, 596 (1982).

[28] F.Lison, P.Schuh, D.Haubrich, D.Meschede, to appear in Phys. Rev. A.

Part II

Translation and Energy Relaxation
in Supersonic Free Jets,
Including Possible Alignment
and Condensation Phenomena

The Physics of Plasma Expansion

D.C. Schram, S. Mazouffre, R. Engeln,
M.C.M.van de Sanden

Eindhoven University of Technology, Department of Physics,
Centre for Plasma Physics and Radiation Technology (CPS),
P.O. Box 513, 5600 MB Eindhoven, The Netherlands.

1 Introduction

Plasma expansion from a hot and dense source, where electric or electromagnetic energy is dissipated, to a low pressure environment, is a very general physical phenomenon which concerns a large variety of objects and covers a broad range of dimensions. The physical issues addressed are relevant to many subjects in science ranging from astrophysical objects [1, 2], like supernovae and solar flares, to small laser spots [3-7] and cathode spots [8] (as in vacuum arc). On intermediate scale, it also includes expansion from thermal plasma sources (DC, RF, microwave torches) for plasma chemistry [9-11], as well as divertor region of Tokamak plasmas [12]. Applications of high density plasmas are laser plasma cutting, welding, heating, and annealing [9], and the generation of VUV and X-ray radiation [13]. For the latter two applications, a high temperature, multiple ionization, and high (laser) power densities are required. Laser produced plasmas are also projected to create modulated plasmas for plasma based free electron lasers. Finally in (vacuum) arc switching, the conductive properties of high density plasmas are used

The expansion from high pressure thermal plasma region to a low pressure region resembles a neutral gas expansion [14], meaning that in first order the particles density, temperature, and velocity, behave similarly to supersonic free jet expansion [15-20]. However, plasma expansion exhibits specific characteristics which make it a richer and more complex process. Firstly, it is a non-equilibrium process [21, 22] and, due to recombination effects, a frozen regime in which all parameters stay constant, can never be achieved even if the expansion would occur into a perfect vacuum. Secondly, the enlarged heat conductivity and viscosity [9] lead to a non isentropic expansion [23]. One should note that three-particle recombination can also disturb the energy transfer mechanisms between charged particles, and therefore create non-adiabatic conditions during the expansion process [24]. Third, the plasma nature gives rise to the generation of fields and currents [24, 25], which in turn influence, or even determine, the currents and thus power dissipation in the hot region. At high power densities, the generated currents and fields may be so strong that the plasma becomes confined by the self-generated magnetic field. Finally, due to plasma–surface interactions, i.e. interactions with dust particles or the vessel wall [26], radicals generated from molecules in the plasma source, decouple from the neutral particles expansion and exhibit anomalous transport properties [27]. These phenomena make the expansion richer, in view of the possible issues to be studied, and also more complex.

We here restrict ourselves to the regime where the plasma remains unmagnetised and singly ionised (low to moderate power densities), in which case the main mass is contained in the neutrals and thus the expansion resembles the usual gas expansion. We will first summarise schematically the creation of the high pressure source plasma by means of simplified mass and energy balances [28]. This analysis gives relations between power density, created pressure, and resulting plasma flow.

Subsequently, we will turn our attention to the expansion from thermal plasma sources. These receive much attention nowadays as efficient particle sources for deposition [29], etching, and surface modification. In these applications the source delivers primary particles as ions and/or radicals, which are used to dissociate injected monomers to deposition or etching precursors. The material and energy efficiencies are of crucial importance and then study of plasma production and expansion process is needed for optimisation. At the same time the study of the intermediate size expansions of thermal plasmas (mm to cm) can serve as a scale model for both the tens of micrometer-sized laser spots and the astrophysical size plasma jets as to some extent they are driven by similar physical mechanisms. Furthermore, an important reason to study plasma flow at an intermediate scale is the possibility of application of a variety of diagnostic techniques, both passive and active, and therefore opening the way to acquire extensive sets of data.

2 Basic physics of source plasma generation

2.1 Plasma formation

Plasma expansion finds its origin in the formation of a high density and high temperature plasma by a concentrated power deposition. This power can be direct current, as in vacuum arc spots, or can be electro-magnetic as in laser spots.

This source plasma is in principle in non-equilibrium; it is ionizing, and the production of new electrons and ions has to be large enough to compensate for the losses of charged particles by diffusion and convection. For efficient plasma production, high densities are to be preferred and a sufficiently high temperature is required. This will lead to a high pressure source and thus to expansion from that source. Note that the expanding plasma is also not in equilibrium, and the temperature is low, however, it is recombining, in contrats with the plasma in the source. As previously mentioned in the introduction, plasma expansion is different from neutral gas expansion in several respects: the source temperature is high and thus heat conduction and viscosity effects cannot be neglected. Moreover, electrons, ions, and radicals are present, and currents can be generated by the pressure gradients.

Therefore it is worthwhile to relate the plasma expansion to the source conditions. This will be done for conditions where the generated currents and fields are still small, the plasma is non-magnetised, and only singly ionised ions are present.

These relations between source and expansion can be obtained by considering the mass and energy balances in the source region, expressing the essential non-equilibrium plasma behaviour. These balances simply state that the ion production has to be sufficient to result in an ion expansion flow and that the dissipated energy has to be sufficient to provide the energy needed to ionize, heat, and expand the flow out of the source region.

In Fig. 1, such a plasma source, characterised by a source radius r_s and length l_s, is schematically depicted. The dissipated power flux is equal to $P/\pi r_\text{s}^2$, where P is the injected power. This figure can serve as a schematic representation of a laser spot, created by a focussed laser beam, or as a cathode spot formed by a contracted current. Other methods of plasma production are CO_2 laser sustained plasma [7], microwave plasma torch [9], plasma guns, inductively coupled plasma [9] and a cascaded arc which will be used as an example in this paper. These different plasma sources and their geometries will be briefly described in the next section.

The main physical mechanism of plasma formation is the dissipation of energy in the source region. As we will see, it follows directly from the mass balance that the electron temperature $\hat{T}_\text{e}$ (expressed in eV) in the source has to be such that the net ion production counterbalances the convective ion loss. In other words, the electron temperature has to be sufficient to provide the ion flow in the expanding plasma.

For a singly ionized plasma, the production per unit volume of ions, and therefore electrons, by electron induced ionization can be written as:

$$\text{ion production} = n_\text{e} n_\text{o} K_\text{ion}(\hat{T}_\text{e}), \qquad (1)$$

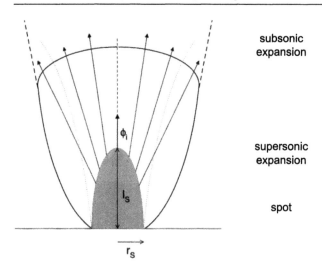

Figure 1: Schematic view of plasma expansion. The created spot has a length l_s and a radius r_s. From the high pressure source (spot), the plasma expands supersonically into a low pressure environment. Collision with the residual background gas results in the formation of a stationary shock wave structure.

where n_e is the electron denisty, n_o the neutral density, and $K_{\mathrm{ion}}(\hat{T}_e)$ is the ionization rate averaged over a thermal electron energy distribution with temperature $\hat{T}_e$. The ionization rate can be written as a product of a nearly constant factor and an exponential factor [30]:

$$K_{\mathrm{ion}}(\hat{T}_e) \simeq C_{\mathrm{ion}} \cdot \exp\left[-\frac{\hat{E}_{\mathrm{ion}}}{\hat{T}_e}\right],$$

$$\text{with} \quad C_{\mathrm{ion}} = 2 \cdot 10^{-14} \sqrt{\hat{T}_e} \left(\frac{\hat{R}_y}{\hat{E}_{\mathrm{ion}}}\right)^2 \quad \text{in m}^3\text{s}^{-1}, \tag{2}$$

where $\hat{R}_y$ ($= 13.6\,\mathrm{eV}$) is the Rydberg energy and $\hat{E}_{\mathrm{ion}}$ the ionisation energy in eV. After being generated, the plasma expands, and thus the ion loss is mainly due to convection away from the source. When recombination is neglected, the ion loss per unit volume is given by:

$$\text{ion loss} \approx n_e/\tau \approx c_s n_e/l_s, \quad \text{with} \quad c_s = 10^4 \sqrt{\hat{T}/A}, \tag{3}$$

where τ is the charged particle life time. This characteristic time can be estimated as the ratio of the source length l_s, and the accoustic velocity c_s. The latter depends on the mass number A, and the heavy particle temperature $\hat{T}$, which is assumed to be equal to the electron temperature $\hat{T}_e$. Temperature equilibrium is a reasonable assumption at high electron density. By equating the ion production to the ion loss, and by using the perfect gas law as an equation of state, $p = n_o kT$, where k refers to Boltzmann's constant, a logarithmic expression can be obtained for the electron temperature:

$$\hat{T}_e = \frac{\hat{E}_{ion}}{\ln[n_o C_{ion} \tau]} \approx \frac{\hat{E}_{ion}}{\ln\left(10 p l_s \sqrt{A}\right) - \ln[\hat{T}]}. \tag{4}$$

The result is that $\hat{T}_e$ is a fraction of the ionization potential, weakly dependent on the product $p l_s \sqrt{A}$. For high pressure plasma sources with relevant value of the product of pressure by spot size, $p \times l_s \sim 100\,\text{mPa}$, we obtain an electron temperature of 1–2 eV. Note, that in the derivation we ignored recombination and the electron and ion contribution to the pressure. However, including these effects does not affect the result appreciably.

The electron density can be estimated from a simplified energy balance, which relates the dissipated energy density $P/\text{Vol} = P/\pi r_s^2 l_s$ to the energy carried away by the produced ions, i.e. the convective energy loss:

$$K_{ion} n_e n_o \left(E_{ion} + \frac{5}{2} kT_e\right) \approx \frac{n_e}{\tau_n} E_{ion} \approx \frac{P}{\xi_P \text{Vol}}, \tag{5}$$

where the mass balance has been used and the kinetic and expansion energies ($\frac{5}{2} kT_e$ term) are neglected with respect to the ionization energy. The correction factor $\xi_P > 1$ represents the other energy losses: energy carried away by neutral particles, radiation, and heat conduction. At high pressures and for moderate power densities (singly ionised plasma $\hat{T}_e$ and $\hat{T}_i$ or both between 1 and 2 eV) this factor is typically $\xi_P = 2$–3. From (5), the ion (and electron) density in the source and the ion flux Φ_i follow:

$$n_e \cong \frac{\tau_n P}{e \xi_P \hat{E}_{ion} \pi r_s^2 l_s} \cong \frac{P \sqrt{A}}{1.6 \times 10^{-15} \xi_P \hat{E}_{ion} . \pi r_s^2 . \sqrt{\hat{T}}}, \tag{6}$$

and

$$\Phi_i = n_e c_s \pi r_s^2 = \frac{P}{1.6 \times 10^{-19} . \hat{E}_{ion} \xi_P}. \tag{7}$$

Typically, a flux of 10^{17} ions $\text{s}^{-1}\text{W}^{-1}$ is obtained. The total heavy particle flow is $\Phi_h = (1/\alpha) \Phi_i$, where $\alpha = n_e/(n_e + n_o)$ is the ionization degree.

The total pressure p is then given by:

$$p = (n_o + n_e + n_i)kT = 10^{-4} \frac{P\sqrt{A.\hat{T}}}{\xi_P.\hat{E}_{\text{ion}}.\pi r_s^2}\left(\frac{1}{\alpha}+1\right). \tag{8}$$

We note that the pressure scales with the inverse of the dimensions of the spot.

From this crude analysis, it is clear that a critical power density is needed to sustain plasma generation. It is also evident that plasma expansion on which we will focus our attention is the natural consequen ce of a concentrated power dissipation in the source region.

The relation between the created plasma expansion flow and the upstream non-equilibrium plasma, and power deposition, has unavoidably as consequence that a strong gradient in electron and ion pressure is generated. This leads not only to flow, but also to current generation. In the most simple form, the electron momentum balance can be reduced to a generalised Ohm's law, which in stationary state reads:

$$\underline{E} = \frac{\nabla p_e}{n_e e} + \frac{\underline{j}}{\sigma} \quad \text{with} \quad \sigma = \sigma^{\text{Coulomb}} = 0.5 \times 10^{-4} \frac{\hat{T}_e^{3/2}}{\ln \Lambda}. \tag{9}$$

Apparently the gradient of the electron pressure gives rise to an additional quasi-potential and currents can be generated, which in stationary state have to be divergence free. The first consequence is that these currents in the source and first expansion can still heat the electrons, therewith rendering the expansion non adiabatic. At high power deposition these currents can become very large, as well as the associated magnetic fields. At these power levels, the plasma is hotter and in higher ionization stages and the generated fields may become so strong that the expansion is modified.

Though we will discuss only singly ionised systems, it is straightforward to generalise the picture to multiple ionization systems. At higher power densities, higher ionization states become accessible; then the temperature must increase for the ion prodution to be sufficient to balance convective losses. The mass balance dictates that the T_e is higher, as it must be still a fraction ($\sim 1/10$) of the then relevant ionization potential ($Z^2 \hat{R}_y$). Electron heat conduction increases strongly with electron temperature (but decreases with Z_i) and substantially higher power densities are needed. Furthermore, the resistivity of the plasma increases with Z_i (as the inverse Bremsstrahlung absorption). For the ion flow, charge neutrality now dictates $Z_i n_i = n_e$ and is modified. As current and field generation become important, their effect on plasma transport has to be included.

For the present analysis, it suffices to illustrate that there is a definite relation between source conditions and plasma expansion. The schematic

	$\hat{T}_e$ (eV)	r_s (m)	l_s (m)	P (W)	p (Pa)	n_e (m^{-3})	Φ_i (s^{-1})
cathode spot	1–2	3.10^{-5}	10^{-4}	10^3	10^6	10^{24}	10^{20}
thermal arc source	1	2.10^{-3}	10^{-3}	10^3	10^4	10^{22}	10^{20}

Table 1: Summary of plasma parameters for cathode spot and thermal arc source when Ar is used as a precursor gas.

picture confirms the relation between the dissipated power density (spot size), the upstream pressure, the ionisation state and the emanating particle flow (ions, electrons and neutrals). Note that the crude analysis was performed for the steady state. For laser spot formation, the laser pulse may be shorter than the expansion time from the hot spot and time dependence has to be included in the analysis. The parameters of a cathode spot and thermal arc heated plasma are summarised in Table 1. In the next section the expansion from the source region will be described.

2.2 Source geometries

In this paper the emphasis will be on the kinetic and transport processes in the plasma expansion from a high pressure source to a low pressure background. We will do so for a particular source geometry: the cascaded arc, which is an example of (sub)atmospheric thermal plasma sources. Still, it is usefull at this point to shortly discuss several plasma source geometries, which are used in plasma chemistry.

The basic source is the DC (or AC) powered plasma gun, which are commonly in use for plasma spraying and for deposition [9]. This source consists of a cathode and an anode nozzle and is characterized by a high current and low voltage. Sometimes extended segmented anode are used to be able to use a larger voltage and thus lower current. A related thermal plasma is the cascade arc source, which consists of cathodes, isolated cascade disks and a nozzle anode, as depicted in Fig. 2a. This source will be discussed later as it is the one used in the further illustration of the expansion characteristics.

A related, but RF powered source, is the inductively coupled plasma (ICP), shown in Fig. 2b. Here RF power with frequencies from 100 kHz to 100 MHz, is induced from the RF load coil to the plasma. This plasma is annular due to the skin effect. RF inductive plasmas need a wide bore for the induction of plasma current and a converging nozzle is needed if supersonic expansion is demanded. These sources are used for spectrochemistry, for deposition and for chemical conversion. Another type of RF source with a small bore, low power and flow is the E-field driven arrangement [31]. Here RF power

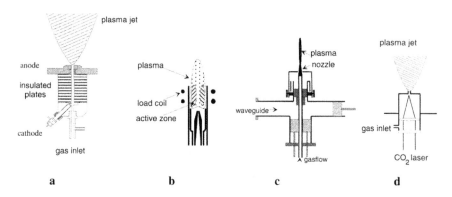

Figure 2: Different kind of plasma sources used in plasma chemistry: (**a**) cascaded arc, (**b**) inductively coupled plasma, (**c**) microwave torch, (**d**) laser sustained plasma.

is fed to a coaxial line of which the inner conductor is used as gas feed. This line ends in the plasma, where the energy is dissipated.

Microwave induced plasmas (MIP) have also appeared in the last years, primarily used for spectrochemistry. A specific popular type is the microwave torch; the so-called "Torche à Injection Axiale", or TIA, is depicted in Fig. 2c. In this case the microwave power is fed to a nozzle, which is the inner conductor of a coaxial line coupled to the waveguide.

Also sources driven by dissipation of optical fields are used. Here a CO_2 laser beam is focussed on a gas target just prior to a nozzle, see Fig. 2d, which connects a high pressure region to a low pressure background. These sources are used also for the study of expansion processes. As mentioned before, CO_2 and other lasers are also employed to create plasmas for X-ray sources (on targets and on droplets) and for laser ablation for deposition (e.g. for high temperature super conductors).

All these plasmas have in common that they combine a concentrated power deposition in the source region, which causes the produced plasma to flow from the source to the treatment area. Hence, plasma expansion always occurs, and in many cases part of the expansion is supersonic. However, the examples for deep expansion (large flow and low pressure) are not so wide spread; the most pertinent examples are the laser plasmas and optical discharges in low pressure, and the thermal plasma which expands from the cascade arc. As for the latter the upstream plasma is well-known and the dimensions of the expansion permit detailed analysis of density, temperature and velocity fields, we will discuss this plasma in more detail.

3 Downstream expansion of thermal plasma sources

The basic phenomena of the expansion of a partially ionised plasma are to some extent similar to those of the expansion of a hot neutral gas, and can thus be described using the supersonic free jet theory [15, 19]. In the expansion of a partially ionised plasma, because of several heating mechanisms, the temperature remains higher than in neutral gas adiabatic expansions. The expansion of a plasma can thus be described as a common density rarefaction because of the supersonic expansion, modified to account for disturbances to an isentropic flow.

In the thermal source plasma the heavy particle temperature (ions and neutrals) is close to the electron temperature despite the inefficient energy transfer from electrons to ions. For a singly ionized plasma, as used in this paper, they are both equal to about 1 eV (as used in this paper), depending slightly on the upstream pressure and the atomic ionisation energy E_{ion}. At the exit of the arc source, the velocity is sonic, as it is usual for Knudsen flow expansions.

Under the assumption that particles originate from a point source and flow along straight stream lines, the theoretical density development along the jet centerline can be determined using the conservation law [32] and one finds:

$$n(z) = n_o \frac{1}{1 + z^2/z_{\text{ref}}^2}, \qquad (10)$$

where z_{ref} is a scaling length determined by the properties in the early expansion and by the source nozzle geometry (it is of the order of the diameter of the source outlet). In the early stage of the expansion process (i.e. when the axial location is less than a few nozzle diameters), this expression differs slightly from the one used in gasdynamics [18]. This modification is connected with the plasma character of the source region with its high temperature, sonic exit velocity, and finite temperature gradients. It allows for a zero first derivative of the density at the source point in order to accommodate for the convective processes in the source, as well as for field generation.

For an adiabatic process, the energy equation can be replaced by the Poisson adiabatic formula [32] which reads:

$$\left(\frac{n}{n_o}\right)^{\gamma-1} = \frac{T}{T_o}, \qquad (11)$$

where γ is the effective value of the specific heat ratio. It is equal to 5/3 for an isentropic flow of a monoatomic gas. In neutral gas expansion, the flow is isentropic in the supersonic domain and the temperature profile can be well described by the Poisson relation [18]. Since temperature can be seen as a measure of the velocity spread (strictly correct if the velocity distribution is Maxwellian), it is usual in compressible fluid dynamics (gas, plasma) to decompose the temperature in two components, one perpendicular to the stream line and the other one parallel. This makes the analysis of temperature less straightforward [15, 18]. In plasma physics, the situation is even more complicated, as one can define several temperatures [9] which may differ because of incomplete coupling. This causes departure from thermal equilibrium in addition to departures from Saha equilibrium because of slow recombination. Therefore, in a plasma the value of γ is different from the monoatomic gas value [23], also because of dissipative effects like recombination or heat conduction. The effect on γ of the presence of charged particles has been calculated by Burm et al. [33] for a thermal plasma and they conclude that $\gamma \approx 1.2$ for a wide range of the ionization degree (0.03–03). Using Thomson scattering [34], which delivers n_e and T_e, a value of $\gamma_e = 1.3$ for the electron gas has been measured [35]. Heavy particle heat conduction keeps the temperature higher than in normal gas expansions. As also from the region behind the barrel shock, heat is conducted back into the expanding flow the temperature minimum occurs before the density minimum, i.e. before the stationary shock front [23]. Using Laser Induced Fluorescence (LIF) γ is found to be 1.3 for atomic radicals [27].

If a quasi-adiabatic relation between the temperature and the density with a lower adiabatic coefficient γ is used instead of the full energy equation, then it follows from the momentum balance that the velocity w_z along the jet axis is given by:

$$w_z = c_{s0} \left[1 + \frac{2}{\gamma - 1} \cdot \left\{ 1 - \left(\frac{1}{1 + z^2/z_{\text{ref}}^2} \right)^{\gamma - 1} \right\} \right]^{1/2}, \quad (12)$$

where c_{s0} is the speed of sound at the source exit. This expression includes the fact that the velocity is sonic at the source outlet. Behind the source exit the velocity increases in a few nozzle diameters, due to conversion of thermal energy gained in the source into kinetic energy by means of collisions. Finally, the flow reaches its terminal values which only depends on γ and on the source temperature. By using a value of 1.3 for γ a final velocity of about three times the sonic exit velocity is obtained.

So far, it has been experimentally established for the supersonic domain of plasma expansions that the density rarefaction, the velocity as well as the temperature, follow roughly the common gas expansion laws, with a smaller

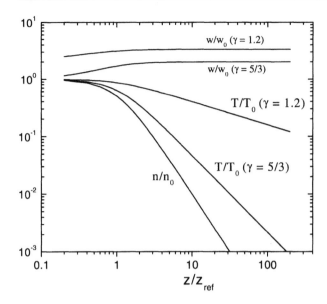

Figure 3: Calculated dependencies of density, temperature and velocity as function of the axial coordinate (source at $z/z_{ref} = 0$). Calculation are presented for both an adiabatic expansion ($\gamma = 5/3$) and a non-adiabatic expansion ($\gamma = 1.2$).

adiabatic exponent, at least for neutrals and charged particles. In Fig. 3, the theoretical behaviour of expansion quantities is depicted as a function of the axial position (relative to the scaling length z_{ref}) for isentropic as well as non-isentropic conditions. However, in the case of radicals, which is treated in the last part of this contribution, the expansion differs strongly from the classical picture [27].

At the end of the supersonic expansion a stationary normal shock is formed, which is the front boundary of the barrel shock structure [19]. Both the position and the thickness of the stationary shock wave depend on the background pressure. In the limit of an extremely low pressure, the shock structure becomes diffuse and vanishes, and it is even possible that the flow never undergoes a supersonic-subsonic transition. However, such a situation is rare in the case of plasma expansion connected to industrial applications, since an important gas flow is used. This large pumping capacity regime is usually not reached for deposition and etching, but offers interesting new possibilities.

The position of the shock front, or more precisely the location of the Mach disk z_M, i.e. the position where the Mach number M becomes 1, depends on the source pressure, the nozzle diameter, and the downstream pressure [14].

For commodity, such a relation can be rewritten in terms of flow (Φ_h in standard $m^3 s^{-1}$), background pressure (p_{back} in Pa), atomic mass number A, and stagnation temperature (T_o in K):

$$z_M^2 = 2.5\,\Phi_h \frac{(1+\gamma_o/2)}{\gamma_o^{1/2}} \frac{T_o^{1/2} A^{1/2}}{p_{back}}. \tag{13}$$

In this formula, z_M is expressed in meters and γ_o is the isentropic exponent in the reservoir. The heavy particle flow Φ_h can be related to the ion flow Φ_i by $\Phi_h = (1/\alpha)\Phi_i$, where α is the ionisation degree.

Throughout the stationary shock front, of which thickness is of the order of one local momentum exchange mean free path, the velocity decreases, and the temperature rises, due to collisions with the standing gas. Therefore, due to the conservation of the forward flux, the density increases abruptly. The Rankine–Hugoniot relations [32] connect such a jump to the Mach number magnitude ahead of the shock wave. In case of a weak shock, the theoretical density profile across the shock can be determined using either hydrodynamic equations, i.e. Navier–Stokes equations to account for viscosity, or the gas kinetic theory, e.g. the Mott–Smith model [36].

Behind the shock wave, the flow is subsonic and the flow pattern is determined by the vacuum vessel geometry. Moreover, diffusion starts to play an important role for ions and radicals, for example.

The plasma expansion, as described above, is nearly similar to the free jet expansion of neutral particles. It appears that charged particles are not completely coupled to the neutrals and there are small differences between the expansion of neutrals and the expansion of charged particles. The electrons and ions have to be taken together, as quasi neutrality forces the electrons to move with the ions. This causes the appearance of an additional (electron) pressure term, leading to a $1 + T_e/T_i$ correction term in the expression of the momentum balance for the ion-electron fluid, and thus to a different shock position. Normally there is a finite coupling between the two fluids, even if in some cases it can only be partial.

The coupling between the fluid of neutrals and charged particles depends strongly on the ionization degree α, and becomes incomplete for $\alpha > 0.1$. Thus, attention has to be paid to the collisional coupling when probing excited states in order to extract information about the expansion of the neutrals in terms of velocity and temperature. The difficulty is due to the recombination in expanding plasmas, as mentioned earlier.

Finally, to conclude this paragraph, it is worthwhile to point out that the flow pattern of a plasma expansion can be more complex than aforementioned due to parasitic effects like vortex formation (especially in the vicinity of the barrel shock) and background gas re-entry into the supersonic region.

Those phenomena determine the general re-circulation pattern which plays a key role in plasma chemistry where reaction times have to be compared with transit time.

4 The argon expanding plasma as a test case

Plasmas expanding from thermal plasma sources have the virtue of a large ion and radical flow capacity and therefore are very suitable for high rate deposition and surface modification. In this study, results from such a system based on a cascade arc source will be described. It may serve at the same time as a model study for expanding plasmas in general, as many characteristics are known from detailed measurements.

The cascade arc consists of a cathode at the upstream side, a stack of isolated cascade plates with a bore of 4 mm, and an anode nozzle. From there, the plasma expands into the deposition chamber, which is held at a low pressure. Typical operating parameters are argon flow $\Phi_h = 50 \times 10^{-6}$ standard $m^3 s^{-1}$, current $I = 50$ A, power 5 kW, downstream background pressure 40 Pa (referred to as standard conditions).

In Fig. 4, the neutral density profile along the jet axis $n_o(z)$ of a pure argon plasma expanding from a cascade arc source is plotted as a function of axial position [11]. The neutral particle density is measured by means of Rayleigh scattering [11, 37, 38]. One observes successively the supersonic expansion with the $1/z^2$ dependence, then the stationary shock front and subsequently the subsonic expansion. In the shock, the density increases of a factor 3–4, as expected. In the subsonic region the neutral density increases slowly, as the temperature decreases and the pressure is constant.

The electron density n_e, also shown in Fig. 4, presents similar behaviour in the supersonic expansion and across the shock, but decreases slowly in the subsonic region by ambipolar diffusion. The data presented there have been measured by means of Thomson scattering [34, 35]. The electron density can be measured using Langmuir probes [23] which, contrary to laser scattering technique, is intrusive. It can be shown that three particle recombination of electrons and ions is negligible for the electron mass balance: the ion flow remains unchanged in atomic plasmas. However, recombination is important for the electron energy balance: it is one of the three heating mechanisms keeping the electrons hotter than in a fully adiabatic expansion. The other two mechanisms are the dissipation by currents generated by the gradient of the electron pressure, and the heat conduction both from the source and from the subsonic side [35]. The direct consequence of this is that the lowest temperature occurs before the shock position [23], as it can be seen in Fig. 5. Apparently the temperature is shocked earlier than the density. As stated before, the temperature decays with a smaller apparent adiabatic exponent

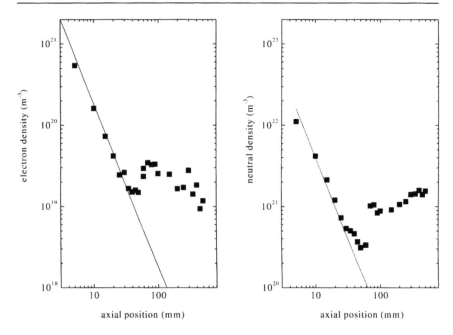

Figure 4: Electron density (*left*) and neutral density (*right*) as function of distance from the source (standard conditions). The line in both graphs shows the $1/z^2$ dependence.

($\gamma = 1.2$–1.4) than in an adiabatic expansion. The temperature of the various particles can also be accessed using diagnostic techniques like emission spectroscopy [39, 40], enthalpy probes [41, 42], Langmuir probes [23], and Laser Induced Fluorescence (LIF) [27, 39].

The velocity of argon neutrals along the beam axis, has been measured by LIF in an Ar–H$_2$ mixture. As can be observed in Fig. 6, the velocity at the exit of the plasma source is already supersonic in this experiment, due to the use of a 45° diverging nozzle. In the supersonic expansion, the velocity increases by about a factor 2, relative to the sonic velocity at the temperature of 0.5 eV at the end of the arc channel, to finally reach a plateau. The velocity decay across the shock front, of which the position is in agreement with (13), can be fully described with the Rankine–Hugoniot relation. The Mach number ahead of the shock is found to be equal to 6 at the aforementioned experimental conditions. In the subsonic domain, behind the shock, the velocity drops because of friction with the non-flowing neutral background gas until it reaches the pump speed. Also shown in Fig. 6 is a fit to the data, deduced from (12), across the supersonic domain. Another possible way to measure gas velocity in plasma jet is to use Pitot tubes [43, 44].

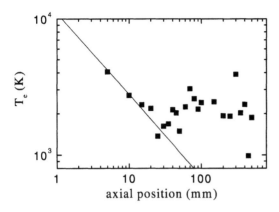

Figure 5: Electron temperature as function of distance from the source, measured with Thomson scattering (standard conditions). The line shows the $(1/z^2)^{\gamma-1}$ dependence, with $\gamma = 1.3$.

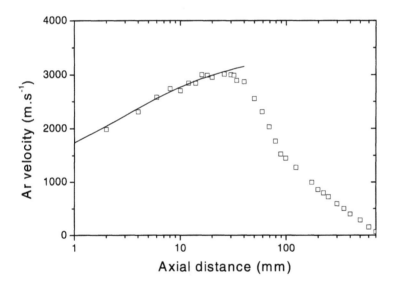

Figure 6: Axial profile of Ar atom drift velocity and comparison (within the supersonic domain) with a theoretical profile obtain using (12).

It appears from other measurements that the shock position depends indeed on the flow and the background pressure as predicted by (13), and also that the first expansion is independent of the background gas pressure, as expected.

5 Excitation processes

5.1 Electronic state distribution in pure argon plasma

The availability of quantitative data on n_e and T_e permits the comparison of excited level densities with quantitative predictions. A pure argon plasma case is discussed in this section. The clearest presentation is to compare the excited state density n_q of level q, with the equilibrium value, i.e. according to Saha equilibrium, n_q^S:

$$n_q^S/g_q = \frac{n_e n_i}{g_e g_i} \cdot \left(\frac{h^2}{2\pi m_e e \hat{T}_e}\right)^{3/2} \exp\left[\frac{\hat{E}_q^{\text{ion}}}{\hat{T}_e}\right]. \tag{14}$$

In this equation, g_q, g_e and g_i are the statistical weight of the qth neutral state, the electrons, and the ions, respectively. The present system is recombining, as excitation from the ground state is impossible because of the low electron temperature in the expansion. Hence all the population of excited levels arises from three particle recombination from the Ar ion ground states and they cascade by de-excitation and radiation [45]:

$$\text{Ar}^+ + e + e \rightarrow \text{Ar}^{**} + e$$
$$\text{Ar}^{**} + e \longleftrightarrow \text{Ar}^* + e \quad \text{and} \quad \text{Ar}^{**} \rightarrow \text{Ar}^* + h\nu.$$

The rate of recombination has been found to be in reasonable agreement with literature. Since in the early expansion the relation between n_e and T_e can be described by a power law: $T_e \propto n_e^{\gamma-1}$, the relative recombination loss decreases with increasing distance and is found to be negligible for the mass balance. However, the recombination is important for the radiation and for the energy balance and it determines the population of the excited states.

The excitation characteristics can be predicted with a collisional radiative model and the experimental results on absolute values of population densities follow the predictions as shown in Fig. 7. In other words, absolute excited state density measurements give information on the plasma parameters, especially high lying levels which are still close to Saha predictions. We also see the typical characteristics of a recombining system: the levels are underpopulated with respect to Saha's law. This expresses the fact that line radiation is a consequence of recombination flow downstream in the excitation space, i.e. line radiation is a sign of the presence of ions. This is in contrast with ionising systems in which levels are overpopulated because of excitation upward from the neutral ground state and line radiation is commonly a sign of neutrals.

Excited states have been measured by emission spectroscopy, which for argon is limited to the $4p$ and higher multiplets. The $4s$ resonant and metastable level population has been measured by absorption spectroscopy [46].

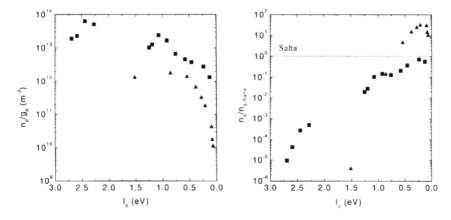

Figure 7: Excited state densities per statistical weight of argon (squares), in pure argon (*left*), and normalized to Saha equilibrium densities (*right*), both at standard conditions and at $z = 20$ mm from the arc outlet. Also shown are the excited state densities of hydrogen (triangles) in a 3% H_2 admixture in argon. Note that in the hydrogen seeded expansion, the argon state densities will be smaller because of lower Ar^+ density.

It appears that the $4s$ level densities closely follow model predictions, in which three particle recombination is balanced with trapped resonant radiation loss. The agreement is satisfactorily considering that radiation trapping is difficult to model precisely. Also it has been established that the $4s$ sub-levels are fully mixed, in other words the sublevels are collisionally coupled by electron collisions. We conclude that excitation in expanding atomic plasmas can be fully understood as a consequence of three particle recombination of atomic ions.

5.2 Addition of molecules to pure argon plasma

If hydrogen is admixed to argon in the arc source, a significant fraction of the molecules is dissociated and thus mainly atomic argon and hydrogen emanate from the source. In contrast with expectations for an atomic gas plasma, a strong additional recombination is observed, as is shown in Fig. 8. This points to a recombination through charge exchange of atomic ions with molecular hydrogen and subsequent dissociative recombination of the resulting molecular ion [47, 48]:

$$Ar^+ + H_2 \rightarrow ArH^+ + H$$
$$ArH^+ + e \rightarrow Ar + H^*.$$

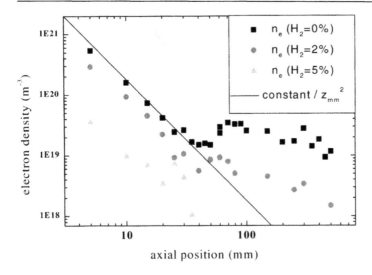

Figure 8: Axial profile of the electron density as determined by Thomson scattering (standard conditions; 0, 2, and 5% H_2 injected in the arc). Note that adding 2 or 5% of hydrogen reduces significantly the electron density.

This loss of ionization is already apparent in the arc source, partly due to narrowing of the plasma channel in the source, but also due to the above mentioned process. This is not in contradiction with substantial dissociation because only a small fraction of molecules in the source is sufficient to explain the rate of electron loss. However, the loss in the expansion is also significant, in particular in the subsonic part after the shock. In addition, dominant red H_α light appears, which is due to the fact that dissociative recombination ends in excited states, as indicated above. In the color pictures shown in Fig. 9, the effect of hydrogen addition is clearly demonstrated. The pure argon plasma jet, shown in Fig. 9a, remains optically visible, whereas the expanding argon/hydrogen mixture, shown in Fig. 9b, radiates mainly the H_α line after some distance, which is a clear sign of molecular induced recombination.

As an example, the excited state densities of atomic hydrogen are sketched in Fig. 7. The plot is also given in terms of Saha densities, which is possible for argon states as the main ion is argon and the ion density is equal to the measured electron density. The hydrogen ion density is unknown and strictly speaking no Saha plot can be made for hydrogen levels in an absolute way. We assumed that the hydrogen ion density is such that, at the ionization limit, the density is in equilibrium. This means that high lying states just below the ionization limit are overpopulated as would be the case for ionising systems. The reason is that the dissociative recombination of formed

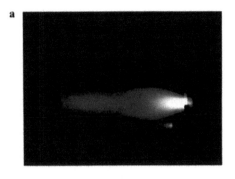

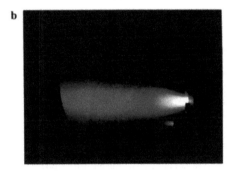

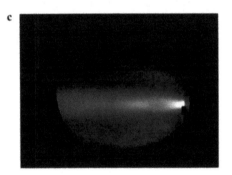

Figure 9: In (**a**), the light emission of a pure argon plasma is shown (Ar flow = 100×10^{-6} standard $m^3.s^{-1}$, $I_{arc} = 75\,A$, $p_{back} = 68\,Pa$). The light consists of line radiation generated by 3-particles recombination and continuum radiation. In (**b**), the light emission is shown for the same conditions when 2% H_2 is admixted in the plasma source. In this case the red H Balmer-α radiation appears. In (**c**), a higher background pressure (110 Pa) is used, showing the larger molecular effect due to the larger H_2 partial pressure. The stationary shock front is closer to the exit of the source, and the plasma jet is narrower. The origin of the blue light at the periphery is unknown, but it may be a sign of the presence of negative ions.

molecular ions (ArH$^+$) populates excited hydrogen states, which makes that the system resembles an ionising one, whereas the argon excited states are still populated by the much weaker three particle recombination. In other words, atomic argon lines point to the existence of argon ions (like in the pure argon case) and atomic hydrogen lines prove the presence of molecular ions and thus of hydrogen molecules. It does not need any further elucidation to say that actinometry needs to be used with care in recombining systems because this technique relies on the assumption of a simple direct excitation process by electron collisions and direct de-excitation by radiative decay. It is interesting to note that also in laser plasma expansion similar remarks could be made as there also the excitation will probably arise more from recombination than excitation from the ground state.

Still the dominance of hydrogen molecules in the subsonic expansion at full dissociation in the source needs to be explained. The reason is simple: hydrogen atoms from the source associate at the downstream vessel wall covered by H atoms, long before they are pumped out in the residence time, and form molecular hydrogen [26, 48]. For the subsonic part, the geometry and nature of the wall, and the recirculation patterns are also important. As wall association is a fast process, the dominant hydrogen particles will be molecules in spite of their full dissociation in the flow.

In Fig. 10 we show the population characteristics of molecular hydrogen as measured by means of Coherent Anti-Stokes Raman Scattering (CARS) spectroscopy in a pure hydrogen plasma [49]. It is evident that there are two features. One is a thermal population of low lying rotational states, for $v = 0$ and $J = 0$ to 5. It has been verified that the density and rotational temperature, which follow from these absolute measurements, yield with good accuracy the measured (partial) H$_2$ pressure. But there is a second non-thermal population with significant densities (a few percent), which may be very important for the downstream chemistry. The question of the origin of excited molecular hydrogen is still under debate. As the rotational-vibrational characteristics are quite independent of axial position, it looks improbable that excited molecules originate from the source. The association at the wall of hydrogen atoms, forming ro-vibrationally excited molecules, which are partially de-excited by collisions, while diffusing inward, looks like a possible explanation [26].

If molecular hydrogen and molecular deuterium are injected into the source, HD molecules are produced twice as efficient as H$_2$ molecules as measured by CARS [50]. This evidence also points to wall association as one of the dominant mechanisms for molecule generation.

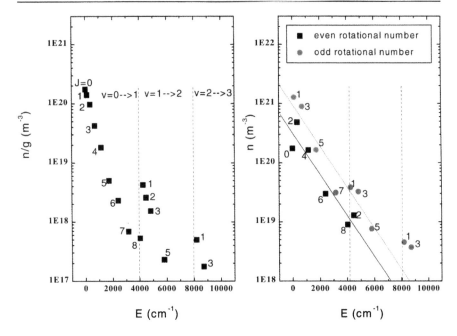

Figure 10: Density per statistical weight (*left*) and total density (*right*) of hydrogen molecules in expanding pure hydrogen plasmas as determined by CARS. The experimental conditions are $I_{\rm arc} = 37.5$ A, hydrogen flow $= 58 \times 10^{-6}$ standard m^3.s^{-1}, $p_{\rm back} = 40$ Pa and axial position $z = 150$ mm.

6 Anomalous radical expansion

In this last section before the general conclusion, we would like to focus on the expansion of radicals. When molecules are seeded either in the plasma source or in the jet, atomic as well as molecular radicals, i.e. unstable particles, are created. For instance, H radical is produced from H_2 molecules and CH from C_2H_2. This kind of very reactive radicals play a key role in chemistry, and thus are of great importance for industrial applications.

Because radicals are uncharged, we could expect that they would expand together with neutrals. Hence, radical expansion should be perfectly understandable in terms of supersonic expansion of a neutral gas [18–20]. However, as we will see, radical expansion is partly decoupled from neutral expansion and does not obey the classical picture of free jet structure.

Plasma jet containing O radicals [51, 52], as well as N radicals [53, 54], have been investigated in the past. Here we propose to look at the transport of ground-state hydrogen atom (H) in an expanding plasma jet generated from an Ar–H_2 mixture.

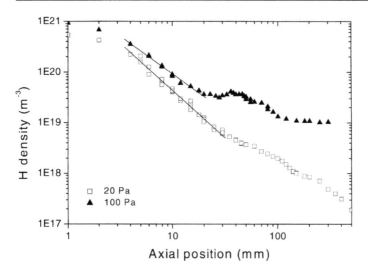

Figure 11: Axial density profile of H ground-state atoms at 20 Pa (*open square*) and at 100 Pa (*solid triangle*) background pressure. The lines are drawn to guide the eye. Contrary to neutral gas expansion, there is no density jump across the stationary shock front, and the sharpness of the density decay in the supersonic domain depends on the background pressure.

In Fig. 11, the H density profile along the jet centerline is depicted for two different background pressures. H is probed by means of Two-photon Absorption Laser Induced Fluorescence (TALIF). Surprisingly, no density jump is measured across the shock front, whereas the velocity is measured to decrease and the temperature development on axis clearly reveals the shock wave structure [27]. The density even decreases further in the subsonic domain. The disappearance of H radicals cannot be due to volume recombination, since the reaction rates are extremely low at the used low pressure ($p < 100$ Pa). Furthermore, the sharpness of the H density drop in the supersonic domain depends on the background pressure, as it can be seen in Fig. 11, which means that H atoms receive information from the region located outside the shock wave.

The clue for this striking disagreement with the classical expansion idea, is given by the role played by surface in the downstream subsonic expansion zone. H radicals associate at the vessel walls to form molecular hydrogen [26] which means that they are almost absent in the ambient background gas. This effect leads to the formation of strong density gradients between the core of the plasma jet and its surroundings, and those gradients are responsible for an outward diffusion of H.

This process, which will be called radical defocusing, is very general and applies to every kind of small mass radical, as for D and N atoms for instance. It is naturally a strong effect for light radicals in a background of heavy neutrals since it is driven by diffusion. To some extent, it could also influence ion transport, and then electron transport, as they can be regarded as radicals, but clear evidence for that effect has not yet been found. Nevertheless, it may be the cause of the relatively low abundance of H^+ ions in a Ar–H_2 plasma flow.

This anomalous transport of radicals has tremendous importance since it means that the confinement of small mass radicals inside a jet, at least at low ambient pressure, is almost impossible. Therefore surface recombination, which leads to the destruction of the radicals, can hardly be avoided, and this can be seen as a loss of the downstream chemical potential.

Recombination of radicals at a surface (that includes cluster, dust, and wall) [26, 55], and the subsequent generation of molecules start nowadays to be considered as a crucial mechanism in the physics of plasma expansion. First, as described above, it may govern to some extend the flow pattern of expanding radicals and thus influence the chemistry. Second, molecules formed at a surface may enter the plasma jet and therefore disturb the transport process, as they can generate new radicals if they react with ions for instance.

Another important issue, which deserves attention, is to investigate experimentally in which quantum state molecules are generated at surfaces exposed to large radical fluxes, since highly excited molecules would again have a large impact on the downstream chemistry.

7 Conclusion

The expansion of a plasma, from a high energy density source, is shown to be a general phenomenon in plasma physics. The produced plasma density and temperature can be estimated from the simplified energy and mass balances. The expansion of the plasma flow from the source into the low pressure background resembles that of neutral gas expansion, at least for neutrals, with minor modifications as a lower apparent adiabatic exponent, the current generation, and the recombination. The expansion of light radicals is shown to differ strongly from the classical expansion, as for instance it is influenced by the presence of surfaces.

The expansion from a thermal plasma source, used for high rate plasma deposition and surface modification, has been analysed with several diagnostics. Therefore, it is used as an example to show the general plasma expansion behaviour. The results confirm the described picture. The excited state population characteristics reflect the recombining character of the expanding plasma. The excited states are underpopulated with respect to Saha equilibrium in atomic plasmas. In the presence of molecules, an anomalously strong recombination is observed, which is proven to be due to charge transfer and dissociative recombination. The rotational and vibrational excitation of hydrogen molecules, as measured by CARS, confirm the dominance of molecules and point to wall association as a source of the molecules. Finally, hydrogen atoms are probed using TALIF spectroscopy. It is indicated that, due to wall-association, hydrogen atoms are scattered out of the expanding plasma. The analysis of the thermal expansion can thus serve as a scale model for expansions at much smaller and much larger scales.

Acknowledgments

This work is financially supported by the Netherlands Technology Foundation (STW) and by the Netherlands Foundation for Fundamental Research on Matter (FOM), partially under FOM-Euratom association agreement, with support from NWO. We thank R.F.G. Meulenbroeks, J.A.M. van der Mullen, and M.G.H. Boogaarts for their contributions concerning the CARS and TALIF experiments. The contributions of M.J.F. van de Sande, H. de Jong and B.F.M. Hüsken are gratefully acknowledged.

References

[1] *Beams and Jets in Astrophysics*, edited by P.A. Hugues (Cambridge University, Cambridge, England, 1991).

[2] J.M. Shull and C.F. McKee, Ap. J. **227**, 131 (1979).

[3] J.C.Miller, R.F.Haglund, *Laser ablation and desorption, Experimental Methods in Physical Sciences* (Academic Press, San Diego, 1998), vol. 30.

[4] Yu. P. Raizer and D.C. Smith, J. Opt. Soc. Am. **70**, 258 (1980).

[5] Yu. P. Raizer, Sov. Phys. Usp. **23**, 789 (1980).

[6] A. Lebéhot and R. Campargue, Phys. Plasmas **3**, 2502 (1996).

[7] J.M. Girard, A. Lebéhot, and R. Campargue, J. Phys. D **26**, 1382 (1993).

[8] E. Hantzsche, Contr. Plasma Phys. **30**, 575 (1990).

[9] M.I. Boulos, P. Fauchais, and E. Pfender, *Thermal Plasmas, Fundamental and Applications* (Plenum Press, 1994), Vol. 1-2.

[10] S.M.Aithal, V.V.Subramaniam, and V.Babu, Plasma Chem. Plasma Proc. **19**, 487 (1999).

[11] M.C.M. van de Sanden, J.M. de Regt, and D.C. Schram, Plasma Sources Sci. Technol. **3**, 501 (1994).

[12] J. Wesson, *Tokamaks* (Clarendon Press, Oxford, 1997).

[13] A. McPherson, B.D. Thompson, A.B. Borisov, K. Boyer, and C.K. Rhodes, Nature **370**, 631 (1994).

[14] H. Askenas and F.S. Sherman, in *Rarefied Gas Dynamics*, edited by J.H. Leeuw (Academic Press, New York, 1966), vol. 2, p. 84.

[15] *Atomic and Molecular Beam Methods*, edited by G. Scoles (Oxford University Press, New York, Oxford, 1988, 1992).

[16] B.B. Hamel and D.R. Willis, Phys. Fluids **9**, 829 (1966).

[17] E.P. Muntz, B.B. Hamel, and B.L. Maguire, AIAA J. **8**, 1651 (1970).

[18] H.C.W. Beijerinck, R.J.F avn Gerwen, E.R.T. Kerstel, J.F.M. Martens, E.J.W. van Vliembergen, M.R.Th. Smits, and G.H. Kaashoek, Chem. Phys. **96**, 153 (1985).

[19] R. Campargue, J. Phys. Chem. **88**, 4466 (1984).

[20] J.M. Girard, *Etude d'un jet supersonique de plasma entretenu par laser*, Ph.D. Thesis, in french, (University of Paris-sud, Orsay, 1994).

[21] S.C. Snyder, A.B. Murphy, D.L. Hofeldt, and L.D. Reynolds, Phys. Rev. E **52**, 2999 (1995).

[22] A.B. Murphy, J. Phys. D: Appl. Phys. **27**, 1492 (1994).

[23] R.B. Fraser, F. Robben, and L. Talbot, Phys. Fluids **14**, 2317 (1971).

[24] M.C.M. van de Sanden, R. van den Bercken, and D.C. Schram, Plasma Sources Sci. Technol. **3**, 511 (1994).

[25] H.J.G. Gielen, *On the electric and magnetic field generation in expanding plasmas*, Ph.D. Thesis (Eindhoven University of Technology, Eindhoven, 1989).

[26] C.T. Rettner and D.J. Auerbach, J. Chem. Phys. **104**, 2732 (1996).

[27] S. Mazouffre, M.G.H. Boogaarts, J.A.M. van der Mullen, and D.C. Schram, Phys. Rev. Lett. **84**, 2622 (2000).

[28] D.C.Schram, I.J.M.M. Raaijmakers, B. van der Sijde, H.J.W. Schenkelaars, P.W.J.M. Boumans, Spectrochimica Acta **38B**, 3369 (1992).

[29] M.C.M. van de Sanden, R.J. Severens, W.M.M. Kessels, R.F.G. Meulenbroeks, D.C. Schram, J. Appl. Phys. **84**, 2426 (1998).

[30] J.A.M. van der Mullen, Physics Reports **191**, 109 (1990).

[31] G. Dinescu, O. Maris, G. Musa, Contrib. Plasma Phys. **31**, 49 (1991).

[32] L. Landau and E. Liftshitz, *Fluid Mechanics* (Pergamon Press, London, 1989).

[33] K.T.A.L. Burm, W.J. Goedheer, D.C. Schram, Phys. Plasmas **6**, 2622 (1999).

[34] S.C. Snyder and R.E. Bentley, J. Phys. D: Appl. Phys. **29**, 3045 (1996).

[35] M.C.M. van de Sanden, J.M. de Regt, and D.C. Schram, Phys. Rev. E **47**, 2792 (1993).

[36] H.M. Mott-Smith, Phys. Rev. **82**, 885 (1951).

[37] S.C. Snyder, L.D. Reynolds, G.D. Lassahn, J.R. Fincke, C.B. Shaw Jr., and R.J. Kearney, Phys. Rev. E **47**, 1996 (1993).

[38] J. Panda and R.G. Seasholtz, Phys. Fluids **11**, 3761 (1999).

[39] L.M. Cohen and R.K. Hanson, J. Phys. D: Appl. Phys. **25**, 339 (1992).

[40] Z. Szymansky, Z. Peradzynski, J. Kurzyna, J. Hoffman, M. Dudeck, M. de Graaf, and V. Lago, J. Phys. D: Appl. Phys. **30**, 998 (1997).

[41] J.R. Fincke, W.D. Swank, S.C. Snyder, and D.C. Haggard, Rev. Sci. Instrum. **64**, 3585 (1993).

[42] M. Rahmane, G. Soucy, and M.I. Boulos, Rev. Sci. Instrum. **66**, 3424 (1995).

[43] O. Chazot, J. Gomes, and M. Carbonaro, AIAA TP **98**, 2478 (1998).

[44] B. Porterie, M. Larini, and J.C. Loraud, J. Therm. Heat. Transfer **8**, 385 (1994).

[45] R.F.G. Meulenbroeks, A.J. van Beek, A.J.G. van Helvoort, M.C.M. van de Sanden, and D.C. Schram, Phys. Rev. E **49**, 4397 (1994).

[46] A.J.M. Buuron, D.K. Otorbaev, M.C.M. van de Sanden, and D.C. Schram, Phys. Rev. E **50**, 1383 (1994).

[47] R.F.G Meulenbroeks; R.A.H. Engeln, C. Box, I. de Bari, M.C.M. van de Sanden, J.A.M van der Mullen, and D.C. Schram, Phys. Plasmas **2**, 1002 (1995).

[48] M.J. de Graaf, R. Severens, R.P. Dahiya, M.C.M. van de Sanden, and D.C. Schram, Phys. Rev. E **48**, 2098 (1993).

[49] R.F.G. Meulenbroeks, R.A.H. Engeln, and D.C. Schram, Phys. Rev. E **53**, 5207 (1996).

[50] R.F.G. Meulenbroeks, D.C. Schram, M.C.M. van de Sanden, and J.A.M. van der Mullen, Phys. Rev. Lett. 76, 1840 (1996).

[51] A. Lebéhot, J. Kurzyna, V. Lago, M. Dudeck, and M. Nishida, Phys. Plasmas **6**, 4750 (1999).

[52] B. Gordiets, C.M. Ferreira, J. Nahorny, D. Pagnon, M. Touzeau, and M Vialle, J. Phys. D: Appl. Phys. **29**, 1021 (1996).

[53] R.W. Bickes Jr., H.R. Newton, J.M. Herrmann, and R.B. Bernstein, J. Chem. Phys. **64**, 3648 (1976).

[54] L. Robin, P. Vervish, and B.G. Cheron, Phys. Plasmas **1**, 444 (1994).

[55] M. Rutigliano, G.D. Billing, and M. Cacciatore, in *Rarefied Gas Dynamics*, edited by R. Brun, R. Campargue, R. Gatignol, J.-C. Leugrand, (Cépadués Edition, Toulouse, 1999), vol. 1, p. 365.

Laser Sustained Plasma Free Jet and Energetic Atom Beam of Pure Argon or Oxygen Seeded Argon Mixture

A. Lebéhot[1], J. Kurzyna[2], V. Lago[1], M. Dudeck[1], R. Campargue[1]
[1] Laboratoire d'Aérothermique
Centre National de la Recherche Scientifique, Orléans, France
[2] Institute of Fundamental Technological Research
Polish Academy of Sciences, Warszawa, Poland

1 Introduction

As emphasized by Schram and co-workers above in this volume, the development of the plasma generation and expansion is relevant to many scientific subjects and a large variety of important applications. This article deals with the physics of the translational and electronic relaxations in plasma free jets, but the results are of potential interest also for the production of plasma flows, as used in ground test facilities for the simulation of spacecraft-atmosphere interactions. This is of importance to study the material degradation and surface modifications due to the very agressive environment in space experiments.
The conditions of the entry into a planetary atmosphere, where a plasma is produced by a shock wave detached in front of the nose of the spacecraft, can be simulated. This is workable in a wind tunnel [1] or in the expansion chamber of a nozzle beam generator, where the surfaces are exposed to a plasma flow of atmospheric species.
As observed during the early missions of the Space Shuttle, another important problem is encountered especially for the satellite surfaces, due to the agressive atomic oxygen in low Earth orbit (LEO). At this altitude (150-300 km), the UV radiation dissociates only O_2 and, consequently, the dominant species are the $O\left(^3P\right)$ atoms and the nitrogen molecules. Thus, the surfaces are exposed to a flux of 10^{13} to 10^{16} $O-atoms.cm^{-2}.s^{-1}$ (depending upon altitude), which are highly reactive at the kinetic energy of 5 eV resulting from the orbital velocity of 8 $km.s^{-1}$ [2]. The observed effects of this bombardment include the production of an optical glow in front of the exposed surfaces. This so-called "shuttle glow" accompanying

the material in the upper atmosphere is not yet well understood. These LEO conditions can be simulated using an oxygen atom beam obtained by skimming the axial part of a plasma free jet generated from a gas mixture containing oxygen [3]. Then, the plasma generator dissociates the oxygen molecules in the nozzle stagnation conditions, and the corresponding high enthalpy accelerates the subsequent O atoms.

Finally, an investigation on the physics of plasma expansion, as described below, is not only of scientific interest, but also can be of great importance to solve serious problems encountered in space experiments.

The results presented in this article concern both:

- the local properties of the plasma free jet (as to be created for entry simulation): electron density and temperature, degree of ionization, degree of dissociation of molecules, electronic excitation of the heavy species (mainly neutral and ionized atoms),

- the properties of the atoms in the final stages of the expansion, as measured in the skimmed beam (of O atoms for LEO simulation): nature of the different species (atoms, ions, molecules), translational properties (velocity distribution, translational temperature), electronic excitation.

Free jet and atom beam with continuum and molecular flow, respectively, are very different media. Nevertheless, the inspection of one of them contributes to the knowledge of the other: the terminal relaxation processes in the free jet define the states frozen in the beam, and the kinetic properties of the atom beam give some insight into the plasma flow, and even its initial stagnation state, through the laws of the isentropic expansion.

Some examples of such investigations are presented below for a laser sustained plasma with pure argon and oxygen-argon mixtures.

2 Experimental Set-up

Plasma flows have been produced by means of various techniques for more than 30 years [4]. Among these techniques, the laser heating by continuous optical discharge (COD) seems to be especially suited to free jet and beam generators for the following reasons: (i) the COD can be operated in a wide pressure range of 1 to 200 bars, which is typically the stagnation pressure range of interest, (ii) the operation of the COD needs no electrode inside the gas and, consequently, yields no gas pollution by material erosion, (iii) the temperature is maximum on the axis of the free jet source, i.e. the centerline part used to generate the atom beam, and (iv) this temperature maximum, as high as $20\,000\ K$, can be achieved very close to the nozzle throat, exactly at the chosen adjustable distance.

2.1 Principle of the continuous optical discharge (COD)

The radiation power of a focused infrared laser beam is absorbed by a gas medium as soon as the power density becomes high enough to initiate the inverse bremsstrahlung process [5]. Then, the free electrons are accelerated up to energies able to excite and ionize atoms and molecules. Thus, the plasma can be produced with a very high accuracy in a well defined wall-less volume limited by strong temperature gradients, and unaffected by any material interaction. The ignition of the plasma is favoured by an auxiliary pulsed laser of high power, which produces the plasma breakdown and then is switched off, as soon as the plasma is maintained steadily by the continuous laser.

2.2 The plasma free jet and the energetic beam

The nozzle beam generator is of the Campargue type [6]: the gas expands through a sonic orifice from a high pressure reservoir into a free jet zone of silence unaffected by the background gas. The beam is extracted from the jet by means of a skimmer. Photographic colored images of the 20 000 K plasma free jet are provided in Fig.1 showing also the skimmer. In the present experiments, the stagnation pressure P_0 is limited below 10^6 Pa by the strength of the ZnSe window through which the laser beams enter the stagnation reservoir (Fig.2). The nozzle is made out of a water cooled brass piece, and the orifice diameter is $D^* = 0.5mm$. The distance x_E between nozzle and skimmer is adjusted by moving axially the assemblage of nozzle reservoir and plasma source. The background pressure in the expansion chamber is maintained, during operation, in the range $0.05 \leq P_1 \leq 0.1$ hPa, by means of Roots pumps with pumping speed of 2000 m^3/h.

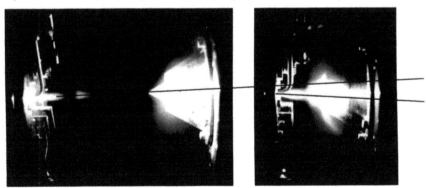

Figure 1: Photographs of an oxygen-argon supersonic free-jet generated from a continous laser sustained plasma at 20 000 K:
a) left: without meaningful skimming, **b)** right: with actual skimming

The beam of the continuous CO_2 laser (OPL 1501 from PRC Corp., with maximum power of 1500 W), operating in the TEM_{00} mode, is driven coaxially and within the hollow beam of the pulsed CO_2 laser (Lumonics TE-820 HP) [7, 8]. The two beams are focused just ahead of the sonic nozzle, inside the stagnation reservoir of the nozzle beam generator. The laser power actually used is here $150 \leq \mathcal{P} \leq 400\ W$ at pressures $P_0 \leq 8.10^5$ Pa.

2.3 Experimental analysis devices

A high resolution time-of-flight (TOF) system (Fig.2), described previously in details [6, 8], is operated on the skimmed atom beam, with a flight path of 4.2 m. Velocity distributions are measured very accurately, for the various species selected by a quadrupole mass spectrometer, in the case of a gas mixture.

Spectral analysis of the light emitted by the plasma free jet is performed with local resolution by using an optical device involving a small monochromator which views narrow slices of the free jet structure (Jobin-Yvon H25, focal length 250 mm, 1200 grooves/mm holographic grating) [8]. Moving a mirror allows a complete scan of the free jet along its diameter to be performed, in order to apply an inverse Abel transform. The emissive power is then restored with the actual axial symmetry of the jet, and the local population density of a number of excited states can be measured.

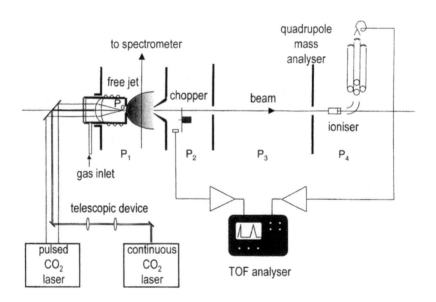

Figure 2: Experimental set-up

3 Results for pure argon

3.1 Translational relaxation

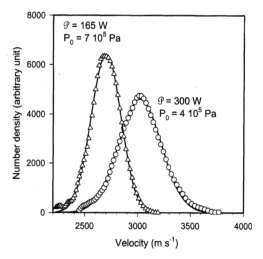

Figure 3: Velocity distributions measured in pure argon beams skimmed from two different plasma free jets, compared with their best Maxwellian fit (full line)

The axial velocity distribution, measured on the atom beam skimmed from the plasma free jet, accounts for the translational properties achieved in the final stages of the expansion. This distribution $f(v_\parallel)$ is found to be closely Maxwellian (Fig. 3), so that, if the speed ratio $S_\parallel = \frac{U}{(2kT_\parallel/m)^{1/2}}$ satisfies the condition $S_\parallel^2 \gg 1/2$, the number density measured on the axis is given by:

$$n(v_\parallel) \propto \frac{1}{(2kT_\parallel/m)^{1/2}} \left(\frac{v_\parallel}{U}\right)^2 exp - \left(\frac{v_\parallel - U}{(2kT_\parallel/m)^{1/2}}\right)^2, \quad (1)$$

where $v_\parallel$ is the longitudinal atom velocity and U is the terminal flow velocity in the free jet, taken as:

$$\frac{1}{2}mU^2 = \frac{\gamma}{\gamma - 1} k (T_0 - T), \quad (2)$$

where γ is the specific heat ratio for the gas of interest, m is the mass of an atom, k is the Boltzmann constant, T_0 is the *stagnation* temperature, and T is the terminal translational temperature, defined as: $\frac{5}{2}kT = \frac{3}{2}kT_\parallel + kT_\perp$ ($T_\parallel$ and $T_\perp$ are the longitudinal and tranverse temperatures, respectively [10]). The specific heat ratio C_P/C_V is taken as the constant $\gamma = 5/3$. This is a rather rough simplification because the value of γ should not be considered as a constant in the wide temperature and pressure ranges encountered in the expansion. Nevertheless, a complete calculation would be very tedious in non-equilibrium conditions. A mean value taken between 1.3 and 1.4 gives sometimes better results with pure argon. Also, it is worth noticing that the meaning of T_0 (here so-called *time-of-flight temperature* [7]) is not straightforward here, where very strong temperature gradients appear upstream of the nozzle throat. T_0 can be simply defined as the temperature at the point located on the nozzle axis, upstream of the nozzle throat, just at the beginning of the gas flow precooling effect (where the Mach number $\mathcal{M} \simeq 0$). This distance is about $x_0/D^* \approx -0.9$, where $\mathcal{M}$ becomes lower than 0.1; upstream of this point, the temperature goes on increasing up to about 20 000 K, in the plasma core [8, 11]. With this definition of T_0, the classical laws governing the supersonic free jet expansion can be applied. Thus, are valid the usual axial laws of the isentropic flow which yield the local properties of the jet, as simple functions of the single local Mach number. Also, the axial Mach number depends only on the distance from the nozzle throat, and can be approximated by polynomial expansions in terms of x/D^* [8, 10].

3.2 Electronic relaxation

3.2.1 Experimental results

The complete wavelength spectra recorded at various distances x/D^* exhibit lines which can be all assigned to the transitions of the ArI system. Neither trace of impurities, nor ionic transitions, are detected within the sensitivity of the experiment.

Nine transitions are examined with special care, in order to evaluate the local population densities of nine electronic levels. For each of them, a complete map is obtained in the free jet and, in each point of this map, the population densities are analysed as a function of the energy of the emitting level, in a Boltzmann diagram. An *excitation temperature* T_{exc} is defined by the equilibrium governing the levels of higher energies (Fig.4). A map of this temperature can then be drawn and the axial values are compared with the results of a one-dimensional relaxation model.

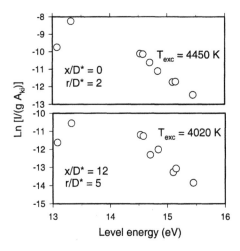

Figure 4: Boltzmann diagrams plotted for the nine levels of argon, at two different points in the plasma free jet, with $\mathcal{P} = 250W$ and $P_0 = 3.5 \times 10^5$ Pa. I is the measured count number after Abel inverse transform, geometrical deconvolution, and correction by the wavelength transmission function of the optical device. g is the degeneracy of the level, A_{ki} is the Einstein coefficient of the transition. T_{exc} is calculated with the upper seven levels.

3.2.2 The one-dimensional relaxation model

The calculation lies upon a collisional-radiative model which connects locally the populations of a number of neutral argon electronic levels to the electron temperature and density [12, 13]. These local parameters can be measured experimentally, generally with electrostatic probes [12], or calculated with a relaxation thermal model [14] which may be reduced to a one-dimensional model [8]. The results of this calculation are compared with the experimental measurements in Fig.5.

The calculated electron temperature T_e is found to be in close agreement with the measured excitation temperature T_{exc}, as expected [13, 15]. The calculated translational temperature of the heavy particles (mainly neutral atoms) is compared with the terminal value corresponding to the Maxwellian distribution which fits the experimental time-of-flight distribution. The system of the collisional-radiative equations is then solved, at given axial distance, with the electron density n_e and temperature T_e calculated at this point. The resulting population densities are compared with the experimental values for a number of states in Fig.6. The calibration factor between calculation and experiment is the same for all sets of curves. The agreement

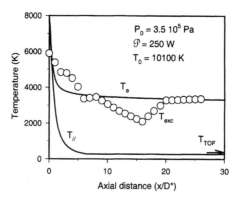

Figure 5: Measured excitation temperature in a plasma free jet of pure argon, compared with calculated electron and translational temperatures (full lines). The measured terminal translational temperature is indicated by an arrow (T_{TOF}).

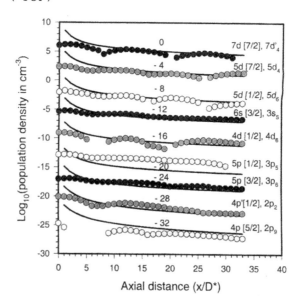

Figure 6: Measured population densities as a function of axial distance in pure argon plasma free jet. For the sake of clarity, the curves are vertically shifted by the amount indicated for each level. The calibration factor between theory and experiment is the same for all states [21].

is quite good, except for the shortest distances, where the lack of geometrical resolution produces an underestimation of the experimental data.

4 Results for oxygen seeded argon mixtures

4.1 Translational relaxation

The plasma is generated in various mixtures containing up to 23 % of oxygen molecules in argon (only the inlet O_2 number concentration is indicated). Generally, the laser power must be increased with the relative concentration of O_2, because the molecule can absorb energy in internal degrees of freedom before being ionized. Thanks to the mass selective detector of the atom beam species, time-of-flight distributions are easily measured separately for oxygen as well as argon atoms. Such distributions are reported in Fig.7, together with the corresponding best fit Maxwellian distributions, for beams skimmed from a plasma jet produced with a mixture of 7% oxygen in argon, $\mathcal{P} = 330\ W$ and a total stagnation pressure $P_0 = 3.5 \times 10^5$ Pa. It appears that the velocity of both components is increased with respect to the velocity of pure argon obtained in the same conditions (Fig.8). It is noteworthy that this gain in velocity is more important than expected from a simple mass effect (Fig.9). This is consistent with a less efficient inverse

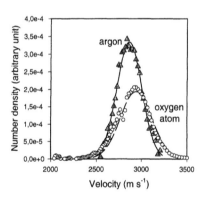
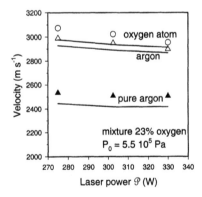

Figure 7: Velocity distributions of argon ($\triangle$) and oxygen atoms ($\circ$) in a beam issued from plasma sustained with 7% oxygen in argon and $P_0 = 3.5 \times 10^5$ Pa, $\mathcal{P} = 330\ W$. The acquisition times are different for each species.

Figure 8: Terminal velocity of Ar and O for the mixture 23% O_2 in Ar, compared with pure argon beam, from plasmas generated with increasing laser power at the same total pressure. Full lines are the calculated results.

bremsstrahlung process which moves the plasma core toward higher laser power density, closer to the nozzle throat. This phenomenon can change largely the effective temperature T_0 [7, 11].

The experimental velocity distributions also give the terminal translational temperature $T_\parallel$ for each species. These data are reported for three different laser powers in Fig.10. The translational temperature of argon is found to be significantly higher than the temperature of oxygen atoms.

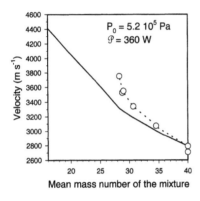

Figure 9: Terminal mean velocity measured in the oxygen atom beam, as a function of the average mass of the mixture. Full line: velocity calculated from T_0 deduced from the pure argon case.

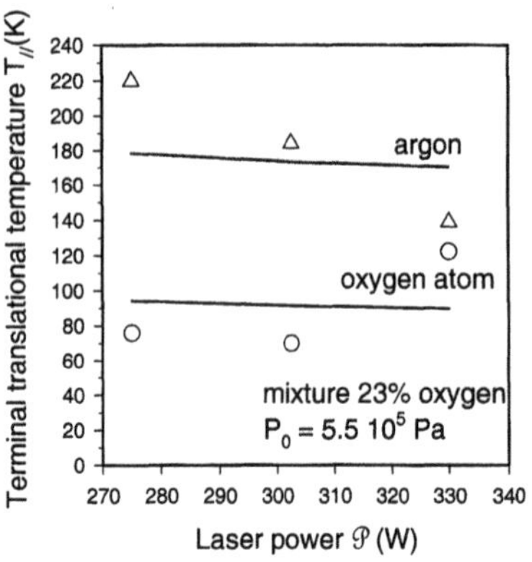

Figure 10: Terminal translational temperature $T_\parallel$ measured in the beam for argon ($\triangle$) and oxygen atoms ($\circ$) in the mixture. Full lines: results of the calculation.

The translational properties of argon and oxygen atoms are also calculated with a model derived from the moment method of the Boltzmann equation, for a binary gas mixture [6, 16]. Such models were very popular in the 60-70's, when great effort was devoted to aerodynamic isotope separation methods [17], and before a better accessibility to Monte-Carlo simulations [18].
It is assumed in these calculations that oxygen is completely dissociated in the expansion and the mixture contains only argon and oxygen atoms. The system of conservation equations is solved using a Runge-Kutta technique. The method suffers of some inconsistency in the assumptions which are necessary to solve a truncated number of equations. Two of them are especially cumbersome: the assumption that hypersonic flow is established before significant non-equilibrium effects occur, and the assumption of a spherically symmetric flow. The hypersonic assumption is required for neglecting the

heat flux terms; it is applied here, in spite of its questionable validity at intermediate distances. The spherical assumption is valid at large distances, where the expansion process may be considered as nearly frozen. As a matter of fact, applying this scheme to the present problem almost prevents any cooling effect to take place, and the kinetic properties are frozen very shortly (mainly $T_\|$). So, the shape of the stream tubes is here calculated with the usual analytical forms of the Mach number at short distances [8, 10]. Beyond a distance of the order of the classical freezing distance [8, 19], the spherical symmetry is then considered as established. With this scheme, the calculated values of the mean velocities for argon and oxygen atoms, as well as the corresponding translational temperatures, are found in satisfactory agreement with the experimental values (Figs. 8, 10). The validity of the method is checked for pure argon and the source at room temperature T_0, all other parameters otherwise equal; the calculated velocity is found very close to the theoretical value (~ 540 instead of $555\ m\ s^{-1}$), while the terminal temperature $T_\|$ is found of the same order of magnitude as the experimental value (3.4 K instead of 1 K). This result seems to be quite reasonable, taking into account a possible condensation effect on the experimental temperature $T_\|$: the increase of collisional cross-sections of clusters may favor the cooling process. In the case of pure argon used with the same pressure and laser conditions as in the mixture, the calculated velocity is also obtained with a rather good accuracy (5 % lower than the experimental value), but the temperature is somewhat higher than the experimental temperature ($\sim 90\ K$ instead of $\sim 70\ K$). Nevertheless, the energy balance given by Eq. 2 is satisfied in any case within a few percent deviation; such a small discrepancy proves the validity of the method.

4.2 Electronic relaxation

Four electronic transitions are selected for each component (Table 1).

The experimental procedure for analysing the population density of the corresponding initial levels, with spatial resolution, is the same as used for pure argon. The experiment is more difficult for mixtures because the emissivity of the plasma jet decreases drastically as soon as oxygen is added into the inlet gas, even in very small amount. This is the reason why only four transitions are considered here, for their intensity and selectivity with respect to the available sensitivity and wavelength resolution (only 0.7 nm).
The collisional-radiative model, previously used for pure argon [8], is extended to the argon-oxygen atom mixture, including electronic excitation of both species by the free electrons and the presence of neutral atoms and residual molecules [21]. The results are compared in Fig.11 with the experimental population densities on the axis of the free jet. The agreement is quite satisfactory and can be still improved by including other processes in

No.	Argon	upper level	eV	g	A_{ki} (10^8 s^{-1})
1	696.5 nm	$2p_2, 4p'[1/2]$	13.33	3	0.067
2	430.0 nm	$3p_8, 5p[5/2]$	14.51	5	0.00394
3	415.8 nm	$3p_6, 5p[3/2]$	14.53	5	0.0145
4	603.2 nm	$5d'_4, 5d[7/2]$	15.13	9	0.0246
	Oxygen				
5	777.3 nm	$3p\,^5P$	10.74	15	0.34
6	436.8 nm	$4p\,^3P$	12.36	9	0.0066
7	645.5 nm	$5s\,^5S^0$	12.66	5	0.071
8	533.0 nm	$5d\,^5D^0$	13.07	25	0.0197

Table 1: Transitions analysed in the plasma free jet of argon-oxygen mixture [20].

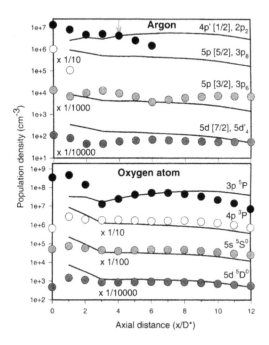

Figure 11: Measured population densities versus the axial distance from the nozzle in argon-oxygen plasma free jets. For the sake of clarity, the curves are vertically shifted by the amount indicated for each level. The calibration factor between theory and experiment is the same for all states (calibration point indicated by an arrow) [21].

the system of equations, such as ion recombination, collisions with van der Waals molecules, etc.

5 Conclusion

A CO_2 laser heated plasma is very appropriate for being used as a source of plasma free jet and energetic atom beam. The continuous optical discharge can be sustained at both high pressure and high temperature (up to 20 000 K at present), in a small volume of gas surrounded by strong temperature gradients. The versatility of the technique results from its flexibility in the optimization of jet and beam parameters, and the possibility of moving axially and precisely the miniplasma in order to use its maximum temperature when the core is close to the nozzle throat. The effective temperature can be deduced accurately from time-of-flight measurements on the atom beam, using the classical laws of the isentropic expansion of neutrals, which can be applied due to the low degree of ionization of the plasma.

Also the characteristics of the atom beam (nature and purity of the species, kinetic properties) are of interest, for a number of applications: the simulation of space environment, the study of surface cathalycity, surface coating, atmospheric chemical processes with energy thresholds, etc. In particular, the fast increase of the atom velocity when decreasing the average mass of the gas mixture makes this type of atom source very interesting, especially for generating fast oxygen atoms (eV range). Furthermore, it appears from the collisional-radiative calculation that the population densities of the electronically excited states are several orders of magnitude lower than the density of the ground state, even for the metastable states. Thus, it may be considered that the ground state O (^3P) is prominently produced in the beam source. The only remaining stray particles are the ground state argon atoms and infra-red photons from the laser beam. As the laser beam is strongly absorbed by the plasma, and also strongly diverging after the nozzle throat, the density of photons becomes negligible in the skimmed atom beam. Consequently, this plasma source can be considered as an interesting tool for the production of energetic atomic species of selected properties with the aim of specific applications.

References

[1] A.T. Schönemann, V. Lago, and M. Dudeck, J. Thermophys. Heat Transfer, **10**, 419, (1996).

[2] See for example: J. Dauphin, in *Looking Ahead for Materials and Processes*, edited by J. de Bossu, G. Briens, and P. Lissac, (Elsevier Sci. Pub., Amsterdam,1987), p. 345; S.L. Koontz, K. Albyn, L.J. Leger, J. Spacecraft, **28**, 315, (1991).

[3] R.B. Cross, Third Annual Workshop on Space Operations, Automation, and Robotics, Houston, 1989-NASA Conference Publication 3059, edited by S. Griffin, p. 269, (1990).

[4] See, for example, T.B. Reed, J. Appl. Phys., **32**, 821, (1961); E.L. Knuth, N.M. Kuluva, and J.P. Callinan, Entropie, **18**, 38, (1967); V.M. Gold'farb, E.V. Il'ina, G.A. Luk'yanov, V.V. Nazarov, N.O. Pavlova, and V.V. Sakhin, High Temp. **14**, 9 , (1976); P.G.A. Theuws, H.C.W. Beijerinck, D.C. Schram, and N.F. Verster, J. Appl. Phys., **48**, 2261, (1977); M.V. Gerasimenko, G.I. Kozlov, and V.A. Kuznetsov, Sov. J. Quantum Electron., **13**, 438, (1983); A. Anders, Contrib. Plasma Phys., **27**, 203, (1987); D.J. Douglas and J.B. French, J. Anal. At. Spectrom., **3**, 743, (1988); B.G. Chéron, L. Robin, and P. Vervisch, Meas. Sci. Technol., **3**, 58, (1992); H. Akatsuka and M. Suzuki, Rev. Sci. Instrum., **64**, 1734, (1993).

[5] Yu.P. Raĭzer, Sov. Phys. Usp., **23**, 789, (1980).

[6] R. Campargue, J. Chem. Phys., **88**, 4466, (1984).

[7] J.M. Girard, A. Lebéhot, and R. Campargue, J. Phys. D: Appl. Phys., **26**, 1382, (1993).

[8] A. Lebéhot and R. Campargue, Phys. Plasmas, **3**, 2502, (1996).

[9] R. Campargue, A. Lebéhot, J.C. Lemonnier, D. Marette, in *Rarefied Gas Dynamics*, edited by S.S. Fisher, (American Institute of Aeronautics and Astronautics, New–York, 1981) p. 823.

[10] D.R. Miller, in *Atomic and Molecular Beam Methods*, edited by G. Scoles, (Oxford University, New-York, Oxford, 1988), p. 14.

[11] J.M. Girard, A. Lebéhot, and R. Campargue, J. Phys. D: Appl. Phys., **27**, 253, (1994).

[12] A. Kimura, M. Minomo, and M. Nishida, in *Rarefied Gas Dynamics*, edited by R. Campargue, (CEA, Paris, 1979), p. 969.

[13] J.A.M. van der Mullen, Phys. Rep., **191**, 109, (1990).

[14] J.J. Beulens, D. Milojevic, D.C. Schram, and P.M. Vallinga, Phys. Fluids, **3**, 2548, (1991).

[15] I.J.M.M. Raaijmakers, P.W.J.M. Boumans, B. van der Sijde, and D.C. Schram, Spectrochim. Acta, **38B**, 697, (1983).

[16] A. Chesneau, R. Campargue, in *Rarefied Gas Dynamics*, edited by O.N. Belotserkovskii, M.N. Kogan, SS. Kutateladse and A.K. Rebrov, (Plenum, New-York, 1985), p. 879.

[17] R. Campargue, J.B. Anderson, J.B. Fenn, B.B. Hamel, E.P. Muntz, and J.R. White, in *Nuclear Energy Maturity*, edited by P. Zaleski, (Pergamon, Oxford, 1975).

[18] G.A. Bird, AIAA J., **8**, 1998, 1970; A.U. Chatwani, in *Rarefied Gas Dynamics*, edited by J.L. Potter, (AIAA, New York, 1977); A.U. Chatwani and M. Fiebig, in *Rarefied Gas Dynamics*, edited by S.S. Fisher, (AIAA, New York, 1981), p. 785.

[19] H.C.W. Beijerinck, N.F. Verster, Physica, **C111**, 327, (1981).

[20] W.L. Wiese, M.W. Smith, and B.M. Miles, *Atomic transition probabilities*, Vol. II, National Stand. Ref. Data Ser., Nat. Bur. Stand. (US), Vol. 22, (1969).

[21] A. Lebéhot, J. Kurzyna, V. Lago, M. Dudeck, M. Nishida, Phys. Plasmas, **6**, 4750, (1999).

Doppler Profiles of the Distribution of O(^{1}D) Relaxing in Ne

Bernie D. Shizgal[1], K. Hitsuda[2] and Yutaka Matsumi[2]
[1] Department of Chemistry, University of British Columbia, Vancouver, Canada
[2] Solar Terrestrial Environment Laboratory and Graduate School of Science, University of Nagoya, Toyokawa, Japan

1 Introduction

The relaxation of hot atoms in a gaseous moderator remains a fundamental problem in rarefied gas dynamics [1, 2]. The rate of relaxation of energetic particles to equilibrium has important applications in diverse fields. In addition to the calculation and measurement of relaxation times, in recent years there have been numerous measurements of the details of the relaxing velocity distribution functions by Doppler spectroscopy [3, 4]. For example, Park et al [5] have investigated the relaxation of H atoms produced in laser dissociation of H_2S at 193 nm. They measured the Doppler profiles of the H atoms at the Lyman-α line during the relaxation by collisions with the rare gases and some molecular moderators. Nan and Houston [6] monitered the velocity relaxation of S(^{1}D) by He, Ar and Xe by measuring the Doppler profile of S(^{1}D) created by pulsed laser photolysis of OCS at 222 nm. Cline et al [7] reported the relaxation of hot I($^2P_{1/2}$) that was produced from photodissociation of of n-C_3F_7I at 266 nm in a background of He.

The distribution functions of O(^{1}D) in the terrestrial atmosphere can be strongly non-Maxwellian owing to the production of these atoms by photodissociation of O_2 and O_3. The relaxation dynamics of O(^{1}D) are important for a characterization of the chemistry of the upper atmosphere. Matsumi et al [8] studied the velocity relaxation of O(^{1}D) produced by photodissociation of O_2 with linearly polarized F_2 laser light at 157 nm. The velocity distributions of O(^{1}D) were determined by Doppler profiles measured by vacuum ultraviolet laser-induced fluorescence. They studied the relaxation of the energy as well as the rate of decay of the anisotropy of the distribution function.

The purpose of the present paper is to study theoretically the relaxation of anisotropic distributions of $O(^1D)$ in Ne with accurate realistic differential elastic scattering cross sections in the Boltzmann equation and to compare with both the measurements and the previous Monte-Carlo simulations.

The experimental arrangement was described in the previous paper [8]. $O(^1D)$ atoms are generated by the photodissociation of O_2 with linearly polarized light at 157 nm. The velocity distributions of $O(^1D)$ were determined from the Doppler profiles measured by vacuum ultraviolet laser induced fluorescence of $O(3s\ ^1D^0 \leftarrow 2p^1D)$ centered at 115.215 nm. The nascent velocity distribution of the O fragments in the center-of-mass frame is of the form

$$f(v,\theta,0) = f^{(0)}(v,0)[1 + \beta(0)P_2(\cos\theta)] \qquad (1)$$

where $f^{(0)}(v,0)$ is the initial isotropic distribution, determined from Doppler spectroscopy, and P_2 is a Legendre polynomial [8]. The initial anisotropy parameter is $\beta(0) = 1.55$ and θ is the angle between the z-axis and the direction of the velocity vector. From a theoretical point of view, this is an ideal experiment for which the initial distribution is known. In particular, the anisotropy of the distribution function contains only a P_2 term in the expansion in Legendre polynomials.

2 Collision Cross Sections

The (Ne,O) system gives rise to to three different singlet potential curves, $^1\Sigma$, $^1\Pi$ and $^1\Delta$, and the detailed radial dependence is taken from the calculations by Langhoff [9] and shown in Fig. 1A. The long range portion of the potentials were fitted to exponential forms from the last data points in the work by Langhoff. The differential scattering cross section, $\sigma(E,\theta)$, is calculated in terms of the quantum mechanical phase shifts, δ_ℓ, for each potential following standard methods. The contribution from each interaction potential is given by the degeneracies of that state. The differential cross section is finite at zero scattering angle unlike the classical cross section, and the small angle cut-off is not necessary [8]. An example of the differential cross section is shown in Fig. 1B. The large forward scattering peak is clearly evident. A large fraction of the total cross section,

$$\sigma_{tot}(E) = 2\pi \int_0^\pi \sigma(E,\theta)\sin(\theta)d\theta, \qquad (2)$$

arises from integration of the differential cross section at small angles. The energy variation of the total elastic cross section and the momentum transfer cross section,

$$\sigma_{mt}(E) = 2\pi \int_0^\pi \sigma(E,\theta)(1-\cos\theta)\sin\theta d\theta, \qquad (3)$$

is shown in Fig. 1C. The momentum transfer cross section is in good agreement with the hard sphere cross section (dashed line) reported earlier [8] with $d = 2.1$. The classical scattering angle versus the impact parameter at $E = 0.1$ eV is shown in Fig. 1D for the Δ potential. The scattering angle can also be calculated from the quantum phase shifts in the semi-classical limit, that is, $\theta(b) = 2d\delta_\ell/d\ell$ with the impact parameter $b = (\ell+\frac{1}{2})/k$ where k is the wavenumber corresponding to relative energy E. The agreement of the classical and the "semiclassical" scattering angle is excellent.

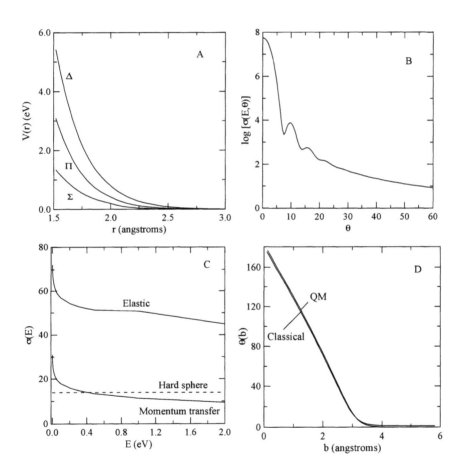

Figure 1: (**A**) Ne-O Interaction potentials; (**B**) Differential cross section at $E = 0.05$ eV; (**C**) Total elastic and momentum transfer cross sections in ^{2}; (**D**) Classical and semiclassical scattering angles at $E = 0.1$ eV

3 The Boltzmann Equation

We consider the nonequilibrium system of O(^{1}D) produced via the photodissociation process. The O density is n, dilutely dispersed in a bath of Ne atoms, denoted by a subscript 1. We assume that $n \ll n_1$, so that only Ne-O collisions need to be considered and the Ne distribution is a Maxwellian at the temperature T, that is,

$$f_1^M(\mathbf{v}) = n_1(m_1/2\pi kT)^{3/2} e^{-m_1 v^2/2kT}. \tag{4}$$

The Boltzmann equation for the time evolution of the O(^{1}D) velocity distribution function is given by

$$\frac{\partial f}{\partial t} = 2\pi \int \int [f' f_1^{M'} - f f_1^M] \sigma(g,\theta) g \sin(\theta) d\theta d\mathbf{v}_1 \tag{5}$$

where $\mathbf{g} = \mathbf{v} - \mathbf{v}_1$ is the relative collision velocity, and θ is the scattering angle in the center of mass frame. The quantities $\mathbf{v}$ and $\mathbf{v}_1$ are the velocities before a collision and $\mathbf{v}'$ and $\mathbf{v}'_1$ are the velocities after a collision.

The Boltzmann equation can be rewritten in the form

$$\frac{\partial f(\mathbf{v})}{\partial t} = \int K(\mathbf{v},\mathbf{v}')f(\mathbf{v}')d\mathbf{v}' - Z(v)f(\mathbf{v}) \tag{6}$$

where the collision frequency is given by

$$Z(v) = \int K(\mathbf{v},\mathbf{v}')f_1^M(\mathbf{v}')d\mathbf{v}'/f^M(\mathbf{v}). \tag{7}$$

The kernel $K(\mathbf{v},\mathbf{v}')$ is well-known [10]. The main objectives, as in the previous work [8], is to determine the time dependent velocity distribution function, the average energy,

$$E_{avg}(t) = \frac{m}{2} \int f(\mathbf{v},t) v^2 d\mathbf{v} \tag{8}$$

and the average anisotropy parameter, that is,

$$\beta_{avg}(t) = \int f(\mathbf{v},t)\beta(v,t)d\mathbf{v} \tag{9}$$

as well as the time dependent Doppler profiles.

In general, $K(\mathbf{v},\mathbf{v}')$ depends only on the magnitudes of the velocities $\mathbf{v}$ and $\mathbf{v}'$, and $\hat{\mu} = \cos\hat{\theta}$, where $\hat{\theta}$ is the angle between $\mathbf{v}$ and $\mathbf{v}'$. We rewrite $K(\mathbf{v},\mathbf{v}')$ as $K(v,v',\hat{\mu})$. The O(^{1}D) distribution function can be written as an expansion in Legendre polynomials,

$$f(\mathbf{v},t) = \sum_{l=0}^{\infty} f^{(l)}(v,t) P_l(\mu), \tag{10}$$

where $\mu = \cos\theta$ and θ is the polar angle between $\mathbf{v}$ and some fixed axis in velocity space. We can then decompose the kernel $K(\mathbf{v}, \mathbf{v}')$ by the ℓth component $K_\ell(v, v')$, where,

$$K_\ell(v, v') = 2\pi \int_{-1}^{1} v'^2 K(v, v', \hat{\mu}) P_\ell(\hat{\mu}) d\hat{\mu}. \tag{11}$$

The $\hat{\mu}$-integration must be done numerically for all but the simplest cross sections [1, 2]. The Boltzmann equation then reduces to the set of uncoupled differential equations

$$\frac{\partial f^{(\ell)}}{\partial t} = \int_0^\infty K_\ell(v, v') f^{(\ell)}(v') dv' - Z(v) f^{(\ell)}(v). \tag{12}$$

The initial distribution function in the experimental work only has $\ell = 0$ and $\ell = 2$ components. Since there is no coupling between ℓ components, the relaxation of each component is independent.

4 Time Dependent Solution of the Boltzmann Equation

We solve the Boltzmann equation, Eq. (12), numerically using the Quadrature Discretization Method (QDM), and also with a Monte-Carlo simulation as described in the previous paper [8] and modified according to the work of Nanbu et al [14]. The QDM has been described in detail in other papers [11, 12]. It is convenient to consider reduced speed defined by $x = v/v_0$ where $v_0 = (2kT_s/m_1)^{1/2}$, and T_s is some arbitrary temperature parameter chosen so as to scale the reduced speed appropriately. The basis of the QDM is to use a set of convenient orthogonal polynomials to approximate the distribution function. All quantities are then evaluated at discrete points (the zeros of the N'th order polynomial), and integrals are evaluated using Gaussian quadrature. Here we use the "speed" polynomials, defined as the set of orthogonal polynomials $B_n(x)$ which satisfy

$$\int_0^\infty w(x) B_n(x) B_m(x) dx = \delta_{nm} \tag{13}$$

where $w(x) = x^2 e^{-x^2}$. With the change of variable to reduced speed, the integration on the right hand side of Eq. (12) can be performed with a quadrature of the form

$$\int_0^\infty w(x) F(x) dx \approx \sum_{i=1}^N w_i F(x_i) \tag{14}$$

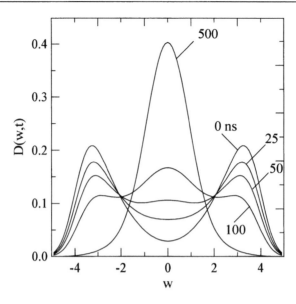

Figure 2: Doppler profiles for the hard sphere cross section at different delay times, w in units of $\sqrt{2kT/m}$, $T = 300K$

where x_i and w_i are the corresponding quadrature points and weights [12]. Eq. (12), written as a quadrature sum (with the ℓ index omitted for clarity), is then of the form

$$\frac{df_i}{dt} = \sum_{j=1}^{N} M_{ij} f_j, \qquad (15)$$

where the subscript i denotes evaluation at the point x_i and

$$M_{ij} = (K_{ij} - Z_i \delta_{ij}) w_j / w(x_j). \qquad (16)$$

In order to conserve particle number accurately, we calculate the collision frequency $Z(x_i)$ using Gaussian quadrature, that is,

$$Z(x_i) = \sum_{j=1}^{N} w_j K(x_i, x_j) f_j^{(M)} / (w(x_j) f_i^{(M)}). \qquad (17)$$

The solution of Eq. (15) is

$$f_i(t) = U_{i1} g_1(0) + \sum_{j=2}^{N} U_{ij} g_j(0) e^{-\lambda_j t} \qquad (18)$$

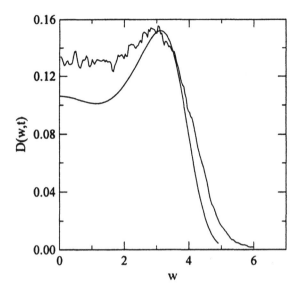

Figure 3: Comparison of the theoretical and experimental profiles at a delay time of 100ns. The position of the maximum and the peak height were adjusted for the comparison.

where $\mathbf{U}$ is a matrix which diagonalizes $\mathbf{M}$, that is, $\mathbf{U}^{-1}\mathbf{M}\mathbf{U} = \mathbf{\Lambda}$ (where $\mathbf{\Lambda}$ is the diagonal matrix of eigenvalues with $\lambda_1 = 0 < \lambda_2 < \lambda_3 < \ldots$), and $\mathbf{g} = \mathbf{U}^{-1}\mathbf{f}$. The term U_{i1} is the i'th component of the eigenfunction with eigenvalue $\lambda_1 = 0$, i.e. a Maxwellian. Thus $\mathbf{f}^{(0)}$ decays to a Maxwellian as $t \to \infty$ with a relaxation time of the order of $1/\lambda_2^{(0)}$. The spectrum of the collision operator determines the detailed time evolution of the distribution functions $\mathbf{f}^{(0)}$ and $\mathbf{f}^{(2)}$.

The time dependent distribution functions are thus determined from the Boltzmann equation and also with a Monte-Carlo simulation as discussed previously [8]. The time dependent Doppler profiles are given by [3, 8],

$$D(w,t) = \frac{1}{2}\int_{|w|}^{\infty}[1 + \beta(v,t)P_2(\cos\Theta)P_2(w/v)]f^{(0)}(v)vdv \qquad (19)$$

where Θ is the angle between the dissociating laser beam and the diagnostic beam. The initial values of $f^{(0)}(v,0)$ and $f^{(2)}(v,0)$ were determined from the experimental setup and reported in the previous paper [8]. The very sharp forward scattering peak in the differential cross section (see Fig. 1B) results in the kernel of the Boltzmann equation to be very narrow when the two speed variables are equal. In the present work, only a hard sphere cross section has been used in the direct solution of the Boltzmann equation. In

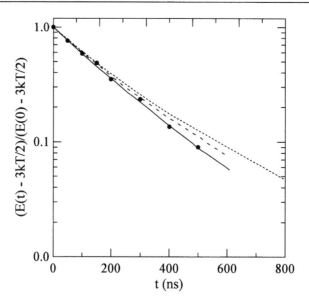

Figure 4: Energy relaxation of O(^{1}D) in Ne: (symbols) experiment; (solid line) Monte Carlo with the ab-initio potentials and classical cross sections; (long dashed line) Monte Carlo with the hard sphere cross section, $d = 2.1$; (short dashed line) Boltzmann equation with the hard sphere cross section

the Monte-Carlo simulations, the classical cross sections for the potentials shown in Fig. 1A were used. For the hard sphere cross section, the solutions for $\ell = 0$ and $\ell = 2$ are given by Eq. (18). The Doppler profiles shown in Fig. 2 at five different delay times, were determined from Eq. (19) with $\beta(v,t) = f^{(2)}(v,t)/f^{(0)}(v,t)$. The experimental profile and the calculated profile are shown for a delay time of 100ns in Fig. 3. The theoretical result has not been convoluted with the diagnostic beam profile [8]. Nevertheless, the agreement, for the hard sphere cross section, is good.

The average energy and the anisotropy parameter given by Eqs. (8) and (9) were calculated for the hard sphere cross section with the Boltzmann equation. They were also calculated with the Monte Carlo simulations for both hard spheres and the classical cross sections for the ab-initio potentials. The results are summarized in Figs. 4 and 5. Figure 4 shows the relaxation of the average energy. The O atoms are initially produced with energies of about five times the thermal energy. The symbols are the experimental measurements, and the solid line is the result of the Monte Carlo simulations with the ab-initio potentials. The long and short dashed curves are the results with the hard sphere cross section and the Monte Carlo and Boltzmann equation results, respectively. The results with the Boltzmann equation give a slower relaxation than with the Monte Carlo simulations.

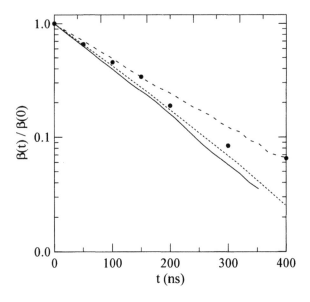

Figure 5: Relaxation of the anisotropy parameter: (symbols) experiment; (solid line) Monte Carlo with the ab-initio potentials and classical cross sections; (long dashed line) Monte Carlo with the hard sphere cross section, $d = 2.1$; (short dashed line) Boltzmann equation with the hard sphere cross section.

Both hard sphere results give a somewhat slower relaxation than the Monte Carlo results with the real potential.

The time evolution of the anisotropy parameter is shown in Fig. 5. The curves refer to the same calculations as described for Fig. 4. The results with the Monte Carlo simulations and the real potentials is somewhat faster than the experimental relaxation behaviour. The hard sphere results obtained with the Boltzmann equation are closer to the experimental results than either Monte Carlo results which are slower than either of the other two calculations. It is important to notice that there is some scatter in the experimental results. Despite some differences in the calculations and with the experiment, the results obtained are in very reasonable agreement.

5 Summary

In this paper, we have presented a solution of the Boltzmann equation with a direct solution method (the QDM) and a Monte Carlo simulation. The results for the energy relaxation and the relaxation of the anisotropy parameter were in good agreement with the experimental results. The multi-domain method introduced by Leung et al [2] is required in order to accurately account for the narrow kernel in the Boltzmann equation that arises because of the sharp forward elastic scattering peak in the quantum elastic cross section. A solution of Boltzmann equation based on this methodology and the actual elastic cross section will be appear in a forthcoming publication [15]. The relaxation of energetic O atoms in the other inert gases will also be reported.

References

[1] B. Shizgal and R. Blackmore, Chem. Phys. **77**, 417 (1983).

[2] K. Leung, G. Arkos, and B. D. Shizgal, Rarefied Gas Dynamics **20**, 118 (1996).

[3] R. J. Gordon, and G. F. Hall, Adv. Chem. Phys. **96**, 1 (1996).

[4] J. L. Kinsey, Chem. Phys. **66**, 2560 (1977).

[5] J. Park, N. Shafer and R. J. Bersohn, Chem. Phys. **91**, 7861 (1989).

[6] G. Nan and P. L. Houston, J. Chem. Phys. **97**, 7865 (1992)

[7] J. I. Cline, C. A. Taatjes and S. R. Leone, J. Chem. Phys. **93**, 6543 (1990).

[8] Y. Matsumi, S. M. Shamsuddin, Y. Sato and M. Kawasaki. J. Chem. Phys. **101**, 9610 (1994).

[9] S. R. Langhoff, J. Chem. Phys. **73**, 2379 (1980).

[10] R. Kapral and J. Ross, J. Chem. Phys. **52**, 1238 (1970).

[11] B. Shizgal and H. Chen, J. Chem. Phys. **104**, 4137 (1996).

[12] B. D. Shizgal, Theochem-J. Molec. Struc. **391**, 131 (1997).

[13] R. Blackmore and B. Shizgal, J. Computat. Phys. **55**, 313 (1984).

[14] K. Nanbu, T. Morimoto and Y. Goto, JSME Int. J. **B36**, 313 (1993).

[15] B. D. Shizgal, K. Hitsuda and Y. Matsumi, J. Chem. Phys. (to be published).

Natural Alignment and Cooling in Seeded Supersonic Free Jets: Experiments and a Quantum Mechanical View

V. Aquilanti, D. Ascenzi, D. Cappelletti
M. de Castro, F. Pirani
Università di Perugia, Perugia, Italy

1 Introduction

A simple and natural source of cold and aligned molecules involves the use of molecular beams seeded with lighter carrier gases: collisions occurring during the supersonic expansion can produce molecules in their lowest rotovibrational states with specific alignment of the rotational angular momentum [1]. Our recent progress on understanding this phenomenon, whose mechanism is qualitatively depicted in Fig. 1, is discussed in this paper. Collisional alignment is by itself an important issue of gas dynamics: after providing the proper background by analyzing the relevant experimental information (section 2), in this paper we present a discussion of the gas dynamics in terms of extensive quantum mechanical calculations of scattering cross sections (section 3) and outline recent applications to scattering experiments providing information on intermolecular forces (section 4).

2 Experimental background

The collisional alignment of molecules in a nozzle expansion under supersonic conditions has been qualitatively anticipated long ago [2] in connection with a transport phenomenon - the Senftleben-Beenakker effect [3] - and was first noticed experimentally for I_2 by Steinfeld and Korving (as reported in Ref. [4]) by a laser induced fluorescence technique. In 1975 Sinha et al. [4] found a significant alignment of Na_2 in supersonic beams of sodium atoms containing dimers, and later similar prominent effects were observed for other alkali-metal dimers [5, 6]. Alignment had been studied for beams of molecular iodine [5, 7] and has been found to be relevant also for a jet of CO in the first excited rotational level [8], for thermal N_2 expanded from a multichannel array [9] and for the molecular ion N_2^+ (in the rotational levels $K=4$ and 10) drifted in He [10].

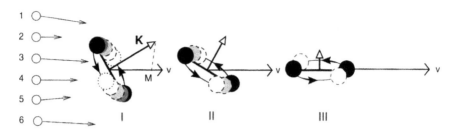

Figure 1: A classical view of acceleration, rotational cooling and alignment in expansions, from high pressure into vacuum, of gaseous mixtures of diatomic molecules diluted in a lighter gas (mono-atomic in this illustration). A diatomic molecule as in I, at the beginning of the expansion, rotates in a plane randomly oriented with respect to the flight direction **v** and experiences during the expansion several collisions by faster atoms of the lighter seeding gas. **K** is the rotational angular momentum, and M its helicity i.e. its projection along **v**. Collisions with impact parameters b of the order of molecular dimension or smaller (e.g. by atoms 3 and 4) lead to acceleration (i.e. exchange of linear momentum and increase in **v**), while those with intermediate b (as by atoms 2 and 5) may also bend the rotational plane, with a net decrease in the helicities M (a bottleneck requiring many collisions, see text); those with large b (as by atoms 1 and 6) will only elastically deflect the molecules, focusing them in the forward direction. After several collisions most molecules (as in II) will have low helicity i.e. will fly edge-on offering target to collisional removal of rotational angular momentum: those which have suffered a sufficiently high number of collisions will be as in III, i.e. fast, very slowly rotating and with a low helicity as can be seen by imagining the (now small) vector **K** distributed uniformly on a plane orthogonal to **v**.

Pullman, Friedrich and Herschbach [11] have extended the study on supersonic beams of I_2 demonstrating the alignment to be facile and large, by seeding the molecules with various gas carriers and probing the fluorescence well downstream. In all of preceding studies, molecules flying edge-on (i.e. rotational angular momentum perpendicular to flow direction) have been found to predominate: Herschbach and co-workers confirmed this finding for iodine but discovered also that, at high source pressure conditions, the sense of rotational alignment can be reversed, favoring broad-side configurations (i.e. rotational angular momentum parallel to flow direction), with a strong dependence on the rotational state probed. This was in line also with a previous report for Na_2 [5]. Later, dependence of alignment on the lateral displacement from the center of a iodine jet has been reported [12].

An extensive investigation of the phenomenon in the early stage of the jet expansion has been more recently carried out for CO_2 [13], confirming some general trends but also showing that some aspects were at variance with previous findings.

From all of these studies emerged the marked influence on the degree of alignment of nozzle geometry and of source pressure conditions. Indeed the collisional alignment was shown to be a strong angular dependent phenomenon, so that making a measurement near the nozzle or well downstream (*i.e.* with a very different sampling angular cone) resulted in different experimental observed behaviors. Also molecules with very different moment of inertia and bond lengths (*i.e.* I_2, Na_2, CO_2) showed a different alignment dependence on the rotational level probed, as an evidence of different propensities in the rotational relaxation.

In 1994 we presented the first evidence of the strong dependence of rotational alignment on final speed v of molecules emerging from supersonic expansions [1] by measuring the variation of paramagnetism of O_2 in continuous seeded beams of molecular oxygen (O_2 is an open-shell species where the electronic spin **S** and the nuclear rotation **K** angular momenta are coupled to give the total angular momentum **J**). Subsequently, measurements of anisotropy effects [14], on collision cross sections for the O_2-Xe system, showed a marked dependence on the velocity of the oxygen molecules within the same supersonic velocity distribution, and confirmed the correlation between molecular alignment and molecular velocity.

Further scattering experiments, performed by using supersonic seeded beams of nitrogen and exploiting anisotropy effects in the measured cross sections, yielded information on the rotational alignment for the case of N_2 and therefore in absence of electronic spin [15]. Comparison of the anisotropy effects measured in the scattering experiments of O_2 and N_2, carried out under similar supersonic seeded expansion conditions, provided further support for the degree of alignment obtained from the paramagnetism measurements, especially those under the extreme velocity and pressure conditions of our experiments.

It is interesting also to note how the measured anisotropy effects are of the same magnitude as in a pioneering work on the scattering of NO, state selected by a hexapolar field and oriented by means of a magnetic field [16]. After our initial reports on this subject, related works using laser probing of N_2^+ drifted in He [17] and of CO in a He seeded supersonic pulsed beam [18] appeared, bringing further evidence of the dramatic dependence of molecular alignment on the final velocity.

3 Quantum Mechanical Analysis

In our work [1] it was shown that O_2 molecules in their rotational ground state $K=1$ where not significantly aligned at low velocities v, while the population of those flying edge-on (*i.e.* with $\mathbf{K} \perp \mathbf{v}$) was found to increase, over those flying broadside ($\mathbf{K} \parallel \mathbf{v}$), with the molecular speed.

A qualitatively similar correlation between degree of alignment and final velocity was found for N_2^+ ($K=15$) ions, drifted in He [10, 17]: for sufficiently low rotational states the alignment can reverse its sign across the velocity profile.

In the case of CO ($J=6$) [18], (for this spin-less molecule J is also the rotational quantum number) molecules with $\mathbf{J} \perp \mathbf{v}$ were found dominant at the lower velocities while those flying broadside (*i.e.* with $\mathbf{J} \parallel \mathbf{v}$) prevailed at intermediate velocities. The maximum polarizations found for O_2 and CO were about the same [1, 18] even if the velocity integrated alignment was lower for the CO case.

The focus on $\mathbf{v}$ as the appropriate quantization axis to describe the polarization arises from the physics of the phenomenon under study - alignment by collision - and from its suitability for perspective applications of scattering experiments with polarized molecules.

The differences in the findings of the various cited experiments have to be interpreted taking into explicit account that each experiment probes a different piece of a complicated mosaic - that of collisional alignment - which is expected to depend on the nature of the investigated system, on the intermolecular potentials and on the number and nature of the collisions which occur during the expansion.

Collisional alignment is the manifestation of hundreds of collisions. The final result depends on propensities of the single collision whose properties are thus fundamental for mechanistic interpretations. A key role is played by the impact parameter, which manifests in the angular dependence of the outcome of collisions.

Recent classical trajectory calculations [19], performed on He–O_2, He–CO and He–CO_2 systems, give a picture of the collisional production of rotationally aligned molecular distributions in a supersonic expansion: this study concerns with the role of the impact parameter, of the collisional energy, of the potential energy surface and of multiple collisions on the final findings. This interesting attempt appears adequate to describe the phenomenon in the first stage, the continuous flow region [12], of the expansion, where excitation and relaxation processes are competitive and a large number of elastic events provide molecular acceleration by momentum transfer. Moreover, in the second stage of the expansion, referred as the transition region [12], collisions become less probable and elastic events still occur but the relative velocity between the carrier and seeded gases is too small to induce rotational excitation: here the relaxation becomes the dominant inelastic process, producing cooling and a further acceleration of the molecules. The behavior of the system is dominated by single collisional events: models based on classical mechanics are not sufficient to obtain a full comprehension of a phenomenon where rotational relaxation plays such a crucial role. This paper focuses on the collisional dynamics which controls this second region of the expansion.

As already authoritatively pointed out [10, 20], close attention must be paid to the quantum differential state-to-state cross sections, which, according to the experiment under consideration, have to be partially or fully integrated, on the angle and on the velocity distribution.

For the atom-molecule collisions of interest here, state-to-state differential cross sections can be classified as *elastic* (when the rotational quantum number K and helicity M are both conserved or when K is conserved but M varies) and *inelastic* (when K varies and M is conserved or when both K and M vary).

Exact quantum mechanical studies of these phenomena are now feasible. Using an accurate interaction potential for O_2-He [21] we have recently performed extensive close-coupling scattering calculations of state-to-state differential cross sections at various collision energies in the range pertinent to the alignment experiments [22].

Some results for O_2 in the rotational state $K=5$ (among the most populated ones in the source, where rotational levels are significantly populated up to $K=21$) and for a center-of-mass collision energy of 6.8 meV (corresponding to a relative velocity of about 600 m/s) are reported in Fig. 2 It is important to note how for high c.m. scattering angles and for particular molecular orientations, the helicity changing and the rotational inelastic cross sections may be comparable to, or even larger than, the pure elastic ones. Further calculations [22] are providing information on dependences on both the collision energies and the involved rotational levels. The motivation of such extensive calculations must be found in the search of propensities and trends in the scattering from an anisotropic potential. Even if the alignment of molecules in a supersonic expansion is the result of many collisions the elucidation of systematic propensities in single collision events may shed light on the alignment mechanisms, explaining the relevant features of the collisional alignment experiments on O_2 molecules [1], particularly the velocity dependence of alignment in supersonic seeded expansions. Moreover the present study can help us to fit into a unified picture other experiments on molecules with mass and rotational level spacings similar to O_2, such as those on seeded CO expansion [18] and drift-tube N_2^+ transport [17].

In our experimental configuration, the beam is analyzed well downstream (about 1 meter after the expansion and with a final defining slit of 0.7 mm). The cone of acceptance is therefore $\leq 10^{-6}$ steradians. This unusually high angular resolution allows the sampling of the molecules that have suffered the largest number of those elastic and inelastic collisions leading to a final

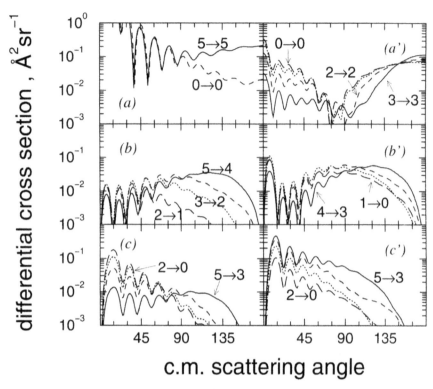

Figure 2: Quantum mechanical state-to-state differential cross sections for the O_2-He system for a center-of-mass collision energy of 6.8 meV and O_2 in the initial rotational level $K=5$. Numbers indicate initial and final helicities M. Panels **a, b, c**: elastic processes (final rotational state, $K=5$). Panels **a', b', c'**: inelastic processes (final rotational state, $K=3$). Among the results not reported, those leading to an increase of helicities are similar to those reported for elastic events but much smaller for inelastic events. Also negligible are events involving rotational and/or helicity jumps larger than ±2.

transversal velocity close to zero.
Small and intermediate angular momentum scattering events enter the laboratory forward component of the differential cross-sections (backward in the center-of-mass, see the figure). Relevant for our experiment are the *inelastic* events, which account for the nearly complete relaxation even from highest populated levels down to $K=1$ [23] (the ground rotational state, which is the one being analyzed). Therefore, these calculations support the cartoon of Fig. 1 (see also [1]): inelastic processes involving small and intermediate

impact parameters are most effective for molecules flying edge-on, rather than for those flying broadside. The latter must bend (see Fig. 2b and 2c) before relaxing or relax but bend at the same time (see Fig. 2b' and Fig. 2c'). Fig. 2 shows that these processes occur sideways in the center of mass and therefore at large impact parameters. For the edge-on flying molecules in high rotational levels, propensity to relax to the ground state, by collisions at small and intermediate impact parameters and involving few helicity changes, is enhanced by the occurrence of a large sequence of inelastic collisions.

This route, which involves large momentum transfer, leads to the observation of high speed, $\mathbf{K} \perp \mathbf{v}$ molecules, in the beam. See also a relevant discussion based on a classical trajectory analysis [20].

In those experiments, which probe not fully relaxed rotational states and which are carried out analyzing a wide molecular beam profile [18, 13], a crucial role is played by the elastic contributions to the differential cross section, to be integrated on the proper range of scattering angles. Indeed for these levels selectivity and role of inelastic collisions are expected to decrease. Molecules as for the example of $K = 5$, when flying broadside may dominate at high final speeds, being accelerated by elastic events in the forward direction for the laboratory frame, or backward in the center of mass (see the case 5→5 in panel a of the figure). We can argue that to this class belong also the molecules which in a drift tube offer the highest resistance to the flow [17].

In conclusion we believe that progress has been made and will be made by explicitly taking into proper account the different experimental observational conditions as a basic requirement to approach a very complex and stimulating problem from complementary viewpoints.

4 Scattering of Aligned Molecules and Measurement of Intermolecular Forces

An important application of molecular beams with controlled rotational alignment is the study of the anisotropy of intermolecular forces [14, 24]. Total integral cross sections for scattering of O_2 on Kr and Xe atoms and on oxygen molecules were obtained with a "hot" effusive molecular beam (which contains fast rotating and randomly oriented O_2 molecules and mainly probe the spherical component of the potential energy surfaces) and with supersonic seeded beams (where molecules are cooled at $K = 1$ rotational level and selectively aligned probe the anisotropy of the potential energy surfaces).

The analysis of the experimental results, based upon close-coupling exact quantum mechanical calculations of the cross sections [14], provides an accurate characterization of the interactions at intermediate and large in-

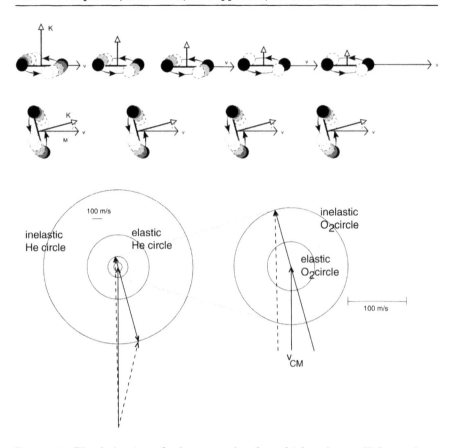

Figure 3: The behavior of *edge-on* molecules which, when collide at short impact parameters, can accelerate and relax because of inelastic events, is compared with the one of *broad-side* targets, which, when do not relax, continue to be accelerated only by elastic collisions. Molecules are backward-scattered in the CM frame which correspond to a forward LAB direction. Typical kinematic conditions for a collisions at $v=342$ m s^{-1} are shown in the Newton diagram. The elastic and inelastic ($K=15 \to 13$) circles are also reported and magnified in the right-hand side for the O_2 case, where the different acceleration due to inelastic and elastic events appears evident.

termolecular distances for the Kr-O_2 and Xe-O_2 systems, the most stable configuration being for perpendicular approach of the rare gas atom, with energies 15.84 meV for Kr and 17.87 meV for Xe, at intermolecular distances of 3.72 and 3.87 Å, respectively [14].
The same technique has been recently applied to the characterization of the $(O_2)_2$ dimer. Analysis of scattering data for the O_2-O_2 system yields a bond

energy in the $(O_2)_2$ dimer of 1.65 kJ·mol^{-1} at an intermolecular distance of 3.56 Å, corresponding to parallel approach [24].
This first complete characterization of the interaction, basic to the theory of weak chemical bonds, and relevant for the dynamics of atmospheric gases, leads also to quantify the (minor) role of spin-spin coupling between isolated molecules, of interest for modeling magnetism in solid O_2 and in O_2 clusters. The splittings among the singlet, the triplet and the quintet surfaces are obtained and a full representation of their angular dependence is reported via a novel harmonic expansion functional form for diatom-diatom interactions [24]. These results indicate that most of the bonding in the dimer comes from van der Waals forces but chemical (spin-spin) contributions in this open-shell–open-shell system are not negligible ($\sim$ 15 % of the van der Waals component of the interaction).

References

[1] V.Aquilanti, D.Ascenzi, D.Cappelletti, F.Pirani, Nature, **371**, 399 (1994); V.Aquilanti, D.Ascenzi, D.Cappelletti, F.Pirani, J.Phys.Chem., **99**, 13620 (1995).

[2] C.J.Gorter, Naturwiss. **26**, 140 (1938).

[3] J.J.M.Beenakker, F.R.McCourt, Ann.Rev.Phys.Chem., **21**, 47 (1970).

[4] M.P.Sinha, C.D.Caldwell, R.N.Zare, J.Chem.Phys., **61**, 491 (1975).

[5] A.G.Visser, J.P.Bekooy, L.K.van der Meij, C.de Vreugd, J.Korving, Chem.Phys., **20**, 391 (1977).

[6] H.G.Rubahn, J.P.Toennies, J.Chem.Phys., **89**, 287 (1988).

[7] W.R.Sanders, J.B.Anderson, J.Phys.Chem., **88**, 4479 (1984).

[8] J.A.Barres, T.E.Gough, Chem.Phys.Lett., **130**, 297 (1986).

[9] H.-P.Neitzke, R.Terlutter, Phys.Rev. B ,**25**, 1931 (1992).

[10] R.A.Dressler, H.Meyer, S.Leone, J.Chem.Phys., **87**, 6029 (1987).

[11] D.P.Pullman, D.R.Herschbach, J.Chem.Phys., **90**, 3881 (1989); D.P. Pullman, B.Friedrich, D.R.Herschbach, J.Chem.Phys., **93**, 3224 (1990); B.Friedrich, D.P.Pullman, D.R.Herschbach, J.Phys.Chem., **95**, 8118 (1991).

[12] H.J.Saleh, A.J.McCaffery, J.Chem.Soc.Faraday Trans., **89**, 3217 (1993).

[13] M.J.Weida, D.J.Nesbitt, J.Chem.Phys., **100**, 6372 (1994).

[14] V.Aquilanti, D.Ascenzi, D.Cappelletti, S.Franceschini, F.Pirani, Phys. Rev.Lett., **74**, 2929 (1995); V.Aquilanti, D.Ascenzi, D.Cappelletti, M.De Castro, F.Pirani, J.Chem.Phys., **109**, 3839 (1998).

[15] V.Aquilanti, D.Ascenzi, D.Cappelletti, R.Fedeli, F.Pirani, J.Phys. Chem. A, **101**, 7648 (1997).

[16] H.H.W.Thuis, S.Stolte, J.Reuss, Chem.Phys., **43**, 351 (1979); H.H.W. Thuis, S.Stolte, J.Reuss, J.J.H.van den Biesen, C.J.N.van den Meijdenberg, C.J.N., *ibid.*, **52**, 211 (1980).

[17] E.B.Anthony, W.Schade, M.J.Bastian, V.M.Bierbaum, S.R.Leone, J.Chem.Phys., **106**, 5413 (1997); E.B.Anthony, Ph.D. Thesis, University of Colorado (Boulder, Colorado, 1998).

[18] S.Harich, A.M.Wodtke, J.Chem.Phys., **107**, 5983 (1997).

[19] J.R.Fair and D.J.Nesbitt, J.Chem.Phys., **111**, 6821 (1999).

[20] D.P.Pullman, B.Friedrich, D.R.Herschbach, J.Phys.Chem., **99**, 7407 (1995).

[21] L.Beneventi, P.Casavecchia, G.G.Volpi, J.Chem.Phys., **85**, 7011 (1986).

[22] V. Aquilanti, D. Ascenzi, D. Cappelletti, F. Pirani, J.Phys.Chem., **103**, 4424 (1999); V. Aquilanti, D. Ascenzi, M. de Castro Vítores, F. Pirani, D. Cappelletti, J.Chem.Phys., **111**, 2620 (1999).

[23] See for instance, N.A.Kuebler, M.B.Robin, J.J.Yang, A.Gedanken, D.R.Herrick, Phys.Rev. A, **38**, 737 (1988); T.Matsumoto, K.Kuwata, Chem.Phys.Lett., **171**, 314 (1990).

[24] V. Aquilanti, D. Ascenzi, M. Bartolomei, D. Cappelletti, S. Cavalli, M. de Castro Vitores, F. Pirani, Phys.Rev.Lett., **82**, 69 (1999); V. Aquilanti, D. Ascenzi, M. Bartolomei, D. Cappelletti, S. Cavalli, M. de Castro Vitores, F. Pirani, J.Am.Chem.Soc., in press (1999).

Nonequilibrium Distributions of Rotational Energies of K_2 Seeded in a Free-Jet of Argon

H. Hulsman
Physics Department, UIA, University of Antwerpen,
B-2610 Wilrijk, Belgium

1 Introduction

In a supersonic free jet the expanding gas cools very fast. As a result, the distribution of molecules among the rotational and vibrational levels develops a non-equilibrium character. Such distributions have been studied earlier by many authors.
The major part of the experimental data has been obtained by studying particles in a molecular beam which had been sampled from a free jet. A large variety of experimental techniques has been used, e.g.: velocity measurements followed by an energy balance [1, 2], laser induced fluorescence measurements [3, 4, 5], infrared absorption spectroscopy [6, 7, 8], Fourier transform emission spectroscopy [9], molecular beam resonance [10], multiphoton ionisation [11], etc. In these cases molecules are studied, which had been sampled from the jet at a large distance from the nozzle exit. Consequently, the results reflect the terminal situation in the jet.
Data on the *evolution* of the internal state distributions in the course of the expansion are more scarce. Rotational distributions in a jet of N_2 have been determined from the spectral analysis of N_2^+ fluorescence, which arises from excitation by an electron beam [12, 13]. Later work, however, has shown [14] that the apparent non-Boltzmann distribution among the rotational levels is (partly or largely) an artifact. Also spontaneous fluorescence measurements on heated CO free jets [15] have explained that some deviations from rotational equilibrium can be due to the penetration of the relatively hot background molecules into the jet. Further, rotational temperatures have been measured in a jet of CO_2 by means of Raman scattering [16] and in various other jets by CARS spectroscopy [17]. In these cases the local densities were relatively high and the reported distributions are Boltzmann distributions. The same conclusions have been obtained by two-foton resonant third harmonic generation in CO free jets [18] as well as in N_2 free jets [19]. Non-Boltzmann rotational distributions in a jet have been observed for CS_2 [20]

and for Na_2 [21] using laser induced fluorescence. All these laser techniques are ideal to probe the free jet, as these are non-intrusive with high spatial resolution and state selective with high spectral resolution.

The analysis of relaxation phenomena in a free jet has been treated in various ways [22]. Two main categories may be distinguished: On the one hand are methods where one assumes the existence of separate rotational, vibrational and translational temperatures, which may differ from each other and from the equilibrium temperature. The relaxation of such temperatures has, e.g., been treated in Refs. [23, 24, 25]. (Such models are known as 'thermal conduction' models). The terminal value of these internal temperatures may also be determined directly using the 'sudden freeze' criterion [4, 24]. Further, the generalized Boltzmann equation has been used [26] to determine the rotational temperatures in a jet together with the 'parallel' and 'perpendicular' temperatures.

On the other hand models have been developed, where the occupations of the internal levels are obtained from the master equation, assuming a model for the state-to-state inelastic cross-sections [2, 27, 28].

The first type of analysis is unsuited for the non-Boltzmann distributions, as observed in molecular beam experiments. It is, however, assumed to be adequate for the first part of the expansion, where $T_{\rm rot}$ and $T_{\rm vib}$ are relatively close to the equilibrium values. The results of the present experiments will show, that this assumption is questionable. We find that the internal state distribution starts deviating from a Boltzmann distribution as soon as the average internal energies differ from the local equilibrium values.

The second type of analysis may be used to obtain information on the state-to-state cross-sections from observed evolutions of the non-Boltzmann distributions in a jet [27, 28]. We hope that the present data will prove to be sufficiently accurate to provide such information.

2 Experimental results

The internal state distribution has been studied for K_2 molecules seeded in argon by means of laser induced fluorescence. The measurements are along the axis of the expansion and extend from the nozzle exit down to a distance where the distribution is essentially 'frozen'. The present experiments are an extension of earlier work on Na_2 under comparable conditions [21]. The motivation for this work is twofold: First, the rotational and vibrational constants of K_2 are significantly smaller then those of Na_2. This means that a larger range of rovibrational levels will be thermally populated, while the relaxation of the internal energies can proceed in smaller quanta. As a result one expects K_2 to be a more classical and more nicely relaxing system. On the other hand it is attractive that for K_2 the results for the largest distances in the jet can be compared with results by Obrebski et al. [29] who studied

II.5 Nonequilibrium Distributions of Rotational Energies in a Free Jet

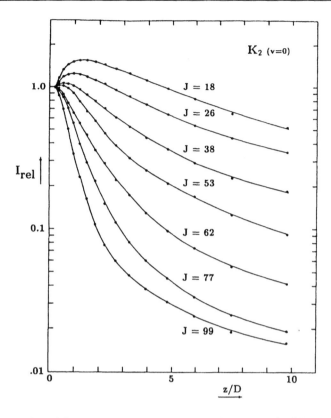

Figure 1: Reduced fluorescence intensities as a function of z/D, the reduced distance from the nozzle exit to the laser beam. Results are for transitions from various rovibrational levels of the electronic ground state of K_2. The stagnation conditions are: $T_0 = 713$ K, $p_0 = 1778$ Pa. The expansion mainly consists of argon (about 80%). The K_2 molecules are only some 0.15% of the mixture, the rest is atomic potassium vapour.

the internal state distributions in a beam. Condensation of potassium in the jet is expected to be negligible under our conditions. [30].

In the experiment we use a single-frequency laser to excite K_2 molecules from a single rovibrational level of the ground state. A narrow beam of laser light intersects the axis of the expansion perpendicularly at a distance z from the exit of the nozzle. When the laser frequency coincides with the central frequency of the transition, the fluorescence mainly originates from a small region around the axis. The intensity of the fluorescence is proportional to the occupation of the rovibrational level at this position. The intensities have been measured as a function of z for a selection of rovibrational levels. Here we will discuss some results for levels in the vibrational ground state

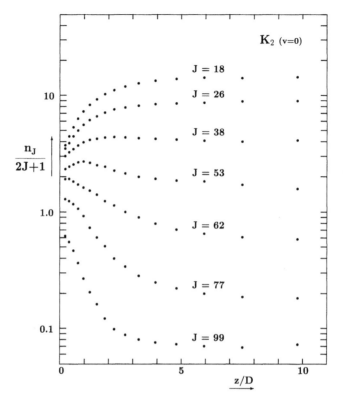

Figure 2: The occupations of various rotational levels along the axis of the expansion as a function of the reduced distance to the nozzle exit.

($v = 0$) with rotational quantum numbers from $J = 18$ to $J = 99$.

The evolutions of the relative intensities as a function of the distance prove to be different for the various levels (Fig. 1): At small distances the intensities for low-energy levels increase, while these decrease for high-energy levels. This reflects the changing occupations of the levels when the gas cools down. At large distances all levels have the same relative decrease of the intensity, which shows that the internal state distribution is frozen.

3 Analysis

In the first part of the expansion, the flow will be close to equilibrium and the local temperatures are fairly well known. In the present experiment the temperature of the nozzle chamber is 713 K. From this, we estimate the local (translational) temperature at the first position of our measurements (at a distance of 0.22 nozzle diameters from the nozzle exit) to be 460 K.

The stagnation pressure is 1778 Pa. The nozzle diameter $D = 515$ μm. Under these conditions the rotational distribution is expected to be close to equilibrium at the position of the first measurement.[1] So, the distribution over the internal states may be estimated for this position in a fairly reliable way.[2] Multiplying the measured ratios of the *intensities* at other positions with the relative populations at $z/D = 0.22$ yields the ratios of the *number densities* of the particles in the various levels at the other positions. The general density of the gas decreases in the course of the expansion. Consequently, these number densities have to be multiplied by a geometrical factor to obtain the *occupations* of the levels which are shown in Fig. 2.

For each position we consider the distribution over the rotational levels (divided by their statistical weights) as a function of the rotational energy. At the first few positions following $z/D = 0.22$ we find that these distributions are Boltzmann distributions to a good approximation: In logarithmic plots like Fig. 3 (Boltzmann plots) one may fit straight lines to the data points. The slopes of these lines give the effective rotational temperatures. These prove to be close to the calculated values of the local translational temperatures. From Fig. 3 one sees, however, that for larger distances the data points no longer fall on straight lines: The levels with higher internal energies are relatively over-populated. (For the largest distance the curvature of the distribution is about the same as that found by Obrebski [29] in his molecular beam experiments). If one tries to determine an effective rotational energy from such distribution by calculating the 'best fitting' straight line, the result depends strongly on the choice of the levels which had been included in the fit.

This is demonstrated in Fig. 4. The ◇ symbols indicate values of the effective temperature which had been obtained using all twelve levels for which data were available. The ▽ symbols were obtained, using only the 4 levels with the lowest rotational energies, while for the △ symbols only the 4 levels with the highest rotational energies have been used. It is apparent that these effective rotational temperatures depend so much on the arbitrary choice of the levels which are considered, that the usefulness may be questioned.

On the other hand, the distributions of Fig. 3 can be used to determine the average rotational energy of the K_2 molecules as a function of the position:

$$E_{\rm rot} = \frac{\Sigma n(J)E(J)}{\Sigma n(J)} \equiv kT_{\rm rot} . \tag{1}$$

In this way one has a definition of $T_{\rm rot}$ which is unequivocal. We have

---

Using the heat-conduction model with a reasonable value for the inelastic cross-section we obtain $T_{\rm rot} = 475\ K$ for the position $z/D = 0.22$.

The results of our analysis do not change, if one varies the assumptions within reasonable limits.

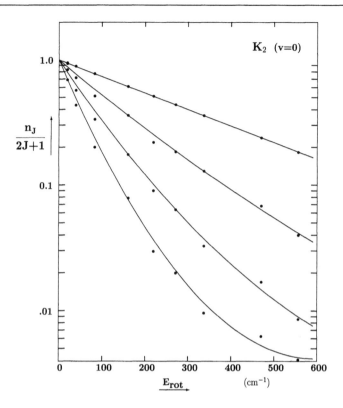

Figure 3: The occupation of rotational levels divided by their statistical weights *vs* the rotational energy for various distances from the nozzle; $z/D = 0.22$, $z/D = 0.98$, $z/D = 2.24$ and $z/D = 9.80$. The lower lines which are more curved, are those for the larger distances. (For easier comparison, the extrapolated occupation of the $J = 0$ level has been put unity).

calculated[3] this rotational temperature from our data. In Fig. 5 its evolution is compared with calculated values for the equilibrium and the translational temperatures. It is seen that it starts deviating very slowly and reaches a limiting value within the region studied.

The behaviour of this experimental $T_{\rm rot}$ may also be compared with results of calculations using the heat-conduction model. It proves that a nearly perfect match may be obtained, using reasonable values for the inelastic cross-section, $(\sigma_{inel} \approx 56 \text{ Å}^2)$.

For the calculation all $n(J)$'s are needed, while the measurements are for a limited number of J's. For the region $J < J_{max}$ we have used a smooth curve fitted to our data. For the extrapolation into the region $J > J_{max}$ we have used a straight line with a slope equal to the slope of the fitted curve in J_{max}. The possible error which is introduced in this way appears to be very small.

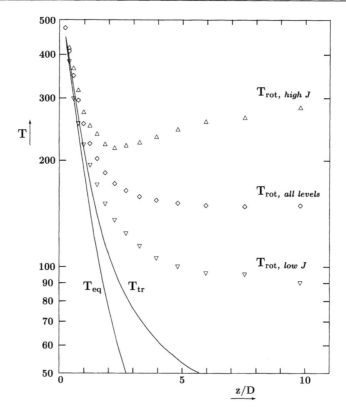

Figure 4: Rotational temperatures for various positions, obtained by fitting Boltzmann distributions to the data (i.e.: straight lines fitted to data points like those of Fig. 3). Different choices of the levels which had been included in the fit: ▽ using 4 levels with rotational energies from 20 to 120 cm^{-1}, △ using 4 levels with rotational energies from 270 to 560 cm^{-1}, ◇ using 12 levels with rotational energies from 20 to 560 cm^{-1}.

4 Conclusion

Generally, the rotational temperature is introduced for two distinctive purposes: Either to describe the distribution of the particles over the internal states, or to describe the energy content of the degree of freedom. For our data on K_2 seeded in argon we find that the relaxation of the energy content can very well be described using the rotational temperature of eq. 1 and a heat-condution model. For the internal state distribution, however, the introduction of T_{rot} appears to be hardly useful.

It is well known, that in molecular beams the internal state distributions are frequently found to be non-Boltzmann. It has, however, been suggested that

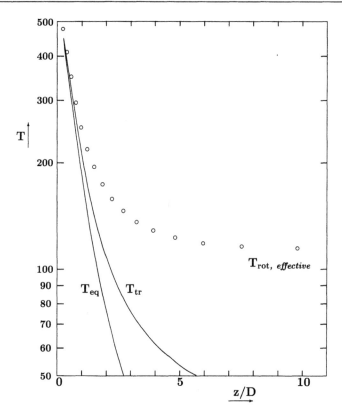

Figure 5: The experimental rotational temperature of eq. 1 as a function of z/D compared to the calculated equilibrium and translational temperatures.

the non-Boltzmann behaviour arises in the last stages of the freezing process and that in earlier stages of the expansion the distribution can be described by a Boltzmann distribution with a characteristic temperature which is different from the local translational temperature. Our data show, that the non-Boltzmann character is already present quite early in the expansion.

In Fig. 6 this is shown some detail: The difference of the two fitted temperatures of Fig. 4, $T_{\rm rot,\ high\ J} - T_{\rm rot,\ low\ J} = \Delta T_{\rm rot}$ is taken as a measure of the uncertainty of the fitted rotational temperatures. The relative importance of this uncertainty is indicated by $\Delta T_{\rm rot}/T_{\rm rot}$, where $T_{\rm rot}$ is the rotational temperature according to eq. 1. In Fig. 6 this relative uncertainty has been plotted vs $T_{\rm rot}$ together with the ratio $T_{\rm rot}/T_{\rm trans}$. One sees, that the uncertainty in the fitted rotational temperature grows roughly in proportion with the deviation of the rotational temperature from the translational temperature. This indicates that the deviations from the equilibrium distribution arise as soon as the energy content of the rotation begins to differ from the

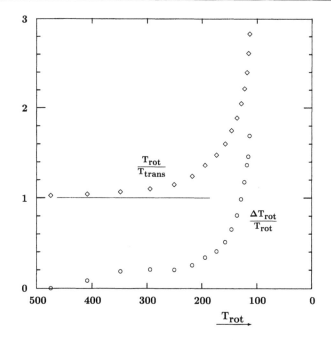

Figure 6: The experimental effective rotational temperature T_{rot} (= the average rotational energy divided by the Boltzmann constant) divided by the calculated translational temperature T_{trans} vs T_{rot} and the uncertainty of the fitted rotational temperature ΔT_{rot} over T_{rot} vs T_{rot}.

energy of the translational degrees of freedom. Therefore one may question the usefulness of the rotational temperature for the description of the internal state distribution in a jet: In our case it proves, that there is never a proper rotational Boltzmann distribution with a characteristic temperature which differs from the local translational temperature.

In the present paper the focus was on a single aspect of our data. A full quantitative discussion and a more thorough analysis of all available data will be published in due course.

Acknowledgment. Expert technical support by Eric De Langhe and Jean-Pierre Huysmans is gratefully acknowledged. We thank Jos Rozema and Randy Brenkers for assistence during the measurements.

References

[1] R.J.Gallagher and J.B.Fenn, J.Chem.Phys. **60**, 3492 (1974).

[2] S.Yamazaki, M.Taki and Y.Fujitani, J.Chem.Phys. **74**, 4476 (1981).

[3] G.M.McClelland, K.L.Saenger, J.J.Valentini and D.R.Herschbach, J.Phys. Chem. **83**, 947 (1979).

[4] G.Gundlach, E.L.Knuth, H.-G.Rubahn and J.P.Toennies, Chem.Phys. **124**, 131 (1988).

[5] P.Zalicki, N.Billy, G.Gouédard and J.Vigué, J.Chem.Phys. **99**, 6436 (1993).

[6] D.Bassi, A.Boschetti, S.Marchetti, G.Scoles and M.Zen, J.Chem.Phys. **74**, 2221 (1981).

[7] T.E.Gough and R.E.Miller, J.Chem.Phys. **78**, 4486 (1983).

[8] N.Dam, C.Liedenbaum, S.Stolte and J.Reuss, Chem.Phys.Lett. **136**, 73 (1987).

[9] S.P.Venkateshan, S.B.Ryali and J.B.Fenn, J.Chem.Phys. **77**, 2599 (1982).

[10] W.L.Meerts, G.Ter Horst, J.M.L.J.Reinartz and A.Dymanus, Chem.Phys. **35**, 253 (1978).

[11] H.Zacharias, M.M.T.Loy, P.A.Roland and A.S.Sudbo, J.Chem.Phys. **81**, 3148 (1984).

[12] P.V.Marrone, Phys.Fluids **10**, 521 (1967).

[13] B.N.Borzenko, N.V.Karelov, A.K.Rebrov and R.G.Sharafutdinov, J.Appl. Mech.Techn.Phys. **17**, 615 (1977).

[14] D.Coe, F.Robben, L.Talbot and R.Cattolica, Phys.Fluids **23**, 706 (1980).

[15] M.A.Gaveau, J.Rousseau, A.Lebehot, R.Campargue and J.P.Martin, in: Rarefied Gas Dynamics (Plenum, New York, 1984) Vol. II, p. 887.

[16] I.F.Silvera and F.Tommasini, Phys.Rev.Lett. **37**, 136 (1976).

[17] P.Huber-Wälchli and J.W.Nibler, J.Chem.Phys. **76**, 273 (1982).

[18] F.Aguillon, A.Lebéhot, J.Rousseau and R.Campargue, J.Chem.Phys. **86**, 5246 (1987).

[19] O.Faucher, F.Aguillon and R.Campargue, J.Chem.Phys. **94**, 4141 (1991).

[20] S.Liu, Q.Zhang, C.Chen, Z.Zhang, J.Dai and X.Ma, J.Chem.Phys. **102**, 3617 (1995).

[21] H.Hulsman, Chem.Phys. **217**, 107 (1997).

[22] R.Campargue, M.A.Gaveau and A.Lebéhot, in *Rarefied Gas Dynamics*, edited by H.Oguchi (University of Tokio Press, Tokio, 1984), Vol II, p551.

[23] R.J.Gallagher and J.B.Fenn, J.Chem.Phys. **60**, 3487 (1974).

[24] C.G.M.Quah, Chem.Phys.Lett. **63**, 141 (1979).

[25] C.E.Klots, J.Chem.Phys. **72**, 192 (1980).

[26] L.K.Randeniya and M.A.Smith, J.Chem.Phys. **93**, 661 (1990).

[27] K.Koura, Phys.Fluids **24**, 401 (1981).

[28] K.Koura, J.Chem.Phys. **77**, 5141 (1982).

[29] A.Obrebski, T.Kaps and U.Cerny, Chem.Phys. **212**, 311 (1996).

[30] A.Obrebski, U.Cerny and T.Kaps J.Chem.Phys. **104**, 1575 (1996).

Rotational and Vibrational Relaxation of Hydrides in Free Jets: HBr and OH

A.E. Belikov[1], M.M. Ahern[2], M.A. Smith[2]

[1] Institute of Thermophysics Russian Academy of Sciences,
Novosibirsk, Russia
[2] Department of Chemistry University of Arizona, Tucson,
AZ, USA

1 Introduction

The study of the rotational and vibrational relaxation of a molecules in a free jet expansion is of interest for some applications: for example, in aerodynamics [1], in crossed-beam experiments regarding rotationally inelastic cross sections [2], and in studies of state-selected chemical and physical reactions within free jets [3], In addition, data on the rates of rotational-translational (RT) energy transfer at the very low collision energies realized in free jets provide a probe of the anisotropy of the intermolecular potential [4], Such studies are particularly sensitive to the long-range attractive portion of the potential which is difficult to investigate at higher collisional energies.

Recent progress in spectroscopic techniques has made possible the measurement of individual rotational level populations allowing general and detailed interpretation of the relaxation. The most accurate description of the relaxation process, under the nonequilibrium distributions which exist both over rotational and translational degrees of freedom in a free jet call for the use of the Waldmann–Snider or Wang–Chang–Uhlenbeck approximations to the generalized Boltzmann equation for the determination of the distribution function, $f_j(v, r, t)$ [5]. Solving these integro-differential equations is a very complicated problem [6].

Using a master equation approach one can numerically solve the set of differential rate equations for the relaxation given the full set of energy dependent rotational relaxation cross sections, $\sigma_{jj'}(E)$ and some model for the collision energy function [7]. Theoretic calculations of the cross sections $\sigma_{jj'}(E)$ run into two difficulties. The principal one is related to the absence of firm data regarding the intermolecular potentials. The second, a technical difficulty, arises from the need to perform very time-consuming calculations for an accurate solution of the scattering problem. To critically evaluate a range of intermolecular interaction models, approximate methods can be useful but need testing as well. Experimental data on relaxation is then quite valuable for addressing this set of coupled problems.

In this work a study was made of the terminal rotational states of HBr in free jets N_2 under varying conditions of P_0D. In a previous work, we have analyzed the HBr/Ar relaxation in terms of a power law relaxation model [8]. Here we present the new expanded data set including N_2 relaxation, and compare the differences in HBr relaxation between these two different expansions. These state-selective results provide valuable data for the study of temperature dependent relaxation cross sections down to the extremely low temperatures within the jet. They also demonstrate different thermal conditions between N_2 and Ar jets possibly inducted by condensation enthalpy in the latter for high values of the stagnation density.

The OH radical is a key intermediate in a wide variety of chemical, atmospheric and astrophysical systems. The inelastic and chemical behavior of OH is important in such diverse environments as high temperature flames and low temperature planetary atmospheres. As a result, the temperature dependence of its inelastic processes over wide ranges of temperature is of both applied and fundamental importance. We have recently reported some aspects of the chemical behavior of OH in the low temperature realm [9]. There have been numerous studies of OH relaxation near 300 K, however, information at low temperatures is lacking [10]. In a free jet, we have observed argon to be a an effective relaxant for both rotation and vibration [11]. As a result, we explore the relaxation efficiency of Ar with the polar OH system to determine the generality of inverse temperature dependence of inelastic processes in the extremes of low temperature.

This study involves the measurement of energy transfer processes occurring within the $A^2\Sigma$ state of the OH molecule. We report the low temperature ($T_{\text{trans}} \approx 5$ K) rate coefficients for vibration/rotation exchange within this manifold upon low temperature single collisions with an Ar buffer. A more thorough study of both the A and X state rate processes for OH and the isotopomer, OD, is in progress [12].

2 Experimental

Our supersonic jet apparatus and diagnostic equipment is a modification of the apparatus used in earlier studies reported from this laboratory for the study of the ion-molecule reactions [13]. The rotational level populations of HBr molecules were determined by REMPI spectroscopy of the HBr using a $(2+1)$ process resonant through the $E\ ^1\Sigma^+(0-0)$ transition at excitation wavelengths near 256.6 nm. An example REMPI spectrum obtained under these conditions is shown in Fig. 1 along with a static 300 K spectrum. From the REMPI spectra obtained at various distances from the nozzle orifice, rotational line intensities for specific j states were obtained. To convert from REMPI line intensities to rotational populations requires the use of the two-photon absorption line strength factors. The complete analysis of population determination and line strength is given in [8].

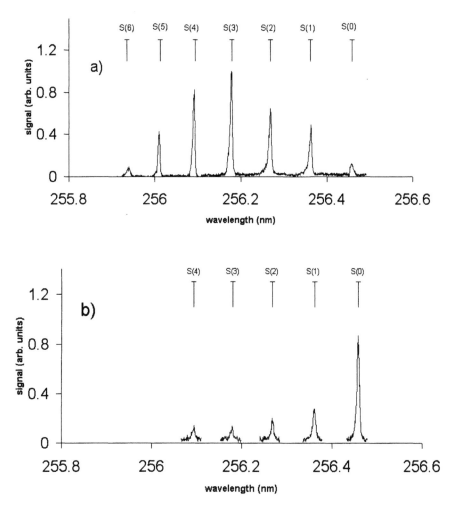

Figure 1: Top: (2 + 1) REMPI spectrum of the S branch of the E $^1\Sigma^+$ state of HBr at 0.5 percent seeded in Ar buffer expanded through a 0.04 cm diameter pulse nozzle orifice with stagnation pressure and temperature of 103 Torr and 298 K, respectively. In the lower spectrum, only the lines were scanned to optimize signal-to-noise for reasonable scan times

For production of jet cooled OH, an expansion of 0.3% water vapor in Ar ($P_0 = 900$ Torr, $d = 0.4$ mm) was expanded through an electrically grounded plate (thickness = 0.3 mm) with a 1 mm hole positioned at a downstream distance of $z/d = 4$. Another plate with a 4 mm hole floated at -600 V was positioned at $z/d = 8$. Self-induced breakdown across the plates generated the OH radicals in the supersonic flow. Since we sample the OH, by laser induced fluorescence, at distances greater than $z/d = 25$, we feel the discharge plates do not present a significant flow perturbation to the central 15 degree sampling cone. This supposition is supported by the fact that the ratio of the mean free path to the aperture diameter (an effective Knudsen number in this case) at the position of each of the apertures is approximately 0.01. This indicates an effective continuum flow regime where the aperture presence would not be expected to seriously effect the central flow streamlines at radial distances less than 10 percent of the aperture diameter. In addition, it indicates that the radicals generated in the plasma contained between the plates undergo many collisions in the subsequent expansion. This argument is further supported by the extremely low internal temperature observed for the OH radical ($T < 15$ K).

The OH was laser excited using a Nd:YAG pumped dye laser frequency doubled to excite the $A^2\Sigma$ state (for example using the $Q_1 + Q_21(1)$ line wavelengths of 307.935 (0-0), 281.997 (1-0) and 288.314 (2-1) nm respectively). Dispersed fluorescence was measured perpendicular to the laser beam. The relaxation rate coefficients are calculated by determining relative populations in different v', N' states through detection of OH fluorescence of A states directly pumped as well as states populated through collisional transfer. An example of of the dispersed spectrum of partially relaxed OH sample is shown in Fig. 2.

Thus as one probes farther down the jet axis, the number of collisions occurring during the average fluorescence lifetime decreases as $1/z^2$. In this way, the fluorescence lifetime serves as the reaction clock for a sample of known density [11].

3 Results and Discussion

3.1 HBr Rotational Relaxation

The rotational energy distributions of HBr molecules were measured on the central axis of the free jets. All of these studies were carried out at a stagnation temperature $T_0 = 295$ K, with two nozzles of diameters $D = 0.36$ and 0.5 mm. Stagnation conditions were varied in the P_0D range from 70 to 550 Torr mm. Since rotational relaxation is a binary collision process, it is expected to scale with P_0D, which is proportional to the inverse nozzle Knudsen number when the nozzle temperature is kept constant [14]. All

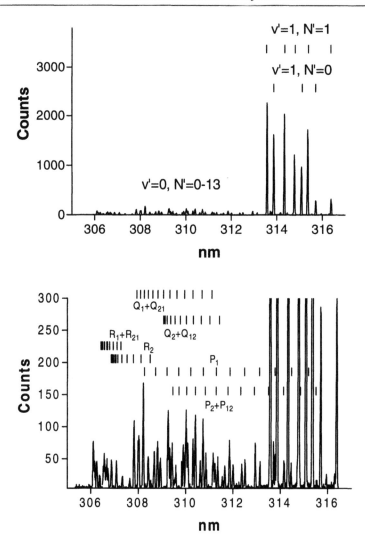

Figure 2: Dispersed fluorescence of the OH X-A transition following excitation of the $Q_1 1(1-0)$ line at a position $z/d = 38$ in an Ar free jet of stagnation pressure 900 Torr and $d = 0.4$ mm. Upper: Low amplification spectrum highlighting rotational relaxation from $N' = 1$ to $N' = 0$ within the OHA($v' = 1$) manifold. Lower: Expanded scale of the same spectrum highlighting emission from OHA($v' = 0, N' = 0 - 13$) induced by vibrational relaxation caused by OHA($v' = 1, N' = 1$) - Ar collisions

measurements are carried out in the far field of flow at distances from the nozzle of $z/D > 80$. It is generally known that as the gas flow moves downstream of the nozzle exit rotational level populations can deviate from equilibrium as the gas generally cools [5]. Finally, as the collision frequency lowers to a point commensurate with the nozzle flow time, the rotational distribution will cease to change ("freeze"). The distance from the nozzle where this occurs, depends on the P_0, (through P_0D) and the efficiency of rotational-translational energy transfer. All of our results are obtained in the area of the flow where the rotational level populations are essentially frozen and so the precise axial position of the measurement is not critically important. For each value of (P_0D) a terminal rotational energy distribution was determined. These energy distributions for all N_2 jet values probed are shown in Fig. 3. Examples of rotational energy distributions of HBr measured in the free jets for two particular expansions are presented in Fig. 4. Distributions similar to these (e.g. indicating strong overpopulation of higher rotational levels for Ar expansions) were observed in all other P_0D values probed (50–550 Torr-mm). We can introduce the notion of a relative population temperature, defined by the slope of the straight line connecting two nearby points in a Boltzmann plot as in Fig. 4. These temperatures increase with rotational quantum number, reaching at $J = 4$ a temperature close to the magnitude in the sonic section of the nozzle ($T_{ex} = 0.75\,T_0$ or 220 K [15] for the heat capacity ratio $= C_p/C_v = 5/3$), which hereinafter does not change. Such behavior is observed for Ar meaning the populations $N_j > 4$ have been frozen already near the exit of the nozzle.

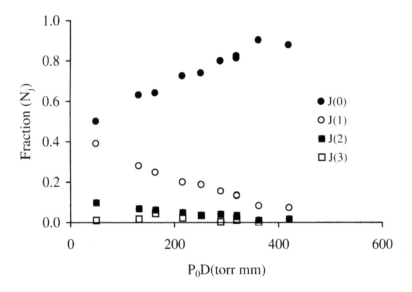

Figure 3: Terminal rotational state distributions for HBr in N_2 as a function of P_0D

This is not surprising, considering the great energy gap between rotational levels of the HBr molecule, which equals 120 K for $j = 5$. What is somewhat more surprising is the fact that this behavior is not seen for N_2 buffers even at the highest observed J values. This would appear to imply that molecular nitrogen is a much greater relaxant even for large energy gaps, but may also be influenced in the case of Ar by Ar condensation enthalpy at the stagnation pressures employed. Such behavior in the case of N_2 might be a result of either the increased anisotropy of the HBr-N_2 interaction potential or the fact that N_2 possesses its own internal rotational bath and in this way can more readily transfer/accept angular momentum through resonant or near-resonant R-R relaxation. The effective temperature observed for HBr is close to that expected of the N_2 buffer itself from both experimental observation and modeling of N_2 rotational relaxation in free jets [15, 16].

The HBr appears to effectively reach equilibrium with the rotational bath of the N_2 buffer prior to any collisional quitting surface for essentially all P_0D values probed in this study. It follows from these results, that information about rotational relaxation at the lowest temperatures could be determined from kinetics of the lowest rotational states in the jet. It should also be noted, that the method of finding the bulk rotational relaxation time is not applicable to the present case, since we are able to measure the populations of the first six lowest levels only. This is not sufficient to obtain a statistically significant average rotational energy of molecules because the sum defined by this average is known to converge only slowly to the limit. For distributions like those in Fig. 4 we need populations up to $j = 12 - 15$ to reach the mean rotational energy with a precision of 5 percent.

In other words, due to the large rotational constant, B, of HBr and therefore the obvious discrete nature of its rotational manifold, especially at prevalent at low temperature, the application of bulk relaxation models is not appropriate. This is particularly true for HBr in Ar where the distribution is non-equilibrium at all pressures, but less so for HBr in the highest density N_2 free jets where the approach to Boltzmann rotational distributions is observed. We have successfully employed a power law model for the temperature dependent $j - j'$ cross sections in HBr-Ar which explains both the non-Maxwellian nature and the higher energy distributions [12], Such a model has not yet been fit to the N_2 data. Such a power law model, with the further simplification of the use of a quitting surface approximation, has been very successful in explaining N_2 rotational relaxation in pure N_2 free jets [16]. It will be interesting to see in the future if a similar model can be used to determine the magnitude and temperature dependence of the HBr-N_2 relaxation cross sections, $\sigma_{jj'}$.

The HBr shows significantly higher energy distributions in the Ar jets with considerable non-Maxwellian character. At nearly all values of P_0D, the HBr distributions in N_2 display a highly Maxwellian character and distribution which appear close to those expected for N_2. It appears that N_2

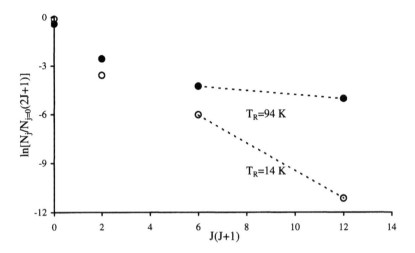

Figure 4: Terminal rotational distributions of HBr in HBr + Ar (solid circles) and HBr + N_2 (open circles) free jets. The value of P_0D is 360 Torr mm ($D = 0.4$ mm) for both jets. Also shown are the asymptotic high N value thermal slopes

is a very effective rotational relaxant over the thermal conditions where collisions remain active in the free jet and is capable of bringing the HBr into near rotational equilibrium.

3.2 OH Relaxation in Ar Free Jets

3.2.1 Rotational relaxation

The rotational relaxation rate coefficients for the OH $A^2\Sigma$ $v' = 0, 1$ and 2 states were measured. The OH $A^2\Sigma$ was prepared by the laser initially in a single N level. The fluorescence from both this initial level and any transfer population in lower N' levels due to Ar collision was monitored through the monochromator by scanning. At close distances from the nozzle, the higher density leads to greater relaxation during fluorescence, causing N' levels emission to grow in. Multiple scans at varying nozzle distances allow a complete progression of relaxation to be observed as a function of jet density. Rate coefficients for population of discrete N' levels are obtained from the integrated rate laws appropriate to relaxation in the free jet [11]. Using the relative product emission intensities extrapolated to far nozzle distance (single collision conditions) allows us to obtain the detailed state-to-state relaxation rate coefficients reported in Table 1.

As can be seen, all rotational transfer rate coefficients are quite large at the very low temperatures present in the free jet. In comparison with the only values reported for 300 K, the rates for OHA$^2\Sigma$(v = 0, N = 2) going to both N' = 1 and 0 show a significant increase of nearly 100% under the jet conditions.

Table 1: Relaxation rate coefficients for OH A$^2\Sigma$ in collisions with Ar (10^{-11} cm^3 s^{-1})

Rotational relaxation from:	T = 300 K	Free Jet
v' = 2, N' = 2		20 ± 4
v' = 2, N' = 1		13 ± 4.5
v' = 1, N' = 2		11 ± 2
v' = 1, N' = 1		6.7 ± 1
v' = 0, N' = 2		$k_{TOT} = 18 \pm 2$
		$k_{2-1} = 12 \pm 1$
		$k_{2-0} = 6.0 \pm 1$
v' = 0, N' = 1		6.6 ± 1
Vibrational relaxation from:		
v' = 2		5.7 ± 2.6
v' = 1	0.41 ± 0.03[a]	3.2 ± 0.8

[a] [19]

3.2.2 Vibrational relaxation

We report vibrational relaxation from the v = 2 to v = 1 and v = 1 to v = 0 A$^2\Sigma$ levels. Comparison of product branching into particular N' levels allows, in principle, conversion of the total vibrational relaxation rate coefficient (summed over all N' levels) to be converted to v, N to v', N' state specific rate coefficients [11]. The total relaxation rates are shown in Table 1. The rates appear to scale with vibration quantum number, v, and appear to be independent of the rotational level pumped. In addition, all rates were observed to be independent, within experimental error, of free jet stagnation pressure from 500 to 1200 Torr (0.4 mm diameter nozzle). Since the degree of Ar clustering should vary considerably in this range, we argue that the bulk of the observed relaxation is due to collisions between OH and Ar monomers.

From the observed rotational temperatures in the OH X state, it appears that rotational relaxation in this manifold is extremely fast, as would be expected from consideration of rotational cooling of a wide variety of molecules in Ar free jets. In addition, the absence of any observable OHX$^2\Pi_{1/2}$(v = 0) in our jets indicates that this low energy excited state relaxes rapidly under the jet conditions. A similar result for the fine structure excited state of NO is well known [17]. It is interesting to note however that fine structure relaxation of the very strong collisions of Xe$^+$ ^{2}P$_{1/2}$ with a variety of rare gases, including Ar, showed no evidence of relaxation in the 1 K limit [18]. Vibrational relaxation of OHX$^2\Pi_{3/2}$(v = 1) by Ar has been found to remain extremely slow at free jet temperatures [11]. This is consistent with the known poor efficiency of Ar as a vibrational relaxant at higher temperatures, but is in contrast to studies of NO$^+$(v = 1) relaxation by Ar near 1 K [17]. In this latter study, Ar was found to relax NO$^+$(v = 1) with near unit efficiency. In the case of OHX$^2\Pi$(v = 1), it would appear that the complex lifetimes remain sufficiently short as to preclude vibrational predissociation of the complex prior to simple adiabatic back-dissociation along the incoming potential surface.

The case of low temperature OH A$^2\Sigma$ rotation/vibration relaxation appears quite different. In the A state, vibrational relaxation by Ar appears to be strongly competitive with rotation, both occurring near the collision limit. The collision rate coefficient can be well approximated using a capture model for complex formation. Employing a two term long range interaction potential including both dipole-induced dipole and simple dispersion terms, one can derive a simple expression for the capture rate coefficient. Using a calculated value (DFT) for the polarizability of the OH A state of 6.9×10^{-31} m^3 and experimental values for all other quantities, we obtain an OH A state-Ar collision capture rate coefficient value near 1.5×10^{-10} cm^3 s^{-1} at 5 K. This value compares favorably with the total relaxation rate coefficients listed in Table 1.

It appears that energy transfer in the OH A state by Ar occurs very efficiently in the limit of low temperature. In the case of the v = 1 state, pure rotational transfer proceeds roughly twice as fast as vibrational transfer. Overall, increases of vibrational relaxation in excess of an order of magnitude can be expected when going from 300 K to temperatures below 10 K. The OH A state-Ar interaction potential at close range has been well determined and is sufficiently shallow (750 cm^{-1}) as to not support vibrational quanta above v = 0 in the OH stretch [20]. The strong inverse temperature dependence observed in vibrational relaxation of free OH is certainly consistent with a complex vibrational predissociation mechanism.

Acknowledgement

This material is based upon work supported by the U.S. Civilian Research and Development Foundation under Award No. RC1-201 and the National Science Foundation under Grant No. CHE 96-23272 and 97-09656.

References

[1] Cattolica R., Robben F., Talbot L., Willis D.R., Phys. Fluids, **17**, 1793 (1974); Boyd I.D., Phys. Fluids, **A2**, 447 (1990).

[2] Faubel M., Kohl K.H., Toennies J.P., J. Chem. Phys., **73**, 2506 (1980); Faubel M.,Kohl K.H., Toennies J.P., Tang K.T., Yung Y.Y., Faraday Discuss. Chem. Sec., **73**, 205 (1982).

[3] Smith M. in: Unimolecular and Bimolecular Reaction Dynamics, eds. Ng C.Y.,Baer T., Fowls I., (John Wiley and Sons, Ltd., New York, 1994), p. 183.

[4] Buck U., Meyer H., LeRoy R.J., J. Chem. Phys. **80**, 5589 (1984); Gianturco F.A., Venanzi M., Faubel M., J. Chem. Phys., **90**, 2639 (1989); Belikov A.E., Sharafutdinov R.G., Chem. Phys. Lett., **241**, 209 (1995).

[5] Marrone P.V., Phys. Fluids, **9**, 521 (1967); Borzenko B.N., Karelov N., Rebrov A.K., Sharafutdinov R.G., J. Appl. Mech. Tech. Phys., **17**, 615 (1976). Anderson J.B. in: Molecular Beams and Low Density Gas Dynamics, ed. Wegner P.P. (Dekker, N.Y., 1974), p. 1; Miller D.R. in: Atomic and Molecular Beam Methods, ed. by Scoles G. (Oxford, N.Y., 1988), Hirschfelder J.O., Curtiss C.F., Bird R.B., Molecular Theory of Gases and Liquids (Wiley, N.Y., 1954).

[6] Toennies J.P., Winkleman K., J. Chem. Phys., **66**, 3965 (1977). Randeniya L.K., Smith M.A., J. Chem. Phys., **93**, 661 (1990). Mazely T.L., Bristow G.R., Smith M.A., J. Chem. Phys., **103**, 8638 (1995).

[7] Koura K., Phys. Fluids, **24**, 401 (1981).

[8] Belikov A.E., Ahern M.M., Smith M.A., Chem. Phys., **234**, 195 (1998).

[9] Atkinson D.B., Jaramillo V.I., Smith M.A., J. Phys. Chem. A, **101**, 3356 (1997). Atkinson D.B., Smith M.A., Rev. Sci. Instrum., **66**, 4434 (1995). Atkinson D.B., Smith M.A., J. Phys. Chem, **98**, 5797 (1994).

[10] Jog A., Meier U., Kohse-Hoinghaus K., J. Chem. Phys., **93**, 6453 (1990). Smith I.W.M., J. Chem. Faraday Trans. 2, **81**, 1849 (1985). Raiche G.A., Jeffries J.B., Crosley D.R., J. Chem. Phys., **92**, 7258 (1990).

[11] Ahern M.M., Smith M.A., J. Chem. Phys., **110**, 8555 (1999).

[12] Steinhurst D.A., Ahern M.M., Smith M.A., J. Chem. Phys., submitted.

[13] Smith M.A., Hawley M. in: Advances in Gas Phase Ion Chemistry, eds. Adams N.G., Babcock L.M., (JAI Press, Greenwich, 1992), p. 167; Hawley M. Mazely T.L., Randeniya L.K., Smith R.S. Zeng X.K., Smith M.A., Int. J. Mass Spectrom. Ion Proc., **80**, 239 (1990).

[14] Miller D.R., Andres R.P., J. Chem. Phys., **46**, 3418 (1967); Poulsen P., Miller D.R. in: Rarefied Gas Dynamics (AIAA, N.Y., 1977), p. 899.

[15] Ashkenas H., Sherman F. in: Rarefied Gas Dyn

[16] Mazely T.L., Smith M.A., J. Phys. Chem, **94**, 6930 (1990). Abad L., Bermejo D., Herrero V.J., Santos J., Tanarro L., J. Phys. Chem. A, **101**, 9276 (1997).

[17] Hawley M., Smith M.A., J. Chem. Phys., **95**, 8662 (1991).

[18] Latimer D.R., Smith M.A., J. Chem. Phys., **101**, 3852 (1994).

[19] Lengel R.K., Crosley D.R., Chem. Phys. Lett., **32**, 261 (1975).

[20] Lester M.I., Loomis R.A., Giancarlo L.C., Chakravarty C., Clary D.C., J. Chem. Phys., **98**, 9320 (1993).

Raman Studies of Free Jet Expansion (Diagnostics and Mapping)

S. Montero, B. Maté, G. Tejeda, J. M. Fernández, A. Ramos
Instituto de Estructura de la Materia, CSIC.
Madrid, Spain

1 Introduction

Raman and Rayleigh diagnostics of free jets have been proposed long ago, but practical applications have been scarce so far [1-5]. The low density of free jets and the intrinsic small cross section of the Raman process contribute multiplicatively, yielding a very weak Raman intensity which is not sufficient for routine work with conventional photomultiplier tubes (PMT's). This severe limitation was overcome in part by means of sophisticated intracavity setups.

Recent instrumental developments, based on a new generation of light detectors (CCD's), along with improvements derived from computer controlled spectrometers, have increased the overall sensitivity of Raman spectroscopy by about two orders of magnitude with respect to photon counting with PMT's [6]. Such gain permits now full exploitation of the appealing posibilities of Raman spectroscopy for jet diagnostics, namely, a) universality, in the sense that all molecular species are detectable, b) spatial resolution of a few μm, c) linear relation between measured intensities and local molecular densities, d) large dynamical range of intensities (over six orders of magnitude), e) wide spectral range (typically from 1 to 6000 cm^{-1}), and f) long term stability (several hours). These capabilities confer Raman spectroscopy great potential for quantitative studies of free jet expansions. In addition to Raman data, Rayleigh data recorded with the same instrumentation complement the information about condensation in the jet.

2 Instrumentation

All examples shown below are based on Raman or Rayleigh spectral data recorded with the non-commercial spectrometer commissioned in our laboratory. This instrument is equipped with a 2360 lines/mm holographic grating as dispersive element, plus a 512×512 pixel CCD detector refrigerated by liquid nitrogen. Scanning and data acquisition are computer controlled. For a laser excitation power of 2 W at 514.5 nm the routine detection limit of the instrument is about 2×10^{13} molecules/cm^3 at a resolution of 1 cm^{-1}. Molecular densities one order of magnitude below that figure of merit can still be measured under nonroutine conditions. A description of this spectrometer has been given in [6, 7].

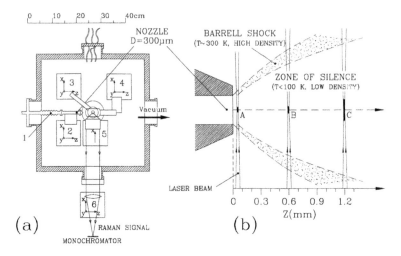

Figure 1: **(a)** Expansion chamber for Raman spectroscopy of jets, **(b)** Recording conditions at $z = 0.05$ mm (A), $z = 0.6$ mm (B), and $z = 1.2$ mm (C) from nozzle.

Raman spectroscopy of jets requires a suited expansion chamber. The one used here, of $42 \times 42 \times 30$ cm^3, manufactured in aluminum, is shown in Fig. 1a. This chamber operates in the stationary flow regime, and is evacuated by a 400 m^3/h Roots pump backed by a rotary pump of 70 m^3/h. A new Roots pump of 1430 m^3/h is being installed. Residual pressures for the jets of the examples discussed below are on the order of $P_r = 40$ Pa.

The nozzle is connected to a gas feeding line (1), and can be moved along the three orthogonal directions of space by remote-controlled microactuators (2). The absolute accuracy of the nozzle position is about ± 10 μm, but relative motions are accurate to ± 1 μm. The dimensions of the flow field covered with present setup are 24 mm (axial), and about 6 mm (radial).

The exciting laser beam, perpendicular to the plane of Fig. 1a, is sharply focused by means of a lens of focal length $f = 35$ mm onto the point (x, y, z) of the jet. The laser beam is multipassed through the focal point in order to increase the Raman signal, which is collected by a composite optical system of aperture $f : 0.95$ and magnification $\times 10$. The optical elements inside the expansion chamber are also mounted on remote–controlled microactuators: for the laser beam (3), for the multipass system (4), and for the collection optics (5), as shown in Fig. 1a.

The focus of the laser beam is a hyperboloid of waist 14 μm. The spatial resolution in the axial direction of the expansion varies between 14 μm and 5 μm, depending on the slit setting of the spectrometer. The spatial resolution in the radial direction is controlled with the readout of the CCD detector by selecting the signal from a narrow track of pixels. Fig. 1b is depicted at scale, to show the actual spatial resolution. At point A, as close to the nozzle as $0.16\ D$, ($D = 300\ \mu\text{m} =$ nozzle diameter) a spatial resolution of 50 μm (radial) and 5 μm (axial) can be attained. Because of the fast decrease of density along the expansion, shown in Fig. 2, and consequently of the Raman signal, such good spatial resolution is only possible in the first two nozzle diameters. At points like C, and beyond, in Fig. 1b a Raman signal-to-noise good enough for quantitative work can only be attained by integration of the signal from a broader track of pixels in the CCD detector. This implies a larger scattering volume 'seen' by the spectrometer and, consequently, a reduction of the spatial resolution along the radial direction of the expansion. In any case, the spatial resolution is good enough to probe the central line of the expansion with no interference from the much warmer and denser barrel shock. This peculiarity confers Raman spectroscopy the capability of studying quantitatively the region very near to the nozzle, hardly accessible to other spectroscopic techniques.

3 Quantitative Measurements

The physical properties of a supersonic jet differ markedly from those of a gas under equilibrium conditions. These properties depend strongly on the stagnation pressure and temperature, on the size and shape of the nozzle, on the residual pressure, and on the composition of the expanded gas. Hence, for most applications it is useful to have a diagnosis of its local properties: Molecular number density, vibrational, rotational and translational temperatures, flow velocity, aggregation state, and geometrical configuration of confining shock waves.

Absolute molecular densities, vibrational and rotational temperatures, and geometrical configuration of confining shock waves are directly obtainable from Raman data, as shown below. Translational temperatures, flow velocities and condensation energy can be deduced from the former data through the principles of conservation of momentum and energy, in combination with the relaxation regime in the jet [8]. Additional information about the aggregation state is obtained from Rayleigh scattering intensities.

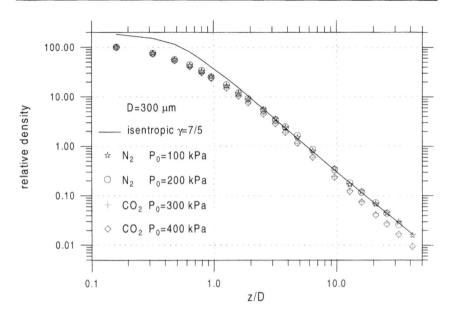

Figure 2: Relative densities in jets of N_2 and CO_2.

Molecular densities are usually measured from the variation of intensity of a strong vibrational band along the jet. An example of this sort of measurement is shown in Fig. 2. Axial relative densities in jets of N_2 expanded under stagnation pressure $P_0 = 100$ and 200 kPa (no condensation), and of CO_2 under $P_0 = 300$ and 400 kPa (strong condensation) are depicted normalized at the first measurement point, $z/D = 0.16$. Before condensation of CO_2 starts ($z/D < 1$) all four expansions are indistinguishable, but the trend of the density ρ versus z/D differs from that of the isentropic relation

$$\rho \propto (1 + (\gamma - 1)M^2/2)^{1/(1-\gamma)},$$

for $\gamma = 7/5$, and Mach number M modelled with the parameters reported in [9]. This discrepancy may be attributed to the nonideal shape of the present nozzle. N_2 obeys the isentropic relation for $z/D > 3$, the density decaying according to z^{-2} within experimental accuracy. On the contrary, CO_2 density decays according to $z^{-\alpha}$, with $\alpha > 2$ depending on the degree of condensation. This suggests a tendency of CO_2 monomers (uncondensed phase) to migrate outwards from the center line of the expansion, while heavier clusters (condensed phase) tend to remain close to it.

Relative densities can be converted to absolute densities comparing the intensity measured at a given reference point of the jet with the intensity measured for the chamber filled with the sample gas at a known pressure

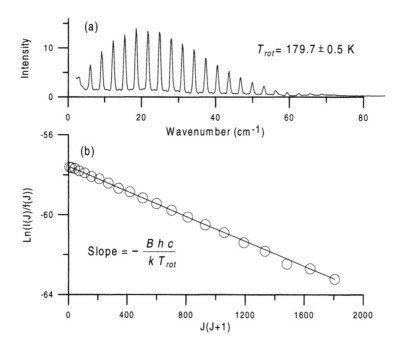

Figure 3: (a) Rotational Raman spectrum of CO_2 expanded under $P_0 = 100$ kPa and $T_0 = 295$ K, recorded at $z/D = 0.16$ from the nozzle (point A in Fig. 1b), (b) Boltzmann plot of the spectrum; $B=0.3902189$ cm^{-1} is the rotational constant of CO_2.

and temperature, usually $P = 100$ hPa and $T = 300 K$. For the range

$$10^{14} < \mathcal{N} < 10^{20} \text{ molecules/cm}^3,$$

the accuracy of absolute number density is on the order of 10 %.

Rotational temperatures can be measured from the relative intensities of rotational lines in the Raman spectrum (non-spherical molecules), or from rotational-vibrational lines (spherical top molecules). For several important species (N_2, O_2, H_2, CO_2, H_2O, CO, NH_3, light hydrocarbons,...) a wide range of temperatures can be measured with accuracy of about 5 %. However, far better accuracy can be attained near the nozzle. For instance, the rotational spectrum shown in Fig. 3a provides, from the slope of its Boltzmann plot on Fig. 3b, a rotational temperature $T_{rot} = 179.7 \pm 0.5$ K. The uncertainty corresponds to ± 1 standard deviation in a robust analysis of the data.

As discussed below (see Nonisentropic Flow), rotational temperatures along the expansion depend strongly on condensation, departing significantly from

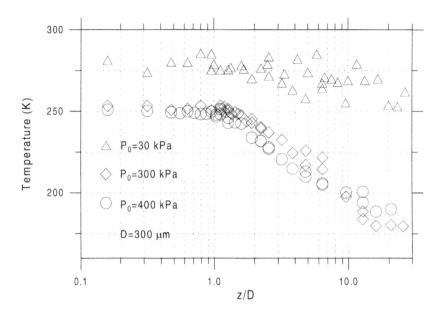

Figure 4: Vibrational temperatures in jets of CO_2.

the prediction of the isentropic model. Further examples of rotational temperature measurements can be seen below.

Vibrational temperatures. Depending on the stagnation temperature the population of some vibrational levels can be monitored along the expansion from the intensity ratio of Stokes to anti-Stokes bands, or from the intensity ratio of hot bands to fundamental bands. In this way, the so called 'vibrational temperature' can be deduced, a useful parameter to account for the vibrational energy. As shown in Fig. 4 this quantity depends on the stagnation pressure (for $z/D < 1$), and on condensation (for $z/D > 1$). For $P_0 = 400$ and $P_0 = 300$ kPa the change of slope at $z/D \approx 1$ may be related to the onset of condensation (see below). On the contrary, for $P_0 = 30$ kPa, where no condensation is expected, vibrational cooling is very mild.

Condensation along the expansion can be monitored efficiently through Rayleigh scattering [5]. In expansions with condensation Rayleigh intensities are strongly enhanced with respect to ideal expansions with no condensation generated under equivalent source conditions. This effect is shown in Fig. 5 for CO_2 expanded at $T_0 = 300$ K through a nozzle of diameter $D = 300$ μm, at source pressures $P_0 = 500, 400, 300, 200$, and 100 kPa. The dashed line represents the ideal isentropic expansion, free of condensation, i. e., strictly proportional to the density of monomers. CO_2 expanded at $P_0 < 20$ kPa, or N_2 expanded at $P_0 < 500$ kPa behave like the ideal case. In

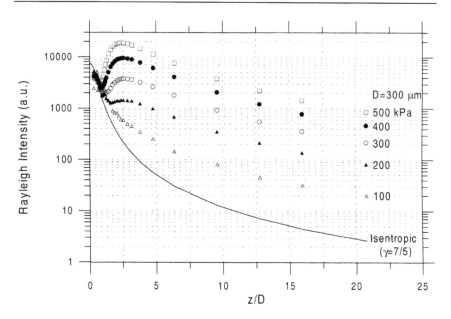

Figure 5: Rayleigh intensity enhancement due to condensation in jets of CO_2.

Fig. 5 the Rayleigh intensities are normalized at $z/D = 1$, immediately before the onset of condensation can be detected. Condensation is noticeable at $z/D > 1$, and for the expansion generated at $P_0 = 500$ kPa, it appears to be completed at $z/D \simeq 9$, since the ratio of Rayleigh intensity to the ideal case remains constant for $z/D > 9$. Conversely, in expansions generated at lower stagnation pressures the condensation process keeps going on still for a few nozzle diameters downstream from $z/D > 9$, in qualitative agreement with the results reported in [5] for N_2O.

4 Raman Mapping

Combining the quantitative capabilities of Raman spectroscopy with the good spatial resolution and the mobility of the nozzle along orthogonal directions, it is possible to map free jets with accuracy. The preliminary results reported in [10, 11] have been improved substantially during the last two years. Maps of absolute densities and rotational temperatures in a jet of CO_2 generated through a nozzle of diameter $D = 300$ μm at source conditions $P_0 = 203$ kPa and $T_0 = 300$ K are shown in Fig. 6.

Each map was generated with the data from 2200 Raman spectra recorded in a grid of points covering the flow field. The map of densities was generated

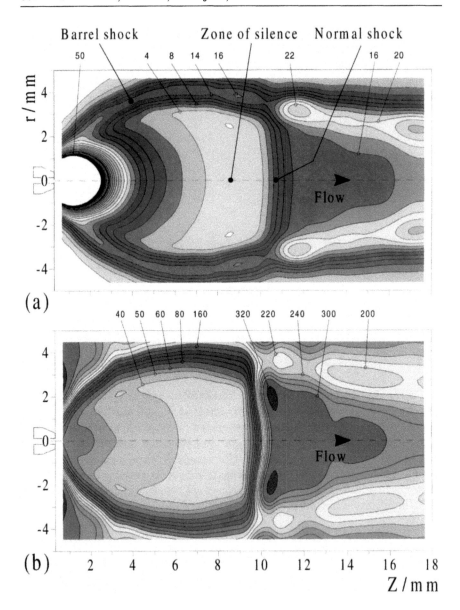

Figure 6: Supersonic jet of CO_2. **(a)** Number density (10^{15} molecules/cm^3), **(b)** Rotational temperatures (Kelvin).

from the vibrational band of CO_2 at 1388 cm^{-1}, and that of rotational temperatures, from the rotational spectra. The data acquisition time was about 40 hours per map, distributed in several working sessions spanning

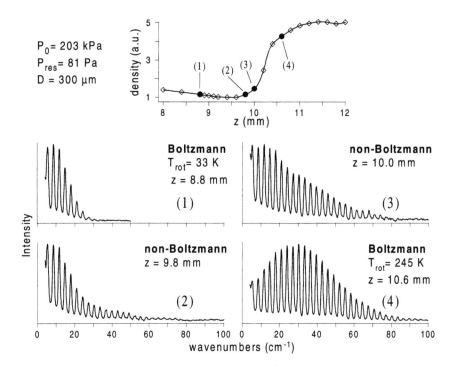

Figure 7: Non-Boltzmann spectra across a normal shock wave of CO_2 located at $z = 10$ mm.

several days. Temperatures in the barrel shock and in the normal shock waves, non–Boltzmann regions of the jet (see below), have been interpolated to allow for the visualization of the flow field.

The maps in Fig. 6 confirm quantitatively the features described qualitatively in the literature [9].

5 Shock Waves

Barrel and normal shock waves surrounding the *zone of silence* of the jet can be distinguished neatly in Figs. 6a and 6b. The structure of these regions is fairly complex, with strong gradients of flow velocity, density and temperature. Rotational temperatures do not obey a simple Boltzmann-like distribution.

Four rotational Raman spectra recorded at representative points of a shock wave are shown in Fig. 7: (1) before, (2) and (3) in between, and (4), after the shock wave. At points (2) and (3) the spectra are markedly non-

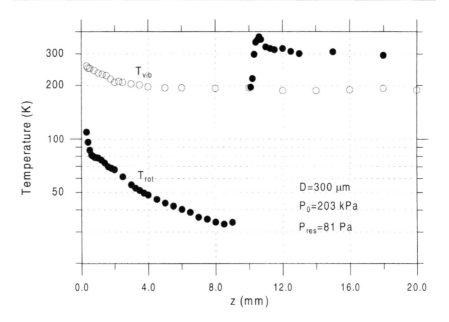

Figure 8: Rotational and vibrational temperatures across a normal shock wave of CO_2 located at $z = 10$ mm.

Boltzmann. Actually the thermal distribution is bimodal, with two rotational temperatures corresponding to two sorts of molecules impinging onto the shock region: The *cold* (and fast) molecules with temperature T_{cold}, and the *warm* (and slow) molecules at temperature T_{warm}. *Cold* molecules are those which have not yet collided with the *warm* and slow molecules at the shock, and are therefore still travelling along z with the velocity and rotational temperature of the zone of silence, approximately 675 m/s and 33 K in the present example. *Warm* molecules are those which have experienced at least one collision with another slow molecule and have become rethermalized. This behaviour is qualitatively similar to that of translational temperatures in monatomic gases, predicted from the solution of the Boltzmann equation for a shock wave [12], and observed from Doppler broadening in electron beam fluorescence experiments [13]. Vibrational temperatures, on the contrary, are insensitive to the shock wave, as can be seen in Fig. 8.

From the simulation of the non-Boltzmann spectra like the ones shown in Fig. 7, in combination with the total absolute number density, it is possible to track the gradual, but very fast, conversion of populations across the normal shock wave and to infer the flow velocity gradient.

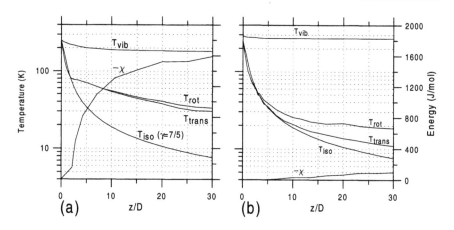

Figure 9: Non-isentropic properties of jets of CO_2 expanded at (a) $P_0 = 203$ kPa, (b) $P_0 = 23$ kPa.

6 Nonisentropic Flow in Molecular Jets

The isentropic model for a perfect gas is an approximation widely utilized to describe a free jet. However, to a greater or lesser degree, real gases tend to depart from this model. This is caused by condensation, and by the contribution of the vibrational degrees of freedom. Vibrational cooling, and condensation, start at the first nozzle diameters of the expansion, depending on source conditions. Both processes do alter the energy balance in the jet and, consequently, the remaining physical properties. So, the general properties of a real jet are conditioned along the first nozzle diameters of the expansion, a region easily accessible to Raman spectroscopy but difficult for most diagnostic techniques.

Combining the Raman data of density, rotational, and vibrational temperatures along the jet by means of the conservation of linear momentum and total energy, and including the rotation-translation relaxation as a constraint, the condensation energy released to the jet, $(-\chi)$, can be deduced [8]. In Fig. 9a the main properties of a jet of CO_2 expanded at $P_0 =200$ kPa are shown. The dashed line stands for the isentropic temperature for $\gamma = 7/5$, and T_{rot}, T_{trans}, and T_{vib}, for rotational, translational and vibrational temperatures, respectively. The energy released to the jet by condensation $(-\chi)$ is highly correlated with the enhancement of Rayleigh signal shown in Fig. 5. Due to the relatively high rate of collisions in this jet, rotational and translational temperatures are almost in equilibrium. For a jet expanded at a stagnation pressure one order of magnitude smaller, shown in Fig. 9b, with condensation in the limit of detectability, the translational temperature is close to the isentropic temperature, and the lag between rotational and translational temperature is about 6 K at $z/D = 30$.

7 Conclusion

The examples presented here show to what extent, and under which experimental conditions, Raman spectroscopy is becoming a mature diagnostics tool to investigate free jets quantitatively at laboratory scale. Work is in progress at our laboratory with the aim of achieving this goal.

References

[1] W.D. Williams and J. W. Lewis, AIAA Journal **13**, 709 (1975).

[2] I. F. Silvera and F. Tommasini, Phys. Rev. Lett. **37**, 136 (1976).

[3] G. Luijks, S. Stolte, and J. Reuss, Chem. Phys. **62**, 217 (1981).

[4] J. W. L. Lewis, 16th International Symposium on Rarefied Gas Dynamics, Pasadena, CA, USA, 1988, Progress in Astronautics and Aeronautics **117**, 107 (1989).

[5] R. G. Shabram, A. E. Beylich, and E. M. Kudriavtsev, 16th International Symposium on Rarefied Gas Dynamics, Pasadena, Ca, USA, 1988, Progress in Astronautics and Aeronautics **117**, 168 (1989).

[6] G. Tejeda, J. M. Fernández-Sánchez, and S. Montero, Appl. Spectrosc. **51**, 265 (1997).

[7] G. Tejeda, Ph. D. thesis, Madrid, 1997.

[8] B. Maté, G. Tejeda, and S. Montero, J. Chem. Phys. **108**, 2676 (1998).

[9] D. R. Miller, in Atomic and Molecular Beam Methods, Edited by G. Scoles (Oxford University Press, New York, Oxford, 1988), p. 14.

[10] G. Tejeda, B. Maté, J. M. Fernández-Sánchez, and S. Montero, Phys. Rev. Lett. **76**, 34 (1996).

[11] B. Maté, Ph. D. thesis, Madrid, 1997.

[12] H. M. Mott–Smith, Phys. Rev. **82**, 885 (1951).

[13] G. Pham–Van–Diep, D. Erwin, and E. P. Muntz, Science **245**, 624 (1989).

Mixed Clusters Produced in Argon-Nitrogen Coexpansions as Evidenced by Two Experimental Methods

E. Fort, A. De Martino, F. Pradère, M. Châtelet and H. Vach
Laboratoire d'Optique Quantique du CNRS,
Ecole Polytechnique, 91128 Palaiseau Cedex, France

G. Torchet, M.-F. de Feraudy and Y. Loreaux
Laboratoire de Physique des Solides (URA 002), Bât. 510,
Université Paris-Sud, 91405 Orsay Cedex, France

1 Introduction

Mixed van der Waals clusters offer unique opportunities to study nucleation and growth phenomena through the dependence of the final cluster sizes, structures and compositions on the mixed cluster preparation method.
Two main techniques are currently available to produce such mixed clusters. In the pickup technique a beam of pure (single-component) clusters obtained by supersonic expansion of a neat gas is made to pass through another gas (the dopant): if the atoms or molecules colliding with the clusters stick on them, mixed clusters with easily controllable composition are eventually obtained. The main shortcoming of this technique is the degradation of beam velocity, divergence and speed ratio when significant amounts of dopant are needed, as for example, for typical chemical stoichiometric proportions. On the other hand, supersonic expansions of gas mixtures can produce mixed clusters with good beam quality, but the cluster nucleation and growth in such expansions are very complex and far from being understood. As a result, the final cluster compositions and sizes are difficult to predict, and the very presence of mixed clusters remains a challenging issue.
In this paper, we present three complementary techniques that provide detailed diagnostics of supersonic expansions involving several species. In Sect. 2 we describe these techniques and show typical raw data obtained with expansions of a mixture composed of 10% argon- 90% nitrogen. Section 3 is devoted to results and discussions.

2 Beam diagnostic techniques

2.1 Beam scattering by a surface or a buffer gas

This technique combines beam scattering by either a solid surface or a buffer gas with angularly resolved mass spectrometric measurements. The experimental setup has been described in detail elsewhere [1], and essentially consists of a Campargue type beam generator [2], equipped with a 0.25 mm sonic nozzle, three differential pumping stages and a UHV chamber containing a graphite sample and rotatable quadrupole mass spectrometer (QMS). With the sample in the beam path, the QMS provides the angular distributions of the species scattered by the surface. Alternatively, with the sample removed from the beam path, the QMS is used to record mass resolved angular profiles of the beam.

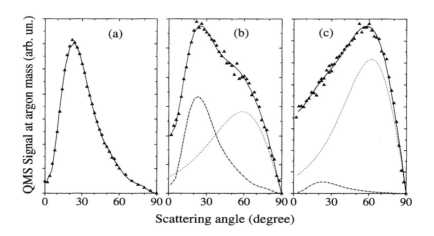

Figure 1: Typical angular distributions of argon monomers scattered off the graphite sample for incoming beams generated by expansion of a 10% argon- 90% nitrogen mixture at different stagnation pressures : (a), 1 bar, signature of incident monomers, (b) 11 bar, incident beam composed of monomers and clusters, (c) 21 bar, incident beam composed of clusters.

Monomers and clusters impinging on the surface yield quite different angular distributions of scattered particles (essentially monomers), as shown in Fig. 1. From these different signatures it is possible to retrieve the proportion of uncondensed particles (monomers) for each species present in the beam [3].

We now turn to the profile measurements carried out in the beam itself. The species eventually detected are not the clusters entering the QMS ion-

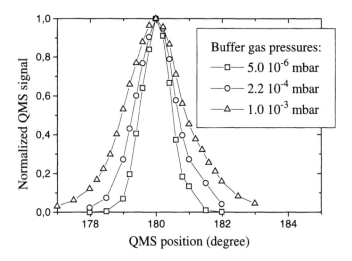

Figure 2: Broadening of the beam profile with buffer gas pressure. Expansion of 10% argon- 90% nitrogen mixture. Detected species : ArN_2 dimers.

ization head (their masses are far beyond the QMS accessible range), but the very small fragments (essentially monomers and dimers) coming from cluster fragmentation on the ionizer meshes. As a result, the QMS is essentially sensitive to the flux of each gas (Ar or N_2), in atoms or molecules, independently of their degree of condensation. Hence the beam composition can be readily obtained from comparison of the signals measured at different masses, via a suitable calibration of the relative sensitivities of the QMS for different gases.

If a buffer gas is placed in the path of a cluster beam, the beam divergence increases due to the cluster bombardment by the buffer gas atoms. The beam broadening with increasing buffer gas pressure can be used to evaluate the average cluster sizes with a typical accuracy of ±50% [4].

In this work, the angular profiles of the beams generated at different stagnation pressures of the argon-nitrogen mixture have been recorded for different pressures of a buffer gas introduced in the last differential pumping stage, at QMS mass settings corresponding to argon monomers, nitrogen monomers, and mixed Ar-N_2 dimers. A typical result is shown in Fig.2.

The very presence of mixed Ar-N_2 dimers in the detected species is a clear evidence of mixed cluster presence in the beam. Moreover, for totally condensed beams, the profiles recorded at the three mass settings have been found to be identical for each value of the buffer gas pressure, indicating that mixed clusters are actually dominant in these beams.

To summarize, beam scattering off the graphite sample yields the monomer percentage of each species, while scattering by a buffer gas shows that fully condensed beams consist essentially of mixed argon-nitrogen cluster and provides reliable values of their average sizes.

2.2 High energy electron diffraction

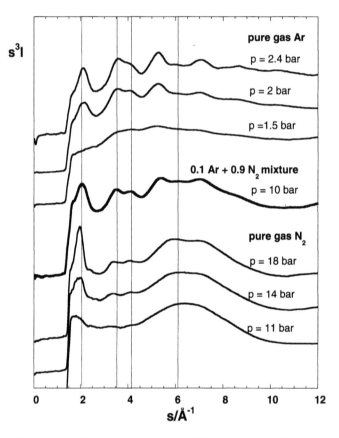

Figure 3: Electron diffraction patterns ($s^3 I$ vs $s = (4\pi/\lambda)\sin(\theta/2)$, with λ electron wavelength and θ diffraction angle). *Upper curves*: pure argon expansion; *lower curves*: pure nitrogen; *middle curve*: 10% argon- 90% nitrogen gas mixture at inlet pressure 10 bar. Vertical lines are provided to facilitate comparison between curves.

In the second technique, the clusters produced during the expansion scatter a high energy (50 keV) electron beam, providing electron diffraction patterns directly related to the Fourier transform of the interatomic distance distribution inside the clusters. The experimental setup used for this technique has been described in detail elsewhere [5]. The supersonic beam of a Campargue type is very similar to the one used in the first technique.

Figure 3 shows diffraction patterns recorded from pure argon (top), pure nitrogen (bottom) and the 10% argon- 90% nitrogen mixture expansions through a 0.2 mm sonic nozzle. From previous studies on pure argon [6] and pure nitrogen [7], the diffraction patterns can be attributed at 1.5 bar for argon and 11 bar for nitrogen to the beginning of condensation, at 2 bar and 14 bar, respectively, to clusters with a polyicosahedral structure and an average size of 30 to 40 atoms. Like nitrogen clusters at 18 bar, argon clusters at 2.4 bar are icosahedral (2 layers) and contain about 50 atoms. It is worthwhile noticing that most of the features present in the pure gas patterns are reproduced in the 10% argon - 90% nitrogen pattern.

3 Results and discussion

We first discuss the nature of the clusters produced by expansion of the mixture. If condensation would create separate Ar_n and $(N_2)_m$ clusters, the corresponding diffraction pattern would be given by some weighted superposition of the patterns recorded with neat argon and neat nitrogen. In spite of the obvious similarities of the curves shown on Fig. 3, this is not the case. For instance, no weighting coefficients can reproduce both the height of the first line, near 2 Å^{-1}, and the plateau near 6 Å^{-1}. More quantitatively, as shown in Fig. 4, the position of the first line is well reproduced by a superposition of pure gas patterns with 35% weight for nitrogen, while for the ratio of the heights of the first and second lines this weight must be close to 70%. The inconsistency of these values thus confirms that mixed clusters are dominant in the beam. The overall aspect of the pattern recorded from the mixture expansion is consistent with some amorphous structure similar to that produced in the case of neat gas expansions. Clearly, further theoretical work is needed to fully retrieve the information about the structure of these clusters and the relative location of argon and nitrogen molecules within the structure.

We now consider the beam and cluster composition. As mentioned earlier, the overall beam composition is readily obtained from the QMS signals at different mass settings. By taking into account the monomer percentages provided by the surface scattering technique, we can extract the cluster composition for partially condensed beams. The results are shown in Fig. 5. Whatever the stagnation pressure, argon concentration within the beam is always larger than in the expanding mixture. Besides, the composition

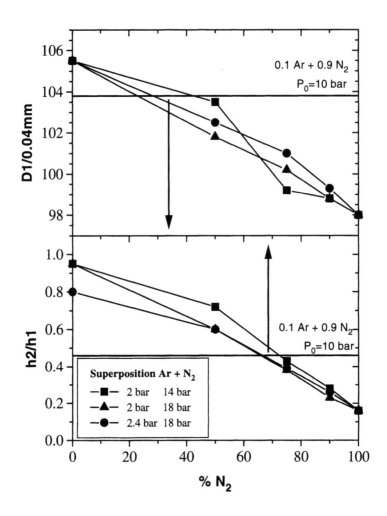

Figure 4: Evolution of the first diffraction line position (D_1) (*top*) and of first (h_1) to second (h_2) line height ratio (*bottom*) vs N_2 pattern percentage for weighted superpositions of pure nitrogen and pure argon diffraction patterns. Experimental values for the mixture pattern are shown as horizontal lines.

undergoes a transition with stagnation pressure from 15% of argon under 7 bar up to about 32% above 13 bar. This transition coincides with the

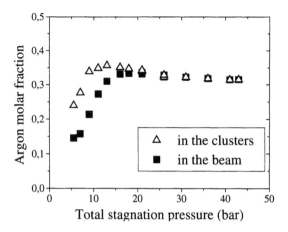

Figure 5: Average composition of the beam and of the mixed clusters vs stagnation pressure for the 10% argon - 90% nitrogen mixture.

condensation which is easily localized from surface scattering density lobes. Most probably nitrogen molecules evacuate the heat of condensation more efficiently than argon atoms. Their sticking probability is lower because of their lower binding energy. Consequently, nitrogen helps argon to nucleate. Note that the role of Mach number focusing effect is difficult to evaluate in presence of nucleation phenomena but it might be crucial in the resulting beam composition.

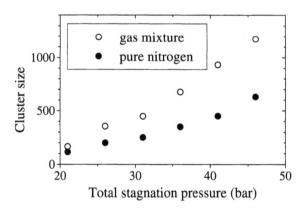

Figure 6: Average cluster size evolution with stagnation pressure, for pure nitrogen and the 10% argon - 90% nitrogen mixture.

Average cluster size versus stagnation pressure is shown in Fig. 6 for both pure nitrogen and argon-nitrogen mixture. Clearly, the clusters obtained with the mixture expansion are larger than with pure nitrogen. Note that this conclusion is still valid if we take into consideration the nitrogen partial pressure for comparison. Hence, the presence of argon helps nitrogen to nucleate.

In this respect, we point out that the best fit for electron diffraction pattern of the mixture at 10 bar (see Fig. 3) by a weighted superposition of pure gas curves is obtained for inlet pressures of 2 bar for argon and 18 bar for nitrogen. These two pressures are higher than the partial pressures in the gas mixture (1 bar for argon and 9 bar for nitrogen), for which no condensation occurs for pure gases.

Thus, we can conclude from both average size measurements and electron diffraction patterns that gas mixing favors nucleation.

4 Conclusion

In summary, by coupling beam scattering on gases and surfaces, and electron diffraction, one can determine on the same system the average cluster size and composition, the amount of uncondensed monomers of each species together with useful information about the structure of the mixed clusters, if any, present in the beam. These techniques are clearly powerful means to investigate heterogeneous nucleation in supersonic beams, as it has been shown here with expansions of a 10% argon - 90% nitrogen gas mixture.

References

[1] F. Pradère, M. Benslimane, M. Château, M. Bierry, M. Châtelet, D. Clément, A. Guilbaud, J.C. Jeannot, A. De Martino, H. Vach, Rev. Sci. Instrum. **65**, 161 (1994).

[2] R. Campargue, J. Phys. Chem. **88**, 4466 (1984)

[3] E. Fort, F. Pradère, A. De Martino, H. Vach, M. Châtelet, Eur. Phys. J. D **1**, 79 (1998).

[4] A. De Martino, M. Benslimane, M. Châtelet, C. Crozes, F. Pradère, H. Vach, Z. Phys. D **27**, 185 (1993).

[5] B. Raoult and J. Farges, Rev. Sci. Instrum. **44**, 430 (1973).

[6] J. Farges, M.-F. de Feraudy, B. Raoult, G. Torchet, J. Chem. Phys. **78**, 5067 (1983).

[7] J. Farges, M.-F. de Feraudy, B. Raoult, G. Torchet, Ber. Bunsenges. Phys. Chem. **88**, 211 (1984).

Part III

Photodissociation Dynamics
and Electronic Spectroscopy in Molecular Beams:
From Simple Molecules to Clusters and Ions

Imaging of State-to-State Photodynamics of Nitrous Oxide in the 205 nm Region of the Stratospheric Solar Window

M.H.M. Janssen, J.M. Teule, D.W. Neyer[1],
D.W. Chandler[1], G.C. Groenenboom[2]
Laser Centre and Department of Chemistry
Vrije Universiteit, Amsterdam, The Netherlands
[1] Sandia National Laboratories
Livermore, CA 94550, USA
[2] Institute of Theoretical Chemistry
University of Nijmegen, Nijmegen, The Netherlands

1 Introduction

Photodissociation experiments provide information about the (anisotropic) molecular forces and the dynamics on the excited state potential energy surfaces involved in the 'half-reaction'. Selection of the quantum states of the parent molecule prior to dissociation as well as knowledge of the internal energy and angular distribution of the photofragments will contribute to a complete understanding of the dissociation dynamics. Two-dimensional ion-imaging detection has proven to be a valuable technique in the study of photofragmentation processes, providing the three-dimensional angular and speed distribution of state-selected fragments [1, 2], which often contains information on alignment effects.

The photodissociation of nitrous oxide, N_2O, has received considerable attention in the past years. Photodissociation around 200 nm forms almost exclusively oxygen atoms in the excited $^1D_2$ state and N_2 in the electronic ground state,

$$N_2O + h\nu \longrightarrow N_2(^1\Sigma_g^+, v, J) + O(^1D_2). \qquad (1)$$

One of the key issues of interest with respect to the global warming is the role of the so-called greenhouse gasses. One of the prominent greenhouse gasses is nitrous oxide [3], which is also involved in the depletion of ozone (O_3) [4]. Per molecule, nitrous oxide has more than 200 times the greenhouse forcing of carbon dioxide. The greenhouse or radiative forcing is the perturbation to the energy balance of the Earth-atmosphere system (in W m^{-2}) following a change in the concentration [3].

The concentration of nitrous oxide in the atmosphere is still increasing and the total biogeochemical source and loss channels and the dynamics between them has been the focus of intense recent research [5]. The loss channels of N_2O in the stratosphere are photolysis (90 %) and photooxidation reactions of $O(^1D_2)$,

$$N_2O + O(^1D_2) \longrightarrow \begin{matrix} 2NO\ (6\%) & (2) \\ N_2 + O_2\ (4\%) & (3) \end{matrix}$$

Reaction (2) is the major source for NO which destroys ozone catalytically. Because of the atmospheric importance of N_2O a detailed understanding of the photolysis (reaction (1), by far the dominant loss channel) is needed. Especially, knowledge of the differences of the photolysis of the various isotopomers of this molecule may provide a solution to solve the various problems associated with the global budget of nitrous oxide.

In this paper two-dimensional ion-imaging experiments are described in which quantum state-selected $N_2(v, J)$ and $O(^1D_2)$ fragment angular distributions are obtained after photolysis of rovibrationally state-selected N_2O. Using the hexapole state-selection technique, the initial quantum state of the parent molecules prior to dissociation is selected. This state-selection technique was previously successfully combined with resonance-enhanced multi-photon ionization (REMPI) and ion-imaging detection in studies on the photodissociation of oriented CD_3I [6]. First results of our experiments on the photolysis of N_2O were reported before [7].

2 Experimental

The experimental apparatus is described briefly. A schematic overview of the total setup is shown in Figure 1. A rotationally cold molecular beam of N_2O is generated by expanding a seeded mixture of N_2O through a pulsed nozzle. The nozzle orifice used was 0.3 mm in diameter and the skimmer, located about 2-4 cm downstream, had a diameter of 1.5 mm. The seeding gasses used were Ar and Kr, with N_2O in concentrations up to 20 %. Vibrationally excited N_2O molecules in the ν_2 bending mode can acquire an angular momentum along the molecular axis by l-type doubling. While travelling through the inhomogeneous electric field of the hexapole, molecules with a positive Stark effect experience a force towards the hexapole axis where the field strength vanishes. In the rotationally cooled N_2O beam, a single (JlM) state can be selected and focussed in the dissociation region, choosing a proper hexapole voltage. Here J is the total rotational quantum number, l is the angular momentum component along the molecular axis and M is the projection of the angular momentum along the (external) elec-

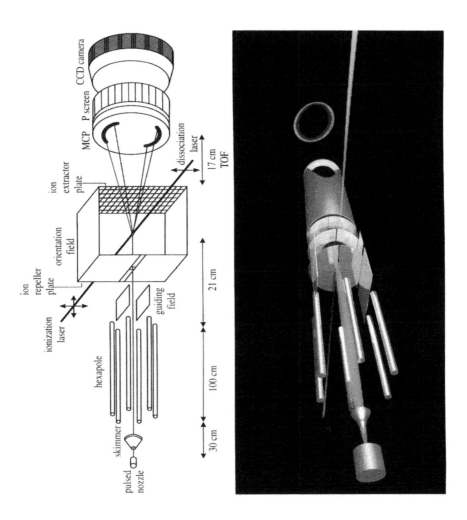

Figure 1: Left panel, original experimental set-up used for the nitrous oxide experiments. Right panel, the recently modified set-up with gridless extraction optics and orientation plates for velocity mapping imaging of oriented molecules.

tric field vector. In the experiments reported here we did not use our guiding and orientation fields downstream of the hexapole to spatially orient N_2O. This means the M quantum number is scrambled in the dissociation region. By photolyzing N_2O and detecting the $O(^1D)$ fragment (using a photomul-

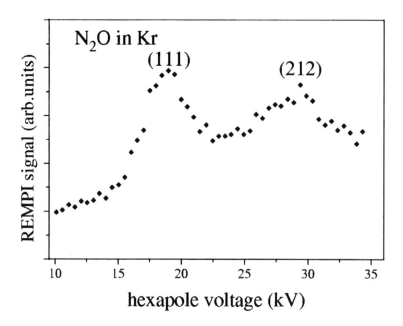

Figure 2: Hexapole focussing spectra, measured by detecting the $O(^1D_2)$ signal as a function of the hexapole voltage. The labels denote the quantum numbers (JIM) of the state focussing in the laser interaction region at the particular voltage.

tiplier tube to collect all phosphorescence from the two-dimensional detector) while scanning the hexapole voltage, a hexapole-focussing spectrum is obtained. Figure 2 presents a focussing spectrum measuring the yield of $O(^1D_2)$ atoms as a function of the hexapole voltage. In the experiments reported here the tripled output from a dye laser ionizes the fragments using (2+1) REMPI with
* 203.7 nm probing $O(^1D_2)$ via the $^1F_3 \leftarrow\leftarrow {}^1D_2$ transition
* 205.4 nm probing $O(^1D_2)$ via the $^1P_1 \leftarrow\leftarrow {}^1D_2$ transition
* 203-204 nm probing $N_2(v, J = 60 - 85)$ via the $a'''^1\Sigma_g^+ \leftarrow\leftarrow X^1\Sigma_g^+$ transition.

In case of the $O(^1D_2)$ detection, the two-photon levels ($^1F_3$ or $^1P_1$) lie above the first ionization threshold [8]. From these states ionization can take place by autoionization, or by absorption of a third photon to a higher ionization continuum. The ions are accelerated by an extraction field perpendicular

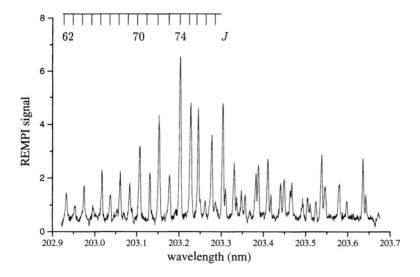

Figure 3: Q-branch REMPI spectrum of N_2 formed in the dissociation of N_2O ($\nu_2 = 1$, J=1, l=1).

to the laser beams into a short TOF tube (17 cm). At the end of the TOF tube the ions hit a mass gated micro-channel-plate (MCP) detector. From the back of the MCP electrons are accelerated onto a fast phosphor screen, which converts the electron spray into light. The fluorescence light is collected with a high quality camera lens and imaged on a charge-coupled-device (CCD) camera. When the total ion yield is measured, the CCD camera is replaced by a photomultiplier tube.

3 Results

3.1 N_2 fragments

Figure 3 presents the (2+1) REMPI-spectrum of N_2 formed in the photodissociation reaction of N_2O ($\nu_2 = 1$, J=1, l=1). A single laser is used to both dissociate N_2O and ionize the N_2 fragment ($v, J = 60 - 85$) via the Q-branch of the $a''^1\Sigma_g^+ \leftarrow\leftarrow X^1\Sigma_g^+$ two-photon resonant transition between 202 and 204 nm. The total N_2 REMPI signal is measured as a function of the wavelength of the ionization laser. In the spectrum a 2:1 intensity ratio for transitions involving even and odd J is observed, which is characteristic of N_2. The N_2 REMPI spectrum in Figure 3 is very similar to the REMPI spectrum obtained by Hanisco and Kummel [9] for the N_2 product from photodissociation of non-state-selected N_2O. The peak in the rota-

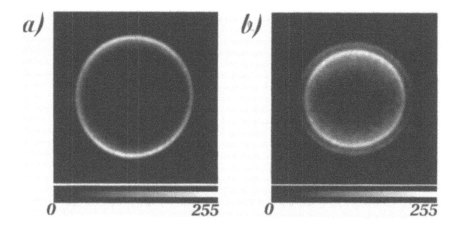

Figure 4: Two-dimensional ion-images of N_2 fragments formed in the photodissociation reaction of N_2O ($\nu_2 = 1$, J=1, l=1) around 203 nm. Images a) and b) show the measured images detecting N_2 photofragments in $J = 70$ and 82, respectively. In image b) two rings are observed. The inner ring corresponds to $N_2(v = 0, J = 82)$, and the outer ring to $N_2(v = 1, J = 63)$, which has less internal energy and therefore higher recoil velocity.

tional distribution lies at $J = 74$, corresponding to $E_{rot} = 1.4$ eV. Although in our experiments the $N_2(J)$ results predominantly from the dissociation of vibrationally excited N_2O ($\nu_2 = 1$), whereas in the experiments of Hanisco and Kummel [9] the $N_2(J)$ results from dissociation of a non-state-selected thermal beam, the rotational distribution is apparently similar in the two cases. This may be partly due to the fact that even in a supersonic beam a large fraction of the N_2 fragments results from dissociation of vibrationally excited molecules. As was estimated by Selwyn and Johnson [10] in gas-cell UV absorption experiments, the photodissociation cross section around 190 nm for N_2O ($\nu_2 = 1$) is about a factor of five larger than for ground state molecules. In Figure 4 ion-images of N_2 photofragments are shown at two different laser wavelengths, detecting rotational levels $J = 70$ and 82. These images (300 × 300 pixels) are obtained by subtraction of a background image (non state-selected) from the image of state-selected N_2O. A background image is measured under exactly the same experimental conditions as the state-selected image, but with zero voltage on the hexapole rods. The intensity of the signal from the direct (non-state-selected) beam is typically less than 20 % of the intensity of state-selected N_2O. Because the laser bandwidth is too narrow to detect all the N_2 fragments independent of their velocity, images are taken by scanning the dye laser wavelength over about 1 cm^{-1}. The intensity of each image in Figure 4 is scaled to the maximum intensity, which is set at 255. The polarization of the laser (used both for photolysis and ionization) is parallel to the detection plane (vertical in Fig. 4). The images show two projected polar caps along the direction of the laser polarization, which is characteristic of a parallel transition. The images exhibit a very narrow velocity distribution because the quantum state-selected $N_2(J)$ corresponds to a single exit channel of the $O(^1D)$ cofragment, and therefore has a very well defined recoil velocity. The images are Abel-inverted to reconstruct the three-dimensional angular distribution from the two-dimensional projections. Since the Q band transition used to probe N_2 is relatively insensitive to alignment the expected angular distribution of each $N_2(J)$ fragment was fitted to the usual angular expression with a J dependent $\beta_{N_2(J)}$

$$I_{N_2(J)} \propto [1 + \beta_{N_2(J)} P_2(\cos\theta)]. \qquad (2)$$

With this procedure for each rotational level studied a β parameter is extracted from the images. Starting with $\beta = 0.96$ for $J = 66$, the anisotropy parameter decreases with increasing N_2 rotation. This can be understood considering the angle between the recoil direction and the transition dipole moment. Higher rotational states in N_2 result from molecules which experienced a larger torque during the dissociation process, in going from the linear ground state configuration to the bent excited state configuration (see Section 4). In these molecules the larger angle between the recoil velocity direction and the direction of the transition dipole moment causes a reduction of the observed β parameter in the laboratory frame. The decreasing trend in β parameter going to higher N_2 rotational levels is also observed in the photodissociation experiments on non-state-selected N_2O by Chandler and coworkers [13]. This similarity between state-selected and non-state-selected N_2O might be partly caused by the substantially higher dissociation cross section for N_2O ($\nu_2 = 1$) molecules. A non-state-selected molecular beam of N_2O molecules still has about 10% population in $\nu_2 = 1$, the vibrational temperature is generally not that much colder than 300 K. Because the cross section for photolysis around 205 nm is an order of magnitude larger for $\nu_2 = 1$ compared to $\nu_2 = 0$, the photofragments observed from photoyusis of a non-state-selected beam are expected to originate for a substantial part from N_2O ($\nu_2 = 1$) molecules.

3.2 $O(^1D_2)$ fragments

Figure 5 shows ion-images of the $O(^1D_2)$ fragments, detected via (a) the $^1P_1 \leftarrow\leftarrow ^1D_2$ transition at 205.4 nm and (b) the $^1F_3 \leftarrow\leftarrow ^1D_2$ transition at 203.7 nm. Again a background image (no hexapole state-selection) is subtracted from the image of state-selected N_2O (ν_2=1,J=1,l=1). The N_2O molecules are photolyzed and ionized using the same laser. Both $O(^1D_2)$

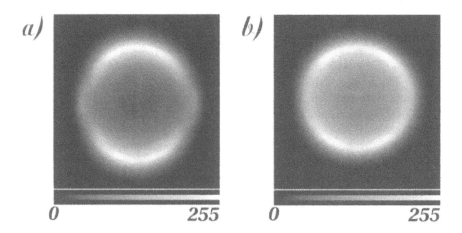

Figure 5: Two-dimensional ion-images of O(^{1}D$_2$) formed in the photodissociation reaction of N$_2$O ($\nu_2 = 1$, J=1, l=1), detected via a) the ^{1}P$_1 \leftarrow\leftarrow ^1D_2$ transition at 205.4 nm and b) the ^{1}F$_3 \leftarrow\leftarrow ^1D_2$ transition at 203.7 nm.

images show a broad translational energy distribution (radius of the image), caused by the variable amount of rotational energy taken up by the N$_2$ fragment. Compared to the ^{1}F$_3 \leftarrow\leftarrow ^1D_2$ image, the ^{1}P$_1 \leftarrow\leftarrow ^1D_2$ image is more peaked along the vertical axis, and shows a dimple at the horizontal axis. The ^{1}F$_3 \leftarrow\leftarrow ^1D_2$ image has a wider distribution. The differences in the O(^{1}D$_2$) images using different detection schemes arise from variations in the detection probabilities for the various m-levels. For analysis, the Abel-inverted images are divided in shells with different radii corresponding to a range of translational energies. Each shell corresponds to a range of rotational levels in the N$_2$ fragment. At first we tried to simulate the recoil distributions using the β parameter obtained from the N$_2$ images, and varying the alignment parameters $A_0^{(2)}$ and $A_0^{(4)}$ to account for angle-dependent detection probability. In this way it was not possible to simulate satisfactorily, especially so for the ^{1}P$_1 \leftarrow\leftarrow ^1D_2$ images. Therefore, we decided to treat the O(^{1}D$_2$) $|m|$-levels independently, assigning a β_m parameter to each m level. Note that we can only deduce populations from our images, and we are insensitive to the sign of m. When only a single m state is populated, the alignment parameter for each m can be easily calculated. We fitted the data to the following expression, I$_{image}$

$$\propto \sum_m n_m [1 + \beta_m P_2(\cos\theta)][1 + a_2 A_{0,m}^{(2)} P_2(\cos\theta) + a_4 A_{0,m}^{(4)} P_4(\cos\theta)],$$

where n_m is the relative m-population. Two restrictions are made in the simulations: (a) the total population in the m-states $\sum_m n_m = 1$ and (b) the overall β parameter equals the β deduced from the corresponding N_2 image. The parameters β_m and n_m are varied till the angular recoil distributions are reproduced well for both the $^1P_1 \leftarrow\leftarrow ^1D_2$ and the $^1F_3 \leftarrow\leftarrow ^1D_2$ images. Although for these fitted data relatively large error bars have to be taken into account, the $|m| = 1$ population shows a clear trend in going from 0.63 for $J = 66$ to 0.12 for $J = 82$. The population distributes among the $m = 0$ and $|m| = 2$ states. The $\beta = 2$ for $m = 0$ in the translational energy range corresponding to low N_2 rotational states decreases to $\beta = 0.3$ for translational energies corresponding to high rotationally excited N_2. In the following section we will discuss these trends in relation to the dissociation dynamics and the excited state potentials involved.

4 Dissociation dynamics

N_2O has 16 valence electrons, and belongs to the $C_{\infty v}$ symmetry group when it is in its electronic ground state $X^1\Sigma^+$ ($4\sigma^2 5\sigma^2 6\sigma^2 1\pi^4 7\sigma^2 2\pi^4$ configuration). Calculations by Hopper [14] show that dissociation using wavelengths around 200 nm occurs via the $2^1A'(^1\Delta)$ state, but the nearby $1^1A''(^1\Sigma^-)$ state can also be involved in the dissociation process. When N_2O is bent, the C_s symmetry group applies and the states are labelled by A' or A''. The $^1\Delta$ and $^1\Sigma^-$ states are bound, but with a much smaller N–N–O bending angle (around 130°). We have recently started high-level ab initio surface calculations as a prelude to quantum mechanical dynamics calculations. In Figure 6 we show a cut through the lowest surfaces relevant at our photolysis wavelength, where the energy is plotted as a function of the N_2O Jacobi scattering angle in the Franck Condon region.

As can be seen there may be two surfaces involved in the photodynamics, a parallel excitation to the $2^1A'$ component of the ($^1\Delta$) surface, and a perpendicular excitation to the $1^1A''$ ($^1\Sigma^-$) surface.

In Figure 7 we show the two-dimensional contour plots of these two surfaces, as a function of scattering angle and N_2-O distance. As is seen the two surfaces are very similar, although some subtle differences are present which may be responsible for different dissociation dynamics. Dynamical wavepacket calculations on the two surfaces will be performed [11] to test the suggestion by Selwyn and Johnston [10] that the $1^1A''$ ($^1\Sigma^-$) surface is predissociative, and is responsible for the observed vibrational structure in the absorption spectrum around 180 nm. As no quantum calculations are present yet, we have tried to infer information on the alignment of the $O(^1D)$ state from a somewhat simplified model. This treatment follows a similar analysis used to interpret orientation effects in reactive scattering of metastable $Ca(^1D_2)$ with oriented CH_3F, CH_3Cl, CH_3Br [12]. We have cal-

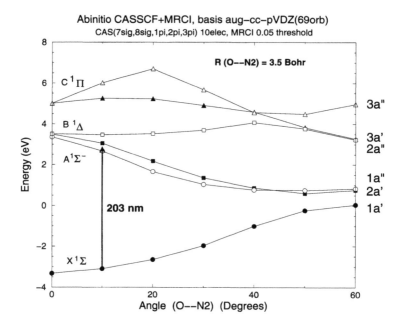

Figure 6: *Ab initio* surface calculation of the most relevant surfaces around 200 nm photolysis. Shown is a one-dimensional cut, with the energy as function of the O–NN scattering angle.

culated the long-range quadrupole-quadrupole interaction between the N_2 and the $O(^1D)$. This interaction gives rise to five electronic surfaces (due to the five m-substates of the D-state). On each of the five surfaces we can calculate as a function of the scattering angle the m-state populations, they simply result from the diagonalization of the interaction matrix and the corresponding eigen vectors. If we now assume a certain initial excitation on the two surfaces involved in the dissociation, the $2^1A'$ and the $1^1A''$ surface, we can calculate the resulting m-state populations. This is shown in Figure 8. The initial excitation to the two surfaces out of the (010) bending wavefunction was estimated from the calculated transition dipole moments as a function of the scattering angle obtained from the *ab initio* calculations. Added to Figure 8 are the m-state populations as obtained from our experiments at two different O-atom speeds, corresponding with different J states of the N_2 cofragment. As can be seen from Figure 8, the trend observed in the experimental results showing a decrease of the $|m| = 1$ population with increasing rotational state J, is represented qualitatively by this simple long-range interaction model. The calculated $|m| = 1$ population in Figure 8 increases again for angles beyond 55°. The populations are mainly (97%) determined by the composition of the eigenvector corresponding to

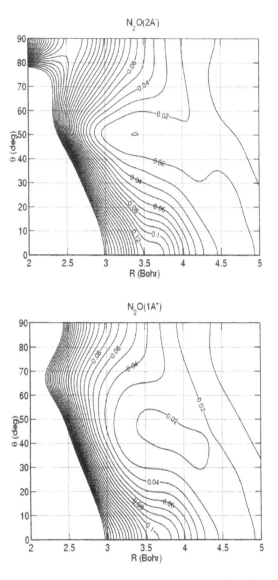

Figure 7: Two-dimensional surface contours of the $2^1A'(^1\Delta)$ and $1^1A''(^1\Sigma^-)$ surfaces. The labels at the contours give the energy in Hartree.

the $2^1A'$ surface. At the linear geometry the $2^1A'$ state has Π symmetry and therefore it has pure $|m| = 1$ character. At the perpendicular geometry this state has B_1 (C_{2v}) symmetry and has again pure $|m| = 1$ character. We

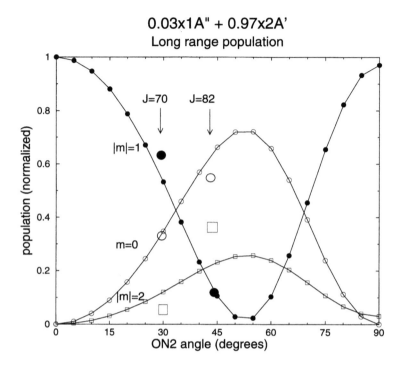

Figure 8: Dependence of the m-state population using the long-range Q-Q interaction, for excitation of 97% to the $2^1A'(^1\Delta)$ surface and 3% to the $1^1A''(^1\Sigma^-)$ surface. Added are the observed m-distributions at two different O-atom speeds correlation to two different N_2 rotational states around J=70 and J=82.

do not want to suggest that the final m-state distribution is determined at an atom molecule separation where the quadrupole-quadrupole interaction is the only important term in the potentials. However, we do think that at a probably more relevant short or intermediate range the symmetries of the states involved are already similar to the symmetries of the states in our long-range model. These ideas need further detailed examination and extensive interpretation of these data will be published elsewhere [11, 15].

5 Conclusions

The two-dimensional ion-imaging detection of photodissociation fragments yields a wealth of information on the angular recoil distribution and the alignment of the fragments. In this study ion-images are presented of both N_2 and $O(^1D_2)$ fragments formed in the photodissociation of rovibrationally single state-selected N_2O around 205 nm. The N_2 REMPI-spectrum reveals high rotational excitation in the N_2 fragment. Furthermore, the N_2 images show a decrease in β-parameter in going to high rotational levels. This is explained considering the angle between the recoil direction and the transition dipole moment. Dissociation occurs via the $2^1A'$ and the $1^1A''$ bent excited state. In going from the linear ground state to a bent excited state, the N_2 fragment experiences a torque causing high rotational excitation. A larger bending angle can explain both the lower anisotropy for high N_2 rotational states, and the decrease in $|m| = 1$ population in the $O(^1D_2)$ fragments. Future experiments are in preparation on state-selected nitrous oxide isotopomers, in order to study the photodynamics in the stratospheric solar window 200-210 nm. These state-to-state photodynamics may provide us with a better fundamental understanding of the role of the bending vibration which may be (partially) responsible for the observed anomalies in the fractionation of nitrous oxide isotopomers in the stratosphere.

Acknowledgements

The authors would like to thank prof. S. Stolte for support and discussions. The research has been financially supported by the councils for Chemical Sciences and Physical Sciences of the Netherlands Organization for Scientific Research (CW-NWO, FOM-NWO).

References

[1] A.J.R. Heck and D.W. Chandler, Annu. Rev. Phys. Chem. **46**, 335 (1995)

[2] D.W. Chandler and D.H. Parker, Advances in Photochemistry **25**, 56 (1999)

[3] J.G. Houghton, L.G. Meira Filho, J. Bruce, H. Lee, B.A. Callander, E. Haites, N. Harris and K. Maskell Eds. *Intergovernmental Panel on Climate Change, Climate Change 1994* (Cambridge Univ. Press, NY, 1995)

[4] World Meteorological Organization, Scientific Assessment of Ozone depletion: 1994, *Global Ozone Research and Monitoring Project-Report 37*, WMO, Geneva, 1995

[5] S.S. Cliff and M.H. Thiemens, Science **278**, 1774 (1997);
T. Rahn and M. Wahlen, Science **278**, 1776 (1997);
J.E. Dore, B.N. Popp, D.M. Karl and F.J. Sansone, Nature **396**, 63 (1998)

[6] M.H.M. Janssen, J.W.G. Mastenbroek and S. Stolte, J. Phys. Chem. A **101**, 7605 (1997)

[7] M.H.M. Janssen, J.M. Teule, D.W. Neyer, D.W. Chandler and S.Stolte, Faraday Discussions **108**, 435 (1997)

[8] S.T. Pratt, P.M. Dehmer, J.L. Dehmer, Phys. Rev. A **43**, 282 (1991)

[9] T.F. Hanisco and A.C. Kummel, J. Phys. Chem. **97**, 7242 (1993)

[10] G.S. Selwyn and H.S. Johnston, J. Chem. Phys. **74**, 3791 (1981)

[11] G.C. Groenenboom, M.H.M. Janssen, work in progress.

[12] M.H.M. Janssen, D.H. Parker and S. Stolte, J. Phys. Chem. **100**, 16066 (1997);
A.J.H.M. Meijer, G.C. Groenenboom and A. van der Avoird, J. Phys. Chem. **100**, 16072 (1997)

[13] D.W. Neyer, A.J.R. Heck and D.W. Chandler, J. Chem. Phys. **110**, 3411 (1999)

[14] D.G. Hopper, J. Chem. Phys. **80**, 4290 (1984)

[15] J.M. Teule, G.C. Groenenboom, D.W. Neyer, D.W. Chandler and M.H.M. Janssen, Chem. Phys. Lett. in print (2000)

The Photodissociation Dynamics of Tropospheric Ozone

G. Hancock, R.D. Johnson, J.C. Pinot de Moira,
G.A.D. Ritchie, P.L. Tyley

Physical and Theoretical Chemistry Laboratory, Oxford University, South Parks Road, Oxford OX1 3QZ, UK

1 Introduction

The wavelength of 310 nm has a special and practical significance in the photodissociation of ozone. It lies within the structured part of the uv absorption spectrum of the molecule in a region which is known as the Huggins bands, and its absorption cross section at this wavelength is about a factor of 100 lower than the maximum absorption in the shorter wavelength Hartley band at 255 nm [1]. Stratospheric ozone acts as a highly efficient uv filter of solar radiation below 300 nm, and thus radiation which reaches the surface of the Earth and can induce photochemistry in the troposphere only has significant intensities at wavelengths above this. In the strong Hartley bands the major fate of ozone molecules following absorption is spin allowed dissociation to produce molecular and atomic oxygen fragments in their first electronically excited singlet states [1]:

$$O_3 + h\nu \rightarrow O(^1D) + O_2(a^1\Delta_g) \tag{1}$$

but this process becomes energetically forbidden at wavelengths above 310 nm for ozone molecules with no internal energy. The photodissociation of ozone is of crucial importance in the troposphere, as the formation of the excited $O(^1D)$ atom leads to production of the OH radical by rapid reaction with water, and this radical, "nature's atmospheric detergent" [2], is responsible for the majority of daytime oxidation processes which remove pollutants from the atmosphere.

The formation rate of $O(^1D)$ from ozone in the troposphere depends upon the product of three factors: first, the solar flux, which increases rapidly with wavelength above ca. 300 nm, secondly, the ozone absorption cross section, which is doing the reverse, and thirdly the quantum yield of ozone to form the excited $O(^1D)$ fragment. In the mid 90s doubt began to be cast on the then accepted view of the quantum yield data [3], namely that the spin allowed process (1) dominated, and that the O(^{1}D) quantum yield dropped rapidly to zero at wavelengths above the 310 nm energetic threshold. Measurements through the threshold region first on the O$_2$(a$^1\Delta_g$) fragment [4–6] and later by several independent groups on the more important O(^{1}D) species [7–14] showed the existence of a long wavelength "tail" on the quantum yield, and that the tail contained two components. The first, between 310 and 320 nm, was temperature dependent, and had a value of about 20% at 298 K. The second, extending out to above 340 nm, was temperature independent at about the 10% level. Although evidence of the long wavelength tail had been published earlier [15, 16] the data had largely been ignored for atmospheric purposes, but its importance was found to be crucial – inclusion of the tail in calculations of tropospheric chemistry has lead to notable upward revisions of the rate of OH production and hence the oxidising capacity of the troposphere [2].

Explanations for the formation of O(^{1}D) above 310 nm have invoked two effects. The first is a highly efficient dissociation of internally excited molecules, particularly those with quanta in the ν_3 stretching mode. Enhanced Franck Condon overlap of the wavefunctions for ground electronic state O$_3$ in the (0,0,1) level and for the dissociative upper ^{1}B$_2$ state predict that despite the small population of molecules in the vibrationally excited level at tropospheric temperatures their increased absorption cross section leads to a marked dissociation yield, which will be temperature dependent as observed experimentally [17, 18]. The second effect is the inclusion of a spin forbidden step, first proposed for the molecular oxygen fragment almost 30 years ago [19]. For the lowest excited singlet molecular and atomic oxygen products two such steps need to be considered, each forming the ground state of the co-fragment and having a threshold at wavelengths longer than 310 nm:

$$O_3 + h\nu \rightarrow O(^3P) + O_2(a^1\Delta_g) \quad \lambda < 611 \text{ nm} \quad (2)$$
$$O_3 + h\nu \rightarrow O(^1D) + O_2(X^3\Sigma_g^-) \quad \lambda < 411 \text{ nm} \quad (3)$$

Indirect evidence for the steps comes simply from the observation of a temperature independent small quantum yield for both O(^{1}D) and O$_2$(a$^1\Delta_g$) fragments at wavelengths well above 310 nm [5, 13, 14], but for a more concrete distinction between the routes for formation, for example, of O(^{1}D) from steps (1) and (3), quantum yield measurements alone are not sufficient. The distinction has recently been possible by the use of molecular dynamics techniques for the measurement of the kinetic energies of the fragments [9, 20–22], and this is the subject of the present article. We shall describe recent results on the molecular beam photodissociation of ozone in the threshold region near 310 nm, detecting channels (1), (2) and (3) through measurements of the kinetic energies of the O(^{1}D) and O$_2$(a$^1\Delta_g$) fragments.

There has been little previous work on the dissociation dynamics of ozone in the weakly absorbing region near 310 nm. Recent work has concentrated on photofragment spectroscopy and ion imaging measurements at 285 nm and below in the strong Hartley band [23–30], with measurements made both on process (1) and its spin allowed analogue to give ground state products (4):

$$O_3 + h\nu \rightarrow O(^3P) + O_2(X^3\Sigma_g^-) \tag{4}$$

Process (4) has a quantum yield of about 0.1 in the Hartley band, and is of interest because of the formation of highly vibrationally excited O_2 in a bimodal distribution, which has been suggested as a possible reactive precursor to ozone formation in the stratosphere [24, 31, 32]. Notable studies at longer wavelengths include early photofragment spectroscopy measurements [33] which yielded a translational anisotropy factor $\beta = 1.18$ for process (1) at 300 nm, and measurements of the $O_2(a^1\Delta_g)$ internal energy distributions by CARS at wavelengths up to 311 nm [34]. More recently measurements of the Doppler profiles of the $O(^1D)$ fragment between 317 and 327 nm have shown the existence of fast species which are consistent with their formation by the spin forbidden channel (3) [9].

2 Experimental

The experimental details have been described in previous publications [20–22] and are briefly summarised here. A pulsed valve was used to introduce a beam of ozone in oxygen (approximately equal mole fractions at a backing pressure of about 80 Torr) into the source region of a modified Wiley–McLaren time of flight (TOF) mass spectrometer [35], where it was crossed by the output of two tunable dye lasers, one to dissociate the ozone, and a second to ionise the $O(^1D)$ or $O_2(a^1\Delta_g)$ fragment by REMPI. The lasers were either excimer or YAG pumped, and produced pulses of 10 or 8 ns duration respectively and were operated at wavelengths which ranged from 204 nm (used to ionise $O(^1D)$ by $2+1$ REMPI via the $^1F_3$ state) to 331 nm (to ionise $O_2(a^1\Delta_g)$ by $2+1$ REMPI via the $3s\sigma_g d^1\Pi_g$ Rydberg state), with photolysis wavelengths predominantly in the range 290–325 nm.

Ions were detected as a function of their time of arrival at a multichannel plate following the laser pulse. Ions formed from neutrals which are initially travelling with their velocity vectors aligned along the TOF axis (which we refer to as the horizonal axis) will show two peaks, corresponding to "early" arrival (those fragments whose velocities point towards the detector) and "late" arrival (those which travel away from the detector, and are turned round by the applied field after ionisation). Fragments with velocities along the vertical axis arrive close to the mid point of these early and late times. The velocity resolution of the present instrument is approximately $50 \, \text{m s}^{-1}$.

3 Results and Discussion

Vibrationally mediated photolysis of ozone

We first describe briefly our early experimental evidence for the importance of the photolysis of vibrationally excited ozone [20]. We concentrate first on the $O_2(a^1\Delta_g)$ fragment, detected by REMPI at 331.5 nm. This wavelength is at the peak of the 2 + 1 REMPI transition excited via the $O_2 d^1\Pi_g(v = 1) \leftarrow\leftarrow a^1\Delta_g(v = 0)$ band, and recent analysis [36] of the complex spectroscopy of this band has confirmed our initial assignment [20] that it corresponds to excitation of $O_2(a^1\Delta_g)$ in low J (J $\sim$ 9).

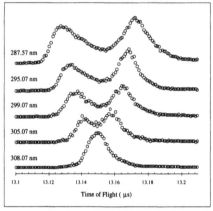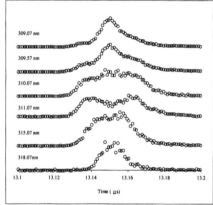

Figure 1: $O_2(a^1\Delta_g v = 0, J \sim 9)$ time of flight profiles for photolysis of ozone at 11 wavelengths between 287.57 and 318.07 nm, taken with the electric vector of both photolysis and REMPI lasers horizontal, i.e., parallel to the TOF axis. The profiles show early and late peaks at short wavelengths, which are unresolved as the available kinetic energy decreases, but then become wider near 311 nm owing to photolysis of internally excited ozone

Figure 1 shows the TOF spectra for photolysis at 11 wavelengths between 287 and 318 nm, passing through the 310 nm threshold, in which the electric vector of the photolysis laser is polarised in the horizontal direction, i.e. along the TOF axis. At the shortest wavelength, 287.57 nm, early and late peaks are clearly seen, and this is caused by the photolysis step being a parallel one, with the fragments departing largely parallel to the transition moment and thus to the laboratory horizontal axis which is coincident with the electric vector. We return to this point later, but note at present that the separation of the early and late peaks gives a (non-linear) measure of the kinetic energy release in the state selected $O_2(a^1\Delta_g)$ fragment, and at 287.57 nm this is entirely consistent with the fragment being formed in process (1).

At longer wavelengths the peaks converge, merging due to the finite resolution of the instrument. However at wavelengths close to the threshold of process (1) this convergence is reversed, and the just separated peaks at 311.07 nm in Fig. 1 imply a higher kinetic energy for the (state selected) fragments than for those formed by photolysis at shorter wavelengths. Internal energy in the ozone is the only possible source of this energy, and the separation of the peaks at 311.07 nm is consistent with this being at a value of $\sim 1000\,\text{cm}^{-1}$. Later measurements of the appearance threshold of $O(^1D)$ by Takahashi et al have also revealed this effect in process (1), and suggested that the internal energy comes from excitation of the ν_3 asymmetric stretch ($1042\,\text{cm}^{-1}$) in ozone [11], thus confirming an observation made some twenty years ago that infrared excitation of the ν_3 mode enhances the uv dissociation of ozone to give $O(^1D)$ [37].

Spin forbidden dissociation of ozone

State selected detection of $O_2(a^1\Delta_g)$ produces more clear cut results than that of $O(^1D)$, as the former species is essentially monoenergetic, whilst the atom can take a range of kinetic energies owing to the internal energy distribution in the diatomic co-fragment. Similar effects of ozone internal energy can still be observed however, and Fig. 2 shows TOF profiles which are now no longer clearly resolved into early and late peaks, but which are broader (and correspond to more kinetic energy) for photolysis at 313 than at 310 nm. We now use the $O(^1D)$ data to illustrate the occurrence of spin forbidden processes at longer wavelengths, in a similar way to that already described for the $O_2(a^1\Delta_g)$ fragment [21, 22]. Figure 3 illustrates the TOF spectra for $O(^1D)$ formed between 319 and 324.8 nm.
What is clear in these examples is that the central peak from $O(^1D)$ with low kinetic energy is gradually subsumed by a broad pedestal of faster fragments which we assign to their formation in the spin forbidden channel (3). In the case of the essentially monoenergetic $O_2(a^1\Delta_g)$ fragment detected from the similar spin forbidden channel (2), well resolved early and late peaks were clearly seen [20, 22], but for $O(^1D)$ the broad distribution of its kinetic energies leads to a smearing of this effect. Figure 3 however shows that at 324.8 nm the expected highest kinetic energy fragment distribution for process (3) (i.e. no internal energy in the $O_2(X^3\Sigma_g^-)$ cofragment) is predicted at the early and late arrival times experimentally observed. We note that Doppler lineshape measurements by LIF of the $O(^1D)$ fragment at wavelengths above 320 nm has also shown the presence of fast atoms which can only be produced by the spin forbidden process (3) [9]. Furthermore, both these Doppler width measurements [9] and our measurements by REMPI [22] indicate that the ratio of fast to slow fragments in the Huggins band region above ca. 320 nm is dependent upon wavelength, showing a general increasing trend with increasing wavelength,

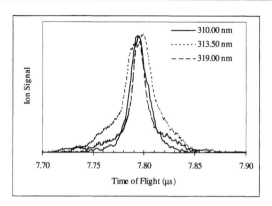

Figure 2: O(^{1}D) time of flight profiles following photolysis of ozone at 310.0, 313.5 and 319 nm, scaled to the same maximum height. The central peak becomes wider for the middle wavelength, again consistent with absorption by internally excited ozone. Evidence of faster fragments from the spin forbidden channel can be seen

but with local maxima appearing at the peaks in the structured Huggins bands.

We add a note of caution in these experiments. Two laser beams are present at relatively high intensities in order to induce multiple photon absorption, and care must be taken to allow for both photolysis of the parent molecule by the REMPI beam, and non-resonant formation of ions by multiple photon ionisation and fragmentation of ozone. This is particularly noticeable for photolysis at the longest wavelengths, and requires data such as those in Figs. 2 and 3 to be taken with the lowest possible laser intensities. Higher intensities produce copious spurious signals, particularly for fragments at low kinetic energies.

Angular distributions

The angular distribution, D[v, θ], of the photofragments' velocities v relative to the electric vector ϵ of the photolysis laser is described in terms of the translational anisotropy factor β

$$D[v, \theta] = (1 + \beta[v]P_2[\cos \theta])/4\pi \qquad (5)$$

where θ is the angle between v and ϵ and P$_2$ is a second degree Legendre polynomial [38]. β can be velocity dependent, as has been found for the O(^{3}P) product of process (4) [29].

Figure 4 shows measurements of the TOF spectra at both horizontal and vertical polarisations of the photolysis laser at 290 nm, with state selected detection of O$_2$(a$^1\Delta_g$) at 331.5 nm. In these experiments the electric vector

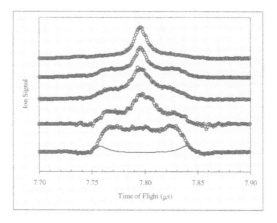

Figure 3: O(^{1}D) time of flight profiles for photolysis of ozone at wavelengths between 319 and 324.8 nm. Fast fragments from the spin forbidden step (3) can be seen to increase in importance as the wavelength is increased. At the longest wavelength the solid line shows a simulation for the fastest possible fragments of process (3), i.e. where all the available energy is partitioned into translational energy of O(^{1}D) and O$_2$(X$^3\Sigma_g^-$)

of the REMPI laser was set at the magic angle of 54.7° to minimise alignment effects in the REMPI process [39, 40]. Simulations are shown with various values of β, from which it can be seen that β can be estimated as 1.3 with a precision of the order of 10%. The fits have still to be improved – the artifact of the overshoot at longer times is taken from early calibrations of the instrument with a thermal source of NO. The value compares well with previous measurements in the Hartley band for photolysis at 248 nm [23, 25], 266 nm [30] and 300 nm [33], but is somewhat outside the range of the measurements at 285 nm ($\beta = 1.6 \pm 0.2$ [25]) and 276 nm ($\beta = 0.71 \pm 0.17$ [28]). However, it is clear that all the measurements are consistent with a parallel absorption step to a state of $^1B_2$ symmetry, followed by a prompt dissociation, for which a value of $\beta = 1.18$ would be calculated [33].

Figure 5 shows the results of the same experiment in which the 290 nm photolysis laser was fixed at the magic angle and the REMPI laser switched between horizontal and vertical polarisation. A difference can be seen, indicating that there is an alignment between the velocity vector of the O$_2$(a$^1\Delta_g$) fragment and its angular momentum vector. The clearest effect is in the upper trace, where enhanced signal is seen for the fastest fragments observed at earliest and latest times, i.e. for those whose velocity vectors lie in the horizontal direction parallel to the TOF axis. REMPI detection favours these species and implies that the transition moment for the REMPI process

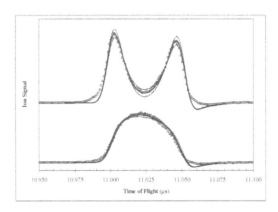

Figure 4: $O_2(a^1\Delta_g v = 0, J \sim 9)$ time of flight profiles for photolysis of ozone at 290 nm with the electric vector horizontal (parallel to the TOF axis, upper trace) and vertical (perpendicular to the TOF axis, lower trace). The REMPI polarisation was set at the magic angle, 54.7°. For each profile three simulations are shown for the spin allowed process (1) taking place with $\beta = 1.1$ (dashed line), 1.3 (solid line) and 1.5 (dot-dashed line), from which it can be seen that the precision of the measurements is of the order of 10%

lies preferentially parallel to the fragments' velocity. For prompt impulsive dissociation of ozone we would expect the $O_2(a^1\Delta_g)$ angular momentum vector J to lie predominantly in a plane perpendicular to its velocity, and thus we predict that the transition moment should be perpendicular to J. Two photon absorption in an O branch of a Π-Δ transition would be expected to give this polarisation dependence: we are as yet not in a position to extract quantitatively the correlations between the velocity and angular momentum vectors because of the unknown polarisation contributions from the intermediate states in the two photon transition, and the possible anisotropy of the ionisation of the aligned $3s\sigma_g d^1\Pi_g$ Rydberg state.

At wavelengths above the 310 nm threshold the anisotropy of the spin forbidden channels (2) and (3) has been investigated. Both processes always yield positive β values. For the $O_2(a^1\Delta_g)$ product these were measured only in single laser experiments, i.e. the same laser used to dissociate ozone and to ionise the fragment [22], and thus measurements in which the laser's plane of polarisation was varied probed the anisotropy of both the dissociation and REMPI steps. Assuming that the former effect dominates, values of $\beta \sim 1.2$ were found in the 320 nm region, similar to those found at shorter wavelengths for the spin allowed process (1) and described above. Formation of the spin forbidden products would be consistent with initial absorption in a parallel transition, followed by rapid (compared with

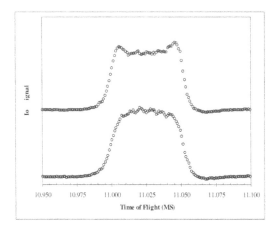

Figure 5: $O_2(a^1\Delta_g$ v = 0, J ~ 9) time of flight profiles for photolysis of ozone at 290 nm with the electric vector of the photolysis laser at the magic angle. In the upper panel the electric vector of the REMPI laser is parallel to the TOF axis, and perpendicular in the lower panel. No alignment in the $O_2(a^1\Delta_g)$ fragment would result in a "top hat" profile: the enhanced detectivity of the fastest fragments seen in the upper trace is consistent with the expected alignment of the angular momentum vector of O_2 perpendicular to the transition moment as explained in the text

a rotational period) crossing to a dissociative triplet state. The initial transition thus could be to a $^1B_2$ state as suggested by the Huggins bands assignments of both Katayama [41] and Joens [42] (where in the latter case the vibronic symmetry of the assigned $^1A_1$ state becomes $^1B_2$ for excitation of odd numbers of quanta in the antisymmetric ν_3 mode). Although our β values do not rule out direct absorption to a triplet state of the correct symmetry, the spectroscopic evidence points to a singlet state being the major carrier of the absorption oscillator strength in this region, with the spin forbidden products formed by intersystem crossing to a dissociative triplet surface. Similar conclusions have recently been reached by TOF measurements of channels (2) and (4) by observations of the $O(^3P)$ fragment, in addition, these studies identified a substantial component of $O_2(b^1\Sigma_g)$ as being formed following photolysis on resolved features in the Huggins band system near 320 nm [43]. Estimates of the β parameter for photolysis at longer wavelengths (327 and 331.5 nm for $O_2(a^1\Delta_g)$, 322 and 324 nm for $O(^1D)$) are subject to larger errors and only qualitative estimates have so far been possible. These show again that the β values are always positive, but indicate a noticeable lowering as wavelength increases, presumably caused by increased lifetime of the molecule in the upper vibronic levels reached by absorption at longer wavelengths.

4 Conclusions

Molecular beam measurements of the dissociation dynamics of ozone have revealed the contribution of several processes to the overall dissociation yield near 310 nm. It has been shown that both vibrationally mediated and spin forbidden photodissociation take place in the formation of $O(^1D)$ at wavelengths above the energetic threshold for its production in the spin allowed step (1). Several groups have now completed measurements of the temperature dependent quantum yields of the $O(^1D)$ fragment [7–14], with good agreement found between results obtained by three different detection methods, all showing a pronounced "tail" in the quantum yield extending above 310 nm. The new data have had a marked effect on calculated values of the $O(^1D)$ production rate, and hence on the oxidising capacity of the troposphere. For example, following the first publication indicating the existence of the tail in the quantum yield of $O_2(a^1\Sigma_g)$, the co-fragment of process (1) [4], Müller et al calculated the effect on the tropospheric $O(^1D)$ yield in the summer in Jülich, Germany. They demonstrated that at least 30% of the noontime flux of the species came from photolysis of ozone in the tail region [44]. More dramatic effects take place at high solar zenith angles and large overhead stratospheric ozone concentrations, such as exist at high latitudes in the winter months. A calculation for the upper troposphere (20 km altitude) with the sun at a zenith angle of 85° indicates that inclusion of the new data increases the calculated $O(^1D)$ production rate by a factor of two [2]. It is pleasing (to the present authors) to note that measurements carried out with molecular beam and laser techniques have made a marked quantitative impact on our understanding of an important part of atmospheric chemistry.

Acknowledgements

This research was supported by the Natural Environment Research Council and by the European Commission Environmental Programme.

References

[1] R.P. Wayne: In: The Handbook of Environmental Chemistry, Vol. 2 Part E, edited by O. Hutzinger, Springer, Berlin (1989) p. 1

[2] A.R. Ravishankara, G. Hancock, M. Kawasaki, Y. Matsumi: Science **280**, 60 (1998)

[3] W.B. DeMore, S.P. Sander, D.M. Golden, R.F. Hampson, M.J. Kurylo, C.J. Howard, A.R. Ravishankara, C.E. Kolb, M.J. Molina: Chemical kinetics and photochemical data for use in stratospheric modeling. Evaluation number 11, JPL publication 94-26 (1994)

[4] S.M. Ball, G. Hancock, I.J. Murphy, S.P. Rayner: Geophys. Res. Letters **20**, 2063 (1993)

[5] S.M. Ball, G. Hancock: Geophys. Res. Letters **22**, 1213 (1995)

[6] S.M. Ball, G. Hancock, F. Winterbottom: J. Chem. Soc. Faraday Disc. **100**, 215 (1995)

[7] W. Armerding, F.J. Comes, B. Schülke: J. Phys. Chem. **99**, 3137 (1995)

[8] K. Takahashi, Y. Matsumi, M. Kawasaki: J. Phys. Chem. **100**, 4084 (1996)

[9] K. Takahashi, M. Kishigami, Y. Matsumi, M. Kawasaki, A.J. Orr-Ewing: J. Chem. Phys. **105**, 5290 (1996)

[10] S.M. Ball, G. Hancock, S.E. Martin, J.C. Pinot de Moira: Chem. Phys. Letters **264**, 531 (1997)

[11] K. Takahashi, M. Kishigami, N. Taniguchi, Y. Matsumi, M. Kawasaki: J. Chem. Phys. **106**, 6390 (1997)

[12] E. Silvente, R.C. Richter, M. Zheng, E.S. Saltzman, A.J. Hynes: Chem. Phys. Letters **264**, 309 (1997)

[13] K. Takahashi, N. Taniguchi, Y. Matsumi, M. Kawasaki, M.N.R. Ashfold: J. Chem. Phys. **108**, 7161 (1998)

[14] R.K. Talukdar, C.A. Longfellow, M.K. Gilles, A.R. Ravishankara: Geophys. Res. Letters **25**, 143 (1998)

[15] J.C. Brock, R.T. Watson: Chem. Phys. **46**, 477 (1980)

[16] M. Trolier, J.R. Wiesenfeld: J. Geophys. Res. D **93**, 7119 (1988)

[17] S.M. Adler-Golden, E.L. Schweitzer, J.I. Steinfeld: J. Chem. Phys. **76**, 2201 (1982)

[18] H.A. Michelsen, R.J. Salawitch, P.O. Wennberg, J.G. Anderson: Geophys. Res. Letters **21**, 2227 (1994)

[19] P.J. Crutzen, I.T.N. Jones, R.P. Wayne: J. Geophys. Res. **76**, 1490 (1971)

[20] S.M. Ball, G. Hancock, J.C. Pinot de Moira, C.M. Sadowski, F. Winterbottom: Chem. Phys. Letters **245**, 1 (1995)

[21] W. Denzer, G. Hancock, J.C. Pinot de Moira, P.L. Tyley: Chem. Phys. Letters **280**, 496 (1997)

[22] W. Denzer, G. Hancock, J.C. Pinot de Moira, P.L. Tyley: Chem. Phys. **231**, 109 (1998)

[23] A.G. Suits, R.L. Miller, L.S. Bontuyan, P.L. Houston: J. Chem. Soc. Faraday Trans. **89**, 1443 (1993)

[24] R.L. Miller, A.G. Suits, P.L. Houston, R. Toumi, J.A. Mach, A.M. Wodtke: Science **265**, 1831 (1994)

[25] M.-A. Thelen, T. Gejo, J.A. Harrison, J.R. Huber: J. Chem. Phys. **103**, 7946 (1995)

[26] J.A. Syage: J. Phys. Chem. **99**, 16530 (1995)

[27] J.A. Syage: J. Phys. Chem. **105**, 1007 (1996)

[28] D.A. Blunt, A.G. Suits: Am. Chem. Soc. Symposium Series **678**, 99 (1997)

[29] R.J. Wilson, J.A. Mueller, P.L. Houston: J. Phys. Chem. **101**, 7593 (1997)

[30] K. Takahashi, N. Taniguchi, Y. Matsumi, M. Kawasaki: Chem. Phys. **231**, 171 (1998)

[31] R. Toumi, P.L. Houston, A.M. Wodke: J. Chem. Phys. **104**, 775 (1996)

[32] P.L. Houston, A.G. Suits, R. Toumi: J. Geophys. Res. **101**, 18829 (1996)

[33] C.E. Fairchild, E.J. Stone, G.M. Lawrence: J. Chem. Phys. **69**, 3632 (1978)

[34] J.J. Valentini, D.P. Gerrity, D.L. Phillips, J.-C. Nieh, K.D. Tabor: J. Chem. Phys. **86**, 6745 (1987)

[35] W.C. Wiley, I.H. McLaren: Rev. Sci. Instr. **26**, 1150 (1955)

[36] B.R. Lewis, S.T. Gibson, J.S. Morrill, M.L. Ginter: J. Chem. Phys. **111**, 186 (1999)

[37] P.F. Zittle, D.D. Little: J. Chem. Phys. **72**, 5900 (1980)

[38] R.N. Zare: Mol. Photochem. **4**, 1 (1972)

[39] M. Mons, I. Dimicoli: Chem. Phys. Letters **131**, 298 (1986)

[40] M. Mons, I. Dimicoli: J. Chem. Phys. **90**, 4037 (1989)

[41] D.H. Katayama: J. Chem. Phys. **71**, 815 (1979)

[42] J.A. Joens: J. Chem. Phys. **101**, 5431 (1994)

[43] P. O'Keeffe, T. Ridley, K.P. Lawley, R.R.J. Maier, R.J. Donovan: J. Chem. Phys. **110**, 10803 (1999)

[44] M. Müller, A. Kraus, A. Hofzumahaus: Geophys. Res. Letters **22**, 679 (1995)

Photodissociation of NO_2 near 225 nm by Velocity Map Imaging

M. Ahmed, D. S. Peterka, A. G. Suits
Chemical Sciences Division
Ernest Orlando Lawrence Berkeley National Laboratory
Berkeley, California 94720, USA

1 Introduction

Since the pioneering work of Busch and Wilson[1] with the inception of Photofragment Translational Energy Spectroscopy (PTS), NO_2 has been one of the most studied molecules in the field of photodissociation dynamics. Recently experimentalists have been using a variety of elegant techniques to unravel the intricate photodissociation dynamics of this model triatomic molecule[2, 3]. A whole body of work has concentrated around the threshold regime (400 nm) for the $O(^3P_j)$ + NO channel. Photodissociation studies beyond the threshold regime have been sparse. These have focussed on recording the internal state distribution of the nascent NO fragment from photolysis at 248 nm utilizing laser-induced fluorescence (LIF)[4] and resonance enhanced multiphoton ionization (REMPI)[5], and photolysis around 226 nm with detection by REMPI[6]. There have also been two reports[7, 8] on the fine- structure population distributions of the oxygen atom produced by photolysis over a range of UV wavelengths. Miyawaki et al. also reported the kinetic energy of the oxygen fragment to be 0.13 eV from analysis of their Doppler profiles and that the energy release did not vary with the photolysis wavelength (355, 337, 266, and 212 nm). Shafer et al.[9] recorded the Doppler spectra for the excited oxygen atom $O(^1D)$ at 205.47 nm and derived a translational energy release which was between 0.11 eV and 0.211 eV, and an anisotropy parameter β of 1.3.

In this paper we present a study of the photodissociation dynamics for NO_2 in the region of 221-226 nm using velocity map imaging[10] (VELMI) in conjunction with REMPI. We chose this region because it allowed us to use a single laser to photodissociate and probe the NO and $O(^3P)$ products using REMPI. Probing the angular and energy distributions of both fragments under state resolved conditions allows for detailed investigations of the product state distributions in the photodissociation dynamics of NO_2.

2 Experimental

The molecular beams apparatus has been described in detail in a recent publication[11]. A 5% mixture of NO_2/O_2 seeded in He was expanded through a Proch-Trickl pulsed valve[12], collimated by a single skimmer and then crossed by a laser beam on the axis of a velocity focusing time-of-flight mass spectrometer. The photolysis and probe light around 221-226 nm was generated by doubling the output of a seeded Nd-YAG pumped dye laser in β- barium borate (BBO), then mixing the resultant UV light with the dye fundamental in a second BBO crystal.

The NO product from NO_2 dissociation was probed using (1+1) REMPI through the $A(^2\Sigma) \leftarrow X(^2\Pi)$ transition (the γ bands). The various spin-orbit states of the O product were probed using (2+1) REMPI through the $3p(^3P_J) \leftarrow \leftarrow 2p(^3P_J)$ transition.

The NO^+ and O^+ ions were accelerated toward a 80-mm diameter dual microchannel plate (MCP) coupled to a phosphor screen and imaged on a fast scan charge-coupled device camera with a integrating video recorder (Data Design AC-101M). Camera threshold and gain were adjusted in conjunction with a binary video look-up table to perform integration of single ion hits on the MCP free of video noise. Images were accumulated while scanning across the Doppler profile of the O atom, since the line width of the laser light was narrower than the Doppler spread. The translational energy and angular distributions were determined from the images by applying the inverse Abel transform and integrating appropriately.

3 Results

Figure1 shows the raw ion-image for the $(^3P_0)$ state of the oxygen atom obtained from photodissociation of NO_2 at 226.23 nm. Similar images were obtained for the $(^3P_2)$ and $(^3P_1)$ spin orbit states of ground state oxygen but there is evidence of atomic orbital alignment for these states[13], which is the subject of a separate study. For those levels, unknown contributions of the orbital alignment can interfere with a direct determination of the anisotropy parameter. For the $(^3P_0)$ state this is not a problem, since there is no total angular momentum and hence no orbital alignment. The shape of the image in Figure1 is from the photofragment angular distribution (the velocity anisotropy), characterized by the β parameter, which is determined to be 1.32 from the fit shown in Figure2(B).

Examining the translational energy distributions of the oxygen atom provides a mapping of the internal state distributions of the NO cofragment because of momentum conservation. Figure2(A) shows such a translational energy distribution for the $O(^3P_0)$ fragment. At 226.23 nm the photon en-

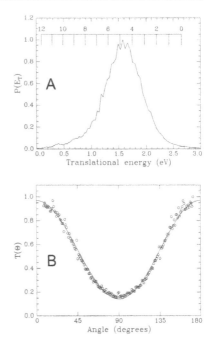

Figure 1: Image of $O(^3P_0)$ from NO_2 dissociation at 226.23 nm

Figure 2: Translational energy (A) and angular distribution (B) of $O(^3P_0)$ derived from the image in the previous figure.

ergy is 5.48 eV. The bond dissociation energy (D_0) for the $NO_2 \rightarrow NO(X^2\Pi)$ + $O(^3P_0)$ channel is 3.11 eV. Internally cold products should show up at 2.37 eV, but the peak in Figure2A is shifted away from this value. The main peak around 1.6 eV corresponds to a range of excited vibrational levels of the NO ($X\ ^2\Pi$) ground state, peaking at v=4 or 5. The comb above the tranlsational energy distribution in Figure2 (A) shows thresholds for indicated vibrational levels in the NO cofragment. Except for v=0, the vibrational distribution for NO obtained from the $O(^3P_0)$ translational energy distributions are in qualitative agreement with the results of McFarlane et al[5]. They obtained highly inverted vibrational distributions of NO peaking at v=5 from NO_2 photodissociation at 248 nm. In contrast to our results, they observed significant population of NO v=0, which could be contributions from NO in their molecular beam or from NO v=5 as discussed below.

Figure3 shows images for various ro-vibrational states of NO. The ground vibrational state images shown in Figure3 (A) and (B) were collected with photolysis at 225.64 and 225.68 nm respectively. The image in Figure3(B) is from the N=16 rotational state probed on the R11, Q21 line. The outer

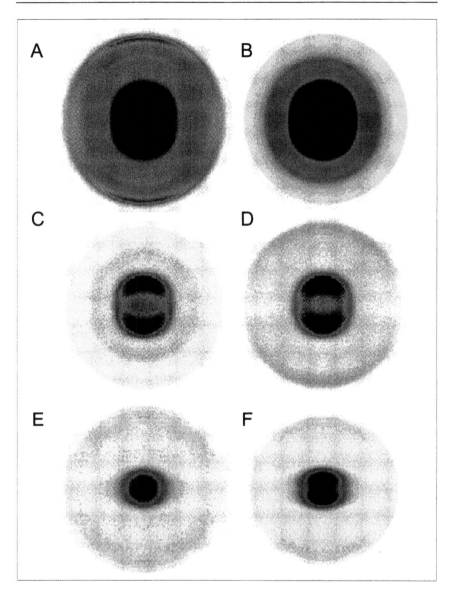

Figure 3: NO$^+$ images on various rotational lines of v=0 (A,B), v=1 (C,D) and v=2 (E,F)

ring in all the images correspond to NO formed in conjunction with O(^3P_J) and the inner ring corresponds to O(^1D). The corresponding translational energies are shown in Figure4. At 225.64 nm (5.49 eV), for NO (v=0), O(^3P) will show up at 2.38 eV (D_0=3.11 eV) while the peak for O(^1D)

will appear around 0.41 eV. This agrees very well with our experimental results as can be seen from the translational energy distributions shown in Figure4(A).

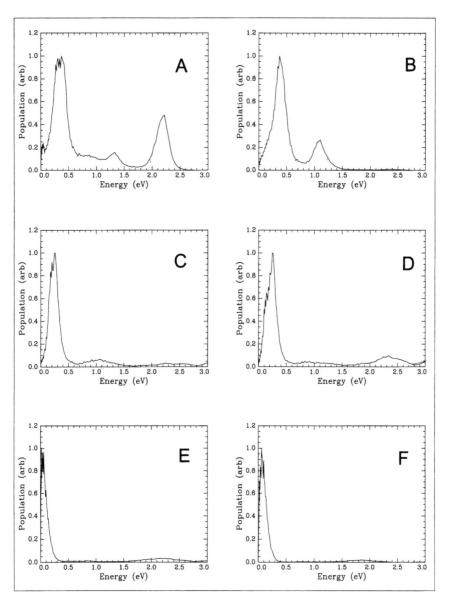

Figure 4: NO$^+$ translational energy distributions for the rotational lines of the previous figure

Comparing Figure4(A) with Figure4(B) shows that for very similar rotational states the branching ratio to the two oxygen atom channels are very different. For NO (v=0,N=16), the ratio for $O(^1D):O(^3P)$ is 1:0.01 while for a nearby line it is 1:0.44. This suggests very different underlying rotational distributions for the two different channels. An alternative but perhaps less plausible explanation is a dramatic change in the dynamics with this small change in photolysis wavelength. Similar results are also seen for results for NO (v=1 and 2). In addition to the two channels mentioned above, we see a ring which is formed with a translational energy release of 1.1 eV. On energetic grounds this ring cannot arise from photodissociation of N_2O_4 (NO dimer) or from HONO (possible contaminant in the NO_2 beam). Kawasaki et al.[14] have shown that photodissociation of N_2O_4 at 193 and 248 nm gives rise to exclusively excited state NO_2 products and they did not find any evidence for NO or O formation.

An examination of the spectroscopy of the γ bands of NO shows that the (4,5) vibrational band has an origin at 225.6 nm. Thus various rotational lines from the (4,5) band can overlap with the (0,0) band which normally is thought to dominate in the 226 nm region. The Frank-Condon factors for the (4,5) and (0,0) bands are 0.057 and 0.162 respectively[15], but we have already seen that NO is formed predominantly in the v=5 level, so its contribution will show up clearly as we scan across the (0,0) band head of NO. This result shows that extracting rotational population distributions for NO (v=0) using the (0,0) transition is fraught with error if NO (v=5) is also formed in the process being examined. This overlapping of bands could account for the large NO (v=0) population observed by McFarlane et al. in the 248 nm photodissociation of NO_2, where they used 1+1 REMPI in the 224.9-227.5 nm region to obtain their NO (v=0) population.

It is apparent from the images shown for NO, that the angular distribution for the $O(^3P)$ component is very different from the $O(^1D)$ component. Figure5 shows the angular distribution for $O(^1D)$, derived from the Abel-transformed NO (v=0) images (the intense inner ring) shown in Figure4(A) and (B). We derive a β of 0.6±0.1 for NO (v=0) formed in conjunction with an excited state oxygen atom. For the NO(v=2) results shown in Figure3(E) it is quite remarkable that the inner ring, corresponding to its formation in coincidence with $O(^1D)$, is completely isotropic, while a nearby line (Figure 3(F)) shows a strongly anisotropic angular distribution. The most likely explanation is simply that this level, so close to the energetic threshold for this process, experiences a long predissociation lifetime.

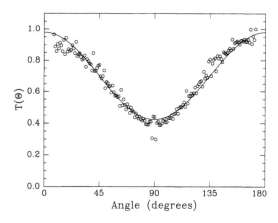

Figure 5: Angular distribution for inner ring of Fig. 3A: NO (v=0) formed in conjunction with O(^1D)

4 Discussion

Figure6 shows the various energy levels for NO_2 that could be accessed with a 226 nm photon, and the energy levels for the dissociation pathways to O and NO. Most studies of NO_2 photodissociation have been performed near the threshold of the O(^3P_j) + NO channel where the NO_2 is primarily excited to the 1($^2B_2$) and/or 1($^2B_1$) electronic states. At 226 nm, in addition to these states, NO_2 can be excited to the 2($^2B_2$) state. A transition to this state is fully allowed from the X ($^2A_1$) ground state; this state is predissociative with a lifetime of about 42 ps at the origin, decreasing to the sub-picosecond regime by the first progression band[16].

Using chemical actinometry techniques, Uselman and Lee[17], showed that upon absorption of a 228.8 nm photon, NO_2 dissociated to form O(^1D) and O(^3P) in equal amounts. They postulated that above 245 nm, where the O(^1D) quantum yield was practically zero, case I predissociation[18] (nonradiative transition to the continuum of another electronic state) dominates. Below 245 nm, case II predissociation (dissociation by vibrational rearrangement to a vibrational manifold with a lower dissociation limit than the ŞdiscreteŤ vibrational manifold, both in the same electronic state) plus case I predissociation was operative. Uselman and Lee suggested that O(^3P) was formed by the case I and O(^1D) by case II mechanisms.

Rubahn et al.[7] carried out an extensive study on the fine-structure population distributions of O(^3P_J) from the photolysis of NO_2 over a range of wavelengths. Compared to 355 nm, they found that at 226 nm, the fine structure distributions had a propensity to populate the 1 and 0 levels.

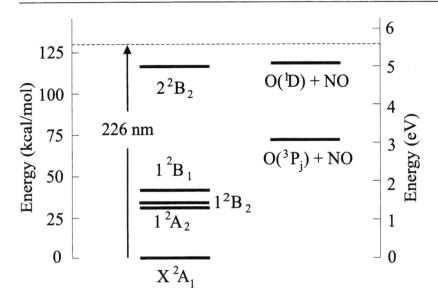

Figure 6: Schematic diagram showing the energy levels for the various relevant states of NO_2 and products.

Miyawaki et al.[8] saw a similar effect for photodissociation at 212 nm. Both groups argued that the opening up of the $O(^1D)$ + NO channel somehow affected the spin-orbit ratios for the ground state oxygen atom.

Since the ground state is totally symmetric ($^2A_1$), the transition dipole will have the same symmetry as the excited state. This means that the transition dipole moment will be perpendicular to the C_2 axis and in the plane of the molecule. Prompt dissociation from this geometry will give rise to $\beta=1.4$. For $O(^3P_0)$, we find $\beta=1.3$, and for the outer component in the various NO images we find similar anisotropy parameters. These results are consistent with prompt dissociation from the $B(^2B_2)$ state. Hradil et al.[19] report a value of 1.2 ± 0.3 for the $O(^3P_2)$ fragment formed in the 355 nm dissociation of NO_2 using photofragment ion imaging. The observation of a lower anisotropy parameter for the $O(^1D)$ channel in our experiments is consistent with a case II predissociation in which the excited state is longer lived compared to case I (the $O(^3P)$ case). The observed angular distribution (the lower β parameter) for $O(^1D)$ could also come from dissociation via a more strongly bent NO_2 excited state geometry.

An explanation for the different ratios for $O(^1D):O(^3P)$ for various NO co-fragments is not forthcoming from the present results. Clearly more experiments should be performed in which separate lasers are used to initate the photodissociation of NO_2 and probe the NO photofragment. Finally experiments are underway to study the $O(^1D)$ + NO channel directly, using REMPI to probe $O(^1D)$ at 205 nm.

Acknowledgments

This work was supported by the Director, Office of Energy Research, Office of Basic Energy Sciences, Chemical Sciences Division of the U. S. Department of Energy under contract No. DE-ACO3-76SF00098.

References

[1] G.E. Busch and K.R. Wilson, J. Chem. Phys. **56**, 3626 (1972).

[2] S. A. Reid, D. C. Dobie, and H. Reisler, J. Chem. Phys. **100**, 4256 (1994).

[3] C.-H. Hsieh, Y.-S. Lee, A. Fujii, S.-H. Lee, and K. Liu, Chem. Phys. Lett. **277**, 33 (1997).

[4] T.G. Slanger, W.K. Bischel, and M.J. Dyer, J. Chem. Phys. **79**, 2231 (1983).

[5] J. McFarlane, J.C. Polanyi, and J.G. Shapter, J. Photochem. Photobiol. A: Chem. **58**, 139 (1991).

[6] L. Bigio, R.S. Tapper, and E.R. Grant, J. Phys. Chem. **88**, 1271 (1984).

[7] H. G. Rubahn, W.J. van der Zande, R. Zhang, M.J. Bronikowski, and R.N. Zare, Chem. Phys. Lett. **186**, 154 (1991).

[8] J. Miyawaki, T. Tsuchizawa, K. Yamanouchi, and S. Tsuchiya, Chem. Phys. Lett. **165**, 168 (1990).

[9] N. Shafer, K. Tonokura, Y. Matsumi, S. Tasaki, and M. Kawasaki, J. Chem. Phys. **95**, 6218 (1991).

[10] A.T.J.B. Eppink and D.H. Parker, Rev. Sci. Instr. **68**, 3477 (1997).

[11] M. Ahmed, D. A. Blunt, D. Chen, and A. G. Suits, J. Chem. Phys. **106**, 7617 (1997).

[12] D. Proch and T. Trickl, Rev. Sci. Inst. **60**, 713 (1989).

[13] M. Ahmed, D. S. Peterka, O. Vasyutinskii, A. S. Bracker, and A. G. Suits, J. Chem. Phys. **110**, 4115 (1999).

[14] M. Kawasaki, K. Kasatani, and H. Sato, Chem. Phys. **65**, 78 (1983).

[15] D. C. Jain and R. C. Sahni, Trans. Faraday Soc. **64**, 3169 (1968).

[16] K. E. J. Hallin and A. J. Merer, Can. J. Phys. **54**, 1157 (1976).

[17] W.M. Uselman and E.K.C. Lee, J. Chem. Phys. **65**, 1948 (1976).

[18] G. Herzberg, *Molecular Spectra and Molecular Structure III, Electronic Spectra and Electronic Structure of Polyatomic Molecules, 2nd Ed.*, Krieger, Coral Gables, Florida, 1991.

[19] V.P. Hradil, T. Suzuki, S. A. Hewitt, P.L. Houston, and B.J. Whitaker, J. Chem. Phys. **99**, 4455 (1993).

The (Ba··FCH$_3$)* Photofragmentation Channels: Dynamics of the Laser Induced Intracluster (Ba··FCH$_3$)* → BaF* + CH$_3$ and Ba* + FCH$_3$ Reaction

S. Skowronek, A. González Ureña*

Unidad de Láseres y Haces Moleculares, Instituto Pluridisciplinar, Universidad Complutense, Paseo Juan XXIII, 1, 28040 Madrid, Spain

1 Introduction

The spectroscopy and dynamics of the intermediate species between reactives and products of chemical reactions, i.e. the transition state region [1], have been the subject of several studies in recent years [1, 2, 3, 4, 5, 6, 7, 8, 9, 10, 11, 12, 13, 14, 15]. There are different experimental approaches to gain insight into this part of the potential energy surface: Zewail's group has probed these short-lived states in real time using femtosecond lasers [4]. Another approach pioneered by Soep and co-workers [5, 6, 7, 8, 9, 10] and by Wittig and co-workers [11, 12] is to study photoinduced chemical reactions in van der Waals molecules, were the reaction surface is accessed directly by laser excitation of the complex. The reactants are oriented and the set of impact parameters is limited by the complex geometry. Soep and co-workers studied the Hg··Cl$_2$ and the Ca··HX (X = halogen) complexes, monitoring the yield of the electronically excited product, while Wittig and co-workers substituted the metal by a molecule as for example in the HI··OCO complex. Unfortunately, the reaction dynamics information obtained from these specific van der Waals approaches where product action spectra are measured, is somewhat blurred by the (normally) unknown spectroscopic factors controlling the laser excitation from the ground (non-reactive) up to the excited (reactive) van der Waals potential.

Polanyi's group [13, 14, 15, 16, 17] extended the van der Waals approach to studies of alkali metal atom "harpooning" reactions with complexes of

---
*To whom Correspondence should be adressed

Na with CH_3Cl, CH_3F and PhF (Ph = phenyl radical). They measured the photodepletion of the parent complexes through time of flight mass spectrometry (TOFMS), rather than the formation of electronically excited products. This provided interesting spectroscopic information about the electronically excited van der Waals potential.

Nevertheless, none of these approaches provided simultaneous information about both parent photodepletion and the product action spectra. This makes it difficult to obtain clear information about the dynamics of the underlying bimolecular reaction.

In the present work TOFMS was used not only to investigate the photodepletion spectrum of the $Ba\cdots FCH_3$ complex, but also to measure the product action spectra, thus gaining a more complete insight into the reaction dynamics of this excited harpooning reaction.

The measurement of the product action spectrum of the BaF reactive channel in the range $16065 - 16340 \, cm^{-1}$, together with the photodepletion spectrum, allowed us to determine the reaction probability of the excited harpooning reaction, e.g.:

$$Ba\cdots FCH_3 + h\nu \to |Ba\cdots FCH_3|^* \to (Ba^+\cdots{}^-FCH_3) \to BaF^* + CH_3 \,,$$

in this interval. On the other hand, both action spectra of the BaF reactive and the Ba non-reactive channels where measured in the range $17795 - 18250 \, cm^{-1}$. The reaction probabilities that have been obtained show an oscillatory behaviour with opposite phase that could be due to interference between the two major photofragmentation channels, i.e. the reactive one, $BaF + CH_3$, and the non-reactive one, $Ba + CH_3F$.

2 Experimental

A detailed description of the experimental apparatus employed in this work has been given elsewhere [3], and only a brief description will be presented here. The weakly bound complex $Ba\cdots FCH_3$ is produced in a laser vaporisation source followed by supersonic expansion. The molecular beam is then probed inside the acceleration region of a linear time of flight mass spectrometer, using the fourth harmonic output of a Nd-YAG laser for ionisation. A tuneable dye laser ($0.08 \, cm^{-1}$ bandwidth), that arrives 10 ns earlier than the UV laser to the beam-laser interaction zone, is used to induce the chemical reaction within the weakly bound complex. The reader is addressed to [3] for further information about the experimental technique and spectroscopic method.

3 Photoionisation Spectra and Photodissociation Channels

Figure 1 displays two time of flight mass spectra. The most intense peaks appearing in the spectra correspond to species containing the ^{138}Ba isotope with the other four most abundant barium isotopes appearing at shorter flight times. The upper trace shows a mass spectrum that results when a mixture of Ba vapour and CH_3F is co-expanded with helium as carrier gas and ionised at 266 nm.

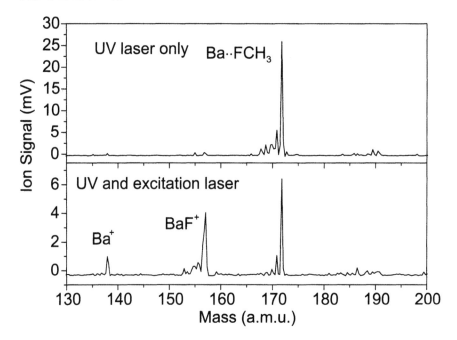

Figure 1: Upper trace: TOF mass spectra of the Ba + CH_3F system when ionised at 266 nm. 1064 nm radiation is used for vaporisation. Strong depletion of the Ba$\cdots$FCH$_3$ complex is observed when the dye laser, tuned to 618.3 nm, interacts with the molecular beam (lower trace). The enhancement of both Ba and BaF signals is clearly noticed as a result of complex photofragmentation.

The lower trace represents the spectrum when the dye laser tuned to 618.3 nm is also allowed to enter the detection chamber and to photoinduce the chemical reaction within the Ba$\cdots$FCH$_3$ complex. Strong depletion of the monomer signal is observed. At the same time, the intensity of the Ba and BaF species increases, suggesting two open channels for the complex fragmentation. Namely:

(a) *Photoinduced charge-transfer reaction*
In this scheme the absorbed photon induces the harpoon reaction, e.g.

$$Ba\cdot\cdot FCH_3 + h\nu \to |Ba\cdot\cdot FCH_3|^\ddagger \to BaF^*(BaF) + CH_3 \qquad (1)$$

leading to ground or electronically excited BaF.
(b) *Breaking of the van der Waals bond*
In this case the products are Ba and CH_3F with the Ba in its excited state, i.e.

$$Ba\cdot\cdot FCH_3 + h\nu \to Ba\cdot\cdot FCH_3^* \to Ba^* + FCH_3 \qquad (2)$$

The breaking of the van der Waals bond with production of Ba in its electronic ground state is, in principle, very unlikely, as it requires a large energy transfer. It would need a large mismatch in energy because of the excited (electronic) energy of the complex has to be channelled into translational energy of the products.

The signals of the Ba and BaF products were recorded as a function of the ionisation laser energy when inducing the depletion at different wavelengths in order to clarify if the products are formed in electronically excited states. Both Ba^+ and BaF^+ signals showed clear single photon dependence when exciting the parent complex at 547 nm [3]. Figure 2 displays the UV energy dependence of the BaF^+ signal at 619 nm. Also in this case the ionisation is a single photon process. These results suggest that the reaction products emerge in an excited state [3]. Results for the Ba channel when depleting at 619 nm does not show such a conclusive evidence for one photon dependence. This may be due to a partial production of fragments in the ground state, which makes it harder to analyse the results.

4 Photodepletion and Product Action Spectra

The parent complex photodepletion spectrum was measured by monitoring the corresponding mass peak in the TOF mass spectrum while tuning the dye laser. Care was taken to keep the excitation laser fluence below the saturation limit [3]. The spectra presented here are the difference between the complex signal without and with excitation laser, i.e. a value proportional to the number of disappearing molecules.

The action spectra of the products were obtained by monitoring the wavelength dependence of the Ba^+ and BaF^+ signals. In this case, the spectra presented show the difference between the signals with and without excitation laser, that is, the number of emerging products.

The dashed line in Fig. 3 (offset to higher intensity values for clarity) shows the photodepletion spectrum obtained by monitoring the parent mass of

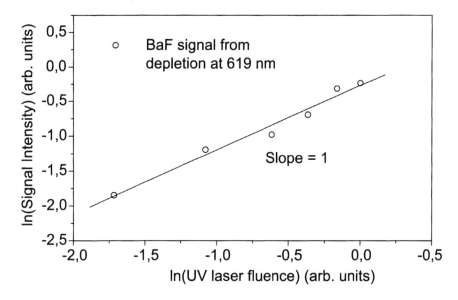

Figure 2: BaF$^+$ product signal as a function of the ionisation laser fluence. The Ba$\cdot\cdot$FCH$_3$ complex fragmentation was induced at 619 nm. The signal shows clear single photon dependence.

the Ba$\cdot\cdot$FCH$_3$ complex in the range 16065 – 16340 cm^{-1}, corresponding to the A-state of the complex. The solid line in Fig. 3 displays the action spectrum for the BaF reaction channel, obtained by the 266 nm ionisation wavelength. A resemblance of the BaF and the Ba$\cdot\cdot$FCH$_3$ photodepletion signals is clearly manifested. However, a closer look reveals an evident shift in their energy dependence. This is more clearly shown in the inset in which the low energy region, near the threshold, of both spectra is displayed. A key feature of this energy dependence is its good energy resolution, determined by the dye laser bandwidth. Indeed, the present experiment provides an energy resolution of ca. 0.01 meV.

The spectra shown in Fig. 3 were obtained averaging several runs. In particular, in the low energy region from 16065 to 16130 cm^{-1}, 100 runs were performed and averaged. A signal to noise ratio of three was estimated for the energy region shown in the inset of this figure.

In the energy region 17795 – 18250 cm^{-1}, corresponding to the B-state of the complex, both BaF$^+$ and Ba$^+$ action spectra were recorded [18]. The product action spectra revealed a similar lack of structure as the parent complex photodepletion spectrum.

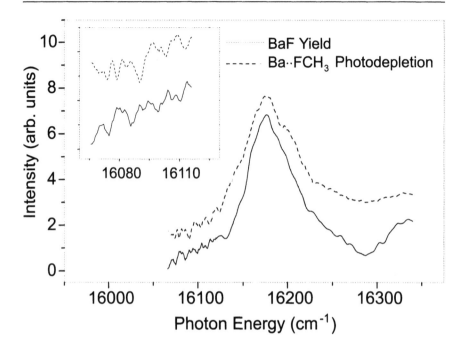

Figure 3: (Dashed line) Photodepletion spectrum of the Ba··FCH$_3$ complex in the range $16065-16340\,\text{cm}^{-1}$. (Solid line) Action spectrum for the BaF reaction channel, i.e. BaF$^+$ signal obtained by the 266 nm ionisation wavelength as a function of the photodepletion photon energy. The inset shows the near threshold part of both spectra for clarity. Notice in the inset the opposite phase of the BaF peaks with respect to those of Ba··FCH$_3$ depletion spectrum.

5 Reaction Probability

Following Soep and coworkers [6, 9] the intensity in the action spectra, $I(\nu)$, can be described as:

$$I(\nu) = c\,|\langle\Psi(G.S)|\Psi(\nu)\rangle|^2 \cdot A(\nu) \tag{3}$$

in which c is a frequency independent factor and the squared term represents a Franck–Condon factor between the complex ground and the excited state wave function at frequency ν. The second factor $A(\nu)$ can be considered as a coupling efficiency to the photofragmentation channel under consideration. Obviously, when several photodissociation channels are opened we have to consider a total coupling factor $A_{\text{tot}}(\nu)$ representing now the total photodissociation probability so as to write the total photodepletion spectrum, I_{ph}, as:

$$I_{\text{ph}} = c \left| \langle \Psi(G.S.) | \Psi(\nu) \rangle \right|^2 \cdot A_{\text{tot}}(\nu). \tag{4}$$

In consequence, we can consider the (unnormalised) reaction probability P_R^i for the i photodissociation channel as given by

$$P_R^i(\nu) = \frac{|\langle \Psi(G.S.) | \Psi(\nu) \rangle|^2 \cdot A^i(\nu)}{|\langle \Psi(G.S.) | \Psi(\nu) \rangle|^2 \cdot A_{\text{tot}}(\nu)} = \frac{A^i(\nu)}{A_{\text{tot}}(\nu)} \tag{5}$$

in which $A^i(\nu)$ stands for the coupling efficiency factor for the i photodissociation channel. Thus, P_R^i can be obtained from the ratio of the two spectra, i.e. the action spectrum of channel i and the total photodissociation spectrum [19]. It is very interesting to point out that the ratio of these two spectra gets rid of the spectroscopic part (Franck–Condon dependence) leaving only the dynamical part contained in the reaction probability.

The energy dependence of the BaF reaction probability in the A-state, and both Ba and BaF reaction probabilities in the B-state, were estimated using Equation (5) from the photodepletion and action spectra shown in Fig. 3 and [18]. The result for the A-state is displayed by the lower trace of Fig. 4. The upper axis of this figure has been adopted to be $E - E_0$ taking into consideration an energy threshold of $E_0 \approx 16065.2 \, \text{cm}^{-1}$ from the photodepletion spectrum. Thus, this abscissa axis represents the excess energy above photodissociation threshold, while the lower axis represents the total energy of the exciting photon. The upper trace of Fig. 4 displays the BaF action spectrum for comparison.

First of all it should be pointed out how the strong maximum in the centre of the $16175 \, \text{cm}^{-1}$ BaF resonance (coincident with the maximum in the Ba$\cdots$FCH$_3$ photodepletion spectrum) has disappeared in the $P_R(E)$ plot. This indicates that the BaF maximum is originated from the Frank–Condon factors of photoexcitation. Conversely, this is not the case for the peak structure observed in the low energy region of the BaF spectrum as it persists in the $P_R(E)$ plot. Certainly, a clear peak structure is noticeable around the red part of the spectrum, i.e. near the threshold of the photodepletion spectrum. Notice that the maxima in $P_R(E)$ cannot be originated from an artefact of the normalisation process of Equation (5) as they are already present in the (experimental) BaF yield. The peaks are spaced by ca. $10.9 \, \text{cm}^{-1}$. This structure could perhaps be attributed to a low energy mode of the van der Waals complex. Since the reaction probability was estimated by eliminating the Franck–Condon dependence these peaks are of dynamical nature, probably related to reactive resonances. A clear explanation for the origin of such peak structure in $P_R(E)$ should wait until good potential energy surfaces for both ground and excited Ba$\cdots$FCH$_3$ become available. Nevertheless, one possibility involves the presence of a low vibrational mode. By analogy with photodepletion and transition

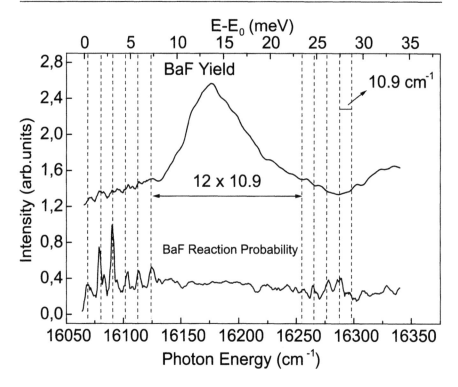

Figure 4: (Upper trace) Same spectrum as in Fig. 3, solid line. (Lower trace) Reaction probability for the BaF photofragmentation channel in the A-state. The ratio of the solid line to the dahed line spectral signals of Fig. 3 is displayed. Notice the energy spacing marked on the different peaks with the aid of vertical dashed-lines.

state ionisation spectra measured by Polanyi and coworkers on the Na$\cdots$FX (X = H, CH_3) system [14, 15, 16, 17] this vibrational progression could be attributed to bending modes in the Ba*$\cdots$$FCH_3$ transition state.

It is also interesting to point out how the peak progression disappears as the excitation energy increases until it becomes almost constant with energy, right over the range at which the photodepletion spectrum reaches it maximum values. In other words, the suggested resonances appear more pronounced at low energy near the photodepletion threshold. In fact this is not surprising at all since it is well known that resonant structures are more pronounced near the saddle point for chemical reactions [20, 21].

The peak structure appears then again near the threshold of the second vibrational state observed in the photodepletion spectrum. The occurrence of the two structures predominantly located near the distinct thresholds clearly suggests a vibrationally adiabatic behaviour for the harpoon reaction. In other words, a non-ergodic behaviour [22] seems to be controlling the reaction dynamics besides the simple electron transfer process. This finding is in principle consistent not only with the expected fast evolution of the transition state for an excited harpoon reaction, but also with the weak nature of the van der Waals bonds which would probably preclude internal vibrational energy redistribution [23] in the excited Ba··FCH$_3$.

Figure 5 displays the Ba and BaF reaction probabilities for the B-state. In both cases the dashed lines represent the experimentally determined reaction probabilities and the solid line a numerical fit to the data. In spite of the small signal to noise ratio (ca. 2), a clear oscillation in $P_R(E)$ is noticeable in both channels. In addition, an opposite phase of the oscillation is clearly manifested. In other words, a maximum in $P_R(E)$ for the Ba channel implies a minimum in $P_R(E)$ for BaF and viceversa. Although the present reaction probabilities data are unnormalised since a quantitative knowledge of the product ionisation efficiencies is not available, their respective yield should, however, show this complementary behaviour as they have been found to be the two major photofragmentation channels. In other words, an increase in BaF yield should be at the expense of the other (Ba) channel yield and viceversa. The solid lines of Fig. 5 represent numerical fits using Equation (6) (see below).

It is interesting to point out an energy spacing of the oscillations in $P_R(E)$ of ca. 400 cm^{-1}. This value may probably reflect some internal degree of motion of the reactive intermediate. In line with the Ar··ClCH$_3$ structure [24] one can consider the ground and first excited Ba··FCH$_3$ state to be a T-shaped complex whose configuration would correlate with the Π symmetry of the asymptotic Ba($^1P_1$) + CH$_3$F. In this view, the next excited state, the B state investigated here, would correlate to the Σ symmetry of the asymptotic Ba($^1P_1$) + CH$_3$F reagents and it should probably adopt a quasi-collinear Ba··F··CH$_3$ configuration. Hence, it is not surprising that some of the internal motions of the quasi-collinear transition state may correlate with those of the products. An example of it could be the observed energy spacing of ca. 400 cm^{-1} in the $P_R(E)$ plot considering that the BaF vibrational frequencies are 435 cm^{-1} and 424 cm^{-1} for the A and B excited states, respectively [25]. As a result the observed peak structure in $P_R(E)$ would reflect the competition between both reactive and non-reactive channels (Equation (2) and (3)) as if it were controlled by the Ba $\rightarrow\leftarrow$ F··CH$_3$ $\rightarrow$ symmetric tension whose value one may expect to be closer to that of free BaF* vibration.

On the other hand, the oscillatory behaviour of the collision energy dependence of the reaction probability constitutes one of the most interesting features in the quantal description of molecular reaction dynamics [26, 27, 28]. For atom-diatom reactions Aquilanti and co-workers [27]

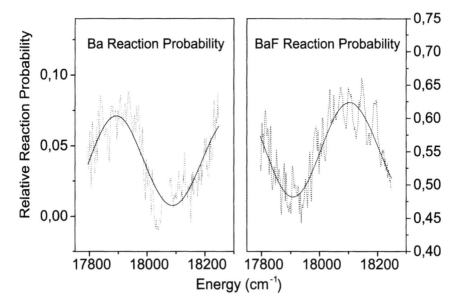

Figure 5: Left: Reaction probability for the Ba photofragmentation channel for the B-state. The ratio of the Product action spectrum to the complex photodepletion spectrum is displayed. (Right) same presentation as in the left Figure, but now for the BaF channel. Notice the clear oscillations with energy in both probabilities and, in addition, their opposite phase. The solid lines in both spectrta represent numerical fits to Equation (6) of the text.

and Hiller et al. [28] deduced the following dependence for $P_R(E)$

$$P_R(E) \approx \sin^2(a + b\sqrt{E_T}) \qquad (6)$$

where a and b are adjustable parameters and E_T the collision energy of the reagents. In this "half collision" reaction what is known is the total energy deposited into the van der Waals reaction. Thus we have adopted the simple hypothesis to consider E_T proportional to $E-E_0$, the excess energy available for reaction in which E is the laser excitation energy and E_0 the photodepletion energy threshold which was taken to be $E_0 = 16100 \text{ cm}^{-1}$ from the photodepletion spectrum. Accordingly, the proportionality constant is considered to be included in b. As displayed in Fig. 5, Equation (6) provides an excellent description of the energy dependence of $P_R(E)$. Table 1 displays the values of the best-fit parameters for Ba and BaF.
Essentially, the oscillatory behaviour inherent to the $\sin^2(E - E_0)^{1/2}$ functionality can be considered to be a reflection of interferences between reactive and non-reactive pathways present in the dynamics of the elementary atom-diatom reaction. In this view, the oscillations would be a manifesta-

Table 1: Best-fit parameters for the adjustment of Equation (6) to the experimental data on $P_R(E)$.

Parameter	Ba	BaF
a (rad)	0.2	1.65
b (rad cm$^{-1/2}$)	0.7	0.7

tion of interferences between the two (observed) channels Equation (1) and Equation (2).

The description provided by Equation (6) for the observed oscillations of both photofragmentation channels is excellent, athough care should be taken in adopting this interpretation until a knowledge of the relevant potential energy surfaces, involved in the photofragmentation dynamics, is available. This semiclassical model describes quite well the oscillations seen in quantum mechanical reactive scattering calculations on the collision energy dependence of the reaction probability for collinear and symmetric heavy + heavy-light atom reactions. A good example of such a system is the I + HI → IH + I reaction [26, 27, 28] in which the vibrationally adiabatic character of the reaction reduces it to a two state symmetric resonance system making possible the interference between the reactive and non-reactive wave. As for the present Ba+FCH$_3$ → (Ba··FCH$_3$)∗ → BaF∗+CH$_3$ system, the lack of the symmetry between both reactive and non-reactive channels does not necessarily imply the absence of the vibrational adiabaticity for the entire photofragmentation process. In fact, the initial vibro-rotational cooling of the van der Waals molecule, as formed in a supersonic beam, together with the high-resolution laser bandwidth ensures a narrow and well-defined initial energy distribution with very few vibro-rotational states populating the excited complex. In addition, the suggested collinear structure of the Ba··FCH$_3$ state plus the fast reaction time of this harpooning reaction [29] may well facilitate a vibrationally adiabatic behaviour for this process. In this view, the occurrence of interference between both photofragmentation channels cannot be ruled out and so the manifestation of oscillations in $P_R(E)$.

6 Concluding Remarks

Results from the photoionisation TOF mass spectra showed that both BaF and Ba are produced (reactive and non-reactive channel). Fragmentation from the excited B-state clearly produced electronically excited Ba and BaF. In the A-state, the BaF product from the reactive channel also emerges in an electronically excited state, while no such clear evidence can be drawn for the Ba (non-reactive) channel. This was attributed to a partial yield of Ba in its ground state.

In addition to the photodepletion spectrum of the parent complex, action spectra of the products were measured. In the A-state, the BaF product spectrum was measured over the $16065-16340\,\mathrm{cm}^{-1}$ energy range. This spectrum showed a similar overall structure as the parent molecule photodepletion spectrum, except in the low energy region. Here, the observed structure showed a clear shift in energy dependence with respect to the photodepletion spectrum. The simultaneous measurement of both spectra allowed us to determine the relative reaction probability $P_R(E)$ of this excited harpooning reaction. A peak structure was found in this spectrum of reaction probability that could reflect resonances associated with excited quasi-bound states of the transition intermediate leading to the reaction products. Moreover, the spacing between peaks in $P_R(E)$ could provide direct information about the structure of the intermediate of the van der Waals reaction.

The action spectra of both the reactive and the non-reactive channel were measured in the $17795-18250\,\mathrm{cm}^{-1}$ energy range, corresponding to the B-state of the Ba··FCH$_3$ molecule. Thus, the relative reaction probability could be determined for each channel from the mentioned spectra. This piece of information is a valuable quantity that provided a clear insight into the reaction dynamics for a van der Waals reaction. Specific conclusions from this result are: (a) an oscillatory behaviour of both Ba* and BaF* reaction probabilities with laser excitation energy. (b) An opposite phase for these two photofragmentation channels which was attributed to consider Ba* and BaF* as the two major photodissociation channels of the Ba··FCH$_3$ complex. (c) The value of the energy spacing in the energy dependence of $400\,\mathrm{cm}^{-1}$ was interpreted as the frequency of the internal motion of the transition state responsible for its fragmentation. In addition, the oscillatory behaviour in the energy dependence of $P_R(E)$ and its satisfactory description by the theoretically deduced $\sin^2(E-E_0)^{1/2}$ functionality constitutes an interesting conclusion for future experimental and theoretical work.

Unfortunately the lack of a reliable potential energy surface for the Ba··FCH$_3$ system precludes a firm acceptation of the conclusions from the reaction probability measurements. Nevertheless, we would like to emphasize that this measurement of the reaction probability from a van der Waals reaction is of major significance in the current field of transition state spectroscopy. As a matter of fact the possibility of carrying out a REMPI spectrum of the (dissociative) Ba··FCH$_3^*$ state either by using femtosecond or (with significantly lower yield) very intense nanosecond pulses, constitutes an interesting objective now in progress. Clearly this type of studies would help to confirm the nature of the observed peak structure in $P_R(E)$.

Acknowledgement

We thank Dr. R. Pereira and J. Jiménez for their collaboration in this project. This work has received financial support from the DGES of Spain (grants PB95/391, PB97/272) and the Ramón Areces Foundation.

References

[1] P.R. Brooks: Chem. Rev., **88**, 407 (1988)

[2] S. Skowronek, R. Pereira, A. González Ureña: J. Chem. Phys., **107**, 1668 (1997)

[3] S. Skowronek, R. Pereira, A. González Ureña: J. Phys. Chem., special issue on Stereodynamics, **101**, 7468 (1997)

[4] A.H. Zewail, Faraday Discuss. Chem. Soc. **91**, 207 (1991) ibid. A.H. Zewail: Science, **242**, 1645 (1988)

[5] C. Jouvet, B. Soep: Laser Chem. **5**, 157 (1985)

[6] C. Jouvet, M. Boiveneau, M.C. Duval, B. Soep: J. Phys. Chem. **91**, 5416 (1987)

[7] B. Soep, S. Abbes, A. Keller, J.P. Visticot: J. Chem. Phys. **96**, 440 (1992)

[8] B. Soep, C.J. Whitham, A. Keller, J.P. Visticot: Farady Discuss. Chem. Soc. **91**, 191 (1991)

[9] A. Keller, R. Lawruszczuk, B. Soep, J.P. Visticot: J. Chem. Phys. **105**, 4556 (1996)

[10] C. Jouvet, M. Boiveneau, M.C. Duval, B. Soep: J. Phys. Chem., **91**, 5416 (1987)

[11] G. Hoffman, Y. Chen, M.Y. Engel, C. Wittig: Isr. J. Chem. **30**, 115 (1990)

[12] C. Wittig, S. Sharpe, R. Beaudet: Acc. Chem. Res. **21**, 341 (1988)

[13] J.C. Polanyi, A.H. Zewail: Acc. Chem. Res. **28**, 119 (1995)

[14] K. Liu, J.C. Polanyi, S. Yang: J. Chem. Phys. **98**, 5431 (1993)

[15] J.C. Polanyi and Ji-Xing Wang: J. Phys. Chem. **99**, 13691 (1995)

[16] M.S. Topaler, D.G. Truhlar, X. Yan Chang, P. Piecuch, J.C. Polanyi: J. Chem. Phys., **108**, 5349 (1998)

[17] M.S. Topaler, D.G. Truhlar, X. Yan Chang, P. Piecuch, J.C. Polanyi: J. Chem. Phys., **108**, 5378 (1998)

[18] S. Skowronek, J.B. Jiménez, A. González Ureña: Chem. Phys. Letter, **303**, 275 (1999)

[19] M. Paniagua, A. Aguado, M. Lara, O. Roncero: J. Chem. Phys., **109**, 2971 (1998)

[20] A. González Ureña: Adv. Chem. Phys., **66**, 213 (1987)

[21] J.M. Alvariño, O. Gervasi, A. Lagana: Chem. Phys. Letters, **87**, 254 (1982)

[22] E.W.-G. Diau, J.L. Herek, Z.H. Kim, A.H. Zewail: Science **279**, 847 (1998)

[23] I. Oref. Science, **279**, 820 (1998)

[24] G.T. Fraser, R.D. Suenram, F.J. Lovas: J. Chem. Phys., **86**, 3107 (1987)

[25] K.P. Huber, G. Herzberg: Molecular Spectra and Molecular Structure IV, Constants of Diatomic Molecules, Van Nostrand Reinhold Company, New York (1979)

[26] V.K. Babamov, V. López, R.A. Marcus: J. Chem. Phys., **18**, 5621 (1983)

[27] V. Aquilanti, S. Cavalli, A. Lagana: Chem. Phys. Letters, **93**, 179 (1982)

[28] C. Hiller, J. Manz, W.H. Miller, J. Römelt: J. Chem. Phys., **78**, 3850 (1983)

[29] V. Stert, P. Farmanara, W. Radloff, F. Noack, S. Skowronek, J. Jimenez, A. González Ureña: Phys. Rev. A, **59**(3), 1727 (1999)

Electronic Spectroscopy and Excited State Dynamics of Aluminium Atom-Molecule Complexes

Paul J. Dagdigian, Xin Yang, Irina Gerasimov, and Jie Lei
Department of Chemistry, The Johns Hopkins University
Baltimore, MD 21218 USA

1 Introduction

There has been considerable interest in the interaction of metal atoms, in both their ground and electronically excited states, with small molecules. In part, this interest stems from a desire to relate properties, such as the electronic absorption spectra, of cryogenic molecular matrices doped with these atoms [1] to the individual 2-body atom-molecule interactions. In our laboratory, we have employed laser fluorescence excitation spectroscopy of weakly bound binary complexes to probe these interactions, as has been done in extensive studies of metal atom-rare gas complexes [2].

We have previously carried out a thorough study of the electronic spectroscopy of complexes of the boron atom with rare gases and the hydrogen molecule [3, 4, 5]. We have recently undertaken the study of complexes of aluminum atoms with small molecules.

Rather than employing electronic transitions of the complex built upon the lowest atomic transition ($4s \leftarrow 3p$ for Al; see Fig. 1), we have exploited detection through excitation to higher molecular states. In an atom-molecule complex, as opposed to a diatomic metal-rare gas complex, there is a high probability that the excited state will decay by nonradiative processes. If these nonradiative pathways involve formation of the Al atom in a lower excited atomic state, then it is still possible to detect the electronic transition by fluorescence excitation since this state will fluoresce. We have detected several Al atom-molecule complexes in this way, through $5s, 4d \leftarrow 3p$ excitation. The following sections present detailed descriptions of our observations and inferences.

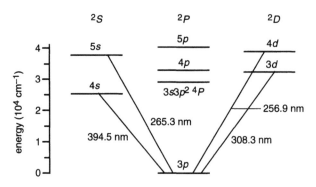

Figure 1: Al atomic energy levels and the lower resonance transitions.

2 Experimental

The experimental apparatus is thoroughly described in previous publications [6, 7, 8, 9], and a brief description is provided here. The chamber housing the molecular beam was evacuated with a water-baffled 10 in. diffusion pump. A mixture of trimethylaluminum (≤ 0.1 atm, Aldrich), the molecule to be complexed (3-80% of the gas mixture), and He or Ar rare gas (total pressure 8-12 atm) was expanded into vacuum through a 0.02 cm diam nozzle orifice by means of a solenoid pulsed valve (General Valve). Al atoms were generated in the early stages of the supersonic expansion by photolysis of trimethylaluminum with a focused, attenuated ArF excimer laser. In our hands, much colder beams were generated through introduction of the atoms into the gas flow by photolysis of a volatile precursor than by laser vaporization of the metal.

Aluminum atoms and Al-containing complexes were detected 1.2 cm downstream of the nozzle by laser fluorescence excitation with the frequency-doubled output of a dye laser. Typical UV pulse energies for the recording of excitation spectra were 5-50 μJ in a 0.2 cm diam beam. The laser-induced fluorescence signal passed through a 1/4 m monochromator and was detected with a photomultiplier. Resolved fluorescence emission spectra were recorded with 250 μm slits installed in the monochromator (0.8 nm spectral resolution).

3 The Al-H_2 Complex

There have been several quantum chemical calculations of Al($3p$) + H_2 potential energy surfaces (PES's) [10, 11, 12]. These predict a T-shaped equilibrium geometry for the van der Waals complex on the lowest-energy $^2B_2$

surface, and there is a significant barrier for formation of the $AlH_2(X^2A_1)$ molecule [13]. Because of the expected small dissociation energy D_0 of the Al-H_2 complex and to motivate the study of this complex, we first sought to generate and detect the very weakly bound AlNe complex, which had not been detected previously. Transitions out of both spin-orbit manifolds of AlNe($X^2\Pi$) to molecular electronic states correlating with the Al $3d$ and $5s$ states have been observed [8]. The dissociation energies of the $X^2\Pi_{1/2}$ and $X^2\Pi_{3/2}$ states were found to be very small (14 and 32 cm^{-1}, respectively) and slightly different because of a transition toward Hund's case (c) coupling.

We have observed transitions from the ground vibronic level of the Al-H_2 complex to bound vibrational levels in the excited AlH_2 electronic state which correlates with the Al($5s$) + H_2 asymptote [5]. This contrasts sharply with our previous observations on the B-H_2 complex. For this species, the B($3s$)-H_2 PES was found to be purely repulsive in the Franck-Condon region [4] and the transition to the electronic states correlating with B($2s2p^2$ ^2D) was found to be broad [9], indicative of rapid reaction in the excited state.

Resonance fluorescence from excited Al($5s$)-H_2 levels was very weak. Rather, these levels decay mainly nonradiatively, and the excitation spectrum was recorded by monitoring emission from the Al($4s$, $3d$) levels, as well as AlH $A^1\Pi - X^1\Sigma^+$ chemiluminescence from chemical reaction within the complex. Figure 2 displays survey spectra, obtained by observing the emission from the different product channels. Prominent in the spectra is a progression of bands in the Al-H_2 van der Waals stretch vibrational mode. The lowest-energy of these bands is assigned as excitation to $v'_{str} = 0$. It was not possible to obtain a reliable estimate of the excited-state dissociation limit. From the displacement of the $v'_{str} = 0$ band from the Al $5s \leftarrow 3p$ atomic transition and the theoretical value for D''_0 obtained by Partridge et al. [10], we obtain $D'_0 = 300^{+160}_{-10}$cm^{-1}.

The action spectra in Fig. 2 also reveal emission when the Al $5s \leftarrow 3p$ atomic line is excited. This emission from lower Al atomic states and AlH($A^1\Pi$) reaction product is the result of bimolecular collisions involving electronically excited Al atoms within the supersonic beam [14].

Comparison of the emission intensities corresponding to the different decay channels for the Al($5s$)-H_2 complex allows us to estimate the excited-state decay lifetime. For example, for the $v'_{str} = 2$ level, relative intensities of the Al $4s \to 3p$, Al $3d \to 3p$, and AlH $A \to X$ emission channels were determined to be 18 : 9 : 1, respectively, ignoring the wavelength dependence of the detection sensitivity of the monochromator/photomultiplier combination. Only an upper bound (<0.04) could be determined for the intensity of resonance fluorescence. If the $5s \to 3p$ radiative decay rate for the complex is equated to the atomic Al($5s$) radiative decay rate, then the Al($5s$)-H_2 de-

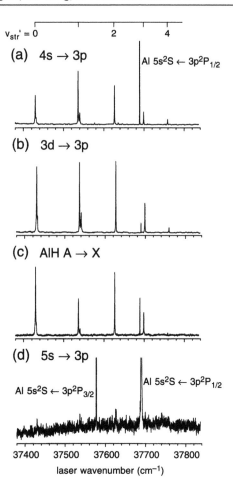

Figure 2: Laser fluorescence excitation spectrum of the Al-H_2 complex. The detected final states are indicated on each panel.

cay lifetime can be estimated to be ≤ 35 ps. This suggests that Lorentzian broadening, with a predicted width of ≥ 0.15 cm^{-1}, should be observable. We also see from Fig. 2 that the product branching ratios are somewhat dependent upon which v'_{str} level is excited.

In analogy with free H_2, the Al-H_2 complex can exist in two nuclear spin modifications. Alexander and Yang [15] found that the dissociation energy of the analogous B-H_2 complex was significantly greater for B-oH_2 than for B-pH_2 (38.6 vs. 27.9 cm^{-1}). This results from the fact that the anisotropy of the B-H_2 PES's is much smaller than the free H_2 rotational spacings so that the molecule can be described as a free rotor within the complex. The

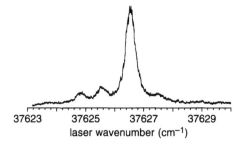

Figure 3: Rotational structure of the transition to the Al(5s)-H$_2$ v$'_{str}$ = 2 level.

oH$_2$ complex involves $j = 1$, which can align itself in the optimum T-shaped geometry, while pH$_2$ ($j = 0$) cannot. As we found for B-H$_2$ [15], we expect that the Al-H$_2$ complexes that we observe predominantly involve the oH$_2$ isotopomer.

Despite the observable Lorentzian broadening, as expected from the above arguments, we have partially resolved the rotational structure of the bands. By analogy with the B(2p)-oH$_2$ complex, whose rotational energy level structure has been studied in detail theoretically by Alexander and Yang [15], we expect that the lowest bend-stretch manifold of the Al(3p)-oH$_2$ complex will be associated with the projection quantum number $P'' = 1/2$. From the calculated equilibrium geometry, we estimate a rotational constant $B'' = 0.86$ cm^{-1}; we also expect a significant parity splitting. Because of the low beam rotational temperature (1-2 K), there will be few rotational levels populated.

Figure 3 presents a scan over the v$'_{str}$ = 2 band. The rotational lines display a Lorentzian broadening of width 0.4 cm^{-1}, in reasonable agreement with the above estimated excited-state decay lifetime. The spectrum will be interpreted in detail with the help of forthcoming calculations on the bend-stretch energies of the Al(3p)-H$_2$ complex by Alexander and Williams [12], based on their *ab initio* calculation of the PES's correlating with the Al(3p) + H$_2$ asymptote.

In order to estimate the internal excitation of the AlH($A^1\Pi$) product from the reactive decay of the Al(5s)-H$_2$ complex, chemiluminescence spectra were recorded by scanning the wavelength of the detection monochromator, operated at a resolution of 0.8 nm, while the excitation laser was tuned to various bands. Figure 4 displays a representative spectrum of the emitting AlH($A^1\Pi$) product. For a $^1\Pi - {}^1\Sigma^+$ electronic transition, the integrated Q-branch intensity should equal the sum of the P- and R-branch intensities. However, it can be seen that the Q branch is quite weak in the spectrum in Fig. 4. The AlH($A^1\Pi$) state is orbitally degenerate, and each rotational

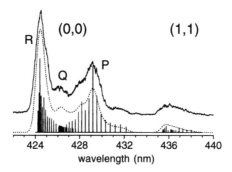

Figure 4: AlH chemiluminescence spectrum of the $A^1\Pi - X^1\Sigma^+$ $\Delta v = 0$ sequence for the reactive decay of the $v'_{str} = 0$ vibrational level of the Al(5s)-H$_2$ complex. Overlaid on the simulated spectrum is the stick spectrum with which the simulated spectrum was generated.

level splits into a closely spaced pair of Λ-doublet levels. The symmetries of these levels with respect to reflection of the electronic wavefunction in the plane of rotation are symmetric (A') and antisymmetric (A''). The Q-branch levels involve emission from the excited levels of A'' symmetry, while P- and R-branch lines involve emission from A' levels.

Although the monochromator resolution was not sufficient to resolve individual AlH rotational lines, it was still possible to perform a band contour analysis to estimate the rotational population distribution [5]. Figure 5 presents the derived rotational state distribution for the spectrum displayed in figure 4. As anticipated by the weakness of the Q branch in figure 4, the populations of the A' levels are much greater than those of the A'' levels. Moreover, the A' populations are peaked at $J = 10$.

The preferred production of A' Λ-doublet levels can be rationalized through symmetry arguments. A single PES, of A' symmetry, emanates from the Al(5s) + H$_2$ reagents. The plane of rotation of the AlH product should lie in the reagent triatomic plane, as the total angular momentum of the complex is small since it was prepared in a supersonic beam. Thus, the preferential formation of A' levels can be explained as a consequence of the conservation of reflection symmetry in the reaction. The actual dynamics are more complicated than this simple picture since the Al(5s) + H$_2$ reagents and AlH($A^1\Pi$) + H products are not connected by an adiabatic PES, but rather this reaction proceeds nonadiabatically. However, the low total angular momentum of the complex dictates that surface crossings will not be easily enabled to A'' PES's by Coriolis couplings, and the nonadiabatic reaction proceeds through crossings with A' PES's.

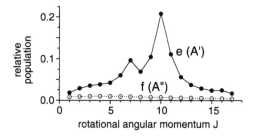

Figure 5: Derived AlH($A^1\Pi$, v = 0) product rotational state distribution for reaction within the $v'_{str} = 0$ vibrational level of the Al(5s)-H$_2$ complex.

4 The Al-N$_2$ Complex

We have also detected the Al-N$_2$ complex with vibrational and partial rotational resolution by excitation to electronic states correlating with the Al(5s, 4d) levels and detection of emission from lower Al atomic levels [16]. In previous work on this complex, Brock and Duncan [17] investigated this complex by photoionization spectroscopy. Quantum chemical calculations by Chaban and Gordon [18] predict that the equilibrium geometry of the Al(3p)-N$_2$ complex is linear and have provided an estimate of the ground-state dissociation energy. The different geometries of the Al-H$_2$ and Al-N$_2$ complex are consistent with the opposite signs of the electric quadrupole moments of the molecules [19].

Figure 6 compares action spectra for excitation of the Al(4d)-N$_2$ state while recording emission from the lower Al $4s, 3d$, and $5s$ atomic levels. Again, a progression in the van der Waals stretch mode dominates the spectrum. For transitions to both the 5s and 4d states, the first member of the progression could not be clearly identified in the spectrum. To assign the v'_{str} quantum numbers, spectra of the Al-^{15}N$_2$ complex were also recorded. This enabled a definite vibrational assignment for Al(5s)-N$_2$, but not for Al(4d)-N$_2$. For the spectra in Fig. 6, we estimate $y = 9 \pm 1$.

It can be seen in Fig. 6 that formation of the Al(5s) level occurs only for vibrational levels $v'_{str} \geq y+3$. This threshold can be used to estimate an upper bound to the ground-state dissociation energy D''_0, through the predissociation process: Al(4d)-N$_2 \rightarrow$ Al(5s) + N$_2$. We obtain $D''_0 \leq 352\pm2$ cm^{-1}, in reasonable agreement with the theoretical estimate $D''_0 = 256 \pm 2$ cm^{-1} [18].

For excitation of several vibrational bands of the Al(5s)-N$_2$ state, partial rotational resolution was achieved. A representative scan is shown in Fig. 7. As the Al-N$_2$ complex is expected to be linear, we compared the experimental scans with simulations of a $^2\Sigma^+ - {}^2\Pi$ transition and a Voigt profile consisting of a Lorentzian component due to nonradiative decay of

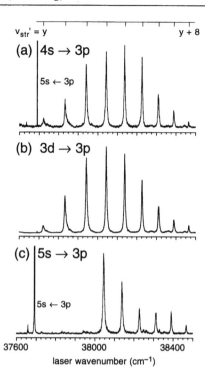

Figure 6: Action spectra for excitation of the Al(4d)-N$_2$ state while monitoring emission from various lower excited Al atomic levels. The labels y etc. denote various excited-state v'_{str} vibrational levels; the absolute vibrational numbering could not be determined.

the complex and a Gaussian profile (FWHM 0.06 cm^{-1}) modeling the dye laser line shape. It can be seen in Fig. 7 that these simulations provide a very good representation of the band contour. For this displayed spectrum, the Lorentzian width Γ was found to equal 0.38 ± 0.05 cm^{-1}, corresponding to a decay rate of 7×10^{10} s^{-1}. The lower-state rotational constant B'' was held fixed at the value appropriate to the equilibrium geometry calculated by Chaban and Gordon [18]. From the derived excited-state rotational constant B', the $5s \leftarrow 3p$ transition is estimated to involve a contraction of the Al-N bond length by 0.6 Å.

There are actually three electronic states ($^2\Delta$, $^2\Pi$, and $^2\Sigma^+$) which emanate from the Al(4d) + N$_2$ asymptote for a linear geometry. The bands displayed in Fig. 6 are assigned to the transition to the most strongly bound, $^2\Delta$, electronic state. The Lorentzian broadening in these bands is significantly greater than in the transition to the Al(5s)-H$_2$ state, and a band contour analysis was not performed.

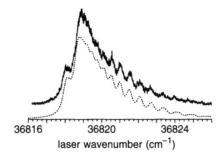

Figure 7: High-resolution scan of the transition to the $v'_{str} = 1$ vibrational level of the Al(5s)-N_2 state. The dotted curve is a simulation of the band, as discussed in the text.

It is interesting to speculate on the mechanism of predissociation in these complexes. The clearest situation is the formation of Al(5s) atoms in the decay of Al(4d)-N_2 since the 5s level lies immediately below 4d. Crossings of Al(4d, 5s)-N_2 PES's cannot take place at large Al-N_2 separations since only one AlN_2 PES arises from Al(5s) + N_2 and this surface is known to support a number of vibrational levels. Rather, the crossings must occur on the inner limb (small Al-N_2 separations). The mechanism for formation of Al(4s, 3d) must be much more complicated since several intervening PES's must be crossed to reach PES's correlating with the final atomic states.

5 The Al-CH_4 Complex

As a final example of an Al atom-molecule complex, we consider Al-CH_4. Figure 8 presents an action spectrum for excitation of the 5s state of this complex [20]. A progression in the van der Waals stretch mode is seen. The widths of the bands are much greater than for the case of Al-H_2 and Al-N_2 and are indicative of an increased nonradiative decay rate and/or more complicated rotational structure. The widths of the corresponding Al-CD_4 bands (not shown) are significantly less, suggesting that H-atom motion is involved in the nonradiative decay. It was not possible to obtain a definitive vibrational assignment for spectra of these complexes and only lower bounds to the excited-state binding energies were derived.

As in the case of Al-H_2, AlH chemiluminescence, as well as emission from lower excited Al atomic levels, was observed. Only in the case of the 4d state was emission of sufficient intensity to record a spectrum. In sharp contrast to the preferential Λ-doublet formation seen for reaction within the Al(5s)-H_2 complex, the AlH($A^1\Pi$) product from reaction within Al(4d)-CH_4 was found to have approximately equal populations of the A' and A'' Λ-doublet

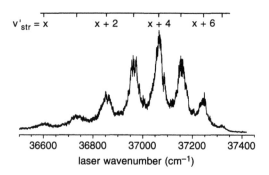

Figure 8: Laser fluorescence excitation spectrum of the Al-CH$_4$ complex, detected by observation of Al $3d \to 3p$ atomic emission.

M	α(M) (Å^3)	D_0(Al(5s)-M)	D_0(Al(4d)-M)
Ne	0.40	81	
Ar	1.64	692	787
Kr	2.48	1216	1429
H$_2$	0.81	300^{+160}_{-10}	
N$_2$	1.74	1218 ± 10	2705 ± 165
CH$_4$	2.20 ($\alpha_{\parallel}$) 2.60	≥ 1489	≥ 1824

Table 1: Binding energies (in cm^{-1}) of Al-M complexes

levels. This is likely the result of the fact that the plane of rotation of the AlH product is not well defined with respect to planes of symmetry in the complex.

6 Discussion and Conclusion

It is of interest to compare the binding energies of various Al-rare gas and Al-molecule complexes. Table 1 presents such a comparison for the excited electronic states correlating with the Al(5s, 4d) atomic levels. For the latter level, the most strongly attractive level ($^2\Delta$ in linear geometry) is considered. The binding energies have been taken from various literature references [5, 8, 16, 20, 21].

Also included in Table 1 are the polarizabilities α of the atomic or molecular ligand [19]. The long-range attractive interaction is usually governed by the dispersion interaction which scales with the polarizabilities of the interacting species. The equilibrium separation, and hence the binding energy, is actually controlled by the balance between the attractive interaction and repulsion between the overlapping electron distributions. There is a strong correlation in Table 1 between the binding energies and the polarizabilities. This suggests that the range at which electron repulsion becomes significant is similar for these complexes. The major uncertainty of the comparison displayed in Table 1 is the uncertainty in the Al-CH$_4$ binding energies, resulting from the lack of definitive vibrational assignment for the spectra of this complex.

For N$_2$, we have included in Table 1 both the isotropic polarizability and the parallel component $\alpha_\parallel$. The latter may be more appropriate for this comparison because of the linear equilibrium geometry of the Al-N$_2$ complex. The geometry is determined from the induction forces since N$_2$ has a permanent, negative quadrupole moment [19].

This study of the electronic spectra of Al-molecule complexes shows the importance of predissociation as a nonradiative decay pathway in such complexes, as compared to complexes involving rare gases. Such decay processes are likely to play an important role in the excited-state decay dynamics of other complexes involving molecules.

Acknowledgment This research was supported by the US Air Force Office of Scientific Research under grant nos. F49620-98-1-0187 and F49620-99-1-0100. The encouragement of Millard Alexander is greatly appreciated.

References

[1] M. E. Fajardo, S. Tam, T. L. Thompson, and M. W. Cordonnier, Chem. Phys. **189**, 351 (1994), and references therein.

[2] For a comprehensive review, see W. H. Breckenridge, C. Jouvet, and B. Soep, in *Advances in Metal and Semiconductor Clusters*, edited by M. A. Duncan (JAI Press, Greenwich, 1995), Vol. 3, p. 1.

[3] P. J. Dagdigian, X. Yang, and E. Hwang, in *Highly Excited States: Relaxation, Reactions, and Structure*, ACS Symp. Ser. edited by A. S. Mullin and G. C. Schatz (American Chemical Society, Washington, 1997), p. 122.

[4] X. Yang, E. Hwang, M. H. Alexander, and P. J. Dagdigian, J. Chem. Phys. **103**, 7966 (1995).

[5] X. Yang and P. J. Dagdigian, J. Chem. Phys. **109**, 8920 (1998).

[6] X. Yang, E. Hwang, P. J. Dagdigian, M. Yang, and M. H. Alexander, J. Chem. Phys. **103**, 2779 (1995).

[7] X. Yang, E. Hwang, and P. J. Dagdigian, J. Chem. Phys. **104**, 8165 (1996).

[8] X. Yang, P. J. Dagdigian, and M. H. Alexander, J. Chem. Phys. **108**, 3522 (1998).

[9] P. J. Dagdigian and X. Yang, Faraday Discuss. **108**, 287 (1997).

[10] H. Partridge, C. W. Bauschlicher Jr. and L. Visscher, Chem. Phys. Lett. **246**, 33 (1995).

[11] G. Chaban and M. S. Gordon, J. Phys. Chem. **100**, 95 (1996).

[12] M. H. Alexander and J. Williams, J. Chem. Phys. (submitted).

[13] G. Chaban, M. S. Gordon, and D. R. Yarkony, J. Phys. Chem. A **101**, 7953 (1997).

[14] D. M. Lubman, C. T. Rettner, and R. N. Zare, J. Phys. Chem. **86**, 1129 (1982).

[15] M. H. Alexander and M. Yang, J. Chem. Phys. **103**, 7956 (1995).

[16] X. Yang, I. Gerasimov, and P. J. Dagdigian, Chem. Phys. **239**, 207 (1998).

[17] L. R. Brock and M. A. Duncan, J. Phys. Chem. **99**, 16571 (1995).

[18] G. Chaban and M. S. Gordon, J. Chem. Phys. **107**, 2160 (1997).

[19] C. G. Gray and K. E. Gubbins, *Theory of Molecular Fluids. Vol. 1: Fundamentals* (Clarendon Press, Oxford, 1984).

[20] I. Gerasimov, J. Lei, and P. J. Dagdigian, J. Phys. Chem. A 103, 5910 (1999).

[21] S. A. Heidecke, Z. Fu, J. R. Colt, and M. D. Morse, J. Chem. Phys. **97**, 1692 (1992).

Electronic Spectra of Cold Polycyclic Aromatic Hydrocarbons (PAH) Cations in a Molecular Beam

Philippe Bréchignac, Thomas Pino, Nathalie Boudin
Laboratoire de Photophysique Moléculaire*, C.N.R.S.,
Université de Paris-Sud, France, Bâtiment 210,
F-91405, Orsay, France
laboratoire associé à l'Université de Paris-Sud

1 Introduction

The Diffuse Interstellar Bands (DIBs) are absorption features observed from ground telescopes throughout the near infra-red and visible region of the electromagnetic spectrum when the line-of-sight of a star intercepts a cloud of interstellar matter. The first observation of two of the strongest of these bands was performed in 1921 by Heger [1], and the proof that they originate from the Interstellar Medium (ISM) was reported by Merrill [2] in 1934. Nowadays at least 200 DIBs have been reported[3],[4], but only a few of them may possibly receive a convincing assignment. Among the most recent and most serious candidates are the C_{60}^+ cation [5], the C_7^- anion [6], and the cations of the polycyclic aromatic hydrocarbons (PAH$^+$s) [7].

The low temperature characteristic of interstellar clouds has motivated a number of spectroscopic studies on species isolated in solid matrices. Application of this matrix isolation spectroscopy (MIS) technique to several relevant ions [8], [9] has provided useful information on their electronic spectra. However, the interactions of the host molecule with its solid environment induce perturbations in the spectra, which are too large to allow detailed comparisons with the DIBs spectra, even in the case of the inert rare gas matrices. In order to match the interstellar conditions, spectra must be taken in the gas phase, when the species are cold and isolated (collision-free). Producing large quantities of cold ions in the gas phase in order to measure absorption directly is very difficult, and alternative routes have to be used.

Recently, a novel experimental technique has been developed in our group specially devoted to access the electronic absorption spectra of cold PAH$^+$

cations [10]-[12]. Indeed PAHs have been suggested to be present in the ISM since 1984 [13], and this suggestion has so far, and particularly in recent years with the ISO satellite flight [14], proved to be very powerful. PAH$^+$ cations do fulfill the constraints relative to the DIBs carrier species (abundance, photostability, intense transitions in the spectral domain of interest)[15]. On the other hand, although polyatomic cations are of fundamental interest, their electronic spectroscopy in the gas phase has been poorly investigated because of intrisic difficulties. In particular the intramolecular dynamics taking place in such large molecular systems is dominated by ultra fast internal conversion which affects the results of multiphoton excitation experiments.

This article presents the principle and some details of the new technique, and illustrates its capabilities through the presentation of original spectroscopic results on some selected PAH$^+$s.

2 The Experimental Technique

In brief, the principle of the experiment consists of inducing a fragmentation of the cation cluster by resonant absorption of laser photons, so that the change in the charge-to-mass ratio can be detected with ultimate sensitivity in a time-of-flight (TOF) mass spectrometer. However, because of their stability, PAH monomers (neutrals as well as cations) need a fairly large excitation energy before they begin to fragment [16]. This requires several laser photons, then high laser flux, and then a number of experimental difficulties, as shown in the tentative work by Syage and Wessel [17] on naphthalene$^+$ and methyl-naphthalene$^+$. They made use of resonant enhanced multiphoton dissociation spectroscopy (REMPDS), and showed that its application to PAH$^+$s is hazardous, because of their high stability. It is possible to overpass these difficulties by studying the photodissociation spectra of the ionic van der Waals (vdW) complexes PAH$^+$-Ar under low laser fluence [10].

Attaching an argon atom to the PAH molecule to form a vdW complex is only a weak perturbation, which can be measured experimentally as it will be explained below. Neutral vdW complexes can be easily formed in a supersonic molecular beam, since the supersonic expansion provides efficient cooling of the species [18]. Thus the cationic complexes PAH$^+$-Ar are produced cold by threshold resonant two-photon two-colour ionisation (R2P2CI) of the PAH-Ar neutrals in the near UV.

The recording of the PAH$^+$ fragment ion signal versus the wavelength of a visible laser provides the photodissociation spectrum of the PAH$^+$-Ar complex cation. This spectrum differs from the spectrum of the bare PAH$^+$ by a small vdW shift. This shift is measured by recording the spectrum of the PAH$^+$-Ar$_2$ vdW (1/1) complex (following standard notation (m/p) as in ref [19]), as the binding sites occupied by the two argon atoms are known to be equivalent ([19], [20]), implying equal shifts induced by each atom.

2.1 Photodissociation Spectroscopy of van der Waals Cationic Clusters

Demonstration of the relevance of the dissociation of cation – rare gas van der Waals (vdW) clusters for spectroscopic investigations has been done on a variety of sytems including metal cations [21], and in particular aromatics[11], [12],[22]-[24]. In this scheme, the ejection of the rare gas (RG) atom from the chromophore results from intramolecular energy transfer within the complex. Vibrational vdW predissociation has been extensively studied in the past decade mainly on neutral species. This has been done by excitation of intramolecular vibrationally excited levels in the electronic ground state S_0 or in the first electronically excited state S_1[25]. Only poor data are available for the upper electronically excited states. A simple reason for that is that the dynamics of these states is often dominated by fast electronic relaxation, mostly internal conversion (IC), so that weakly bound complexes dissociate easily due to the large amount of vibrational energy made available to the system (of the order of the energy gap between the initially excited state and the acceptor state).

This property is used in our experiment. In PAH$^+$ cations the doublet states accessible by absorption of near infrared or visible photons from their ground state D_0 are lying rather close to one another in energy, and IC is expected to be very efficient. Thus, recording the electronic absorption spectrum can be done by monitoring the fragment signal versus the wavelength of a time-delayed laser, since the appearance of the bare fragment ion is a signature of the absorption of one photon by the parent cluster ion. The photodissociation spectrum reflects directly the absorption spectrum of PAH$^+$-Ar, since the quantum efficiency is equal to one for the ejection of the argon atom[12].

2.2 Ion Preparation and Detection Scheme

It is a well known problem that ion internal temperature may affect the results of a given experiment, and cooling devices are often used behind ion sources. The combination of (1+1') R2PI with pulsed lasers and a molecular beam offers a reliable source of cold molecular ions for spectroscopic investigations. Indeed, this technique enables the production of cations in a few (or a single) vibrational state(s) of their electronic ground state and preserves the rotational cooling of the neutrals in the beam. This is an efficient way to avoid spectral congestion in the cation spectra which would result from a preparation of the cation in many initial states. Furthermore, the supersonic expansion used to generate the molecular beam also allows

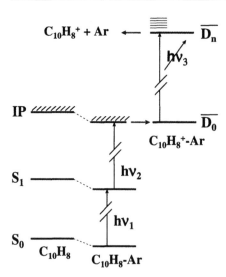

Figure 1: Energy levels diagram (here for naphthalene) showing the principle of the experimental method : two photons are used to prepare the cold van der Waals complex cation, the third one photodissociates this cation.

to form vdW clusters, and the optical selectivity of the (1+1') R2PI process permits ionization of a selected neutral system.

The procedure is shown in Figure 1. A first photon excites the neutral clusters in the S_1 state in a well-defined vibronic level (below the first predissociative state), a second photon ionizes the excited clusters to just above the ionization threshold to produce vdW cluster ions in their ground electronic state D_0. Then a third laser probes the cationic clusters to investigate their photodissociation spectrum.

The fragment PAH$^+$ time-of-flight (TOF) signal resulting from the dissociation of the cluster ion has to be separated from the signal of residual PAH$^+$. This is achieved by delaying the dissociation laser, as earlier done by Bieske et al [22]. Effectively, by using a static field in the extracting chamber, the PAH$^+$ ions produced via the delayed photodissociation step gain a lower total kinetic energy in the acceleration chamber than the background ions. This is because the parent ions have already started to drift when the fragments are formed. Hence the fragment ions have a longer time-of-flight (TOF). Indeed, by separating the two steps in time (ionization and dissociation) by a few hundreds of nanoseconds (ns), the difference in the TOF's is of the order of one hundred ns. Thus the fragment signal can be detected on a zero-background, giving a sensitivity of detection only determined by the efficiency of ion-collection and detection by the mass-analyzer, which is close to unity.

2.3 Measurement of the Absolute Absorption Cross-Section

According to the procedure presented in the previous section, the mass spectrum gives the absolute parent to fragment ratio, f, at a defined wavelength and intensity of the photodissociation laser. It can easily be shown that the number of remaining parents after the path of a laser pulse through the interaction volume depends exponentially on the radiant energy in this pulse, E_{laser} (in mJ), according to :

$$\frac{N_p(E_{laser})}{N_p(E_{laser}) + N_f(E_{laser})} = \exp(-\frac{\lambda}{hc \cdot \pi R^2}\sigma(\lambda) E_{laser}) = \frac{1}{1+f} \quad (1)$$

where λ is the fixed laser wavelength, $N_p(E_{laser})$ and $N_f(E_{laser})$ are the measured parent and fragment signals, πR^2 the measured area of the laser beam cross section interaction with the parent cations. All these parameters can be measured. The last parameter is the cross section of the process : $\sigma(\lambda)$. Thus, by recording mass spectra at a fixed wavelength for a large set of dissociation laser intensities, the absolute value of this cross section can be derived.

It is of interest to remark that the absolute photodissociation cross section can be measured without having to know the number of cations involved in the measurement. Our error bars only depend on experimental parameters which can be measured accurately, namely : E_{laser} and the laser beam radius R, providing thus a very safe determination of this cross section. This is a major point because the knowledge of the oscillator strength is of fundamental interest for the astrophysical problem.

2.4 Set–up

Most of the so-called "ICARE" molecular beam apparatus used in this study has already been described in previous publications [26]. PAH vapors were produced in the heated sample chamber maintained at constant temperature, seeded in a mixture of $\simeq 25\%$ Ar in Ne with a backing pressure of 1 to a few bars. The gas was expanded through a 0.9 mm orifice of a heated pulsed solenoid valve, with a pulse length of $\simeq 250$ μs. A home-made time-of-flight mass spectrometer with a total length of about 30 cm, set perpendicular to the molecular beam, provides a mass resolution of about 300 at an average mass of 200 a.m.u..

The ionizing lasers consist of two nanosecond pulsed dye lasers pumped by the same YAG laser operating at 10 Hz and 532 nm. The laser used for the dissociation step was an optical parametric oscillator (MOPO-730, Spectra-Physics, 0.2 cm^{-1} bandwidth, 5 ns pulse duration) operating at 10 Hz, and delivering photons from 430 to 690 nm using signal output, and from 730 to 2000 nm using the idler output. All these lasers were synchronized with the pulsed molecular beam via a four channel digital delay/pulse generator (Stanford Research System, model DG535) combined with a home-made pulse generator system. All the beams were directed to the center of the extraction zone. The ion current collected by a MCP detector is fed into a digital oscilloscope (Lecroy 9361) connected to a PC via IEEE bus. All the experiments were driven via Labview programs, consisting essentially in monitoring the fragment and parent currents versus the photodissociation laser wavelength.

3 The Spectra

3.1 Naphthalene$^+$

The naphthalene molecule (Nap = $C_{10}H_8$) is the smallest of the PAHs family. Prior to this work the Nap$^+$ cation has been studied in detail by photoelectron spectroscopy[27] (PES), and by absorption spectroscopy in the near-infrared to the ultra-violet range in rare gas matrices[28], [29]. Semi-empirical calculations have also been performed for the interpretation of the experimental spectra[30], [31]. Cavity Ring Down Spectroscopy (CRDS) has recently been successfully applied to this cation [32], produced by a discharge in a supersonic slit jet, but only two bands in the vicinity of the origin of the second electronic transition $D_2 \leftarrow D_0$ were measured. We present here the electronic spectrum of Nap$^+$ ($C_{10}H_8^+$) in the whole visible range [12].

The spectroscopy of the neutral Nap ($C_{10}H_8$) and its vdW complexes has already been investigated in other groups and is well documented in the literature[33], [20]. The first electronic transition $S_1(\tilde{A}^1 B_{3u}) \leftarrow S_0(\tilde{X}^1 A_g)$ originates at 32 020 cm^{-1}. This electronic transition has a low oscillator strength ($f = 0.002$) and the most intense band of the spectrum, 10 times stronger than the origin, is a vibronic transition localized at + 436 cm^{-1} above the electronic origin, assigned to the $\bar{8}$ vibrational mode. From a study of the vdW clusters, this excited state was found to be below the dissociation threshold of $C_{10}H_8$–Ar [20]. Thus, we choose this $\bar{8}^1$ state in S_1 as the intermediate resonant state for the first stage of production of the cationic complexes. Its band is shifted, relatively to the transition of the bare molecule, by -14.5 and -29 cm^{-1} to excite the $C_{10}H_8$–Ar and $C_{10}H_8$–Ar$_2$ (1/1) respectively. The wavelength of the ionizing laser was carefully chosen to excite just above the IP (a few tens of cm^{-1}) of each species, in agreement with the ZEKE measurements by Vondrák et al[34], giving a vdW shift per argon atom of -85 cm^{-1} relatively to the first IP of $C_{10}H_8$, localized at 65 687 cm^{-1}. The latter laser (MOPO system) was then scanned continuously in the visible region for the investigation of the $C_{10}H_8^+$–Ar$_{1,2}$ spectra.

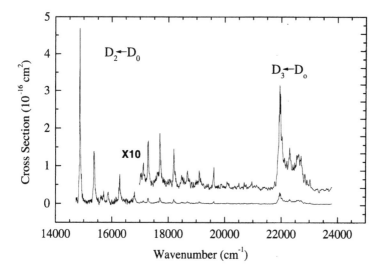

Figure 2: Electronic spectrum of the naphthalene cation in the visible

The observed visible absorption spectrum of the $C_{10}H_8^+$–Ar is shown in Figure 2. The first vibronic system originating at 14 863 cm^{-1}(strongest peak) is identified as the $^2B_{2g}(D_2) \leftarrow X^+ {}^2A_u(D_0)$ electronic transition. The second vibronic system originating at 21 959 cm^{-1} (strongest peak in this range) is identified as the $^2B_{3g}(D_3) \leftarrow X^+ {}^2A_u(D_0)$ electronic transition. These assignments are done by comparison with previous work by PES, MIS measurements and semi-empirical calculations. The spectrum looks very similar to the one obtain in rare gas matrix by Andrews et al [28] and by Salama et al[29], but an important difference is that no background is seen in our results, and the simulated spectrum obtained by Negri et al[31] is in agreement with our gas phase spectrum.

A fit of all the bands of the $D_2 \leftarrow D_0$ electronic transition was performed in order to evaluate (to a few cm^{-1} accuracy) the position of their centers, assuming lorentzian lineshapes with a common FWHM. The values of the frequencies, as well as their assignments, comparison with the values in a Ne matrix reported by Salama et al [29] and with theoretical calculations by Negri et al [31] can be found in reference [12]. In the case of the $D_3 \leftarrow D_0$ electronic transition, no clear assignment could be performed because of the large widths of the features (FWHM = 90 cm^{-1}, as compared to 20 cm^{-1} for the $D_2 \leftarrow D_0$ transition).

Also the bands in the photodissociation spectra of the Nap$^+$-Ar and of the Nap$^+$-Ar$_2$ clusters are lorentzian in shape and their widths are the same within experimental uncertainty for a given electronic transition. This is an indication for excited state lifetime broadening. From the absolute values

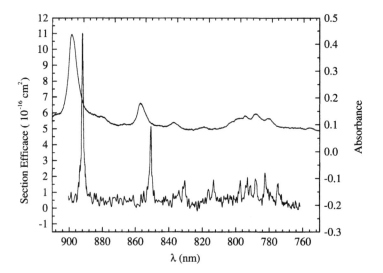

Figure 3: Electronic spectrum of the phenanthrene cation in the near infrared. Vertical scales : left for lower curve (gas phase spectrum), right for upper curve (matrix spectrum).

of the FWHM's the non radiative transition rates can be extracted : 4.10^{12} s^{-1} for the $D_2(0)$ state, and 1.7×10^{13} s^{-1} for the $D_3(0)$ state.
A question then arises : what is the non radiative mechanism responsible for such short lifetimes? The vibrational predissociation of the vdW cluster can be excluded, since its associated dynamical time cannot be shorter than the typical vdW vibrational period, i.e. $\simeq 10^{-12}$ s. Then it must be concluded that a non radiative process within the Nap$^+$ chromophore is involved. This process can be identified as internal conversion (IC) since the D_1 state of the same spin multiplicity is not far in the energy scale.

3.2 Phenanthrene$^+$

The phenanthrene molecule (Ph = $C_{14}H_{10}$) is the non linear PAH made with three 6-membered rings. In this case the energy of the first UV photon was set to 29 313 cm^{-1} to excite the ground vibrational state in the S_1 electronic state of Ph-Ar, red shifted by 13 cm^{-1} from the monomer band [35]. The energy of the second UV photon was tuned to about 34 000 cm^{-1} (slightly above the ionization threshold) to excite the vdW complex from S_1 to the ground electronic state D_0 of the Ph$^+$-Ar cation, without vibrational excitation. The third near infrared photon, delivered by the idler beam generated in the OPO laser system, was scanned from 900 nm to 760 nm while recording the intensities of the fragment and parent peaks in the mass spectrum and keeping the energy of the laser pulse relatively small enough (of the order of a few microjoules per pulse) to avoid undesired effects. The result is shown by the bottom trace in Figure 3.

The upper trace is a reproduction of the Ne matrix spectrum obtained by Salama et al [36], and assigned to the $D_2 \leftarrow D_0$ transition in the Ph^+ cation. The two spectra exhibit similar patterns. But the comparison emphasizes the effect of the solid phase environment of the cation on the positions of the bands and more importantly on their widths : the matrix broadening is of the order of 80 cm^{-1} to be compared with the gas phase widths of about 16 cm^{-1}. Consequently the fine structure of the bands between 770 and 800 nm is well resolved, which was not the case in the matrix spectrum. The accurate positions and vdW shifts of all the bands can be found in reference [11]. Note that the bands were found lorentzian in shape, like in Nap^+, which leads to the non radiative transition rate for IC from the $D_2(0)$ state : 3.10^{12} s^{-1}.

3.3 Phenylacetylene$^+$

The phenylacetylene (PA = C_8H_6) molecule is not a proper PAH, but its interest in our context resides in the fact that it is the major fragment in the photofragmentation of the naphthalene cation and an intermediate species in PAH formation reactions. From the experimental point of view, it offers a favorable accidental circumstance : the ionization potential is lying at an energy which is only slightly larger than twice the energy of the $S_1 \leftarrow S_0$ transition [37]. Consequently it can be ionized with a single frequency-doubled pulsed dye laser with little excess energy deposited in the ion. From the Threshold PhotoElectron Spectroscopy (TPES) data of Dyke et al. [38] under (1+1') R2PI laser excitation, one expects that, with the UV laser set at 35 860 cm^{-1} ($S_1 \leftarrow S_0$ origin band of PA-Ar), the PA^+-Ar complex is formed predominantly (70%) in its ground vibrational state, and minorily (30%) in the vibronic state involving one quantum in the intramolecular mode 13, which lies at 460 cm^{-1} in the D_0 state. The latter population of PA^+-Ar may be expected to undergo vibrational predissociation on a nanosecond timescale, if the binding energy of the argon atom (not accurately known) is smaller than this value.

Figure 4 shows the visible absorption spectrum recorded by monitoring the PA^+ fragment intensity. The band of strongest intensity near 17 800 cm^{-1} is assigned to the origin of the $D_3 \leftarrow D_0$ transition on the basis of the PES study by Maier and Turner[39]. It exhibits an interesting fine structure, as shown by the insert in the figure, which involves vdW states and will be commented upon in a forthcoming paper[40]. The band to the red of the origin is assigned to the 13_1^0 hot band. From its intensity it seems that the D_0 13_1 state of PA^+-Ar has partially predissociated during the 300 ns time

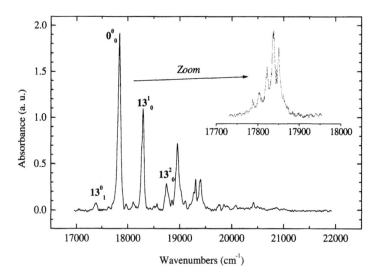

Figure 4: Electronic spectrum of the phenylacetylene-argon complex cation in the visible. The insert shows a zoom of the strongest band structure.

delay. The progression involving the same intramolecular mode in the D_3 state is easily recognized, and a few more modes can be seen.

Concerning the bandshape, we note a complex structure in this case. And if one wishes to recognize a lorentzian bandshape, it can only be found in the individual components of the fine structure, which are about 7 cm^{-1} wide. An upper limit of the non radiative rate of 1.3 ×10^{12} s^{-1} is associated. But no clear conclusion can be drawn without interpretation of the whole pattern[40].

4 Absolute Cross Sections and Oscillator Strengths

The absolute values of the photodissociation cross section have been measured carefully following the procedure described in Section 2.3 for the origin bands of the electronic transitions of Nap$^+$ and Ph$^+$. From these calibrations and the knowledge of the spectra, we can derive in each case the oscillator strength by integration of the spectrum over each electronic transition as a whole, using the standard formula :

$$\int \sigma(\lambda)d\lambda = \frac{e^2}{4\varepsilon_0 m_e c^2}\lambda^2 f \qquad (2)$$

We obtained for Nap$^+$ $f_2 = 0.052\pm0.005$ for the $D_2 \leftarrow D_0$ and $f_3 = 0.010\pm0.005$ for the $D_3 \leftarrow D_0$ transitions respectively. The latest value can be slightly underestimated since we did not measure the whole spectrum of Nap$^+$ (edge of the MOPO wavelength range). For Ph$^+$ we obtained $f_2 = 0.15\pm0.05$ for the $D_2 \leftarrow D_0$ transition. Theoretical results [30], [31] for these oscillator strengths are in agreement with our values, but previous evaluations in direct absorption MIS spectroscopy experiments gave much smaller values[29], [36]. It seems to show that the density of cations in the matrix samples has been overestimated in these studies. These numerical values are very useful from the astrophysical point of view. They will be used to estimate the upper limits of abundance of these cations in the ISM, in a careful examination of astronomical spectra.

5 Conclusion

It has been shown in this review that the recently developed method of photodissociation of vdW ionic complexes of the type PAH$^+$-RG$_n$ (RG = rare gas), selectively prepared by (1+1') R2PI in a molecular beam, is a very powerful technique to obtain the electronic absorption spectra of such large cations in the gas phase. This task had remained unfeasible until recently, mainly because of the high stability of these molecular species, combined to a fast intramolecular relaxation. The adequate use of the properties of vdW clusters has proven to be very useful. In particular the measurement of the vdW shift induced by the RG partner has been measured for each transition, and its additivity in the case of clusters with two RG atoms has been checked in several cases. This allows to recover the exact absorption wavelengths for the bare PAH$^+$ cations.

The astrophysical motivation for these studies was the Diffuse Interstellar Bands mystery. The possibility to compare these spectra with interstellar ones is a reality now, and not only frequencies but also bandshapes are provided by this work. The applied goal of this research activity, i.e. providing to the astrophysical community the proper database for searching for the possible presence of these species in interstellar clouds, is going to be gradually fulfilled. As a matter of fact, up to now, the quality of the astronomical spectra may well be the limiting factor for the detection of specific PAH$^+$ cations in space. But such new data is capable to motivate significant improvements also on that side.

Acknowledgment : The financial support from the CNRS Programme National PCMI (Physique et Chimie du Milieu Interstellaire) is gratefully acknowledged. The authors wish to thank Pascal Parneix for very helpful discussions.

References

[1] M. L. Heger, Lick. Obs. Bull. **10**, 146 (1921).

[2] P.W. Merrill, Publ. Astron. Soc. Pac. **46**, 206 (1934).

[3] G.H. Herbig, Ann. Rev. Astrophys. **33**, 19 (1995) and references therein.

[4] P. Jenniskens, F.X. Désert, Astron. Astrophys. Suppl. Ser. **106**, 39 (1994).

[5] (a) B.H. Foing, P. Ehrenfreund, Nature **369**, 296 (1994); (b) B.H. Foing, P. Ehrenfreund, Astron. Astrophys. **317**, L59 (1997).

[6] M. Tulej, D.A. Kirkwood, M. Pachkov, and J.P. Maier, Astrophys. J. **506**, L69 (1998).

[7] See the General Discussion starting on page 217 of the Faraday Discuss. **109** (1999).

[8] F. Salama in "Solid Interstellar Matter : The ISO Revolution", Les Houches No **11**, C. Joblin, L.D'Hendecourt and A. Jones eds, EDP Sciences, Les Ulis, France (1999).

[9] F. Salama, E.L.O. Bakes, L. J. Allamandolla, A.G.G.M. Tielens, Astrophys. J. **458**, 621(1996) and references therein.

[10] Bréchignac Ph., 1998, Faraday Discusssion **109**, General Discussion, 237.

[11] Ph. Bréchignac and T. Pino, Astron. Astrophys. **343**, L49 (1999).

[12] T. Pino, N. Boudin and Ph. Bréchignac, J. Chem. Phys. **111**, 7337 (1999).

[13] (a) A. Léger, J. L. Puget, Astron. Astrophys. **137**, L5 (1984); (b) L. J. Allamandolla, A.G.G.M. Tielens, J.R. Barker, Astrophys. J. **290**, L25 (1985); (c) A. Léger, L. D'Hendecourt, N. Boccara, Ed., Polycyclic Aromatic Hydrocarbons and Astrophysics, Reidel (1987).

[14] See the special issue on the first results of the ISO misssion: Astron. Astrophys. **315-2** (1996).

[15] A. Léger, in "The Diffuse Interstellar Bands", A.G.G.M. Tielens and T.P. Snow, p 363 (1995).

[16] C. Lifshitz, Int. Rev. in Phys. Chem. **16**, 113 (1997).

[17] J. Syage and J.E. Wessel, J. Chem. Phys. **87**, 3313 (1987).

[18] C. Moutou, L. Verstraete, Ph. Bréchignac, S. Piccirillo, A. Léger, Astron. Astrophys. **319**, 331 (1997).

[19] See S. Douin, P. Hermine, P. Parneix, Ph. Bréchignac, J. Chem. Phys. **97**, 2160 (1992).

[20] T. Troxler and S. Leutwyler, J. Chem. Phys. **95**, 4010 (1991).

[21] J.E. Reddic and M.A. Duncan, J. Chem. Phys. **110**, 9948 (1999).

[22] E.J. Bieske, M.W. Rainbird and A.E.W. Knight, J. Phys. Chem. 94, 3962 (1990).

[23] H. Piest, G. von Helden and G. Meijer, J. Chem. Phys. **110**, 2010 (1999).

[24] H. Piest, G. von Helden and G. Meijer, Astrophys. J. Lett. **520**, L75 (1999).

[25] See for instance the volume "Structure and dynamics of Van der Waals Complexes", J.Chem.Soc. Faraday Trans. **96** (1994).

[26] S. Douin, P. Parneix, F.G. Amar, Ph. Bréchignac, J. Phys . Chem. A **101**, 122 (1997).

[27] W. Schmidt, J. Chem. Phys. **66**, 828 (1976).

[28] L. Andrews, B.J. Kelshall and T.A. Blankenship, J. Phys. Chem. **86**, 2916 (1982)

[29] F. Salama, L. J. Allamandolla, J. Chem. Phys. **94**, 6964 (1991).

[30] O. Parisel, G. Berthier and Y. Ellinger, Astron. Astrophys. **266**, L1 (1992).

[31] F. Negri and M.Z. Zgierski, J. Chem. Phys. **100**, 1387 (1994).

[32] D. Romanini, L. Biennier, F. Salama, A. Kachanov, L.J. Allamandola and F. Stoeckel, Chem. Phys. Lett. **303**, 165 (1999).

[33] M. Stockburger, H. Gatterman, and W. Klusmann, J. Chem. Phys. **63**, 4519 (1975).

[34] T. Vondrák, S. Sato, K. Kimura, Chem. Phys. Lett. **261**, 481 (1996).

[35] T. Troxler, R. Knochenmuss, S. Leutwyler, Chem. Phys. Lett. **159**, 554 (1989).

[36] F. Salama, C. Joblin, L.J. Allamandola, J. Chem. Phys. **101**, 10252 (1994).

[37] P.D. Dao, S. Morgan and A.W. Castleman, Chem. Phys. Lett. **111**, 38 (1984).

[38] J.M. Dyke, H. Ozeki, M. Takahashi, M.C.R. Cockett and K. Kimura, J. Chem. Phys. **97**, 8926 (1992).

[39] J.P. Maier and D.W. Turner, J.C.S. Faraday II **69**, 196 (1973).

[40] N. Boudin, T. Pino and Ph. Bréchignac, to be published.

Structure and Dynamics of van der Waals Complexes by High Resolution Spectroscopy

G. Pietraperzia[1], M. Becucci[1], I. López-Tocón[1]
Ph. Bréchignac[2]
[1] LENS and Dip. Chimica, Univ. Firenze, Firenze, Italy
[2] LPM-CNRS, Univ. Paris-Sud, Orsay Cedex, France

1 Introduction

Molecular beams represent an extraordinary tool for the study of molecular properties [1]. The expansion of a gas through a small nozzle into a vacuum chamber has a number of consequences which have been extensively exploited to setup different experiments. The adiabatic expansion process results in a very efficient cooling of the sample. An experiment carried out on a sample cooled in a supersonic expansion allows to observe molecular aggregates bound by weak interactions, or metastable species. It is well known that, increasing the distance from the nozzle, the frequency of molecular collisions enormously decreases: the mean free path, especially in a molecular beam experiment with differentially pumped vacuum chambers, can be as large as a few meters. Therefore, if during the first moments of the expansion some molecular aggregate bound by weak forces is formed, it can survive long enough to be observed. In a conventional gas cell a similar aggregate would be immediately broken by collisions. The sample preparation in a supersonic molecular beam makes easier its spectroscopic study. The spectrum can be highly simplified because most of the molecules are in their ground vibrational state and then the hot bands are very weak. Also only a few rotational states are significantly populated and the rotational contour becomes much simpler. The experimental spectral resolution can be readily improved in a molecular beam experiment because contributions to both homogeneous and inhomogeneous broadening are strongly reduced. The number of collisions taking place during the effective experimental time drops virtually to zero at a distance from the nozzle larger than a few nozzle diameters. Then the collisional broadening, a homogeneous contribution to the linewidth, is completely removed. If the free supersonic expansion is collimated with a skimmer, the resulting molecular beam has a very limited divergence. Then, in a spectroscopic experiment where a narrow band, well collimated, laser beam crosses perpendicularly the molecular beam, the Doppler broadening (an inhomogeneous contribution to the linewidth) can be strongly reduced. The combination of these techniques allows to obtain a resolving power as good as 10^8 for UV excitation [2]. Many different detection schemes can be readily setup for the study of different properties of van der Waals complexes, such as laser induced fluorescence (LIF) for visible-ultraviolet excitation, multiphoton ionization (MPI), direct absorption, Fourier transform infrared spectroscopy (FTIR) or optothermal detection [3, 4].

We are mainly involved in high resolution electronic spectroscopy experiments on van der Waals complexes of aromatic molecules. Typically we are able to observe by LIF the vibronic components of electronic $S_1 \leftarrow S_0$ transitions under conditions of complete rotational resolution. The resolution and the assignment of the rovibronic transitions allow the determination of both ground and excited state rotational constants, as well as the measure of the linewidths (correlated to the lifetime of the excited states). This offers the unique opportunity to gain access at the same time to the structure and to the dynamics of these systems. The aim of this contribution is to report on the detailed description of the van der Waals complexes of aromatic molecules with rare gas atoms, as it can be achieved by high resolution spectroscopy. In particular we will show some evidence of correlation between the dynamics and the *effective* structure in the different excited states.

The 1:1 van der Waals complexes formed between aromatic molecules and rare gas atoms offer a key system for the understanding of the weak binding interactions responsible for the solvation process and other phenomena in many different fields. The molecular chromophore can be easily excited with near UV lasers and favorable Franck-Condon factors allow to excite several different vibronically excited states. Then, exciting the corresponding transitions in the complex, it is possible to put a large amount of energy in a weakly bound system and to study how the relaxation processes are affected by the presence of new predissociative relaxation channels.

2 Experimental technique

Our molecular beam-laser spectrometer has been described in detail in previous reports [5]. Briefly, it consists of a first vacuum chamber with the nozzle, pumped by a 7000 l s^{-1} diffusion pump backed by a 250 m^{3}hour^{-1} mechanical booster and rotary pump combination system. The jet is collimated by a 400 μm diameter conical skimmer (Mod.2, Beam Dynamics Inc.) placed 10 mm downstream of the nozzle. The resulting molecular beam enters a second high vacuum chamber, pumped by a 2000 l s^{-1} baffled diffusion pump to 10^{-4} Pa. The nozzle assembly is made of stainless steel. It has a sample compartment and can be provided with different diameter circular nozzles, with diameter ranging from 50 to 200 μm. The nozzle assembly can be heated up to 80° C to enhance the sample vapor pressure and to prevent sample condensation and clogging of the nozzle. The formation of binary clusters depends on the rate of three body collisions out of the nozzle, Z_3. It can be demonstrated that Z_3 depends on P_0 and T_0, pressure and temperature of the stagnation state, and d, the nozzle diameter, according to the expression [1]

$$Z_3 \propto P_0^2 d/T_0^2 \qquad (1)$$

Because the cluster formation goes with the square of the pressure and linear with the nozzle diameter, taking into account the available pumping speed (which limits the gas flow, proportional to $P_0 d^2$), the highest efficiency in cluster formation can be reached working with small diameter nozzles and high pressures. We have found that X-Ne (X=aniline, 4-fluoroaniline, styrene, 4-fluorostyrene) 1:1 complexes are best formed using pure neon as the carrier gas, a 50 μm diameter nozzle and 500 KPa pressure (limited by the pumping speed). A cluster efficiency up to 10% has been obtained working on the aniline-neon complex. The laser system is based on a single mode ring dye laser, frequency stabilized against a reference cavity, pumped by a large frame argon ion laser. The emission of the dye laser is sent to a second resonant ring cavity, which contains a BBO crystal for second harmonic generation. We obtained conversion efficiency up to 10% generating UV radiation at 288 nm with 1 W of injected power. A small portion of the fundamental wave is used for calibration: the working wavelength is measured in a wavelength meter with a 0.02 cm^{-1} accuracy against a reference He-Ne laser. Furthermore during the laser scan the relative frequency shift is calibrated on the transmission peaks of a temperature and pressure stabilized 150 MHz free spectral range étalon. The UV laser radiation enters and exits the vacuum chamber through two fused silica (UV grade) windows aligned at the Brewster angle. The power of the transmitted beam is measured for normalization of the LIF signal. The laser beam, focused using a 1 m focal length lens, crosses the molecular beam 28 cm downstream the nozzle and the resulting fluorescence emission, collected with two spherical mirrors, is imaged on a photomultiplier by means of two lenses. The LIF signal is acquired with photon counting technique and stored in a personal computer. The personal computer is also used to drive the laser scan and to acquire the calibration signals.

An iterative simulation and fitting procedure, based on a modified version of the ASYROT Fortran code by Sears [6], allows to assign the rotational part of the spectrum and to determine ground and upper state rotational constants. The experimental spectrum of the $S_1 \leftarrow S_0$ 0_0^0 transition of the aniline-neon complex and its best fit simulation are reported in Fig. 1.

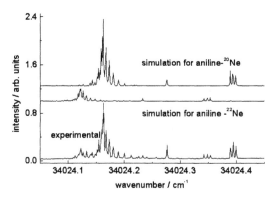

Figure 1: Experimental and simulated spectrum of the $S_1 \leftarrow S_0$ 0_0^0 transition of the aniline-neon van der Waals complex: an expanded view around the band center.

3 Structural determination from high resolution data

We have studied a few different van der Waals complexes between rare gas atoms and small aromatic molecules, such as aniline [7, 8] and 4-fluorostyrene [9]. Here we will discuss in some detail recent results obtained for the excitation of several $S_1 \leftarrow S_0$ vibronic bands of the aniline-neon complex. The first step in this kind of research is a complete characterization of the isolated molecules. Normally it is a good approximation to assume that their structures are not affected by the complex formation because of the large difference in strength between intra- and intermolecular forces. Under this assumption, we can transfer directly any information available on the isolated chromophore molecule to the complex. As a starting point for the study of the aniline-X (X = neon, argon) complexes we have recorded the high resolution electronic excitation spectra of aniline for different vibronic excitations. The spectra of the 0_0^0, $6a_0^1$, I_0^2 and 1_0^1 were completely assigned and the upper state rotational constants were determined with a 0.1 MHz accuracy [5, 10]. The inversion (I) vibration of aniline has been widely studied both in the S_0 and in the S_1 state. It is a low frequency, *large amplitude motion,* similar to the well known *"umbrella"*-like inversion of ammonia. This vibration was roughly described as the motion of the amino hydrogens above and below the aromatic ring plane. This allowed to build potential energy curves for the inversion vibration which well reproduce the spacings of the observed inversion levels deduced from the spectra

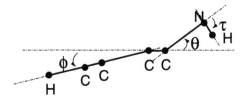

Figure 2: Schematic representation of aniline (C_s symmetry) projected in the symmetry plane (distances and angles not in scale). Some of the geometrical parameters used in the flexible model are also shown.

of normal aniline and of deuterated isotopomers [11, 12]. It is recognized that aniline is non planar in its S_0 state: microwave and infrared data showed that the inversion barrier is ~ 600 cm^{-1} and the equilibrium out-of-plane angle (τ_0) is about 46° degrees. In the S_1 state the barrier lowers practically to zero, according to the analysis of the vibrational spacings in the inversion progression. It was also demonstrated that this description of the inversion vibration is not accurate enough to predict the changes in the rotational constants in the different inversion excited states [13]. We have recently demonstrated [14] that it is possible to accurately reproduce, at the same time, both the vibrational spacings and the changes in the rotational constants in the progression of a large amplitude vibration by the use of a *flexible model* for the description of the molecular motion [15] supported by *ab initio* quantum calculations. The inversion potential is modeled as a function of the inversion angle τ, described in Fig. 2, according to the following expression

$$V(\tau) = a\tau^2 + b\tau^4 \tag{2}$$

In our model the molecular flexibility is introduced by allowing some other internal coordinates to change as a function of the inversion angle τ:

$$\begin{aligned}
l(\tau) &= l_0 + l_1 (\tau/\tau_0)^2 \\
\theta(\tau) &= \theta_1 (\tau/\tau_0) \\
\phi(\tau) &= \phi_1 (\tau/\tau_0) \\
\alpha(\tau) &= \alpha_0 + \alpha_1 (\tau/\tau_0)^2
\end{aligned} \tag{3}$$

l and α are respectively the CN bond length and the HNH angle, the other parameters are described in Fig. 2. The l_0 and α_0 values are referred to the planar (C_{2v}) molecular conformation (where θ_0 and ϕ_0 are zero by symmetry).

We have calculated the equilibrium structure of aniline in the S_1 electronic state by the CIS method with a 6-31G* basis set using the GAUSSIAN 94 package. This approach has already been found reliable in similar cases [14].

It came out that aniline in S_1 is slightly non planar. Then we repeated the calculation imposing a C_{2v} symmetry and we obtained the barrier height for inversion and a picture of the changes in the molecular structure between the two conformations. The optimized molecular structure in C_{2v} symmetry and its changes on going to the equilibrium non planar conformation are reported in Table 1 together with the relative energies. These results suggested the choice of the set of parameters allowed to change with the application of the flexible model.

During the flexible model calculations the potential parameters were freely allowed to change and the resulting values of vibrational frequencies and the changes of the planar moments of inertia from the S_1 ground to the I^2 level ($\Delta M_{ii} = M_{ii}(v=2) - M_{ii}(v=0)$; $i = a, b, c$) were compared with the experimental data [11, 8]. We prefer to compare the experimental and calculated changes in the molecular structure using the planar moments of inertia, defined as

$$M_{aa} = \frac{\hbar}{8\pi}\left(-\frac{1}{A} + \frac{1}{B} + \frac{1}{C}\right) = \sum_i m_i a_i^2 \qquad (4)$$

and similarly for the b and c axes, and not directly the rotational constants because the former ones directly represent the mass distribution along the principal axes of inertia.

The same calculation of the vibrational frequencies and of the changes in the planar moments of inertia for excitation of the amino inversion was also performed assuming a *rigid* planar conformation for the aniline frame: only the amino hydrogens were moved above and below the plane during the inversion. The results of the flexible and rigid models are reported in Table 2. It can be clearly seen that the spacings of the vibrational levels is well reproduced in both cases but only the flexible model is able to describe the change in the planar moments of inertia. The change in the mass displacements along the c axis, represented by the ΔM_{cc} term, is not yet very accurate.

The accurate determination of the rotational constants of the isolated aniline and of the complex in the corresponding states allows to gain information on the relative position of the two partners in the complex from the definition of the inertia tensor. A good choice for the reference frame to write the inertia tensor of the complex is the center of mass of the complex with the axes parallel to the principal inertia axes of the isolated aniline [7]. It can be shown that, with this choice, the inertia tensor can be expressed as

$$\left\{\begin{array}{ccc} k/A_0 + \mu\left(Y^2 + Z^2\right) & -\mu XY & -\mu XZ \\ -\mu XY & k/B_0 + \mu\left(X^2 + Z^2\right) & -\mu XZ \\ -\mu XZ & -\mu YZ & k/C_0 + \mu\left(X^2 + Y^2\right) \end{array}\right\}$$

$$= \left\{ \begin{array}{ccc} k/A & 0 & 0 \\ 0 & k/B & 0 \\ 0 & 0 & k/C \end{array} \right\} \qquad (5)$$

where A_0, B_0, C_0 and A, B, C are respectively the rotational constants (cm^{-1}) of aniline and its complex with the rare gas, $k = (h/8\pi^2 c)$ (cm^{-1} amu Å^2) and X, Y, Z are the Cartesian coordinates (Å) of the rare gas atom with respect to the aniline center of mass. The coordinates of the rare gas atom can be obtained by repeated eigenvalues calculations of the inertia tensor over a set of reasonable rare gas positions. Only those positions giving a good agreement between experimental and calculated rotational constants of the complex were retained and averaged with a weighting factor inversely proportional to the difference between the two values. The results obtained with this treatment however must be taken with some caution because of the loose nature of the binding in the complex. It has been shown by quantum calculations that the amplitude of the zero point motion of a rare gas atom above the aromatic ring surface can be as large as 0.5 Å[16].

We must also take into account that we obtain these coordinates from the inertia tensor which, for a fixed geometry, can be expressed as given by (5), i. e. the I_{xx} component is proportional to Y^2 and Z^2. Therefore, in a system like a van der Waals complex where the atoms are subjected

	C_{2v}	Δ		C_{2v}	Δ
bond lengths (Å)					
C_1C_2	1.4217	-0.0005	C_2C_3	1.4108	-0.0001
C_3C_4	1.4074	0.0004	C_2H	1.0744	-0.0001
C_3H	1.0730	0.0000	C_4H	1.0755	-0.0001
C_1N	1.3435	0.0044	NH	0.9936	0.0012
internal angles (deg.)					
$C_1C_2C_3$	118.53	0.04	$C_2C_3C_4$	119.36	-0.01
$C_2C_1C_6$	122.02	-0.03	$C_3C_4C_5$	122.21	-0.04
C_1C_2H	119.68	-0.03	C_2C_3H	120.45	0.03
C_2C_1N	118.99	0.01	C_3C_4H	118.90	0.01
C_1NH	121.20	-1.43	HNH	117.60	-1.30
dihedral angles (deg.)					
C_3C_2-C_1C_6	0.00	-0.05	C_2C_3-C_4C_5	0.00	0.03
C_4C_3-C_2H	180.00	-179.15	C_1C_2-C_3H	180.00	-180.22
C_2C_3-C_4H	180.00	-180.25	τ_0	0.00	18.3
energy (cm^{-1})	0.0	-17.5			

Table 1: CIS/6-31G* results on the S_1 state of aniline: C_{2v} structural parameters and their changes (Δ) in going to the C_s symmetry. The relative energy of the two molecular conformations are also given.

		ab initio	rigid model	flexible model
l_0	(Å)	1.3435	1.3435	1.3428
l_1	"	0.0044	-	0.0045
θ_1	(deg.)	1.3	-	2.1
ϕ_1	"	0.04	-	0.8
α_0	"	117.6	117.6	117.8
α_1	"	-1.30	-	-1.26
τ_0	"	18.3	-	19.5
a	(cm^{-1} deg.$^{-2}$)	-	0.06	0.06
b	(cm^{-1} deg.$^{-4}$)	-	0.00010	0.00014
		experimental		
I_0^1	(cm^{-1})	333	333	337
I_0^2	"	760	753	757
I_0^3	"	1224	1230	1224
ΔM_{aa}	(amu Å^2)	0.164	-0.95	0.14
ΔM_{bb}	"	0.030	0.22	-0.04
ΔM_{cc}	"	0.313	-0.12	0.12

Table 2: Parameters and results for the flexible model calculation of the inversion progression in S_1

to large amplitude motions, the inertia tensor will depend on the averaged values of the squares of the rare gas atom coordinates. These values of the coordinates determined from the analysis of the inertia tensor can be taken as the root mean square values, *i. e.*

$$\langle X^2 \rangle^{1/2} = \left(\langle X \rangle^2 + (\Delta X)^2 \right)^{1/2} \quad (6)$$

where ΔX is the mean deviation from the equilibrium value of the X coordinate. A direct interpretation of experimental values of the rare gas atom position is then really difficult because the two quantities, equilibrium position and mean displacement, can take values of the same order of magnitude. As it was suggested by Meerts and coworkers [17], this analysis carried on symmetric molecules can give a direct measure of the mean displacement of the rare gas atom if its equilibrium position is in a symmetry plane. In general, if the mean deviation is much smaller than the equilibrium value of the coordinate it is possible to safely take the experimental value as a measure of the structure of the complex. It was already demonstrated [16] for the aniline-neon complex that the rare gas atom is positioned in the symmetry plane of aniline at a distance of ∼3.4 Å and the mean deviation around this position is ∼0.1 Å Then our evaluation of the distance of the rare gas atom from the ring plane would reflect directly the equilibrium distance. Instead

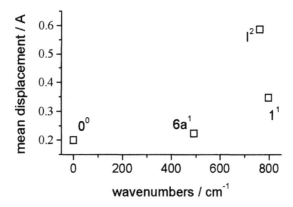

Figure 3: Mean displacement of the neon atom from the symmetry plane of aniline in the different vibronic states of S_1. The x axis is the vibronic energy with its origin set to the 0_0^0 transition.

the experimental value for the distance of the neon atom from the aniline symmetry plane is a measure of the mean displacement of the neon atom from this plane. In Fig. 3 we report the mean displacement of the rare gas atom from the aniline symmetry plane in the different vibronic levels of the S_1 state. The average distance between the rare gas atom and the aromatic ring is 3.400(5)Å in S_0 and 3.375(5)Å in S_1. In conclusion it should be noted that this "structural" analysis of the information contained in the rotational constants already includes dynamical effects due to the extent of the vibrational wavefunction.

4 Dynamics of van der Waals complexes in the S_1 state

High resolution spectroscopy gives a direct access to the lifetimes of the excited states by the measure of the linewidths of the transitions in the eigenstates resolved spectra, since in the vibronic excitations considered here the common 0_0 ground state has a practically infinite lifetime. We are mainly interested in the dynamics induced by the formation of the van der Waals complex. We account for the dynamics of the corresponding excited states in the isolated aromatic molecule assuming, as usual, that the same relaxation processes exist both in the isolated molecule and in the complex. Then it is possible to describe the relaxation processes in the complex as resulting from the simultaneous existence of additional pathways for the

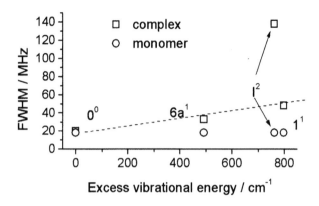

Figure 4: FWHM for the different excited states of aniline and its complex with neon. The x axis is the vibronic energy with its origin set to the 0_0^0 transition.

energy decay. The new channels available for the complex are associated to the increase of the density of states involving the new *external* vibrations (the displacements of the neon atom with respect to the aniline frame) and, in principle would lead to IVR (intramolecular vibrational relaxation) or VP (vibrational predissociation).

Figure 4 shows the linewidths observed for the individual rovibronic transitions of the corresponding S_1 vibronic bands in the isolated molecule and in the complex. It can be noticed that the 0^0 excited state in both systems has the same lifetime. This supports our hypothesis that both the ISC (intersystem crossing from S_1 to triplet states) and IC (internal conversion from S_1 to S_0) relaxation processes are not affected by the complex formation. The relaxation dynamics introduced by the complex formation then take place directly in the S_1 electronic state. In any case it derives, in terms of frequency resolved experiments, from the mixing of the zero order internal (aniline-like) vibrational eigenstate with the external vibrational eigenstates. Then it could be expected that an increase in the linewidth for a given excited state would correspond to a larger contribution of the external zero-order vibrational wavefunctions to the description of the true molecular eigenstate. The position of the rare gas atom that we determine from the analysis of the inertia tensor could be of some interest in the understanding of the dynamics of the excited states. We already pointed out the effective nature of these coordinates: they contain both the contribution from the equilibrium position and the mean displacement of the atom. If no coupling occurs between the internal and external vibrations no change

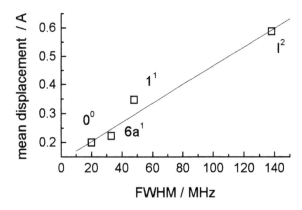

Figure 5: Correlation between the mean displacements of the rare gas atom from the aniline symmetry plane and the observed linewidths.

in the effective coordinate describing the mean displacement of the neon atom from the aniline symmetry plane is expected. If some mode mixing occurs then a decrease in the lifetime of the excited states is expected and the excitation of the vibronic bands, corresponding to internal vibrations of aniline, will have some partial character of the external vibrations. Figure 5 shows the correlation existing between this large amplitude motion and the observed linewidth measured in the different excited states: it is a clear proof of the existence of mode mixing between the internal and external vibrational coordinates.

The results reported in Figs. 4 and 5 show another important feature of the relaxation dynamics in the aniline-neon van der Waals complex, that is likely to be general for similar complexes: the relaxation rate seems to be mode specific. Similar results were reported for instance on the aniline-argon complex [18, 13]. It can be seen that the regular trend of the increase of the relaxation rate with the vibrational energy available in the S_1 state is not followed by the I^2 excited state, which exhibits an anomalously large relaxation rate. Indeed this is the only out-of-plane mode among the series of vibronic states we studied, and its nature of large amplitude, anharmonic motion has been clearly demonstrated in Section 3. Then it is quite reasonable to think that the inversion mode can have a good coupling with the external vibrations. At present we are working on theoretical quantum calculations trying to elucidate the nature of this coupling.

The support from MURST and EU (under Contract Nr. ERBMGECT 950017) is kindly acknowledged. One of us (I.L.T.) acknowledges the Spanish Ministerio de Educacion y Ciencia for the grant PF97/28482592. The authors express their gratitude to the colleagues involved at different stages of this research: N.M.Lakin, E.Castellucci, E.R.Th.Kerstel and W.Caminati.

References

[1] G.Scoles ed., Atomic and Molecular Beams Methods (Oxford University Press, New York, Oxford, 1988-1992).

[2] W.A.Majewski and W.L.Meerts, J. Mol. Spectr. **104**, 271 (1984).

[3] J.M.Hollas and D.Phillips eds., Jet Spectroscopy and Molecular Dynamics (Blackie Academic and Professionals, London, 1995).

[4] D.W.Pratt, Annu. Rev. Phys. Chem. **49**, 481 (1998).

[5] E.R.Th.Kerstel, M.Becucci, G.Pietraperzia and E.Castellucci, Chem. Phys. **199**, 263 (1995).

[6] T.J.Sears, Comp. Phys. Comm., **34**, 123 (1984).

[7] M.Becucci, G.Pietraperzia, N.M.Lakin, E.Castellucci and Ph.Bréchignac, Chem. Phys. Lett. **260**, 87 (1996).

[8] M.Becucci, N.M.Lakin, G.Pietraperzia, E.Castellucci, Ph.Bréchignac, B.Coutant and P.Hermine, J. Chem. Phys. **110**, 9961 (1999).

[9] N.M.Lakin, G.Pietraperzia, M.Becucci, E.Castellucci, M.Coreno, A.Giardini-Guidoni and A. van der Avoird, J. Chem. Phys. **108**, 1836 (1998).

[10] E.R.Th.Kerstel, M.Becucci, G.Pietraperzia and E.Castellucci, J. Mol. Spectr. **177**, 74 (1996).

[11] J.M.Hollas, M.R.Howson, T.Ridley and L.Halonen, Chem. Phys. Lett. **98**, 611 (1983).

[12] R.A.Kydd and P.J.Krueger, Chem. Phys. Lett. **49**, 539 (1977).

[13] W.E.Sinclair and D.W.Pratt, J. Chem. Phys. **105**, 7942 (1996).

[14] M.Becucci, E.Castellucci, I.López-Tocón, G.Pietraperzia, P.R.Salvi and W.Caminati, J. Phys. Chem. **103**, 8946 (1999).

[15] R.Meyer, J. Mol. Spectr. **76**, 266 (1979).

[16] P.Parneix, N.Halberstadt, Ph.Bréchignac, F.G.Amar and A. van der Avoird, J. Chem. Phys. **98**, 2709 (1993).

[17] W.L.Meerts, W.A.Majewski and W.M. van Herpen, Can. J. Phys. **62**, 1293 (1984).

[18] E.R.Bernstein, Annu. Rev. Phys. Chem. **46**, 197 (1992).

Femtosecond Pump-Probe Experiments with a High Repetition Rate Molecular Beam

W. Roeterdink, A.M. Rijs, G. Bazalgette, P. Wasylczyk[1],
A. Wiskerke, S. Stolte, M. Drabbels and M.H.M. Janssen

Laser Centre and Depeartment of Chemistry, Vrije Universiteit
Amsterdam, The Netherlands
[1]Physics Department, Warsaw University
Warsaw, Poland

1 The laser system

This paper describes the first results in the field of femto chemistry obtained by our group at the VUA. A high repetition rate (1 KHz) molecular beam machine and laser system have been used. The laser system was supplied by Spectra Physics and consists of a chirped amplified titanium sapphire system. Its source, formed by a titanium sapphire oscillator (Tsunami) produces 70 fs pulses with a repetition rate of 82 MHz and an energy of approximately 10 nJ per pulse. The output is tunable between 760 nm and 820 nm. A homebuilt autocorrelator was used to determine the pulse duration of the oscillator. The autocorrelator is based on a Michelson-type interferometer with a Hamamatsu photodiode as detector[1]. The semi conductor detector has a bandgap exceeding the energy of one photon of 800 nm, so a two photon process is needed to generate a signal. The measured pulse duration of 70 fs (FWHM) shown in Fig.1 is in agreement with the specified value of the Tsunami.

These femtosecond oscillator pulses are amplified in a regenerative amplifier (Regen). To avoid damage to the amplifier and to improve overlap between the pump laser and the oscillator pulses, the oscillator pulses are stretched approximately 10^4 times in the time domain. This is done by introducing a linear chirp. The pulses are then amplified 10^5 times in energy by the Regen. Subsequently, these pulses are compressed to give 130 fs pulses of 1 mJ at a repetition rate of 1 kHz. At 800 nm, the output bandwidth of the amplifier is 8 nm. The 130 fs pulse duration of the amplifier was determined with a homebuilt second harmonic single shot autocorrelator [2] [3]. A single shot monochromator was constructed to measure the spectral shape of the

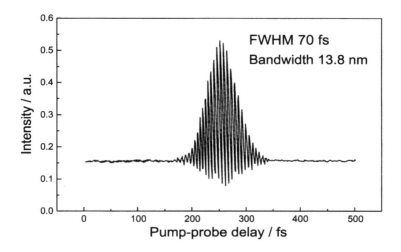

Figure 1: Interferometric autocorrelation transient of the tsunami.

oscilator laser pulse as well as the amplified laser pulse.
The amplified pulses are used to pump an optical parametric amplifier (OPA), giving a continuously tunable output over the range 1150 to 2000 nm. The signal can be doubled to 50 μJ with a central wavelength of 600 nm and a bandwidth of 16 nm (FWHM) and a pulse duration of 160 fs (FWHM). Further compression with a four prism setup shortens these pulses to 90 fs (FWHM). The alignment of the prism compressor is easily done with the help of a homebuilt FROG (Frequency Resolved Optical Grating) which enables us to minimize the linear chirp [4].

2 Test experiments on molecular iodine

To test the laser system pump-probe experiments have been carried out in an iodine cell. The potential curves of I_2 of interest are shown in Fig.2. A pump pulse, employing 30 μJ of varying wavelength in the range of 633 nm to 570 nm, prepares a coherent wave packet in the B-state of I_2 by exciting a manifold of vibrational levels. The B-state is probed by a second pulse, delayed (0-20 ps) with respect to the pump. The remainder of the 800 nm pump beam from the amplifier behind the OPA is doubled to 400 nm and used as a probe, taking the coherently prepared wave packet into an upper fluorescent state (the E-state). The resulting interferogram (Fig.3) resembles earlier findings [5]. Its Fourier transform reproduces the vibrational frequencies of the B-state of iodine. High resolution spectroscopic data from Luc et al. [6] are used to determine the wavelengths of

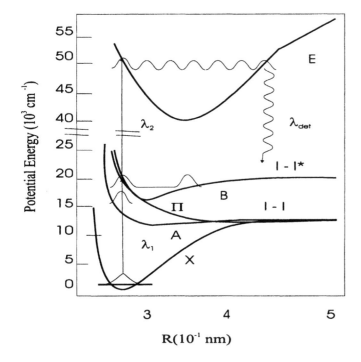

Figure 2: Potential energy surfaces in iodine, adapted from [5].

the vibrational levels. The Franck-Condon factors are taken from [7] and assuming a thermal distribution, gives a maximum population in the J=58 rotational level. The two observed peaks are attributed to a quantum beat between $\nu'=17$ and $\nu'=18$ (92.5 cm^{-1}) and between $\nu'=18$ and $\nu'=19$ (94.3 cm^{-1}), which compares favorably with the literature values 92.68 cm^{-1} and 94.52 cm^{-1}, and are in accordance with the specified resolution 0.1 μm of the delay line.

3 Molecular beam machine

A new machine has been developed to enable femtosecond experiments with our laser system in a molecular beam (Fig.4). The source chamber is pumped by a 5000 l/s diffusion pump with a water cooled baffle. A seeded molecular beam is produced in a fast piezo pulsed valve [8], yielding gas pulses with a repetition rate of 1 kHz and a pulse width in the range of

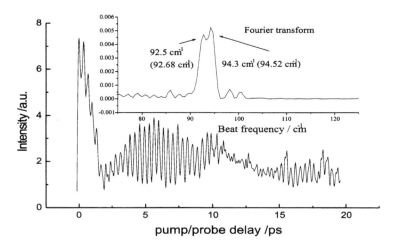

Figure 3: Pump-probe transient of the B-state of iodine. The number in the brackets give the value found by [6]

50–200 µs. If required, LIF measurements can be performed directly behind the piezo valve, in the source chamber, were the number density is high.
The molecular beam is doubly skimmed before entering the ion optic chamber. The pressure in the source under normal operating conditions is 1×10^{-4} mbar, whereas in the ion optics chamber it is 1×10^{-6} mbar.
Currently a ion imaging set up is under construction, which will employ the velocity map imaging technique with electrostatic lenses [9]. The advantage of this technique is that the ion image is not a convolution of the spatial profile of the molecular beam with the recoil distribution, because all molecules with the same velocity are imaged on the same point of the MCP. Another advantage of the velocity imaging technique is the absence of extraction grids, which enhances the resolution of the image. In our initial experiments a conventional ion time of flight technique is used, we report femtosecond pump probe experiments on I_2 and CD_3I, via different REMPI schemes.

4 Lifetime of the B state of CD_3I

The first electronically excited band in CD_3I is the A state, which is a fast dissociative band around 266 nm with a bandwidth of 100 nm. In Fig 5, in which the electronic level scheme of CH_3I is depicted, the spectral size of the A-band continuum is indicated with a block. There are two available dissociation channels: $CD_3 + I\ (^2P_{3/2})$ leading to iodine in the electronic

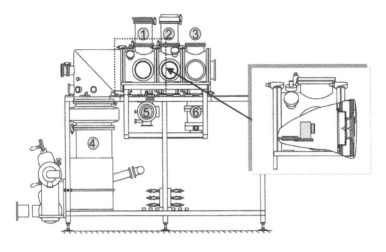

Figure 4: Molecular beam machine.
1: Source chamber 2: REMPI chamber
3: For future experiments 4: 5000 ls^{-1} diffusion pump
5,6 : turbomolecular pumps

ground state and $CD_3 + I$ ($^2P_{1/2}$) in which iodine is formed in the excited spin orbit state. These dissociative potentials intersect the higher lying B-state and cause its predissociation. With the help of ion imaging the branching ratios between the two dissociation channels can be determined. Excited CD_3I can be described as a Rydberg atom, the ion core has a $^2E_{3/2}$ symmetry which can be attributed to the lower spin orbit component and a $^2E_{1/2}$ to the higher spin orbit component. These two states of the ion core are coupled with the outer s-electron which leads to the $^2E_{3/2}[2]$ and $^2E_{3/2}[1]$ states for the lower spin orbit component and $^2E_{1/2}[1]$ and $^2E_{1/2}[0]$ for the upper spin orbit level. The number in the bracket gives the projection of the total angular momentum onto the symmetry axis of the molecule. Lifetime measurements on the $^2E_{3/2}[2]$ and its vibronic bands have been performed by several groups with different techniques. Zeigler et al. [10] reported a study on the B state dynamics using Raman and hyper Raman excitation profiles and depolarisation dispersions. They measured a lifetime of 1.2 ps for the vibrationless band origin of the B-state of CD_3I.

In comparison Syage and Steadmann [11] have measured jet cooled absorption line widths for various excited levels of the B state of CH_3I. These linewidths have been transformed to a lower limit of the lifetime, and led to a result for the 2_0^1 of 830 fs, for the 3_0^1 of 920 fs (the C-I stretch), for the 6_0^1 of 670 fs and the band origin 870 fs, which suggest the presence of decreased

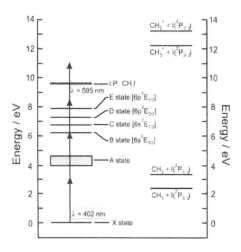

Figure 5: Level scheme of CD$_3$I.

dissociation rate upon excitation of the C-I stretch. Direct pump probe lifetime measurements of the B state of CD$_3$I and CH$_3$I, by Baronavski and Owrutsky have been carried out in a cell employing a (1+1) REMPI scheme with pump pulses around 192-205 nm and probe pulses of 256-273 nm. These experiments yielded a lifetime of the band origin of 1.84 ps for CD$_3$I [12].

In our current set up we have also measured the lifetime of the B state of CD$_3$I. Our study is carried out in a supersonically cooled beam of molecules. Presently a (2+2) REMPI scheme is used, in which two photons of 401 nm are used to excite to the B state and the B state is probed with two photons around 600 nm, which ionizes the parent molecule. Apparently, fragmentation of the parent ion or ionization of the neutral fragments does not take place at these colours. Preliminary results detecting the parent ion mass as a function of the delay between pump and probe pulses are shown in Fig.6. The part of the transient around 0 fs shows the correlation function between the pump and probe pulses each with a duration of 160 fs. The tail at the right part of the transient reflects the lifetime of the B state. Fitting these raw data with a single exponential decay, a lifetime of 1.2 ps has been extracted. The beatings in the transient are believed to originate from a coherent superposition between the $^2E_{3/2}[2]$ 0^0 state and the $^2E_{3/2}[1]$ 6^1 state, yielding a energy difference of 149 cm^{-1}. We have also applied a (2+1) REMPI scheme, two photons of 401 nm were used to excite the B state and one photon of 266 nm, the third harmonic of 800 nm, was used to ionize the B state. Type I second harmonic generation

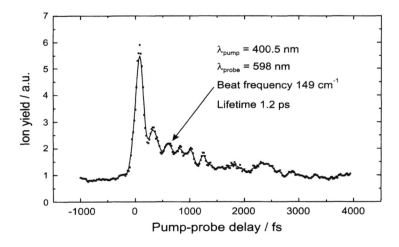

Figure 6: Pump-probe transient of the B-state of methyliodide.

is used to generate 400 nm photons. Type II mixing is used to generate 266 nm. The first crystal has a thickness of 0.5 mm which gives a delay of several hundreds of femtoseconds between the second harmonic and the fundamental, which causes a bad overlap of the 800 nm and 400 nm pulses in the mixing crystal. To avoid this problem the mixing crystal is rotated around the ϕ angle. Powers up to 60 μJ are obtained. In this way the 400 nm beam consists of two orthogonally polarized components. In this scheme the majority of the fragments turn out to be produced by one 266 nm photon absorption in the A state followed by prompt photodissociation, giving rise to the autocorrelation peak alone.

Acknowledgements

Financial support from NWO/CW and NWO/FOM enabled the construction of our femtolaser apparatus. We are grateful to the University for a special laboratory for femtosecond laser experiments. Paul Griffiths is acknowledged for help with LaTeX, and for preparation of some of the figures. WR also acknowledges his special Ph.D student position of the young chemist program of NWO/CW.

References

[1] D.T. Reid, M. Padgett, C. McGowan, W.E. Sleat and W. Sibbet, Optics Lett. **22**, 223 (1997)

[2] A. Brun, P. Georges, G. Le Saux and F. Salin, J. Phys. D: Appl. Phys. **24**, 1225 (1991)

[3] J. Jansky, G. Corradi, Optics Comm. **23**, 293 (1977)

[4] R. Trebino, K.W. DeLong, D.N. Fittinghoff, J.N. Sweetser, M.A. Krumbugel, B.A. Richman and D.J. Kane, Rev. Sci. Instrum. **68 – 9**, 3277 (1997)

[5] R.M. Bowman, M. Dantus and A.H. Zewail, Chem. Phys. Lett **161**, 297 (1989)

[6] S. Gerstenkorn and P. Luc, J. Physique **46**, 355 (1985),

[7] J. Tellinghuisen, J. Quant. Spectrosc. Radiat. Transfer. **19**, 149 (1978)

[8] D. Gerlich, Private communication.

[9] A.T.J.B. Eppink and D.H. Parker, Rev. Sci. Instrum. **68 – 9**, 3477 (1997)

[10] D.J. Campbell and L.D. Ziegler, Chem.Phys. Lett. **201**, 159 (1993)

[11] J.A. Syage and J. Steadmann, J.Chem.Phys. **54**, 7343 (1990)

[12] A.P. Baronavski and J.C. Owrutsky, J.Chem. Phys. **108**, 3455 (1998)

Part IV

Dynamics of Elementary Molecular Collisions and Femtochemistry: Theory, Real-Time Probing, and Imaging of Crossed-Beam Reactions

Femtochemistry: Recent Progress in Studies of Dynamics and Control of Reactions and Their Transition States*

Ahmed H. Zewail

Arthur Amos Noyes Laboratory of Chemical Physics,
California Institute of Technology, Pasadena, CA 91125, USA

1 Introduction and Perspectives

This Centennial Issue of *The Journal of Physical Chemistry* marks not only its achievements since birth in 1896 but also an era of great scientific contributions to the field of chemical dynamics. Pioneering work in the studies of molecular reaction dynamics, which continued over a century, led to new methods of experimentation and to new concepts. Letokhov [1] has identified "three waves" in the development of molecular dynamics, depending on the time scale. In this account we focus on the third wave characterized by femtosecond time resolution. Chemistry occurring on this time scale, femtochemistry, is microscopic, on the length scale of a bond, allowing us to address the nature of transition states and their control, a subject also started in the first part of this century.

---

*Revised and reprinted with permission from J. Chem. Phys. 1996, 100, 12701. Copyright 1996 American Chemical Society.

Editorial Note. The 1999 Nobel Prize in Chemistry has been awarded to A. H. Zewail for his impressive accomplishments in molecular reaction dynamics, reaching the time resolution of femtosecond, the fastest for chemical reactions. Thus, Zewail's Femtochemistry has created the impact and explosion of experimental research in molecular reaction dynamics. Important concepts and perspectives in Femtochemistry, with applications in studies of dynamics and control of chemical reactions in different systems and phases, are highlighted and discussed in this feature article to the Centennial Issue of The Journal of Physical Chemistry. Some new opportunities and developments are also discussed. We are glad and honored to include this new edition in our volume, expected for a large audience. We recommend also his recent references [122–132] and his Nobel lecture summarized in the J. Phys. Chem., Volume 104, page 5660 (2000) article and detailed in the forthcoming Les Prix Nobel book.

Roger Campargue, Editor

How do reactions proceed and what are their rates? Arrhenius [2] gave the first description of the change in rates of chemical reactions with temperature and formulated (1889) the familiar expression for the rate constant,

$$k = A\exp(-E_\mathrm{a}/kT) \qquad (1)$$

which, as Arrhenius acknowledges, has its roots in van't Hoff's (1884) equations [3]. Besides the value of (1) in the well-known plots of "$\ln k$ vs $1/T$" to obtain the energy of activation E_a. Arrhenius introduced a "hypothetical body", now known as the "activated complex", a central concept in the theory of reaction rates – the reaction, because of collisions or other means, proceeds only if the energy is sufficient to exceed a barrier whose energy describes the nature of the complex. Various experimental data for different temperatures T were treated with (1), giving E_a and the preexponential factor A.

A few years after Arrhenius' contribution, Bodenstein (1894) [4] published a landmark paper on the hydrogen/iodine system, which has played an important role in the development of gas-phase chemical kinetics, in the attempt to understand *elementary* reaction mechanisms. In the 1920s, Lindemann (1922) [5] and Hinshelwood (1926) [6] and others developed, for unimolecular gas-phase reactions, elementary mechanisms with different steps, defining activation, energy redistribution, and chemical rates. By 1928, the Rice–Ramsperger–Kassel (RRK) theory was formulated, and Marcus, starting in 1952, blended RRK and transition state theories in a direction which brought into focus the nature of the initial and transition-state vibrations in what is now known as the RRKM theory of chemical kinetics [7].

The rate constant, $k(T)$, described above, does not provide a detailed molecular picture of the reaction. This is because $k(T)$ is an average of the microscopic, reagent-state to product-state rate coefficients over all possible encounters. These might include different relative velocities, mutual orientations, vibrational and rotational phases, and impact parameters. A new way was needed to describe, with some quantitative measure, the process itself of chemical reaction: how reagent molecules approach, collide, exchange energy, sometimes break bonds and make new ones, and finally separate into products. Such a description is the goal of molecular reaction dynamics [8]. For some time, theory was ahead of experiment in the studies of microscopic molecular reaction dynamics. The effort started shortly after the famous publication of the Heitler–London quantum-mechanical treatment (1927) of the hydrogen molecule [9]. One year later (1928), at Sommerfeld's Festschrift (60th birthday), London [10] presented an approximate expression for the potential energy of triatomic systems, e.g., H_3, in terms of the Coulombic and exchange energies of the "diatomic" pairs. In 1931 Henry Eyring and Michael Polanyi [11], using the London equation, provided a semiempirical calculation of a potential energy surface (PES) of the $H + H_2$

reaction describing the journey of nuclei from the reactant state of the system to the product state, passing through the crucial transition state of activated complexes. The birth of "reaction dynamics" resulted from this pioneering effort, and for the first time, one could think of the PES and the dynamics on it. Figure 1 gives typical reaction paths for different systems, and Figure 2 reproduces some of the early results of the theoretical studies of the dynamics and the time scale for elementary reactions – *in those days, often expressed in atomic units of time!*

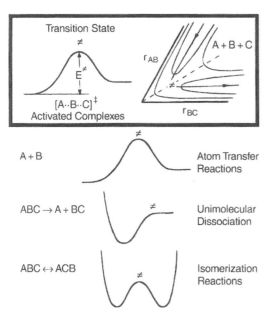

Figure 1: Schemes of the reaction path in different classes of reactions, generically labeled by three bodies where A, B, and C represent atoms or molecules

No one could have dreamed in the 1930s of observing the transient molecular structures of a chemical reaction, since the time scale for those *far from equilibrium* activated complexes in the transition state was estimated to be less than a picosecond (ps). Such a time scale was rooted in the theory developed for the description of reaction rates, starting with the work of Polanyi and Wigner (1928) [12]. In 1935, Eyring [13] and, independently, Evans and Polanyi [14] formulated *transition-state theory*, which gave an explicit expression for Arrhenius' preexponential factor:

$$k = (kT/h)(Q^{\ddagger}/Q) \exp(-E_{\rm a}/kT) \tag{2}$$

Quantum statistical mechanics was used in deriving the expression, defining the partition function Q of the reactant and $Q^{\ddagger}$ of the activated complex.

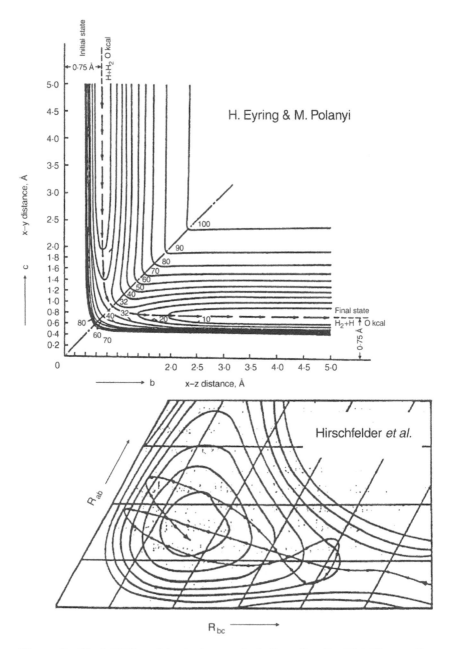

Figure 2: First PES and trajectory calculations for the H + H$_2$ reaction: see text

The theory made four assumptions [13, 14], including equilibration of the activated complexes, and gave an analytical formula for the rate constant with a "frequency" for the passage through the transition state, kT/h. This frequency factor typically corresponds to $\sim 10^{-13}$ s, *the time scale of molecular vibrations*. Kramers' (1940) [15] classic work modified the preexponential factor to include friction, but the description of the transition state still is similar. From a classical mechanical point of view, this estimate of time scale is consistent with knowledge of the velocities of nuclei and distance changes involved in a chemical reaction: For a velocity of 1 km/s and a distance of 1 Å, the time scale is about 100 femtoseconds (fs). Molecular dynamics simulations have shown a range for the time scales, ps to fs, depending on the reaction.

Experimentally, an important stride was made by the development of flash photolysis by Norrish and Porter (1949) [16] and the relaxation method by Eigen (1953) [17] to study chemical reaction intermediates with millisecond to microsecond lifetimes. Molecular collisions occur on a much shorter time scale, and new approaches were needed to examine the dynamics of single collisions. Significant contributions were made by Herschbach [18], Lee [19], and Polanyi [20] in the development of crossed molecular beams and chemiluminescence techniques to study such collisions and provide deeper understanding of the reactive processes.

A large body of results has accumulated over the past three decades involving the characterization of such "*before*" (reactant) and "*after*" (product) observables and their relationship to the dynamics on semiempirical or ab initio PESs. The time scale was deduced from knowledge of the angular distribution of products utilizing what is known as Herschbach's osculation model of the collision complex. In addition to the large number of crossed molecular beams and chemiluminescence studies, an expanding literature of crossed molecular beam laser results has probed dynamics via careful analyses of product internal energy (vibrational and rotational) distributions and steady-state alignment and orientation of products (see the article by Zare and Bernstein) [21].

To probe the transition-state region more directly, different methodology had to dawn. Polanyi's approach [20] of transition-state (TS) spectroscopy was an important trigger for many studies. Emission, absorption, scattering, and electron photodetachment are some of the novel methods presented for such "time-integrated" spectroscopies. The key idea was to obtain, as Kinsey [22] puts it, *short-time dynamics from long-time experiments*. With these spectroscopies in a CW mode, a distribution of spectral frequencies provides the clue to the desired information regarding the distribution of the TS over successive configurations and potential energies. Recently, this subject has been reviewed by Polanyi and the author. In section 2, we give a few examples from the review and discuss the general definition of the TS. The resolution in time of the elementary dynamics (femtochemistry) [24] offers an opportunity to observe a molecular system in the continuous process of its evolution from reactants to transition states and then to products. Important to such transformations, as detailed elsewhere [24], are three fundamentals of the dynamics, namely (1) intramolecular vibrational energy redistribution (IVR), (2) reactant-state to product-state rates, i.e., $k(E)$, in contrast to $k(T)$, and (3) the nature of transition states. Generally, the time scale for IVR is on the order of picoseconds, the rates are tens of picoseconds and longer, and the "lifetime" of transition-state species is picoseconds to femtoseconds. Because the phenomena are ultrafast in nature, the sensitivity in femtochemistry is enhanced by orders of magnitude [25]. The scope of applications now spans many systems and in different phases, including biological structures [23–30], as illustrated in Figure 1.

Several important advances have been part of the evolution toward the fs resolution. The mode locking of lasers and the development of dye lasers [31] were essential to the generation of ps pulses and in the applications to chemical and biological systems in the 1970s. Significant contributions in the area of ps chemistry were made by Eisenthal's group which studied in liquids chemical processes such as collision-induced predissociation, the cage effect, and proton and electron transfer; as discussed below, these primary processes have been studied in femtochemistry. The groups of Rentzepis, Hochstrasser, Kaiser, and others, with the ps resolution of the 1970s, played major roles in the studies of nonradiative processes in solutions; these include internal conversion, intersystem crossing, and vibrational relaxation and dephasing [32–34]. The success of Shank and his colleagues in developing the CPM laser in 1981 made it possible to generate fs pulses [34]. Also, the generation of a continuum of ultrashort pulses by Alfano and Shapiro [32] provided the ability to tune the wavelength. The latest titanium:sapphire laser development by Sibbett in 1991 [32] has taken the technology into a revolutionary road. Finally, the issues concerning problems of sensitivity and spatial and temporal considerations, especially for femtochemistry in molecular beams, have been addressed [24]. Essentially all detection schemes have now been implemented: *laser-induced fluorescence, multiphoton ionization mass spectrometry, photoelectron and ZEKE detection, stimulated-emission pumping, and absorption* (for reviews see ref [23]).

On the theoretical side, a major step forward was made when Heller [35] reformulated the time-dependent picture for applications in spectroscopy, and Kinsey and Imre [36] described their novel dynamical Raman experiments in terms of the wave packet theory. The progress was greatly helped by advances made in the theoretical execution and speed of computation by Kosloff [37] and subsequently by many others. The groups of Imre [38] and Metiu [39] did the first "exact" quantum calculations of femtochemical dynamics, and these studies proved to be invaluable for future work. There is a parallelism between the experimental diversity of applications in different areas (Figure 1) and the impressive theoretical applications to many experiments and systems. Manz, who has played a significant role in this field, has given an overview to the recent progress made in the two volumes he edited with Wöste [23].

The coming section 2 introduces general definitions and the CW spectroscopic advances made in probing the TS. Section 3 is concerned with concepts in femtochemistry and section 4 with applications to elementary and complex reactions of various types and classes. In section 5, we discuss some new opportunities, the development of ultrafast electron diffraction and reaction control, and in section 6 we conclude the paper.

2 The Transition State: Definition and Spectroscopy

For an elementary reaction of the type

$$A + BC \rightarrow [ABC]^\ddagger \rightarrow AB + C \qquad (3)$$

the whole trip from reagents to products involves changes in internuclear separation totaling ~ 10 Å. If the atoms moved at $10^4 - 10^5$ cm/s, then the entire 10 Å trip would take $10^{-12} - 10^{-11}$ s. If the "transition state", $[ABC]^\ddagger$, is defined to encompass *all configurations of ABC significantly perturbed from the potential energy of the reagents A + BC or the products AB + C*, then this period of $1 - 10$ ps is the time available for its observation. To achieve a resolution of ~ 0.1 Å, the probe time window must be $10 - 100$ fs. The above definition of the transition state follows the general description given by Polanyi et al. [40], namely, the full family of configurations through which the reacting particles evolve en route from reagents to products. This description may seem broad to those accustomed to seeing the TS symbol, $\ddagger$, displayed at the crest of the energy barrier to a reaction. As stated in reference [40], even if one restricts one's interest to the overall rates of chemical reactions, one requires a knowledge of the family of intermediates sampled by reagent collisions of different collision energy, angle, and impact parameter. The variational theory of reaction rates further extends the range of TS of interest, quantum considerations extend the range yet further, and the concern with rates to yield products in specified quantum states and angles extends the requirements most of all. A definition of the TS that embraces the entire process of bond breaking and bond making is therefore likely to prove the most enduring.

In CW spectroscopies, the first experiments involved alkali metal atomic exchange reactions, studied at pressures orders of magnitude lower than those at which collisional pressure broadening would normally be observed. Nonetheless, there was evidence of far-wing emission [41] and absorption [42] extending ~ 100 nm away from the alkali metal atom line center. This was

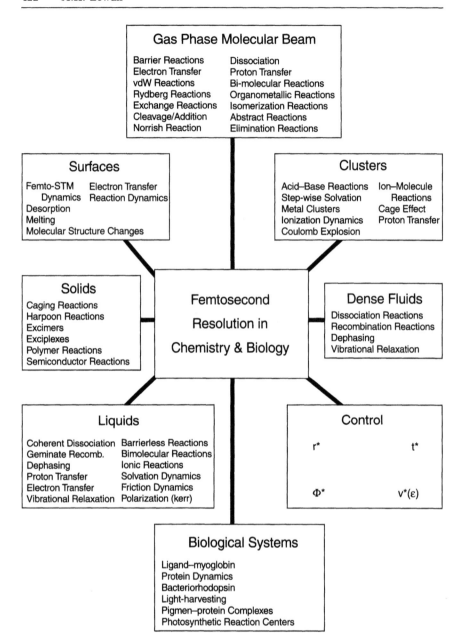

Figure 3a: Schematic indicating the different phases studied with femtosecond resolution, and the area of control studied by spatial (r^*), temporal (t^*), phase (Φ^*), or potential energy (V^*) manipulation. (b, bottom)

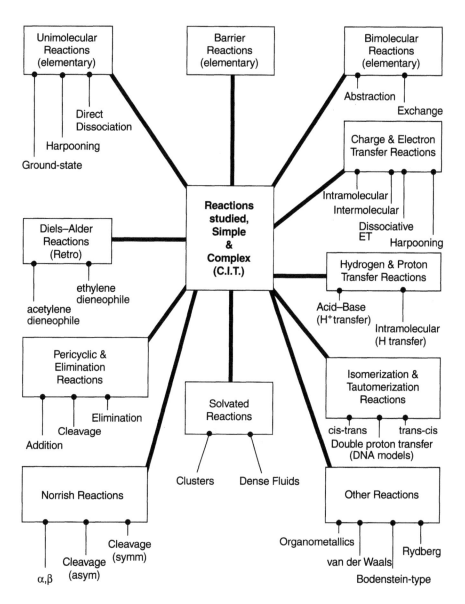

Figure 3b: Schematic indicating the reactions so far studied at Caltech (see text)

indicative of the occurrence of a strong collision [43, 44] in this case between the products of reactions (4) and (5) [41, 42]:

$$F + Na_2 \rightarrow NaF + Na^* \quad (4)$$
$$K + NaCl + h\nu \rightarrow KCl + Na^* \quad (5)$$

In the case of reaction (4) [41], wings to the red and the blue of the sodium D-line evidenced themselves in emission. The integrated intensity of the wing emission was $\sim 10^{-4}$ that of the sodium D-line emission (lifetime $\sim 10^{-8}$ s), giving a total lifetime for the $[FNa_2]^\ddagger$ of $\sim 10^{-12}$ s. For reaction (5) [42], laser absorption by $[KClNa]^\ddagger$, tens of nanometers to the red of the sodium D-line and also a little to the blue, was detected by recording the resulting D-line wavelength.

A limitation in both of these early experiments, as for line-broadening experiments in general, was the sparsity of structure in the wings. Early calculations [44] had shown that structure should be present, corresponding to the marked variation in density of TS species along the reaction coordinate. In recent work [45] on the exchange reaction K + NaBr, Brooks' group has found evidence of structure centered 20 nm to the red of the sodium D-line, when one component of the line, $D_2(^2P_{3/2} \rightarrow {}^2S_{1/2})$, is monitored but not when D_1 is the product. This structure in the TS spectrum was attributed to absorption of radiation by $[KBrNa]^\ddagger$ reacting by way of an electronically excited transition state.

Kinsey and co-workers [36] have demonstrated in an impressive fashion that rich structures can be obtained if one starts $[ABC]^\ddagger$, on its path to products from well-defined TS configurations. The free species (the TS) was formed in a narrow range of configurations by laser excitation of a bound molecule (methyl iodide). During its stay in the free state the dissociating species $[CH_3 \cdots I]^{*\ddagger}$ emitted to quantum states of the ground electronic state, as determined by the Franck–Condon overlap for successive configurations as the TS dissociated to products.

In the examples being cited, the TS has been accessed (directly or indirectly) from a bound state that defines the internuclear separations. A powerful method for TSS that embodies this principle is to "complex" the reagents or their precursor prior to reaction. Since the complexed species are a few angstroms apart, following excitation by light the reaction proceeds from a TS to products. In the first variant on this approach, the complex was a van der Waals cluster. The method being described was pioneered by Soep and co-workers using, for example, complexes Hg··Cl$_2$ or Ca··HCl [46]. Pulsed irradiation triggered reaction, by electronically exciting the metal atom. This approach has been extended to the classic alkali metal atom "harpooning" reactions, complexed as Na··XR (XR = organic halide) by Polanyi's group in a series of papers [47].

A significant development is due to Wittig and co-workers [48], who replaced the metal atom in the complex by a molecule that on photolysis aimed a photofragment along a preferred direction, with a somewhat restricted range of impact parameters, at the remainder of the complex. Thus, the complex IH··OCO when irradiated with 193 nm UV released an H atom that reacted with the CO_2 end of the complex to yield OH whose internal energy was probed. This type of "spectroscopy starting in the TS" has analogy in studies of "surface-aligned photochemistry" [49] where the surface is a "fixed" reagent.

Among the most informative experiments using complexes to arrange the reagents in a TS configuration have been those in which the complex existed only as a negatively charged ion. This approach introduced by Neumark [50] is a form of emission spectroscopy. What is emitted, however, is not photons but electrons. The emission is triggered by light of fixed wavelength which transforms the stable complex ABC^- to the labile TS $[ABC]^{\ddagger}$, resulting in the emission of electrons whose translational energies have imprinted on them the preferred energies of $[ABC]^{\ddagger}$. These energies are vibrational modes approximately orthogonal to the reaction coordinate and also resonances corresponding to quantum "bottlenecks" along the reaction pathway. The work began with the observation of vibrational modes in the TS of hydrogen-transfer reaction, e:g., I + HI.

3 Concepts in Femtochemistry

In this section, we highlight some concepts governing the dynamics of motion at the atomic scale of time resolution and discuss the underlying fundamentals of reaction dynamics in real time. Perhaps the most significant concepts emerging from developments in femtochemistry are the following: (1) the concept of *time scales*, in relation to vibrational and rotational motions (snapshots), and *localization* with de Broglie wavelength reaching the atomic scale for motion; (2) the concept of coherence (for *state, orientation*, and *nuclear wave packet* dynamics) and single-molecule trajectory of motion, instead of ensemble averaging; (3) the concept of physical (spreading and dephasing) and chemical (nuclear motions) time scales and their impossible separability in the spectra of complex systems; (4) the concept of probing transition states and intermediates directly in real time; (5) the concept of controlling reactivity with ultrashort light pulses. These concepts are outlined in Figure 4 as a summary.

The idea of a wave group was introduced (1926) by Schrödinger [51] in order to make a natural connection between quantum and classical descriptions. The use of wave groups (or wave packets) in physics, and certainly in chemistry, was limited to a few theoretical examples in the applications of quantum mechanics. The solution of the time-dependent Schrödinger

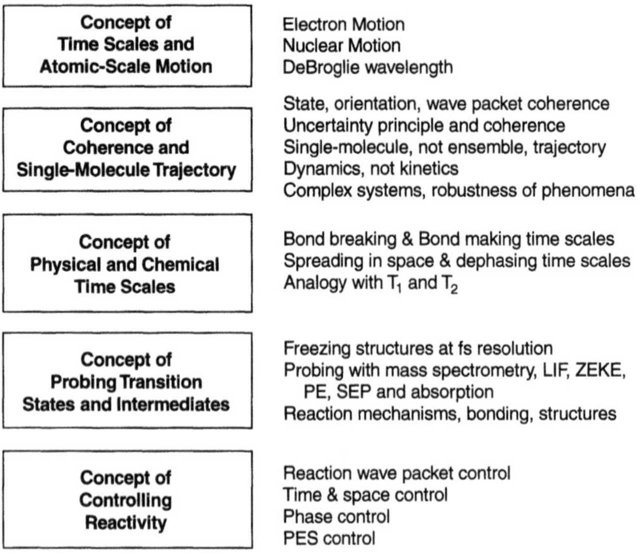

| Concept of Time Scales and Atomic-Scale Motion | Electron Motion
Nuclear Motion
DeBroglie wavelength |

| Concept of Coherence and Single-Molecule Trajectory | State, orientation, wave packet coherence
Uncertainty principle and coherence
Single-molecule, not ensemble, trajectory
Dynamics, not kinetics
Complex systems, robustness of phenomena |

| Concept of Physical and Chemical Time Scales | Bond breaking & Bond making time scales
Spreading in space & dephasing time scales
Analogy with T_1 and T_2 |

| Concept of Probing Transition States and Intermediates | Freezing structures at fs resolution
Probing with mass spectrometry, LIF, ZEKE, PE, SEP and absorption
Reaction mechanisms, bonding, structures |

| Concept of Controlling Reactivity | Reaction wave packet control
Time & space control
Phase control
PES control |

Figure 4: Summary of some concepts in femtochemistry discussed in text

equation for a particle in a box or a harmonic oscillator and the elucidation of the uncertainty principle by superposition of waves are two of these examples. However, essentially all theoretical problems are presented as solutions in the time-independent frame picture. In part, this practice is due to the desire to start from a quantum-state description. But, more importantly, it was due to the lack of *experimental* ability to synthesize wave packets.

Even on the ps time scale, molecular systems reside in eigenstates, and there is only one evolution, the change of *population* with time from that state. Hence, with the advent of ps spectroscopy, employed in numerous applications in chemistry and biology (section 1), one is mainly concerned with kinetics, not dynamics. On the femtosecond (fs) time scale, an entirely new domain emerges. First, a wave packet can now be prepared, as the temporal resolution is sufficiently short to "freeze" the nuclei at a given internuclear separation. Put in another way, the time resolution becomes shorter than the vibrational (and rotational) motions such that the wave packet is prepared, highly localized with a de Broglie wavelength of ~ 0.1 Å. Second, this synthesis of packets is not in violation of the uncertainty principle, as the key here is the *coherent* preparation of the system [24]. It has been shown in, for example, a two-atom system (iodine) that because its width is < 0.2 Å, the wave packet oscillates spatially and executes distance changes between 2 and 5 Å depending on the energy. The preparation and probing are done coherently, and only as such can one see the bond stretch and compress and the molecule rotate [24]. For both nonreactive and reactive systems, the same picture applies.

Third, because of this coherent synthesis, the transition from *kinetics* to *dynamics* is made, as one is able to monitor the evolution, at the atomic resolution of motion, of single-molecule trajectory and not an ensemble-averaged behavior. There is a fourth important point: Ehrenfest's classical limit [52] is actually reached on this time scale for molecular systems. The spreading of the wave packet was found to be negligible, contrary to anticipation, and we now experimentally know why [24]. Theoretically, by considering the motion of a Gaussian packet in free space, we find that the wave packet disperses, but not significantly on the fs time scale. The dispersion time is given by

$$\tau_d = 2m\{\Delta R^2(t=0)\}\hbar^{-1} \qquad (6)$$

and is on the ps time scale for molecular systems (m is the mass and R is the separation) [24]. Note that $\Delta R(t=0)$ relates to the momentum ΔP by the uncertainty relationship. Thus, the fs resolution provides the required $\Delta P(\Delta E)$, which in turn gives the sub-angstrom ΔR resolution in complete accord with the uncertainty principle.

As mentioned before, the key is the *coherent* nature of the experiment at preparation and probing and the time resolution, which is *shorter* than the vibrational and rotational times of the motion. A temporal image of the dynamics becomes real in time. We can speak of the motions in the same way that we conceive of them. The beauty of wave packets lies in their simplicity and direct visualization of the dynamics on the time scale of theory and experiment. The nature of wave packets in molecular systems is dictated by the forces of interactions for bound or unbound motion. For bound systems, the structural detail of eigenstates is reflected in the wave packet motion. In fact, because of this relation, one can obtain the potential from direct observation of the wave packet coherence [24]. The observed vibrational dynamics are separated from the rotational dynamics because of the difference in their time scales and also because of their distinct vectorial polarization characteristics.

For reactive, unbound systems, the wave packet motion describes the processes of bond breaking and/or bond making. For these chemical and biological changes the realization of dynamics from spectra, mentioned above for bound systems, fails. All reactions involve transition states and many possess structural intermediates, and to study these transient structures, one must "isolate" them in time. In other words, the initial or final state of the reaction is not directly related to intermediate changes along the reaction path. This concept can perhaps be appreciated by relating back to the importance of flash photolysis in isolating radical intermediates which have no relationship in their spectra to the precursors from which they are made. The point will be illustrated by numerous examples in section 4 where reactions with intermediates have very broad initial spectra and the dynamics of the intermediate have no relationship to such spectra. Another serious problem has to do with the complexity of the spectra, due to inhomogeneous broadenings in complex molecular systems. A third complication comes about from dephasing, even in homogeneous systems.

In general, we must consider two time scales: the "chemical time scale", which measures the dynamics of bond breaking/bond making, and the "physical time scale", which reflects the process of dephasing or spreading of the wave packet. Dephasing of wave packets can be caused by a very small motion (< 0.05 Å) on a steep repulsive potential during bond breakage [53] or by avoided crossings [54]. In both cases, the initial absorption spectra will be severely broadened, though not due to chemical dynamics. This point has analogy in the T_1 and T_2 language of describing dynamics [55].

The ability to probe the motion offers new opportunities for controlling chemical reactions with the concept [56] of using ultrashort pulses for the manipulation. Several schemes for control utilizing the quantum nature of the level structure have emerged, and this new direction of research has begun experimental fruitation as reviewed recently by Wilson [57] and by Rice [58]. Our own efforts have focused on timing of wave packets and on the development of multiple-pulse techniques (see section 5).

4 Elementary and Complex Reactions Studied

Applications to different classes of reactions in different phases are numerous in many laboratories around the world; here, we limit ourselves to some examples studied by the Caltech group [24–26] (more recent references are listed here). The specific examples given are chosen to span classes of reactions which display different structures and dynamics in the transition state region. A summary is given in Figure 1b.

4.1 Unimolecular Reactions

4.1.1 Direct Dissociation

The first of these femtochemistry studies dealt with unimolecular reactions. The experiments were performed on the elementary reaction

$$\text{ICN} \to [\text{I}\cdot\cdot\text{CN}]^{*\ddagger} \to \text{I} + \text{CN} \tag{7}$$

to define the meaning of bond breakage. When tuning to the perturbed CN fragment absorption, the transients exhibited a buildup and a decay characteristic of the transition states. On-resonance absorption of the free CN fragment gave a rise delayed from $t = 0$ by $\tau_{1/2} = 205 \pm 30$ fs. The time for bond breaking depends on the nature of the PES. In the simple case of a one-dimensional PES, the clocking time is directly related to the repulsion length scale (L) between the fragments:

$$\tau_{1/2} = L/v \ln(4E/V_f) \tag{8a}$$

where V_f is the final potential energy probed (with $E \gg V_f$). The potential is repulsive, $V(R) = E \exp[-(R - R_i)/L]$. The above formula defines the fundamentals of bond breaking: dissociation of the bond depends on the terminal values of the potential and velocity; the time is determined by the energy of recoil ($E = 1/2\mu v^2$) and steepness of the potential (length parameter L). The characteristic time $\tau_{1/2}$ in this case is simply the time for the potential to drop to a value equal to V_f. For the ICN experiment, L was deduced to be ~ 0.8 Å ($v = 0.0257$ Å/fs). The transition state survives for only ~ 50 fs or less depending on the region probed on the PES. This is because its lifetime is given by

$$\tau^{\ddagger} = \Delta V(R^{\ddagger})/[v(R^{\ddagger})|F(R^{\ddagger})|] \tag{8b}$$

and, as in (8a), v is the velocity, but now at $R^{\ddagger}$ of the transition state. F is the force and ΔV is the energy window around $R^{\ddagger}$.

Theoretically, these observations were reproduced with classical-mechanical as well as quantum-mechanical treatments of the dissociation. Even kinetic equations can describe the general behavior of the transients, but, of course, not as a wave packet motion. In general, however, the PESs are more complex and the data are better inverted to obtain the shape. Such a procedure has been developed and applied to the ICN reaction.

To learn about the reaction trajectory to different final CN rotational states, $\tau_{1/2}$ was measured for different angular momentum states. The fs alignment was also measured. From these experiments information was obtained on the magnitude of the torque between I and CN (angular part of the PES) and on the extent to which coherence was lost among the different product states. Other systems we have studied for this class of reactions are Bi_2, CH_3I, ClO_2, and C_6H_5I. Soep and colleagues studied the interesting system of alkyi nitrites (see Mestdagh et al. in ref [1]).

4.1.2 Covalent-to-Ionic Resonance along the Reaction Coordinate

One example that illustrates the methodology of femtosecond transition-state spectroscopy (FTS), the concept of localization, and the resonance dynamics along the reaction coordinate comes from studies [24] of the dissociation reactions of alkali halides. For these systems, the covalent (M + X, where M denotes the alkali atom and X the halogen) potential and the ionic ($M^+ + X^-$) potential cross at an internuclear separation larger than 3 Å.

The electronic structure along the reaction coordinate therefore changes character from being covalent at short distances to being ionic at larger distances. The reaction occurs by a harpoon mechanism, has a large cross section, and is central to mechanisms described by curve crossings (ET, S_N2, etc.).

In studying this system, the first fs pulse takes the ion pair M^+X^- to the covalent "bonded" MX potential at a separation of 2.7 Å. The activated complexes $[MX]^{*\ddagger}$, following their coherent preparation, increase their internuclear separation and ultimately transform into the ionic $[M^+\cdots X^-]^{*\ddagger}$ form. With a series of pulses, delayed in time from the first one, the nuclear motion through the transition state and all the way to the final $M + X$ products can be followed. The probe pulse examines the system at an absorption frequency corresponding to either the complex $[M\cdots X]^{*\ddagger}$ or the free atom M.

Figure 5 gives the observed transients of the NaI reaction for the two detection limits. The resonance along the reaction coordinate describes the motion of the wave packet when the activated complexes are monitored at a certain internuclear separation. The steps describe the *quantized, coherent buildup* of free products, with separations matching the resonance period of the activated complexes. The complexes do not all dissociate on every out-bound pass, since there is a finite probability that the I atom can be harpooned again when the Na$\cdots$I internuclear separation reaches the crossing point at 7 Å. The complexes survive for about 10 oscillations before completely dissociating to products. When the position of the crossing point is adjusted by a change in the difference in the ionization potential of M and the electronegativity of X (e.g., switching from NaI to NaBr), the survival of the complex changes (NaBr complexes, for example, survive for only one period).

The results in Figure 5 illustrate some important additional features of the dynamics. The sequential recurrences are damped, in a quantized fashion, due to *bifurcation* of the wave packet at the crossing point of the covalent and ionic potentials, as shown in the quantum calculation given in the figure. In fact, it is this damping which provides important parts of the dissociation dynamics, namely the reaction time, the probability of dissociation, and the extent of covalent and ionic characters in the bond. These observations and their analyses have been made in more detail elsewhere, and other systems have been examined similarly. Numerous theoretical studies of these systems have been made to test quantum, semiclassical, and classical descriptions of the reaction dynamics and to compare theory with experiment.

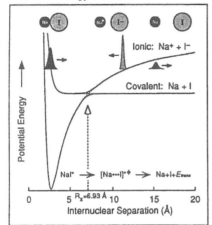
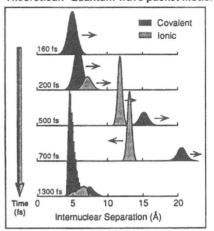
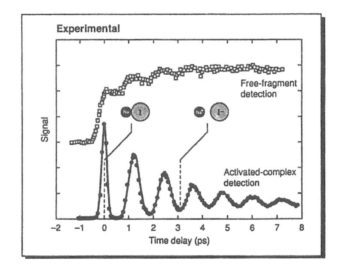

Figure 5: Femtosecond dynamics of dissociation (NaI) reaction. Bottom: experimental observations of wave packet motion, made by detection of the activated complexes [NaI]*‡ or the free Na atoms. Top: potential energy curves (left) and the "exact" quantum calculations (right) showing the wave packet as it changes in time and space. The corresponding changes in bond character are also noted: covalent (at 160 fs), covalent/ionic (at 500 fs), ionic (at 700 fs), and back to covalent (at 1.3 ps). Adapted from the following: Rose, T. S.; Rosker, M. J.; Zewail, A. H.: J. Chem. Phys. **88**, 6672 (1988); **91**, 7415 (1989). Engel, V.; Metiu, H.; Almeida, R.; Marcus, R. A.; Zewail, A. H.: Chem. Phys. Lett. **152**, 1 (1988)

4.1.3 Ground-State Reactions

For this class of reactions and with similar methodology we have studied the following systems and obtained the rates:

(i) $H_2O_2 \to 2OH$ (vibrationally initiated)

(ii) $NCNO \to NC + NO$

(iii) $C_2H_2O \to CH_2 + CO$

4.2 Barrier Reactions: Saddle-Point Transition State

Our first observation of the transition state in barrier reactions (with a saddle point) was made in the ABA (IHgI) system, which is the simplest for understanding the dynamics on a multidimensional PES: $[ABA]^\ddagger \to AB + A$. This system is the half-collision of the A + BA full collision (see Figure 6); it involves one symmetrical stretch (Q_s), one asymmetrical stretch (Q_a), and one bend (q), and there is a barrier along the reaction coordinate. In the same system, the coherent interference of reaction channels in the barrier descent motion, termed *coherence in products*, was first discovered [24].

The "lifetime" of the transition state over a saddle point near the top of the barrier is the most probable time for the system to stay near this configuration. It is simply expressed, for a one-dimensional reaction coordinate (frequency ω) near the top of the barrier, as $\tau^\ddagger = 1/\omega$. For values of $\hbar\omega$ from 50 to 500 cm^{-1}, $\tau^\ddagger$ ranges from 100 to 10 fs. In addition to this motion, one must include the transverse motion perpendicular to the reaction coordinate, with possible vibrational resonances, as discussed below.

For the IHgI system, the activated complexes $[IHgI]^{*\ddagger}$, for which the asymmetric (translational) motion gives rise to vibrationally cold (or hot) nascent HgI, were prepared coherently at the crest of the energy barrier (Figure 6). The barrier descent motion was then observed using series of probe pulses. As the bond of the activated complexes breaks during the descent, both the *vibrational motion* (~ 300 fs), of the separating diatom, and the *rotational motion* (~ 1.3 ps), caused by the torque, can be observed (Figure 6). These studies of the dynamics provided the initial geometry of the transition state, which was found to be bent, and the nature of the final torque which induces rotations in the nascent HgI fragment. Classical and quantum molecular dynamics simulations show the important features of the dynamics and the nature of the force acting during bond breakage. Two snapshots are shown in Figure 6.

The force controls the remarkably persistent (observed) coherence in products, a feature which was unexpected, especially in view of the fact that all trajectory calculations are normally averaged (by Monte Carlo methods) without such coherences. Only recently has theory addressed this

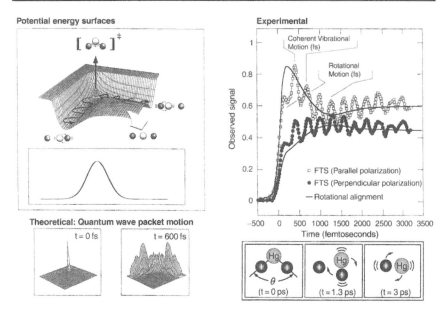

Figure 6: (a, left) Potential energy surfaces, with a trajectory showing the coherent vibrational motion as the diatom separates from the I atom. Two snapshots of the wave packet motion (quantum molecular dynamics calculations) are shown for the same reaction at $t = 0$ and $t = 600\,\text{fs}$. (b, right) Femtosecond dynamics of barrier reactions, IHgI system. Experimental observations of the vibrational (femtosecond) and rotational (picosecond) motions for the barrier (saddle-point transition state) descent, $[\text{IHgI}]^{*\ddagger} \to \text{HgI}(\text{vib}, \text{rot}) + \text{I}$, are shown. The vibrational coherence in the reaction trajectories (oscillations) is observed in both polarizations of FTS (femtosecond transition-state spectra). The rotational orientation can be seen in the decay of FTS (parallel) and buildup of FTS (perpendicular) as the HgI rotates during bond breakage (bottom). Adapted from the following: Dantus, M.; Bowman, R. M.; Gruebele, M.; Zewail, A. H.: J. Chem. Phys. **91**, 7437 (1989). Gruebele, M.; Roberts, G.; Zewail, A. H.: Philos. Trans. R. Soc. London, A **332**, 223 (1990)

point [59–61] and emphasized the importance of the transverse force, i.e., the degree of anharmonicity perpendicular to the reaction coordinate. The same type of coherence along the reaction coordinate was found for reactions in solutions [62–64], in clusters [65–68], and in solids [69], offering a new opportunity for examining solvent effects on reaction dynamics in the transition-state region.

Even more surprising was the fact that this same phenomenon was also found to be robust and common in biological systems, where wave packet motion was found in the twisting of a bond, e.g., in rhodopsin and bacteriorhodopsin [70], in the breakage of a bond, e.g., in ligand–myoglobin [71] systems, in photosynthetic reaction centers [72], and in the light-harvesting antenna of purple bacteria [73, 74]. The implications, e.g., as to the global motion of the protein, are fundamental to the understanding of the mechanism (coherent vs nonstatistical energy or electron flow), and such new observations are triggering numerous theoretical studies in these biological systems (see, e.g., ref [75]).

For the wave packet motion in dissociation and barrier reactions, discussed above, there have been recent studies of the same (or similar) systems but in solutions, and the results are striking. Sundström and co-workers [76] have observed the wave packet motion for the twisting process in a barrierless isomerization reaction in solutions. The findings give a direct view of the motion and examine the nature (coherent vs diffusive) of the coupling to the solvent during the reaction. This ICN-type behavior indicates the persistence of coherence along the reaction coordinate and provides the time scales for intramolecular motion and solvation. Hochstrasser's group [77] has shown that for HgI_2 in ethanol solutions the HgI is formed in a coherent state, similar to the observation we made in the gas phase. Their study is rich with information regarding solvated wave packet dynamics, relaxation in the solvent, and the effective PES. Of particular interest is the study of solvent-induced relaxation of nascent fragments.

4.3 Bimolecular Abstraction and Exchange Reactions: Ground-State Transition States

Ground state, bimolecular reactions represented an opportunity to elucidate the dynamics of collision complexes directly in real time and to address the concept of time scales ($\tau^\ddagger$, τ_{vib}, τ_{rot}) in relation to the reaction mechanism, usually described by extreme two limits involving electronic (stripping) or nuclear energy redistribution (complex mode). Furthermore, the results could then be quantitatively compared with ab initio potential and dynamical calculations on ground state surfaces to improve our understanding of these abundant reactions. Real-time clocking of abstraction reactions was first performed on the $I - H/CO_2$ system for the dynamics on the *ground-state* PES [24]:

$$H + CO_2 \rightarrow [HOCO]^\ddagger \rightarrow CO + OH \qquad (9)$$

Two pulses were used, the first to initiate the reaction and the second delayed to probe the OH product. The decay of $[HOCO]^\ddagger$ was observed in the buildup of the OH final fragment. The two reagents were synthesized in a van der Waals complex, following the methodology discussed in section 2. The PES along the reaction coordinate and results (at one energy) are shown in Figure 7. The results established that the reaction involves a *collision complex* and that the lifetime of [HOCO]‡ is relatively long, about a picosecond.

Wittig's group [78] has recently reported accurate rates with subpicosecond resolution, covering a sufficiently large energy range to test the description of the lifetime of [HOCO]‡ by an RRKM theory. Recent crossed-molecular beam studies of OH and CO, from the group in Perugia, Italy [79], have shown that the angular distribution is consistent with a long-lived complex. Vector correlation studies by the Heidelberg group [80] addressed the importance of the lifetime to IVR and to product-state distributions. The molecular dynamics calculations (with ab initio potentials) by Clary, Schatz, and Zhang [81] are also consistent with such lifetimes of the complex and provide new insight into the effect of energy, rotations, and resonances on the dynamics of [HOCO]‡. This is one of the reactions in which both theory and experiment have been examined in a very critical and detailed manner. For exchange reactions, the fs dynamics of bond breaking and bond making was examined in the following system [24]:

$$Br + I_2 \rightarrow [BrII]^{\ddagger} \rightarrow BrI + I \qquad (10)$$

The dynamics of this $Br + I_2$ reaction (Figure 7) was resolved in time by detecting the BrI with the probe pulses using laser-induced fluorescence. The reaction was found to be going through a sticky (tens of picoseconds) collision complex. More recently, McDonald's group [82] has monitored this same reaction, using multiphoton ionization mass spectrometry, and found the rise of I (and I_2) to be, within experimental error, similar to the rise of BrI (Figure 7). They proposed a picture of the PES for the dynamics. With molecular dynamics simulations, comparison [24] with the experimental results indicated the trapping of trajectories in the [BrII]‡ potential well; the complex is a stable molecular species on the picosecond time scale. Gruebele, et al. (see ref [24]) drew a simple analogy between collision ($Br + I_2$) and half-collision ($h\nu + I_2$) dynamics based on the change in bonding and using frontier orbitals to describe it. More recently, S. Yabushita (private communication; see also ref [115] discussed below) has considered the effect of spin–orbit coupling on the PES and found evidence for a conical intersection involving the two spin–orbit channels. We plan further MD studies on this surface.

Other bimolecular reactions of complex systems, such as those of benzene and iodine and acid–base reactions, will be discussed below. Currently, we are examining the inelastic and reactive collisions of halogen atoms with polyatomics (e.g., CH_3I). Other groups at NIST and at USC have studied a new class of reactions: $O + CH_4 \rightarrow [CH_3OH]^{\ddagger} \rightarrow CH_3 + OH$ (ref 83) and $H + ON_2 \rightarrow HO + N_2$ or $HN + NO$ (ref [84]).

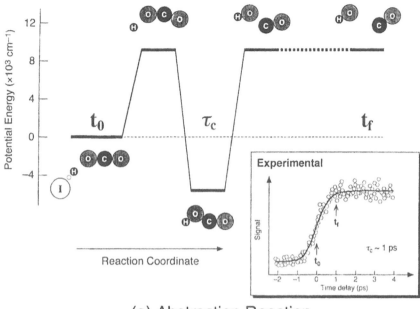

(a) Abstraction Reaction

Figure 7a: Femtosecond dynamics of abstraction (H + CO_2 → OH + CO) and exchange (Br + I_2 → BrI + I) reactions. The PES along the reaction coordinate, and the observed rise of the OH from the breakup of the collision complex [HOCO]‡; lifetime $\tau_c \sim 1$ ps. The corresponding structures are noted with emphasis on three snapshots τ_0, and t_f (final) at the asymptote region. Adapted from the following: Scherer, N. F.; Khundkar, L. R.; Bernstein, R. B.; Zewail, A. H.: J. Chem. Phys. **87**, 1451 (1987). Scherer, N.F.; Sipes, C.; Bernstein, R. B.; Zewail, A. H.: J. Chem. Phys. **92**, 5239 (1990). Ionov, S. I.; Brucker, G. A. Jaques, C.; Valachovic, L.; Wittig, C.: J. Chem. Phys. **99**, 6553 (1993). Sims, I. R.; Gruebele, M.; Potter E. D.; Zewail, A. H.: J. Chem. Phys. **97**, 4127 (1992). Wittig, C.; Zewail A. H.: In *Chemical Reactions in Clusters*, Bernstein, E., Ed.; Oxford University Press: Oxford, 1996

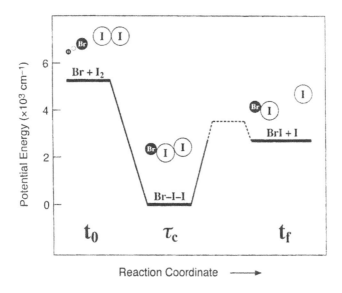

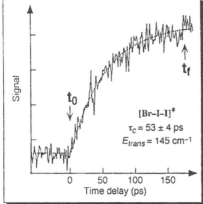

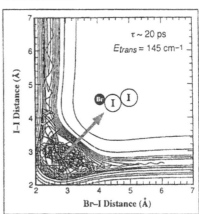

(b) Exchange Reaction

Figure 7b: Similar to part (a) but for the exchange reaction. Here, $\tau_c = 53$ ps, and as shown in theoretical molecular dynamics the [BrII]‡ complex is very long lived (see text).

4.4 Isomerization Reactions

For systems with a large number of degrees of freedom (N), the situation is more complex. First, perpendicular to the reaction coordinate, there is now $(N-1)$ possible motions. Second, the wave packet may suffer fast spreading as its structure involves a large number of modes. The isomerization of diphenylethylene (stilbene) is an example of such a reaction with 72 modes:

$$\underset{(cis)}{\overset{H}{\underset{\phi}{>}}C=C\overset{H}{\underset{\phi}{<}}} \longrightarrow \left[\text{Twisted}\right]^{*\dagger} \longrightarrow \underset{(trans)}{\overset{H}{\underset{\phi}{>}}C=C\overset{\phi}{\underset{H}{<}}}$$

The reaction coordinate is usually described by a single motion about the double bond (torsional angle θ). The molecule at the excited cis configuration is essentially unbound in the θ coordinate but, in principle, is bound along all the other coordinates; a saddle-point transition state is defined.

The wave packet was initially prepared by a fs pulse (Figure 8). The temporal evolution was then probed by resonance multiphoton ionization [24]. The transient exhibits an exponential decay (reaction time) with a superimposed oscillatory pattern. The PES and trajectory in Figure 8 illustrates the molecular changes and corresponding structures. In the twisting of the double bond there are at least three angular coordinates, out of the 72 modes, directly involved: The C_e-C_e torsional angle, θ reaction coordinate, the $C_e - C_e - C_{Ph}$ in-plane bending angle α, and the $C_e - C_{Ph}$ torsional angle ϕ.

To compare with experiments, molecular dynamics calculations were made by solving equations of motion, starting from $t = 0$ and continuing until isomerization is complete in selected coordinates. Two time scales are involved: one for the initial dephasing of the packet (tens of fs) and the other for the relatively slower (hundreds of fs) nuclear dynamics of twisting. The reaction coordinate involves the θ as well as α and ϕ coordinates, as shown for the reaction trajectory and structures displayed in Figure 8. The total reaction time toward twisting is ~ 300 fs.

This type of coherent twisting motion is now evident in other systems [62, 76, 85]. Remarkably, this same resonance behavior of the motion in N-dimensional space was also observed for stilbene in solutions by the Hochstrasser group [85]. The phenomenon is also evident in biological systems, e.g., isomerization of rhodopsins [70], the key primary event of vision. In a recent work, we have also examined the trans-to-cis twisting dynamics on the fs time scale, covering an energy range of 9000 cm^{-1} and elucidating the nature of barrier crossing, different from the barrierless cis, of interest for many years [24]. The new study separates the influence of IVR and

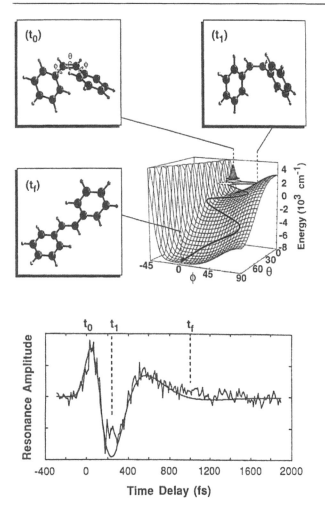

Figure 8: Femtosecond dynamics of isomerization (stilbene) reaction. Bottom: experimental observation of the twisting (decay) and resonance (oscillation) motions depicted on the PES (middle). The trajectory shown on the PES describes the changes in the molecular structure, and three snapshots at different times display the corresponding structures. Adapted from the following: Pedersen, S.; Bañares, L.; Zewail, A. H.: J. Chem. Phys. **97**, 8801 (1992); work to be published

alignment and provides their effect on the microcanonical rates at different energies in the isolated molecule and under stepwise solvation in solvent clusters of hexane and ethane.

4.5 Pericyclic Addition and Cleavage and Elimination Reactions: TS and Orbital Symmetry

For more than a century [86], one of the most well-studied addition/cleavage reactions, both theoretically and experimentally, is the ring opening of cyclobutane to yield ethylene or the reverse addition of two ethylenes to form cyclobutane (Figure 9). Such is a classic case study for a Woodward–Hoffmann description of concerted reactions. The reaction, however, may proceed directly through a transition state at the saddle region of an activation barrier, or it could proceed with a diradical intermediate, beginning with the breakage of one σ-bond to produce tetramethylene, which in turn passes through a transition state before yielding final products. The concept of time scale, therefore, besides being important to the definition of diradicals as stable species, is crucial to the nature of the reaction mechanism: a concerted one-step process vs a two-step process with an intermediate. The fundamental issues discussed here are encapsulated in the following questions: Given me time scale of the nuclear motion, what is meant by *concertedness*, and what does *simultaneous* bond breakage or formation really mean?

Experimental and theoretical studies have long focused on the possible existence of diradicals and on the role they play in affecting the processes of cleavage, closure, and rotation. The experimental approach is based primarily on studies of the stereochemistry of reactants and products, chemical kinetics, and the effect of different precursors on the generation of diradicals. The time "clock" for rates is internal, inferred from the rotation of a single bond, and is used to account for any retention of stereochemistry from reactants to products. Theoretical approaches basically fall into two categories: those involving thermodynamical analysis of the energetics (enthalpic criterion) and those concerned with semiempirical or ab initio quantum calculations of the (PES) describing the motion of the nuclei.

Real-time studies of these reactions should allow one to examine the nature of the transformation and the validity of the diradical hypothesis. The Caltech group [87] reported direct studies of the fs dynamics of the transient diradical structures. The aim was at "freezing" the diradicals in time, in the course of the reaction. Various precursors were used to generate the diradicals and to monitor the formation and the decay dynamics of the reaction intermediate(s). The parent (cyclopentanone) or the intermediate species was distinctly identified using time-of-flight mass spectrometry; the CO leaves in less than 100 fs, allowing for the prompt preparation of the intermediates. The concept behind the experiment and some of the results are given in Figure 9.

The mass spectra obtained at different fs time delays show the changes of the intermediates. At negative times there is no signal present. At time zero, the parent mass (84 amu) of the precursor cyclopentanone is observed, while the intermediate mass of 56 amu is not apparent. As the time delay

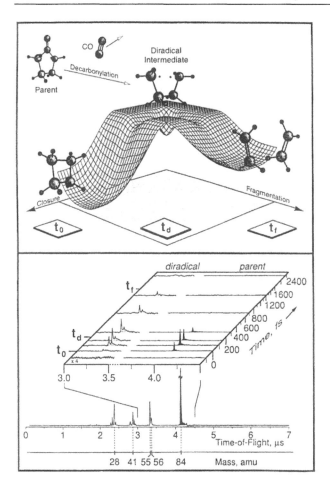

Figure 9: Femtosecond dynamics of addition/cleavage reaction of the cyclobutane–ethylene system. Bottom: experimental observation of the intermediate diradical by mass spectrometry. Top: the PES showing the nonconcerted nature of the reaction, together with three snapshots of the structures at t_0 (initial), t_d (diradical), and t_f (final). The parent precursor is also shown. Adapted from the following: Pedersen, S.; Herek J. L.; Zewail, A. H.: Science (Washington, D.C.) **266**, 1293 (1994); work to be published

increases, a *decrease* of the 84 mass signal was observed and, for the 56 mass, first the *increase* and then *decrease* of the signal. The 56 mass corresponds to the parent minus the mass of CO. Its dynamics directly reflect the nature of the transition-state region.

Considering the dynamics of nuclei near the top of the barrier, it is impossible at these velocities to obtain the observed time scales if a wave packet is moving translationally on a "flat", *one-dimensional* surface. For example, over a distance of 0.5 Å, which is significantly large on a bond scale, the time in the transition-state region would be $\sim$ 40 fs. The reported (sub)-picosecond times therefore reflect the involvement of other nuclear degrees of freedom.

In the original publication [87], the rates were related to the stability of the diradical species. By varying the total energy and using different substituents, these studies gave evidence that the diradical is a stable species on the global PES. The approach is general for the study of other reactive intermediates in reactions and since then has been extended to cover other classes of reactions. We have completed studies of trimethylene and the isotopically substituted species. Without the fs-resolved mass spectra it would have been impossible to observe the evolution of the parent reagents and the dynamics of the intermediate (see section 3).

Using the same techniques, elimination reactions were clocked [24] in order to address a similar problem: the nature of two-center elimination by either a *one-step* or *two-step* process. The reaction of interest is

$$F_2XC-CXF_2 \longrightarrow F_2C=CF_2 + 2X$$
(ethanes) (ethylenes)

where the 2X (in this case 2I) elimination leads to the transformation of ethanes to ethylenes. The questions then are as follows: Do these similar bonds break at the same time or consecutively with formation of intermediates? What are the time scales?

Methyl iodide, whose C–I nonbonding to antibonding orbital transition is at $\sim$ 2800 Å, is known to undergo fragmentation with the formation of iodine in both spin–orbit states (I and I*). The CH_3 fragment produced is vibrationally excited. In both I and I* channels, the iodine atoms rise in a time of less than 0.5 ps. For $I - CF_2 - CF_2 - I$, the situation is entirely different. The bond breakage, which leads to elimination, is consecutive, nonconcerted: a primary bond breakage ($\sim$ 200 fs), similar to methyl iodide, and a much slower (32 ps) secondary bond breakage. The dynamics of the prompt breakage can be understood by applying the theoretical techniques mentioned above for coherent wave packet motion in direct dissociation reactions [24], but what determines the dynamics of the slower secondary process?

The observation of a 30 ps rise, by monitoring I, indicates that, after the recoil of the fragments in the primary fragmentation, the total internal energy in the intermediate $[CF_2I--CF_2]^{\ddagger}$ is sufficient for it to undergo secondary dissociation and produce I in the ground state. From the photon energy (102 kcal mol^{-1}) used in the experiment, the C–I bond energy (52.5 kcal mol^{-1}), and the mean translational energy, the internal energy of the fragment was found to be comparable with the activation energy. The (30 ps)$^{-1}$ represents the average rate for this secondary bond-breaking process (barrier 3 – 5 kcal mol^{-1}).

When the total energy decreased in these experiments, the decay became slower than 30 ps. This energy dependence of the process could be understood considering the time scale for IVR and the microcanonical rates, $k(E)$ at a given energy, in a statistical RRKM description. Dynamics of consecutive bond breakage is common to many systems and also relevant to the mechanism in different classes of reactions discussed in the organic literature.

4.6 Diels–Alder Reactions

The Diels–Alder addition reaction of 1,3-butadiene, the simplest diene, with ethylene, the simplest dieneophile, is a prototype case. It has been central to a volume of experimental and theoretical work and has stimulated debate since 1929 (see ref [88]). Houk [88] has discussed the nature of the transition state and considered the secondary isotope effect and ab initio calculations to address the issue of concertedness discussed above.

For this class of reactions we examined norbornylene (retro-Diels–Alder), the product in the addition reaction of ethylene to cyclopentadiene:

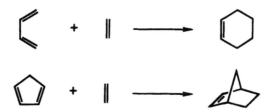

Similarly, we have studied norbornadiene, the product for addition to acetylene and for which there exist theoretical and experimental studies of the electronic structure (for an excellent article, see ref [89] by Roos et al.).

In the fs-resolved mass spectrometry, we have observed the temporal dynamics of the parent and the intermediates. For example, for norbornylene, the parent mass at 94 amu decays to intermediates in 160 fs, while the intermediate mass at 66 amu builds up and decays with a 220 fs time constant. The system was energized by the initial fs pulse, which, as in the case of the diradical system (section 4.6), deposits ~ 8 eV of energy (Rydberg/valence region) [89]. Several points can be made. First, it is clear that the decay of the parent (160 fs) is an order of magnitude faster than the *inertial rotation* (~ 1 ps) about a C – C bond. Thus, the *stereochemistry* is retained.

Second, the fact that we are isolating an intermediate with a finite lifetime indicates the presence of a transient structure which undergoes rearrangement to the final products of cyclopentadiene and ethylene. Third, the "arrows of bookkeeping" of electron shifts is only meaningful if the separation of time scales for electron and nuclear motions is established.

This class of reactions is rich in the questions to be asked, and we are continuing further studies of the transition-state region and its crucial role in mechanisms. The work by Horn et al. [90]. discussed here will be published with details and comparisons with other systems.

4.7 Charge-Transfer Reactions: TS and Harpoon Mechanisms

An approach which makes it possible to directly study the transition-state dynamics of charge-transfer (CT) reactions was recently reported [91]. The entire system is prepared on a reactive potential energy surface and in a well-defined impact geometry. To define the zero of time, we start from the van der Waals (vdW) configuration in a molecular beam, similar to other real-time studies discussed above in section 4.3. A fs pulse induces the CT. We then follow the dynamics of the transition state using probe pulses which monitor either the transition state or the final products of the reaction, using mass spectrometry. The system of interest is the bimolecular reaction of benzene (Bz) with iodine, and similar derivatives:

$$Bz \cdots I - I \rightarrow [Bz^+ \cdots I^- \cdots I^-]^{*\ddagger} \rightarrow Bz \cdot I + I \qquad (11)$$
$$\text{reactants} \quad \text{transition state} \quad \text{products}$$

This system (see Figure 10) is unique in many aspects of the structure and dynamics and has historic roots for nearly 50 years since the seminal work by Hildebrand and Mulliken. Mixing of benzene and iodine results in a new color, new absorption spectrum, and a new theory.

Mulliken [92] attributed the strong absorption band of the system to the excitation of the ground-state complex to the CT state with the aromatic molecule acting as the electron donor and the iodine as the acceptor, i.e., $Bz^+ \cdot I_2^-$. Several spectroscopic and theoretical studies have predicted that the $Bz \cdot I_2$ ground state has a C_{6v} axial structure with the I–I bond being perpendicular to the benzene molecular plane. The heat of formation of this complex in the gas phase was determined by spectrometric methods to be on the order of 2 – 3 kcal/mol, and our ab initio calculations support these values.

The product we monitor is the I atom using fs-resolved mass spectrometry. The other product is the BzI species. The initial fs pulse prepares the system in the transition state of the harpoon region, i.e., $Bz^+ I_2^-$. The iodine atom is liberated either by continuing on the harpoon PES or by electron transfer

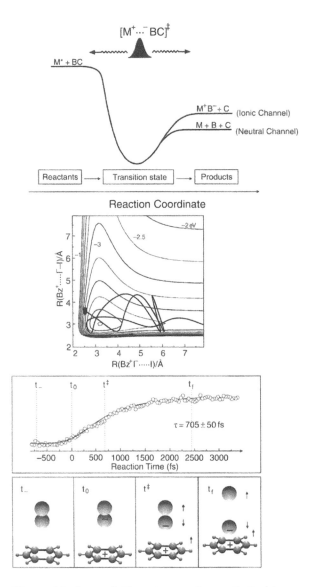

Figure 10: (a, top) Generic reaction path for charge-transfer reactions with both channels of harpooning and electron transfer indicated. Molecular dynamics of the Bz/I_2 bimolecular reaction is shown at the bottom (see text). (b, bottom) The observed transient for the Bz/I_2 reaction (I detection) and the associated changes in molecular structure. References are given in text. Note that we observe the two channels of the reaction, shown in (a), with different kinetic energies and rises of the I atom (see text)

from iodine (I_2^-) to Bz^+ and dissociation of neutral I_2 to iodine atoms. We have studied the fs dynamics of these channels (Figure 10) by resolving their different kinetic energies and temporal behavior. The mechanism for the elementary steps of this century-old reaction is now clear, and in more recent work we also studied the effect of solvation in clusters and in solutions. The observed fs dynamics of this dissociative CT reaction is related to the nature of bonding. Upon excitation to the CT state, an electron in the HOMO of benzene (π) is promoted to the LUMO of $I_2(\sigma^*)$. Vertical electron attachment of ground-state I_2 is expected to produce molecular iodine anions in some high vibrational levels below the dissociation limit. In other words, after the electron transfer, the I–I bond is weakened but not yet broken. While vibrating, the entire I_2 and benzene begin an excursion motion within the Coulombic field, and the system proceeds from the transition-state region to final products. Additionally, an electron may return to the benzene cation, leaving I_2 on a dissociative potential. The resulting neutral Bz–I then loses the I atom. The apparent 750 fs reaction time actually is made of two components: a fast one ($\sim$ 200 fs), describing the back electron transfer, and a slow one ($\sim$ 1 ps), describing both the ionic channel and the secondary Bz–I dissociation. This is an important conclusion pertinent to dissociative CT reactions in solutions (see reviews by Eberson [93] and Savéant [94]), to CT surface reactions [95], and to future transition-state studies of surface-aligned, photoinduced reactions [40]. To give more insight into the molecular dynamics in the transition-state region, we performed classical trajectory calculations.

The results, which we detailed elsewhere, reveal that the transition state of CT reactions can be studied at well-defined impact geometries. The dissociative CT reaction of benzenes with iodine occurs with an elementary harpoon/electron-transfer mechanism. The time scales for the CT and for the product (I) formation define the degree of concertedness and, as reported elsewhere [91], are significant to the recent elegant studies in condensed media by Wiersma and colleagues and by Sension [96]. So far we have studied the electron donors of benzene, mesitylene, and cyclohexane, and we plan extension to other systems.

The above CT systems represent the case for *intermolecular* electron transfer. We have also examined *intramolecular* electron-transfer systems and studied the influence of IVR and geometrical changes. This work is detailed elsewhere [24].

4.8 Proton (Hydrogen Atom)-Transfer Reactions

4.8.1 Acid–Base Reactions

A key point in the measurements of proton-transfer rates is the ability to induce an acid–base reaction on a relatively short time scale ("pH jump") and to follow the change with time. Fundamental to the proton-transfer reaction are the energetics of the initial and final states and the dynamical structure of the solvent. With fs and ps time resolution, it is possible to study the elementary processes of structural changes due to solvent organization.

One of the systems studied in this group is 1-naphthol (referred to here as AH or 1-NpOH) solvated, in a stepwise manner, by ammonia, water, and piperidine. The pK_a of 1-NpOH in solution has been determined by other groups to be 9.4 in the ground state and 0.5 ± 0.2 in the excited state. This large change in pK_a by the excitation makes it possible to induce the acid–base reaction on a very short time scale. In the solution phase there have been many studies, and together with work on the spectroscopy and kinetics, they are reviewed in ref [97].

The system of interest has the unique acid–base structure displayed in Figure 11 and determined by rotational coherence and by high-resolution spectroscopies (see ref [97]). It is a bimolecular reaction in solution. In solvent cluster cages, the acid–base reaction involves the following elementary steps:

$$A^*H \cdot \cdot B_n \leftrightarrow A^{*-} \cdot \cdot \cdot H^+ B_n \rightarrow A^{*-} H^+ \cdot \cdot B_n \quad (12)$$
$$\text{acid} - \text{base unequilibrated} \quad \text{equilibrated}$$

The above steps define the processes involved as the system changes from the unequilibrated to the equilibrated structure. Because of the weak hydrogen bond, the zero of time can be defined, similar to other studies of bimolecular reactions in real time (see section 4.3), and the diffusion-controlled processes are eliminated. The reversible process indicated above depends on a finite number of solvent molecules around the solute, miscible or immiscible, and has analogy in bulk studies of acid reactions.

The PES is normally thought of as a simple double well. This is because the hydrogen bond is relatively weak, and the configurations A^*H and H^+B_n are much different. For the ion-pair product state, the shape of the potential energy surface is modified due to Coulombic interaction and the solvent cage effect. If tunneling is the dominant mechanism, then the time for transfer will depend on the nature of the potential (barrier height and width), which is strongly dependent on the intermolecular distance of the solute and the solvent (here, the O–N distance). The rearrangement of solvent molecules is also expected to affect the potential energy surface in both static and dynamical ways.

In these finite-sized clusters, we reported real-time ps and fs studies of solvation involving the proton transfer [97]. The solute/solvent cluster size changes with n, the number of solvent molecules, being 1 to 21. The occurrence of proton transfer on the ps time scale was observed for the ammonia clusters at a critical number with $n = 3$, while for piperidine $n = 2$.

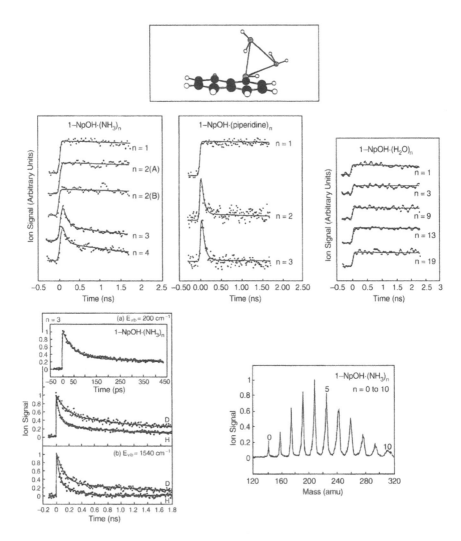

Figure 11: Molecular structure of the acid 1-naphthol with two water solvent molecules, as determined by rotational coherence spectroscopy. The ps and fs transients for three solvents are shown, together with the mass spectra obtained for finite-sized clusters of naphthol in ammonia. References are given in text

The water clusters show no sign of short time scale (picosecond) dynamics for $n = 1 - 21$, indicating no evidence of proton transfer in these solvent clusters. To understand the nature of the transfer and the role of structural changes, we made the following studies: (1) accurate measurements of the transient decay and its form (biexponential, etc.); (2) the isotope effect; (3) the vibrational energy dependencies; (4) the effect of the number and type of solvent molecules. From these results, which have been detailed in ref [97], we proposed a simple model which takes into account deprotonation by tunneling process, protonation, and solvent reorganization.

The time scales for the tunneling phenomena, which we shall address in section 4.9, and solvent reorganization were obtained, relating the dynamics to the nature of the solvent cage structure and its finite size. Important insights were gained. For example, we were able to explain why 1-naphthol is a stronger acid than 2-naphthol and why a certain isomer of 1-naphthol·$(NH_3)_n$ is reactive while others are not. The barrier to proton transfer is due to a crossing of a *covalent* reactant state and a *Coulombic* ion-pair state. The characteristic proton-transfer times of $50 - 100$ ps are well reproduced in the theoretical model of tunneling between two states.

It is suggested that the overall dynamics of the transfer is governed by the interplay between the energetics and the solvent effective dielectric screening which determines the strength of the Coulombic interactions of the ion pairs. By considering the rates of deprotonation and protonation and the cluster size dependence, we concluded that the change in free energy with the structural changes shifts the equilibrium toward the acid as the cluster size increases. The critical value found for the number of solvent molecules ($n = 3$ for NH_3; $n = 2$ for piperidine; $n > 21$ for H_2O) for proton transfer elucidates the key role of the local structure on proton transfer, a central point to the argument made in bulk studies by Eigen, Robinson, and others (see ref [97]). We are currently involved in further studies of the molecular dynamics under these stepwise solvation conditions in these and other acid–base reactions.

4.8.2 Intramolecular Hydrogen Atom Transfer

As discussed above, bond-breaking and bond-making dynamics involve the redistribution of electrons between "old" and "new" bonds or the transfer of an electron (a harpooning reaction) when the nuclei are in an appropriate configuration. In a whole class of reactions, loosely called proton-transfer reactions, the key to the breaking or making of bonds is hydrogen atom (H) motion or proton (H^+)-transfer dynamics, each of which also occurs on the ps to fs time scale. The generic description of these reactions involves a reaction coordinate of the type $O_a - H \cdots O_b$, where the light hydrogen nucleus is between two heavy oxygen atoms. With H moving (or transferring) between O_a and O_b, the $O_a - H$ bond is broken and a new one

$(H - O_b)$ is formed. This phenomenon, which may involve neutral H motion or zwitterion (H^+O^-) formation, is abundant in organic photochemistry and proton-transfer spectroscopy.

Even under collisionless conditions, the motion may not be that simple. The motion of hydrogen on the ps or fs time scale may be localized, or it may involve nuclear motions with a simultaneous redistribution of electrons in many bonds. The nature of bonding and electronic charge distribution, as dictated by symmetry rules, frontier orbitals, or the nodal pattern of the wave function, determines the reaction pathway, while IVR plays a role if the nuclei have enough time to change their positions in the course of the reaction.

Femtosecond time resolution is ideally suited for probing the dynamics in such reactions and for initiating the reaction from a localized (nuclear) wave packet [24]. In the gas phase and under collisionless conditions, studies of the initial, intermediate, and final states associated with the H motion could reveal the dynamics in real time. The packet may find its way directly, or it may search through other modes for the reaction coordinate, analogously with direct and complex mode reactions.

A prototype large molecular system exhibiting hydrogen transfer is methyl salicylate (MS), whose structure (Figure 12) is either ketonic (before the transfer) or enolic (after the transfer). Based on the dual emission spectrum of MS in solution, which was first investigated in 1924, the idea of proton transfer to form a zwitterionic species was proposed by Weller in 1956. Subsequently, a voluminous literature has discussed MS and similar systems (see ref [24]).

A double-well potential along a reaction coordinate corresponding to the two forms and responsible for the dual fluorescence is an attractive description and could simplify the problem. However, in MS, the validity of the double-well model, which has received considerable, attention, was not proven. For example, if a double-well potential description is correct, the tunneling time in the isolated molecule is expected to be $\sim 10^{-7}$ s. The spectroscopy at 4.2 K is not consistent with such a double-well behavior. Picosecond time-resolved studies of isolated MS in a molecular beam by this group (and also in solution by other groups) have failed to resolve the dynamics of the transfer, indicating that such H motion occurs on a time scale of less than 10 ps. The time scale of the dynamics and the mechanism depends on the path of the motion and the nature of the reaction coordinate.

The fs dynamics of hydrogen atom transfer in the gas phase under collisionless conditions was studied by a depletion-technique, variant of FTS (Figure 12). We explored the dynamics from the early fs times to the ps time scale, where the hydrogen-transferred species undergoes nonradiative decay. Within 60 fs, the wave packet in MS evolves to cover all configuration space along the reaction coordinate. We observed no deuterium isotope effect; this is consistent with ultrafast hydrogen movement. However, the subsequent

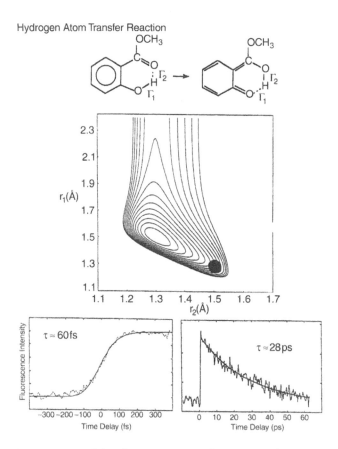

Figure 12: Molecular structure of methyl salicylate and the fs dynamics of intramolecular hydrogen atom transfer. See text. The wave packet position and the two coordinates are displayed. Adapted from: Herek, J. L.; Pedersen, S.; Bañares, L.; Zewail, A. H.: J. Chem. Phys. **97**, 9046 (1992)

redistribution of energy in the new form is on the ps time scale (Figure 12). The results indicate that intramolecular bond–electron rearrangement involves the molecular framework (nuclear motion and IVR); this process is similar to the case of $A + BC$ reactions where bond breaking and bond making occur simultaneously. The potential of the motion is highly asymmetric along the $O - H \cdots O$ reaction coordinate. This potential asymmetry, common to many of these hydrogen atom-transfer reactions, leads to wave packet motion on a timescale only comparable with the half-period of the low-frequency modes, but slower than that of the OH reaction coordinate. A description of the dynamics is shown in Figure 12.

4.9 Tautomerization Reactions: DNA Models

The above picture of the dynamics of a single hydrogen bond can be extended to more complex systems. Multiple hydrogen bonds commonly lend robustness and directionality to molecular recognition processes and supramolecular structures. In particular, the two or three hydrogen bonds in Watson–Crick base pairs bind the double-stranded DNA helix and determine the complementarity of the pairing. Watson and Crick pointed out, however, that the possible tautomers of base pairs, in which hydrogen atoms become attached to the donor atom of the hydrogen bond, might disturb the genetic code, as the tautomer is capable of pairing with different partners. But the dynamics of hydrogen bonds in general, and of this tautomerization process in particular, are not well understood.

Recently, we reported observations of the fs dynamics of tautomerization in model base pairs (7-azaindole dimers) containing two hydrogen bonds [98]. Because of the fs resolution of proton motions, we were able to examine the cooperativity of formation of the tautomer (in which the protons on each base are shifted sequentially to the other base) and to determine the characteristic time scales of the motions in a solvent-free environment. The first step was found to occur on a time scale of a few hundred femtoseconds, whereas the second step, to form the full tautomer, is much slower, taking place within several picoseconds; the time scales are changed significantly by replacing hydrogen with deuterium. These results elucidate the molecular basis of the dynamics and the role of quantum tunneling.

The molecular structures and some of the transients are shown in Figure 13. There are two possible mechanisms of double proton transfer in these model base pairs: a stepwise transfer from the base-pair structure (b-ps) to the tautomer structure (ts) through an intermediate (zwitterionic) structure (is) or a direct cooperative transfer of (b-ps) to (ts). We have studied the fs transients for the fully undeuterated (NH, NH, CH) pair, at two different vibrational energies, E, and for the isotopic species. For the 236 amu mass species, the decay was fit to a biexponential function giving decay times of $\tau_1 = 650$ fs and $\tau_2 = 3.3$ ps when $E = 0$. On the other hand, when the vibrational energy content became ~ 1.5 kcal/mol, these rates changed significantly, giving $\tau_1 = 200$ fs and $\tau_2 = 1.6$ ps. This drastic change reflects the presence of a reaction barrier.

The observed decays for $E = 0$ (and at higher energies) give the rates at which the base-pair structure is changing with time due to proton transfer. The fact that the initial tautomerization is on the fs time scale, when the total vibrational energy is zero, indicates that the proton-transfer motion is *direct* and does not involve the entire vibrational phase space of the pair. The implication is that the motion can be described as "localized" in the coordinate of $N - H \cdots : N$. Furthermore, the two decay components indicate the presence of the intermediate structure, which reflects the two-step motion in the transfer.

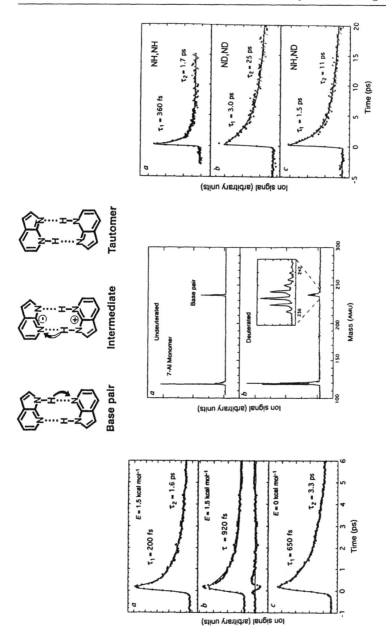

Figure 13: Molecular structures of the base-pair model of 7-azaindole. The fs transients and the mass spectra for protonated and deuterated pairs are shown. References are given in text

The rate of tautomerization can be related to a simple model describing the transformation of the (b-ps) to (is) by considering the motion of the proton in a double-well potential, with the system in either the N − H or the NH$^+$ configuration of the two moieties. In this model, the rate is given by the tunneling expression

$$k = \nu \exp[-\pi a \sqrt{2mU_0}] \tag{13}$$

where ν is reaction coordinate frequency (N − H), $a_0 = \hbar a$ is the half-width of the energy barrier, and U_0 is its height. m is the effective mass of the particle. Using the measured k and taking ν for the N − H of 2800 cm^{-1} and $a_0 = 0.27$ Å, we obtained $U_0 = 1.2$ kcal/mol. This picture assumes that the distance between the two nitrogens is fixed. On relaxing this condition and averaging over the stretch motion of the N⋯N centers at $E = 0$, we again obtained $U_0 \approx 1.3$ kcal/mol. For self-consistency we have repeated these two types of calculations and obtained satisfactory agreement for the effect of isotope substitution and for the excess vibrational energy. We have also made similar studies on the deuterated structures and observed the decrease in rates due to quantum tunneling ($E = 0$). The phenomenon could be general to biological systems and is relevant to Löwdin's description of quantum genetics (see ref [98] for more details).

With the equivalence of the two hydrogen bonds in the static structure of the molecule it is interesting to ask, *what is the nature of the process which leads to the dynamical structures?* We proposed the following picture. Because the time scale of the proton motion is observed to be relatively short, compared to the energy redistribution, the "reaction center" involves primarily the N − H and N⋯N motions. The time scale of the proton motions, however, is longer than or comparable to the changes in the electronic distribution upon excitation and the nuclear vibrational motions of the N − H and N⋯N stretches. This distinction in time scales allows for the asymmetric motion of one of the protons, and because one moiety is excited, the proton ultimately transfers leading to the (is). A consequence of this transfer is a stability for the second N − H motion and a higher barrier toward TS formation. The N⋯N stretch is ~ 120 cm^{-1} and the N − H is ~ 2800 cm^{-1}, giving 280 and 12 fs, respectively. Therefore, on the time scale of 0.5 to 10 ps, typical reaction times, the "asymmetric reaction coordinate" for the two particles is established. Very recent ab initio calculations by Douhal et al. support this proposed model.

The process of mutation by tautomerization is similar to the excited-state process described here. If a "misprint" induced by a tautomer takes place during replication, then an error is recorded. Because reaction path calculations of DNA base pairs show similar potential energy characteristics to those discussed here, we anticipate being able to explore the relevance of tautomerization dynamics to mutagenesis. In this area, we are currently examining these and other systems, also in solutions.

4.10 Norrish Reactions: TS in Concerted and Stepwise Mechanisms

Norrish type I reactions of ketones have been among the most extensively investigated areas in photochemistry. Such reactions have provided good model systems to address some important issues such as the dissociation mechanism, coupling of electronic states, nascent product-state distributions, and concertedness of reactions with multi-bond-breaking events. They have also been used as a convenient source of hydrocarbon free radicals, widely used in organic syntheses, and are important combustion species themselves.

The acetone molecule, because of its relative simplicity, has been of particular interest in numerous photochemical studies and for the investigation of dissociation at many different energies. At relatively low energies ($\lambda \geq 266$ nm), the major dissociation channel is the cleavage of one C – C bond which is in the α-position to the carbonyl group, producing CH_3CO and CH_3 radicals. The CH_3CO radical formed does not have enough internal energy to surmount the barrier to further dissociation. However, at high energies ($\lambda \leq 193$ nm), where Rydberg states are excited, both C – C bonds adjacent to the carbonyl group of acetone dissociate, giving two CH_3 radicals and CO as final products with a quantum yield of nearly unity. This breakage of the two chemically equivalent bonds in the reaction of $(CH_3)_2CO \rightarrow 2CH_3 + CO$ has been intensively studied as a model system for resolving a fundamental issue, namely, do the two bond-breaking events occur in a *concerted* or in a *stepwise* manner?

The distinguishing criterion between the concerted and stepwise mechanisms has in the past been defined by using the internal molecular clock, the calculated rotational period ($\sim$ ps) of the intermediate. From product-state distribution and alignment data the lifetime of the intermediate has been deduced to be longer (or shorter) than the rotational period, inferring the mechanism of the reaction to be stepwise (or concerted). This definition of concertedness is not fundamental, and the direct measurement of the intermediate lifetime with fs resolution is a key to the resolution of the issue of concertedness and synchronicity.

Femtosecond-resolved mass spectrometry was invoked to answer these questions and to determine the elementary mechanism. For acetone [99], real-time dynamics of the dissociation, when excited to the (n–4s) Rydberg state, shows two elementary steps for the two C – C bonds with distinctly different time scales for the *primary* and *secondary* α-bond breakage:

$$(CH_3)_2CO^* \rightarrow CH_3CO\cdot^\dagger + CH_3\cdot \quad (50 \text{ fs}) \tag{14}$$

$$CH_3CO\cdot^\dagger \rightarrow CH_3\cdot + CO \quad (0.5 \text{ ps}) \tag{15}$$

The experimental results indicate that the concertedness of the reaction should be judged from the dynamical time scale for the actual nuclear motions of the intermediate or transition states along the reaction coordinate [99]. *The two elementary steps shown above differ in their time scales by an order of magnitude, and yet both are faster than the rotational period.* Therefore, the reaction mechanism would have been assigned as concerted, which is clearly not the case. Castleman's group (ref [1]) has shown slower, and also nonconcerted, dynamics at lower energies and studied novel processes in clusters.

For asymmetric α-cleavage, the problem is very different because the strengths of the two C – C bonds are not identical, the vibrational phase space is distinct, and there are multiple reaction coordinates. With time resolution, the evolution of the cleavage could be followed, and with mass resolution, the transient intermediates formed along different pathways can be positively identified. Of particular interest to study [99] were the asymmetric ketones, R_1COR_2, where R_1 and R_2 are CH_3 and C_2H_5, and for them the C – C bonds are of a different nature. As shown in Figure 14, this picture has roots in transition-state spectroscopy where, in this case, the preparation is into a quasi-bound state and the clocking involves two channels.

For the asymmetric cleavage, the primary and secondary breakage also occurs in a *stepwise* mechanism with two distinct time scales. The time for the primary C – C bond cleavage is slower for methyl ethyl ketone and diethyl ketone than for acetone. This increase of the time constant for the primary α-cleavage with increasing number of atoms in the molecule is the result of the increased dimensionality of the potential energy surface where the wave packet motion is significant in coordinates other than the C – C reaction coordinate. We found a correlation between the time constants and the number of degrees of freedom, although the impulse in the C – C bond is caused by the σ^* repulsion.

The secondary C – C bond cleavage dynamics of the intermediates is governed by the internal energies obtained during the primary C – C bond-breaking event. We considered the energy partitioning during the primary C – C bond cleavage using the impulsive, simple statistical, and RRKM models. The key is the time scale of the cleavage in relation to IVR time scales. We have also examined the primary and secondary isotope effect on the dynamics of these Norrish reactions. Currently, we are exploring new directions for studying the dynamics of IVR and α-cleavage at energies reaching 100 kcal/mol above the bond energy in the ground state.

4.11 Other Reactions

With the same approach, we have studied other classes of reactions and these include (i) reactions of van der Waals complexes, (ii) reactions of organometallics, (iii) reactions of the Bodenstein type, (iv) reactions of Rydberg states, and (v) reactions under solvation (dense fluids and clusters).

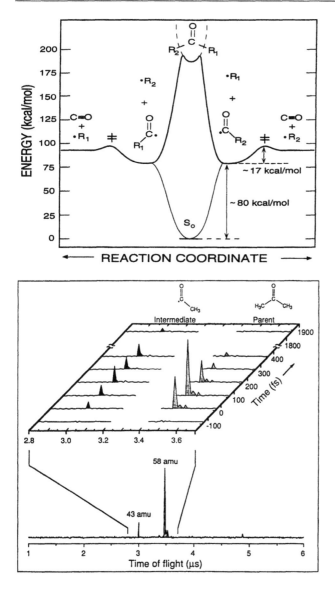

Figure 14: (a, top) PES for Norrish type I reactions, indicating the established stepwise mechanism; see text. (b, bottom) The fs-resolved mass spectra displaying the parent and intermediate temporal dynamics. References are given in text

More details regarding these studies by our group can be found in refs [24–26] and in ref [28].

5 New Opportunities

5.1 Ultrafast Diffraction and Molecular Structures

For the studies of complex molecular structures with ultrashort time resolution two approaches have been developed in this laboratory. One of them, *rotational coherence spectroscopy*, has been reviewed recently in ref [23]. The second is UED. Over the past 60 years, ever since the pioneering work by H. Mark and R. Wierl, gas-phase (continuous beam) electron diffraction has become a powerful tool for studying the *static* nature of molecular structures. To capture the *dynamics* of structures in transition, ultrafast time resolution of the change must be introduced into the diffraction. As discussed above, it has been possible to probe such changes with fs spectroscopy and fs-resolved mass spectrometry by exposing individual molecules and reactions and studying their elementary nuclear motion. In 1991, we proposed the extension of FTS to employ ultrafast electron diffraction (UED) for the study of the structure of complex systems. We reported our first successful UED from beams of isolated molecules (CCl_4, I_2, CF_3I), opening the door to new studies of molecular and chemical changes on the ps time scale and below [24].

The Caltech apparatus is composed of a femtosecond laser, a molecular beam assembly, an electron pulse (15 keV) and lens system, and a newly designed computerized CCD camera for immediate visualization of the diffraction patterns in two dimensions. Conceptually, the idea is simple. However, for UED, sensitivity, timing, and space-charge effects (i.e., temporal dispersion of electrons due to their repulsion) are some of the barriers that had to be overcome. For example, it was difficult to predict whether ultrafast diffraction could be obtained from a beam of molecules, since the sensitivity would be several orders of magnitude lower than that of solid film experiments. Additionally, the electron pulses had to be generated with minimum space-charge dispersion in order to obtain the desired temporal resolution. Finally, the electron pulse had to be time delayed relative to the initiation pulse, so that changes in the diffraction pattern may be followed.

In our apparatus, we use a laser to create, through the photoelectric effect, the ultrashort electron pulse from a photocathode. A second laser is used to initiate a chemical change. The temporal resolution and detection sensitivity were achieved using a new CCD camera design, capable of single electron detection and a high electric field near the photocathode (30 kV/cm).

Figure 15 shows a typical diffraction pattern recorded on our CCD camera and presents the radial distribution functions for some of the systems studied. The radial distribution functions, which yield the internuclear distances, were obtained by a sine Fourier transform of the molecular scattering intensities. Data for standard CCl_4, for example, gave the characteristic diffraction pattern as was evident from comparison with theoretical calcu-

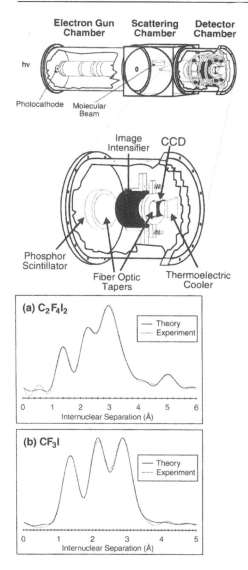

Figure 15: (a, top) Ultrafast electron diffraction apparatus, second generation, built recently at Caltech. The first generation machine is detailed elsewhere: Williamson, J. C.; Zewail, A. H.: J. Phys. Chem. **98**, 2766 (1994). Dantus, M.; Kim, S. B.; Williamson, J. C.; Zewail, A. H.: J. Phys. Chem. **98**, 2782 (1994) . (b, bottom) Some representative examples of the radical distribution function obtained with the UED apparatus. See text and ref [24]. The new apparatus is the work of C. Williamson, H. Frey, and H. Ihee, from this group, to be published

lations based on known scattering amplitudes [24]. In the radial distribution function, peaks were found at 1.76 and 2.89 Å corresponding to the C – Cl and Cl⋯Cl internuclear distances, respectively. For calibration, diffraction rings from Al foil (70 Å) were observed to give the characteristic lattice constant. In Figure 15, we display results for $C_2F_4I_2$ and CF_3I.

To demonstrate the feasibility of studying reactions, we have used the electron beam and the initiation laser to reveal the diffraction of CF_3 radical from the dissociation of CF_3I [24]. More recently, we have time-resolved the structural changes of CH_2I_2 when breaking the C – I bond, and this work is in the process of publication. In our apparatus, the time resolution was measured (by streaking techniques) to be ∼ 1 ps, only limited by the laser (fs) and electron (ps) pulses. Currently, we are studying these and other systems using this second new apparatus displayed in Figure 15.

In recent publications [24], we detailed the experimental and theoretical analyses and outlined new features for probing molecular structure changes with coherent dynamics. Just as with FTS, we also included the time-dependent alignment in UED and introduced a methodology for obtaining both ground- and excited-state structures. Wilson's group [100] has shown the generation of ps X-ray pulses to study reactions, and with these two new sources we expect an exciting new direction. We are also collaborating with Russian colleagues (A. Prokhorov and M. Schelev) on new electron sources for 50 fs resolution.

5.2 Reaction Control

In a 1980 paper [56] we proposed an approach for controlling the outcome of a chemical reaction – the use of ultrashort pulses. With the ability to probe the dynamics on the fs time scale, we more recently decided to explore new avenues for this idea. Here, we discuss our effort in the study of population and reaction yield control in elementary systems.

5.2.1 Control by Pulse Sequences

Population control of the wave packet motion in isolated iodine on the bound B state was demonstrated using fs pulse sequences [24]. The experiments involved the introduction of a "second dimension" to the traditional two-pulse (FTS) scheme. A three-pulse sequence was used; the first two pump pulses prepared the B state, and the third pulse probed the resulting motion. It was shown that this simple sequence of pulses can *build up* wave packet population on the B state surface with *well-defined* and controllable phase difference.

Delaying the second pump pulse by τ_1 from the first introduces a phase shift or angle, θ, completely determined by τ_1 and the frequencies of the wave packet ω_{ij}. One, therefore, has a control over the preparation process.

On the other hand, controlling the probe wavelength allowed us to observe the *in-phase* and *out-of-phase* motion of the packet, similar to the IVR case observed in anthracene [24]. The sequence used was $\lambda_1 - \tau_1 - \lambda_1 - \tau_2 - \lambda_2(\lambda_2^*)$, where λ_1 refers to the wavelength of the pump pulses and $\lambda_2(\lambda_2^*)$ refers to the probing pulse wavelength. Such fs pulse sequences are reminiscent of the sequences we introduced in the 1970s for coherent transients [24], but the pulse phase is not prescribed. Theoretically, the observations can be modeled by using the density matrix formalism or simple perturbation theory, but the idea is intuitive. Hartke and Manz [101] have studied the dynamics of the wave packet under controlled conditions of preparation and made a video for the motion picture illustrating the temporal dynamics.

5.2.2 Control of Unimolecular Reactions

In this area of control, the first quantum wave packet calculations by Rice, Tannor, and Kosloff (for review see ref [58]) showed how a wave packet can be manipulated according to the timing of pulses and the potentials governing the motion. Several theoretical schemes for achieving control have been advanced and reviewed in recent articles by Rice [58] and by Wilson [57]. On the experimental side, progress has been made in shaping techniques for ps and fs pulses and in extending the generation of phase-coherent pulses from nanosecond to ps to fs regimes. With these pulses, there are numerous possibilities for controlled excitation such as selective control of population, locking of dynamics, and interferometry.

For control of a chemical reaction, our first experimental demonstration was made [24] on a bimolecular reaction (following section). With fs pulses, Gerber's group has shown control of the electronic state prepared in Na_2 depending on the intensity [102]. They also demonstrated that, through multiphoton ionization at fs delays, the ratio of the $Na^+ + Na$ or Na_2^+ channels can be altered.

For unimolecular reactions, we studied the control of the dissociation of NaI [103]. The aim was to affect the branching of products by intercepting the wave packet in the transition-state region. Control was achieved with the following pulse sequence: the first to pump (i.e., prepare) the system on a reactive surface and the second to take some fraction of the dissociative system, via stimulated absorption or emission, to another potential energy surface. A third laser pulse is used to probe the system and monitor the change. The perturbation induced by the intermediate pulse effectively removes population from the reactive channel at a prescribed internuclear separation, thereby decreasing the product yield along the initially dissociative path.

5.2.3 Control of Bimolecular Reactions

For reaction yield control, we invoked the concept of wave packet timing [24]. As mentioned before, the critical stage in a chemical reaction, the progression through the transition states, occurs in less than a ps. The coherence and very short duration of fs pulses make them ideal for the possible influence of the reaction during this stage or when the wave packet is localized at a given configuration. Using two sequential coherent laser pulses, we could control the reaction of iodine molecules with xenon atoms to form the product XeI. The yield of product XeI was shown [24] to be modulated as the delay time between the pulses was varied, reflecting its dependence on the nuclear motions of the reactants.

The reaction involves the translational motion of Xe toward I_2, the nuclear vibrational motion of iodine, and the formation of the transition state by Xe harpooning the I_2 through electron transfer:

$$Xe + I - I^* \to [Xe^+ \cdot\cdot I^- \cdot\cdot I]^{*\ddagger} \to XeI^* + I \qquad (16)$$

To exploit the nuclear motions on a fs time scale, we used two pulses separated by a time delay to pump and control the $Xe + I_2$ reaction. The first pulse, λ_p, created a well-defined, coherent wave packet in an intermediate state, the B state of iodine. Following this preparation, the wave packet could move back and forth with a well-defined period as the internuclear separation in the iodine molecule varies between 2.5 and 5 Å. Some femtoseconds after the pump pulse, a second pulse λ_c was sent to lift this wave packet above the reaction threshold for reaction with Xe.

These results demonstrated that the product yield of a chemical reaction (A + BC) can be controlled temporally if the motion of the wave packet toward (direct or indirect) formation of the transition state is timed using pump-and-control coherent fs pulses. The methodology should have applications to other bimolecular (and unimolecular) reactions and to other types of molecular collisions. Because the mechanism of the collision pair of XeI_2 is not yet established, we are continuing further studies of this system, but for different initial states of the reactants. With this first success in controlling reactions with fs pulses (the above-mentioned work and the work of Gerber's group [102]), we hope to extend these ideas to other molecular systems. There are also a number of experimental possibilities based on the theoretical strategies developed by Rice, Tannor, and Kosloff [58], by Manz and Paramonov [104], and by Wilson [57].

6 Concluding Remarks

In a *Faraday Discussion* review paper, five years ago, several new directions to the field were proposed [105]. Many of these proposed studies of new systems and of advancing additional methodologies have, for the most part, been accomplished. However, the advances made have naturally given birth to new thoughts, and we mention a few here.

It is now possible to photodetach a negative ion to form the neutral on the fs time scale and then ionize to the positive ion (see the paper by Wöste, Berry, and colleagues in ref [1]). This process of "negative, neutral, positive" adds to the ability to study dynamics in clusters. It is also possible to study in real time the dynamics of ions and ion–molecule reactions. A new approach in this direction is the study of Coulomb explosion and ion–molecule reactions by the Castleman group [106]. Our own effort in this area will still focus on the negative ion as a source of a neutral launching to examine transition-state dynamics of ground-state reactions [105].

The range of schemes to probe transition-state dynamics is impressive, and new schemes are possible [40]. The fs probing and detection are realized in various ways. These include, for probing, the methodology of laser-induced fluorescence (LIF), absorption, mass spectrometry with multiphoton ionization (MPI), photoelectron kinetic energy and ZEKE, and stimulated emission pumping (see ref [1]). Nonlinear four-wave-mixing (FWM) techniques, such as degenerate FWM (DFWM), provide an additional and significant probing method, especially for generalization of absorption techniques in gas-phase reaction dynamics (Figure 16). Only recently has this approach been introduced to study uni- and bimolecular reactions [107], and we expect future applications, including extension to using the impressive pulse schemes advanced recently by the group of Wiersma [108]. Another area to be exploited is the use of intense pulses to probe by Coulomb explosion wave packet dynamics at distances of more than 100 Å, as demonstrated nicely by Corkum's group [109].

One exciting development for the studies of organic and inorganic complex reactions (see section 4) will continue to explore other reactions with intermediates and transition states hitherto unobserved. This area is rich with unresolved questions, and we plan many extensions, including new systems and new utilization of fs-resolved mass spectrometry. The coupling with the detection [110, 111] methodology of electron kinetic energy will surely enhance the range of applications; in very large systems, the ultrashort time resolution becomes critical for channel switching [112]. The same approach should be expanding in the studies of surface femtochemistry, especially with the recent development of STM and other probing techniques with fs resolution (see ref [1]). There is no doubt that the breakthrough technological advance of the Ti:sapphire laser will continue the expansion of femtochemistry, but now with ease even for the non-laser experts!

Have the experiments reached the so-called "chemistry time scale"? In 6 fs, the nuclear motions are indeed those which characterize chemical reactions and molecular dynamics. For reactions, this time corresponds, typically, to a motion of ca. 0.1 Å, and it is shorter than any TS lifetime (deduced by

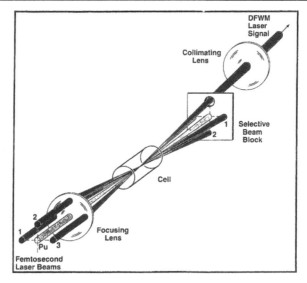

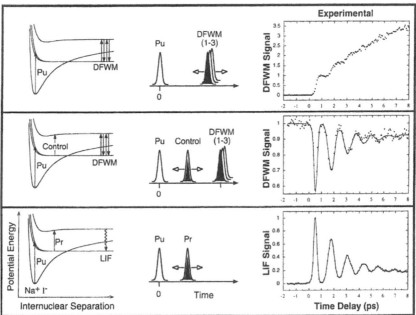

Figure 16: (a, top) Schematic of the different beams invoked for fs degenerate-four-wave-mixing studies of uni- and bimolecular reactions, (b, bottom) Some results obtained with the technique for the dissociation reaction of NaI and comparison with results of LIF. References are given in text

classical and quantum calculations). The 6 fs duration represents the state of the art in laser pulse generation. The energy uncertainty ΔE, although *not* a problem for the preparation even in bound systems (see section 3), should be compared with bond energies; ΔE is 0.7 kcal/mol for a 60 fs pulse and 7 kcal/mol for a 6 fs pulse. So does it help to shorten the pulse further? One may say not, only because the energy uncertainty of subfemtosecond pulses (attosecond regime) is very large (100 as = 420 kcal/mol) compared to bond energies. As illustrated here and elsewhere [24], despite the broadening, the pulse can still be used for coherent preparation, and this "avoids" the issue of the uncertainty principle.

While sub-femtosecond pulses' resolution, which is reaching feasibility [113], may be considered outside the "limit of chemistry", they may prove useful for electron motion and valency. As discussed before [105], for example, based on Pauling's simple description of the bonding in H_2^+, the electron will hop between the two nuclei in a time of 2 fs (about an order of magnitude longer than the orbital motion about the nucleus of the electron in the hydrogen atom) because of the 50 kcal/mol resonance energy. The localization length of the initial packet must be on the atomic or molecular scale, still orders of magnitude larger than the scale of femtophysics (1 fm). Perhaps attosecond pulses would allow us to see such the process of electron valency (in H_2^+, benzene Kekulé structures, etc.) in real time, just as fs pulses make it possible to expose the (nuclear) dynamics of the chemical bond.

But there is, at least for now, an exciting area where *electron dynamics* can be studied on the fs time scale. In atoms, the electron orbit period scales with n^3 (n^5 with lm perturbation). For $n = 10$ the period is in the fs regime. The coupling to the core is attractive Coulombic but the centrifugal (because of angular momentum) is repulsive r^{-2}, and the potential has a barrier (depending on l) for preventing the penetration to the core. For elliptical orbitals, the electron spends most of the time at the outer region, and thus the survival probability is related to the period. In molecules, for large n (large l), the electron does not couple as efficient to the core because of the anisotropy of the orbit and the size. For lower n, the time scale is 10^{-10} s or shorter, and the coupling to the core (electron-ion collision) produces electronic, vibrational, and rotational excitations in the core. The dynamics of such inelastic and chemical bond changes are very interesting. We have made studies [114] of the $n = 5, 6$ of CH_3I, and CD_3I and the results (see Figure 17) are significant in showing the nuclear wave packet motion, coupled to the electron motion on similar time scales. Ab initio PESs of these heavy-element systems are now becoming available [115]. Much work in this area is expected and to cover larger molecules. The analogy with Rydberg atoms [116] is of fundamental interest.

In section 5, we discussed the new direction for reaction control with fs pulse timing, pulse shaping, and pulse sequencing. Future directions will include the control of the effect of the geometric phase and IVR, e.g., in Na_3 [117],

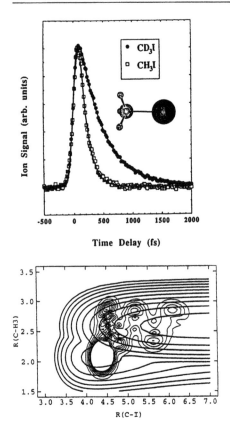

Figure 17: Rydberg-state fs dynamics of methyl iodide, CH_3I and CD_3I, dissociation reaction. The calculated wave packet trajectory is also shown. Adapted from: Janssen, M. H. M.; Dantus, M.; Guo, H.; Zewail, A. H.: Chem. Phys. Lett. **214**, 281 (1993). Guo, H.; Zewail, A. H.: Can. J. Chem. **72**, 947 (1994)

the onset of chaotic behavior [118], and the control of rotationally selected packets [119] and dissociation [120]. Clearly, it is now possible to probe and elucidate the nature of ultrafast phenomena (Figure 18) in chemistry and biology. With the energy landscape established, control schemes will continue to develop at atomic resolution and with increased molecular complexity [121].

Acknowledgment

This work was supported by grants from the National Science Foundation and the U.S. Air Force Office of Scientific Research. We have received two

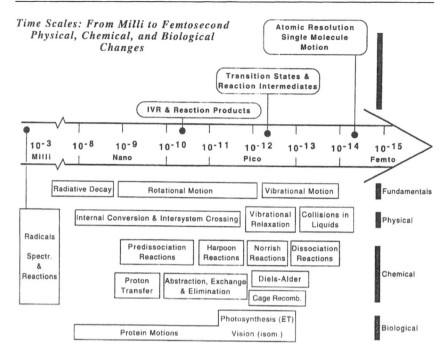

Figure 18: Time scales and scope of ultrafast phenomena in physical, chemical, and biological changes. On the upper side of the arrow of time we display the fundamental elements for the dynamics of the chemical bond, defining the scales for IVR, transition states, and single-molecule motion in femtochemistry. The time scales for the vibrational and rotational motions are shown. Below the arrow of time, examples of studies of physical, chemical, and biological changes are given

referee reports for this article, and wish to thank both referees for their very thorough reading of the manuscript and for the helpful suggestions. The work presented in this review was the result of the dedicated efforts by members of the Caltech group, past and present. They made possible the story told here, and I hope they will find it as exciting as their research in Femtoland!

References

[1] Femtochemistry: Ultrafast Chemical and Physical Processes in Molecular Systems; Chergui, M., Ed.; World Scientific: Singapore, 1996

[2] Arrhenius, S.: Z. Phys. Chem. (Leipzig) **4**, 226 (1889)

[3] van't Hoff, J. H.: In Etudes de Dynamiques Chimiques, F. Muller and Co.: Amsterdam, 1884; p 114 (translation by T. Ewan, London, 1896)

[4] Bodenstein, M.: Z. Phys. Chem. (Munich) **13**, 56 (1894); **22**, 1 (1897); **29**, 295 (1899)

[5] Lindemann, F. A.: Trans. Faraday Soc. **17**, 598 (1922)

[6] Hinshelwood, C. N.: The Kinetics of Chemical Change in Gaseous Systems; Calendron: Oxford, 1926 (second printing, 1929; third printing, 1933); Proc. R. Soc. London 1926, A113, 230.

[7] See, e.g.: Laidler, K. J.: Chemical Kinetics, 3rd ed.; Harper Collins: New York, 1987

[8] Levine, R. D.; Bernstein, R. B.: Molecular Reaction Dynamics and Chemical Reactivity; Oxford University Press: Oxford, 1987 and references therein

[9] Heitler. W.; London, F.: Z. Phys. **44**, 455 (1927)

[10] London, F.: Probleme der Modernen Physik, Sommerfeld Festschrift 1928; p 104

[11] Eyring, H.; Polanyi, M.: Z. Phys. Chem. **1312**, 279 (1931); **B72**, 279 (1931) Polanyi, M.: Atomic Reactions, Williams and Norgate: London, 1932

[12] Polanyi, M.; Wigner, E.: Z. Phys. Chem., Abt. A **139**, 439 (1928)

[13] Eyring, H.: J. Chem. Phys. **3**, 107 (1935) See also: J. Chem. Phys. **3**, 492 (1935)

[14] Evans, M. G.; Polanyi, M.: Trans. Faraday Soc. **31**, 875 (1935); **33**, 448 (1937)

[15] Kramers, H.: A. Physica (Utrecht) **7**, 284 (1940)

[16] Norrish, R. G. W.; Porter, G.: Nature **164**, 658 (1949) Porter, G.: In The Chemical Bond: Structure and Dynamics; Zewail, A. H., Ed.; Academic Press: Boston, 1992; p 113

[17] Eigen, M.: Discuss. Faraday Soc. **17**, 194 (1954); In Techniques of Organic Chemistry; Interscience: London, 1963; Vol. VIII, Part II. Eigen, M.: Immeasurable Fast Reactions. In Nobel Lectures (Chemistry); Elsevier: Amsterdam, 1972; p 170 and references therein

[18] Herschbach, D. R.: Angew. Chem., Int. Ed. Engl. **26**, 1221 (1987) and references therein

[19] Lee, Y. T.: Science **236**, 793 (1987) and references therein

[20] Polanyi, J. C.: Science **236**, 680 (1987) and references therein

[21] Zare, R. N.; Bernstein, R. B.: Phys. Today **33**, 11 (1980)

[22] Johnson, B. R.; Kinsey, J. L.: In ref [23].

[23] Manz, J., Wöste, L., Eds. Femtosecond Chemistry; VCH Verlagsgesellschaft: Weinheim, 1994

[24] Zewail, A. H.: Femtochemistry–Ultrafast Dynamics of the Chemical Bond; World Scientific: Singapore, 1994; Vols. I and II. Most of the Caltech publications in this field, to be referenced here (1976–1994), are collected in these two volumes. These references will not be detailed here, and only newer references will be given for the sake of reducing this section of references

[25] Zewail, A. H.: In ref [23].

[26] Zewail, A. H.: J. Phys. Chem. **97**, 12427 (1993)

[27] Manz, J., Caleman, A. W., Jr., Eds. Femtosecond Chemistry Special Issue. J. Phys. Chem. **48**, 97 (1993)

[28] Wiersma, D., Ed. Femtosecond Reaction Dynamics; Royal Netherlands Academy of Arts and Sciences; North-Holland: Amsterdam, 1994

[29] Zewail, A. H., Ed. The Chemical Bond: Structure and Dynamics; Academic Press: Boston, 1992; p 223

[30] Beddard, G.: Rep. Prog. Phys. **56**, 63 (1993)

[31] See: Dye Lasers: 25 Years, Topics in Applied Physics; Stake, M., Ed.; Springer-Verlag: Berlin, 1992; Vol. 70

[32] Martin, J.-L., Migus, A., Mourou, G. A., Zewail, A. H., Eds. In Ultrafast Phenomena VIII; Springer-Verlag: New York, 1992 and the previous volumes in the series

[33] See the review of the work at that time: Laubereau, A.; Kaiser, W.: Annu. Rev. Phys. Chem. **26**, 83 (1975) See articles by: Eisenthal, K.; Hochstrasser, R.; Kaiser, W.: In Ultrashort Laser Pulses; Kaiser, W., Ed.; Springer-Verlag: Berlin, 1988

[34] Shank, C. V.: In ref [33]. Fleming, G. R.: Chemical Applications of Ultrafast Spectroscopy; Oxford University Press: Oxford, 1986

[35] Heller, E. J.: Acc. Chem. Res. **14**, 368 (1981)

[36] Imre, D.; Kinsey, J. L.; Sinha, A.; Krenos, J.: J. Phys. Chem. **88**, 3956 (1984) Johnson, B. R.; Kittrell, C.; Kelly, P. B.; Kinsey, J. L.: J. Phys. Chem. **700**, 7743 (1996)

[37] Kosloff, R.: J. Phys. Chem. **92**, 2087 (1988) and references therein

[38] Williams, S. O.; Imre, D. G.: J. Phys. Chem. **92**, 6636 6648 (1988)

[39] Engel, V.; Metiu, M.; Almeida, R.; Marcus, R. A.; Zewail, A. H.: Chem. Phys. Lett. **152**, 1 (1988) Engel, V.; Metiu, H.: Chem. Phys. Lett. **755**, 77 (1989); J. Chem. Phys. **90**, 6116 (1989)

[40] Polanyi, J. C.; Zewail, A. H.: Acc. Chem. Res. **28**, 119 (1995) and references therein

[41] Arrowsmith, P.; Bartoszek, F. E.; Bly, S. H. P.; Carrington, T., Jr.; Charters, P. E.; Polanyi, J. C.: J. Chem. Phys. **73**, 5895 (1980) Arrowsmith, P.; Bly, S. H. P.; Charters, P. E.; Polanyi, J. C.: J. Chem. Phys. **79** 283 (1983)

[42] Hering, P.; Brooks, P. R.; Curl, R. F.; Judson, R. S.; Lowe, R. S.: Phys. Rev. Lett. **44**, 698 (1980) Brooks, P. R.; Curl, R. F.; Maguire, T. C.: Ber. Bunsen-Ges. Phys. Chem. **86**, 401 (1982) Brooks, P. R.: Chem. Rev. **88**, 407 (1988)

[43] For a review, see: Gallagher, A.: In Spectral Line Shapes; Bumett, K., Ed.; De Gruyter: Berlin, 1982; Vol. 2, p 755

[44] Polanyi, J. C.; Wolf, R. J.: J. Chem. Phys. **75**, 5951 (1981)

[45] Barnes, M. D.; Brooks, P. R.; Curl, R. F.; Harland, P. W.; Johnson, B. R.: J. Chem. Phys. **96**, 3559 (1992)

[46] Jouvet, C.; Soep, B. Chem. Phys. Lett. **96**, 426 (1983) Soep, B.; Whitham, C, J.; Keller, A.; Visticot, J. P.: Faraday Discuss. Chem. Soc. **91**, 191 (1991)

[47] Liu, K.; Polanyi, J. C.; Yang, S.: J. Chem. Phys. **96**, 8628 (1992); **98**, 5431 (1993) Polanyi, J. C.; Wang, J.-X; Yang, S. H.: Isr. J. Chem. **34**, 55 (1994) Polanyi, J. C.; Wang, J. X.: J. Phys. Chem. **99**, 13691 (1995)

[48] Hoffman, G.; Oh, D.; Chen, Y.; Engel, Y. M.; Wittig, C.: Isr. J. Chem. **30**, 115 (1990) For a recent review see: Wittig, C.; Zewail, A. H.: In Chemical Reactions in Clusters; Bernstein, E., Ed.; Oxford: New York, 1996

[49] Polanyi, J. C.; Williams, R. J.: J. Chem. Phys. **88**, 3363 (1988) Polanyi, J. C.; Rieley, H.: In Dynamics of Gas-Surface Interactions; Rettner, C. T., Ashford, M. N. R., Eds.; Royal Society of Chemistry: London, 1991; Chapter 8, p 329

[50] Metz, R. B.; Bradforth, S. E.; Neumark, D. M.: Adv. Chem. Phys. **81**, 1 (1992) Neumark, D. M.: Acc. Chem. Res. **26**, 33 (1993) Weaver, A.; Metz, R. B.; Bradforth, S. E.; Neumark, D. M.: J. Phys. Chem. **95**, 5558 (1988)

[51] Schrödinger, E.: Ann. Phys. **79**, 489 (1926)

[52] Ehrenfest, P.: Z. Phys. **45**, 455 (1927)

[53] Mukamel, S.: Principles of Nonlinear Optical Spectroscopy; Oxford: New York, 1995. Beswick, J. A.; Jortner, J.: Chem. Phys. Lett. **168**, 246 (1990)

[54] Seidner, L.; Stock, G.; Domcke, W.: J. Chem. Phys. **103**, 3998 (1995)

[55] Zewail, A. H.: Acc. Chem. Res. **13**, 360 (1980)

[56] Zewail, A. H.: Phys. Today **33**, 27 (1980)

[57] Kohler, B.; Krause, J. L.; Raksi, F.; Wilson, K. R.; Yakovlev, V. V.; Whitnell, R. M.; Yan, Y.: Acc. Chem. Res. 1995,28,133 and references therein

[58] Rice, S. A.: Proc. XX Solvay Conf. 1996, xx, xxx and references therein

[59] Manz, J.; Reischl, B.; Schroeder, T.; Seyl, F.; Wartmuth, B.: Chem. Phys. Lett. **198**, 483 (1992) Burghardt, I.; Gaspard, P.: J. Chem. Phys. **100**, 6395 (1994)

[60] Ben-Nun, M.; Levine, R. D.: Chem. Phys. Lett. **203**, 450 (1993)

[61] Metiu, H.: Faraday Discuss. Chem. Soc. **97**, 249 (1991) and earlier references therein

[62] For a collection of recent examples, see: Ultrafast Phenomena VIII; Martin, J. L., Migur, A., Mourou, G. A., Zewail, A, H., Eds.; Springer-Verlag: New York, 1993

[63] See: Ruhman, S.; Fleming, G.; Hochstrasser, R.; Nelson, K.; Mathies, R.; Shank, C.; Barbara, P.; Wiersma, D.: In Femtosecond Reaction Dynamics; Wiersma, D. A., Ed.; Royal Netherlands Academy of Arts & Sciences; North-Holland: Amsterdam, 1994

[64] Banin, U.; Bartana, A.; Ruhman, S.; Kosloff, R.: J. Chem. Phys. **107**, 8461 (1994) and references therein

[65] Baumert, T.; Gerber, G.: Isr. J. Chem. 1994, 34,103 and references therein. Also: Gerber et al. in ref [23]

[66] Papanikolas, J. M.; Vorsa, V.; Nadal, M. E.; Campagnola, P. J.; Gord, J. R.; Lineberger, W. C.: J. Chem. Phys. **97**, 7002 (1992)

[67] Liu, Q.; Wang, J.-K; Zewail, A. H.: Nature (London) **364**, 427 (1993). Potter, E. P.; Liu, Q.; Zewail, A. H.: Chem Phys. Lett. **200**, 605 (1992) Wang, J.-K; Liu, Q.; Zewail, A. H.: J. Phys. Chem. **99** 11309, 11321 (1995)

[68] Kühling, H.; Kobe, K.; Rutz, S.; Schreiber, E.; Wöste, L.: J. Phys. Chem. **98**, 6679 (1994)

[69] Zadoyan, R.; Li, Z.; Ashjian, P.; Martens, C. C.; Apkarian, V. A.: Chem. Phys. Lett. **218**, 504 (1994) Apkarian, V. A.; Martens, C. C.: In ref [1].

[70] Wang, Q.; Schoenlein, R. W.; Peteanu, L. A.; Mamies, R. A.; Shank, C. V.: Science **266**, 422 (1994) and references therein. Dexheimer, S. L.; et al.: Chem. Phys. Lett. **188**, 61 (1992)

[71] Zhu, L.; Sage, J. T.; Champion, P. M.: Science (Washington, D.C.) **266**, 629 (1994)

[72] Martin, J. L.; Breton, J.; Vos, M. H.: In ref [63]. See also: Vos, M. H.; et al.: Nature (London) **363**, 320 (1993)

[73] Chachisvilis, M.; Pullerits, T.; Jones, M. R.; Hunter, C. N.; Sundström, V.: Chem. Phys. Lett. **224**, 345 (1994)

[74] Bradforth, S. E.; Jimenez, R.; Van Mourik, F.; van Grondelle, R.; Fleming, G.R.: J. Phys. Chem. **99**, 16179 (1995)

[75] Gehlen, J. N.; Marchi, M.; Chandler, D.: Science (Washington, D.C.) **263**, 499 (1994) and references therein

[76] Yartsev, A.; Alvarez, J.-L; Åberg, U.; Sundström, V.: Chem. Phys. Lett. **243**, 281 (1995)

[77] Pugliano, N.; Palit, D. K.; Szarka, A. Z.; Hochstrasser, R. M.: J. Chem. Phys. **99**, 7273 (1993) Pugliano, N.; Szarka, A. Z.; Gnanakaran, S.; Triechel, M.; Hochstrasser, R. M.: J. Chem. Phys. **103**, 6498 (1995)

[78] Ionov, S. I.; Brucker, G. A.; Jaques, C.; Valachovic, L.; Wittig, C.: J. Chem. Phys. **99**, 6553 (1993)

[79] Alagia, M.; Balucani, N.; Cassavecchia, P.; Stranges, D.; Volpi, G. G.: J. Chem. Phys. **98**, 8341 (1993) See also: J. Chem. Soc., Faraday Trans. **91**, 575 (1995)

[80] Jacobs, A.; Volpp, H.-R.; Wolfrum, J.: Chem Phys. Lett. **218**, 51 (1994)

[81] Clary, D. C.; Schatz, G. C.: J. Chem. Phys. **99**, 4578 (1993) Hernandez, M. I.; Clary, D. C.: J. Chem. Phys. **101**, 2779 (1994) Zhang, D. H.; Zhang, J. Z. H.: J. Chem. Phys. **103**, 6512 (1995)

[82] Wright, S. A.; Tuchler, M. F.; McDonald, J. D.: Chem. Phys. Lett. **226**, 570 (1994)

[83] van Zee, R. D.; Stephenson, J. C.: J. Chem. Phys. **102**, 6946 (1995)

[84] See: Wittig, C.: In ref [1].

[85] Raftery, D.; Sension, R. J.; Hochstrasser, R. M.: In Activated Barrier Crossing; Fleming, G. R., Hänggi, P., Eds.; World Scientific: Teaneck, NJ, 1993; p 163. Szarka, A. Z.; Pugliano, N.; Palit, D.; Hochstrasser, R. M.: Chem. Phys. Lett. **240**, 25 (1995)

[86] Berson, J.: Science (Washington, D.C.) **266**, 1338 (1994) and references therein

[87] Pedersen, S.; Herek, J. L.; Zewail, A. H.: Science (Washington, D.C.) **266**, 1359 (1994); work to be published

[88] Storer, J. W.; Raimondi, L.; Houk, K. N.: J. Am. Chem. Soc. **116**, 9675 (1994) and references therein

[89] Roos, B. O.; Merchán, M; McDiarmid, R.; Xing, X.: J. Am. Chem. Soc. **116**, 5927 (1994)

[90] Horn, B.; Herek, J.; Zewail, A. H.: to be published

[91] Cheng. P. Y.; Zhong, D.; Zewail, A. H.: J. Chem. Phys. **103**, 5153 (1995); Chem. Phys. Lett. **242**, 368 (1995); J. Chem. Phys., in press

[92] Mulliken. R. S.: J. Am. Chem. Soc. **72**, 610 (1950); **74**, 811 (1952) Mulliken, R. S.; Person, W. B.: Molecular Complexes: A Lecture and Reprint Volume; Wiley-Interscience: New York, 1969

[93] Eberson, L.: Acta Chem. Scand. **B36**, 533 (1982). Eberson, L.: Electron Transfer Reactions in Organic Chemistry; Springer-Verlag: Berlin, 1987

[94] Savéant, J. M.: Adv. Phys. Org. Chem. **26**, 1 (1990)

[95] Dixon-Warren, St. J.; Jensen, E. T.; Polanyi, J. C.; Zu, G. Q.; Yang, S. H.; Zeng, H. C.: Faraday Discuss. Chem. Soc. **91**, 451 (1991)

[96] Lenderink, E.; Duppen, K.; Wiersma, D. A.: Chem. Phys. Lett. **211**, 503 (1993); J. Phys. Chem. **700**, 7822 (1996) Pullen, S.; Walker II, L. A.; Sension, R. J.: J. Chem. Phys. **103**, 7877 (1995)

[97] Breen, J. J.; Peng, L. W.; Willberg, D. M.; Heikal, A.; Cong, P.; Zewail, A. H.: J. Chem. Phys. **92**, 805 (1990) Kim, S. K.; Breen, J. J.; Willberg, D. M.; Peng, L. W.; Heikal, A.; Syage. J. A.; Zewail, A. H.: J. Phys. Chem. **99**, 7421 (1995) and references therein

[98] Douhal, A.; Kim, S. K.; Zewail, A. H.: Nature (London) **378**, 260 (1995); work to be published

[99] Kim, S. K.; Pedersen, S.; Zewail, A. H.: J. Chem. Phys. **103**, 477 (1995). Kim, S. K.; Zewail, A. H.: Chem. Phys. Lett. **250**, 2790 (1996) Kim, S. K.; Guo, J.; Baskin, J. S.; Zewail, A. H.: J. Phys. Chem. **100**, 9202 (1996)

[100] Wilson, K.: In ref [1].

[101] Hartke, B.: Chem. Phys. Lett. **175**, 322 (1990) Hartke, B.; Kolba, E.; Manz, J.; Schor, H. H. R.: Ber. Bunsen-Ges. Phys. Chem. **94**, 1312 (1990)

[102] Baumert, T.; Engel, V.; Meier, C.; Gerber, G.: Chem. Phys. Lett. **200**, 488 (1992) Baumert, T.; et al. In ref [23]

[103] Herek, J. L.; Materny, A.; Zewail, A. H.: Chem. Phys. Lett. **228**, 15 (1994)

[104] Manz, J.; Paramonov, G. K.; Polásek, M; Schütte, C.: Isr. J. Chem. **34**, 115 (1994) Dohle, M.; Manz, J.; Paramonov, G. K.: Ber. Bunsen-Ges. Phys. Chem. **99**, 478 (1995)

[105] Zewail, A. H.: Faraday Discuss. Chem. Soc. **97**, 207 (1991)

[106] Castleman, A. W., Jr. In ref [1]; see also: Castleman, A. W., Jr.; Wei, S.: Annu. Rev. Phys. Chem. **45**, 685 (1994) and references therein. Snyder, E. M.; Wei, S.; Purnell, J.; Buzza, S. A.; Castleman, A. W., Jr.: Chem. Phys. Lett. **248**, 1 (1996); Chem. Phys. **207**, 355 (1996)

[107] Motzkus. M.; Pedersen, S.; Zewail, A. H.: J. Phys. Chem. **100**, 5620 (1996)

[108] Nibbering, E. T. J.; Wiersma, D. A.; Duppen, K.: Chem. Phys. **183**, 167 (1994) Also in: Coherence Phenomena in Atoms and Molecules in Laser Fields; Bandrauk, A. D., Wallace, S. C., Eds.; Plenum: New York, 1992

[109] Stapelfeldt, H.; Constant, E.; Corkum, P. B.: Phys. Rev. Lett. **74**, 3780 (1995)

[110] Fischer, I; Villeneuve, D. M.; Vrakking, M. J. J.; Stolow, A.: J. Chem. Phys. **102**, 5566 (1995); also in ref [1]; Chem. Phys. **207**, 331 (1996)

[111] See articles in refs [1, 23, 24] and the recent work by V. Engel, T. Baumert, and G. Gerber, (Ultrafast Phenomena Proceedings; Springer-Verlag: New York, 1997)

[112] Weinkauf, R.; Aicher, P.; Wesley, G.; Grotemeyer, J.; Schlag, E. W.: J. Phys. Chem. **98**, 8381 (1994)

[113] Hänsch, T. W.: Opt. Commun. **80**, 71 (1990) Farkas, G.; Tom, C.: Phys. Lett. A **168**, 447 (1992) Kaplan, A. E.: Phys. Rev. Lett. **73**, 1243 (1994) Vartak, S. D.; Lawandy, N. M.: Opt. Commun. **120**, 184 (1995) Corkum, P.: Opt. Photon. News 1995, May issue, p 18 and references therein

[114] Janssen, M. H. M.; Dantus, M.; Guo, H.; Zewail, A. H.: Chem. Phys. Lett. **214**, 281 (1993) Guo, H.; Zewail, A. H.: Can. J. Chem. **72**, 947 (1994)

[115] Yabushita, S.: American Chemical Society Abstracts (Physical Chemistry), No. 165, New Orleans meeting, 1996; also private communication

[116] Garraway, B.; Suominen, K.-A.: Rep. Prog. Phys. **58**, 365 (1995) Yeazell, J. A.; Stroud, C. R., Jr.: Phys. Rev. A **43**, 5153 (1991)

[117] See the papers from the group of Manz, Wöste, and Gerber in ref [1] Also: Schön, J.; Köppel, H.: Chem. Phys. Lett. **231**, 55 (1994) Reischl, B.: Chem. Phys. Lett. **239**, 173 (1995)

[118] Burghardt, I.; Gaspard, P.: J. Phys. Chem. **99**, 2732 (1995) Somlóz, J.; Tannor, D. J.: J. Phys. Chem. **99**, 2552 (1995)

[119] Papanikolas, J.; Williams, R.; Kleiber, P.; Hart, J.; Brink, C.; Price, S.; Leone, S.: J. Chem. Phys. **103**, 7269 (1995)

[120] See the work on NO_2, by double resonance, of Wittig's group in ref [1]

[121] Ball, P.: Designing the Molecular World; Princeton University Press: Princeton, 1994. von Baeyer, C. Taming the Atom; Random House: New York, 1992

[122] Pedersen, S.; Herek, J.L.; Zewail, A.H.: Science **266**, 1359 (1994)

[123] Polanyi, J.C.; Zewail, A.H.: Accounts of Chemical Research (Holy-Grail Special Issue), **28**, 119 (1995)

[124] Douhal, A.; Kim, S.K.; Zewail, A.H.: Nature **378**, 260 (1995)

[125] Cheng, P.Y.; Zhong, D.; Zewail, A.H.: J. Chem. Phys. **105**, 6216 (1996)

[126] Zewail, A.H.: Femtochemistry & Femtobiology (V. Sundström, ed.). World Scientific (Singapore 1998)

[127] Williamson, J.C.; Cao, J.; Ihee, H.; Frey, H.; Zewail, A.H.: Nature **386**, 159 (1997)

[128] Baskin, J.S.; Chachisvilis, M.; Gupta, M.; Zewail, A.H.: J. Phys. Chem. **102**, 4158 (1998)

[129] Diau, E. W.-G.; Abou-Zied, O.; Scala, A.A.; Zewail, A.H.: J. Am. Chem. Soc. **120**, 3245 (1998)

[130] Chachisvilis, M.; Garcia-Ochoa, I.; Douhal, A.; Zewail, A.H.: Chem. Phys. Lett. **293**, 153 (1998)

[131] Diau, E. W.-G.; Herek, J.L.; Kim, Z.H.; Zewail, A.H.: Science **279**, 847 (1998)

[132] Zewail, A.H.: Nobel lecture, summarized in the J. Phys. Chem. **104**, 5660 (2000); detailed version in the forthcoming Les Prix Nobel book

Femtosecond Spectroscopy of Collisional Induced Recombination: A Simple Theoretical Approach

V. Engel[†]
Department of Chemistry
Technical University of Denmark
Lynby, Denmark

[†] Permanent Address:
Institut für Physikalische Chemie
Universität Würzburg
Würzburg, Germany

1 Introduction

The interaction of molecules with ultrashort laser pulses leads to the preparation of an ensemble in a non-stationary state, i.e. a wave packet. This allows for the detection of quantum mechanical motion if appropriate experimental schemes are employed [1]. Femtosecond pulses have been used to monitor internal motion or fragmentation dynamics of gas-phase molecules [2, 3, 4, 5] as well as temporal changes in the liquid and solid phase [6, 7]. In what follows we discuss the perturbation of coherently excited molecules by collisions. Although femtosecond spectroscopy of molecular species has been carried out in molecular beams, to our knowledge no experiment combining femtosecond excitation with the crossed molecular beam technique has been performed up to date. On the other hand, several experiments investigated molecules excited with ultrashort laser pulses in high pressure environments [8, 9, 10, 11, 12].
Let us be more specific and discuss the following scenario:

1. A femtosecond (pump) excitation with an electromagnetic field $E_1(t)$ is performed on AB molecules:

$$AB + E_1(t) \rightarrow (AB)^{(1)}. \tag{1}$$

2. A collision with an atom C takes place:

$$(AB)^{(1)} + C \rightarrow ((AB)^{(1)} + C)_{out}. \tag{2}$$

3. A second short laser pulse (probe) interacts with AB:

$$(AB)^{(1)}_{out} + E_2(t) \rightarrow (AB)^{(2)} \tag{3}$$

and a measurement involving $(AB)^{(2)}$ is performed.

The above separation of the laser excitation and collision processes is based on a separation of time-scales. Here we consider the case where the mean collision time τ_c is much longer than the temporal widths of the laser pulses τ_p. This is fulfilled in the example we will discuss below, where τ_c can be estimated to be in the picosecond domain whereas τ_p is shorter than 100 femtoseconds.

Several approaches can be adopted in a description of such processes:

Reduced density matrix formalism. Since measurements are performed on AB whereas the C atom is not observed, one can set up the equations for the reduced density matrix of the AB system coupled to the "bath" C. A difficulty arises in solving these equations since the system-bath coupling which occurs during the scattering in general is quite strong which prohibits a perturbative treatment of the interaction.

Time-dependent wave packets. The pump excitation is described by solving the time-dependend Schrödinger equation for the molecule interacting with the pump laser. Likewise the scattering of an incoming wave packet (describing the atomic motion) from AB is treated. The outgoing scattering state for the combined AB+C system is used as an initial state for the computation of the probe transition and the time-resolved signal.

Mixed quantum-classical dynamics. The AB molecule and its interaction with the pump- and probe-pulses is treated quantum mechanically whereas the motion of C is described using the laws of classical mechanics. In a mean field approach the quantum system moves in a time-dependent potential where the trajectory of the atom C enters as a parameter. Simultaneously the classical particle experiences a potential containing the expectation value of the potential energy with respect to the quantum degrees of freedom. In choosing this approach one has to keep in mind that the limits of the mixed method are not always very well defined.

Classical dynamics. All degrees of freedom are treated using classical trajectories. The wave packet which is prepared through the pump interaction is represented by a swarm of classical trajectories which are employed to calculate pump-probe signals. To obtain good statistics many molecular dynamics simulations have to be performed. Naturally this method cannot account for purely quantum mechanical effects.

Classical-statistical model. As in the full classical treatment the AB wavepacket motion is represented by an ensemble of trajectories. The scattering events are included by a calculation of a collision probability at every timestep. The decision if a collision occurs is taken by a Monte-Carlo procedure. The energy transfer in case of a collision is calculated and a new orbit with modified energy is started. An average over many events leads to signals to be compared to experiment.

The above list is far from being complete. Below we will outline a simple classical-statistical method (Sec. 2) which then is applied to I_2–Ar scattering in Sec. 3. A short conclusion is presented in the final section.

2 Classical-statistical approach to the pump-probe spectroscopy of collision systems

As outlined in the Introduction here the AB wave packet is represented by a swarm of classical trajectories. One way to choose these orbits is to calculate the wave packet prepared by the pump pulse using quantum mechanics and sample the initial conditions from its coordinate and momentum distribution. The trajectories are then integrated in time. At each time-step a collision probability for a fixed density of AB molecules is determined as

$$P_c(\Delta t) = \Delta t \, |\vec{v}(A, BC)| \, \rho(C) \, \sigma(AB, C). \tag{4}$$

Hence it is assumed that the probability for a collision to occur is proportional to the time-step Δt, the relative velocity $|\vec{v}(A, BC)|$ between the centre-of-mass motion of AB and the C motion, the atomic density $\rho(C)$ and the A+BC scattering cross section $\sigma(AB, C)$, for details see Ref. [14]. Briefly, a calculation involves the following steps:

1. An AB trajectory is chosen.

2. The translational momenta $\vec{p}_{AB}, \vec{p}_C$ for AB and C are selected from Maxwell-Boltzmann distributions determined by the experimental conditions and the collision probability P_c is calculated according to Eq. (4).

3. A random number s is chosen from the unit interval. If $P_c < s$ the trajectory is integrated unperturbed. If $P_c > s$ a collision occurs.

The collision leaves the center-of-mass velocity of the collision partners unchanged as well as the magnitude of their relative velocity. The orientation of the relative velocity after the collision is chosen at random. Since an isotropic distribution of the rare gas atoms is assumed this amounts to an average over all impact parameters [15]. Then we calculate the momentum transfer along the AB axis and the trajectory is integrated with the modified kinetic energy.

4. An ensemble of many trajectories which have undergone collisions at different times is used to calculate the pump/probe signal. The latter is assumed to be propotional to the number of trajectories which, at a particular delay-time, are localized within the spatial window, where a resonant probe excitation occurs.

The above scheme has been applied to several systems [14, 16, 17, 18] and seems to provide a reliable method for the calculation of pump-probe signals.

3 An example: I_2 - rare gas collisions

Recently the collision induced recombination of I_2 in rare-gas environments was investigated using femtosecond pump-probe spectroscopy [11]. The excitation scheme is displayed in Fig. 1: the molecule is excited from its electronic ground state (X) to the electronically excited B-state. Since the carrier frequency (ω_1) of the pump-pulse exceeds the dissociation limit of the B-state, fragmentation is initiated. The probe excitation (ω_2) induces a transition to an ion pair state (E) and the pump-probe signal consists of the total fluorescence from this state detected as a function of delay time. The excitation is only effective if the B-state wave packet is in the region where excitation is most effective (the Franck-Condon window). This region is centered around the point where the difference between the potentials in the B- and E-state equals the laser frequency ω_2 as indicated in Fig. 1.

If one studies pure gas phase iodine molecules the pump-probe signal consists of a single peak with a maximum at the time when the wave packet passes the Franck-Condon window. Regard now the same excitation process in a surrounding of atoms C. If a collision takes place at times when the outgoing wave packet is localized close to the B-state potential well, it is possible that the I+I relative motion looses kinetic energy. This loss might be large enough so that, as a consequence of the collision, stable I_2 molecules are built. The existence of this *caging* process was verified in the experiments which showed that in helium and argon surroundings the pump-probe signals exhibit additional peaks which stem from the caged I_2 molecules [11].

Complete quantum calculations were performed and gave very good agree-

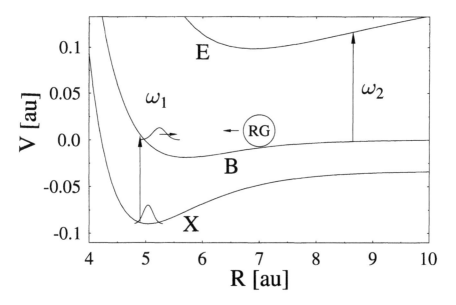

Figure 1: Pump-probe excitation scheme of I_2: the pump transition (ω_1) consists of a transition from the electronic ground state (X) to the excited B-state. Dissociation takes place unless a collision with an atom (RG) occurs. The probe excitation (ω_2) to the E-state is effective if the wave packet is located in the Franck-Condon window between 8 and 9 au.

ment with experiment [19]. The experiment detects dissociating molecules and those which are formed by collisions shortly after the pump excitation. Thus, the average to be taken over collision times (or initial positions of incoming rare gas atoms) has to be performed in a small time interval only. Nevertheless, a quantum calculation of long time pump-probe signals is essentially not possible owing to limited computer time.

Opposite to the above approach the classical-statistical method (Sec. 2) is able to predict pump-probe signals even for long times without any difficulties. It was applied to the present system and produces excellent agreement with experiment [18]. The method allows us to examine the density of trajectories $\rho(R,t)$ describing the I_2 relative motion at any time. This density is depicted in Fig. 2 for the case of a 50 bar Ar environment. The pump-probe signal at a given time is proportional to the density located in the Franck-Condon window around a bond length of $R=8.5$ au at the respective time. By taking a cut through the figure around this value of R one finds a large peak at short time corresponding to directly dissociating molecules. At later times there arises a background which carries regular structures

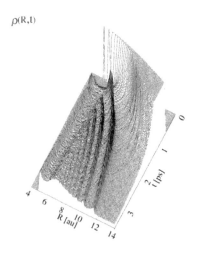 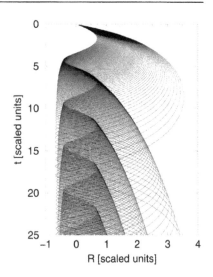

Figure 2: Radial distribution function $\rho(R,t)$ of I_2 trajectories obtained for collisions with Ar at 50 bar pressure.

Figure 3: Motion of bound-state trajectories in Morse-potential exhibiting caustic lines. Scaled units are used for distance and time.

appearing one after the other in the Franck-Condon window. This is precisely what is seen in the experiment.

Recently we performed full-dimensional molecular dynamics simulations of the above described experiments and found pump/probe signals which are almost identical to those obtained from the simple classical-statistical model [20]. It could be shown that the characteristic structures in the radial distribution functions are *induced by single collisions and do not depend on pressure*. Their origin can be explained as follows. Short after the pump excitation, the collisions prepare caged molecules having bond-lengths within a small interval but, owing to the different efficiency of the energy transfer, within in broad distribution of internal enrgies. The time-evolution of the ensemble then is that of neighbouring orbits originating from a small volume in phase space. If multiple collisions do not occur the trajectories evolve unperturbed in time. As time goes along two respective trajectories which differ only slightly in the value of their initial momentum will intersect each other. The points of intersection of many trajectories are bounded by a tangent curve which in optics (where the trajectories correspond to different light rays) is known as a caustic [21]. On these lines (which do, in general, not coincide with classical turning points) an enhanced density can be found. Fig. 3 illustrates this behaviour. Several trajectories are shown which start at the same distance but with different momenta. For the illustration we used the motion in a scaled Morse-potential where all paramteres where set to 1.

As an important result one arrives at the conclusion that pump-probe signals arising from the caged molecules reflect properties of the classical motion underlying the dynamical behaviour of the quantum system.

4 Conclusion

Several possibilities to describe the real-time spectroscopy of collision processes were outlined. We introduced a simple classical-statistical approach which can be used to describe complex systems since it only requires the classical treatment of the observed molecules. Here the environment enters in form of a random perturbation. Although we described experiments performed at high pressures, the interesting structures appearing in the pump-probe signals are due to single collisons. Thus they should be observable in molecular beam experiments which allow for a much better characterization of the scattering events compared to the situation in gas cells.

Acknowledgments

Financial support by the Deutsche Forschungsgemeinschaft is gratefully acknowledged. The author thanks the Technical University of Denmark at Lyngby for the opportunity to participate in its program for visiting professors. Stimulating discussions with H. Dietz, V. A. Ermoshin and C. Meier are acknowlegded.

References

[1] A. H. Zewail, *Science* **242**, 1645 (1988).

[2] A. H. Zewail, *Femtochemistry, Vols.1,2* (World Scientific, Singapore, 1994).

[3] *Femtosecond Chemistry*, J. Manz, L. Wöste, Eds. (VCH, Weinheim, 1995).

[4] *Femtochemistry*, M. Chergui, Ed. (World Scientific, Singapore, 1996).

[5] *Femtochemistry and Femtobiology*, V. Sundström, Ed. (World Scientific, Singapore, 1997).

[6] *Ultrashort Laser Pulses*, W. Kaiser, Ed. (Springer, Heidelberg 1993).

[7] *Ultrafast Phenomena X, XI*, Springer Series in Chemical Physics, Nos. **62, 63**, (Springer, Heidelberg).

[8] C. Lienau and A. H. Zewail, J. Phys. Chem. **100**, 18629 (1996).

[9] A. Materny, C. Lienau, and A. H. Zewail, J. Phys. Chem. **100**, 18650 (1996).

[10] Q. Liu, C. Wan, and A. H. Zewail, J. Phys. Chem. **100**, 18666 (1996).

[11] C. Wan, M. Gupta, J. S. Baskin, Z. H. Kim, and A. H. Zewail, J. Chem. Phys. **106**, 4353 (1997).

[12] G. Knopp, M. Schmitt, A. Materny, and W. Kiefer, J. Phys. Chem. A **101**, 4852 (1997).

[13] M. Dantus, R. M. Bowman, and A. H. Zewail, Nature **343**, 737 (1990).

[14] H. Dietz and V. Engel, J. Phys. Chem. A **102**, 7406 (1998).

[15] L. C. Woods, *An Introduction to the Kinetic Theory of Gases and Magnetoplasmas* (Oxford University Press, New York, 1993).

[16] H. Dietz, G. Knopp, A. Materny, and V. Engel, Chem. Phys. Lett. **275**, 519 (1997).

[17] H. Dietz and V. Engel, Theo. Chem. Acc. **100**, 199 (1998).

[18] H. Dietz and V. Engel, J. Chem. Phys. **110**, 3335 (1999).

[19] C. Meier, V. Engel, and A. J. Beswick, Chem. Phys. Lett. **287**, 487 (1998).

[20] V. A. Ermoshin, C. Meier, and V. Engel, J. Chem. Phys. (submitted).

[21] M. Born and E. Wolf, *Principles of Optics* (Pergamon Press, New York, 1959).

Direct Observation of Collisions by Laser Excitation of the Collision Pair

J. Grosser[1], O. Hoffmann[1], F. Rebentrost[2]
[1] Institut für Atom- und Molekülphysik
Universität Hannover, Germany
[2] Max-Planck-Institut für Quantenoptik
Garching, Germany

1 Introduction

Experimental studies of atomic or molecular collisions are usually indirect. Typically, an experimental control of the velocity vectors and the quantum states of the collision partners before and after the collision is achieved, but the collision event itself remains uncontrolled and unobserved. The considerable knowledge which we possess about the mechanisms of collisions is based for the greatest part on the discussion and analysis of such indirect experiments. The optical excitation of the collision complex has often been considered as a possibility to study collisions in a more direct way [1]. In particular the process

$$A + M + h\nu \rightarrow A^* + M,$$

in which an atom A is excited by a photon while it is close to a collision partner M, has been experimentally realized for a considerable number of cases, see for example [2]. The process is usually denoted as an "optical collision". Optical collision studies provide indeed a basically new approach to the study of the A+M collisional dynamics, as demonstrated repeatedly. However, optical collisions were investigated in the past only in gas cell experiments. One gains a more direct access to the collision process in this way, but one looses at the same time all the advantages provided by beam techniques.

The study of optical collisions with beam methods faces two major difficulties: First, because the typical particle densities in beams are considerably smaller than those in gas cells, the expected signal level is very low. Second, a differential detection of the collision products is difficult, because the excited product atoms decay typically on a nanosecond time scale. The first beam experiment on the optical excitation of atom-atom collision pairs was reported by our group in 1994 [3]. We attacked the intensity problem by using an optimized beam arrangement and a high density target beam, and we solved the detection problem by the application of a novel and efficient detection scheme. Since then, the step from gas cell to beam techniques has turned out to provide numerous new possibilities to study the properties of collision pairs, to observe details of a collision, and even to control collisions. For the future, such exciting prospects as the geometric observation of the transition state of a chemical reaction or the geometric time resolved observation, practically a motion picture, of a collision seem within reach.

2 Optical Collisions

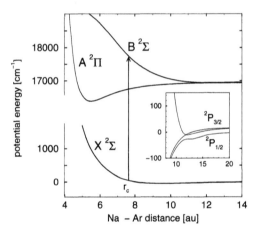

Figure 1: The Na+Ar potential curve system. The ground state potential converges to Na(3s)+Ar, the excited ones to Na(3p)+Ar. The *inset* shows an enlarged view of the nonadiabatic coupling region at the convergence of the $A^2\Pi$ and $B^2\Sigma$ states. Zero corresponds to the Na(3s)+Ar asymptote in the main figure and to $Na(3p)^2P_{1/2}+Ar$ in the *inset*

We study optical collisions of Na atoms,

$$Na(3s) + M + h\nu \to Na(3p) + M,$$

with atomic and molecular targets M. The photon energy $h\nu$ is detuned from the Na and target resonance energies. Figure. 1 demonstrates for the case of an atomic target that the optical transition can occur only when the colliding atoms are at a distance r_c where

$$h\nu = V_e(r_c) - V_g(r_c). \tag{1}$$

V_g and V_e are the ground and excited molecular potentials. The quantity r_c is the Condon radius and depends on the frequency. In place of the frequency we usually indicate the detuning $(\nu - \nu_{res})/c$, where ν_{res} corresponds to the $Na(3s)^2S \to (3p)^2P_{1/2}$ transition. In a simple case as in Fig. 1, the detuning selects a single molecular state, which is excited at a single value for r_c. Figure 2 shows a classical trajectory picture of the collision. The Condon radius is indicated by the circle. The transition can occur only where a trajectory intersects the circle. When, as in our experiments, the velocity directions before and after the collision are fixed, there are only few possible trajectories. In the case of Fig. 2, which is typical for positive detuning, there are two possibilities only: In one case, the optical transition occurs at the first passage through r_c, in the other one at the second passage. Of course, optical transitions are possible at the other intersections; but, travelling under the influence of other potentials, the particles would experience other forces in such a case, and the trajectories would lead to other deflection angles. The transition points are denoted as Condon points, the correponding vectors $\mathbf{r}_1$ and $\mathbf{r}_2$, which connect the target and projectile atoms in the moment of the transition, as Condon vectors.

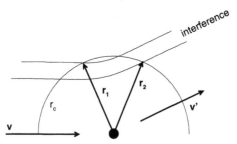

Figure 2: The geometric characteristics of an optical collision: The classical trajectories $\mathbf{r}(t)$ for the vector $\mathbf{r}$ connecting the atoms, the Condon radius r_c, and the Condon points and Condon vectors $\mathbf{r}_{1,2}$ which indicate where the optical transition occurs on the trajectory (calculations for NaAr at a detuning of 120 cm^{-1})

3 Theory

For a theoretical treatment, the molecular potential curves are required. For atom-atom optical collisions, we perform a full quantum mechanical calculation on this basis [4]; nonadiabatic couplings due to spin-orbit and rotational interactions are fully taken into account. We use only one significant approximation, the limit of low laser intensity, which is however adequate for the actual experimental conditions. The results are expected to be correct, provided the potentials and couplings are. The quantum calculations are followed by a convolution procedure which accounts for angular and velocity resolution effects. The quantum approach is used whenever a quantitative comparison between experiment and theory is desired; nonadiabatic effects are treated exclusively by this approach. Alternatively, the concept of Fig. 2 can be used to formulate a semiclassical theory [4]. The semiclassical expression for the cross section is

$$\sigma = |\Sigma f_i|^2$$
$$f_i = \text{const } \sqrt{b_i \cdot db_i/(\sin\theta\, d\theta)}\; v_{\text{rel}}^{-1}\; \exp(i\Phi_i)\; (\mathbf{e}\cdot\langle g|\mathbf{d}|e\rangle_i), \qquad (2)$$

where b is the impact parameter, θ the scattering angle, v_{rel} the radial component of the relative velocity, and Φ_i a classical phase expression. $\mathbf{e}$ is the polarization vector of the radiation and $\langle e|\mathbf{d}|g\rangle$ the electronic transition dipole moment, calculated at the Condon point. The summation goes over all trajectories. For the Na + rare gas systems at positive detuning, we deal with a Σ-Σ transition; under these conditions, the electronic transition dipoles are parallel to the corresponding Condon vectors. The semiclassical description is generally in good agreement with the full quantum results. This means in particular that the underlying notions, i.e. classical trajectories, Condon radius and Condon vectors, and interference provide a description which is close to correct. Even though the simple semiclassical theory considered here does not incorporate any nonadiabatic interactions, it is very useful for qualitative discussions.

4 The Beam Experiment

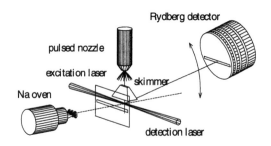

Figure 3: The experimental set-up used to study optical collisions with atomic beams and a differential detector

The experimental set-up [3, 5, 6] is shown in Fig. 3. We use two particle beams and two laser beams. The first ("excitation") laser provides the photons for the optical collision process. The differential detection of the short lived Na(3p) collision products is achieved by a transfer to a long lived Rydberg state

$$\text{Na}(3p)^2 P_{1/2,3/2} + h\nu' \rightarrow \text{Na}(nl), \quad n = 20\ldots 40,$$

a method first applied by Welge and coworkers [7]. The photons are provided by the second ("detection") laser. The lasers are pulsed, but the pulse duration is long compared to the collision time; the detection transition occurs long after the collision, at least on the statistical average. The Rydberg atoms travel to a distant rotatable detector, where they are registered via field ionization. Due to their large linear dimension, the Rydberg atoms undergo numerous collisions with the background gas on their way to the detector. However, these collisions occur with very large impact parameter. They therefore affect the motion and the quantum state of the Rydberg electron, but not the motion of the Na$^+$ ion which forms the core of the Rydberg atom. The velocity vector of a Rydberg atom remains virtually unchanged on its way to the detector. The velocity vector $\mathbf{v}'_{Na}$ determined from the time and direction of flight of the Rydberg atom is therefore practically identical to that of the Na(3p) collision products immediately after the collision [6].

We use a Na beam with a broad distribution of initial velocities $\mathbf{v}_{Na}$. For collisions with atomic targets, the knowledge of the Na velocity $\mathbf{v}'_{Na}$ after the collision is sufficient to determine all velocity vectors before and after the collision by simple kinematic calculations. For molecular targets (Sect. 9), the possible transfer of rotational or vibrational energy makes the calculation ambiguous. Relative velocity vectors were obtained in this case neglecting rotational and vibrational energy transfer; this is expected to cause no significant error under the present conditions. An additional velocity analysis, which will resolve the problem ultimately, is under construction.

Table 1 collects a number of characteristic data of the experiment. Note in particular the high density of the target beam, which is achieved by working with a distance of only 11 mm between the nozzle and the scattering volume, and the large detection efficiency for the excited atoms.

Beams	Na beam: thermal, $3 \cdot 10^{11}$ atoms/cm^3
	Target beam: supersonic, $v/\Delta v = 10$, 10^{15} atoms/cm^3
	Excitation laser: pulsed, 15 ns, 0.3 mJ
	Detection laser: pulsed, 15 ns, 0.1 mJ
Scattering volume	1 mm × 1 mm × 10 mm
Detector	at a distance of 70 mm, effective area 3 mm × 30 mm
	Estimated detection efficiency for Na(3p) atoms: 5%
Typical counting rate	1 event per laser pulse

Table 1: Characteristic data of the experimental set-up

5 Differential Cross Sections: Probing the Potentials

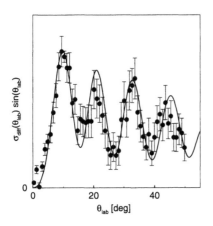

Figure 4: Typical differential cross section for the optical collision at positive detuning. Symbols: experiment, line: theory (Na+Ne, detuning 240 cm^{-1}, v'_{Na}=1125 m/s)

Figure 4 shows a typical result for a differential cross section. The oscillations are caused by interference between different classical paths. The interference pattern and its variation with the external parameters form a sensitive probe of the potential curve system. We were able to propose an improved B$^2\Sigma$ curve for NaKr on this basis [8]. Theoretical potentials for NaNe and NaAr were reported by Kerner and Meyer [9]. Figure 4 contains a theoretical curve calculated with these potentials. The good agreement was achieved without any adjustable parameters, except for the intensity scaling of the experimental results, and indicates a high accuracy of the theoretical results. We performed systematic measurements of differential cross sections and a systematic comparison between experiment and theory. We estimate on this basis that the theoretical NaNe potentials are correct within ± 10 cm^{-1} (1.2 meV).

6 Final State Analysis: Probing Nonadiabatic Processes

By tuning the detection laser correspondingly, collision products in the Na(3p) substates $^2P_{1/2}$ and $^2P_{3/2}$ can be detected separately. Figure 5 shows experimental results for the $^2P_{1/2}$ fraction of the Na(3p) population. The data provide direct insight into the nonadiabatic processes which occur during the collision [10]. The data refer to a positive detuning, corresponding to the excitation of the molecular B$^2\Sigma$ state. Under conditions of strictly adiabatic behaviour (i.e. no transitions between different electronic states), the collision should populate exclusively the $^2P_{3/2}$ state, see the inset in Fig. 1. The considerable population of the $^2P_{1/2}$ component is caused by nonadiabatic transitions, which occur at atomic distances between 10 and 20 au. The experimental data represent directly the nonadiabatic transition probability. The theoretical curves in the figure were obtained using the Kerner-Meyer potentials as input for the quantum calculations. The good agreement indicates once more the accuracy of these potentials; changes by few cm^{-1} in the coupling region can result in significant changes in the calculated $^2P_{1/2}$ fraction. In contrast to the differential cross section, which probes the potentials typically below 10 au, the present data probe the system in the nonadiabatic region, that is beyond 10 au. By varying the polarization of the detection laser, it is possible to determine also the alignment and orientation of the products. This technique will give access to other nonadiabatic coupling mechanisms, e.g. the decoupling of the electronic angular momentum from the rotating molecular axis [11].

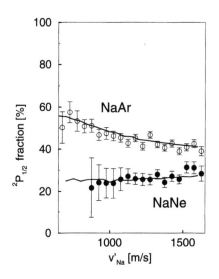

Figure 5: The $^2P_{1/2}$ fraction of the Na(3p)^{2}P population. Symbols: experimental data, lines: theory (Na+Ne,Ar, detuning 360 cm^{-1}, $\theta_{lab} = 10.8°$)

7 Polarization Experiments: Collision Photography

The graphs in the left hand column in Fig. 6 show the variation of the signal intensity with the polarization of the excitation laser. The data form geometric images of the collision pair [12]. As seen from (2), the polarization enters into the intensity by the scalar product with the electronic transition dipole moments. Therefore, polarization experiments probe the directions of the transition dipoles at the Condon points. In the case of Fig. 6 we work at a positive detuning and therefore deal with a $^2\Sigma$-$^2\Sigma$ transition. The transition dipole for a given Condon point is then parallel to the Condon vector. This means for instance that one obtains the largest signal from a single trajectory when the polarization is parallel to the Condon vector. For the actual situation with signal contributions from two or more trajectories, it is often sufficient to consider an incoherent superposition. It is easily

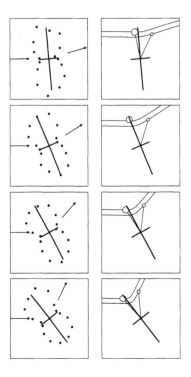

Figure 6: NaKr polarization results, 120 cm^{-1}. *Left*: The intensity variation of the optical collision signal as a function of the linear polarization direction in the form of polar diagrams; the dots are the experimental data, the bars visualize the largest and smallest intensities. The *boxes* represent different scattering angles, the arrows show the relative velocity directions before and after the collision. *Right*: The bars show the same experimental results as before, the other symbols are theoretical data. The curves are the classical trajectories for the optical collision. The circles indicate the two Condon vectors, their diameters represent the weight of the corresponding signal contribution

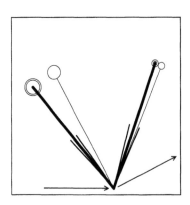

Figure 7: The separate Condon vectors. Massive lines with circles: experimental results for the Condon vectors and their weight factors; the experimental error margins are indicated as well. Thin lines with circles: the corresponding theoretical data (NaKr, detuning 80 cm^{-1}, relative velocity 1125 m/s, center-of-mass scattering angle 27°)

shown that the direction of the largest intensity then represents a weighted average over the directions of the Condon vectors. This is confirmed by the comparison between experimental and theoretical data in the right hand column of Fig. 6. Note in particular, how the calculated Condon vectors turn more and more to the left as the scattering angle increases; this is completely reproduced by the experimental data. Polarization data obviously provide a straightforward possibility to visualize the geometric properties of the collision.

It is possible to see even more details, for instance as follows. Every single trajectory can be switched off by choosing the polarization at right angles to the Condon vector. This results in a vanishing contrast of the angular interference pattern and permits the separate determination of both Condon vector directions. The idea has been realized, a result is shown in Fig. 7. Figure 6 and even more Fig. 7 contain the same sort of information as an everyday photography; we therefore speak of a "collision photography".

8 Polarization Experiments: Control of Collisions

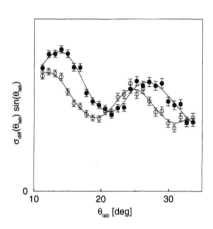

Figure 8: Angular distributions with left and right circular polarization of the excitation laser. Experimental points with polynomial fits (NaKr, detuning 288 cm^{-1}, v'_{Na}=1140 m/s)

Figure 8 shows that the interference pattern is different with left and right circular polarization of the excitation laser. This is a striking example for the control of the collision by coherent (here polarization) properties of the excitation laser. It can be shown from (2) that both the contrast of the oscillation pattern and the angular position of the maxima can be given arbitrary values using elliptic polarization.

The ability to switch every single trajectory off at will, already mentioned above, forms another example for the control of a collision by polarization.

9 Experiments with Molecular Targets

Collisions involving molecules are a very active field of research. Much more than for atom-atom collisions, the potentials of these systems are often not known with sufficient accuracy, the accurate treatment even of the adiabatic dynamics is far from trivial, and nonadiabatic behaviour might turn out to be the rule rather than an exception [13, 14]. The new techniques can provide new insight into all these problems. As a first application, Fig. 9 shows polarization data as Fig. 6, but now for molecular targets. Clearly, because the optical transition occurs during the collision, the data show geometric properties of the collision complex as before. The details of the interpretation are different, however. Strictly speaking, the data represent the alignment tensor of the electronic transition dipole moments [12]. In practice, the optical transition is essentially from a Na(3s) to a Na(3p) orbital; the large axis represents then the preferred (average) direction of the 3p orbital in the moment of the transition. Fig. 9 shows striking similarities and dissimilarities between the different collision systems. Obviously, the data contain nontrivial geometric information about the collision complex. For three of the four molecular targets, there exist clear directions of preference, which are similar to the NaKr case discussed above. This seems to indicate that the electronic structure of the collision pairs NaN_2, NaCO, and even NaC_3H_8 is similar to that of the NaKr collision pair. We conclude that the collisional molecules possess repulsive states which ressemble a Σ state, that is, the Na(3p) orbital points towards the target molecule. In contrast, the dissimilarity of the CO_2 result with the other data seems to indicate a basically different electronic structure of the $NaCO_2$ molecule.

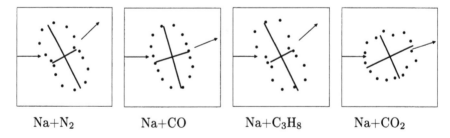

$Na+N_2$ $Na+CO$ $Na+C_3H_8$ $Na+CO_2$

Figure 9: Polarization dependence of the signal for different molecular targets. The dots show the signal intensity as a function of the excitation laser polarization in polar diagrams, the bars represent the conditions of smallest and largest intensity, the arrows show the relative velocities before and after the collision.

10 Conclusion

Optical collision experiments with beams and angle resolved detection provide both new qualitative and quantitative methods for the study of atom-atom and atom-molecule collisions:

- By an analysis of interference structures, the new method allows the test or the determination of molecular potentials. This applies to situations which are usually not covered by spectroscopy, e.g. repulsive potentials and large atomic distances. The accuracy competes with that of spectroscopic potential determinations.

- The new technique permits the selective population of a single molecular state of the collision complex. Together with a final state analysis, this allows a direct and detailed study of nonadiabatic processes.

- Collision photography: Polarization experiments provide geometric images of the collision complex. Images showing few details (Figs. 6 and 9) have been obtained for atomic as well as molecular targets. For atomic targets, the analysis of interference structures makes more details of the collision geometry accessible.

- Different schemes for the control of collisions by the polarization of the incident radiation have been demonstrated.

Besides applications to other diatomic systems, future developments and applications are expected to follow two main lines. First, the new techniques will more and more be applied to atom-molecule collisions. This will yield information about the electronic structure of the collision complex, see Sect. 9. Detailed studies of typical molecular nonadiabatic mechanisms will be performed. It will be possible to observe the different collision geometries of rotationally elastic and inelastic collisions, and, eventually, the transition state geometry of chemically reactive collisions. Second, the present techniques can be combined with femtosecond pump-probe methods. This will provide tools for the time resolved geometric observation of a collision, or in other words, for the production of a motion picture of the collision event [15].

References

[1] P.D.Kleiber, W.C.Stwalley, and K.M.Sando, Annu. Rev. Phys. Chem. **44**, 13 (1993)

[2] W.Behmenburg, A.Makonnen, A.Kaiser, F.Rebentrost, V.Staemmler, M.Jungen, G.Peach, A.Devdariani, S.Tserkovny, A.Zagrebin, and E.Czuchaj, J. Phys. B **29**, 3891 (1996)

[3] J.Grosser, D.Gundelfinger, A.Maetzing, and W.Behmenburg, J. Phys. B **27**, L367 (1994)

[4] F.Rebentrost, S.Klose, and J.Grosser, Eur. Phys. J. D **1**, 277 (1998)

[5] J.Grosser, D.Hohmeier, and S.Klose, J. Phys. B **29**, 299 (1996)

[6] S.Klose, Ph. D. thesis, Hannover, 1996

[7] L.Schnieder, W.Meier, K.H.Welge, M.N.R.Ashfold, and C.M.Western, J. Chem. Phys. **92**, 7027 (1990)

[8] J.Grosser, O.Hoffmann, S.Klose, and F.Rebentrost, Europhys. Lett. **39**, 147 (1997)

[9] C.Kerner, Ph. D. thesis, Kaiserslautern, 1995; C.Kerner and W.Meyer (to be published).

[10] J.Grosser, O.Hoffmann, F.Schulze Wischeler, and F.Rebentrost, J. Chem. Phys. **111**, 2853 (1999)

[11] J.Grosser, Z. Phys. D **3**, 39 (1986)

[12] J.Grosser, O.Hoffmann, C.Rakete, and F.Rebentrost, J. Phys. Chem. A **101**, 7627 (1997)

[13] D.R.Yarkoni, Rev. Mod. Phys. **68**, 985 (1996)

[14] L.J.Butler, Annu. Rev. Phys. Chem. **49**, 125 (1998)

[15] J.Grosser, Comments At. Mol. Phys. **21**, 107 (1988)

Counterpropagating Pulsed Molecular Beam Scattering

Henning Meyer
The University of Georgia
Athens, GA 30605-2451, USA

1 Introduction

Over the past years great progress has been made in the characterization of molecular processes at the microscopic level.[1] In particular, the dynamics of a variety of photodissociation processes has been studied at the quantum state resolved level.[2] These advances have been made possible through the application of ever more sophisticated laser spectroscopic techniques; often combined with molecular beam techniques.[3] In the study of bimolecular collision processes, progress has been much more difficult to achieve. The main reason is the difficulty in combining laser spectroscopic techniques with the typical crossed beam set-up[4] while at the same time achieving single collision condition and sufficiently high signal levels. These problems are especially severe if angular resolved product distributions are to be detected employing pulsed lasers with inherently low duty cycle. Final state resolved integral cross sections have been determined using infrared absorption[5], laser induced fluorescence[6, 7] or resonance enhanced multiphoton ionization (REMPI).[8, 9] Combining a conventional crossed molecular beam set-up with laser induced fluorescence detection, Gentry et al. were able to measure state resolved differential cross sections for the scattering of NO from Ar.[10] A very elegant method based on Rydberg atom tagging has been developed by Schnieder et al. [11] Its application to the reaction H+H_2 and its isotopic variants has demonstrated excellent sensitivity and angular resolution. Alternatively, the technique of two-dimensional ion imaging[12] has been applied recently to crossed molecular beam scattering.[13, 14]

In this contribution we review the method of counterpropagating pulsed molecular beam scattering. This new approach to the experimental study of bimolecular collision processes enables us to determine state-resolved integral and differential scattering cross-sections. Depending on the applied spectroscopic detection scheme, even information on the collision induced angular momentum alignment becomes available. The potential of this method is illustrated with recent results on the scattering of NO from Ar, with special emphasis on the detection of collision induced angular momentum alignment.

2 Counterpropagating pulsed molecular beam scattering

Instead of attempting to incorporate the spectroscopic detection into an existing crossed molecular beam set-up, we chose to design the molecular beam scattering arrangement around the laser detection. In our experiments, we achieve state specific detection through REMPI. Therefore, the state probed is selected by the frequency of the detection laser while the subsequent ion detection step is used to access information about the direction of the final velocity vector enabling the detection of state and angle resolved product distributions. In the case of the scattering of randomly oriented molecules, the collision system exhibits cylindrical symmetry around the direction of the initial relative velocity vector. The direction of the final velocity vector is uniquely specified by the scattering angle θ with respect to the direction of the initial relative velocity vector. Since, in an ideal experiment, the azimuthal angle ϕ is redundant, the scattered flux should be integrated over ϕ. In this way, a high signal level can be achieved without giving up information on the scattering dynamics.

It is for this reason that we chose the counterpropagating pulsed molecular beam geometry. The initial relative velocity vector now coincides with the direction of the molecular beams. It is interesting to note that the cylindrical symmetry of the distribution of scattered molecules is preserved even if the molecules are initially aligned or oriented with respect to the beam directions. The integration over the azimuthal direction can be accomplished in different ways. We chose the ion time-of-flight (TOF) method in combination with a drift field arrangement along the molecular beam direction. With this method, the component of the final velocity vector onto the direction of the initial relative velocity is measured. In contrast to other methods, the scattering angle is thus determined through a measurement in velocity space rather than in real space. This feature has important implications for the simultaneous optimization of signal levels and angular resolution.

In conventional crossed molecular beam scattering experiments, the scattering angle is determined by moving a detector around the volume defined as the intersection of the two molecular beams. In order to achieve good angular resolution, the size of the scattering volume must be minimized resulting in small signal levels. Similar problems are faced when the method of two-dimensional ion imaging is used in combination with a crossed molecular beam set-up. In this method, the velocity component perpendicular to the field direction is determined by measuring the spatial distance of the impact position of an ion from the center of the detector.

On the other hand, ion TOF analysis requires good time resolution. The combination of short laser pulses for the product ionization and the use of fast ion detectors allow the accurate measurement of the TOF. Consequently, this method takes advantage of pulsed laser systems despite their inherent poor duty cycle. Furthermore, there is no real limitation on the size of the scattering volume as long as the initial velocity vectors are well defined and a uniform ion detection is guaranteed. The ultimate limitation is determined by the requirements of achieving single collision condition. Since we use relatively short molecular beam pulses, single collision condition can always be achieved by limiting the overlap of the two pulses. Also, the size of the volume probed by the laser is not very critical. Its extension in the direction of the acceleration field is accounted for by using an ion TOF set-up under space focusing conditions.

In conclusion, using a counterpropagating beam arrangement, we are able to realize the following major advantages over conventional crossed molecular beam set-ups:

- State-specific product detection.

- Detection within the scattering volume.

- Integration over the (redundant) azimuthal angle.

- Parallel detection of angle resolved product distributions.

The resulting high signal levels allow us to employ higher order multiphoton processes for the state specific product detection. In this way, the number of systems which can be studied is increased substantially. Furthermore, the molecular beam axis coincides with the initial relative velocity vector and therefore with the quantization axis in the collision frame. Consequently, the counterpropagating molecular beam geometry is ideally suited for the study of collision induced alignment through the polarization dependence of the scattered intensity.

3 Experimental Set-up

Details of the experimental set-up have been described previously.[15, 16] Briefly, the scattering apparatus consists of two differentially pumped vacuum chambers. The two molecular beam pulses are generated in the source chamber evacuated by an $11000 l/s$ diffusion pump. Nested into the source chamber is the scattering chamber which houses the ion TOF and detection system. This chamber is pumped by a $3000 l/s$ diffusion pump achieving operating pressures better than 10^{-6} mbar. Molecular beam pulses enter the scattering chamber through skimmers in the walls facing the molecular beam sources. The primary beam is generated with a home built piezo electric valve while the target beam originates from a commercial beam source. The molecular beam pulses show a gaussian time profile with a FWHM of about 40 μs to 65 μs depending on the type of seed gas (usually Ne or Ar) used. Excellent translational and rotational cooling has been observed for both types of sources. It is important to realize that these pulses have a typical length of about 4-7cm. Therefore, we have complete control over the overlap of the two pulses without interference from wall collisions.

In a typical scattering experiment, the pulses are overlapped in the region of a drift field arrangement. It consists of an acceleration field, a field free region, an electrostatic mirror and a microchannel-plate detector (MCP). The electrostatic mirror deflects the ions towards the MCP detector which is mounted off-axis. In order to allow for the passage of the target beam pulse, the mirror has a hole of 10mm diameter. Using wire mesh of high transmission (>90%) to cover the various holes and openings, all field and field free regions are well defined.

The two counterpropagating and partially overlapping beam pulses are intersected by the laser under right angle. Depending on the type of multiphoton process employed, the probe laser radiation is focused onto the molecular beam using lenses with varying focal lengths (typical 300mm to 500mm). The direction of linear laser polarization is controlled using a set of double fresnel rhombs mounted on a motorized stage. The laser radiation is generated with two Nd:YAG laser pumped dye laser systems. The visible output of the dye lasers is frequency doubled providing us with ultraviolet light tunable from 210nm to 370nm. All components of the experiment are controlled by a PC; e.g. timing of the various pulses, frequency and direction of polarization of the UV light, and data acquisition.

To demonstrate the principle of the scattering experiment, we display in Fig. 1 raw TOF spectra recorded for the scattering of NH_3 from the rare gases He, Ne, and Ar at the indicated collision energies. NH_3 is detected through a (2+1) REMPI process with its B state acting as the resonance enhancing state. At the indicated laser frequency the rotational state $|JK\epsilon\rangle = |43-\rangle$ is probed. The symmetric top quantum numbers, J and K, refer to the total angular momentum and its projection onto the molecule fixed symmetry axis. ϵ characterizes the symmetry of the wavefunction. Due to incomplete rotational cooling, a small but noticeable thermal rest population is detected for this particular quantum state. In the absence of the target beam, this population is responsible for the single background peak observed at 10.62μs. This peak provides a convenient reference for the arrival time of undeflected NH_3 molecules. In order to determine a TOF spectrum representing the pure scattering contribution, TOF spectra are measured with

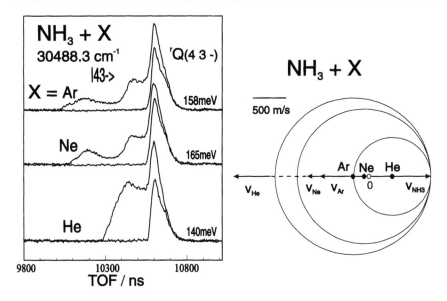

Figure 1: TOF spectra and Newton diagrams for the scattering of NH_3 from He, Ne, and Ar. See text for details.

and without target beam. When calculating the difference spectrum, the small depletion of the rest population due to the scattering (usually less than 10%) is taken into account. In this particular experiment, the drift field is oriented in such way that NH_3^+-ions representing backward scattering events arrive at the detector at earlier times. For all TOF spectra, the same electric field parameters are used enabling the direct comparison of the spectra.

In the case of He scattering, a very narrow distribution is observed while the width of the distribution for scattering from Ne is almost doubled. Compared to the latter, the TOF spectrum for NH_3+Ar is only slightly increased in width. All three spectra show peaks associated with backward scattering. For scattering from Ne and Ar, an additional peak in the forward direction can be distinguished.

The general structure of these spectra reflects the different collision kinematics. In the right part of Fig.1, the velocity or Newton diagram for the various systems is displayed. The center-of-mass (c.m.) velocity is indicated by black dots labeled with the target atom. For the light NH_3 − He system the c.m. is shifted towards the tip of the laboratory NH_3 velocity vector resulting in a small radius of the Newton sphere. In contrast, the radius of the sphere for scattering from Ne is nearly doubled while for Ar scattering an additional small increase is observed in perfect agreement with the observed TOF spectra.

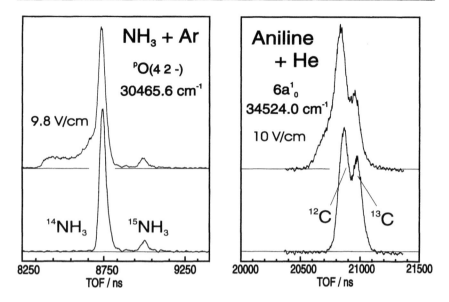

Figure 2: TOF spectra for the scattering of $NH_3 + Ar$ and *aniline* + He (top spectra). The spectra in the bottom part represent background TOF spectra with only the primary beam present. At the indicated laser frequencies, REMPI signals due to the indicated isotopes are detected.

In comparison to standard TOF mass spectrometer of the Wiley-McLaren type [17], we use only weak acceleration fields with a strength of less than $10 V/cm$. Nevertheless a mass resolution is achieved which is sufficient to distinguish different isotopes of NH_3 or aniline as can be seen in Fig. 2.

4 Resonance Enhanced Multiphoton Ionization: (n+m) REMPI

An important aspect of this experiment is the extraction of state populations from measured intensities. We employ different REMPI detection schemes distinguished as (n+m)-REMPI. After the absorption of n photons, the molecule reaches an excited state while the subsequent ionization process requires the absorption of at least m photons. In the following, we summarize the rotational state dependence for different multiphoton processes: one-photon, non-resonant two- and three-photon absorption. Note that here the term non-resonant absorption describes a situation in which no intermediate state can be reached with a single photon. In the theoretical description of the absorption process, it is important to allow for a product ensemble whose rotational angular momentum distribution is anisotropic.

In particular, we assume an aligned ensemble which results from the scattering of initially unpolarized molecules. For simplicity, we assume linearly polarized laser light of a single color with the laser beam propagating in a direction perpendicular to the molecular beam axis. Furthermore, the intensity of the employed laser field is restricted to sufficiently small values which allows a semiclassical treatment using nth order time dependent perturbation theory. In the following, we discuss the detection of an aligned ensemble of symmetric top molecules whose energy levels are characterized by the quantum numbers of the total angular momentum and its projection onto the molecular symmetry axis, J and K, respectively. Note that the quantum number K can take on positive and negative values. The extensions to more complex molecules, e.g. open shell species, asymmetric top molecules or molecules with internal rotation, are straight forward but rather tedious.[18, 19, 20]

In the following, we assume that molecules, initially prepared in a rotational state $|J_i K_i M_i\rangle$, are collisionally excited to some rotational state $|J_f K_f M_f\rangle$. The population of different magnetic sublevels belonging to the energy level $E(J_f, K_f)$ is described by the diagonal elements $\rho_{M_f M_f}^{J_f K_f}$ of the corresponding density matrix. The individual matrix elements are expanded in terms of state multipole moments $T_0^{(Q)}(J_f K_f)$ as defined in Ref. [21] :

$$\rho_{M_f,M_f}^{J_f K_f} = \sum_Q (-)^{J_f - M_f} \sqrt{2Q+1} \begin{pmatrix} J_f & J_f & Q \\ M_f & -M_f & 0 \end{pmatrix} T_0^{(Q)}(J_f, K_f) \quad (1)$$

If the ensemble is aligned, only even rank multipole moments are allowed. In the spectroscopic detection scheme, these final states are probed by exciting the molecule to an intermediate state whose rotational part is described by $|J_e K_e M_e\rangle$. The intensity for a non-resonant multiphoton process depends on the direction of the laser polarization as well as the angular momenta involved. In the case of one-photon absorption, we find the following expression for the intensity:

$$I = (2J_f + 1)(2J_e + 1) \sum_{Q=0,2} (-)^{J_f + J_e} \sqrt{2Q+1}\ T_0^{(Q)}(J_f, K_f)$$
$$\times P_Q(\cos\beta) \begin{pmatrix} 1 & 1 & Q \\ 0 & 0 & 0 \end{pmatrix} \begin{Bmatrix} 1 & J_f & J_e \\ J_f & 1 & Q \end{Bmatrix} \quad (2)$$
$$\sum_q \begin{pmatrix} J_e & 1 & J_f \\ K_e & -q & -K_f \end{pmatrix}^2 |\langle n_f v_f | R_q^{(1)} | n_i v_i \rangle|^2$$

The terms in brackets, (...) and {...}, are regular Wigner 3j-symbols and 6j-symbols.[22] The properties of the first 3j-symbol restrict contributions to the signal from only the first even multipole moments with Q=0,2. The contributions due to these moments are modulated with a Legendre polynomial $P_Q(\cos\beta)$ of the same degree. Here, β specifies the angle of the linear laser polarization with the axis of cylindrical symmetry, i.e. the molecular beam axis. Finally, the intensity is determined by the rovibronic transition probability as given by the factor in the last line of Eq. 2. The vibronic transition matrix elements define the spherical tensor components of the transition moment vector. Depending on the particular system, parallel ($q = 0$) or perpendicular transitions ($q = \pm 1$) are possible.

In the case of non-resonant two-photon absorption, the vibronic transition matrix elements involve either a zeroth rank tensor or the components of a second rank tensor.[23] The former is responsible for rotational Q-branches ($\Delta J = 0$) while the latter gives rise to rotational transitions with $\Delta J = 0, \pm 1$, and ± 2. If the transition is mediated by a second rank tensor component, the intensity will depend also on the alignment moments $T_0^{(2)}$ and $T_0^{(4)}$.

$$I = (2J_f + 1)(2J_e + 1) \sum_{Q=0,2,4} (-)^{J_f+J_e} \sqrt{2Q+1}\; T_0^{(Q)}(J_f, K_f)$$

$$\times \sum_{j,j'=0,2} \sqrt{(2j+1)(2j'+1)} \begin{pmatrix} 1 & 1 & j \\ 0 & 0 & 0 \end{pmatrix} \begin{pmatrix} 1 & 1 & j' \\ 0 & 0 & 0 \end{pmatrix}$$

$$\times P_Q(\cos\beta) \begin{pmatrix} j & j' & Q \\ 0 & 0 & 0 \end{pmatrix} \begin{Bmatrix} j & J_f & J_e \\ J_f & j' & Q \end{Bmatrix} \quad (3)$$

$$\times \sum_{q,q'} \begin{pmatrix} J_e & j & J_f \\ K_e & -q & -K_f \end{pmatrix} \begin{pmatrix} J_e & j' & J_f \\ K_e & -q' & -K_f \end{pmatrix}$$

$$\langle n_f v_f | R_q^{(j)} | n_i v_i \rangle \langle n_f v_f | R_{q'}^{(j')} | n_i v_i \rangle^*$$

A special situation arises when the two-photon transition is carried by both types of tensors. In the case of an isotropic ensemble, the spectrum will be made up of two independent contributions. On the other hand, if an anisotropic ensemble is excited, additional terms due to higher order multipole moments contribute in the form of cross terms between the different types of tensor components. As a consequence, the spectrum for an isotropic ensemble can be dominated completely by the contribution of the zeroth rank tensor while a strong polarization dependence can be observed for the anisotropic case.[24]

The expression for three-photon absorption is similar to the one given in Eq. 2.[25] Now, the transition can be carried by the components of a first rank or a third rank tensor. In the latter case, even state multipole moments with $Q \leq 6$ can contribute to the signal. If the transition is carried by the first rank tensor only, the three-photon absorption spectrum becomes identical to a one-photon spectrum.

$$I = (2J_f + 1)(2J_e + 1) \sum_{Q=0,2,4,6} \sqrt{2Q+1} T_0^{(Q)}(J_f, K_f) P_Q(\cos\beta)$$

$$\times \sum_{j,j'=1,3} \sqrt{(2j+1)(2j'+1)} \begin{pmatrix} j & j' & Q \\ 0 & 0 & 0 \end{pmatrix} \begin{Bmatrix} j & J_f & J_e \\ J_f & j' & Q \end{Bmatrix}$$

$$\times (-)^{J_f + J_e} \sum_{j_3=0,2} \sqrt{2j_3+1} \begin{pmatrix} 1 & 1 & j_3 \\ 0 & 0 & 0 \end{pmatrix} \begin{pmatrix} j_3 & 1 & j \\ 0 & 0 & 0 \end{pmatrix} \quad (4)$$

$$\times \sum_{j_3'=0,2} \sqrt{2j_3'+1} \begin{pmatrix} 1 & 1 & j_3' \\ 0 & 0 & 0 \end{pmatrix} \begin{pmatrix} j_3' & 1 & j' \\ 0 & 0 & 0 \end{pmatrix}$$

$$\times \sum_{q,q'} \begin{pmatrix} J_e & j & J_f \\ K_e & -q & -K_f \end{pmatrix} \begin{pmatrix} J_e & j' & J_f \\ K_e & -q' & -K_f \end{pmatrix}$$

$$\langle n_f v_f |^{(j_3)} R_q^{(j)} | n_i v_i \rangle \langle n_f v_f |^{(j_3')} R_{q'}^{(j')} | n_i v_i \rangle^*$$

For a n-photon process, we can summarize the polarization dependence of the detected intensity in the following compact form:

$$I(\beta) = C \sum_{q=0}^{n} B^{2q}(J_f, J_e) T_0^{(2q)}(J_f, K_f) P_{2q}(\cos\beta) \quad (5)$$

The coefficients $B^{2q}(J_f, J_e)$ depend on the involved angular momenta as well as the type of multiphoton process employed (i.e. the particular tensor component carrying the transition).

5 Data Analysis

Another important aspect of any scattering experiment is the extraction of c.m. cross sections from the experimentally determined data. This data analysis has to take into account the scattering kinematics as well as the finite resolution of the apparatus. Ideally a direct inversion of the experimental data is desirable. Unfortunately, the uncertainty in the velocity and angular distributions of the molecular beam pulses as well as their time profiles must complicate any successful inversion attempt. Therefore, we chose a forward convolution procedure based on an apparatus function. This function must be calculated from the distribution functions characterizing the molecular beam pulses. It takes into account the scattering kinematics, the finite laser detection volume, and the flux-to-density transformation.

For the calculation of the apparatus function, it is important to make a clear distinction between the scattering volume and the detection volume. In this experiment, the actual scattering angle is determined by the projection of the final velocity vector onto the axis of cylindrical symmetry. Because the velocity component is measured through the TOF of the ion towards the detector, the size of the scattering volume can be maximized without compromising the angular resolution. The major contribution of the detected intensity is due to scattered flux generated outside of the detection volume. It depends critically on the relative time delays between the molecular beam pulses and the laser pulse.

The contribution of a small volume element ΔV_s to the flux Δj_{scatt} scattered towards a detector volume element ΔV_d is given by the following expression:

$$\Delta j_{scatt} = \frac{1}{r_d^2}\frac{\Delta N_{scatt}}{\Delta\Omega\Delta t} = \frac{1}{r_d^2}n_p n_t g \frac{d\sigma}{d\omega}\frac{d\omega}{d\Omega}\Delta V_s \tag{6}$$

Here, we assume that the distance r_d between the two volume elements is larger than the size of the detector element. n_p and n_t describe the densities of the primary beam and the target beam in the volume element ΔV_s, respectively. Note that they depend on the location of the volume element as well as the relative timing of the pulses. The differential c.m. cross section is denoted $\frac{d\sigma}{d\omega}$ while the Jacobi factor describing the c.m. to lab. transformation is given by $\frac{d\omega}{d\Omega}$. The factor g, denoting the relative velocity of the particles, completes the flux to density transformation. Combining the two equations, we find for the number of particles scattered into the detector volume element:

$$\Delta N_{scatt} = n_p n_t g \frac{d\sigma}{d\omega}\frac{d\omega}{d\Omega}\Delta V_s \Delta t \Delta\Omega \tag{7}$$

The time interval Δt is determined by the volume illuminated by the detection laser. Assuming a spherical probe laser volume element with radius r_p, we find $\Delta t = \frac{2r_p}{v_f}$ where v_f is the final velocity of the product molecule. For the case of a spherical volume element, we find the solid angle element $d\Omega = \pi(\frac{r_p}{r_d})^2$.

The total number of particles deflected by the c.m. scattering angle θ is found by integrating Eq. 7 over all space. Naturally, only those positions contribute to the integral where both beam densities do not vanish at the appropriate times. To see how the time profile is incorporated into the calculation, we assume that the laser is fired at time $T = T_0$. In order for a product molecule to reach a particular detection volume element, it must have been deflected by an angle which is uniquely defined by the relative position of the two volume elements and the initial velocity vectors of both particles. Assuming a pair of velocity vectors, we can calculate the collision kinematics for a specific inelastic process. In particular, we can calculate the final velocity of the product molecule for the laboratory deflection angle. Using the distance between the two volume elements, we then calculate the time ΔT it takes the molecule to reach the detection volume element. Consequently, the scattering event must have taken place at time $T = T_0 - \Delta T$. The contribution of this particular element of the scattering volume to the overall signal is weighted with the beam densities at this location at time $T = T_0 - \Delta T$.

If the volume element ΔV_s partially overlaps the detection volume element ΔV_d, it is necessary to perform an additional integration over the finite detector solid angle. Using Eq. 6, we define the apparatus function $G(v_{fz}, \theta)$.

$$G(v_{fz}, \cos\theta) = \int d^3r d^3v_p d^3v_t d\Omega n_p(\vec{r}, \vec{v}_p, T) n_t(\vec{r}, \vec{v}_t, T) |\vec{v}_p - \vec{v}_t|$$
$$\times \frac{d\omega}{d\Omega} \delta(v_{fz} - \tilde{v}_{fz}) \delta(\cos\theta - \cos\tilde{\theta}) \qquad (8)$$

The quantities $\tilde{v}_{fz}$ and $\cos\tilde{\theta}$ are determined by the kinematics for a given set of variables $\vec{r}, \vec{v}_p, \vec{v}_t$, and Ω. The Dirac δ-functions restrict contributions to only those values of the variables which yield the desired values for v_{fz} and $\cos\theta$.

The scattered intensity as a function of the final velocity component can now be written as a convolution of the apparatus function with the differential cross section.

$$\delta N_{scatt}(v_{fz}) = \int d\cos\theta\, G(v_{fz}, \cos\theta) \frac{d\sigma}{d\omega}(\cos\theta) \qquad (9)$$

A typical apparatus function for the elastic scattering of NO from Ar is shown in Fig. 3. The strong correlation of the projection v_{fz} of the final velocity vector with $\cos\theta$ is clearly seen. We also notice the increased probability for detection of NO molecules scattered into the forward or backward directions. This property reflects the fact that scattering events occurring on the molecular beam axis far away from the detection volume can only contribute to the signal when they result in final velocity vectors parallel to the symmetry axis; i.e. forward or backward scattering events. Typically the main contributions originate from regions within $\pm 4mm$ of the detection volume.

If a narrow bandwidth laser source is used for the REMPI detection, it might not cover the complete Doppler profile of the transition. In this case also the dependence on the off-axis velocity component must be taken into account.

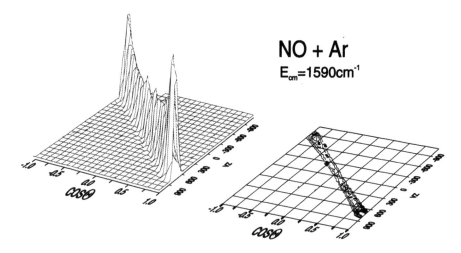

Figure 3: Apparatus function $G(v_{fz}, \cos\Theta)$ for the elastic scattering of NO from Ar as a function of the velocity projection v_{fz} and $\cos\Theta$.

Similarly, the use of an electrostatic mirror in our experiment breaks the cylindrical symmetry, causing a small dependence on the azimuthal angle. Instead of extending the simulation of the apparatus function to cover also the dependence on the off-axis velocity component, we use the most probable Newton diagram to determine the absolute value of the final velocity vector for a given velocity component v_{fz}. Once v_f is known, we can calculate either the Doppler profile or include the ϕ-dependence of the ion TOF.

In order to determine the differential cross section, the integral in Eq. 9 is replaced by a sum over well defined intervals for which the differential cross section is assumed to be constant. Their values are determined in a least squares fit procedure. As can be seen in the right part of Fig. 3, we expect a constant resolution for the variable $\cos\theta$. In the present set-up, the resolution is limited by the aberrations introduced by the electrostatic mirror. This effect is most severe for $\cos\theta = 0$, but negligible for forward and backward scattering. Depending on the kinematics of the collision system and the employed electric fields, 6 to 12 intervals are sufficient to reproduce a measured TOF spectrum indicating a resolution $d\cos\theta = 0.15 - 0.30$.

6 Collision-induced Alignment

In Section 4, we presented the dependence of the rotational linestrength on the laser polarization for an ensemble of cylindrical symmetry. On the other hand, the dependence on the scattering direction was not included explicitly. In order to describe the polarization dependence for the case of angle resolved product distributions, we relate the expression of the scattered intensity to the state multipole moments of the scattering amplitude. Throughout this section, we assume the direction of the incoming relative velocity as the quantization axis for the initial as well as the final rotational angular momentum. Because the symmetry axis of the counterpropagating beam arrangement coincides with this natural choice of the quantization axis, it is especially suitable to detect any dependence of the state resolved cross section on the magnetic quantum numbers.

In an ideal experiment, a single initial quantum level characterized by the rotational angular momentum quantum number J_i is prepared. Assuming single collision condition, the density of product molecules in a specific final state $|J_f M_f\rangle$ moving with a final velocity vector (direction $\hat{R}$) is proportional to the differential scattering cross section $\frac{d\sigma}{d\omega}(J_i M_i \to J_f M_f; \hat{R})$.

$$\rho_{M_f M_f}^{(J_f)}(\theta) = \frac{\rho_0^{(J_i)}}{2J_i + 1} \sum_{M_i} \frac{d\sigma}{d\omega}(J_i M_i \to J_f M_f; \hat{R}) \tag{10}$$

The M-dependent differential cross section is calculated from the scattering amplitude $f_{J_i M_i \to J_f M_f}(\hat{R})$:

$$\frac{d\sigma}{d\omega}(J_i M_i \to J_f M_f)(\hat{R}) = \frac{k_{J_f}}{k_{J_i}} |f_{J_i M_i \to J_f M_f}(\hat{R})|^2 \tag{11}$$

Here k_{J_i} and k_{J_f} refer to the wavevectors of the incoming and the outgoing waves, respectively. In the close coupling formulation, the scattering amplitude is expressed in terms of the elements of the T-matrix $T_{J_i L, J_f L'}^J$ [27, 28]:

$$f_{J_i M_i \to J_f M_f}(\hat{R}) = \frac{2\pi i}{(k_{J_i} k_{J_f})^{1/2}} \sum_{J,L,L'} (2J+1) i^{L-L'} \sqrt{\frac{2L+1}{4\pi}} T_{J_i L, J_f L'}^J$$

$$\times \begin{pmatrix} L' & J_f & J \\ M_i - M_f & M_f & -M_i \end{pmatrix} \begin{pmatrix} L & J_i & J \\ 0 & M_i & -M_i \end{pmatrix} Y_{L', M_i - M_f}(\hat{R}) \tag{12}$$

Following Alexander and Davis[26], we introduce the state multipole moments of the scattering amplitude:

$$f_{J_i J_f}^{Qq}(\hat{R}) = \sum_{M_i M_f} (-)^{J_f - M_f} \sqrt{2Q+1} \begin{pmatrix} J_f & J_i & Q \\ M_f & -M_i & -q \end{pmatrix}$$

$$\times f_{J_i M_i \to J_f M_f}(\hat{R}) \tag{13}$$

with the inverse relationship:

$$f_{J_iM_i \to J_fM_f}(\hat{R}) = \sum_{Q,q}(-)^{J_f-M_f}\sqrt{2Q+1}\begin{pmatrix} J_f & J_i & Q \\ M_f & -M_i & -q \end{pmatrix} f_{J_iJ_f}^{Qq}(\hat{R}) \qquad (14)$$

In a next step, we determine the state multipole moments of the ensemble of particles scattered into the direction $\hat{R}$.

$$T_0^{(\tilde{Q})}(J_f,\hat{R}) = \sum_{M_f}(-)^{J_f-M_f}\sqrt{2\tilde{Q}+1}\begin{pmatrix} J_f & J_f & \tilde{Q} \\ M_f & -M_f & 0 \end{pmatrix} \rho_{M_fM_f}^{J_f}(\hat{R}) \qquad (15)$$

Substituting Eqs. 10, 11, and 14 into Eq. 15, we arrive at the desired relationship between the state multipole moments of the population and of the scattering amplitude:

$$T_0^{(\tilde{Q})}(J_f,\hat{R}) = \frac{\rho_0^{(J_i)}}{2J_i+1}\frac{k_{J_f}}{k_{J_i}}\sum_{M_iM_fQ'q'Q''q''}(-)^{M_f-J_f}\sqrt{2Q'+1}\sqrt{2Q''+1}$$

$$\times \begin{pmatrix} J_f & J_f & \tilde{Q} \\ M_f & -M_f & 0 \end{pmatrix}\begin{pmatrix} J_f & J_i & Q' \\ M_f & -M_i & -q' \end{pmatrix}\begin{pmatrix} J_f & J_i & Q'' \\ M_f & -M_i & -q'' \end{pmatrix} \qquad (16)$$

$$\times \sqrt{2\tilde{Q}+1}f_{J_iJ_f}^{Q'q'}(\hat{R})\left[f_{J_iJ_f}^{Q''q''}(\hat{R})\right]^*$$

The sum over the quantum numbers M_i and M_f is performed analytically, and, after some algebra, we obtain the result:

$$T_0^{(\tilde{Q})}(J_f,\hat{R}) = \frac{\rho_0^{(J_i)}}{2J_i+1}\frac{k_{J_f}}{k_{J_i}}\sum_{Q'qQ''}\sqrt{2Q'+1}\sqrt{2Q''+1}(-)^{\tilde{Q}-q}$$

$$\times \begin{pmatrix} Q' & Q'' & \tilde{Q} \\ -q & q & 0 \end{pmatrix}\begin{Bmatrix} Q' & J_f & J_i \\ J_f & Q'' & \tilde{Q} \end{Bmatrix}\sqrt{2\tilde{Q}+1}f_{J_iJ_f}^{Q'q}(\hat{R})\left[f_{J_iJ_f}^{Q''q}(\hat{R})\right]^* \qquad (17)$$

Since the two scattering amplitude terms, which now depend on the same quantum number q, enter Eq. 17 as a product, the dependence on the azimuthal angle ϕ disappears. The results simplify even further if we assume $J_i = 0$ for the initial state. In this case, we find for the state multipole moments of the final state angular distribution:

$$T_0^{(\tilde{Q})}(J_f,\theta) = \frac{\rho_0^{(J_i)}}{2J_i+1}\frac{k_{J_f}}{k_{J_i}}\sum_q(-)^{J_f-q}\sqrt{2\tilde{Q}+1}$$

$$\times \begin{pmatrix} J_f & J_f & \tilde{Q} \\ q & -q & 0 \end{pmatrix}|f_{0J_f}^{J_fq}(\theta)|^2 \qquad (18)$$

A compact expression for the state multipole moments of the scattering amplitude was derived by Alexander and Davis[26]:

$$f^{Qq}_{J_i J_f}(\hat{R}) = 2\pi i \sqrt{\frac{2Q+1}{k_{J_i} k_{J_f}}} \sum_{L,L'} i^{L-L'} \sqrt{\frac{2L+1}{4\pi}} Y_{L'q}(\hat{R}) \begin{pmatrix} L & L' & Q \\ 0 & q & -q \end{pmatrix}$$

$$\times \sum_J (2J+1)(-)^{J+J_f} T^J_{J_i L, J_f L'} \begin{Bmatrix} J_i & J_f & Q \\ L' & L & J \end{Bmatrix} \quad (19)$$

For the case $J_i = 0$ and $Q = J_f$, this expression reduces to the one for the scattering amplitude given in Eq. 12. As expected, the state multipole moments of the ensemble of scattered particles become identical to the corresponding moments of the state-resolved differential cross sections:

$$T_0^{(\tilde{Q})}(J_f, \theta) = \frac{\rho_0^{(J_i)}}{2J_i+1} \frac{k_{J_f}}{k_{J_i}} \sum_{M_f} (-)^{J_f - M_f} \sqrt{2\tilde{Q}+1} \begin{pmatrix} J_f & J_f & \tilde{Q} \\ -M_f & M_f & 0 \end{pmatrix}$$

$$\times \frac{d\sigma}{d\omega}(00 \to J_f M_f; \theta) \quad (20)$$

Having determined the differential state multipole moments of the cross section, we can now combine the results of Sects. 4 and 5 to find an expression for the polarization dependence of the TOF spectra. The M_f dependent scattering signal corresponding to a specific velocity projection v_{fz} is given by:

$$\delta N^{scatt}_{J_f M_f}(v_{fz}) = \int d\cos\theta G_{J_f}(v_{fz}, \cos\theta) \frac{d\sigma}{d\omega}(00 \to J_f M_f, \cos\theta) \quad (21)$$

The apparatus function does not depend on the magnetic quantum numbers because of the cylindrical symmetry of the experimental set-up. Using Eqs. 20 and 21, we calculate the state multipole moments $N_0^{(Q)}$ of the intensity scattered into the directions θ corresponding to a final velocity component v_{fz}:

$$N_0^{(Q)}(J_f, v_{fz}) = \int d\cos\theta G_{J_f}(v_{fz}, \cos\theta) T_0^{(Q)}(J_f, \theta)) \quad (22)$$

Once the state multipole moments are known, we can derive an expression for the polarization dependence of the final velocity distribution using Eq. 5:

$$I(\beta, v_{fz}) = C \sum_{q=0}^{n=1,2,or3} B^{2q}(J_f, J_e) N_0^{(2q)}(J_f, v_{fz}) P_{2q}(\cos\beta)$$

$$= C \int d\cos\theta G_{J_f}(v_{fz},\cos\theta) \sum_{q=0}^{n} T_0^{(2q)}(J_f,\theta)) \quad (23)$$

$$\times P_{2q}(\cos\beta)B^{2q}(J_f,J_e)$$

Eq. 23 provides us with two possible schemes for the extraction of the multipole moments of the differential cross section from the recorded polarization dependence of the TOF spectra. First, the comparison with Eq. 6, suggests the definition of a polarization dependent cross section:

$$\frac{d\sigma}{d\omega}(\cos\Theta,\beta) = \sum_{q=0}^{n} T_0^{(2q)}(J_f,\theta))P_{2q}(\cos\beta)B^{2q}(J_f,J_e) \quad (24)$$

These polarization dependent cross sections are extracted from the original TOF spectra in a least squares fit procedure. In a second step, the multipole moments are calculated from these polarization dependent cross sections.[29, 30, 31] Alternatively, we can calculate directly the multipole moments of the TOF spectra and use Eq. 22 to determine the moments of the differential cross sections in a least squares fit.

For a one-photon resonance, it is possible to extract the monopole moment (the degeneracy averaged cross section) and the quadrupole moment by measuring TOF spectra for $\beta = 0°$ and $\beta = 90°$ and by taking the appropriate linear combinations of $I(\beta,v_{fz})$. For the case of two-photon resonances, the unambiguous extraction of the various multipole moments is complicated by the presence of a possible hexadecapole moment. Nevertheless, it is possible to extract the first two non-vanishing moments, by measuring TOF spectra at those angles β for which the fourth order Legendre polynomial vanishes. Unfortunately under these conditions, the polarization effect is substantially reduced. A special situation arises for rotational lines with $\Delta J = \pm 2$, i.e. lines of O-branches and S-branches. In this case, it can be shown that the contribution of the hexadecapole moment is very small. Consequently, TOF spectra, recorded for $\beta = 0°$ and $\beta = 90°$, can be used to determine the degeneracy averaged cross section and the quadrupole moment.

7 Application to NO+Ar

In this section, we illustrate the different aspects of the scattering experiment with recent results on the NO+Ar scattering. Due to the limited scope of this article, a complete presentation and analysis will be given in a separate publication.[32]

The primary beam is generated by expanding a mixture of 5% NO in Ne at a backing pressure of 1.8 bar. For the target beam, pure Argon is expanded at a pressure of 1.5 bar. The velocity distributions of the two beams

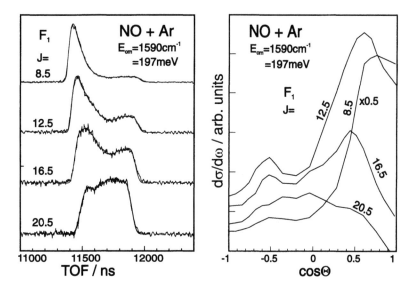

Figure 4: Degeneracy averaged TOF spectra and differential cross sections for the scattering of NO from Ar. Final states belonging to the F_1 manifold are probed for the indicated angular momenta J.

are characterized by most probable velocities of 780 m/s and 710 m/s, respectively. An average collision energy of $1590 cm^{-1} (197 meV)$ results. The velocity spread (FWHM) is determined to be 15% for both beams. At the achieved level of rotational cooling in the molecular beam expansion, we find more than 90% of the NO molecules in the lowest rotational level of the F_1-component, i.e. the lower spin-orbit component. For the subsequent analysis, we assume equal populations for the two parity states labeled e ($\epsilon = +1$) and f($\epsilon = -1$). Therefore, we can determine only final state resolved cross sections averaged over the two initial parity states. The timing of the beam pulses ensures that only the leading parts overlap. Under these experimental conditions, we detect a depletion of 5.5%.

TOF spectra are recorded on various rotational lines for different laser polarizations. Due to the small lambda doubling in NO, several rotational branches are completely overlapped. Since the polarization behavior depends strongly on the rotational branch, it is very difficult to extract reliable alignment data from TOF spectra recorded on these rotational lines. The problem can be avoided by detecting NO on the branches $S_{21}(\epsilon = -1), O_{11}(\epsilon = +1), S_{22}(\epsilon = -1)$, and $O_{12}(\epsilon = +1)$. In all other cases, the individual linestrengths have to be considered. For example, we also recorded TOF spectra of scattered NO molecules on lines of the R_{11}-branch which coincides with the Q_{21}-branch. Since the latter is character-

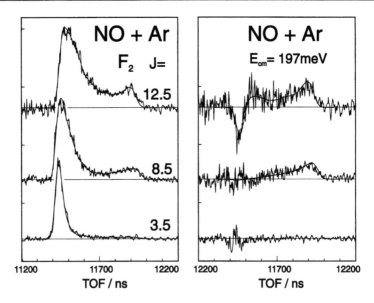

Figure 5: TOF spectra for fine structure changing collisions. Left part: Monopole moments (degeneracy averaged), right part: Quadrupole moments.

ized by extremely small linestrengths, the spectra can be analyzed assuming a polarization dependence characteristic to a pure R-line. TOF spectra for fine structure changing collisions are recorded on lines of the O_{12}-branch. In each case, TOF spectra are measured with and without target beam. Although the signal levels are considerably reduced in comparison with the ones found for the NO+He system[29], a satisfactory signal-to-noise ratio was achieved by accumulating less than 5000 laser shots.

Typical TOF spectra, recorded on lines of the R_{11}-branch, are displayed in Fig. 4. Lines of this branch probe NO levels of the lower fine structure component with a symmetry quantum number $\epsilon_f = -1$. In order to avoid problems due to a possible hexadecapole moment, TOF spectra have been recorded at $\beta = 30°$ and $\beta = 70°$, i.e. the approximate zeros of the fourth order Legendre polynomial. Since, in this case, the polarization effect due to the quadrupole moment is reduced substantially, we determine only the monopole moment, i.e. the degeneracy averaged cross section. In a first step, we calculate monopole moments by taking the appropriate linear combination of the measured TOF spectra. The resulting spectra are displayed in the left part of Fig. 4. Degeneracy averaged differential cross sections are determined in a least squares fit as described in Section 4 and are shown in the right part of Fig. 4. To ensure the correct relation between cross sections for the excitation of different final states, we normalize the TOF spectra and

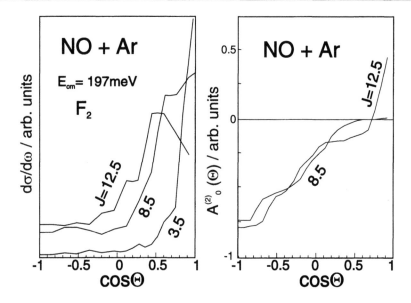

Figure 6: Differential cross section. Left part: Degeneracy averaged cross section $\frac{d\sigma}{d\omega}(J_f, \cos\Theta)$. Right part: Quadrupole moment $A_0^{(2)}(J_f, \cos\Theta) = C_{gz}T_0^{(2)}(J_f, \cos\Theta)$. See text for details.

scale the extracted differential cross sections with the appropriate overall population differences $\delta N_{J_f}^{scatt}$. The latter are determined from frequency dependent spectra recorded in a standard baseline subtraction mode.[29] To a first approximation, the $\delta N_{J_f}^{scatt}$ can be considered proportional to the corresponding integral cross sections.

For the fine structure conserving collisions, we find that, with increasing value for the final angular momentum, the scattered intensity shifts from forward scattering to large angle scattering. The c.m. cross sections show the typical rotational rainbow structure in the form of a maximum shifting towards larger scattering angles with increasing angular momentum transfer ΔJ. A second maximum is clearly visible for some of the higher J-states. As pointed out by Bontuyan et al.[13], the double peak structure in the cross section for fine structure conserving collisions reflects the heteronuclear nature of the NO molecule.

TOF spectra for levels of the excited fine structure component F_2 are recorded on various lines of the O_{12}-branch for $\beta = 0°$ and $\beta = 90°$. These spectra exhibit a strong polarization dependence which allows us to determine the monopole and quadrupole moments of the recorded spectra. The results for several rotational states are displayed in Fig. 5. While monopole spectra exhibit a behavior similar to the one found for fine structure conserving collisions, quadrupole spectra show a very different angular dependence. For small values of J_f, the quadrupole moment is almost undetectable. It increases dramatically for final states with large J_f. We also notice that it changes from negative values to positive values in going from flight times corresponding to forward scattering to ones corresponding to backward scattering. Both types of spectra are fitted using the procedure outlined above. The resulting c.m. cross sections (monopole and quadrupole moments) are displayed in Fig. 6. For the quadrupole moment, we use here the more familiar definition, $A_0^{(2)}$, of Greene and Zare.[22] The factors $C_{gz}(J_f)$ for the conversion between the quadrupole moments of Refs. [21] and [22] have been reported, for example, in Ref. [33].

Comparing the cross sections for fine structure conserving and changing collisions, we find that the integral cross sections are smaller for Ω-changing collisions. For the same value of J_f, we find very similar forms of the differential cross sections, although substantially larger energy transfers are involved in the case of fine structure changing collisions. These findings indicate that the magnitude of the integral cross sections is governed by the energy gap while the angular dependence of the differential cross sections is mainly controlled by the angular momentum gap. The results, obtained for collision induced alignment so far, can be summarized in the following way: The quadrupole moment of the differential cross section is clearly angle dependent. It also depends sensitively on the angular momentum of the final state. In the limit of large angular momentum transfer (with the differential cross section peaked in the backward direction), the quadrupole moment changes from negative values at large scattering angles to positive values for forward scattering. These experimental findings are consistent with the assumption of a sudden collision which causes the rotational angular momentum vector to be preferentially aligned perpendicular to the direction of the linear momentum transfer.[34]

8 Conclusion

Counterpropagating pulsed molecular beam scattering is a very promising approach to study experimentally molecular collision processes at the truely microscopic level. Naturally, the successful experiment requires the availability of an efficient REMPI detection scheme. In this case, state resolved integral and differential cross sections become experimentally accessible. The realized large signal levels allow the application of higher order multiphoton processes for the product detection, thus increasing the number of systems which can be studied. Finally, the signal levels allow us to detect, for the first time, collision induced alignment under single collision condition.

Acknowledgment

This work was supported by a grant from the National Science Foundation (CHE-9707670).

References

[1] Levine R.D., and Bernstein R.B., *Molecular Reaction Dynamics and Chemical Reactivity*, Oxford University Press, Oxford, 1987.

[2] Liu K., and Wagner A.(Edts.), *The Chemical Dynamics and Kinetics of Small Radicals*, World Scientific, Singapore, 1995.

[3] Dai H.L., and Field R.W. (Edts.), *Molecular Dynamics and Spectroscopy by Stimulated Emission Pumping*, World Scientific, Singapore, 1995.

[4] Buck U. Adv. Phys. Chem. **30**, 313(1975).

[5] Chapman W.B., Schiffman A., Hutson J.M., and Nesbitt D.J. J. Chem. Phys. **105**, 3497(1996).

[6] Butz K.W., Du H., Krajnovich D.J., Parmenter C.S. J. Chem. Phys. **87**, 3699(1987).

[7] Screel K., Schleipen J., Eppink A., and ter Meulen J.J. J. Chem. Phys. **99**, 8713(1993).

[8] Seelemann Th., Andresen P., Schleipen J., Beyer B., and terMeulen J.J., Chem. Phys. **126**, 27(1988).

[9] Lin A., Antonova S., Tsakotellis A.T., and McBane G.C., J. Phys. Chem. **103**, 1198(1999).

[10] Jons S.D., Shirley J.E., Vonk M.T., Giese F.G., and Gentry W.R. J. Chem. Phys. **97**, 7831(1992).

[11] Schnieder L., Seekamp-Rahn K., Liedeker F., Steuwe H., and Welge K. Faraday Discuss. Chem. Soc. **91**, 259(1991).

[12] Chandler D.W. and Houston P.L. J. Chem. Phys. **87**, 1445(1987).

[13] Bontuyan L.S., Suits A., Houston P.L., and Whitaker B.J., J. Phys. Chem. **97**, 6342 (1993).

[14] Chandler D.W. this volume.

[15] Meyer H., J. Chem. Phys. **101**, 6686(1994).

[16] Meyer H., J. Chem. Phys. **101**, 6697(1994).

[17] Wiley W.C., and McLaren I.H., Rev. Sci. Instrum. **26**, 1150(1955).

[18] Meyer H., J. Chem. Phys. **107**, 7721(1997).

[19] Meyer H., Chem. Phys. Lett. **262**, 603(1996).

[20] Kim Y., Fleniken J., and Meyer H., J. Chem. Phys. **109**, 3401(1998).

[21] Blum K., *Density Matrix Theory and Applications*, Plenum Press, New York, 1981.

[22] Zare R.N., *Angular Momentum*, Wiley, New York, 1987.

[23] McClain W.M. and Harris R.A., in *Excited States* ed. by Lim E.C., Vol. 3, Academic, New York, 1977.

[24] The two-photon spectrum of the H state of NO is dominated by a zeroth rank tensor component. Nevertheless, a minor second rank tensor component is responsible for a strong polarization dependence of the scattered intensity for the NO+He system.

[25] Meyer H., J. Chem. Phys. **102**, 3110(1995).

[26] Alexander M.H. and Davis S.L., J. Chem. Phys. **78**, 6754(1983).

[27] Arthurs A.M., and Dalgarno A., Proc. R. Soc. Lond. A **256**, 540(1960).

[28] Stolte S. and Reuss J. in *Atom-Molecule Collision Theory: A Guide to the Experimentalist* ed. by Bernstein R.B., Plenum, New York, 1979.

[29] Meyer H., J. Chem. Phys. **102**, 3151(1995).

[30] Meyer H., Mol. Phys. **84**, 1155(1995).

[31] Meyer H., J. Phys. Chem. **99**, 1101(1995).

[32] Meyer H., manuscript in preparation.

[33] Meyer H., Chem. Phys. Lett. **230**, 510(1994).

[34] Khare V., Kouri D.J., and Hoffmann, D.K. J. Chem. Phys. **74**, 2275(1981).

Ion Imaging Studies of Chemical Dynamics

D.W. Chandler[1], J.R. Barker[2], A.J.R. Heck[3],
M.H.M. Janssen[4], K.T. Lorenz[1], D.W. Neyer[1],
W. Roeterdink[4], S. Stolte[4], L.M. Yoder[2]

[1] Sandia National Laboratory, Livermore CA, 94550
[2] University of Michigan, Ann Arbor
[3] Utrecht University, Sorbonnelaan 16, 3584 CA Utrecht, The Netherlands
[4] Free University of Amsterdam, De Boelelaan 1083, 1081 HV Amsterdam, The Netherlands

1 Introduction

Chemical dynamics is the exacting study of individual details of chemical reactions. This involves studying both intramolecular and intermolecular motion of energy and mass. A new tool in the study of chemical dynamics is ion imaging. Unimolecular photodissociation [1, 2, 3], reactive scattering [4, 5] and inelastic scattering [6, 7] are areas that have been investigated over the last several years using ion imaging. Many of these experiments were performed at the Chemical Dynamics Visitors Laboratory at Sandia National Laboratories. Two-dimensional imaging techniques allow the measurement of the velocity of a state-selectively photoionized product of a unimolecular dissociation or a bimolecular interaction. In the last few years there have been significant improvements in these techniques with the further development of velocity mapping and the accelerated application of the techniques to the study of bimolecular reactions. In this paper we present an overview of some experiments that have been performed at Sandia that highlight the present experimental capabilities available.

Ion imaging is a technique to measure the velocity of a laser ionized cation or a photoelectron. Velocity mapping, first demonstrated by Eppink and Parker [8, 9] is a variation of the "Ion Imaging" technique demonstrated by Chandler and Houston [10] in 1987 and it has improved the velocity resolution of the ion imaging technique. This improvement is obtained by incorporating an electrostatic lens system into the time-of-flight path in order to focus the ions as they travel from the ionization region to the detector. The lenses are adjusted in such a way as to project the charged particle onto the detector so that only the velocity and not the original position where the charged particles are created determines the postition on the detector that the particle impact. The velocity of the particle is mapped onto the detector. The biggest impact of this is on bimolecular scattering experiments where products are ionized by a laser beam over a large volume and the blurring that would be associated with this is eliminated by the velocity mapping lens arrangement.

A typical ion imaging apparatus used for the study of unimolecular dissociation reactions consists of a single, skimmed molecular beam directed along the axis of a time-of-flight mass spectrometer. Along the time-of-flight axis is a electrostatic lens to focus the ions onto the position sensitive ion detector. This geometry eliminates from the image the velocity spread along the propagation axis of the molecular beam but not the smaller velocity spread perpendicular to the molecular beam propagation axis.

The addition of velocity mapping to the unimolecular dissociation apparatus has allowed us to improve the velocity resolution so that we have measured kinetic energies of molecules with as little as $10 \, \text{cm}^{-1}$ of translational energy. At 1 eV translational energy the resolution is on the order of 30 meV ($240 \, \text{cm}^{-1}$). This improved velocity resolution has allowed us to observe the electron recoil from H_2 following 2 + 1 REMPI [11], resolve vibrational structure in photoelectron images from the photoionization of D_2 [11, 12] and measure the low energy recoil associated with Pyrazine/Ar dimer dissociation [13]. The incorporation of velocity mapping into the crossed molecular beam apparatus has made it possible to obtain raw data images that are almost directly comparable to differential scattering cross sections.

For the study of bimolecular reactions two different geometries have been employed. A traditional crossed-molecular-beam apparatus having two, skimmed molecular beams intersecting at the ionization region of a time-of-flight mass spectrometer has been used by Suits and co-workers [4]. Laser ionization occurs at the intersection region of the beams and the ions are projected out of the plane of the molecular beams through an electrostatic lens toward a position sensitive ion detector.

A second geometry has been used for the study of reactive scattering following the work of Dr. Welge's group in Bielefeld [14]. In this geometry two, parallel molecular beams are produced a short distance from each other. Photolysis of a radical precursor in one beam generates a reactive species that flies into the other beam by virtue of the recoil energy it obtained in the photodissociation event. Reaction is detected by laser ionization of the product and the ions are projected along the axis of the molecular beams toward the ion detector. For both experimental arrangements it is important to project the ions perpendicular to the relative velocity vector of the collision. Because scattering is cylindrically symmetric about this axis the three-dimensional scattering distribution can be generated from a

single image using an inverse Abel transform [15].
The data presented here were taken on an apparatus having a crossed-molecular-beam arrangement. The study of bimolecular scattering in the crossed-molecular-beam apparatus is hampered by the fact that the scattering is typically a quasi-continuous process (hundreds of microseconds) compared to the laser beam duration (nanoseconds). Therefore, the laser interacts with the products that have just scattered as well as products that scattered at earlier times. The scattering products produced several microseconds before the laser ionization pulse arrives that scatter into the laser beam volume will be preferentially detected. Therefore, in order to extract the true differential cross-section from the raw data one must normalize the data for this instrument function effect. For this we use a forward convolution program that generates the apparatus detectivity function. The program takes into account the velocity spreads of the molecular beams, the overlap of the laser beam with scattering products, the ion optics, the velocity of the collision products and the geometry of the apparatus. Despite the averaging associated with the experiment the angular resolution is sufficient for observation of structure in the quantum-state-selected rotationally inelastic scattering of HCl with Ar and CO with Ne.

2 Unimolecular Photodissociation

We have studied the photodissociation of N2O after excitation at ~ 200 nm [3]. The Q branch of the $a''^1\Sigma_g \leftarrow\leftarrow X^1\Sigma_g$ two-photon transition was used to obtain the velocity mapped images of the rotationally hot N_2 that is produced in the photodissociation. The N_2 is formed in rotational states between $J = 50$ an $J = 90$ in the ground and first vibrational states. The angular distribution, designated by the β parameter, changes with J; for $J =\sim 50 \beta =\sim 1.0$ and for $J =\sim 90 \beta =\sim 0$ (isotropic distribution). An image of $N_2(J = 74)$ is shown in Fig. 1C. The angular distribution of the concurrently formed O atoms must be a weighted average of the N_2 images. The $O(^1D_2)$ images, however, do not have this appearance. Analysis of these images shows that the $O(^1D_2)$ atoms are formed with an aligned $O(^1D_2)$ orbital. The alignment can be described by the m_j populations of the $O(^1D_2)$. In order to quantify the alignment of the $O(^1D_2)$ atoms several two-photon transitions are used to ionize the $O(^1D_2)$ atoms. The $^1P_1 \leftarrow\leftarrow {}^1D_2$ two-photon transition near 205.4 nm and the $^1F_3 \leftarrow\leftarrow {}^1D_2$ transition near 203.7 nm were used to detect the $O(^1D_2)$ atoms (Figs. 1A and 1B).

We determined that the $O(^1D_2)$ is primarily formed in the $m_j = 0$ and $m_j = 1$ states with very little population in the $m_j = 2$ state. The population of these m_j states is found to be velocity dependent, implying that different rotational states of the N_2 form in a correlated manner with specific m_j state distributions of the O($^1D_2$). This implies that more than one dissociation pathway is active in the molecule. The detailed modeling of these data show that two electronic states are involved in the initial excitation of the N$_2$O. Both the $1^1A''(^1\Sigma^-)$ (perpendicular transition) and the $2^1A'(^1\Delta)$ (parallel transition) states are energetically accessible with 200-nm excitation. These two states lead to different but overlapping distributions of correlated N$_2$(J) and O($^1D_2$) pairs. Measurement of the alignment of the O(^{1}D) atom gives insight into which initially excited N$_2$O states correlate to specific N$_2$(J) and O(^{1}D) pairs. The higher rotational states, J > 75, seem to be produced exclusively from the $2^1A'(^1\Delta)$ state and the lower rotational states, 48 < J < 75, appear to be produced by both the $1^1A''(^1\Sigma^-)$ and the $2^1A'(^1\Delta)$ states in an approximately 40% to 60% ratio. A similar conclusion was made from the imaging study of Suzuki and co-workers [16].

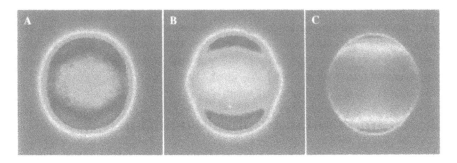

Figure 1: A) Image of O($^1D_2$) obtained using the $^1P_1 \leftarrow\leftarrow {^1D_2}$ two-photon transition near 205.4 nm. B) Image of O($^1D_2$) obtained using the $^1F_3 \leftarrow\leftarrow {^1D_2}$ transition near 203.7 nm. C) Image of N$_2$ (J = 74) using the $a''^1\Sigma_g \leftarrow\leftarrow X^1\Sigma_g$ two-photon transition near 205 nm

Because of the improved resolution velocity mapping has provided we are able to measure the recoil energy distribution of triplet pyrazine fragments following vibrational predissociation of pyrazine/Ar van der Waals clusters [13]. The pyrazine/Ar dimer has a slightly shifted 0^0_0 band for the first excited singlet electronic state, S$_1$. Excitation on this dimer origin transition generates an electronically excited dimer having no vibrational and little rotational energy. The electronically excited pyrazine quickly undergoes intersystem crossing to form a metastable triplet molecule having about 4000 cm^{-1} of vibrational excitation. Some of the vibrational energy is transferred to the argon atom causing dimer dissociation. The triplet electronic character of the pyrazine monomer is left unchanged. Because the triplet state of the monomer is long lived (> 500 ns) it can be photoionized after a small time delay ($\sim$ 50 ns) using 205-nm light. In Fig. 2 we show the image of pyrazine taken after excitation on the origin band of the pyrazine monomer (Fig. 2B) and after excitation on the origin band of the pyrazine/Ar dimer (Fig. 2A). The increased velocity spread is due to recoil from dimer dissociation.

By measuring the magnitude and shape of the recoil of the ionized triplet pyrazine products with ion imaging we determined directly the vibrational-to-translational (V-T) energy-transfer distribution of this process. The recoil probability distribution is found to be a monotonically decreasing function of transferred energy, with an average transfer of $\sim 95\,\mathrm{cm}^{-1}$. The V-T energy-transfer distribution cannot be fit to a simple exponential energy gap law. A surprisal analysis model assuming a momentum constraint and access to only two translational degrees of freedom does a good job at parameterizing the data. The present analysis assumes that the energy is completely randomized in the cluster during dissociation, but this may not be the case, as the low frequency van der Waals modes are expected to be more strongly coupled to the lower frequency out-of-plane pyrazine modes than to the high frequency modes. Further insight will be gained by using more complete models and by performing experiments on other molecules at widely varying vibrational energies. Molecules that include a free rotor, to dramatically change the state densities, will also provide important insight.

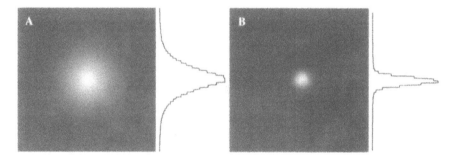

Figure 2: A) Image of the pyrazine ions produced following the ~ 330-nm photolysis of the pyrazine/argon dimer. B) Image of the pyrazine ions produced following the ~ 330-nm excitation of the pyrazine monomers in the molecular beam

A study of the multiphoton dissociation of D_2 has benefited from the improved velocity resolution and the ease of obtaining photoelectron images that velocity mapping affords. We have studied the 532-nm, non-resonant ionization and dissociation of D_2 [11]. Intensities of up to $1 \times 10^{14}\,\mathrm{W/cm^2}$ are used to ionize D_2 forming D_2^+ and then to subsequently photodissociate the D_2^+ via one-, two-, or three- photon absorption. In Fig. 3A is shown a photoelectron image and a photofragment image of D^+ at a laser intensity of approximately $5 \times 10^{13}\,\mathrm{W/cm^2}$. The ability to obtain high-resolution photoelectron images allows us to observe several fundamental photophysical phenomena.

At 5×10^{13} W/cm^2 intensity the ponderomotive potential of the laser beam has shifted the ionization potential of D_2 above the energy of seven 532-nm photons. The few electrons that remain at this energy are distorted into a stripe at the center of the image. Structure still remains in the image at velocities that correspond to eight-photon ionization. Furthermore, in Fig. 3A the electrons are observed to be distorted from their expected circular appearance into an oval as a result of the shape of the ponderomotive potential. Several above threshold ionization peaks are observed in the images. Analysis of these features yields the vibrational distribution of D_2^+ that is produced at this light intensity. Peaks corresponding to production of $v = 3$ and 6 are prominent and marked in the image. Nine-photon ionization (two-photons above threshold ionization, ATI) is also easily seen in the image having the same vibrational distribution as the eight-phonon ATI. The angular distribution of the images shows an undulating intensity pattern that has not been previously observed. There are several possible explanations for this pattern [17].

Figure 3B shows the photofragment image of the D^+ taken at this same

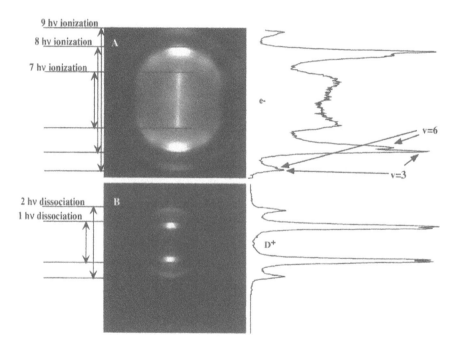

Figure 3: A) Photoelectron image following the 532-nm photolysis of D_2 at a laser intensity of 5×10^{13} W/cm^2. B) Photofragment image of D^+ following the 532-nm photodissociation of D_2 at a laser intensity of 5×10^{13} W/cm^2

laser intensity. One-photon dissociation of the D_2^+ is observed to be predominantly from the $v = 7$ and 8 vibrational levels as determined from the recoil velocity of the D^+. These fragments are very localized along the laser beam polarization axis due to alignment of the D_2 in the strong laser field. The two-photon dissociation signal appears at larger radii and has a broader angular distribution. This unusual observation is attributed to the fact that the two-photon absorption by D_2^+ to the repulsive $2p\sigma u$ state is parity forbidden. Fragments appearing at the two-photon energy have undergone three-photon absorption and as they are falling apart they encounter a curve crossing near 4 Bohr at which point they emit a photon and proceed on to products having the energy of a two-photon dissociation process [18]. The scattering at this curve crossing is seen in the broadened angular distribution of the D^+ fragment.

3 Bimolecular Scattering

The first images of quantum-state-selected Newton spheres from a new crossed molecular-beam machine have been obtained. With the new crossed molecular beam imaging apparatus preliminary data sets for the rotationally inelastic scattering systems HCl/Ar (Fig. 4), CO/He, CO/Ne, NO/D2, NO/He, and HCl/Kr have been obtained. These systems were chosen due to the availability of good potential energy surfaces to compare with and the ability to perform state-selected REMPI on the collision partner. They represent a range of important inelastic scattering systems. HCl has a large dipole, CO has a very small dipole and NO has two nearly degenerate "ground states" (lambda doublets) as do all radical species. We have been successful at using both $1 + 1$ REMPI detection on NO as well as $2 + 1$ REMPI detection of CO and HCl.

The apparatus consists of two, pulsed, doubly skimmed molecular beams intersecting (90°) at the opening of a time-of-flight mass spectrometer containing velocity-mapping ion optics and a position-sensitive ion detector. In the typical experiment, the molecule of choice for inelastic scattering is seeded in a noble gas in one of the beams. About 5% dilution is typical. The pulsed molecular beams are timed to intersect and the inelastic scattering product is detected via resonantly enhanced multiphoton ionization (REMPI). The laser beam is oriented so that the beam bisects the two molecular beams and lies in the plane of the molecular beams. For instance, in Fig. 4 the laser beam would be traveling vertically along the center-of-mass (C-of-M) velocity vector. With this geometry the ionization laser overlaps many Newton spheres of scattered product. The velocity mapping ion optics arrangement projects the ions onto the detector in such a manner that blurring due to the extended source of ions is minimal.

In Fig. 4 is presented a data set of quantum-state-selected Newton spheres for the HCl/Ar system recorded with the new crossed molecular-beam, ion-imaging machine. In this initial experiment, a molecular beam of HCl seeded in Ar crosses a molecular beam of neat Ar.

The rotational temperature of the HCl is approximately 5 K as measured by REMPI spectroscopy. At this temperature there is about 97% population in $J = 0$ and about 3% population in the $J = 1$. The Ar collides with the cold HCl causing translational-to-rotational energy transfer producing HCl in all rotational states between $J = 1$ and $J = 6$. Individual rotational states are ionized by 2 + 1 resonance-enhanced multiphoton ionization through the E state of HCl. These ions are velocity-map focused and projected onto a position-sensitive ion detector located above the plane of the molecular beams. The image is a ring with an intensity pattern indicative of the scattering probability into a particular angle. Because the HCl is seeded in Ar, and the mass of HCl is approximately that of Ar and the backing pressures for the two valves is equal the relative velocity vector of the collision lies on a horizontal line at the center of the images. This is shown in Fig. 4. The scattering angle is easily obtained from inspection of the images.

Low J HCl is preferentially forward scattered (with respect to the HCl's initial velocity) and high J is preferentially back scattered. The collisional

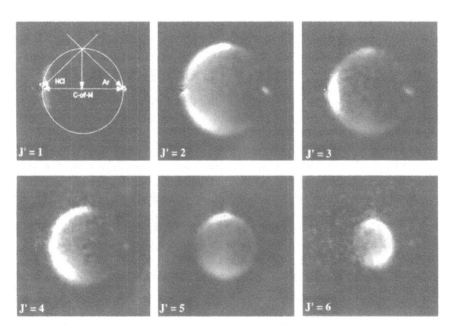

Figure 4: Image of HCl ($v = 0$, $J = 1, 2, 3, 4, 5, 6$) following collision of HCl ($v = 0$, $J = 0, 1$) with Ar atoms in a crossed molecular beam apparatus. 2 + 1 REMPI detection is through the E state of HCl

energy is the same for all images of Fig. 4 ($\sim 500\,\text{cm}^{-1}$) but the translational energy available for scattering decreases with J due to the amount of rotational energy tied up in the molecule. For the most part these data are consistent with calculated differential cross sections as calculated using Jeremy Hutson's intermolecular HCl/Ar potential [19, 20] and his MOLSCAT program. However there are clear discrepancies at the higher J states and at small scattering angles.

4 Conclusions:

The ion imaging technique in combination with molecular beam technology has shown itself to be a valuable tool for the study of chemical dynamics. The multiplex advantage of sampling fragments moving at all angles with each laser pulse and the quantum state resolution obtained by using tunable laser for ionization makes the technique powerful. Unimolecular dissociation dynamics is typically studied in a single molecular beam apparatus with an imaging detector along the axis of the molecular beam. With this arrangement the maximum velocity resolution for the product species is obtained. Resolution of 30 meV at 1 eV translational energy is easily obtained. Higher resolution is obtained for slower moving fragments.

Bimolecular reactions are studied in a crossed molecular beam arrangement. The same advantages of multiplex detection and laser spectroscopic internal state resolution apply here as well. Velocity mapping has made it possible to use the laser to ionize fragments over a large laboratory space and not observe significant blurring of the data. Problems still exist in extracting quantitative differential cross sections from the data but qualitative information (forward versus back scattering, the position of rotational rainbows and the speed of the recoil) is read directly from the images.

References

[1] A.J.R. Heck, D. W. Chandler.: Ann. Rev. of Phys. Chem., **46**, 335 (1995)

[2] T. Droz-Georget, M. Zyrianov, H. Reisler, D.W. Chandler: Chem. Phys. Lett. **276**, 316 (1997)

[3] D.W. Neyer, D.W. Chandler: J. Chem. Phys. **110**, 3411 (1999)

[4] M. Ahmed, D.S. Peterka, A.G. Suits: Chem. Phys. Lett. **301**, 372 (1999)

[5] T.N. Kitsopoulos, M.A. Buntine, D.P. Baldwin, R.N. Zare, D.W. Chandler: Science **260**, 1605 (1993)

[6] P.L. Houston: J. Phys. Chem. **100**, 12757 (1996)

[7] A.G. Suits, L.S. Bontuyan, P.L. Houston, B.J. Whitaker: J. Chem. Phys. **96**, 8618 (1992)

[8] A.T.J.B. Eppink, D.W. Parker: Rev. of Sci. Instrum. **68**, 3477 (1997)

[9] A.T.J.B. Eppink, D.H. Parker: J. Chem. Phys. **109**, 4758 (1998)

[10] D.W. Chandler, P.L. Houston: J. Chem. Phys. **87**, 1445 (1987)

[11] D.W. Chandler, D.H. Parker: Velocity Mapping of Multiphoton Excited Molecules, John Wiley and Sons, New York (1999)

[12] D.W. Chandler, D.W. Neyer, A.J.R. Heck: Proc. of SPIE **3271**, 104 (1998)

[13] Yoder L.M., Barker J.R., Lorenz K.T., and D.W. Chandler: Chem. Phys. Lett. **302**, 602 (1999)

[14] L. Schnieder, K. Seekamprahn, E. Wrede, K.H. Welge: J. Chem. Phys. **107**, 6175 (1997)

[15] R.N. Strickland, D.W. Chandler: Applied Optics **30**, 1811 (1991)

[16] T. Suzuki, H. Katayanagi, Y.X. Mo, K. Tonokura: Chem. Phys. Lett. **256**, 90 (1996)

[17] T. Zuo T., A.D. Bandrauk, P.B. Corkum: Chem. Phys. Lett. **259**, 313 (1996)

[18] A. Zavriyev, P.H. Bucksbaum: Moleculaes in Laser Fields, Dekker, New York (1993)

[19] A.E. Thornley, J.M. Hutson: Chem. Phys. Lett. **198**, 1 (1992)

[20] K.M. Atkins, J.M. Hutson: J. Chem. Phys. **105**, 440 (1996)

Orientation of Parity-Selected NO and Its Steric Asymmetry in Rotational Energy Transfer Collisions[‡]

M.J.L. de Lange, S. Lambrechts, J.J. van Leuken[§]
M.M.J.E. Drabbels, J. Bulthuis, J.G. Snijders*, S. Stolte
Department of Physical Chemistry and Laser Center,
Vrije Universiteit,
de Boelelaan 1083, 1081 HV Amsterdam, The Netherlands
*Department of Chemical Physics and Materials Science Center,
Rijksuniversiteit Groningen,
Nijenborgh 4, 9747 AG Groningen, The Netherlands

1 Introduction

In order to study stereodynamic effects in nonreactive scattering using molecular beams, it is important to be able to orient the (molecular) collision partner. All common methods of orientation for the gas phase make use of a uniform electric field. The conventional method, which is restricted to polar symmetric top molecules or molecules behaving as such, is to select the molecules in a definite rotational state and subsequently orient them in a preferential direction by a uniform electric field of moderate strength. The second method, brute force orientation [2], introduced in 1990, is based on applying a strong uniform electric field, so that orientation by Stark mixing of different rotational states occurs. A comparison of the advantages and disadvantages of the two methods is given in Ref. [3]. More recently a

---
[‡]Part of this contribution has been published in Ref. [1]
[§]Present address: Philips Medical Systems, P.O. Box 10.000, 5684 PC Best, The Netherlands
[¶]present address: Laboratoire de Chimie Physique Moléculaire (LCPM), École Polytechnique Fédérale de Lausanne (EPFL), CH-1015 Lausanne, Switzerland

beam of oriented metastable CO molecules was produced by populating a single M-sublevel in the presence of an electric field using a narrow bandwidth laser [4].

Ideally, the orientation distribution function should be exactly known. In practice, knowledge of this distribution function is obtained in an indirect way, for instance by making use of the symmetry in the collision geometry and preparing orientations that are related in a well-defined way (e.g. head- or tail orientation) [5]. To be able to relate the reactive cross section to the Legendre moments of the reactant orientation it is necessary to express the steric opacity function also as a Legendre series [6].

When all relevant molecular constants are known, the exact shape of the orientation distribution function can be calculated, but in order to be useful for the interpretation of experimental results it is necessary that the relevant experimental parameters, specifically the local value of the orienting electric field, be known. Therefore, it is of interest to have a method of measuring the molecular orientation in situ in an independent way [7].

Such a direct method is available if the oriented molecule can be selected in a single (rotational) state and if transitions that are parity forbidden in zero orienting field become allowed owing to mixing of different parity states in nonzero field. Open shell molecules, such as NO, fulfill this condition. OH is another example, which was recently the subject of a thorough scattering study by Schreel and ter Meulen [8].

NO can be conveniently selected by an electrostatic hexapole in the upper component of the Λ-doublet in its ground rotational state, which has a negative parity [9]. In the presence of an electric (orientation) field, the NO molecule is oriented to a degree that depends on the degree of mixing with the opposite parity Λ-doublet state. If precautions are taken to avoid saturation, the decrease and increase of selected optical transitions from the oriented state can be measured with high accuracy by laser induced fluorescence (LIF), and from this the values of the mixing coefficients of the two Λ-component functions are obtained. The absolute value of the orientation follows directly, without recourse to the magnitude of the orienting field. Reversely, the degree of orientation can be used to calculate the magnitude of the electric field if the relevant molecular constants that determine the Stark splitting are known, as is the case for NO.

2 Theory

For $^2\Pi$ molecules with a large spin-orbit splitting, like NO in its electronic ground state, the interaction between the two spin-orbit states is weak and as a consequence the rotational states can in first approximation be written in a HundÕs case (a) basis. In this letter we concern ourselves with the $J = \frac{1}{2}$ Λ-doublet of the $^2\Pi_{1/2}$ ground state of NO. For this level the interaction with the $^2\Pi_{3/2}$ state is absent and parity defined rotational states are then given by the Hund's case (a) wavefunctions as

$$\left|JM\Omega\epsilon\right\rangle = \frac{1}{\sqrt{2}}\left[\left|JM+\Omega\right\rangle + \epsilon\left|JM-\Omega\right\rangle\right] \quad (1)$$

Here J is the total angular momentum of the NO molecule, Ω is the magnitude of the projection of J on the internuclear axis, M is the projection of J on the space fixed axis, and $\epsilon = \pm 1$ defines the symmetry of the rotational state. States with $\epsilon = +1$ are labeled e whereas states with $\epsilon = -1$ are labeled f. The total parity of the rotational states is by convention [10] related to this e/f symmetry according to $(-1)^{J-\epsilon/2}$. In a first approximation the two parity states are degenerate. Interactions with other electronic states can lift this degeneracy, giving rise to the so-called Λ-splitting. For the $J = \frac{1}{2}$ doublet of the $^2\Pi_{1/2}$ state the magnitude of this Λ-splitting equals, $W_\Lambda = 0.0119$ cm^{-1} [11, 12].

In order to study parity breaking caused by an applied electric field, molecules with a well-defined parity have to be selected. This is realized by using an electrostatic hexapole state selector which selects only the NO molecules in the $J = \frac{1}{2}^-$ (the superscript $-$ indicates f-symmetry) state. When these molecules are placed in an electric field, parity is broken and the resulting rotational wavefunctions are superpositions of the field free e- and f-symmetry wavefunctions given by Eq. (1). For the state selected $J = \frac{1}{2}^-$ upper Λ-doublet level (U) the resulting wavefunction is given by

$$\left|J\pm|M|\Omega U\right\rangle = \alpha(E)\left|J\pm|M|\Omega\,\epsilon=-1\right\rangle \pm \beta(E)\left|J\pm|M|\Omega\,\epsilon=+1\right\rangle \quad (2)$$

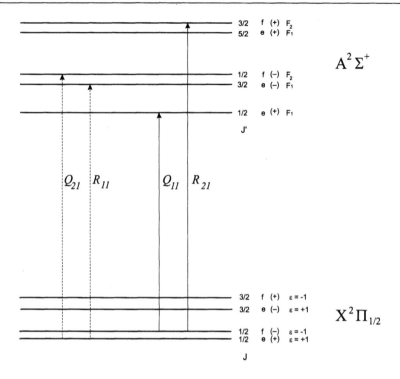

Figure 1: Energy diagram and symmetry labels of the lowest rotational levels in the $X\,^2\Pi_{1/2}$ and $A\,^2\Sigma^+$ states of NO. It should be noted that Λ-doubling in the electronic ground state and the spin-rotation splitting in the excited state are not drawn to scale. The solid lines indicate the zero field allowed transitions from the state selected $J = \frac{1}{2}^-$ (f-symmetry) level. The dashed lines indicate transitions that contribute by mixing of the wavefunctions due to the electric field.

In these equations $\alpha(E)$ and $\beta(E)$ are the electric field dependent mixing coefficients given by

$$\alpha(E) = \sqrt{\frac{1}{2} + \frac{1}{2\sqrt{1 - E_{\text{red}}^2}}} \qquad \beta(E) = \sqrt{\frac{1}{2} - \frac{1}{2\sqrt{1 - E_{\text{red}}^2}}} \qquad (3)$$

where E_{red} is the absolute value of the reduced electric field strength defined by $E_{\text{red}} = 2W_{\text{Stark}}/W_\Lambda$, i.e. the Stark energy relative to the

Λ-splitting. The Stark energy is given by the well-known expression

$$W_{\text{Stark}} = -\langle \boldsymbol{\mu} \cdot \boldsymbol{E} \rangle = \mu E \langle \cos\theta \rangle_{\max} = \mu E \frac{\Omega M}{J(J+1)} \quad (4)$$

Here E is the static electric field strength and μ is the permanent electric dipole moment, which in case of the NO molecule equals $\mu = 0.158$ D [13], and θ is the angle between the electric field vector and the molecular axis and taken from O to N. Note that this is opposite to the direction of the dipole moment. Clearly, when $E_{\text{red}} \to 0$ we have $\alpha(E) = 1$ and $\beta(E) = 0$ and we recover the unoriented $|J\Omega M\,\epsilon = -1\rangle$ state. On the other hand, if $E_{\text{red}} \to \infty$ we have $\alpha(E) = \beta(E) = 1/\sqrt{2}$ and the totally oriented states, adiabatically evolving from the selected f-state become

$$\left| J \pm |M| \,\Omega\, \text{U} \right\rangle \;\Rightarrow\; \pm \left| J \pm |M| \pm \Omega \right\rangle \quad (5)$$

The average orientation of the NO molecule, for which hyperfine effects may be neglected [14], can be expressed as a function of the electric field strength in terms of the mixing coefficients and the maximum degree of orientation, $\langle \cos\theta \rangle_{\max} = 1/3$, according to

$$\langle \cos\theta \rangle_E = 2\alpha(E)\beta(E)\langle \cos\theta \rangle_{\max} \quad (6)$$

The degree of orientation can thus be calculated once the mixing coefficients are known. These mixing coefficients can be determined from transition intensities in an excitation spectrum. In the present experiment the $\gamma(0,0)$ band of NO is used. Energy levels of the $X\,^2\Pi$ and $A\,^2\Sigma^+$ states, along with the transitions that are relevant for this experiment, are given in Fig. 1. The solid lines in the figure indicate the zero field allowed transitions, $Q_{11}(\frac{1}{2})$ and $R_{21}(\frac{1}{2})$, originating from the prepared $J = \frac{1}{2}^-$ (f- symmetry) state, whereas the dashed lines indicate the transitions that become allowed by the electric field, $R_{11}(\frac{1}{2})$ and $Q_{21}(\frac{1}{2})$, originating from the $J = \frac{1}{2}^+$ (e- symmetry) level. The spin-rotation splitting in the $A\,^2\Sigma^+$ state of NO is the order of 0.01 cm^{-1} and as a consequence the R_{11} and Q_{21} cannot be resolved. The mixing coefficients can now be directly determined from the intensity ratio between the various transitions provided the transitions are not saturated. Since the sum of the Hönl-London factors for the transitions starting from the two Λ-doublet components are equal the following relation holds

$$\frac{\alpha(E)^2}{\beta(E)^2} = \frac{1-\beta(E)^2}{\beta(E)^2} = \frac{I_{Q_{11}(\frac{1}{2})} + I_{R_{21}(\frac{1}{2})}}{I_{R_{11}(\frac{1}{2})+Q_{21}(\frac{1}{2})}} \qquad (7)$$

The right part of the equation gives the ratio between the measured intensities of the transitions originating from the prepared $J = \frac{1}{2}^-$ state and the transitions, that cannot be resolved, from the admixed $J = \frac{1}{2}^+$ state.

3 Experimental

The experimental setup is schematically shown in Fig. 3.
A beam of rotationally cold NO molecules is produced by expanding 3 atm. of a 20% NO/Ar mixture into a diffusion-pumped vacuum chamber through a pulsed valve with an orifice diameter of 0.8 mm. After passing two conical electroformed skimmers the molecular beam enters a 1.66 meter long hexapole that focuses NO molecules in the $J = \frac{1}{2}^-$ (f-symmetry) state into the detection region, approximately 3 meters from the source. In the separately pumped scattering chamber the NO molecular beam is crossed at right angles with the secondary Ar beam. The NO and Ar beams ate both pulsed with repetition rates of 10 Hz and 5 Hz respectively. An electric orientation field can be applied at the detection region by applying a voltage (up to 20 kV) pairwise across four parallel metal rods, of 1.6 mm diameter. The 10 cm long rods are positioned in a square of 10 mm width (center to center of the rods) placed at an angle of 45° with respect to the molecular beam. Analysis of this electric field configuration shows that the electric field in the region of the molecular beam has a value that is 0.668 of that expected for a flat plate capacitor with a 10 mm gap and has an inhomogeneity less than 5%. At this field region the molecular beam is crossed at an angle of 45° by a laser beam propagating in the direction of the electric field. The direction of the orientation field will be directed along or opposite to the relative velocity of the incoming NO-Ar.
To resonantly excite the NO molecules to the $A\,^2\Sigma^+$ ($v = 0$) state a pulsed dye laser (Spectra Physics PDL 3) is employed that is pumped by the third harmonic of a Q- switched Nd:YAG laser (Spectra Physics

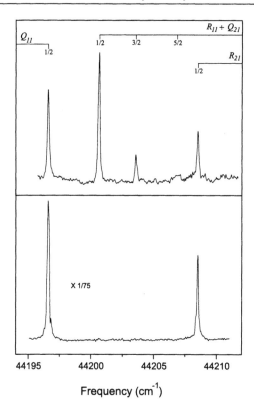

Figure 2: Upper panel: LIF spectrum of the $A\,^2\Sigma^+$ $(v = 0) \leftarrow X\,^2\Pi_{1/2}$ $(v = 0)$ transition in NO recorded with no voltage applied to the hexapole state-selector. Lower panel: The same spectrum recorded with a voltage applied to the hexapole that focuses NO molecules in the $J = \frac{1}{2}^-$ (f- symmetry) state at the detection region. Due to the focussing the intensity of the transitions starting from this level is increased by a factor of about 150.

GCR-150). The output of the dye laser, operated with Coumarine 47, is frequency-doubled in a BBO crystal to yield radiation at 226 nm with a pulse energy of 0.5 mJ and a bandwidth of 0.09 cm^{-1}.
In order to avoid saturation of the $A\,^2\Sigma^+$ $(v = 0) \leftarrow X\,^2\Pi_{1/2}$ $(v = 0)$ transitions the frequency-doubled output of the laser is attenuated to a μJ level using a Glan-Laser polarizer. An unfocussed laser beam of 2 mm diameter is used. The corresponding fluence of less than 0.1 mJ/cm^2 is well below the saturation fluence for this transition [15].

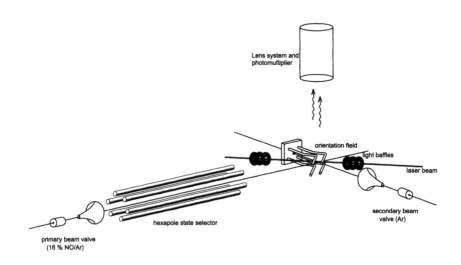

Figure 3: Schematic representation of the experimental setup.

The laser-induced fluorescence (LIF) from the $A\,^2\Sigma^+$ ($v = 0$) state back to the various vibrational levels of the electronic ground state is imaged by a quartz lens system onto a photomultiplier. In order to reduce the scattered 226 nm light from the laser, a dichloromethane liquid cut-off filter is placed in front of the photomultiplier tube. The signals from the photomultiplier are processed by a digital oscilloscope (LeCroy LS140) and a boxcar averager (SRS 250) interfaced with a PC.

4 Results and discussion

The experimental setup has been described in detail in Ref. [1].
As already mentioned, it is necessary to start with molecules that are in one specific quantum state. In the present experiment an electrostatic hexapole is used for this state selection. The upper panel of Fig. 2 shows the LIF spectrum with no voltage applied to the hexapole. Clearly, rotational levels higher than the $J = \frac{1}{2}$ are also populated in the molecular beam. From the observed intensity distribution a rotational temperature of about 4K can be deduced. The

lower panel shows an excitation spectrum where a voltage is applied to the hexapole that focuses the $J = \frac{1}{2}^-$ (f- symmetry) at the detection region and defocuses other quantum states. The transitions originating from the selected $J = \frac{1}{2}^-$ state, $Q_{11}(\frac{1}{2})$ and $R_{21}(\frac{1}{2})$, have increased by a factor of about 150 whereas the other transitions have decreased in intensity. As a result the spectrum consists of only two strong transitions. From the observed intensities the purity of the beam is determined to be better than 99%.

The state selected NO molecules can now be oriented by an electric field. Fig. 4 shows the recorded LIF spectra at various values of the applied electric field. At zero field only the $Q_{11}(\frac{1}{2})$ and $R_{21}(\frac{1}{2})$ transitions are observed starting from the prepared state. The 2:1 intensity ratio of the two peaks, measured as the ratio of the surface underneath each peak, reflects the Hönl-London factors for these transitions and is a clear indication that the transitions are not saturated by the laser powers used. When an electric field is applied the $Q_{21}(\frac{1}{2})$ and $R_{11}(\frac{1}{2})$ transitions become allowed by the mixing of the zero field wavefunctions, as can be seen in Fig. 4. The intensity of these two transitions with respect to the other zero field transitions, $Q_{11}(\frac{1}{2})$ and $R_{21}(\frac{1}{2})$, can be used to determine the mixing coefficients using Eq. (7). As can be seen the degree of mixing at the maximum obtainable electric field of 13.3 kV/cm (corresponding to an applied voltage of 20 kV) is only moderate, $\beta(E)^2 = 0.27$. In contrast to the mixing of the wavefunctions, the degree of orientation, given by Eq. 6, already approaches its limiting value within 90% at this electric field strength. Using the known molecular constants [11, 12, 13] of the NO molecule, the degree of mixing, $\beta(E)^2$, and orientation, $2\alpha(E)\beta(E)$, can be calculated. These results are in excellent agreement with the experimental values.

As mentioned in the introduction, the degree of mixing can be used to determine the local electric field value. These field values have been used as input in a least-squares fitting routine that determines the correction factor for the present electric field geometry with respect to a flat plate capacitor. The correction value found in this way equals 0.664 ± 0.018 and is in perfect agreement with the calculated value of 0.668. This clearly shows the accuracy of this method.

As mentioned before, great care was taken to avoid saturating the molecular transitions. The reason for this is that the strength of the

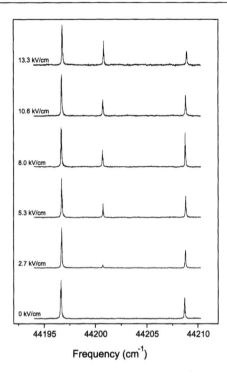

Figure 4: Excitation spectrum of the $A\,^2\Sigma^+$ ($v = 0$) ← $X\,^2\Pi_{1/2}$ ($v = 0$) transition in NO recorded at various values of the applied electric field. At zero field only the two zero field allowed transitions originating from the state selected upper Λ-doublet component are observed. At the higher electric field values, transitions originating from the other Λ-doublet component contribute due to mixing of the zero field wavefunctions.

transitions depends on the applied electric field. Consequently, the saturation behavior of the various transitions is also electric field dependent. This poses no problem if the absolute cross sections, laser power and spectral distribution of the laser are known. However, in practice the laser has an unknown multi-mode spectral distribution within its bandwidth that fluctuates from shot to shot. This makes that the mixing coefficients and thus the degree of orientation cannot be determined. It is therefore necessary that saturation of the molecular transitions is avoided in order to determine the degree of orientation.

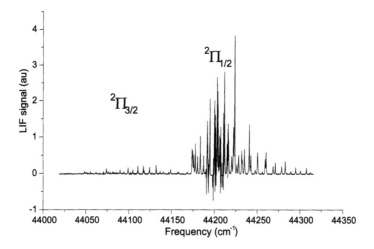

Figure 5: Overview of the $\gamma(0,0)$ LIF signal of scattered NO molecules as a function of laser frequency. The LIF signal reflects the density of scattered NO molecules. The negative signals appearing for very strongly populated low rotational levels of the $^2\Pi_{1/2}$ state reflect an experimental artifact induced by an overload of the digitized voltage converter at the PC-input.

In conclusion we have presented a method that allows the determination of the degree of orientation using spectroscopic techniques. The method has been successfully demonstrated on the NO molecule and can in principle be applied to a large class of molecules.

In our earlier work [16] on collisions of state selected NO with Ar atoms the sensitivity of the measurements was limited by scattered laser light. Recently, we succeeded in increasing the sensitivity of our setup considerably by placing a pair of light baffles along the path of the laser beam. The light baffles strongly reduce the background signal due to the scattered laser light. Fig. 5 shows the S/N that can now be obtained, which compares very favorably with our earlier work [16]. The strength of multiplet changing inelastic collisions, $^2\Pi_{3/2} \leftarrow\, ^2\Pi_{1/2}(J=\frac{1}{2})$, is encouraging for future experimental investigation of the steric asymmetries for these transitions.

The steric asymmetry for the NO-Ar collisions can be defined as the difference between the inelastic collision cross section with the N-end in front and the one with the O-end in front divided by the sum of these cross sections,

$$S = \frac{\sigma_{NO} - \sigma_{ON}}{\sigma_{NO} + \sigma_{ON}} \quad (8)$$

This steric asymmetry has been measured as a function of the final rotational state J, for spin-orbit conserving collisions at a collision energy of 442 cm^{-1}. In addition close coupling scattering calculations of the steric asymmetry have been carried out. The experimental and theoretical results are compared in Fig. 6.

This figure shows that the steric asymmetry alternates its sign with $\Delta J = 1$ and this sign alternation is independent of the parity of the final rotational state. Theory nicely agrees with experiment in this respect, although the absolute magnitude of the steric asymmetry is overestimated by the theory. Also for the highest J values that can be attained with the given collision energy, theory agrees with

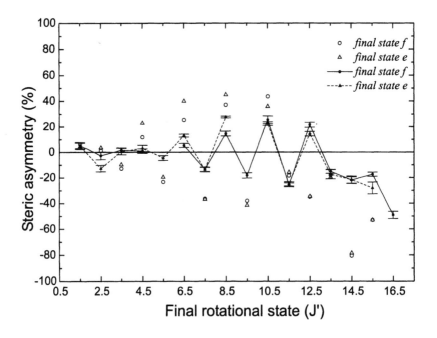

Figure 6: Experimental and theoretical values of the steric asymmetries of fine structure conserving collisions of NO and Ar. Note that the steric asymmetry appears to depend strongly on the J' value and only weakly on the parity of the final state. The state with $J' = 31/2$ is the highest energetically allowed rotational level.

experiment in that the steric asymmetry remains negative for $J >$ 25/2. Clearly, in this region a collision at the O-end of NO is more effective for rotational excitation than is a collision at the N-end. The agreement between experiment and theory regarding the sign and magnitude of the steric asymmetry appears to be encouraging for coming to an understanding of the NO-Ar potential surface. Further work is in progress to elucidate these steric phenomena.

Acknowledgments

This research was made possible the financial support of Scheikundig Onderzoek Nederland (SON) and Fundamenteel Onderzoek der Materie (FOM). M. Drabbels gratefully acknowledges the support of the Dutch Royal Academy of Sciences (KNAW).

References

[1] M.J.L. de Lange, J.J. van Leuken, M.M.J.E. Drabbels, J. Bulthuis, J.G. Snijders, and S. Stolte, Chem. Phys. Lett. **294** 332 (1998).

[2] H. J. Loesch and A. Remscheid, J. Chem. Phys. **93**, 4779 (1990); B. Friedrich and D. R. Herschbach, Z. Phys. D., **18**, 153 (1991).

[3] J. Bulthuis, J. J. van Leuken and S. Stolte, J. Chem. Soc., Faraday Trans. **91**, 205 (1995).

[4] M. Drabbels, S. Stolte and G. Meijer, Chem. Phys. Lett. **200**, 108 (1992).

[5] D. H. Parker and R. B. Bernstein, Annu. Rev. Phys. Chem. **40**, 561 (1989).

[6] S. Stolte, Ber. Bunsenges. Phys. Chem. **86**, 413 (1982).

[7] J. J. van Leuken, Ph.D. Thesis, Vrije Universiteit, Amsterdam (1994).

[8] K. Schreel and J. J. ter Meulen, J. Phys. Chem. A **101**, 7639 (1997).

[9] F. H. Geuzenbroek, M. G. Tenner, A. W. Kleyn, H. Zacharias and S. Stolte, Chem. Phys. Lett. **187** , 520 (1991).

[10] J. M. Brown, J. T. Hougen, K. P. Huber, J. W. C. Johns, I. Kopp, H. Lefebvre-Brion, A. J. Merer, D. A. Ramsay, J. Rostas and R. N. Zare, J. Mol. Spectr. **55**, 500 (1975).

[11] W. L. Meerts and A. Dymanus, J. Mol. Spectr. **44**, 320 (1972).

[12] C. Amiot, R. Bacis and G. Guelachvili, Can. J. Phys. 56, 251 (1978).

[13] R. M. Neumann, Astrophys. J. **161**, 779 (1970).

[14] M. G. Tenner, E. W. Kuipers, W. Y. Langhout, A. W. Kleyn, G. Nicolasen and S. Stolte, Surf. Sci. **236**, 151 (1990).

[15] D. C. Jacobs, R. J. Madix and R. N. Zare, J. Chem. Phys. **85**, 5469 (1986).

[16] J. J. van Leuken, F. H. W. van Amerom, J. Bulthuis, J. G. Snijders, and S. Stolte, J. Phys. Chem. **99**, 15573 (1995).

Collisional Energy Dependence of Insertion Dynamics: State-Resolved Angular Distributions for $S(^1D) + D_2 \to SD + D$

S.-H. Lee, K. Liu

Institute of Atomic and Molecular Sciences (IAMS) Academia Sinica, Taipei, Taiwan 106

1 Introduction

The past few years witness a remarkable achievement that has been made through the state-resolved differential cross section measurement in our better understanding of the dynamics for the two prototypical three-atom reactions, $H + H_2$ [1] and $F + H_2$ [2, 3, 4, 5]. Extension to more than three-atom systems, such as Cl + hydrocarbons [6, 7], further highlights the need of the state-resolved angular distribution in shaping our conceptual understanding about chemical reactivity. Those are the typical direct abstraction reactions. For an indirect complex-forming reaction, conventional wisdom often invokes statistical treatments in describing its reactivity [8]. Could the state-resolved differential cross section reveal the "hidden" dynamics for this type of reaction? And with which could further insights into its reactivity be gained? These are the questions we have been trying to address over the past few years [9, 10, 11].

Insertion is an important indirect reaction pathway, which is characterized by a simultaneous one-bond rupture and two-bonds formation process in forming the intermediate complex. The reaction product is then produced upon the subsequent decomposition of the complex. Previous experiment has suggested [10] that the reaction of $S(^1D) + H_2$ (and its isotopomers) is an excellent system for detailed study of insertion dynamics over a wide collision energy range. A recent high quality ab initio characterization of the potential energy surfaces (PESs) for this reaction [12] confirms this notion. While the lowest, insertive type PES ($^1\Sigma^+$) is barrierless, the other abstractive type PES's ($^1\Pi$ and $^1\Delta$) exhibit collinear barriers $\gtrsim 42$ kJ/mol. Thus, the low energy collisions of this reaction is totally dominated by insertion only. Compared to the analogous, better-known reaction of $O(^1D) + H_2$, both reactions involve deep wells (377 (494) kJ/mol with respect to the SH+H(OH+H) product channel); but the overall exoergicity of S(^{1}D)+H$_2$ is substantially lower (28.8 kJ/mol versus 181 kJ/mol). In the case of S(^{1}D) + D$_2$ → SD + D(ΔH° = −25.9 kJ/mol), a detailed differential cross section measurement at 22.2 kJ/mol has also been reported [11]. It was found that the angular distribution is roughly forward-backward (f-b) peaking. A strong coupling between the product angular and speed distributions was noted. Phase-space theory (a statistical theory for a chemical reaction involving a loose complex) gave a fair description about product translational energy distribution, though significant discrepancies were noted for angular and angle-specific speed distributions.

Reported here is the measurement of the differential cross section of this reaction at a lower collision energy of 9.6 kJ/mol. Comparison with the previously reported results at 22.2 kJ/mol [11] offers us an opportunity to explore the collisional energy dependence of the detailed insertion dynamics, which is the subject of the present report.

2 Apparatus and Methodology

The experiment was carried out in a rotating-sources, crossed-beam apparatus described previously [11, 13, 14]. In brief, a skimmed S(^{1}D) beam was generated by laser photolysis of CS$_2$ ≲ 0.5%CS$_2$ seeded in He, 15 atm.) at 193 nm near the throat of a pulsed valve. The subsequent supersonic expansion confined and translationally cooled the S(^{1}D) beam which then collided with the target D$_2$ beam from the second pulsed valve.

The three dimensional (3-D) velocity distribution of the D-atom product was interrogated by the Doppler-selected TOF method [9, 14, 15]. As illustrated in Fig. 1, to measure a 3-D product velocity distribution $I(v_x, v_y, v_z)$, the Doppler-shift technique is applied to select a subgroup of the D-atom products with $v_z \pm \delta v_z$ via the (1 + 1) resonance-enhanced multiphoton ionization (REMPI) detection scheme. Rather than collecting all those ions from the REMPI detection process as a single data point in the conventional Doppler-shift technique, those Doppler-selected ions are further dispersed both spatially (in v_x) and temporally (in v_y). A slit placed in front of the detector restricts only those ions with $v_x \approx 0$ to be detected, and the v_y distribution of those v_x- and v_z- selected ions is then measured by a high resolution ion TOF technique. Since both the Doppler slice and the ion TOF measurement are essentially in the center-of-mass (c.m.) frame and the v_x- component associated with the c.m. velocity vector is small and can be largely compensated experimentally, this highly multiplexed measurement maps out directly the desired c.m. distribution $(d^3\sigma/v^2 dv d^2\Omega)$ in a Cartesian velocity coordinate $(d^3\sigma/dv_x dv_y dv_z)$. This is to be contrasted with the conventional neutral TOF technique (either in the universal machine [2] or by the Rydberg-tagging approach [1]) for which the laboratory-to-c.m. transformation must be performed, or with the 2-D ion imaging technique [16] which involves a 2-D to 3-D back transformation.

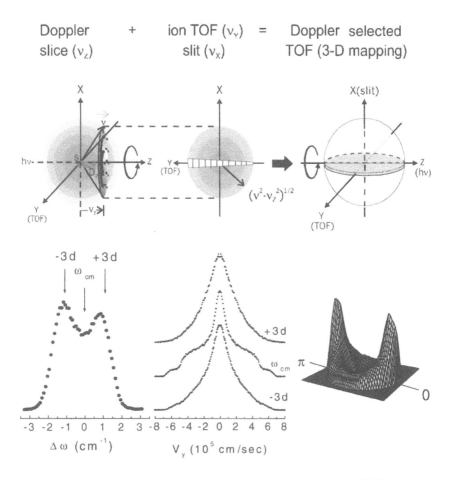

Figure 1: Illustration of the basic idea of the Doppler-selected TOF method. The lower panels show the actual data for the D-atom product from the reaction of $S(^1D) + D_2(r)SD + D$ at $E_c = 22.2\,\text{kJ/mol}$

To take advantage of the cylindrical symmetry of the product distributions around the initial relative velocity axis ($\hat{V}_z$) in a typical crossed-beam scattering experiment, the probe laser is always directed along $\hat{V}_z$ in this approach. Experimentally, this is readily achieved with the present rotating-sources machine.

3 Global View of the Collisional Energy Dependences

As exemplified in Fig. 1, the TOF measurements were performed for a total of about 24 equally spaced Doppler selections to cover the entire profile for each Ec. By stacking those data together (after the L-α doublet complication is removed [15]), the entire 3-D velocity distribution can be mapped out, which is depicted in Fig. 2. A strong coupling of the product angular and speed distribution is readily observed. For example, the step structures are quite prominent for sideward-scattered products, but they merge together as the scattering angle shifts toward the backward direction.

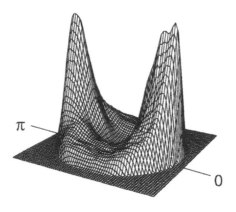

$d^3\sigma/dvd^2\Omega$ representation

$E_c = 9.6$ kJ/mol

Figure 2: D-atom c.m. velocity-flux contour, $d^2\sigma/dvd(\cos\theta)$, from the reaction of S(^{1}D)+D$_2$ at $E_c = 9.6$ kJ/mol. The contours are constructed directly from a total of about 24 slices of the Doppler-selected TOF measurements. The contour for $E_c = 22.2$ kJ/mol is depicted in Fig. 1 for comparison

The results of a global analysis of the contour for $E_c = 9.6$ kJ/mol are presented in Fig. 3, along with the previously reported ones at 22.2 kJ/mol [11]. The product translational energy distributions ($d\sigma/dE_t$) are rather broad. At this level of detail, the vibrational structure is either absent (9.6 kJ/mol) or barely discernible (22.2 kJ/mol). The fraction of the average kinetic energy release $\langle f_t \rangle$ increases from 0.43 to 0.48 with the increase in collision energies. Thus, the amount of increase in initial kinetic energy seems largely channelled into the product translational energy ($\Delta \mathbf{T} \to \Delta \mathbf{T}'$ – a behavior normally referred to the Polanyi's rule [17] for a direct abstraction reaction!

We will return to this point later.

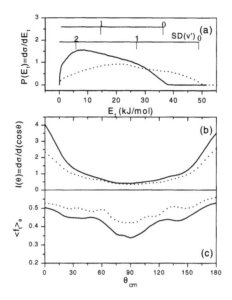

Figure 3: Comparisons of (a) the product translational energy distributions, (b) the product c.m. angular distributions, and (c) the fraction of the average kinetic energy release for $E_c = 9.6$ kJ/mol (solid lines) and 22.2 kJ/mol (dotted lines). The results for the two energies are normalized to their integral cross sections [10]. Also marked in (a) are the onsets of the vibrational states of the SD product for the two cases

The product c.m. angular distributions $(d\sigma/d(\cos\theta))$ are displayed in Fig. 3 (b), and both cases show pronounced f/b peaking and nearly symmetric distributions. It implies that as the insertion complex decomposes, the initial orbital angular momentum $\mathbf{L}$ (or equivalently the total angular momentum $\mathbf{J}$ in the present case) is preferentially disposed into the final orbital angular momentum $\mathbf{L}'$: , i.e., $\mathbf{L} \approx \mathbf{J} \approx \mathbf{L}'$ [18]. This high specificity of the orbital angular momentum disposal $(\mathbf{L} \to \mathbf{L}')$ is the global picture. It will be shown later that the detailed angular momentum disposal is actually far richer than that.

Depicted in Fig. 3(c) are the fractions of the average translational energy release as a function (c.m., which gives a quantitative measure of the coupling of the product angular and speed distributions as alluded to early. In both cases, a distinct dependence of $\langle f_t \rangle$ on $\theta_{c.m}$ is seen. Comparing the two $\langle f_t \rangle_\theta$ distributions reveals a roughly constant upward-shift in $\langle f_t \rangle_\theta$'s with the increase in E_c. As was shown previously [11], a qualitatively similar dependence is also predicted by phase-space theory (PST). The deviations from the PST prediction, both in the magnitudes of the $\langle f_t \rangle_\theta$'s and in the striking oscillatory features, are the manifestations of the hidden dynamics of this reaction.

4 Angle- and State-Specific Analysis

As was presented previously [11], a more informative way to reveal the detailed dynamics afforded by the present direct 3-D mapping approach is to examine the angle-specific kinetic energy distribution $P(E_t; \Delta\theta)$ over a limited range of $\Delta\theta$. Without losing generality, Fig. 4 illustrates this point for every 15° angular segment. To explore the collisional energy dependence, we have converted the $P(E_t; \Delta\theta)$ distribution into $P(E_{int}; \Delta\theta)$, where $E_{int} = E_{total} - E_t$.

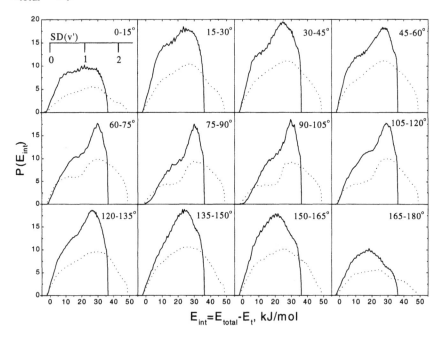

Figure 4: Angle-specific internal energy distributions of the SD product over every 15° angular range, converted from the contours shown in Fig. 2. The solid (dotted) lines are for $E_c = 9.6(22.2)$kJ/mol. The results for the two cases have been normalized to their integral cross sections [10]

Two important conclusions can be drawn from the comparison of these angle-specific distributions at the two energies. First, the vibrational structures of the SD product are clearly seen, especially in the sideward directions. Since the total reaction cross section decreases with the increase in initial collision energy [10], the integrated areas of $P(E_{\rm int}; \Delta\theta)$'s become smaller for $E_c = 22.2$ kJ/mol. Nevertheless, judging from the shape of each of these distributions, the product vibrational state distributions also alter with collision energies. (Note that since the distributions are not deconvoluted for the finite experimental resolution, the spread in the initial collision energy (about 10% of E_c) extends the tail of the distribution into the negative $E_{\rm int}$ region. Also note that the finite slit width in the TOF measurement [15] amplifies the $P(E_t)$ distribution near $E_t \approx 0$, consequently the $P(E_{\rm int})$ distribution is distorted in the highest $E_{\rm int}$ region.)

Second, at a given E_c the internal energy distributions of the SD product are significantly different for different c.m. scattering angles. It is the ramification of the product angle-speed coupling mentioned earlier.

These qualitative observations can be put forward semiquantitatively through the product state analysis. Figure 5 illustrates the vibrational state partitioning for three angular segments. Although such a partitioning is never unique, the essential features described below remain unaltered. The resulted angle-specific energy disposal and the vibrational branchings are summarized in Fig. 6. As is seen, the average vibrational excitations remain about the same ($\sim 13\%$) at the two E_c's, thus the observed $\Delta \mathbf{T} \to \Delta \mathbf{T'}$ is mainly at the expense of the average rotational excitation. In terms of the vibrational state branching, both cases display an intriguing behavior: the sideward scattered products exhibit more vibrational excitation, and the angular dependence of the ratio $P_{v\neq 0}/P_{v=0}$ becomes more pronounced with the increase in E_c. These observations suggest that different types of complexes and/or mechanisms must be involved for the formation of different vibrational states of SD. One plausible explanation lies on the correlation between the initial impact parameter and the type of the complex, as proposed previously [11]. Theoretical work along this line is currently in progress.

The shape of the vibrational "band" exemplified in Fig. 5 also contains the information about the rotational distribution. The recoil energy E_t and the product rotational energy E_r for vibrational state v' are related by

$$E_t = E_{\rm total} - E_v - E_r$$

For a fixed c.m. angle, the "classical" rotational state distribution is

$$P(j') = P(E_t)dE_t/dj' = 2B_{v'}(j' + 1/2)P[E_{\rm total} - E_v - B_v j'(j' + 1)],$$

where $B_{v'}$ is the rotational constant for vibrational state v'. Note that only the essential features in the product state distributions are sought here, the nature of the energy level structure of the open-shell species (SD($^2\Pi$)) is then neglected.

Figure 7 displays the 3-D plots showing how the rotational state distributions for the two dominant vibrational states vary with the c.m. angles,

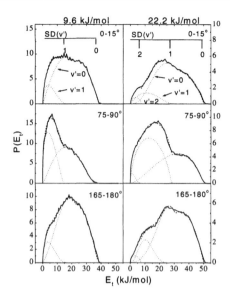

Figure 5: Partition of the angle-specific translational energy distributions into the SD vibrational state manifolds, exemplified by three different angular segments for each case

and how they vary with the initial collision energies. As is seen, the product state-specific angular distributions are strikingly different: while a pronounced f/b peaking distribution is evident for the $v' = 0$ state, a nearly isotropic one is displayed for $v' = 1$! (Note that an isotropic angular distribution for a given (v', j') state means $d\sigma_{v'j'}/d\cos\theta = $ constant, which will appear as sideward peaking in the $d\sigma_{v'j'}/d\theta$ representation due to the $\sin\theta$ term in the integration over all azimuthal angles.) The global angular distribution shown in Fig. 3(b) corresponds to the result of the summation of these (v', j')-resolved distributions according to their cross sections. For a reaction which forms a collisional complex, the exact shape of the c.m. angular distribution of products is determined by the angular momentum disposal [18], even if the complex is short-lived. In general, a f/b peaking angular distribution corresponds to $\mathbf{L} \approx \mathbf{L}'$ and an isotropic distribution implies $\mathbf{L} \approx \mathbf{j}'$. It is quite remarkable that both extreme types of directional correlations in angular momentum disposal are revealed here for different product vibrational states in the very same reaction. Furthermore, the differences in the angular contours for $v' = 0$ and $v' = 1$ apparently become more pronounced at the higher collision energies. To our knowledge, this is the first time that such a dramatic contrast in product state-specific angular distribution is revealed for any complex-forming reaction.

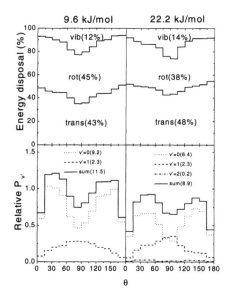

Figure 6: Summary of the angle-specific energy disposals (the upper panels) and the angle-specific vibrational branchings (the lower panels) for the two collision energies. The numbers in the parentheses give the corresponding angle-integrated values

5 Conclusion

The doubly differential cross section $(d^2\sigma/dvd\cos\theta)$ for the reaction $S(^1D)+D_2$ at $E_c = 9.6\,\mathrm{kJ/mol}$ was mapped out by a newly-developed technique called Doppler-selected TOF method. This new technique is simple and robust, yet its resolution is sufficiently high to observe the vibrational structures directly and to infer the shape of the rotational energy distribution. Detailed analysis procedure was presented. Compared to the previous results at $E_c = 22.2\,\mathrm{kJ/mol}$, the dependence of the detailed insertion dynamics on the initial collision energy is then revealed. As shown in Figs. 3-5 and summarized in Figs. 6 and 7, clearly there is rich dynamical information underneath the statistical treatment to be uncovered for an indirect complex-forming reaction. Theoretical efforts, both the ab initio characterization of the interaction potential and the dynamics calculation, are currently underway. We hope that the work presented here will provide the stimulus for more detailed investigations and for better understanding of insertion reactions in general.

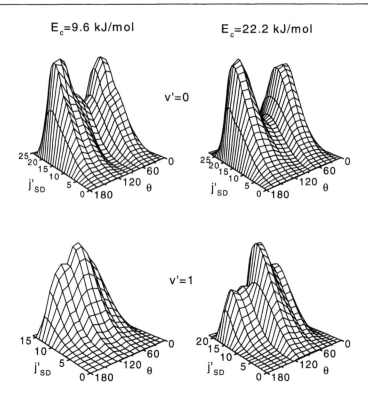

Figure 7: The 3-D representations of the product ro-vibrational state-specific angular distributions $(d\sigma_{v'j'}/d\theta)$. See text for details

Acknowledgments

This work was supported by the National Science Council of Taiwan and the Chinese Petroleum Corporation.

References

[1] L. Banares, F.J. Aoiz, V.J. Herrero, M.J. D'Mello, B. Niederjohann, K. Seekamp-Rahn, E. Wrede, L. Schnieder: J. Chem. Phys. **108**, 207 (1998)

[2] D.M. Neumark, A.M. Wodtke, G.N. Robinson, C.C. Hayden, Y.T. Lee: J. Chem. Phys. **82**, 3045 and 3067 (1985).

[3] M. Baer, M. Farrbel, B. Martinez-Haya, L.Y. Rusin, U. Tappe, J.P. Toennies: J. Chem. Phys. **108**, 9694 (1998)

[4] G. Dharmasena, K. Copeland, J.H. Young, R.A. Lasell, T.R. Phillips, G.A. Parker, M. Keil: J. Phys. Chem. A **101**, 6429 (1997)

[5] D.E. Manolopoulos: J. Chem. Soc. Faraday Trans. **93**, 673 (1997)

[6] W.R. Simpson, T.P. Rakitzis, S.A. Kandel, A.J. Orr-Ewing, R.N. Zare: J. Chem. Phys. **103**, 7313 (1995)

[7] S.A. Kandel, T.P. Rakitzis, T. Lev-on, R.N. Zare: J. Chem. Phys. **105**, 7550 (1996) ; ibid **107**, 9392 (1997)

[8] R.D. Levine, R.B. Bernstein: Molecular Reaction Dynamics and Chemical Reactivity, Oxford University Press, New York, (1987)

[9] Y.-T. Hsu, K. Liu: J. Chem. Phys. **107**, 1664 (1997)

[10] S.-H. Lee, K. Liu: Chem. Phys. Lett. **290**, 323 (1998)

[11] S.-H. Lee, K. Liu: J. Phys. Chem. A **102**, 8637 (1998)

[12] D. Simah, B. Hartke, and H.-J. Werner: J. Chem. Phys. **111**, 4523 (1999); A.M. Mebel (private communication).

[13] R.G. Macdonald, K. Liu: J. Chem. Phys. **91**, 821 (1989)

[14] Y.-T. Hsu, K. Liu, L.A. Pederson, G.C. Schatz: J. Chem. Phys. **111**, 7921 (1999)

[15] J.-H. Wang, Y.-T. Hsu, K. Liu: J. Phys. Chem. A **101**, 6593 (1997)

[16] A.J.R. Heck, D.W. Chandler: Ann. Rev. Phys. Chem. **46**, 355 (1995)

[17] J.C. Polanyi: Science **236**, 680 (1987)

[18] W.B. Miller, S.A. Safron, D.R. Herschbach: Discuss. Faraday Soc. **44**, 108 (1967)

High Resolution Translational Spectroscopic Studies of Elementary Chemical Processes

X. Liu[1], J.J. Lin[1], D.W. Hwang[1,3], X.F. Yang[1,4],
S. Harich[1], X. Yang [1,2]*

[1] Institute of Atomic and Molecular Sciences, Academia Sinica,
Taipei, Taiwan
[2] Department of Chemistry, National Tsing-Hua University,
Hischu, Taiwan
[3] Department of Chemistry, National Taiwan University,
Taipei, Taiwan
[4] Department of Chemistry, Dalian University of Technology,
Dalian, China

1 Introduction

Nascent product quantum state distribution in chemical reactions can now be measured routinely using many laser-based techniques such as laser-induced fluorescence (LIF), resonance-enhanced multiphoton ionization (REMPI) etc. For example, in the $O(^1D) + H_2 \rightarrow OH(^2\Pi) + H$ reaction, the OH radical products are frequently detected using LIF through the A ← X transition. Quantum state distribution of nascent chemical products, however, only carries part of the product information in a molecular beam experiment. Information on the angular distributions of reaction products is at least as important to the understanding of the whole picture of reactions. In reality, quantum state resolved differential cross section measurements could provide the most detailed mechanistic information on a chemical reaction, and also the most stringent test for a quantitatively accurate theoretical picture for this process.

Product translational spectroscopy in various forms has been used to measure the product angular distributions and angular resolved product translational energy distributions in both unimolecular and bimolecular reaction

---

*Corresponding author. E-mail address: xmyang@po.iams.sinica.edu.tw

processes. However, quantum-state resolved differential cross sections are hard to measure, to say the least. This is largely due to the limited translational energy resolution of various translational spectroscopic techniques and poor molecular beam condition under which crossed molecular beam experiments are routinely carried out. Recent development of the H atom Rydberg "tagging" TOF technique [1] has provided us an extremely powerful tool for measurement of state resolved differential cross sections for both unimolecular and bimolecular reactions with unprecedented translational energy resolution and extremely high sensitivity. This technique has been applied successfully to the studies of the important benchmark reaction $H+D_2 \to HD+H$ recently [2] and many important unimolecular dissociation processes [3].

Since H atom products from chemical reactions normally do not carry any internal energy excitation with its first excited state at ~ 10 eV, the high resolution translational distribution of the H atom products directly reflects the quantum state distribution of its partner product. For example, in the photodissociation of H_2O in a molecular beam condition,

$$H_2O + h\nu \to OH(\alpha, v, N) + H,$$

where α, v, N represent the electronic, vibrational and rotational states of the OH product. Total energy and linear momentum in the photodissociation process should be conserved:

$$E_{int}(H_2O) + E_{h\nu} - D_0(H-OH) = E_{int}(OH) + E_{trans}(OH) + E_{trans}(H)$$
$$m_{OH}E_{trans}(OH) = m_H E_{trans}(H).$$

Normally, in a molecular beam condition, $E_{int}(H_2O)$ (0 is not difficult to achieve. Therefore,

$$E_{int}(OH) = E_{h\nu} - D_0(H-OH) - (1+m_H/m_{OH})E_{trans}(H)$$

Since the photon energy, $E_{h\nu}$, and the dissociation energy, $D_0(H-OH)$, are constants and already known, the measured laboratory (LAB) H atom product translational energy ($E_{trans}(H)$) distribution can be easily converted into the center-of-mass (CM) OH product internal energy ($E_{int}(OH)$) distribution. With sufficient high translational energy resolution for the H atom product, the OH quantum state distribution can be determined for this photodissociation process by simply measuring the H atom product TOF spectra. OH product angular distribution can also be determined from the H atom product angular distribution since H and OH always run in opposite directions, ie, $\theta_{OH} = -\theta_H$. Similarly, these considerations also apply to the case of bimolecular reactions such as the $O(^1D)+H_2 \to OH(^2\Pi)+H$ reaction with only slight modifications in the calculations. In a crossed molecular beam study of a bimolecular reaction, translational energy resolution is normally limited by the spread in the collision energy rather than the intrinsic resolution of the Rydberg tagging method itself. In the photodissociation, higher translational energy resolution can usually be achieved by detecting H atom products at the perpendicular direction of the molecular beam.

In this article, we will present our recent experimental investigations on the unimolecular dissociation of the important H_2O molecule and the related bimolecular reaction $O(^1D) + H_2 \rightarrow OH + H$ using the H atom Rydberg "tagging" TOF technique. Through these studies, detailed dynamical information can be extracted experimentally for these important systems.

2 Experimental Methods

The Rydberg "tagging" TOF technique for the hydrogen atom was developed in the early 1990s by Welge and coworkers [1]. The detailed experimental methods used to study the crossed beam $H + D_2 \rightarrow HD + H$ reaction have been described in great details before [2]. Detailed descriptions of this technique used for studying molecular photodissociation can also be found in the [3]. The central scheme of this technique is the two step efficient excitation (see Fig. 1) of the H atom to its high Rydberg levels ($n = 35 \sim 90$) without ionizing the H atom product directly. Figure 2 shows a spectrum of Rydberg transitions of the H atom from the $n = 2$ level. These high Rydberg H atoms are known to be long lived to millisecond time scale in a small electric field ($\sim 20\,\text{V/cm}$). The enhancement of Rydberg H atom lifetime in a small electric field is likely caused by mixing of l quantum number in the Rydberg H atom [4, 5]. These quite long-lived neutral H atoms allow us to measure the time-of-flight spectrum of the neutral H atom chemical product with extremely high translational energy resolution (as high as 0.1 % in translational energy has been achieved). The neutral Rydberg H atoms are also quite easy to detect using field ionization. The extremely high translational energy resolution can be achieved by minimizing the physical sizes of the "tagging" region and the field ionization region which ionizes those Rydberg H atoms before reaching the multichannel plate (MCP) ion detector.

The excitation of the ground state H atom product ($n = 1$) is made by the following two step excitation scheme:

$$H(n = 1) + h\nu(121.6\,\text{nm}) \rightarrow H(n = 2) \tag{1}$$

and

$$H(n = 2) + h\nu(365\,\text{nm}) \rightarrow H^*(n \sim 50) \tag{2}$$

The 121.6 nm VUV light used in the first step excitation is generated using a two photon resonant $(2\omega_1 - \omega_2)$ four wave mixing scheme in the Kr gas cell. $2\omega_1(212.5\,\text{nm})$ is resonant with the Kr (4p-5p) transition [6]. ω_1 is

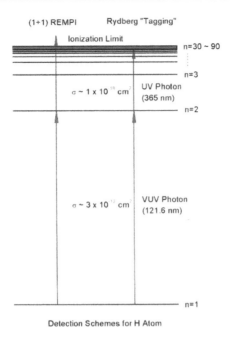

Figure 1: Detection schemes for H atoms. Rydberg tagging technique is slightly different from the $(1+1)$ REMPI detection scheme in which H atom products are directly ionized, while Rydberg tagging only pumps the H atom to the high Rydberg states

generated by doubling a dye laser pumped by a Nd:YAG (355 nm) laser, while ω_2(845 nm) is the direct output of a dye laser pumped by the second harmonic of the same YAG laser. During the experiment, a few mJ of 212.5 nm and 845 nm laser light are generally used. The efficiency of the VUV generation can be enhanced by adding the Ar gas as the phase matching medium in about a 3 : 1 ratio between Ar and Kr. By generating about 50 µJ of the 121.6 nm laser light, the first step can be easily saturated since this transition has a huge excitation cross section (3.0×10^{-13} cm^2). Following the first step VUV excitation, the H atom product is then sequentially excited to a high Rydberg state with $n \approx 50$ using a 365 nm light, which is generated by doubling a dye laser pumped by the same YAG laser. These two excitation lasers have to be overlapped very well in space and time. The neutral Rydberg H atom then flies a certain TOF distance to reach a MCP detector with a fine metal grid (grounded) in the front. After passing through the grid, the Rydberg tagged H atom products are then immediately field-ionized by the electric filed applied between the front plate of the Z-stack MCP detector and the fine metal grid. The signal

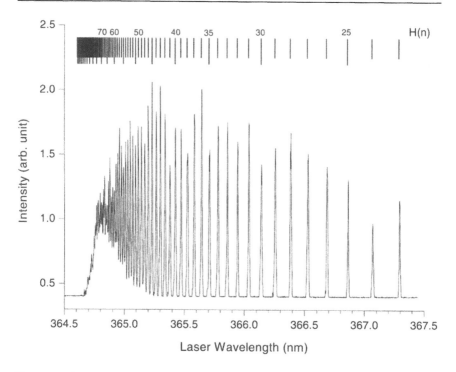

Figure 2: H atom Rydberg transitions from the $n = 2$ level to the higher n states

received by the MCP is then amplified by a fast pre-amplifier, and counted by a multichannel scaler.

Two types of experiments, unimolecular photodissociation (H_2O) and crossed beam bimolecular reaction ($O(^1D)+H_2 \to OH+H$), have been carried out in this work. In the photodissociation experiments, dissociation dynamics at two photolysis wavelengths (157 nm and 121.6 nm) were studied. At 157 nm excitation, H_2O is excited to the $^1B_1$ surface; while at 121.6 nm, H_2O is excited to the $^1B_1$ surface. The experimental setup used to study the H_2O photodissociation is described by a simple schematic shown in Fig. 3. A fluorine laser (Lambda Physik) is used as the photolysis laser source for the photodissociation of H_2O at 157 nm; while the same laser light to pump the H atom Lymann a transition is also used as the photolysis laser for the 121.6 nm photodissociation. The 157 nm laser light is unpolarized, while the 121.6 nm laser light is polarized. The polarization direction of the 121.6 nm photolysis light can be changed by rotating a polarization of the 845 nm laser using a rotatable half-waveplate for angular anisotropy measurement. In the H_2O photodissociation, a molecular beam of H_2O was generated by expanding a mixture of H_2O and Ar at a stagnation pressure of 600 Torr through a 0.5 mm diameter pulsed nozzle. The mixture of H_2O and Ar was made by bubbling Ar through the water sample at room temperature. The rotational temperature of the H_2O molecules in the molecular beam is about ~ 10 K. The detector in the photodissociation experiment is fixed at the perpendicular direction of the molecular beam. The molecular beam and the photolysis laser beam are all perpendicular to each other.

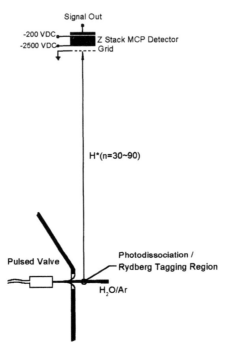

Figure 3: Experimental setup for studies of H_2O photodissociation

The experimental setup used for crossed beam experiments is slightly different from that used for photodissociation studies. In the crossed beam study of the $O(^1D) + H_2 \rightarrow OH + H$ reaction, two parallel molecular beams (H_2 and O_2) were generated with similar pulsed valves. The $O(^1D)$ atom beam was produced by the 157 photodissociation of the O_2 molecule through the Schumann–Runge band. The $O(^1D)$ beam was then crossed at 90° with the H_2 molecular beam. The H atom products were detected by the Rydberg "tagging" TOF technique with a rotatable MCP detector.

3 Unimolecular Dissociation of H_2O

3.1 H_2O on the $\tilde{A}^1B_1$ Surface: A Direct Dissociation

Photodissociation of H_2O in the VUV region is a very important topic in atmospheric [7] and instellar chemistry [8]. Dissociation on the first electronic excited states $\tilde{A}^1B_1$ surface of H_2O is also an excellent case for direct dissociation that can be studied from the first principle without too much complication. During the last decade or so, photodissociation of H_2O at 157 nm from the $\tilde{A}^1B_1$ state has been extensively studied both theoretically [9, 10, 11, 12, 13, 14, 15] and experimentally [16, 17, 18, 19, 20, 21, 22, 23, 24, 25, 26, 27, 28]. Modern laser spectroscopic techniques can now easily be used to detect individual quantum states of the OH radical product making direct experimental measurements of OH product ro-vibrational distributions possible. Theoreticians have also made significant progresses in understanding of unimolecular and bimolecular reactions during the last decade or so. More accurate potential surfaces and better dynamical calculations are more accessible than ever before. Theoretical predictions of the product quantum state distributions are more accurate than ever.

Good agreements have been found between theory and experiment in the rotational distributions of the OH(v) radical product from H_2O photodissociation at 157 nm [7]. However, there have been significant discrepancies between theory and experiment on the exact vibrational state distribution of the OH product from the photodissociation of H_2O at 157 nm. Recently, we have measured the translational energy distribution of the H atom product from H_2O photodissociation at 157 nm. As pointed out in the introduction, the number density of H atom product excited to the Rydberg states in the interaction region is directly proportional to the number density of total H products in the photodissociation volume. Therefore, the TOF spectrum of the H atom products will provide us with a most direct measurement of the OH product vibrational state distribution with different rovibrationally excited OH products detected with exactly the same efficiency.

Figure 4 shows the translational energy spectrum of the H atom product from photodissociation of H_2O in the molecular beam. Five rovibrational features have been observed in the spectrum. These features can be easily assigned to the vibrational excited OH (v = 0, 1, 2, 3, 4) products from the photodissociation of H_2O at 157.6 nm. Rotational structures with rotational quantum number N larger than 5 are partially resolved. The relative vibrational state populations of the OH product were obtained by simply integrating all signals for each rovibrational structure in Fig. 4. The relative vibrational distributions obtained from the experimental measurements in this work are listed in Table 1. Previous experimental and theoretical results are also shown in the table. All results shown in Table 1 are plotted in Fig. 5 for comparisons.

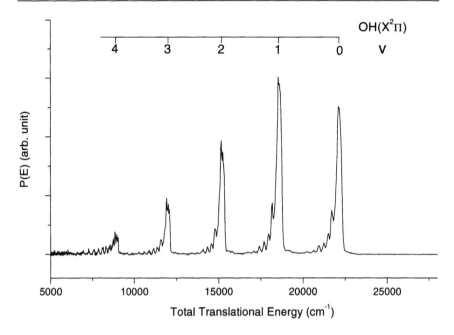

Figure 4: The total translational energy distribution of H_2O photodissociation at 157 nm. The peaks correspond to the different ro-vibrationally excited OH products

Table 1: Relative vibrational state populations of the OH(X, v) products from photodissociation of H_2O at 157.6 nm

Ref	v = 0	v = 1	v = 2	v = 3	v = 4	v = 5	Remarks
[16]	1.0	0.96	0.15	–	–	–	LIF, 300 K
[26]	1.0	0.56	0.10	0.02	0.003	$< 3 \times 10^{-4}$	LIF, flow cell
[8]	1.0	0.89	0.63	0.41	0.28	0.24	3D trajectory
[14]	1.0	0.81	0.53	0.34	0.18	0.10	3D ab initio
This work	1.0	1.11	0.61	0.30	0.15	0	R tagging, beam

From Table 1 and Fig. 5, it is quite obvious that the relative OH product vibrational state distribution obtained in this work is significantly different from those measured using laser induced fluorescence (LIF), especially at the higher vibrational states. The relative vibrational population obtained here for v = 4 are as high as 50 times larger than that obtained by previous LIF measurements. This large discrepancy indicates that the vibrational distributions of OH obtained by previous LIF measurements may have serious errors. Since the previous LIF measurements of the OH vibrational distributions were done with great care, these large errors are not likely caused by human errors. Instead, these errors are likely due to the LIF schemes used for the OH LIF detection. This fact will have certain significant impacts since the LIF detection scheme is widely used to measure the product OH vibrational state distributions in the studies of many chemical reactions, indicating that one must be very careful when using LIF to measure vibrational distributions of the OH($^2\Pi$) reaction product.

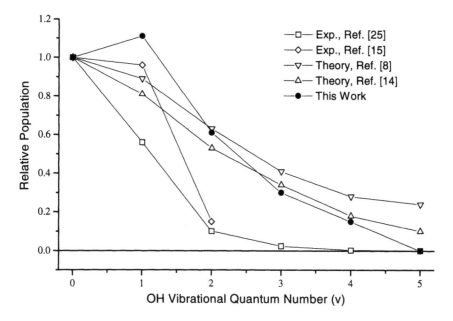

Figure 5: The OH product vibrational state distributions from H_2O photodissociation at 157 nm obtained from this work are compared with previous experimental and theoretical results

Generally speaking, LIF is a nice and sensitive detection method for molecular species. However, there are a few problems associated with LIF when one uses this technique to measure relative product vibrational populations. Firstly, in order to determine the relative population distributions in many vibrational states, one normally could not cover all vibrational states with one dye tuning range. This will likely cause larger propagation errors. Secondly, saturation effects may also cause significant errors when one calculates relative populations using relative LIF intensities. When converting relative LIF intensities to relative populations, one needs information that is usually not available, such as transition dipole moments etc. Since there is also propagation errors, this error can be amplified. Predissociation of the excited electronic state in the LIF scheme can also cause serious problems.

And it is well known that OH predissociates in the A $^2\Sigma$ state. This could be a major source of errors in measuring the vibrational distribution of OH($^2\Pi$) because the effect of predissociation is hard to account for quantitatively in many cases.

In comparison with previous theoretical calculations, the agreement of our experimental results with both theoretical results is rather good, indicating previous theoretical calculations should be generally all right. However, there are still obvious quantitative differences between the experimental results and theoretical calculations. From our estimation, uncertainties in the experimental results should be less than 10%. Therefore the discrepancies are more likely resulting from inaccuracies in the theoretical calculations, especially the potential energy surfaces used in the previous theoretical works. More accurate surfaces are available already at the moment, it is therefore interesting to recalculate the OH vibrational distribution for H_2O photodissociation at 157 nm to see whether better quantitative agreement between theory and experiment can be achieved for this important system.

3.2 H_2O on the $\tilde{B}^1A_1$ Surface: Dynamical Interference

Dissociation of H_2O at 121.6 nm through the B state has also been a topic of extensive experimental and theoretical studies because of its fundamental importance and interesting dynamics [29, 30, 31, 32, 33, 34, 35, 36, 37, 38, 39, 40, 41]. Photodissociation of H_2O from the B state is much more complicated than the A state since it involves much more complicated potential energy surfaces and multiple conical intersections. Energetically there are three possible channels, which produce H atom product, at 121.6 nm excitation:

$$H_2O + h\nu(121.6\,\text{nm}) \rightarrow OH(X^2\Pi) + H \quad (1)$$

$$\rightarrow OH(A^2\Sigma) + H \quad (2)$$

$$\rightarrow O(^1D) + 2H(\text{triple dissociation}) \quad (3)$$

Previous experimental studies show that the OH(X) ground state products are extremely rotationally excited. Experimental results also indicate the existence of both channel (2) and channel (3). Due to the limited resolution and sensitivity of the previous experimental studies, however, many details of the photodissociation process remain unresolved, such as the product angular anisotropy distributions, the effect of parent rotation on different dissociation channels, and the dynamics of the triple dissociation process etc. Recently, we have carried out very detailed experimental studies on the photodissociation of H_2O at 121.6 nm using the experimental technique described above. The translational energy distribution of the H atom products have been measured with very high translational energy resolution, product angular anisotropy has also been determined. Many interesting dynamical phenomena have been unraveled for this process.

Figure 6 shows the product translational energy distributions from the H_2O photodissociation at 121.6 nm when the photolysis laser polarization is either perpendicular or parallel to the detection axis. The experimental translational energy spectra were measured in a molecular beam condition in which the rotational temperature of H_2O is about 10 K. From these two spectra, the total product translational energy distribution at the magic angle can be determined, which is shown in Fig. 7. According to the well-known spectroscopic data for the OH radical, nearly all sharp features can be assigned to the OH product ro-vibrational states. Above $\sim 7500\,\text{cm}^{-1}$, most of the observed features are due to the OH ground electronic state products while below $\sim 7500\,\text{cm}^{-1}$, most of the sharp features are due to the OH $A^2\Sigma$ state products.

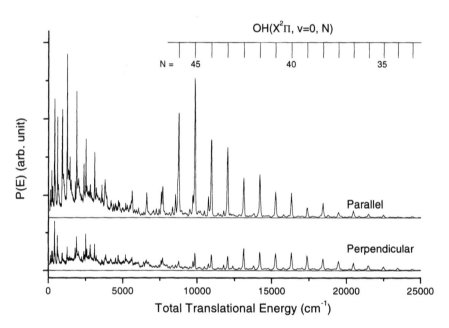

Figure 6: The translational energy distributions of H_2O photodissociation at 121 nm obtained with polarized lasers. The upper trace was acquired with the photolysis laser polarization parallel to the detection direction, while the lower trace was obtained with the photolysis laser polarization perpendicular to the detection direction

From Fig. 6, it is obvious that the major OH products in the ground electronic state are in the v = 0 level. Vibrationally excited OH products are also clearly present, their contributions to the total dissociation process, however, are quite small. The high sharp peaks above $\sim 7500\,\mathrm{cm}^{-1}$ can be mostly assigned to the rotationally excited OH(X) products at v = 0 of the ground electronic state with the highest peak N = 45, which corresponds to about 4 eV rotational excitation. This extremely high rotational excitation of the OH product is clearly caused by a peculiar dissociation pathway for H_2O dissociation on the B surface [42, 43]. As one of the OH bond stretches out, the HOH molecule becomes linear before the OH bond finally ruptures through the conical intersections. This would provide a large angular momentum kick to the OH product, thus producing very high rotationally excited OH radicals. Another intriguing observations from the TOF spectra in Fig. 6 is the clear oscillation behavior in the OH (X $^2\Pi$, v = 0) product rotational distribution with the odd N levels of OH(X, v = 0) have more population than the even N levels when the photolysis laser polarization is parallel to the detection axis. The oscillation pattern observed is very reproducible in the experiment. This oscillation was attributed to the dynamical interference through the two conical intersection dissociation pathways, H–OH and OH–H, on the B surfaces by Dixon et al. [44]. This idea has been tested with wave-packet calculations on realistic H_2O potential energy surfaces [44].

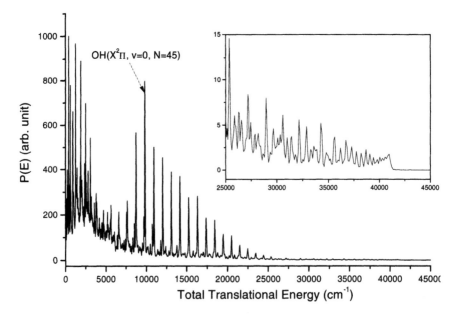

Figure 7: The total product translational energy distribution of H_2O photodissociation at 121.6 nm (magic angle). This distribution is obtained from the two distributions shown in Fig. 6

Since the product recoil velocity has a definite orientation with respect to the molecular coordinate system, this photoselection of molecules can result in an anisotropic angular distribution for the photofragments, which is conventionally described in the following way,

$$I(E, \theta) = I_0(E)(1 + \beta P_2(\cos\theta))/4\pi,$$

where θ is the angle between the observation direction (in lab frame) and the **E** field, $P_2(\cos\theta)$ is the second Legendre polynomial, and β is the anisotropy parameter which describes the angular distribution of photofragments. From the translational energy distributions (Fig. 6) taken at the perpendicular and parallel schemes, the product angular anisotropy parameters can also be determined using two points (perpendicular and parallel). Figure 8 shows the OH rotational dependence of the anisotropy parameters for OH at v = 0 level in the ground electronic state. It is very clear that the anisotropy parameter increases as the OH rotational excitation. This phenomenon can be described using the conical intersection dissociation pathway, indicating photodissociation on the B state mainly occurs through the conical intersection between the B and X states of H_2O. Assuming a distance of R_c between H and OH where dissociation occurs, the anisotropy parameters for rotationally excited OH products can then be calculated using the following formula (see detailed derivation of this formula for NH_3 photodissociation in reference [45]):

$$\beta(N) = A_0 \left\{ \frac{3N(N+1)\hbar^2}{2\mu R_c^2 [E_{avail} - E_{int}(OH)]} - 1 \right\}$$

where A_0 is a factor which accounts for the blurring of anisotropy due to parent rotational excitation. In Fig. 8, the calculated anisotropy parameters are also shown in comparison with the experimental results assuming $A_0 = 1.0$ and $R_c = 2.2$ Å. The results from the simple model calculations are in rather good agreement with the experimental results, which indicates that the dynamical picture presented above should be quite reasonable.

In the expanded plot of the translational energy distribution from 0 to 6000 cm^{-1} (Fig. 9), all the sharp line can be assigned to the OH($A^2\Pi$) products. In addition to these sharp lines, broad peaks and a underlying continuum are also present. These structures are believed to be from the dissociation of water clusters (3) after investigations based on stagnation pressure dependence and concentration dependence. It is quite interesting that clear structures have been observed in the photodissociation of the water clusters, these structures are likely due to the small clusters such water dimer or water-Ar complex. Further investigations are needed in order to understand fully these broad structures. Work in this direction is under way. Extensive studies on the dissociation of the water isotomopers at 121.6 nm have also been performed. Based on the above experimental

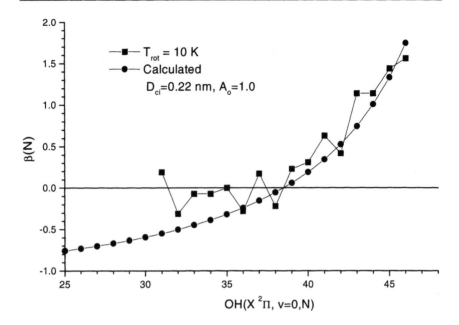

Figure 8: Rotational dependence of the angular anisotropy parameter (β) for the OH($X^2\Pi$, v = 0, N) products from H_2O 121.6 nm photodissociation

results, it is our belief that constructing a quantitative theoretical picture of the H_2O photodissociation at 121.6 nm is now possible with accurate potential energy surfaces.

4 $O(^1D) + n-H_2(^1\Sigma) \rightarrow OH(X^2\Pi, v, N) + H(^2S)$: The Insertion Mechanism

The reaction of $O(^1D) + H_2$ plays a significant role in atmospheric [46] and combustion chemistry [47]. The rate of this reaction has been found to be nearly gas kinetic. [48, 49, 50, 51, 52, 53, 54, 55, 56, 57, 58, 59, 60, 61, 62, 63]. This reaction is also a well-known bench mark system for an insertion type chemical reaction. Extensive experimental and theoretical studies have been carried out in order to elucidate the dynamics of this reaction [64, 65, 66, 67, 68, 69, 70, 71, 72, 73]. Very recent experimental studies show that at the collision energy above $\sim$ 1.8 kcal/mol, an elusive abstraction mechanism also becomes possible [74, 75, 76]. At energies below 1.8 kcal/mol, however, this reaction remains the most well known example of an insertion type reaction. Recently, we have studied this reaction using the H atom Rydberg tagging TOF technique. The experiment was carried

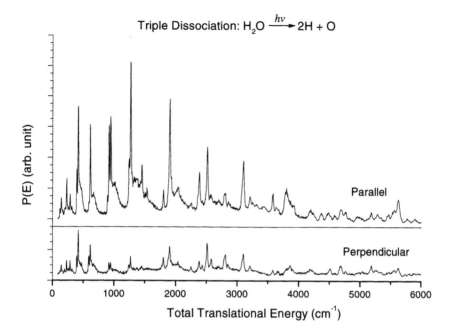

Figure 9: The product translational energy distributions in the 0 to 6000 cm^{-1} energy range. All the sharp peaks can be assigned to the OH(A$^2\Sigma$, v, N) products, while the broad peaks and the underlying continuum are attributed to the dissociation of water clusters

out at collision energy of $\sim$ 1.4 kcal/mol which is significantly below the 1.8 kcal/mol barrier for the abstraction mechanism inferred by previous experimental studies. Therefore this investigation would provide an ideal case for the pure insertion type reaction. The purpose of the experiment is to measure the rotationally resolved differential cross sections for this reaction to provide theoreticians a solid test ground for this model system. For the last few decades, experimental research on chemical reaction dynamics has provided extremely insightful information on mechanisms of a range of important chemical reactions. However, accurate measurements on fully rotational state resolved differential cross-sections have been carried out only for a couple of reactions so far: H + D$_2$ → HD + H [2] and F + H$_2$ → HF + H [77] due to the difficulties and the many limitations in this type of measurements. These reactions are probably the most thoroughly studied systems so far, which have provided excellent examples for understanding the reaction dynamics at the state-to-state level. Through the studies of O(^{1}D) + H$_2$ → OH($^2\Pi$) + H in this work, we hope to establish this reaction as the true model system for the insertion type reaction at this low energy, and provide a solid test ground for many concepts involving the insertion type at the state-to-state level.

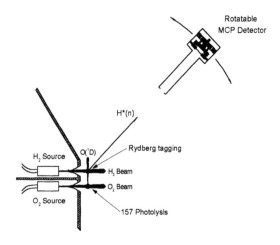

Figure 10: A simple schematic of the experimental setup for studying the crossed beam $O(^1D) + H_2 \to OH + H$ reaction

Figure 10 shows a simple scheme of the experimental setup used for the crossed molecular beam studies of the $O(^1D) + H_2 \to OH + H$ reaction. As described above, the $O(^1D)$ beam is generated by photolysis of O_2 in a molecular beam using a 157 nm F_2 laser. This process produces one $O(^1D)$ atom, one $O(^3P)$ atom which has very small reaction cross-section with H_2. The $O(^1D)$ beam is then crossed with a normal H_2 molecular beam which is produced by a pulsed valve cooled down to liquid nitrogen temperature. This is to reduce the energy uncertainties of the collision energy by minimizing the beam velocity spread. The Newton diagram of the reaction performed in this experiment is shown in Fig. 11. The velocity of the $O(^1D)$ beam has been measured to be 2050 m/s, while the velocity of the H_2 beam is 1350 m/s. Normal hydrogen (n-H_2) is used in this experimental study. It contains both p-H_2 and o-H_2 with a ratio of p-H_2 (mainly $J = 0$): o-H_2 (mainly $J = 1$) = 1 : 3 in the molecular beam. The energetic limits of the H atom products that correspond to the different OH vibrational states are described by the different circles in the Newton diagram.

Time-of-flight spectra of the H atom products have been measured at many laboratory angles. Figure 12 shows the TOF spectra at three laboratory angles: $\Theta_L = 117.5°$, $90°$, $-50°$. The laboratory angles at which the TOF spectra are shown are indicative of forward ($-50°$), backward ($117.5°$) and sideward scattering ($90°$). From Fig. 12, it is quite clear that that these spectra consist of a lot of sharp structures. These sharp structures clearly correspond to the OH quantum state resolved products, indicating that these measured TOF spectra have indeed achieved rotational state resolu-

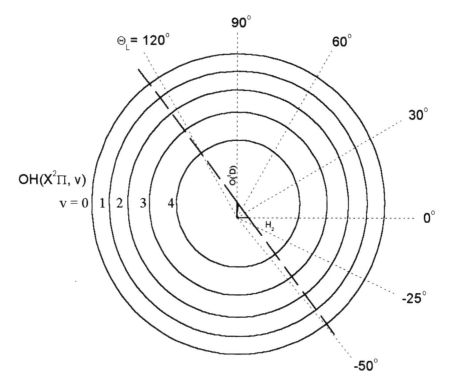

Figure 11: The Newton diagram for the crossed beam O(^{1}D)+H$_2$ → OH+H reaction

tion for the O(^{1}D) + n − H$_2$ → OH($^2\Pi$) + H reaction. From these TOF spectra, the product translational energy distributions at different center-of-mass (CM) scattering angles of the title reaction can be obtained.

Figure 13 shows the translational energy distributions obtained from the experimental measured TOF spectra. Since the center-of-mass angles are not constant for a certain laboratory angle at different product (H) velocity (see Newton diagram in Fig. 11), each of the translational energy distributions obtained from the three different laboratory angles contains information in a range of CM angles. However, the three translational energy distributions still carry the basic information on forward, backward and sideward scattering.

From these distributions, it is hopeful that information on ro-vibrational state distributions of the OH product can be obtained from reasonable simulations. The low rotational OH(X, v = 0) products, which correspond to the cutoff energy near 46 kcal/mol, are clearly more pronounced at the forward and backward directions than at the sideward scattering. The over-

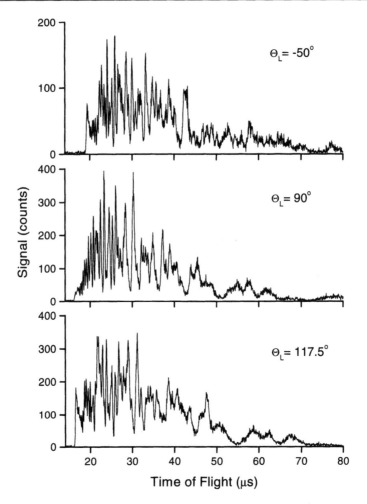

Figure 12: The time of flight spectra at three different laboratory angles for the H atom products obtained from the crossed beam $O(^1D)+H_2 \rightarrow OH+H$ reaction using the Rydberg tagging TOF technique

all features at backward and forward direction scattering are more similar in comparison with the sideward scattering. The total signal for backward scattering is, however, clearly larger than the forward backward scattering. This observation is due to the experimental instrument function. More detailed modeling is required to resolve this issue. The detailed symmetry between the forward and the backward scattering in state-to-state details is, however, not so clear at the moment. This will be definitely an interesting test case that can provide us the most detailed information on this issue.

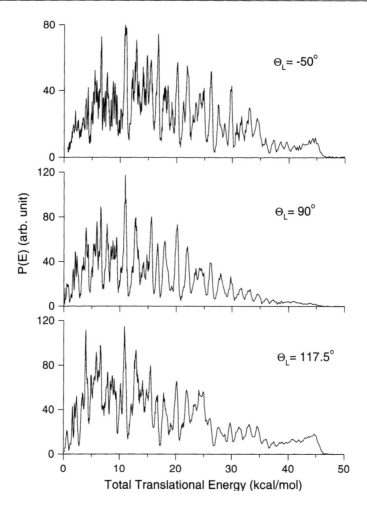

Figure 13: The total product translational energy distributions converted from the TOF spectra shown in Fig. 12

It is necessary to point out here that the data presented here are very preliminary, much detailed analyses are required in order to answer this type question on more solid ground. Nevertheless, we have provided an excellent example of quantum state resolved reactive scattering for an important insertion type reaction. Through more thorough experimental and theoretical investigations, quantitative understandings of this important reaction will be possible.

5 Concluding Remarks

In this article, we have shown a few examples of quantum state resolved reactive scattering studies of both unimolecular and bimolecular reactions using the H atom Rydberg "tagging" TOF technique. Detailed dynamical information can be learned from these well-controlled experimental investigations. Photodissociation of H_2O at both 157 nm and 121.6 nm has been studied in great details. From the studies of H_2O at 157 nm, the OH vibrational state distribution was measured. Out results indicate the inaccuracies using LIF technique to measure the OH vibrational state distributions. Photodissociation of H_2O at 121.6 has provided an excellent dynamical case of complicated, yet direct dissociation process. These experimental investigations have provided a solid ground to test the accuracy of the currently available H_2O potential energy surfaces and to improve these surfaces. The $O(^1D) + H_2 \rightarrow OH + H$ reaction has also been studied using the Rydberg tagging TOF method. Rotational state resolved differential cross-sections have been measured for this reaction for the first time. These measurements will provide the most detailed experimental results for this prototype insertion reaction.

Acknowledgments

This work is supported by National Science Council, Academia Sinica, the China Petroleum Co. We are very grateful for the helpful advice and support from Prof. Yuan T. Lee and Prof. Kopin Liu. The authors also wish to thank Prof. Richard N. Dixon for many insightful discussions.

References

[1] L. Schnieder et al.: J. Chem. Phys. **92**, 7027 (1990)

[2] L. Schnieder et al.: Science **269**, 207 (1995); L. Schnieder et al.: J. Chem. Phys. **107**, 6175 (1997)

[3] M.N.R. Ashfold, D.H. Mordaunt, S.H.S. Wilson, Advances in Photochemistry, Volume 21, edited by D.C. Neckers, D.H. Volman and G. v. Bunau, John Wiley & Sons, Inc, 217 (1996)

[4] W.A. Chupka: J. Chem. Phys. **98**, 4520 (1993)

[5] A. ten Wolde et al.: Phys. Rev. A **40**, 485 (1989)

[6] J.P. Marangos et al.: J. Opt. Soc. Am. B **7**, 1254 (1990)

[7] R.P. Wayne: Chemistry of Atmospheres, 2nd ed., Oxford Science, Oxford, 1991.

[8] P. Andresen, G.S. Ondrey, B. Titze: Phys. Rev. Lett. **50**, 486 (1983)

[9] H. Guo, J.N. Murrel: Mol. Phys. **65**, 821 (1998)

[10] S. Hennig, V. Engel, R. Schinke: Chem. Phys. Lett. **149**, 455 (1988)

[11] K. Kuhl, R. Schinke: Chem. Phys. Lett. **158**, 81 (1989)

[12] R. Schinke, V. Engel, V. Staemmler: J. Chem. Phys. **83**, 4522 (1985)

[13] P. Andresen, R. Schinke: In: Molecular Photodissociation Dynamics, edited by M.N.R. Ashfold, J. Baggot, Royal Society of Chemistry, London (1987)

[14] D. Imre, J. Zhang: Chem. Phys. **139**, 89 (1989)

[15] V. Engel, R. Schinke, V. Staemmler: Chem. Phys. Lett. **130**, 413 (1986); J. Chem. Phys. **88**, 129 (1988)

[16] P. Andresen, G.S. Ondrey, B. Titze, E.W. Rothe: J. Chem. Phys. **80**, 2548 (1984)

[17] P. Andresen et al.: J. Chem. Phys. **83**, 1429 (1985)

[18] D. Hausler, P. Andresen, R. Schinke: J. Chem. Phys. **87**, 3949 (1987)

[19] R.L. Vander Wal, F.F. Crim: J. Chem. Phys. **93**, 5331 (1989)

[20] R.L. Vander Wal, R.J. Scott, F.F. Crim: J. Chem. Phys. **92**, 803 (1990)

[21] R.L. Vander Wal et al.: J. Chem. Phys. **94**, 3548 (1991)

[22] R. Schinke et al.: J. Chem. Phys. **94**, 283 (1991)

[23] R.L. Vander Wal, J.L. Scott, F.F. Crim: J. Chem. Phys. **94**, 1859 (1991)

[24] M. Brouard, S.R. Langford, D.E. Manolopoulos: J. Chem. Phys. **101**, 7458 (1994)

[25] A.U. Grunewald, K.-H. Gericke, F.J. Comes: Chem. Phys. Lett. **133**, 501 (1987)

[26] K. Mikulecky, K.-H. Gericke, F.J. Comes: Chem. Phys. Lett. **182**, 290 (1991)

[27] D.F. Plusquellic, O. Votava, D.J. Nesbitt: J. Chem. Phys. **101**, 6365 (1994)

[28] D.F. Plusquellic, O. Votava, D.J. Nesbitt: J. Chem. Phys. **107**, 6123 (1997)

[29] H.J. Krautwald et al.: Faraday Discuss. Chem. Soc. **82**, 99 (1986)

[30] D.H. Mordaunt, M.N.R. Ashfold, R. Dixon: J. Chem. Phys. **100**, 7360 (1994)

[31] K. Weide, R. Schinke: J. Chem. Phys. **87**, 4627 (1987)

[32] R. Dixon: Mol. Phys. **85**, 333 (1985)

[33] A. Hodgson et al.: Mol. Phys. **54**, 551 (1985)

[34] M.P. Docker, A. Hodgson, J.P. Simons: Mol. Phys. **57**, 129 (1986)

[35] A. Hodgson: Faraday Discuss. Chem. Soc. **82**, 190 (1986); L.J. Dunne: ibid. **82**, 190 (1986); J.N. Murrel, ibid. **82**, 191 (1986); M.N.R. Ashfold, R.N. Dixon, ibid. **82**, 193 (1986)

[36] E. Segev, M. Shapiro: J. Chem. Phys. **77**, 5604 (1982)

[37] L.J. Dunne, J, Guo, J.N. Murrel: Mol. Phys. **62**, 283 (1987)

[38] M. Brouard, M.P. Docker, J.P. Simons: Faraday Discuss. Chem. Soc. **82**, 188 (1986)

[39] N. Shafizadeh et al.: Chem. Phys. Lett. **152**, 78 (1988)

[40] L.J. Dunne, J.N. Murrel, J.G. Stamper: Chem. Phys. Lett. **112**, 497 (1984)

[41] B. Heumann et al.: Chem. Phys. Lett. **166**, 385 (1990)

[42] D.H. Mordaunt, M.N.R. Ashfold, R.N. Dixon: J. Chem. Phys. **100**, 7360 (1994); R.N. Dixon: J. Chem. Phys. **102**, 301 (1995)

[43] R. Schinke, V. Engel, V. Staemmler: J. Chem. Phys. **83**, 4522 (1985)

[44] R. N Dixon et al.: Science **285**, 1249 (1999)

[45] D.H. Mordaunt, M.N.R. Ashfold, R.N. Dixon: J. Chem. Phys. **104**, 6460 (1996)

[46] J.G. Anderson: Ann. Rev. Phys. Chem. **38**, 489 (1987), and references therein

[47] G. Dixon-Lewis, D.J. Williams: Comprehensive Chem. Kinet. **17**, 1 (1977)

[48] G. Paraskevopoulos, R.J. Cvetanovic: J. Am. Chem. Soc. **91**, 7572 (1969)

[49] R.J. Donovan, D. Husain, L.J. Kirsch: Chem. Phys. Lett. **6**, 488 (1970);

[50] R.F. Heidner III, D. Husain: Int. J. Chem. Kinet. **5**, 819 (1973);

[51] M.J.E. Gauthier, D.R. Snelling: J. Photochem. **4**, 27 (1975)

[52] L.J. Stief, W.A. Payne, R.B. Klemm: J. Chem. Phys. **62**, 4000 (1975)

[53] J.C. Tully: J. Chem. Phys. **62**, 1893 (1975)

[54] J.A. Davidson et al.: J. Chem. Phys. **64**, 57 (1976)

[55] J.A. Davidson et al.: J. Chem. Phys. **67**, 5021 (1977)

[56] P.H. Wine, A.R. Ravishankara: Chem. Phys. Lett. **77**, 103 (1981)

[57] A.M. Pravilov, V.N. Pauk, S.E. Ryabov: Kinet. Catal. **22**, 1109 (1981)

[58] P.J. Ogren et al.: J. Phys. Chem. **86**, 238 (1982)

[59] W.B. DeMore et al.: Chemical kinetics and photochemical data for use in stratospheric modeling. Evaluation number 9, JPL Publication 90-1, Pasadena, CA., 1990, pp1

[60] R. Atkinson et al.: Evaluated kinetic and photochemical data for atmospheric chemistry. Supplement IV. IUPAC subcommittee on gas kinetic data evaluation for atmospheric chemistry, J. Phys. Chem. Ref. Data **21**, 1125 (1992)

[61] Y. Matsumi et al.: J. Phys. Chem. **97**, 6816 (1993)

[62] S. Koppe et al.: Chem. Phys. Lett. **214**, 546 (1993)

[63] T. Laurent et al.: Chem. Phys. Lett. **236**, 343 (1995)

[64] P.A. Whitlock, J.T. Muckerman, E.R. Fisher: J. Chem. Phys. **76**, 4468 (1982)

[65] R. Schinke, W.A. Lester, Jr.: J. Chem. Phys. **72**, 3754 (1980)

[66] S.W. Ransome, J.S. Wright: J. Chem. Phys. **77**, 6346 (1982)

[67] P.J. Kuntz, B.I. Niefer, J.J. Sloan: J. Chem. Phys. **88**, 3629 (1988)

[68] G.C. Schatz et al.: J. Chem. Phys. **107**, 2340 (1997)

[69] J.E. Butler et al.: Chem. Phys. Lett. **95**, 183 (1983); J.E. Butler et al.: J. Chem. Phys. **84**, 5365 (1986)

[70] R.J. Buss et al.: Chem. Phys. Lett. **82**, 386 (1981)

[71] K. Tsukiyama, Katz, R. Bersohn: J. Chem. Phys. **83**, 2889 (1985); Y. Matsumi et al.: J. Phys. Chem. **96**, 10622 (1992)

[72] M.S. Fritzcharles, G.C. Schatz: J. Phys. Chem. **90**, 3634 (1986); L.J. Dunne: Chem. Phys. Lett. **158**, 535 (1989); A.J. Alexander, F.J. Aoiz, M. Brouard, J.P. Simons, ibid. **256**, 561 (1996)

[73] J.K. Badenhoop, H. Koizumi, G.C. Schatz: J. Chem. Phys. **91**, 142 (1989); T. Peng et al.: Chem. Phys. Lett. **248**, 37 (1996)

[74] D.-C. Che, K. Liu: J. Chem. Phys. **103**, 5164 (1995)

[75] Y.T. Hsu, K. Liu: J. Chem. Phys. **107**, 1664 (1997)

[76] Y.-T. Hsu, J.-H. Wang, K. Liu: J. Chem. Phys. **107**, 2351 (1997)

[77] D.M. Neumark et al.: J. Chem. Phys. **82**, 3045 (1985); D.M. Neumark et al.: J. Chem. Phys. **82**, 3067 (1985); M. Faubel et al.: Chem. Phys. **207**, 227 (1996); M. Baer et al.: J. Chem. Phys. **110**, 10231 (1999)

Reaction Kinetics in Uniform Supersonic Flows at Very Low Temperatures

B. R. Rowe, A. Canosa, C. Rebrion-Rowe
Laboratoire Physique des Atomes, Lasers, Molécules et Surfaces
UMR 6627 du CNRS, Université de Rennes I
35042 Rennes Cedex, France.

1 Introduction

There is a strong fundamental interest in the study of molecular collisions under conditions of very low temperatures or energies. For example, for reactive collisions, the effect of very small energy barriers or the influence of rotational or spin-orbit states on the reactivity can be highlighted. One of the strongest motivations however, comes from modern astrophysics and is related to the fascinating problem of the birth of stars and planets from the gravitational collapse of molecular cloud cores [1]. Molecular clouds are the densest part of the interstellar medium and typical conditions in a dense core are, a kinetic temperature of around 10K and a concentration of molecular hydrogen of 10^4-10^5 cm^{-3}. In the collapse of a rotating cloud core, a circumstellar disk is naturally generated in which planets, comets and various objects can be formed by accretion and this process is certainly not restricted to our solar system. If dust is present in various parts of the interstellar medium as grains of sub-micron size, its concentration in dense clouds makes them opaque to the harsh UV and visible radiation field of the neighbouring stars. With the striking developments of radioastronomy in the last few decades, it is, however, now possible to track the physical conditions prevailing inside these clouds mainly through the rotational spectra (most often in the sub-millimetre and millimetre range) emitted by polar molecules. Spin-orbit state transitions are also widely used and, most recently, infrared observations from space (with ISO) have yielded a great wealth of new information.

The observations have allowed more than one hundred molecules to be detected in the interstellar medium, and most especially in molecular clouds. Some of these molecules, such as those with highly unsaturated carbon chains, are quite exotic. This raises the question of their formation in conditions of extremely low temperature and pressure which look at first glance to be quite unfavourable to molecule formation. Besides this fundamental question of how molecules are formed, it is now understood that the physics and chemistry of the interstellar medium are highly linked and that the study of the chemistry of molecular clouds is important for our understanding of the process of the evolution of matter in the universe towards the formation of planets and of life. This field of science, which includes the study of the physics and chemistry of dense and diffuse clouds, and of planetary and cometary atmospheres is now known as "molecular astrophysics" or "astrochemistry".

It is recognised that, in dense clouds, the chemistry is initiated by cosmic ray ionisation followed by a rich variety of processes, either in the gas phase (ion-molecule reactions, dissociative recombination of electrons with molecular ions, radical-molecule reactions) or on the surface of grains. The most recent numerical models of interstellar chemistry [2] now include hundreds of species and thousands of reactions. Such models are highly demanding in laboratory data obtained in the relevant temperature range.

2 Experimental techniques at very low temperatures or energies

To design an experiment able to study reactions or relaxation kinetics at a temperature well below that of liquid nitrogen is a considerable challenge. Since the early 1980s several experiments have been built for this purpose most being restricted to the study of ion-molecule reactions, due to the fact that ions can be easily manipulated by electromagnetic fields and detected with a very high efficiency. An example is the cryogenically cooled RF ion trap of Gerlich and coworkers [3]. Such techniques are restricted generally to hydrogen as the molecular reactant at the lowest temperatures, due to heterogeneous condensation of other neutral species onto the walls of the apparatus. They provide rate coefficients under conditions of thermodynamic equilibrium, (at least as far as the kinetic energy is concerned).

The experimental methods that do not use cryogenic cooling rely on supersonic expansions under rarefied conditions to produce low temperatures, thus avoiding heterogeneous condensation. In fact merged beam techniques, in which the cross section for a process is measured as a function of very low centre-of-mass energy [4] belong to this class of method, if at least one of the beams is formed by supersonic expansion. A free jet flow reactor has also been used by the group of M. Smith at Tucson to provide extremely interesting results for the kinetics of ion-molecule reactions at temperatures close to 0 Kelvin [5]. The first results on ion-molecule reactions down to very low temperatures (8K), however, were obtained by the author and his coworkers in the early 1980s at the Laboratoire d'Aérothermique de Meudon, using the so-called CRESU technique (a French acronym for "Cinétique de Réactions en Ecoulements Supersoniques Uniformes" which means "Kinetics of Reactions in Uniform Supersonic Flows").

The idea behind the CRESU technique grew out of the recognition that the uniform supersonic flow obtained at the exit of a Laval nozzle, under rarified conditions, is an ideal flow reactor in which one could perform chemical physics studies at extremely low temperatures. Such flows have been of course, widely used in rarefied wind tunnels for aerodynamic studies [6] but curiously were ignored by the chemical physics community. Before reviewing the most recent developments of the technique it is interesting to compare the free jet, the merged beams and the uniform flow from various point of views:

- Energy resolution : both the free jet flow reactor and the CRESU techniques are clearly recognised as thermal methods which produce measurements of rate coefficients as a function of temperature. This is particularly true for the CRESU in which a local thermodynamic equilibrium exists at least for the translation and rotation of the various species in the supersonic flow [7,8]. For the free jet, the situation is more complicated and the measured rate coefficient is related to non-equilibrium distributions even for translation. A careful analysis however [5], allows one to assign a given "true" temperature to a measured rate coefficient in a number of cases. It has to be understood that the situation for energy resolution is not any better for the merged beam technique. In an ideal experiment, the cross section (E) would be measured as a function of the centre of mass kinetic energy E with a very small spread in energy E. This information is more fundamental than a rate coefficient which can be derived, using the formula:

$$k(T) = \frac{1}{kT} \left(\frac{\gamma}{\pi\mu kT}\right)^{\frac{1}{2}} \int E\,\sigma(E)\,exp\frac{-E}{kT}\,dE$$

In fact at very low centre of mass energy (E around 1 meV and below), the energy spread E in a merged beam experiment is at least of the same order of magnitude as E. Therefore the energy resolution is not better than in a truly thermal experiment and to deconvolute in order to obtain a thermal rate coefficient, the energy and angular distributions of the beams have to be known. Due to this problem it is sometimes difficult to obtain a reliable rate coefficient for extremely low temperatures from such data.

- Collision regime : here is in fact the main difference between a CRESU and a merged beam experiment. Merged beams are used under molecular free conditions and collisions are very sparse. The cross sections are obtained from the detection of the reaction products formed in the rare events occurring in the overlapping zone of the beams. In the CRESU experiments the large density ensures that the flow remains in the continuous regime and a local thermodynamic equilibrium, at least for translation and rotation (hydrogen excepted in some cases) exists throughout the flow.

Using the CRESU technique a large body of data has been obtained on ion-molecule reactions down to 8 K and this has reinforced our confidence in the theoretical models of capture, as shown in Figure 1 for the case of the reaction of the N^+ ion with ammonia. Two apparatuses using stationary flows are now used world-wide, one in Rennes, (France) and the other at the University of Birmingham (UK) in the group of I.W.M. Smith [9]. Recently an interesting extension has been devised by M. Smith which uses uniform supersonic pulsed flows [10]. The technique has been shown to be very versatile and is now used for studies of radical-molecule reactions, electron attachment, relaxation and nucleation processes at very low temperatures and some of our most recent results in these fields will be reviewed in the next paragraphs. These experiments have been performed using a variety of nozzles which allow us to perform experiments under many pressure and buffer gas conditions. It has in fact been shown that the same nozzle can be used not only with different reservoir temperatures, but also with different buffer gases or buffer gas mixtures.

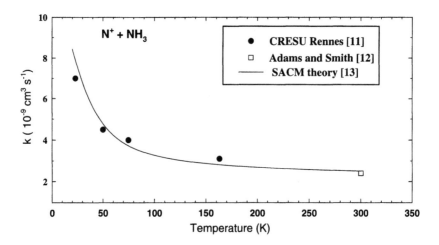

Figure 1: Rate coefficient of the reaction $N^+ + NH_3$ as a function of temperature. Experimental results are compared with a theoretical model.

3 Radical - molecule reactions

The database concerning reactions of neutral radicals with molecules below 200 K was very sparse in the early 1990s with virtually no results below liquid nitrogen temperature. In a collaboration with the Birmingham group of I.W.M. Smith, we have implemented the powerful PLP-LIF (Pulsed Laser Photolysis - Laser Induced Fluorescence) technique in one of the chambers of our CRESU apparatus, dedicated to this kind of experiment. A twin facility now exists at the School of Chemistry at Birmingham. Reactions of the radicals CN, CH, OH, Al, Si, C_2H and C (the last two at Birmingham) have now been studied with various molecules. The experiments have been described in detail in several papers (for example see [7,8]) and only some essential features will be described below. A sketch of the Rennes experimental chamber is shown in Figure 2.

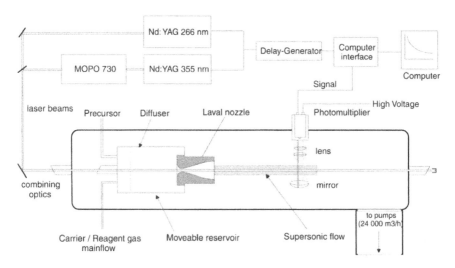

Figure 2: Sketch of the CRESU apparatus for the study of neutral - neutral reactions.

Inside the chamber, the Laval nozzle is attached, by means of a quick mounting flange, onto a reservoir which can be moved along the chamber axis. The buffer gas, either argon, nitrogen or helium, is introduced into the chamber together with a precursor of the radical to be studied and a neutral reactant gas. The buffer and reactant flow rates are controlled and measured by Tylan flow controllers and the various pressures measured using capacitance gauges. Two pulsed laser beams are made co-linear with the axis of the supersonic jet. The time widths of the laser pulses are shorter than 10 ns and the delay between the two pulses can be varied precisely between 0 and several milliseconds. The first beam, either a quadrupled or tripled Nd-Yag laser (266 or 355 nm), photolyses the precursor forming the neutral radical which is detected downstream of the nozzle exit, using Laser Induced Fluorescence (LIF) techniques, the emitted photons being detected by a photomultiplier connected to an optical collection system. Fluorescence of the radical is induced by the second laser beam which is generated by a tuneable OPO (MOPO 730 of Spectra-Physics).

When the photolysis can be accomplished without focusing the laser beam (i.e. when it is not necessary to use multiphoton processes) then the reactant radical is created with a uniform concentration along the flow axis and its temporal behaviour as found by varying the delay between the two laser pulses for a fixed distance between the photomultiplier and the nozzle exit, is solely linked to the kinetics of the studied reaction. For example in the case of aluminium atoms reacting with oxygen:

$$Al + O_2 \longrightarrow AlO + O$$

The LIF signal is proportional to the aluminium atom concentration [Al] which has the simple temporal dependence :

$$[Al] = [Al]_0 exp^{-k_1[O_2]t}$$

$[Al]_0$ is a constant due to the initial uniform concentration of aluminium atoms created by the photolysis of trimethylaluminium at 266 nm and molecular oxygen is in large excess, yielding pseudo first-order kinetics for the reaction. From the exponential decrease of the LIF signal it is therefore possible to deduce a pseudo first-order rate coefficient $k_1[O_2]$. A plot of this coefficient as a function of O_2 concentration yields the bimolecular rate coefficient k_1.

When it is necessary to focus the laser beam in order to photolyse the radical precursor, the initial density of the reactant radical is no longer uniform along the flow. This is, for example, for the case of silicon atoms formed by the photolysis of tetramethylsilane at 266 nm. Under this condition, it is necessary to adjust the distance between the focal point and the photomultiplier for each different time delay. Figure 3 and 4 show respectively, the temperature dependence of the reaction of molecular oxygen with aluminium and silicon atoms. The inverse temperature dependence of reaction (1) is well reproduced by capture theory [14]. In several other cases however, it has been found that the theory has difficulty in reproducing the observed temperature dependence for neutral reactions down to very low temperatures.

The finding that radical-molecule reactions often exhibit a strong inverse temperature dependence has shed a new light on the cold chemistry of space and modellers have now been forced to include neutral chemistry in their models to a much larger extent [15].

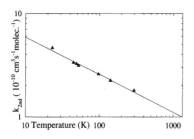
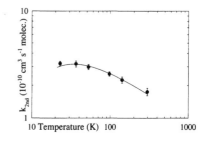

Figure 3: Rate coefficient for the reaction Al + O_2 as a function of temperature.

Figure 4: Rate coefficient for the reaction Si + O_2 as a function of temperature.

4 Electron attachment

There is strong evidence that Polycyclic Aromatic Hydrocarbons (PAHs) are present in interstellar clouds and it has been suggested [16] that they could attach electrons, therefore leading to the formation of massive anions. This would have strong implications, not only for the chemistry of the interstellar medium but also for the physics of star formation, through the correlation between the magnetic field and the conductivity of the tenuous interstellar plasma.

There exist a number of techniques which allow electron attachment to be studied at very low electron energy. These include Threshold Photo-Electron (TPE) as well as Rydberg Atom (RA) techniques. In the latter case, very high electron energy resolution (a few microelectronvolts) can be achieved. In most of these experiments however, the attaching neutral gas is held at room temperature.

We have recently devised a new experiment incorporating a Langmuir probe into the CRESU apparatus, which allows us to measure the rate coefficient β for attachment processes, whether dissociative or not [17]. In order to cool the electrons to the temperature of the flow, helium or nitrogen, at a sufficient pressure, have been used. An important difference, compared to some other techniques, is that the internal states of the reactant neutrals are in local thermodynamic equilibrium, especially low lying vibrational states. This is illustrated in Figure 5 which shows the value of $\beta(T)$ as measured by different experiments for the process :

$$CF_3Br + e \longrightarrow Br^- + CF_3$$

The CRESU results are in excellent agreement with those of the FALP technique which is also a thermal experiment, but the difference between these

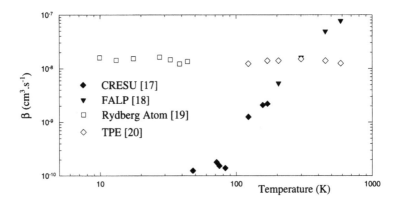

Figure 5: Rate coefficient for the dissociative attachment to CF_3Br vs electron temperature.

results and those obtained with other techniques is striking. It has been shown that this behaviour is completely linked to the difference in vibrational temperatures (which is equal to room temperature for the Threshold Photo-Electron and Rydberg Atom techniques) in the various experiments [17]. This demonstrates clearly that it is necessary to cool down, not only the electrons, but also the attaching molecules in order to obtain results relevant to interstellar chemistry. We have recently obtained our first results for electron attachment to anthracene which shows no temperature dependence down to 70 K [21].

5 Spin-orbit relaxation

Spin-orbit relaxation is an important process in the interstellar medium as it participates in the cooling of the clouds. This is particularly true for carbon cations and oxygen and carbon atoms. In many experiments also, at very low temperature or energy, it is important to know precisely if the spin-orbit states are well relaxed. The PLP-LIF technique, used for the study of radical-molecule reactions, can also be applied to this kind of process. As an example, we have recently studied the spin-orbit relaxation of aluminium atoms in argon at 53 and 137 K [22]. In the photolysis of trimethylaluminium, both $P_{\frac{3}{2}}$ and $P_{\frac{1}{2}}$ states are formed in nearly equal abundance. The $P_{\frac{3}{2}}$ state then relaxes rapidly towards its equilibrium value due to collisions with argon. As in the case of chemical reactions, the LIF signal corresponding to a transition from this state, follows an exponential decrease versus time which yields a pseudo first-order rate coefficient for the relaxation process. It can be shown that the rate coefficient derived from the LIF decay of the $P_{\frac{3}{2}}$ state is actually the sum of the relaxation rate coefficient k_r and the excitation rate coefficient k_e. Various nozzles corresponding to different argon densities for the same flow temperature have been used and this allows the determination of the binary rate coefficient to be made. A plot of this coefficient versus temperature is shown in Figure 6 and compared with theory. Experiments concerning the reaction of aluminium atoms with oxygen have then been conducted for reaction times, sufficient to ensure complete relaxation.

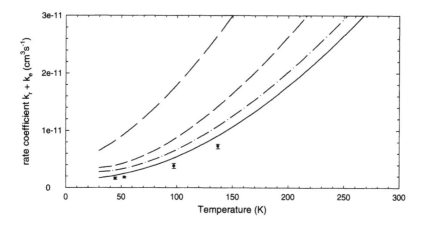

Figure 6: Total (relaxation and excitation) rate coefficient for the equilibrium recovery of Al ground state fine structure. Experimental data set with their error bars are compared to calculations carried out using different potentials for the AlAr complex [22].

6 Dimer formation

At very low temperatures, it is expected that the free enthalpy change in the formation of slightly bound neutral dimers, becomes negative or, in other words, that there is no longer an entropy barrier to nucleation. In the CRESU experiment, there is indirect evidence of the onset of homogeneous nucleation when using large flow rates of highly condensable species. For example, this can be seen in the non-linearity of the plot of the pseudo first-order rate coefficient versus neutral reactant concentration for a given reaction [8]. It has also been found that the dimer of CCl_2F_2 attaches electrons much more efficiently than the monomer although it was not possible to determine quantitatively the amount of monomers that undergo association [17]. There is a strong fundamental interest in measuring the rate coefficient for neutral dimer formation as it is the first step involved in homogeneous nucleation. We have recently studied the kinetics of benzene dimer formation between 15 and 123 K. Figure 7 shows the behaviour of the benzene LIF signal (i .e. of the true monomer concentration) as a function of the concentration deduced from the benzene flow rate in a buffer flow of helium at 36 K. The departure from a linear plot is a signature of the formation of dimers and, at larger flow rates, of clusters of higher masses. It is possible, by working under conditions where this process begins only at the nozzle exit, to measure the rate coefficient for the process:

$$C_6H_6 + C_6H_6 + He \longrightarrow (C_6H_6)_2 + He$$

A $T^{-2.8}$ temperature dependence for the termolecular rate has been found along with evidence that the measurements have been conducted close to the low pressure limit [23]. A temperature dependence of this kind is normal for association reactions.

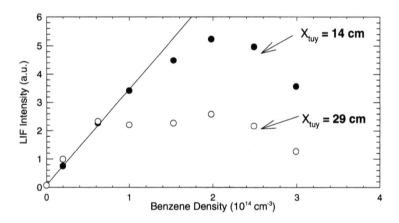

Figure 7: LIF intensity of the benzene transition $6_0^1\ 1_0^1$ as a function of the benzene concentration deduced from the experimental benzene flow rate. Data are obtained for two different nozzle positions with respect to the LIF detection area

7 Conclusion

In this presentation, we hope to demonstrate by the variety of the reactions that have been examined, the versatility of the CRESU technique for the study of physico-chemical reactions at very low temperatures. This versatility proves to be very useful when directed towards the problems presented by astrochemistry, namely the extremely diverse chemical reactions seen in the interstellar medium and in cold planetary atmospheres. From a fundamental standpoint, the results obtained concerning reaction kinetics in this unique temperature range, allow us to expand our knowledge of theoretical models for collision processes and chemical reactivity.

Ackowledgements

The authors would like to thank J.B.A. Mitchell for assistance in the preparation of this manuscript.

References

[1] J.S.Lewis, *Cosmic abundances Matter*, AIP conference proceedings **183**, 17 (1989).

[2] H.H.Lee, R.P.A.Bettens and E. Herbst, Astron. Astrophys. Suppl. Ser. **119**, 111 (1996).

[3] D.Gerlich and S.Horning, Chem. Rev. **92**, 1509 (1992).

[4] D.Gerlich, XII International Symposium on Molecular Beams, Perugia, Italy, edited by V. Aquilanti (1989), p 37.

[5] M.Hawley, T.L.Mazely, L.K.Randemiya, R.S.Smith, X.K.Zeng and M.Smith, Int. J. Mass. Spectrom. Ion. Proc. **80**, 239 (1990).

[6] J.M.Owen and F.S.Sherman, University of California,Technical Report HE 150-104, (1952).

[7] B.R.Rowe, G.Dupeyrat, J.B.Marquette and P.Gaucherel, J. Chem. Phys. **80**, 4915 (1984).

[8] I.R.Sims, J.L.Queffelec, A.Defrance, C.Rebrion-Rowe, D.Travers, P.Bocherel, B.R.Rowe and I.W.M.Smith, J. Chem. Phys. **100**, 4229 (1994).

[9] D.Chastaing, P.L.James, I.R.Sims and I.W.M.Smith, Faraday Discuss. **109**, 165 (1998).

[10] D.B.Atkinson and M.Smith, Rev. Sci. Instr. **66**, 4434 (1995).

[11] B.R.Rowe, A.Canosa and V.Le Page, Int. J. Mass Spectrom. Ion Proc. **149/150**, 573 (1995).

[12] N.G.Adams, D.Smith and J.F.Paulson, J. Chem. Phys. **72**, 288 (1980).

[13] J.Troe, J. Chem. Phys. **122**, 425 (1985).

[14] S.D.Le Picard, A.Canosa, D.Chastaing, D.Travers, B.R.Rowe and T.Stoecklin, J. Phys. Chem. A **100**, 14928 (1998).

[15] E.Herbst, H.H.Lee, D.A.Howe and T.J.Millar, Monthly. Not. Roy. Astron. Soc. **268**, 335 (1994).

[16] A.Dalgarno and R.A.McCray, Astrophys. J. **181**, 95 (1973).

[17] J.L.Le Garrec, O.Sidko, J.L.Queffelec, J.B.A.Mitchell and B.R.Rowe, J. Chem. Phys. **107**, 54 (1997).

[18] E.Alge, N.G.Adams and D.Smith, J. Phys. B **17**, 3827 (1984).

[19] A.Kalamarides, R.W.Marawar, X.Ling, C.W.Walter, B.G.Lindsay, K.A.Smith and F.B.Dunning, J. Chem. Phys. **92**, 1672 (1990).

[20] S.H.Alajajian, M.T.Bernius and A.Chutjian, J. Phys. B **21**, 4021 (1988).

[21] T.Mostefaoui, C.Rebrion-Rowe, J.L.Le Garrec, J.B.A.Mitchell and B.R.Rowe, Faraday Discuss. **109**, 71 (1998).

[22] S.D.Le Picard, B.Bussery, C.Rebrion-Rowe, P.Honvault, A.Canosa, J.M.Launay and B.R.Rowe, J. Chem. Phys. **108**, 10319 (1998).

[23] S.Hamon, S.D.Le Picard, A.Canosa, B.R.Rowe and I.W.M.Smith, J. Chem. Phys. (March 2000).

Ion–Neutral Collisions in Beam Experiments. Production of ArN^{n+} $(n = 1, 2)$ in the Reaction of Ar^{n+} with N_2

Paolo Tosi and Davide Bassi

INFM and Dipartimento di Fisica
Università di Trento, I-38050, Povo, Trento, Italy

1 Introduction

Ion-neutral collisions at low energies (below about 10 eV) play a central role in the physics and chemistry of low density plasmas and ionized gases. Interstellar clouds, planetary ionospheres, plasmas used for the deposition of thin films and gas discharges are relevant examples of systems where ion-neutral collisions are particularly important. At low energies, possible collision processes include chemical reactions, charge transfer, excitation and quenching of internal energy states. In this paper we will consider a particular class of collisions, the so-called "atom transfer" reactions *i.e.*:

$$X^+ + Y_2 \longrightarrow XY^+ + Y$$

These reactions are particularly relevant for the synthesis of molecular ions. For example, the formation of NH^+ in the collision of N^+ with H_2 is the first step of the reaction chain leading to the synthesis of ammonia in dense interstellar clouds [1, 2].

Reactions of argon cations and dications with molecular nitrogen have been extensively studied in recent years. However, most of these studies are limited to the investigation of the charge-transfer channels [3, 4]:

$$Ar^+ + N_2 \longrightarrow Ar + N_2$$
$$Ar^{2+} + N_2 \longrightarrow Ar + N_2^{2+} \longrightarrow Ar + 2\,N^+$$

Very little information are available [5] on the energy dependence of the endoergic nitrogen-transfer reactions leading to the production of the molecular ions ArN^{n+} $(n = 1, 2)$ *i.e.*:

$$Ar^{n+} + N_2 \longrightarrow ArN^{n+} + N$$

Recent results concerning these reactions are presented in this paper.

2 Experimental

Two different experimental setups have been used. The first one – hereafter called cell apparatus – is a differentially pumped version (see Fig. 1) [6] of an ion-molecule reaction (IMR) tandem mass spectrometer previously used in our laboratory.[7] The ion beam is guided by means of a radio-frequency octopole and reacts with the neutral molecules contained into a room-temperature scattering cell. In the crossed-beam apparatus the scattering cell is replaced by a supersonic molecular beam which crosses the ion beam at 90°.[1] The nitrogen supersonic beam is produced by expansion through a 10^{-2} cm diameter nozzle cooled at 150 K.

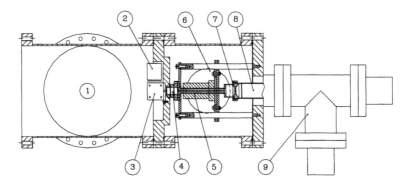

Figure 1: Schematic view of the cell apparatus: 1) 1000 l/s turbomolecular pump, 2) ion source, 3) magnetic-sector mass spectrometer, 4) ion optics, 5) octopole ion guide and scattering cell, 6) 250 l/s turbomolecular pump, 7) ion optics, 8) quadrupole mass spectrometer, 9) ion detector. In the crossed-beam apparatus the scattering cell is replaced by a supersonic molecular beam.

The cell apparatus provides high sensitivity, but unfortunately the influence of the random motion of the target gas produces a significant dispersion of the collision energy. The effective cross section, measured at the nominal collision energy E_{cm0}, results from the convolution of the reaction cross section $\sigma(E_{cm})$ – where E_{cm} is the collision energy – with the energy distribution $f(E_{cm}, E_{cm0})$ of reactants:

$$\sigma_{eff}(E_{cm0}) = \int_0^\infty \sqrt{\frac{E_{cm}}{E_{cm0}}} f(E_{cm}, E_{cm0}) \sigma(E_{cm}) dE_{cm} \qquad (1)$$

If T is the temperature of the scattering cell, and assuming that $E_{cm0} \gg k_B T$ the energy distribution function $f(E_{cm}, E_{cm0})$ is centered around E_{cm0}, with a FWHM given by:

$$\Delta E_{cm0} \simeq \sqrt{11.1 \frac{m_I}{m_I + m_N} k_B T E_{cm0}} \qquad (2)$$

where m_I and m_N are the ion and the neutral mass, respectively. As an example, for the reaction of Ar^+ with N_2, ΔE_{cm0} is about 1 eV at room-temperature and at the collision energy of 6 eV.

A much better energy resolution may be achieved using the crossed-beam apparatus[1]. Unfortunately, the crossed-beam apparatus is much less sensitive with respect to the cell machine, and measurements may be very time consuming.

The combined use of the two apparatuses enable us to develop a very effective measurement strategy. Ion-molecule reactions are first investigated using the high-sensitivity cell apparatus. The molecular beam apparatus is then used to measure the cross-section at a limited number of collision energies, studying in more detail the energy regions where a better resolution is required. This allow us to save time, obtaining detailed information only when required.

3 Results

3.1 $Ar^+ + N_2 \longrightarrow ArN^+ + N$

Although this reaction is a minor process in the argon-nitrogen ion chemistry, its characterization is an essential step towards a more complete understanding of the reaction dynamics of a relatively simple triatomic system as $(Ar-N_2)^+$. Moreover, the production of ArN^+ is of particular interest since its ground state $X^3\Sigma^-$ is characterized by a substantial bond between N^+ and Ar in spite of the generally assumed chemical inertness of rare gases. An open problem concerns the dissociation energy of ground state ArN^+ ions. *Ab initio* calculations give 2.08 eV and 2.13 eV, whereas experimental estimates range from 0.5 eV to 2.3 eV (see [8, 9] and references therein). We have investigated the energy dependence of the reaction:

$$Ar^+ + N_2 \longrightarrow ArN^+ + N$$

and the symmetric charge-state reaction:

$$N_2^+ + Ar \longrightarrow ArN^+ + N$$

in the energy range from 5 to 45 eV. Both reactions show an endoergic behavior, which corresponds to well defined thresholds in the energy dependence of the cross-section.

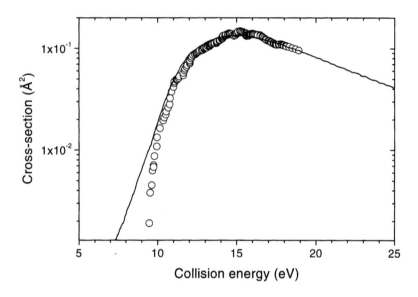

Figure 2: Effective cross-section for the reaction $Ar^+ + N_2 \longrightarrow ArN^+ + N$ as a function of the nominal collision energy E_{cm0}. Crossed-beam data are shown by means of open points, while beam-cell data are shown by means of the solid line.

In Fig. 2 we compare the results obtained in the cell experiment (line) with those obtained in the crossed-beam experiment (points).[9] A clear deviation is observed at collision energies below about 10 eV. It is due to the different energy resolution of the two experiments. Crossed-beam results show – as expected – a sharper reaction onset. From the analysis of experimental results we obtain an estimation of the energy threshold which is 8.2±0.5 and 8.9 ± 0.2 eV for the cell and the crossed-beam measurements, respectively. These values are in substantial agreement with the results of a previous experiment[5]. A comparison with the reaction $N_2^+ + Ar \longrightarrow ArN^+ + N$ and an analysis of the diabatic state correlation diagram show that the reaction $Ar^+ + N_2$ produces ArN^+ essentially in its first excited state $A^3\Pi$, while the reaction $N_2^+ + Ar$ should produce ground state ArN^+ ions.[9] This fact explains the discrepancies between different estimates of the dissociation energy of ArN^+ obtained in previous studies.

3.2 $Ar^{2+} + N_2 \rightarrow ArN^{2+} + N$

In the past few years a growing interest in the properties of diatomic dications has stimulated experimental and theoretical research. Many of these species are thermodynamically unstable due to the Coulomb explosion $XY^{2+} \rightarrow X^+ + Y^+$. Nevertheless they may be observable molecules, due to the presence of barriers towards the fragmentation channel. In this case, one deals with the special situation of a stable electronic state whose dissociation limit lies below the potential well minimum. The most celebrated example is He_2^{2+}, which was first described theoretically by Pauling.[10]
One of the reasons of interest for such states concerns the possibility of energy storage. Once these species are formed, it should be possible to trigger energy release either via unimolecular fragmentation or via exothermic reactions with other molecules. In fact, He_2^{2+} has been proposed as a source of propulsive energy with performances which far exceed those of all known propellants.[11]
Molecular dications with rare gas ligands received considerable theoretical attention in recent years as they often display surprisingly strong bonds despite the well known chemical inertness of neutral rare gas atoms.[12, 13] Many of them have been theoretically predicted to be stable molecules, but their experimental identification is limited to only a few species, because of the difficulty to generate stable neutral precursors that can subsequently be ionized. Here we show the production of ArN^{2+} in a chemical reaction, i.e. the atom transfer process $Ar^{2+} + N_2 \rightarrow ArN^{2+} + N$. It is worth noting that the process studied in the present work converts a double-charge atomic ion into a molecular dication. This result is unusual as collisions of dications with molecules generally produce atomic ions via dissociative electron capture from molecular gas. In fact, previous measurements of the reaction of Ar^{2+} with N_2 observed only N^+ as product ion, probably originating from dissociation of an excited state of N_2^{2+} formed by double–charge transfer with Ar^{2+}.[4]
Fig. 3a shows the effective cross section for the reaction $Ar^{2+} + N_2 \rightarrow ArN^{2+} + N$ as a function of the nominal collision energy.[14] At the lowest energies the cross section is roughly constant, while above about 2 eV the production of ArN^{2+} starts to rise. After a maximum at about 10 eV a rapid fall-off is observed at higher energies. The overall shape of the cross section versus the collision energy indicates that probably different reaction mechanisms are operative.
Wong and Radom,[13] have calculated the ground state of ArN^{2+} obtaining values for the dissociation energy, $D_0(ArN^{2+}) = -4.67$ eV, and for the dissociation barrier, $D_0^*(ArN^{2+}) = 0.25$ eV. By using the theoretical results listed above and the appropriate values for the heats of formation ($\Delta H_{f,0}(Ar^+) = 15.76$ eV, $\Delta H_{f,0}(N^+) = 19.413$ eV, $\Delta H_{f,0}(Ar^{2+}) = 43.38$ eV, $\Delta H_{f,0}(N) = 4.88$ eV) we get $\Delta H_{f,0}(ArN^{2+}) = 39.84$ eV and $\Delta E = 1.34$

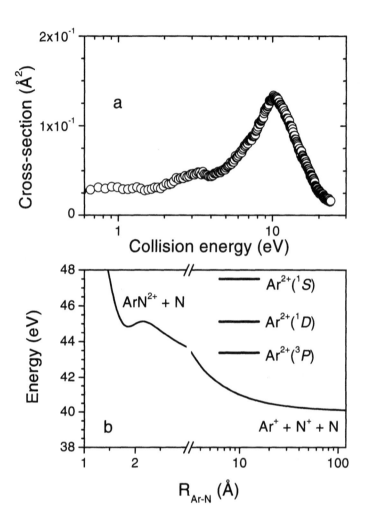

Figure 3: (a) Effective cross section for the reaction $Ar^{2+} + N_2 \rightarrow ArN^{2+} + N$ as a function of the nominal collision energy. (b) Schematic representations of the reaction energetics. The zero energy level is referred to as $Ar + N_2$.

eV, where ΔE is the reaction endothermicity. While this result indicates that the reaction should be endothermic by 1.34 eV, we observe products also at lower energies. This discrepancy is likely due to the possible presence of metastable Ar^{2+} ions in our experiment. As the primary ion beam is produced by electron bombardment of Ar, all the three low lying states $Ar^{2+}(^3P)$, $Ar^{2+}(^1D)$, and $Ar^{2+}(^1S)$ are likely to be produced according to their statistical weights $9:5:1$. The energy difference between the ^1D and the ^3P ground state is 1.74 eV, whereas the ^1S lies 4.12 eV above the ^3P (see Fig. 3b). This suggests that the reactivity observed at the lowest energies may be due to the fraction of metastable states present in our Ar^{2+} beam, and in particular to the ^1D component. The ^1S state should not react because N_2 has Σ symmetry whereas ArN^{2+} has Π symmetry, and therefore there is no way to conserve the symmetry going from reactants to products. A definitive answer on the role played in this reaction by the first low-lying states requires further studies making use of state selected ions.

In conclusion, our experiment suggests that ion-molecule reactions may be used for producing molecular dications containing rare gases in an relatively simple way. This method has been recently extended to the synthesis of other molecular dications like – for example – ArC^{2+}.[15, 16]

Our guided-beam method opens new prospects for a systematic investigation of molecular dications. In particular, it may be used for investigating molecular dications using both low energy collision-induced dissociation (CID) and photodissociation techniques. Moreover, it will be interesting to explore the possibility of accumulating molecular dications into an ion trap. This could allow the development of miniaturized devices in which the release of energy, stored at molecular level, can be controlled by triggering the Coulomb explosion.

Acknowlegments. This work has been partially supported by MURST, contract 98-02623099.

References

[1] P. Tosi, O. Dmitriev, D. Bassi, O. Wick, and D. Gerlich, J. Chem. Phys., **100**, 4300 (1994).

[2] R. Tarroni, P. Palmieri, A. Mitrushenkov, P. Tosi, and D. Bassi, J. Chem. Phys., **106**, 10265 (1998).

[3] P. Tosi, O. Dmitriev, D. Bassi, Chem.Phys.Letters, **200**, 483 (1992).

[4] H. Störi, E. Alge, H. Villinger, F. Egger, and W. Lindinger, Int. J. Mass Spectrom. Ion Phys., **30**, 263 (1979).

[5] G.D. Flesch, and C.Y. Ng, J. Chem. Phys., **92**, 2876 (1990).

[6] W. Lu, P. Tosi, and D. Bassi, J. Chem. Phys., **111**, 8852 (1999).

[7] D. Bassi, P. Tosi, and R. Schlögl, J. Vac. Sci .Technol. A, **16**, 114 (1998).

[8] L. Broström, M. Larsson, S. Mannervik, and D. Sonnel, Chem. Phys., **94**, 2734 (1991).

[9] P.Tosi, R.Correale, W. Lu, and D.Bassi, J. Chem. Phys., **110**, 4276 (1999).

[10] L. Pauling, J. Chem. Phys. **1**, 56 (1933).

[11] C.A. Nicolaides, Chem. Phys. Letters, **161**, 547 (1989).

[12] G. Frenking, W. Koch, D. Cremer, J. Gauss, and J.F. Liebman, J. Phys. Chem., **93**, 3397 (1989).

[13] M.W. Wong, and L.Radom, J. Phys. Chem., **93**, 6303 (1989).

[14] P. Tosi, R. Correale, W. Lu, S. Falcinelli, and D. Bassi, Phys. Rev. Letters, **82** 450 (1999).

[15] P. Tosi, W. Lu, R. Correale, and D. Bassi, Chem. Phys. Letters, **310** 180 (1999).

[16] W. Lu, P. Tosi, and D. Bassi, J. Chem. Phys., (in press).

Crossed Beams and Theoretical Study of the (NaRb)$^+$ Collisional System

A. Aguilar[1], J. de Andrés[1], T. Romero[1], M. Albertí[1],
J.M. Lucas[1], J.M. Bocanegra[1], J. Sogas[1], F.X. Gadea[2]

[1] Departament de Química Física. Universitat de Barcelona,
Barcelona. Spain
[2] Laboratoire de Physique Quantique, I.R.S.A.M.C.,
Université Paul Sabatier, Toulouse, France

1 Introduction

Ion-atom collisions are one of the most assiduously studied topics in Chemical Dynamics, since their relative simplicity make their study easy to set up and carry out. When both colliding species are alkali atoms this interest becomes still greater, not only because their dynamics can be studied using simple models such as those involving pseudopotentials, but also because species in excited electronic states are readily formed by the collisions (either by direct excitation or via electron exchange processes) whose fluorescent decay emissions can furnish useful information and offer the perspective of such interesting technical applications as lasers and gas-discharge lamps.

While beam-chamber experiments were very easy to perform on these systems, it was the development of the crossed beams technique that finally permitted their study in carefully selected conditions to obtain true cross-section data. Foremost in this field were Aquilanti *et al* whose work included the determination of collision cross-sections for a number of symmetrical and unsymmetrical collision systems, being able in some cases to resolve the fluorescent emissions into their J-components, and even to study polarisation fractions accurately enough to determine the branching ratios between M_J states (for a full review of papers in this field see [1] and references mentioned there).

In order to develop and improve on these studies, this Group has built a crossed molecular beam machine with which a series of studies have been carried out, using Na atoms as a target [2, 3], both extending the energy ranges previously studied and trying to probe further into the fine structure of the species formed. Continuing on these lines we report here on exper-

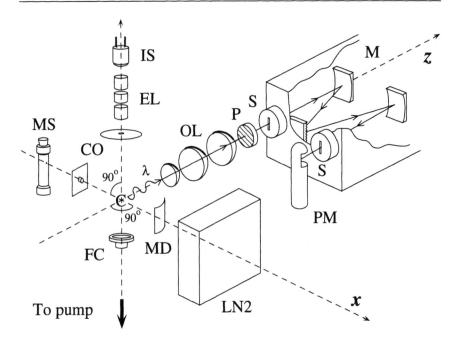

Figure 1: Experimental setup diagram. Letters CO indicate the beam collimators. For the rest see text.

iments involving the (NaRb)$^+$ quasimolecule, which has been already the object of some of our studies [4, 5].

2 Experimental setup

Since a full description of the crossed molecular beam machine can be found in [1], only a brief outline will be given here. Experiments have been carried out in a single chamber machine (see Fig 1) pumped by a rotative-diffusive train. In order to reduce background gas as much as possible, the pumps were overdimensioned in relation to chamber size, the target beam generator was surrounded by a water-cooled sleeve (not shown) to condense as much metal vapor as possible and, after crossing, this beam was directed to a liquid nitrogen trap (LN2). In these conditions it has proved possible to work with Rb beams with background pressures on the order of $10^{-6} - 10^{-7}$ mbar.

Neutral beams are generated by metal vapor effusion trough a thin-walled hole in a combined oven/heating element (MS). Sodium beams are generated simply by heating the metal, but the rather hazardous use of Rb has been obviated by loading the oven with a mixture of rubidium chloride and

Table 1: Electron exchange processes

Process	Emission (Å)
$Na^+(^1S_0) + Rb(5^2S_{1/2}) \rightarrow Na(3^2P_{1/2,3/2}) + Rb^+(^1S_0)$	5889.9-5895.9
$Na^+(^1S_0) + Rb(5^2S_{1/2}) \rightarrow Na(4^2D_{5/2,3/2}) + Rb^+(^1S_0)$	5688
$Na^+(^1S_0) + Rb(5^2S_{1/2}) \rightarrow Na(5^2D_{5/2,3/2}) + Rb^+(^1S_0)$	4983
$Na^+(^1S_0) + Rb(5^2S_{1/2}) \rightarrow Na(6^2D_{5/2,3/2}) + Rb^+(^1S_0)$	4669
$Na^+(^1S_0) + Rb(5^2S_{1/2}) \rightarrow Na(7^2S_{1/2}) + Rb^+(^1S_0)$	4751
$Rb^+(^1S_0) + Na(3^2S_{1/2}) \rightarrow Rb(5^2P_{3/2,1/2}) + Na^+(^1S_0)$	7800.2-7947.6
$Rb^+(^1S_0) + Na(3^2S_{1/2}) \rightarrow Rb(6^2P_{3/2,1/2}) + Na^+(^1S_0)$	4201.8

barium which, once heated, produces Rb according to the process:

$$RbCl(s) + Ba(s) \rightarrow BaCl(s) + Rb(g)$$

This procedure has been proved by fluorescence and surface ionisation (MD) measurements to produce stable, pure, high density beams. Oven temperature (800 K) was selected empirically as that giving the best signal-to-noise ratio. Ions are produced by thermoelectric effect heating an eucryptite wafer, accelerated by an electric field and focussed by an einzel lens system (EL). A series of tests proved that the resulting beam did not change in geometry on varying collision energy. Beam intensity was determined using a Faraday cup (FC).

The fluorescent emission produced by the decay of the excited species formed in the collisions is focussed by an optical system (OL), analysed by a 50 cm monochromator (M) and measured by a photomultiplier (PM). A series of baffles and light traps (not shown) prevented "noise" (black body emission from the hot beam generators) from masking the signal. A rotatable polariser (P) can be installed in the optical path as necessary to measure polarisation fractions.

In order to give absolute cross-sections the eucryptite was replaced by an electron-emitting device and the results given by Chen et al [6] for Rb - e⁻ collisions were used to obtain a normalisation constant. In Fig. 2 it can be seen how both collisional systems show a similar behaviour.

3 Experimental results and discussion

3.1 Total absolute cross sections

Fluorescent emissions attributable to electron exchange processes detected with enough intensity to allow the study of their collision energy dependence are summarized in Table 1. Also, direct electronic excitation processes observed are shown in Table 2.

For all of these processes, absolute cross-section dependence on collision

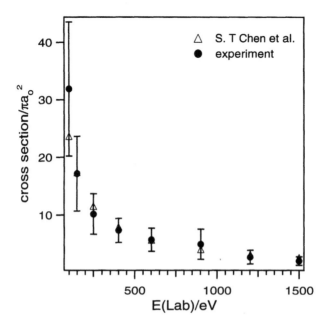

Figure 2: Calibration curve for an Rb beam. Comparison with data for $Rb(5s) + e^- \to Rb(5p) + e^-$

Table 2: Electronic excitation processes

Process	Emission (Å)
$Na^+(^1S_0) + Rb(5^2S_{1/2}) \to Na^+(^1S_0) + Rb(5^2P_{1/2,3/2})$	7800.2-7947.6
$Rb^+(^1S_0) + Na(3^2S_{1/2}) \to Rb^+(^1S_0) + Na(3^2P_{3/2,1/2})$	5889.9-5895.9
$Rb^+(^1S_0) + Na(3^2S_{1/2}) \to Rb^+(^1S_0) + Na(7^2S_{1/2})$	4751
$Rb^+(^1S_0) + Na(3^2S_{1/2}) \to Rb^+(^1S_0) + Na(6^2D_{5/2,3/2})$	4669
$Rb^+(^1S_0) + Na(3^2S_{1/2}) \to Rb^+(^1S_0) + Na(5^2D_{5/2,3/2})$	4983
$Rb^+(^1S_0) + Na(3^2S_{1/2}) \to Rb^+(^1S_0) + Na(4^2D_{5/2,3/2})$	5688

energy was determined, an example being shown on Fig. 4. While excited Na atoms are readily formed either via direct electronic excitation or electron capture processes, excited Rb atoms are not produced by electron exchange. It must also be pointed out that, while many of the Na excited species mentioned above decay to the $Na(3^2P_{1/2,3/2})$ level, their absolute cross-section values were low enough in all cases as to give a negligible cascade effect in the formation of this state.

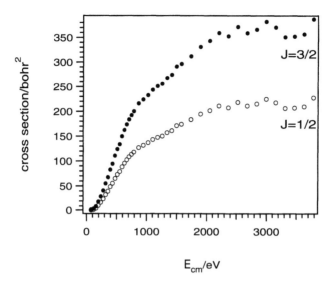

Figure 3: State-to-state excitation cross section for the formation of Na(3^2P_J) by collisions between Rb and Na$^+$

3.2 State-to-state resolved collision cross-sections

In all the collision processes for which the wavelengths of both J-states are given in Tables 1 and 2, it was possible to improve resolution until both signals became separated so that the branching ratio between both states could be determined by simple peak area measurements. Their energy dependence for Na($3^2P_{1/2,3/2}$) formation by charge transfer is shown in Fig. 3. It can be seen that its value is very close to 2, with no dependence on energy, giving an statistical distribution between both J-states.

The same result was obtained when this state was formed via electron exchange. This can be justified on the basis of the small spin-orbit splitting of the Na atom, a supposition confirmed by the fact that, for Rb($5^2P_{1/2,3/2}$) formation, while the distribution becomes statistical at high collision energies, $J = 1/2$ tends to be populated in preference to $J = 3/2$ at the lower end of the energy range (below 800 eV approx.).

3.3 Polarisation effects and magnetic cross-sections

Measurable polarisation effects were observed for all the transitions described above. According to theory, however, only the 3/2-1/2 ones ought to show this behaviour, so the anisotropy shown by 1/2-1/2 transitions was attributed to distortions caused by the monochromator grids, and their values

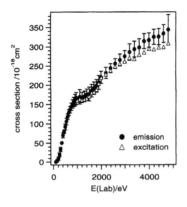 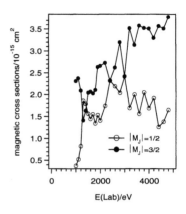

Figure 4: Total emission cross section for Na($3^2P_{3/2,1/2}$) formation in Na + Rb$^+$ collisions

Figure 5: Magnetic cross sections for Na($3^2P_{3/2}$) formation in Rb + Na$^+$ collisions

were used as correction factors to calculate the true values for the 3/2-1/2 processes. Using the polarisation fraction measurements, the total emission cross-section values were corrected to obtain the true excitation ones, but in all cases the obtained values were nearly identical to the original ones (see Fig. 4).

The magnetic cross sections for both M_J states as a function of collisional energy for Na($3^2P_{3/2}$) formation by electron exchange are given in Fig. 5. Although the magnetic cross-sections seem to show some structure between 1300 and 1400 eV, this behaviour can be considered as a result of the experimental uncertainty at these low collision energies. Conversely, above 3100 eV, the $M_J = 3/2$ sublevel is populated more readily than the $M_J = 1/2$ one, as was to be expected since at these high collision energies, polarisation fractions have negative values. The same behaviour was observed when the exit channel was an electronic excitation one, and it contrasts with that exhibited by Rb($5^2P_{3/2}$) for which the $M_J = \pm 1/2$ sublevels nearly always show a higher population than the $M_J = \pm 3/2$ ones.

4 Computational Study

4.1 Ab initio calculations

Since the 60's much effort has been devoted to the interpretation of the salient features of experimental cross sections in inelastic atomic collisions. Although nice explanations about excitation and charge transfer have been given based on qualitative profiles of potential energy curves, accurate ones seem to be needed to reproduce experimental results quantitatively. It is also well known how much the success of any dynamical calculation depends on the choice of appropiate electronic wave functions to be used as a basis. In the present case, the (NaRb)$^+$ quasimolecule has been treated as a pseudomonoelectronic system. NonÐempirical pseudopotentials derived by Durand and Barthelat [7, 8] have been used in order to replace sodium ($1s2s2p$) and rubidium ($1s2s2p3s3p3d4s4p$) cores. Moreover, relativistic parameters derived by Pavolini et al [9] for the Rb atom have been also included. Potential energy curves have been then calculated using a ROHF technique. Due to the high polarizability of the atoms involved in this calculation it has been necessary to take into account the core-valence correlation. This has been done adding a polarization potential (CPP) in the Fock operator. The CPP formalism used here has been the one proposed by Foucrault *et al* [10], itself a modification of the Müller *et al* one [11]. This means that the cutoff is a step function which includes a l-dependent parameter in order to improve the result of the method when applied to the excited states.

The core-core repulsion has been also corrected from the pure coulombic value $1/R$. This has been necessary first because we are working with large cores and, second, because the collision energies we are dealing with allow the colliding particles to approach each other nearer than 1 Å. Thus, the core-core repulsion has been calculated as the energy difference between NaRb^{2+} and Na$^+$ and Rb$^+$ in an all-electron calculation with minimal basis set:

$$E_{\text{core-core}} = E[NaRb^{2+}] - E[Na^+] - E[Rb^+]$$

The valence electron has been described by extended Gaussian Type Orbital (GTO) basis sets: a $[7s, 4p, 5d, 1f/6s, 4p, 4d, 1f]$ basis set for Rb [9] and a $[7s, 6p, 5d, 2f/5s, 5p, 4d, 2f]$ for Na [12].

Figure 6 shows potential energy curves for the ground state and some excited ones of the (NaRb)$^+$ system. The extended GTO basis set has allowed us to calculate with high reliability 29 potential curves corresponding to 44 molecular states. Computational results show that the experimental (Na + Rb$^+$) entrance channel correlates with the 1 $^2\Sigma^+$ molecular ground state of the system while the (Rb + Na$^+$) does to the first electronic excited one. In Fig. 6 it can be seen that some relevant features of the curves are due to crossings or avoided crossings. This is the case for example of the 1 and 2 $^2\Sigma^+$ molecular states at about 15 a.u.. These define regions where different nonÐadiabatic transitions can eventually take place leading to the formation of the collision products that have been detected in our crossed beams experiments.

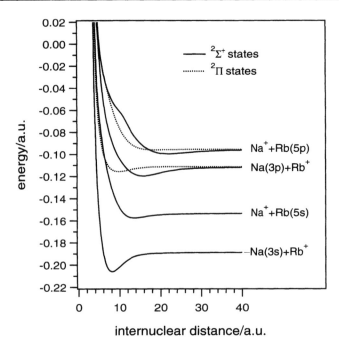

Figure 6: Potential energy curves for some relevant collision channels in the (NaRb)$^+$ collisional system

4.2 Dynamical calculations

Although simple two-state models like the Landau-Zener one [13] can provide good empirical fittings of the cross-section energy dependence, they do not give much information about the behaviour of the systems when the number of states which can potentially be involved in the process is large enough at the considered collision energy. This is the situation for the collisional (NaRb)$^+$ system reported in this paper where, even at the lowest experimental collision energy, a manifold of electronic states can be populated, playing then an important role in the excitation and charge tranfer processes.

From the above mentioned ab initio calculations [14] performed previously by our group on the (NaRb)$^+$ system and using the effective hamiltonian metric proposed by Gadea et al [15], radial couplings given by $\langle \phi_j | \partial/\partial R | \phi_i \rangle$ between two electronic ϕ_i, ϕ_j states of the same symmetry ($\Sigma - \Sigma$ or $\Pi - \Pi$) have been obtained by finite increment techniques. Some of these radial couplings are shown in Fig. 7. It can be appreciated there how the couplings spread out over all the range of relevant internuclear distances R between the Na and Rb cores and also how one given state is coupled to the

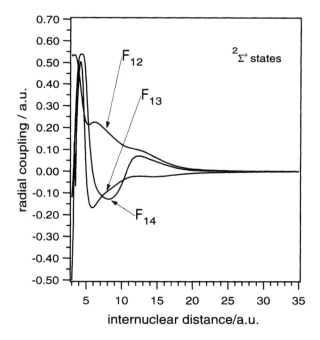

Figure 7: Radial couplings between the ground and the first three excited Σ^+ states as a function of the internuclear distance

other electronic states of the proper symmetry. These considerations show the inadequateness of any simple two states models which assume that the coupling is localised at well defined values of the internuclear separation, as the Landau-Zener model does.

Dynamical calculations for systems where electronically nonadiabatic transitions play an important role can be done by using as a basis one of both adiabatic or diabatic set of electronic states. In the former case (adiabatic representation) the Born-Openheimer electronic hamiltonian matrix is diagonal, the electronic excitation processes being caused by the nuclear motion. Radial couplings $\langle \phi_j | \partial/\partial R | \phi_i \rangle$ (or rotational ones for states of different symmetry) become of crucial importance in this representation, and can be obtained from the knowledge of the different electronic wave functions in a range of R values. In the second case (diabatic representation) the Born-Openheimer electronic hamiltonian matrix is not diagonal, the electronic excitation being interpreted as caused by the interaction between the different electronic states which give non zero off-diagonal matrix elements. Diabatic states can be obtained from the adiabatic ones by means of an unitary matrix transformation and vice versa but there are many ways about how such a transformation can be performed.

In order to carry out dynamical calculations we have performed an unitary adiabatic to diabatic transformation of the original adiabatic states obtained in the previous section. Such a transformation has been done by ensuring that radial couplings are everywhere zero along the range of variation for R (0.50 Å$\leq R \leq$ 20.00 Å) so that any electronic excitation produced along the collision can be seen as caused by a purely electronic interaction between the manifold of electronic sates involved in the diabatization procedure [16]. Taking into account that our experimental measurements allow us to obtain state-to-state cross sections (i.e. from initial states with $J = 1/2$ to final states with $J = 1/2$ or $J = 3/2$) it would be convenient to carry out dynamical calculations considering the spin-orbit interaction for the different diabatic states. Such a spin-orbit effect has been included in the way usual in many calculations, i.e. an empirical one based on the spectrally observed atomic spin-orbit splittings [17]. In this approximation the magnitude of the spin-orbit interaction is assumed to be independent of the internuclear separation and the corresponding values are introduced as off-diagonal matrix elements in the diabatic representation of the electronic hamiltonian. In our particular case and according to the experimental measurements we have only included the spin-orbit interaction correction for those molecular states correlating with the asymptotic excited atomic states Na*($3^2P_{1/2,3/2}$) and Rb*($5^2P_{1/2,3/2}$). This means that instead of the two potential energy curves $^2\Sigma^+$ and $^2\Pi$ asymptotically correlating with the atomic states $^2P_{1/2,3/2}$ of Na and Rb atoms, the inclusion of the spin-orbit interaction makes this set of curves to split into three curves where the relevant quantum number (Ω) is the projection on the internuclear axis of the total electronic angular momentum. The new spin orbit molecular states $|1/2, \Omega = \pm 1/2\rangle$, $|3/2, \Omega = \pm 1/2\rangle$, $|3/2, \Omega = \pm 3/2\rangle$ correlate with the asymptotic atomic states with $J = 1/2$ and $3/2$ respectively, and transitions to those states leading to Na* or Rb* excited atoms can be induced by collision from the entrance channel $|1/2, \Omega = \pm 1/2\rangle$ in both Na$^+$+ Rb and Rb$^+$+ Na experiments.

Dynamical calculations for the colliding (NaRb)$^+$ system have been performed in the framework of the hemiquantal mechanics [18] using a computer code implemented by us in our computational facilities. This method requires moderate computational time and hopefully should lead to a proper description of the process, the accuracy of the results being limited by the accuracy of the potential energy curves and couplings used in the dynamical study. Calculations have been performed in an Apollo HP 735 work station type computer. In our first approach to the dynamical calculation for the (NaRb)$^+$ system we have considered only electronic transitions induced by the collision process between states of the same symmetry then focusing our attention on transitions from the entrance channels $|1/2, \Omega = \pm 1/2\rangle$

The basic asumption of the hemiquantal mechanics in our atom-atomic ion collision process is that the internuclear coordinate R is treated classically while the electronic evolution of the colliding system is treated by solving the time dependent Schrödinger equation. To this end the complete electronic wave function of the collision system is expanded on a basis set of diabatic molecular states obtained from the ab initio calculations, the expansion coefficients $a_i(t)$ being a function of time. The number of electronic diabatic states included in the total electronic wave function expansion was taken to be equal to twelve including the properly calculated spin-orbit molecular states for both previously considered exit channels.

Hemiquantal trajectories require the resolution of the set of coupled Hamilton's equations as well as the time dependent Schrödinger equation for a given set of initial conditions, namely orbital angular momentum and collision energy, the initial electronic state being selected according to the experimental entrance channel. In the present theoretical study we focus our attention on the direct electronic excitation process $Na(3^2S_{1/2}) + Rb^+ \rightarrow Na^*(3^2P_{1/2}) + Rb^+$.

At the end of a hemiquantal trajectory $(t \rightarrow \infty)$ performed at a fixed initial impact parameter (l) and fixed collision energy (E), the transition probability from the initial i electronic state to a given final f state $P^l_{f \leftarrow i}(E)$ is given by:

$$P^l_{f \leftarrow i}(E) = |a_f(l, \infty)|^2$$

where the $a_f(t)$ coefficients are allowed to be complex. Total inelastic cross section at fixed energy for direct electronic excitation of Na atom from their ground electronical state i to the first excited state f with $J = 1/2$ $(Na^*(3^2P_{1/2}))$ is given by :

$$\sigma_{f \leftarrow i}(E) = \frac{\pi}{2\mu E} \sum_{l=0}^{l_{max}} (2l+1)|a_f(l, \infty)|^2$$

where μ is the reduced mass of the colliding system and l_{max} the maximum orbital angular momentum leading to non zero probability for the excitation processes considered.

5 Results and discussion

In figure 8 the computed total excitation cross sections for the process $Na(3^2S_{1/2}) + Rb^+ \rightarrow Na^*(3^2P_{1/2}) + Rb^+$ are shown together with the experimentally measured values. As it can be seen from the picture, calculated cross sections are in very good agreement with the experimental ones and the shape of the computed excitation function also agrees with the experimental determined behaviour for the transition leading to the excited $J = 1/2$ spin-orbit state of the Na atoms.

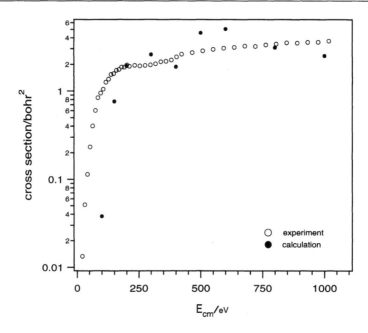

Figure 8: Computed and experimental cross sections for the electronic excitation process $Rb^+(^1S_0) + Na(3^2S_{1/2}) \to Rb^+(^1S_0) + Na(3^2P_{3/2,1/2})$

As it was mentioned above in computing state-to-state transition probabilities for a fixed initial state, dynamical calculations show that, at the end of a trajectory, there is a finite non-zero probability for the population of the different electronic states included in the expansion of the total electronic wave function, this one being function of the collision energy. A detailed analysis of the behaviour of a given excitation probability $P^l_{f \leftarrow i}(E)$ along the collision process shows that such a probability changes continuously along the trajectory and, consequently, a non-defined region of the configuration space seems to produce the main contribution to the excitation process considered. The time evolution of the state probabilities and also their asymptotic values (the excitation probabilities), are moreover strongly influenced by the orbital angular momentum of the trajectory. This analysis of the evolution of the excitation probability $P^l_{f \leftarrow i}(E)$ along a collision process proves the limitations of the simple two states models. Although good empirical fittings of the excitation function can be obtained using such simple models they cannot provide the proper description of the detailed microscopic mechanism involved in a particular process.

Although both the calculated cross section values and their energy dependence are in quite good agreement with the experimentally measured values, a less than perfect coincidence between both values has been obtained. Independently of the dynamical method used in the study of a non adiabatic collision process, the accuracy of the computed inelastic total cross sections is strongly influenced by the number of electronic states included in the total electronic wave function expansion. The number of states included in the present calculation seems to be enough to give a good description of the excitation channel studied, but nevertheless an expansion on a larger set of molecular states has to be performed to reach an improved convergence of the computed cross section values with the experimental ones. This means that larger computer memory facilities and longer computational times will be required and work is in progress in this direction, which will also be extended to the study of other inelastic processes like those leading to the charge transfer from both $Na + Rb^+$ and $Rb + Na^+$ entrance channels.

Acknowledgements

This work has been supported by the Spanish DGICYT projects PB91-0553, PB94-0909 and PB97-0919; the EU Human Capital Mobility Program through the Structure and Reactivity of Molecular Ions under cont. No. CHRX-CT93-0150. Thanks are also due to the Catalan Comissionat Per a Universitats i Recerca (FI grant FI/94-1034 and 1996SGR-00040 project).

References

[1] A. Aguilar, M.Albertí, M. Prieto, J. de Andrés, M. Gilibert, X . Giménez, M. González, J.M. Lucas, R. Sayós, A. Solé., Chem. Phys. Letters, **220**, 267 (1994)

[2] J. de Andrés, M. Prieto, T. Romero, J.M. Lucas, M. Albertí, A. Aguilar, Chem. Phys. Letters, **238**, 338 (1995)

[3] T. Romero, J. de Andrés, M. Albertí, J.M. Lucas, A. Aguilar, Chem. Phys., **209**, 217 (1996)

[4] T. Romero, J. de Andrés, M. Albertí, J. Sogas, J.M. Lucas, J M. Bocanegra, A. Aguilar, Chem. Phys. Letters. **281**, 74 (1997)

[5] T. Romero, J. de Andrés, M. Albertí, J.M. Lucas, A. Aguilar, Chem. Phys. Letters. **272**, 271 (1997)

[6] S.T. Chen, A.C. Gallagher Phys. Rev. A **17**, 571 (1978)

[7] Ph. Durand, J.C. Barthelat Theoret. Chim. Acta **55**, 43 (1980)

[8] Ph. Durand, J.C. Barthelat, M.Pelissier. Phys. Rev. A **21**, 1773 (1981)

[9] D. Pavolini, J. P. Daudey, T. Gustaffson, F. Spiegelman J. Phys. B **22**, 1721 (1989)

[10] M. Foucrault, Ph. Millie, J.P. Daudey. J. Chem. Phys. **96**, 1257 (1992)

[11] W. Müller, W. Meyer. J. Chem. Phys. **80**, 3331 (1984)

[12] S. Magnier, Ph. Millié, O. Dulieu and F. Masnou-Seeuws, J. Chem. Phys. **98**, 7113 (1993)

[13] E.E. Nikitin. Theory of Chemical Elementary Processes in Gases. Springer, Berlin 1978

[14] T. Romero, J. de Andrés, M. Albertí, J.M. Lucas, J. Rubio, J.P. Daudey, A. Aguilar, Chem. Phys. Letters., **261**, 583 (1996)

[15] F.X. Gadea, M. Pellissier. J. Chem. Phys. **93**, 545 (1990)

[16] T. Romero, A. Aguilar and F.X. Gadea. J. Chem. Phys. **110**, 6219 (1999)

[17] J. S. Cohen and B. Schneider, J. Chem. Phys. **61**, 3230 (1974)

[18] L.L. Halcomb and D.J. Diestler, J. Chem. Phys. **84**, 3130 (1986)

Part V

Clusters and Nanoparticles: Diffraction,
Size Selection, Polarizability, and Fragmentation
by Photodissocation, Electron Impact,
and Atom Collision

Diffraction of Cluster Beams from Nanoscale Transmission Gratings

L.W. Bruch[1], W. Schöllkopf[2], J.P. Toennies[2]
[1] Dept. of Physics, University of Wisconsin, Madison, USA
[2] Max-Planck-Institut für Strömungsforschung
37073 Göttingen, Germany

1 Introduction

Since van der Waals clusters produced in molecular beam expansions have a broad size distribution, experimental techniques are needed to select clusters of well defined sizes. A size selective detection of clusters is not possible with conventional mass spectrometers because of cluster fragmentation and dissociation in the ionization process. Recently our group has demonstrated that it is possible to select and analyze light clusters by diffracting them from a free-standing nanostructured transmission grating [1, 2]. Since this technique depends on the wave nature of the particles it has the advantage of being essentially non-destructive. This feature has been demonstrated with small clusters of ^{4}He and allowed for the unambiguous detection of He$_2$ and He$_3$ [1, 2]. With a predicted binding energy of only $1.3\,\text{mK}\,(=1.1\cdot 10^{-7}\,\text{eV})$ and a mean nuclear distance of about 55 [3] the helium dimer is thought to be the weakest bound diatomic molecule in the ground state. The bond is sufficiently weak to support an Efimov state in He$_3$ [4]. Efimov states are special long range three particle bound states which occur only if the corresponding two particle system is sufficiently weakly bound.

The non-destructive detection of small clusters opens up a wide variety of new experiments some of which have been reported earlier [2, 5]. In the present report we would like to focus on two recent developments. In Sect. 3 of this article we use the diffraction technique to investigate the dependence of the helium dimer, trimer, and tetramer mole fractions on the source pressure reflecting the onset and the first stages of condensation. The cluster formation in the free-jet expansion is described by a kinetic model that gives insight into the deviations from equilibrium during the expansion and that predicts the main features of the observed pressure dependence.

Recently it has been established that the cluster diffraction intensities are affected by the finite size i.e. by the respective molecular wave function of the cluster [5, 6] opening up opportunities to measure the molecular wave functions. Initially we had discounted the effect of long range dispersion forces, but recently we have observed that these forces can affect the diffraction intensities significantly. Hence, an extraction of the size information out of the diffraction data is only possible if the latter effect is understood quantitatively. Therefore, a systematic study of the effect by diffraction of He, Ne, Ar, and Kr atom beams is presented in Sect. 4 of this paper. A detailed evaluation of the observed diffraction intensities allows to determine the attractive part of the respective atom–surface van der Waals potential quantitatively [7].

2 Experimental Technique

The experimental setup is shown schematically in Fig. 1. The cluster beam is produced by expanding pure gas from a source chamber through a thin walled, 5 µm wide orifice into high vacuum. The temperature T_0 and the pressure P_0 inside the source can be varied from 4 to 300 K and 0 to 200 bar, respectively. After passing through the skimmer the beam is collimated by two 10 µm wide, 5 mm tall slits to reduce the angular divergence before it impinges on the transmission grating. The grating is made out of silicon nitride and has a period of 100 nm, a nominal slit width of 50 nm, and a thickness of about 90 nm. It was made by Tim Savas and Henry I. Smith from MIT, USA, using the method of achromatic interferometric lithography [8]. The far-field diffraction pattern is measured by precisely rotating the mass spectrometer detector by ϑ around an axis parallel to the grating slits. With the 25 µm wide, 5 mm tall entrance slit of the detector a high angular resolution of $\approx 7 \times 10^{-5}$ rad (FWHM of the molecular beam) is achieved.

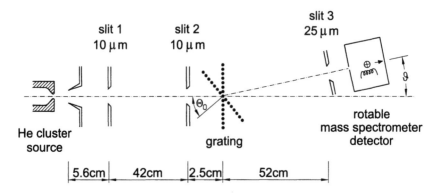

Figure 1: Important dimensions of the experimental setup used in the diffraction of atom or cluster beams by a 100-nm-period transmission grating.

3 Diffraction of Small Helium Clusters

The mass selection of the small clusters is based on the fact that clusters and atoms in the molecular beam have a narrow velocity spread of about $\Delta v/v \approx$ 1% with the same mean velocity determined mainly by T_0. Therefore, the de Broglie wavelength of a particular cluster He_N is inversely proportional to its mass $N \cdot m_{He}$. In the Fraunhofer limit the diffraction angles are thus inversely proportional to the mass leading to a separation of the clusters. The technique is essentially non-destructive since only those clusters which do not hit the grating bars contribute to the coherent diffraction pattern.

For each source temperature T_0 there is a range of source pressures P_0 for which an intense signal of dimers and trimers is observed: at low pressures pure atomic beams and for large pressures beams containing large clusters and microdroplets are formed. Thus in the intermediate range small clusters are formed by homogeneous nucleation which gradually increases with increasing pressure. Hence, the diffraction method opens up the possibility to observe onset and evolution of condensation in helium expansions. Once the resolution is further improved to detect larger clusters well beyond the present limit of $N < 10$, tests of the classical theories of homogeneous nucleation which date back to Gibbs (1876) will become possible.

In Fig. 2 helium cluster beam diffraction patterns at $T_0 = 12\,\text{K}$ are shown for various source pressures P_0 between 0.5 and 6 bar. In each case a section of the diffraction pattern from $\vartheta = -1$ to $+4.5\,\text{mrad}$ is shown on a logarithmic intensity scale including the zeroth order peak in the forward direction and the first order diffraction peaks of the atoms, dimers, trimers, and tetramers as indicated. The data were acquired with the mass spectrometer set to detect the He^+-Ion, which is the favourable fragment-ion for dimers and trimers [2]. The intensities of the dimers, trimers, and tetramers vary significantly with increasing P_0 until eventually larger clusters are formed as indicated by widening of the flank of the zero order peak. Thus, the data provide stepwise snapshots of the evolution of the condensation occuring in the free jet expansion. The dependencies of the helium cluster mole fractions on P_0 as evaluated from many such measurements are shown in Fig. 3. With increasing pressure first dimers then trimers and then tetramers are formed. The fraction of dimers shows a maximum when the trimer fraction increases, while the latter, in turn, reach a maximum when the tetramer fraction increases.

To understand this peculiar behavior a kinetic model of rate equations which are integrated numerically from the source orifice to steady state final conditions is used for each source pressure to describe the cluster formation in the expansion. In these preliminary simulations only three reactions involving atoms, dimers, trimers, and tetramers are taken into account.

$$\text{He} + \text{He} + \text{He} \;\rightleftharpoons\; \text{He}_2 + \text{He}, \tag{1}$$

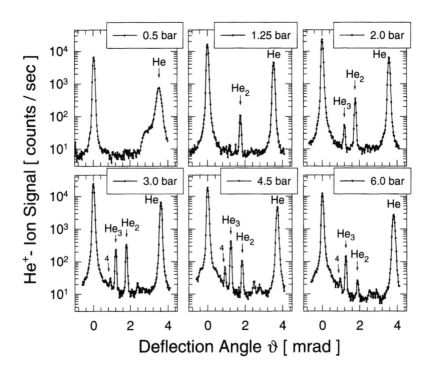

Figure 2: Helium cluster beam diffraction patterns at $T_0 = 12$ K for various source pressures P_0 showing the stepwise rise and fall of cluster concentrations.

$$\text{He} + \text{He} + \text{He}_2 \rightleftharpoons \text{He}_3 + \text{He}, \qquad (2)$$
$$\text{He}_2 + \text{He}_3 \rightleftharpoons \text{He}_4 + \text{He}. \qquad (3)$$

Unfortunately, little is known about the cross sections for these reactions. Due to the very low temperatures in the final stages of the expansion ($\ll 1$ K) and the large de Broglie wavelengths and large scattering lengths ($a \approx 100$) quantum effects are expected to play an important role. For the reaction (1) and (2) the backward rate cross sections were calculated as a function of temperature by assuming it to be equal to the energy dependent integral collision cross section of the atoms approximated by effective range theory or to the geometric cross section of the respective cluster derived from its molecular wave function, whichever is larger. For 10^{-3} K these cross sections become as large as $7 \times 10^{4\,2}$. For the third reaction a constant stripping cross section of $500^{\,2}$ is assumed to calculate the forward rate constant. The other rate constants, backward or forward, are then determined

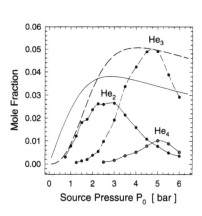

Figure 3: Small helium cluster mole fractions at $T_0 = 12\,\text{K}$ as function of source pressure P_0 according to experiment (*points*) and kinetic model (He_2 *continuous line*, He_3 *dashed line*).

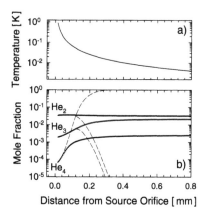

Figure 4: (**a**) Calculated temperature as a function of the distance from the orifice for a helium expansion from $T_0 = 12\,\text{K}$, $P_0 = 1.5\,\text{bar}$. (**b**) Evolution of the cluster mole fractions with distance assuming either local equilibrium (*dashed lines*) or freezing according to the kinetic model (*continuous lines*).

from the equilibrium constant via partition functions, in which it is assumed that the dimer possesses only one bound state with a negligible binding energy of $-1\,\text{mK}$ and the trimer possesses the ground state of $-125\,\text{mK}$ and one excited state of an also negligible energy [4]. The temperature T as a function of the distance r from the source orifice is computed assuming an adiabatic expansion via [9]

$$\left(\frac{T_0}{T}\right)^2 / \sqrt{\frac{T_0}{T} - 1} = \left(\frac{r}{r^*}\right)^2 \left(\frac{4}{3}\right)^2 \sqrt{3}, \qquad (4)$$

where r^* denotes the sonic radius of 0.7 times the nozzle diameter. This leads to a $T \propto r^{-\frac{4}{3}}$-law in the range where the acceleration has ceased and the beam velocity has reached its final value.

Using these calculated rate constants the rate equations of cluster and atomic densities are then integrated numerically over r until the freeze-out occurs accounting for the above reactions and for the density decrease due to the expansion. In Fig. 4 the evolution of the mole fractions of He_2, He_3, and He_4 according to the kinetic model (continuous lines) are compared to equilibrium (dashed lines) for $T_0 = 12\,\text{K}$ and $P_0 = 1.5\,\text{bar}$. Of course, equilibrium leads to the eventual condensation into the most strongly bound tetramer. The kinetic model, however, reveals the gradual deviation from equilibrium already at about 0.1 mm from the orifice, where the temperature is less than 100 mK and the atomic density is in the range of 3×10^{23} particles/m^3, and subsequent freezing-out at about 0.3 mm.

As shown in Fig. 3 the freeze-out mole fractions as a function of P_0 calculated without any fit parameter reproduce the main features of the measurements, i.e. the successive rise of the cluster mole fractions, the occurrence of a maxima in the dimer and trimer mole fractions, and the correct shape of the curves. However, there are still interesting deviations especially at low P_0 where in the experiment there are no dimers observed below 0.75 bar. This deviation at small P_0 might be related to a freeze-out of the temperature before the mole fractions freeze out so that Eq. (4) is not appropriate over the whole range of r. In addition, for $P_0 = 0.5$ bar a shoulder in the first order diffraction peak is clearly observed (Fig. 2) which might be caused by fast atoms or very slow dimers, the occurrence of either is not yet understood and might play a role in the early stages of cluster formation.

4 Diffraction of Rare Gas Atoms

To investigate the influence of the van der Waals interaction between the particles of the beam and the silicon nitride grating bars on the diffraction patterns atomic beams of He, Ne, Ar, and Kr as well as molecular beams of CH_3F and CHF_3 were diffracted under normal as well as off-normal incidence. The latter experiments are carried out by rotating the grating by Θ_0 as shown in Fig 1. As can be seen in Fig. 5 in addition to the anticipated decrease of the diffraction angles with increased mass of the atom, which is inversely proportional to $\sqrt{m}$, there is a strong change in the relative diffraction intensities, which is especially evident for the ratio of the 2nd to 3rd orders of atomic diffraction intensities which change drastically from He to Kr. This behavior deviates from classical optics where, according to Kirchhoff's theory in the Fraunhofer limit, the diffraction pattern is determined solely by the Fourier transform of the grating.

The diffraction intensities can be understood in detail by taking into account the van der Waals interaction between the rare gas atoms and the inner walls of the grating bars [7]. This interaction introduces a phase shift the effect of which can be described in terms of a complex effective width of the grating slits. Its real part decreases from He to Kr, for which the interaction is strongest. As in classical optics the effective slit width determines the envelope function which becomes broader with decreasing real part of the slit width. The attractive part of the atom-surface potential has been determined quantitatively from the effective slit widths [7] by accounting for the trapezoidal geometry of the grating bars, which is derived from atom beam

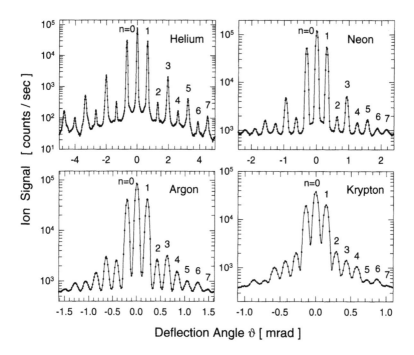

Figure 5: Diffraction patterns of atomic beams of He, Ne, Ar, and Kr at $T_0 = 300\,\text{K}$ from the same grating at normal incidence showing significant changes in the relative diffraction intensities.

transmission measurements [10] and confirmed by electron microscopy.

5 Conclusions

The diffraction of helium clusters by a transmission grating provides a unique tool to observe the evolution of homogeneous nucleation in rare gases, which is initiated by the successive formation of dimers and trimers. The corresponding dependencies of dimer and trimer mole fractions on the source pressure representing the first stage of condensation can be understood by a kinetic model which includes the unusually large, quantum mechanical cross sections of helium at low temperatures. The model leads to a qualitative agreement with the experiment, however, it does not correctly predict the absence of cluster formation at small source pressures. This discrepancy is presumably correlated with the relatively low terminal speed ratios of the atoms observed under these experimental conditions which correspond to a freeze-out of the temperature that is not included in the present calculations.

Further, the diffraction patterns of He, Ne, Ar, and Kr atom beams indicate that the van der Waals interaction of the particles with the grating bars leads to a significant change in the diffraction intensities with increasing strength of the potential. Calculations which take into account this interaction are in excellent agreement with the measurements and allow to determine the attractive part of the interaction quantitatively [7]. The quantitative understanding of the effect of the van der Waals interaction on the diffraction intensities provides the prerequisite for extracting the cluster size information out of diffraction patterns of helium clusters [5, 6].

Acknowledgment

The authors are indebted to Tim Savas and Hank Smith, MIT, for providing the diffraction gratings, Jim Anderson for his contributions to the kinetic model and to Gerhard Hegerfeldt, Thorsten Köhler, and Robert Grisenti for their theoretical work on rare gas atom diffraction patterns.

References

[1] W. Schöllkopf and J.P. Toennies, Science **266**, 1345 (1994).

[2] W. Schöllkopf and J.P. Toennies, J. Chem. Phys. **104**, 1155 (1996).

[3] F. Luo, G. Kim, G.C. McBane, C.F. Giese, and W.R. Gentry, J. Chem. Phys. **98**, 9687 (1993).

[4] B.D. Esry, C.D. Lin, and C.H. Greene, Phys. Rev. A **54**, 394 (1996).

[5] W. Schöllkopf and J.P. Toennies, 17th Int. Symposium on Molecular Beams, Univ. Paris XI, Orsay, France, Book of Abstracts, 143 (1997).

[6] G.C. Hegerfeldt and T. Köhler, Phys. Rev. A **57**, 2021 (1998).

[7] R.E. Grisenti, W. Schöllkopf, J.P. Toennies, G.C. Hegerfeldt, and T. Köhler, Phys. Rev. Lett. **83**, 1755 (1999).

[8] T.A. Savas, S.N. Shah, M.L. Schattenburg, J.M. Carter, and H.I. Smith, J. Vac. Sci. Technol. B **13**, 2732 (1995).

[9] J.B. Anderson, in *Molecular Beams and Low Density Gasdynamics*, edited by P.P. Wegener (Dekker, New York, 1974).

[10] W. Schöllkopf, J.P. Toennies, T.A. Savas, and H.I. Smith, J. Chem. Phys. **109**, 9252 (1998).

Electron Impact Fragmentation of Size Selected Ar_n (n=4 to 9) Clusters

P. Lohbrandt[1], R. Galonska[1], H.-J. Kim[1], M. Schmidt[2],
C. Lauenstein[1], U. Buck[1]

[1] Max-Planck-Institut für Strömungsforschung Bunsenstr. 10
37073 Göttingen, Germany

[2] Fakultät für Physik, Universität Freiburg, H.-Herderstr. 3
79104 Freiburg, Germany

1 Introduction

One of the exciting features of supersonic molecular beams is the possibility to form clusters in the adiabatic expansion zone. This effect has attracted a lot of interest due to the fact that these aggregates of atoms or molecules exhibit properties between those of the free molecule and the condensed state with its solid and liquid phases. The major problem, which has to be overcome experimentally, is that in the expansion always a distribution of cluster sizes is generated, which itself is only a portion of the beam formed by monomers. Early experiments were performed by varying the expansion parameters such as backing pressure or nozzle orifice, to change the maximum of the size distribution [1]. Other experiments are carried out with ionic clusters which can be easily separated by mass filters [2]. For neutral clusters special selection techniques have to be applied.
In principle, the problem could be solved by a size specific detection method. The most commonly used method for this purpose, however, the ionization and the subsequent mass selection in a mass spectrometer, is hampered by the ubiquitous fragmentation during the ionization process. It is caused by the energy released into the system as the clusters change from their neutral to their ionic equilibrium structure. This excess energy then leads to evaporation of neutral subunits, and thus fragmentation occurs [3, 4, 5]. In any case, a simple mass spectrum does not at all characterize the neutral cluster distribution.
Nevertheless, one possibility is to use *special ionization techniques* for a certain class of molecules. A well known example is the two-color resonant two-photon ionization of aromatic molecules [6, 7]. In this case the neutral

and ionic equilibrium structures are similar and by carefully adjusting the ionization energy near the threshold region, fragmentation is avoided. In addition, the first step, the excitation of an electronically excited state can be made size specific, so that a very reliable method results. This is, however, restricted to molecules with suitable electronic transitions. Similar considerations hold for the detection of metal clusters.

Another technique for labelling specific neutral cluster sizes is the charge exchange with ionized cluster ions which are selected by a mass filter [2]. This technique is only feasible for specific aggregates which survive the processes of charge exchange without excitation or dissociation.

A more universal method is the collision process with atoms [8]. Because of the momentum transfer in the scattering process, the different cluster sizes are dispersed into different angular ranges according to their size. It turns out that by moving the detector to a specified angle, a subsequent discrimination against larger cluster sizes is achieved. In this way detection angles can be found were only clusters with sizes smaller than a specific size are detected. By the use of a second analysing technique like time of flight (TOF) analysis for the velocity or mass selection after ioniziation, size specific information can be obtained. This technique has been very sucessfully used in the fragmentation analysis of various atomic and molecular cluster systems [4, 5]. In addition, this method has been applied in a variety of infrared dissociation experiments of molecular clusters as a function of the cluster size [9, 10]. In the latter case, the mass filter was sufficient for the final discrimination, since usually a small fraction was detected at the nominal mass. In the former case with TOF analysis, we have the disadvantage that the size selection is measured in the detector, that is after a possible interaction so that measurements like product state distributions with size selected species cannot be carried out.

In this article we will present a new setup which uses the scattering process for prelabelling and a directly downstream placed velocity selector to achieve a complete discrimination against all cluster sizes but one. For this purpose a new apparatus is installed with two crossed supersonic beams, a single disc velocity selector with variable resolution and a quadrupole mass spectrometer with electron bombardment ionization. A typical application for using this machine are photodissociation experiments of chromophore molecules embedded in single sized clusters, a series of experiments which we already started [11, 12].

As a first example for a successful running of this apparatus we will present the fragmentation analysis of Ar_n clusters up to $n = 9$. This system was among the first ones which had been studied by the scattering method [8] and it clearly gave evidence for the severe fragmentation process occuring in and predicted for this type of weakly bound clusters [3]. In these first experiments fragmentation probabilities were measured up to $n = 6$. For $n \geq 3$, however, the monomer channel could not be detected because of

background problems caused by the remaining monomers in the beam.
We will compare the new experimental results of the present experiment with recent hemi-quantal calculations of the fragmentation probabilities [13] and similar results obtained in experiments of the photofragmentation of argon cluster ions Ar_n^+ [14]. In this way we will explain the general experimental findings of the preference for the dimer Ar_2^+ channel, although the equilibrium calculations of the structure tell us that the core ion in these clusters is at least a trimer [15, 16]. Finally we will try to estimate the amount of energy which stays in the clusters upon electron bombardment and, based on these results, predict the fragmentation of very large clusters using statistical considerations which are not valid for small clusters.

2 Method of cluster separation

The full separation of the different cluster sizes is achieved by two consecutive steps. The first one is the scattering of the cluster beam by a rare gas beam. In this experiment a pure Ar beam was scattered by a He beam. This is visualized in Fig. 1 which shows the Newton diagram for this system. It documents clearly the different angular ranges for the different cluster sizes. At the given detection angle Θ three cluster sizes (n=1,2,3) will contribute to the scattered beam. But the Newton diagram shows as well that these contributions have different velocities in the laboratory system. The corresponding velocity distribution of the scattered beam are also presented in Fig. 1. A velocity selector which is placed directly behind the scattering center can be adjusted in a way that all particles which have velocities corresponding to the shaded areas are blocked out of the beam. The transmitted particles, however, are then trimers only. This principle of operation is possible for all cluster sizes. The limiting condition is the finite velocity and angular resolution in the scattering experiment, which finally makes it impossible to separate the different clusters. Nevertheless it is possible with this method to get a beam of neutral clusters of one size only for sizes up to $n = 9$ as is shown here for the system Ar_n scattered from He. Slowing the argon beam, for instance, by cooling the nozzle should increase the angular range of the different clusters in a way that even larger cluster are separable. In fact, we have already achieved the selection of $(CH_3CN)_{13}$ [17] and $(HBr)_{15}$ [12] by changing the scattering beam from helium to neon, which leads to larger deflection angles.

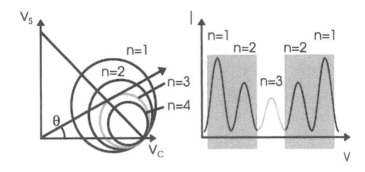

Figure 1: Schematic velocity vector diagram for the size selection of the cluster $n = 3$ by angular and velocity selection.

3 Apparatus

3.1 General concept

The apparatus described is a new setup specially designed for the investigation of size separated clusters. Some restrictions were taken into account for a versatile, easy to modify apparatus. The scattering chamber and the following section containing the velocity selector were designed with minimum distances to achieve high beam intensities. Therefore the maximum detectable scattering angle is limited to 30°, which is sufficient because all systems can by exchange of the carrier gas or the scattering partner adjusted in their Newton diagram to fit this limitation. All other vacuum chambers like the beam chambers and the detector chamber, are universal to fit many purposes. The installed quadrupole chamber can be exchanged by an existing reflectron. A bolometer- or matrixcryostat chamber are easily fitted into the system.

3.2 Vacuum system

A horizontal cross section is shown in Fig. 2. The two cylindrical beam chambers (1,3) are mounted on the walls of the nearly quadratic scattering chamber (2). The primary beam chamber (1) has a diameter of 500 mm which allows us to install various beam sources. The secondary beam chamber (3) has only a diameter of 350 mm, since here always simple rare gas sources are built in. Both sources are mounted in cones which are optimized for high pumping speed in the beam chambers and in the scattering chamber as well. With this arrangment and the skimmers used, the distance between the cluster beam source and the scattering center is 90 mm, whereas the value for secondary beam is 42 mm.

The beam chambers are pumped by diffusion pumps (6000 l/s) and a com-

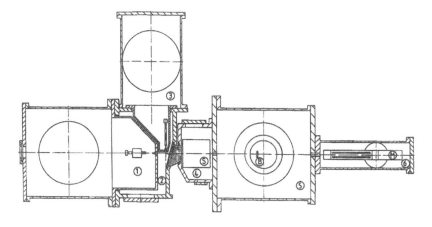

Figure 2: Top view of the molecular beam apparatus.

bination of a rootsblower (1000 m³/h or 500 m³/h) and a mechanical pump (175 m³/h) as forepumps. The idle pressure is 3×10^{-6} mbar and with full gas load it reaches 1×10^{-3} mbar for both source chambers. The scattering chamber is evacuated by a diffusion pump (2700 l/s) and a mechanical forepump (37 m³/h). The whole assembly of this three chambers is mounted on a big ball bearing in a way which allows us to rotate it around the scattering chamber. The connection to the fixed selector chamber is made by a special bellow which is very flexible and allows us a twisting under vacuum conditions of close to 40°. The following selector chamber (4) is only 200 mm long to minimize the distance between scattering chamber and the place where experiments with the completly size selected cluster beam can be performed. This chamber holds not only the velocity selector, which will be described in more detail in the next section, but also a pseudo-random time-of-flight chopper which allows us to carry out fast and easy checks of the direct and scattered primary beam. This chamber is pumped by a small diffusion pump (100 l/s) and a forepump (16 m³/h) due to the fact that only a small gas load has to be pumped and a wide connection exists to the scattering chamber. The pressure obtained in this chamber system is 3×10^{-7} mbar under working conditions. In the present configuration a bolometer chamber (5) is next, which serves for improvement of the angular resolution and as a differential pump stage for the detector. The last chamber which holds the detecting system (6) reaches pressures of better than 8×10^{-10} mbar and with a filled liquid nitrogen trap 4×10^{-10} mbar. The other chambers are also equiped with liquid nitrogen traps, which are used in the experiment to reduce the background.

3.3 Velocity selector

A mechanical velocity selector is used to finally select out of the different sizes scattered into one angle the one size wanted. Since for an optimal performance a variable resolution is required, a single rotating disk, whose rotation axis can be changed with respect to the beam direction, was constructed [18]. This is a special case of the usual slotted disk arrangement. Using the nomenclature of Ref. [19], the angle α between the rotation axis of the disk and the beam is given by $\tan\alpha = (\phi r)/L$, where ϕ is the angular shift, r the radius, and L the length of the disk. The resolution $R = \Delta v/v$ is given by

$$R = \frac{lL\tan\alpha}{L^2\tan^2\alpha - 0.25l^2}, \qquad (1)$$

where l is the slit width. This formula is based on the value $\gamma = l/(L\tan\alpha)$. By varying α, the resolution is easily changed. The only further condition which has to be fulfilled is that the transmitted velocity $v_0 = \omega L/(\phi\cos\alpha) = (\omega r)/\sin\alpha$ also depends on α. Here ω the frequency of the disk. The actual design parameters are as follows: The disk with a radius r=47 mm is L=10 mm thick and rotates with a frequency of maximal 800 Hz. The beam passes one of the 785 slits with a width of l=0.2 mm. In order to get a resolution of R=0.045, the disk has to be tilted by 24° and the velocity which can be reached is 580 m/s, just a little bit above the required velocity of the scattered argon clusters.

4 Fragmentation analysis

In this experimental arrangement the determination of the fragmentation probabilities upon electron impact ionization is straightforward. Once the cluster is size selected, the intensities of measured mass spectrum give the result.

The argon cluster beam is generated by an adiabatic expansion from stagnation pressures of 1.3 bar through a nozzle of 100 µm diameter. The helium beam for deflection is produced by the same type of expansion under high stagnation pressure of 37 bar through an orifice of 30 µm diameter. Both beams are at room temperature of 303 K. According to their measured properties, the corresponding Newton diagrams and the limiting angles Θ_n are calculated: These are 9.1, 7.3, 6.1, 5.2, 4.6, and 4.1° for the size $n = 4$ to $n = 9$, respectively. Now measurements are carried out close to these angles with the velocity selector set to the corresponding intensity maxima at about 550 m/s with a resolution of 0.045. In case that this resolution is not sufficient, experiments with 0.03 were carried out and small contributions from the next larger clusters were corrected based on these results and a Monte-Carlo simulation of the distribution. Finally, the transmission of the quadrupole mass filter has to be calibrated by measurements with a gas of known fragmentation probabilities.

We note that in the scattering process with the helium atoms the cluster is internally excited. The energy transfer is estimated to be about 16 to 18 meV in good agreement with calculations using the impulsive model. This amount of energy is not sufficient to dissociate the clusters collisionally from the trimer upwards and will not influence the results. In addition, quantum calculations which will be decribed in the next section, revealed that internal excitations in the range of 10 to 25 meV do not influence the final results [13].

The results are presented in Tab. 1 and Fig. 3. The dominant channel for all measured sizes up to $n = 9$ is the dimer ion Ar_2^+. The remarkable fact is that for $n = 3$ and $n = 4$ no intensity was observed at other masses than those of dimer and monomer, the parent masses included. These results agree with the tendency observed in the previous measurement [8] of the dimer ion channel which gave the probabilities 1.00 for $n = 4$, 0.98 for $n = 5$, and 0.95 for $n = 6$. Note that in this experiment the monomer channel Ar^+ was not measured. We also remark that the monomer and dimer channel do not exhibit a clear tendency with cluster size, while the trimer channel gradually increases.

exp	$n=2$	$n=3$	$n=4$	$n=5$	$n=6$	$n=7$	$n=8$	$n=9$
$k=1$	0.40	0.30	0.56	0.29	0.42	0.22	0.22	0.34
$k=2$	0.60	0.70	0.44	0.67	0.54	0.71	0.73	0.51
$k=3$	–	0.00	0.00	0.04	0.04	0.07	0.05	0.15
cal		$n=3$	$n=4$					
$k=1$		0.40	0.53					
$k=2$		0.60	0.47					
$k=3$		0.00	0.00					

Table 1: Fragmentation probabilities of Ar_n clusters in Ar_k^+ products. Experimental results from Ref. [8] ($n = 2, 3$) and this work ($n = 4 - 9$). Calculations from [13] ($n = 3, 4$).

5 Interpretation and Discussion

5.1 Dynamical calculations

In the first molecular dynamics (MD) calculations the formation of the Ar_2^+ was assumed and an appreciable boiling off of Ar atoms was found [20]. Similar results were obtained for Ne clusters using realistic potential models in an adiabatic treatment [21]. The first non-adiabatic calculation was

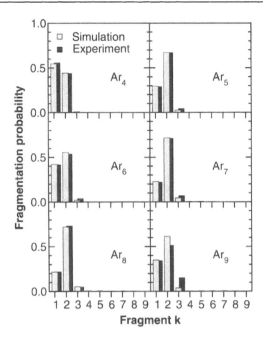

Figure 3: Fragmentation probabilties of Ar_n clusters; black: experiment; shaded: simulation for the determination of the internal cluster energy as described in the text.

carried out for Ar_3 using a DIM (diatomics in molecules) potential of the first three adiabatic states and a classical path surface-hopping model [22]. Here, mainly Ar_2^+ ions were found in contrast to the experimental results where, in addition, also monomer Ar^+ ions were observed. Very recently a hemi-quantal calculation with a complete DIM basis set of nine potentials was performed for Ar_3 and Ar_4 [13]. Here the electronic problem is treated quantum mechanically, while the nuclear motion is described classically under the action of a quantum averaged potential. The results for the fragmentation probability of $n = 3$, 0.40, 0.60, 0.00, and $n = 4$, 0.53, 0.47, 0.00, corresponding to the fragment channels k=1, 2, 3, respectively, are in good agreement with experiment. They are also presented in Table 1. The final discrepancies in the order of a few percent might be caused by the approximations (DIM potentials without spin-orbit coupling, classical motions of the nuclei) still made in the calculations and/or the experimental uncertainties in the same error range. Importantly, all detailed features of the experiment are reproduced in the calculations. Although for neutral trimers the stable ion is Ar_3^+, it cannnot be reached dynamically and the system decays into Ar_2^+ and Ar^+ by the coupling of all nine potential curves. If only a few states are coupled, this leads to incorrect branching ratios. In case of the tetramer neither Ar_4^+ nor Ar_3^+ are reached and the system ends up in the same final channels as are observed for the trimer. Considering the binding energies of an Ar atom in the dimer (1.2 eV) and the trimer (0.2 eV), the result is not surprising. There should, however, be a transition to the situation in which the trimer ion Ar_3^+ is stabilized by the evaporation of neutral atoms. The calculations also revealed that Ar_3^+ is first produced from neutral pentamers, again in agreement with the experimental results.

5.2 Comparison with Cluster Ions

The results of the present experiments are the only ones available for neutral clusters. Thus a comparison is only possible with the much more detailed results which have been obtained for the fragmentation of Ar_n^+ cluster ions after photoexcitation. A special case is the trimer ion Ar_3^+ which decays in fast and slow, charged or neutral monomer fragments [23, 24, 25]. The complete process including the kinetic energy distributions of the neutral and ionic fragments as well as the amount of slow fragment ions as function of the excitation wavelength has been calculated [26] and is in good agreement with the data. Larger clusters have been investigated by measuring the kinetic energy release of the ionic and the neutral fragments [27, 28, 29] and the fractions of argon monomer and dimer ions [30]. The picture which can be derived from all these investigations is that following the absorption of a photon a fast neutral atom is ejected and the remaining energy is dissipated through the loss of further neutral atoms with low kinetic energies. The time scale for the first ejection is fast [21, 31] and the kinetic energy reaches values of 0.4 eV up to cluster sizes of $n = 25$ at an excitation energy of 2.3 eV [28].

The direct comparison with the results of Ref. [30] for 580 nm (2.1 eV), which gives for $n = 5$ the probability of 0.8 for the dimer ($k = 2$) and 0.2 for the monomer ($k = 1$) ion, shows the same trend as is obtained in the present experiments. This continues for $n = 6$ with a similar result and for $n = 4$ with a smaller amount of dimer ions. What is definitly different is the lack trimer ion contributions in the ion experiments which we observe as increasing fraction with increasing cluster size. We have to keep in mind that in the ion experiments a well defined photon energy is applied to dissociate the prepared ion, while in our experiments electrons with much higher energy are used to first ionize and then dissociate the clusters. In a further experiment Ar_3 was ionized by photons and the kinetic energy release of Ar_2^+ was measured in a coincidence experiment [32]. At an excess energy of 1.1 eV a value of 0.4 eV was obtained which could be reproduced with a statistical model. Later a dynamical calculations based on a direct autoionization mechanism came to the same result on a much shorter time scale [31]. The results of this section clearly demonstrate that direct mechanisms dominate the fragmentation of small argon clusters in

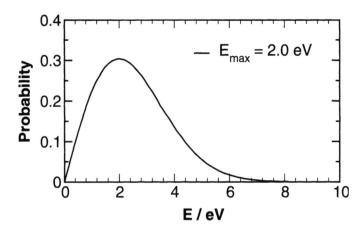

Figure 4: Distribution of the energy left in the cluster after electron impact ionization.

the first steps and that statistical concepts are only of limited value.

5.3 Determination of the Internal Energy of the Cluster

One of the crucial parameters in the somewhat unspecific ionisation by electron impact is the amount of energy which is left in the cluster. It is well known from coincidence measurements of electron ionization of large molecules that only a small fraction of the typical electron energy of about 70 eV is found in the molecule [33]. It is a distribution between 0 and 4 eV. A direct estimation from the calculated binding energies of an Ar atom in Ar_n^+ gives for the $n = 9$ about 2.0 eV before Ar^+ is reached. Similar results are obtained from the hemi-quantum calculations [13] in which for $n = 3$ the range of the different DIM-potential surfaces involved extends from 14.5 to 16.0 eV.

In order to confirm these energetic considerations in a more quantitative way, we have carried out calculations which take into account the kinetic energy release of the evaporating atoms. The cluster of size n is considered to be a system with the initial energy E which is stored in the cluster by the ionization process. Then the cluster decays with the rate constant $k(n, E, D)$ where D is the energy necessary for the dissociation of an atom. We start the process under the assumption of monomer evaporation and calculate after each step the energy balance according to $E^* = E - D - \epsilon$ where ϵ is the kinetic energy taken with the evaporating atom. The process is finished, if i) the available energy E^* is not large enough for a further dissociation, or ii) the monomer ion is already reached. In the size range considered here the results do not depend on the special form of the rates. Thus we take those from statistical theories [34], while D is obtained from reliable calculations [15]. What is crucial is the kinetic energy ϵ of the evaporating atoms. In our calculations they are taken from molecular dynamics simulations [35] which agree with quasi exact calculations based on phase space theory. The special form is taken from measurements of the decay of ionic clusters [36] and is obtained from fits to the high energy tail of the results of [35]. In the calculation ϵ is randomly selected out of this distribution in each fragmentation step. In order to obtain a statistical mean value, we average over 5000 trajectories. By comparing the results of these calculations with the experimental framentation probabilities, we determined the initial energies. It turned out that the experimental data with three fragment channels could only be reproduced by assuming a distribution of initial energies of the type $P(E) = E/(E_{max}^2)exp[-E^2/(2E_{max}^2)]$ with E_{max} = 2.0 eV. The calculated values are also shown in Fig. 3. The resulting distribution is presented in Fig. 4.

Although it is quite obvious from the results of the last section that exact predictions of the fragmentation probabilities can only be obtained from dynamical calculations, we used this simplified method only for the determination of the internal energy distribution of the clusters.

5.4 Statistical considerations

Based on the result of the last section, we are now able to predict the fragmentation pattern of larger clusters using the statistical theory. It will be valid, if at all, for cluster size $n \geq 50$. We calculate the final averaged cluster size $\overline{N} = \sum c_k k / \sum c_k$ where c_k is the probability that the fragment channel k is reached. We start the dissociation process at a given size n under the assumption that the averaged energy left into the system is given by the distribution of Fig. 4 with E_{max}=2.0 eV. The result is shown in Fig. 5. At n=100 the predicted number is still $\overline{N}$=72. The effect gradually diminishes and disappears at n=400. These values are larger than those obtained by similar considerations but with assumed values for E from which then ϵ is calculated [37]. We think that our values are more realistic for a reasonable but still simple correction. For a correct description of the process nearly exact calculations are necessary similar to those which are available for $n = 3$ and $n = 4$ [13] and which take into account the complete dynamics.

6 Conclusions

A new type of universal machine for the investigation of size selected molecular clusters is used to analyze the fragmentation pattern of argon clusters

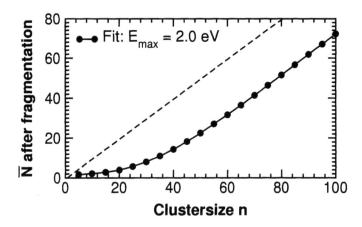

Figure 5: Final average cluster size $\overline{N}$ after electron impact ionization as function of the initial size n.

after the electron impact ionization with 70 eV electrons. The complete size selection is achieved by the combination of two well known techniques. First, the cluster beam is deflected by the scattering process with a He beam. Secondly, a velocity selector with a variable resolution is used to complete the separation. The results clearly indicate that for Ar_n clusters with $n \leq 9$ the preferred fragment channel with th eexception of the tetramer is the dimer ion.

For $n=3$ and 4 hemi-quantal calculations are available which are in good agreement with the experimental values. This clearly demonstrates that, although the chromophore ion is Ar_3^+, the main fragment channel which is reached dynamically is Ar_2^+. From the pentamer onwards also the trimer ion channel Ar_3^+ is detected. Based on an extended calculation of the energy balance, we have determined the energy which is released into the cluster during the ionization process by reproducing the experimental results. It is a distribution with a peak at 2.0 eV and a tail which extends to 6 eV. For the interpretation of the behaviour of large clusters, we have predicted a correlation between the original size and the average size of a fragmented cluster using the internal energy distribution as input for a statistical calculation. We acknowledge the support by the Deutsche Forschungsgemeinschaft in SFB 357.

References

[1] O.F. Hagena, Rev. Sci. Instrum. **63**, 2374 (1992).

[2] *Clusters of Atoms and Molecules*, edited by H. Haberland (Springer, Berlin, 1994).

[3] H. Haberland, Surf. Sci. **156**, 305 (1985).

[4] U. Buck, J. Phys. Chem. **92**, 1023 (1988).

[5] U. Buck, in *The Chemical Physics of Atomic and Molecular Clusters*, edited by G. Scoles (North-Holland, Amsterdam, 1990), p. 543.

[6] K. O. Börnsen, L. H. Lin, H. L. Selzle, E. W. Schlag, J. Chem. Phys. **90**, 1299 (1989).

[7] S. Leutwyler, J. Bösinger, Chem.Rev. **90**, 489 (1990).

[8] U. Buck and H. Meyer, J. Chem. Phys. **84**, 4854 (1986).

[9] U. Buck, Adv. At. Mol. and Opt. Phys. **35**, 121 (1995).

[10] U. Buck, *Advances in Molecular Vibrations and Collision Dynamics*, edited by J.M. Bowman and Z. Bačić (JAI Press, Stamford, 1998), p. 127.

[11] R. Baumfalk, U. Buck, C. Frischkorn, S.R. Gandhi and C. Lauenstein, Chem. Phys. Lett. **269**, 321 (1997).

[12] R. Baumfalk, U. Buck, C. Frischkorn, S.R. Gandhi and C. Lauenstein, Ber. Bunsenges. Phys. Chem. **101**, 606 (1997).

[13] A. Bastida, N. Halberstadt, J.A. Beswick, F.X. Gadéa, U. Buck, R. Galonska, C. Lauenstein, Chem. Phys. Lett. **249**, 1 (1996).

[14] H. Haberland in *Clusters of Atoms and Molecules*, edited by H. Haberland (Springer, Berlin, 1994), p. 392.

[15] P.J. Kuntz and J. Valldorf, Z. Phys. D **8**,195 (1988).

[16] T. Ikegami, T. Kondow, and S. Iwata, J. Chem. Phys. **98**, 3038 (1993).

[17] U. Buck and I. Ettischer, Faraday Discuss. Chem. Soc. **97**, 215, 1994.

[18] G.O. Este, B. Hilko, D. Sawyer, G. Scoles, Rev. Sci. Instrum. **46**, 223 (1975).

[19] C.J.M. van den Meijdenberg, *Atomic and Molecular Beam Methods*, edited by G. Scoles (Oxford, New York, 1988), Vol. I, p. 345.

[20] J.J. Saenz, J.M. Soler, and N. Garcia, Surf. Sci. **156**, 121 (1985).

[21] P. Stampfli, Z. Phys. D **40**, 345 (1997).

[22] P.J. Kuntz and J.J. Hogreve, J. Chem. Phys. **95**, 156 (1991).

[23] J. T. Snodgrass, C.M. Roehl, and M.T. Bowers, Chem. Phys. Lett. **159**, 10 (1989).

[24] T. Nagata, J. Hirokawa, T. Ikegame, T. Kondow, and S. Iwata, Chem. Phys. Lett. **171**, 433 (1990).

[25] H. Haberland, H. Hofmann, and B. von Issendorf, J. Chem. Phys. **103**, 3450 (1995).

[26] A. Bastida and F.X. Gadéa, Z. Phys. D **39**, 325 (1997).

[27] C. A. Woodward and A.J. Stace, J. Chem. Phys. **94**, 4234 (1991).

[28] J.A. Smith, N.G. Gotts, J.F. Winkel, R. Hallet, C. A. Woodward, and A.J. Stace, J. Chem. Phys. **97**, 397 (1992).

[29] B. von Issendorf, H. Hofmann, and H. Haberland, J. Chem. Phys. **111**, 2513 (1999).

[30] T. Nagata, J. Hirokawa, and T. Kondow, Chem. Phys. Lett. **176**, 526 (1991).

[31] A. Bastida, N. Halberstadt, J.A. Beswick, and F.X. Gadéa, J. Chem. Phys. **104**, 6907 (1996).

[32] K. Furuya, K. Kimura, and T. Hirayama, J. Chem. Phys. **97**, 1022 (1992).

[33] H. Ehrhardt and F. Linder, Z. Naturforsch. **22**, 444 (1967).

[34] M. Jarrold, in *Clusters of Atoms and Molecules*, edited by H. Haberland (Springer, Berlin, 1994), p. 163.

[35] S. Weerasinghe and F.G. Amar, Z. Phys. D **20**, 167 (1991).

[36] P. Sandler, T. Peres, G. Weismann, and C. Lifshitz, Ber. Bunsenges. Phys. Chem. **96**, 1195 (1992).

[37] R. Karnbach, M. Joppien, J. Stapelfeldt, J. Wörmer, and T. Möller, Rev. Sci. Instrum. **64**, 2838 (1993).

Static Dipole Polarizability of Free Alkali Clusters

Ph. Dugourd[1], E. Benichou[1], R. Antoine[1], D. Rayane[1],
A.R. Allouche[1], M. Aubert-Frecon[1], M. Broyer[1],
C. Ristori[2], F. Chandezon[2], B. A. Huber[2], C. Guet[2]

[1] Laboratoire de Spectrométrie Ionique et Moléculaire,
CNRS and Université Lyon 1, Villeurbanne, France
[2] Service des Ions, des Atomes et des Agrégats,
CEA Grenoble, Grenoble, France

1 Introduction

To what extent do clusters made of only a few alkali-metal atoms behave as small metal particles ? The static electric dipole polarizability, α_0, is a basic observable for discussing this question since it is very sensitive to the effectiveness of the delocalization of valence electrons, as well as to the structure and shape [1, 2]. Despite numerous investigations of alkali metal clusters, until very recently, polarizability measurements were only available for sodium clusters and for selected sizes of potassium clusters [3]. Nothing was known about lithium clusters. However, the static and dynamic response of lithium clusters to electric fields is in many respects the most interesting and puzzling. The polarizability of lithium atom is abnormally large as compared to the value of the bulk and to other alkali atoms. Moreover, several experiments have shown that the optical response of lithium clusters is significantly redshifted as compared to the classical prediction for a finite metallic sphere [4, 5]. This has been back up to non local effects in electron- ion interactions [6, 7, 8]. For a metallic sphere, the Mie frequency is directly related to the polarizability of the sphere,

$$\omega_M^2 = \frac{e^2 N}{m \alpha_{cl}}. \tag{1}$$

A direct experimental determination of the electric polarizability is therefore crucial both for understanding the size evolution and for the interpretation of the optical response.

In this article, we present the first measurement of lithium cluster polarizabilities [9]. We have also measured values for sodium clusters. The article

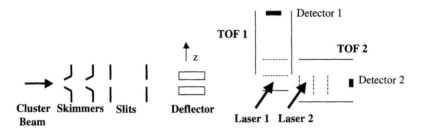

Figure 1: Schematic of the experiment

is organized as follows. Experiment and experimental results are presented in Sect. 2 and Sect. 3. In Sect. 4, we compare the experimental values to results of Density Functional Theory (DFT) calculations and we discuss the size evolution. Consistency with optical spectra is discussed in Sect. 5 in the frame of non local pseudopotential calculations.

2 Experiment

The polarizability measurements are made by deflecting the cluster beam in a static inhomogeneous transverse electric field. A schematic of the experiment is given in Fig.1. Lithium and sodium clusters are produced in a supersonic beam. Metal vapor (0.1 bar pressure) is co-expanded with argon (3 bars pressure) through an aperture of 100 μm diameter. To prevent clogging, the nozzle temperature is slightly hotter compared to the oven temperature. It is close to 1350 K. The collinear part of the beam is extracted by a skimmer and collimated by two 0.35 mm slits. The distance between the two slits is 1 m. The collimated beam passes between the two cylindrical pole faces of a 15 cm long deflector. A difference of potential of 30 kV can be applied between the two pole pieces which are 1.7 mm apart. With the electric field magnitude along the z-axis denoted by F, the force acting on the passing cluster is

$$f_z = \alpha F \frac{dF}{dz}. \qquad (2)$$

The deflection is measured 1 m out of the deflector. Clusters are ionized by a low flux laser ($\lambda = 308$ nm or $\lambda = 266$ nm) and are subsequently mass selected in a time of flight (TOF) mass spectrometer. The set of voltages applied in the TOF is adjusted so that the arrival time at the detector is

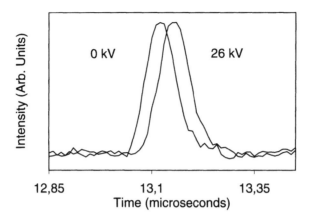

Figure 2: Arrival time distributions for Na_2 with and without deflecting field.

sensitive to the ionization position. The polarizability is proportional to the measured deflection in the z-direction: $\Delta z = K f_z/(Mv^2)$, where M is the mass of the cluster and v its velocity. The constant K is a geometrical factor which actually does not need to be precisely known since it cancels when one takes the ratio of the polarizability of a given cluster and the well-known polarizability of the sodium atom. The velocity is determined with a coaxial TOF mass spectrometer. Velocity measurements are described in details in Ref. [10]. The main source of error is due to the velocity measurement. The precision on the measured polarizabilities is estimated to be $\pm 10\ \%$ for clusters and $\pm 4\ \%$ for monomers and dimers. The relative precision between clusters with neighboring masses is much better.

Figure 2 shows an example of arrival time distributions measured for Na_2 with and without deflecting field. A shift in the arrival time of 34 ns is observed between the two distributions. This shift corresponds to a deviation of the beam of 0.3 mm and a polarizability of 32.8 Å^3.

3 Experimental results and discussion

Figure 3 shows the absolute static polarizability (per atom) for lithium and sodium clusters. For lithium, the values measured for the atom and for the dimer (24.3 Å^3 and 32.8 Å^3, respectively) are in agreement with published data (24.3 Å^3 [11] and 34.0 Å^3 [12, 13], respectively). A sharp decrease in the polarizability per atom by about a factor of 2 from the monomer to the trimer is observed. For larger sizes, $n \geq 4$, the polarizability per atom is

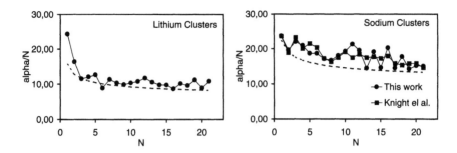

Figure 3: Static dipole polarizability per atom of lithium and sodium clusters as a function of the number of atoms in the cluster (Å^3). For sodium, experimental values of Knight et al. [3] are also plotted. Experimental values are compared to the value calculated for classical metallic spheres (dashed lines)

slowly decreasing. Small oscillations are superimposed on the average trend, especially for $n \geq 15$, where one observes a marked odd-even alternation.

For sodium clusters, our values are compared to the values obtained by Knight et al. [3] (Fig.3). The average evolution is similar in both experiments. A very good agreement is observed for small sizes ($n \leq 9$). For larger sizes, the small odd-even alternation that was observed by Knight et al. is strongly enhanced in our data. The most striking difference between sodium and lithium is that no sharp decrease is observed for small sizes in sodium. The polarizability per atom decreases slowly with the cluster size.

In the classical limit, the polarizability of a metallic sphere is directly related to the volume of the sphere. The dashed lines in Fig.3. correspond to the polarizabilities calculated for finite metallic spheres,

$$\alpha = (N^{1/3}r_s + \delta)^3, \qquad (3)$$

where r_s is the Wigner-Seitz radius (1.75 Å and 2.12 Å, for lithium and sodium respectively) and δ is the electronic spillout (0.75 Å and 0.69 Å) [2]. The Li-Li bond is shorter and stronger than the Na-Na bond. As a consequence, the electronic density is higher in lithium. In the finite sphere model, this leads to different limits for the polarizability per atom for lithium and sodium, in agreement with experimental results for large sizes. However, the simple metallic model cannot be used for very small sizes and cannot explain the different behavior observed between sodium and lithium. In the next section, we focus on small sizes (n=1-8). We give results of DFT calculations in this size range and discuss the specific evolution in electronic properties observed for lithium clusters.

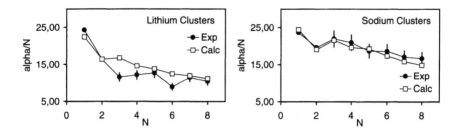

Figure 4: Calculated (DFT) and experimental static polarizabilities for lithium and sodium clusters

	Exp.	T DFT	T Pseudo	T [8]	T [4]	T [17]
Li_8	10.4	11.1	10.6	10.3	12.4	12.1
Na_8	16.8	14.9	16.5			

Table 1: Measured static dipole polarizability of sodium and lithium octamers (Exp.), compared to different theoretical estimates (T). DFT and Pseudo estimates correspond to the calculations decribed in section 4 and 5, respectively.

4 Comparison with DFT calculations

DFT polarizabilities have been obtained for the lowest-energy structures of lithium and sodium clusters (n=1-8). Calculations are described in details in Ref.[10]. Briefly, the values given in this article have been obtained with the non local Perdew-Wang 91 functional [14] and the Gaussian basis sets of Sadlej and Urban [15]. We used the GAUSSIAN 94 Package[16]. Calculated values are plotted in Fig.4 with our experimental values. Polarizabilities which are very sensitive to delocalized orbitals are difficult to compute. The overall agreement observed in Fig.4 is very good. Calculations are able to reproduce both the smooth evolution in sodium cluster polarizabilities and the sharp decrease in lithium cluster polarizabilities. For sodium, there is a perfect match between experimental and theoretical values. For lithium, the values calculated for Li_3 and Li_6 are higher than experimental values. For octamers, which have close electronic structures, several values are available in the literature [4, 6, 9, 17]. Calculated values are given in Table 1. In particular, our values obtained with non local pseudopential calculations [7, 9] are in good agreement with the experimental values. Thess calculations and the effects of non- locality are discussed in the next section.

To understand the sharp decrease in polarizability observed for small lithium clusters, we have analyzed the electronic density maps obtained with the

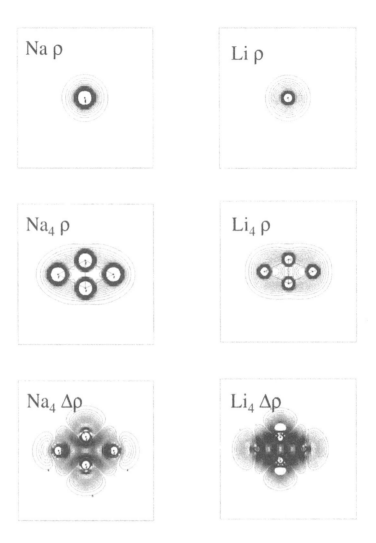

Figure 5: Total electronic densities (ρ) for atoms and tetramers and differential electronic densities ($\Delta\rho = \rho(Cluster) - \rho(Atoms)$) for tetramers

DFT calculations. Total electronic density maps for atoms and tetramers and differential electronic density maps for tetramers are plotted in Fig.5. First, for atoms, while the numbers of electronic inner shells are different,

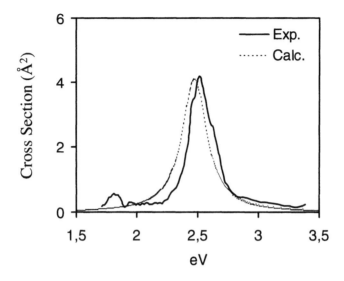

Figure 6: Comparison of experimental photoabsorption cross section of Li_8 (solid line) and calculated transition (dashed line). The experimental data are from Ref. [4]. The experimental curve has been scaled to the theoretical values by a factor 3.2. Calculated transitions have been broadened with a 0.25 eV width Lorentzian.

the volume occupied by the $2s$ electron in lithium atom is close to the volume occupied by the $3s$ electron in sodium atom. This explains the similitude of atom polarizabilities. For tetramers, Fig.5 shows that the metallic bond in lithium clusters induces a strong contraction of the valence electronic cloud : there is a transfer of electrons from the exterior to the center of the cluster and a rapid increase in the electronic density. Thus, the specific evolution observed for lithium cluster polarizability is due to the onset of the metallic bonding. For sodium, the contraction of the electronic cloud is much smaller and the size evolution is much smoother.

5 Consistency with optical spectra

Figure 6 shows the optical photoabsorption spectrum of Li_8 clusters, measured some years ago [4]. The spectrum is dominated by a strong resonance at about 2.5 eV, just as in Na_8 clusters [18, 19]. Though the density of lithium is larger than the density of sodium by about 80%. This immediately rules out a simple interpretation of the observed resonance in Li_8 in terms of the simple Mie theory (Eq. 1), which works well for sodium clusters. The latter will naturally predict the dipole resonance in Li$_8$ to lie above that of Na$_8$ by about 35%, i.e. around 3.5 eV. Such a strong red shift of the experimental resonance with respect to the Mie-Drude model prediction has also been systematically observed in charged lithium clusters[5, 20]. To summarize, for sodium clusters, static and dynamic responses are close to the response expected for a finite metallic sphere. For lithium, the static response is close to the response expected for a metallic sphere while the dynamic response is in total disagreement with a so simple model. In this part, we try to explain with simple physical arguments this apparent contradiction for lithium clusters.

Pseudopotential calculations allow a thorough understanding of the electronic properties of lithium clusters. These calculations are described in details in previous publications [7, 9]. Since lithium has only s core electrons, s-wave delocalized electrons cannot scatter deep in the core region due to Pauli repulsion, whereas p-wave electrons do not suffer such a repulsion. This means that a physically sound pseupotential for lithium should have an $l = 0$ component different from the $l \neq 0$ components. Note that for sodium, which has s and p core electrons, nonlocal effects are expected to be much weaker. Non local effects lead to a substantial red-shift of the giant dipole resonance. In the case of Li$_8$, the theory actually yields a single optical transition at 2.47 eV in perfect agreement with the mean position of the experimental resonance [4]. The calculated theoretical displacement may be phenomenologically understood in terms of effective mass. The Mie frequency ω_M should be replaced by $\omega = \omega_M/\sqrt{m^*/m}$, where m^* is an effective electronic mass. The effective mass for bulk lithium is close to $m^* \simeq 1.4m$[7].

As already mentioned in Section 4, the value calculated for the polarizability of Li$_8$ with this model (Table 1, Pseudo value) is in perfect agreement with the experimental value.

To relate the static and dynamic response with Eq. 1, one has to take into account a second effect of the nonlocality. There is a violation of the Thomas-Reiche-Kuhn sum rule. This sum rule ensures that the total sum of oscillator strength is exactly given by the number of valence electrons, but the rule only holds when the Hamiltonian is local. Approximating the non locality with a constant effective mass, m^*, the dipole sum rule is multiplied by the ratio m/m^* such that the effective number of active electrons is $N^{\text{eff}} \simeq 0.77N$ in lithium clusters. Assuming that all the oscillator strength is concentrated in a single transition, in (1), N should be replaced by N^{eff}. The electric dipole polarizability can then be written as :

$$\alpha_0 = \frac{e^2 N^{\text{eff}}}{m\omega^2}, \qquad (4)$$

where N^{eff} is the reduced strength and ω is the calculated resonance frequency $[\omega^2 = \omega_M^2(m/m^*)]$. One can see that, within the present approximations (single transition, m^* constant, ...), the nonlocal effects in (4) acting both on the oscillator strength and on the frequency cancel out. This explains why experimental polarizabilities are in close agreement with those of the finite metallic sphere, while dynamic polarizabilities are in clear disagrement.

6 Conclusion

We have measured the static dipole polarizability of lithium and sodium clusters. Values measured for sodium clusters are in agreement with previous experiments. The polarizabilities of sodium and lithium atoms are almost equal. For sodium clusters, a smooth decrease in the polarizability per atom is observed as the cluster size increases. For lithium, a sharp decrease in polarizability is observed from lithium atom to size 3-4. This decrease illustrates in a striking manner the onset of the metallic bonding in lithium clusters. Finally, we have shown in the frame of non local pseudopotential calculations that the dynamic and static responses are consistent and that it is necessary to take into account non-local interactions to correctly describe lithium cluster electronic properties.

References

[1] K.D. Bonin and V.V Kresin, *Electric-dipole polarizabilities of atoms, molecules and clusters* (World Scientific, Singapore, 1997)

[2] W.A. de Heer, Rev. Mod. Phys. **65**, 611 (1993)

[3] W.D. Knight, K. Clemenger, W.A. de Heer, and W.A. Saunders, Phys. Rev. B **31**, 2539 (1985)

[4] J. Blanc, V. Bonacic-Koutecky, M. Broyer, J. Chevaleyre, Ph. Dugourd, J. Koutecky, C. Scheuch, J.P. Wolf, and L. Wöste, J. Chem. Phys. **96**, 1793 (1992)

[5] C. Bréchignac, Ph. Cahuzac, F. Carlier, and J. Leygnier, Phys. Rev. Lett. **70**, 2036 (1993)

[6] L Serra, G.B. Bachelet, V.G. Nguyen, and E. Lipparini, Phys. Rev. B **48**, 14708 (1993)

[7] S.A. Blundell and C. Guet, Z. Phys. D **33**, 153 (1995)

[8] J.M. Pacheco and J.L. Martins, J. Chem. Phys. **106**, 6039 (1997)

[9] E. Benichou, R. Antoine, D. Rayane, B. Vezin, F.W. Dalby, Ph. Dugourd, M. Broyer, C. Ristori, F. Chandezon, B.A. Huber, J.C. Rocco, S.A. Blundell, and C. Guet, Phys. Rev A **59**, R1 (1999)

[10] R. Antoine, D. Rayane, A.R. Allouche, M. Auber-Frecon, E. Benichou F.W. Dalby, Ph. Dugourd, M. Broyer, and C. Guet, J. Chem. Phys. **110**, 5568 (1999)

[11] R.W. Molof, H.L. Schwartz, T.M. Miller, and B. Bederson, Phys. Rev. A, **10**, 1131 (1974)

[12] R.W. Molof, T.M. Miller, H.L. Schwartz, B. Bederson, and J.T Park, J. Chem. Phys. **61**, 1816 (1974)

[13] V. Tarnovsky, M. Bunimovicz, L. Vušković, B. Stumpf, and B. Bederson, J. Chem. Phys. **98**, 3894 (1993)

[14] J.P. Perdew and Y. Wang, Phys. Rev. B **45**, 13244 (1992)

[15] A.J Sadlej and M. Urban, J. Mol. Struct. (THEOCHEM) **80**, 234 (1991)

[16] Gaussian 94, Revision B.3, M.J. Frisch, G.W. Trucks, H.B. Schlegel, P.M.W. Gill, B.G. Johnson, M.A. Robb, J.R. Cheeseman, T. Keith, G.A. Petersson, J.A. Montgomery, K. Raghavachari, M.A. Al-Laham, V.G. Zakrzewski, J.V. Ortiz, J.B. Foresman, C.Y. Peng, P.Y. Ayala, W. Chen, M.W. Wong, J.L. Andres, E.S. Replogle, R. Gomperts, R.L. Martin, D.J. Fox, J.S. Binkley, D.J. Defrees, J. Baker, J.P. Stewart, M. Head-Gordon, C. Gonzalez, and J.A. Pople (Gaussian, Inc., Pittsburgh PA, 1995)

[17] A. Rubio, J.A. Alonso, X. Blase, L.C. Balbas, and S.G. Louie, Phys. Rev. Lett. **77**, 247 (1996). The values given in this article should be divided by a factor 3, X. Blase and A. Rubio, private communication.

[18] K. Selby, M. Vollmer, J. Masui, V.V. Kresin, W.A. de Heer, and W.D. Knight, Phys. Rev. B **40**, 5417 (1989)

[19] S. Pollack, C.R.R. Wang, and M.M. Kappes, J. Chem. Phys. **94**, 2496 (1991)

[20] Ch. Ellert, Ph.D Thesis, University of Freiburg, 1995

[21] P. Blaise, S.A. Blundell, and C. Guet, Phys. Rev. B **55**, 15856 (1997)

Photodissociation of Water Clusters

K. Imura[1,2], M. Veneziani[1], T. Kasai[2], R. Naaman[1]

[1] Department of Chemical Physics, Weizmann Institute
of Science Rehovot, 76100, Israel
[2] Department of Chemistry, Graduate School of Science,
Osaka University, Toyonaka, Osaka 560 Japan

1 Introduction

The effect of cluster formation on the photophysical properties of molecules is important from the perspective of understanding the influence of "solvent" molecules on electronic structure of the constituent molecules. It may also be an important factor in understanding the absorption properties of species in the atmosphere. It is well known that cluster formation can introduce profound changes in the spectroscopy of atoms and molecules that are incorporated into the cluster. In the past it was shown that in the water-ozone complex there is a red-shift in the absorption of ozone and its absorption cross section at 355 nm increases by about two orders of magnitude compared to that of the isolated ozone [1]. In the case of $(O_2)_2$ it was found that oxygen atoms are produced upon excitation at wavelength well below the dissociation threshold of the isolated O_2 molecule, due to concerted formation of O_3 [2].

Cluster induced processes in water complexes are of particular interest, due to their role in atmospheric photophysics. In the past several years it became clear that solar absorption by water clouds is anomalously large [3, 4] when compared to theoretical estimates. Although water clusters and clusters of water with other molecules are predicted to be present in the atmosphere [5, 6, 7], current models only include the absorption of water monomers. The photodissociation of water monomers due to absorption in the $^1B_1 \leftarrow {}^1A_1$ transition has been investigated intensively, both experimentally and theoretically [8, 9]. The first band of the featureless $^1B_1$-$^1A_1$ transition is characteristic of repulsive surfaces and comprises a broad continuum absorption, whose onset is at about 190 nm and it extends to 140 nm.

Since spectral shifts in the cluster absorption spectrum can explain some of the discrepancies between observations and theory, experimental and theoretical studies have been performed on water clusters absorption in the infrared-visible region [6] and in the UV [10]. Most relevant to the present work is reference [8] in which the UV spectra of the $^1B_1 \leftarrow {}^1A_1$ transition was calculated for small water clusters, $(H_2O)_n$, for $n = 2$–6. Few studies have been performed on the photochemistry of water clusters in the UV spectral region. Recently, a state selective (hence, size selective) photodissociation of the $Ar - H_2O$ complex was performed and the OH product state distribution was monitored [11].

Here we report results on the effect of cluster formation on the photodissociation of water molecules due to excitation at 193 nm and 212.8 nm.

2 Experimental

The experimental system has been described in details before [12]. Briefly, the molecular beam was produced by expanding a mixture of He and H_2O, whose reservoir temperature was kept at 298 K (32 mbar of water vapor), through a 10 Hz pulsed nozzle (General Valve, 0.8 mm diameter). The pressure in the reaction chamber was kept below 5×10^{-5} Torr. Photodissociation of water was induced either at 193 or at 212.8 nm. 193 nm radiation was generated by an Lambda Physik (Compex 102) ArF excimer laser that delivered around 30 mJ and whose beam was gently focused. 212.8 nm radiation was generated by frequency mixing of the second and third harmonics of a Quanta-Ray, PRO 220 YAG laser in a BBO crystal. The energy of the final 212.8 nm output was 6 mJ. When using this frequency the laser beam was focused with a 30 cm focus lens in the scattering region. Following the photodissociation laser, with a delay of 150 ns, a frequency doubled dye laser output (Spectra-Physics, PDL-2 operated with DCM and pumped by second harmonic of a GCR 150 Quanta Ray YAG laser) was used to probe the OH products. The photolysis and the probe laser beams were counter propagated though two buffer arms and intersected the molecular beam at about 8 mm downstream from the nozzle.

Water and water clusters absorptions were monitored by detecting the laser induced fluorescence of the $OH(^2\Pi)$ fragment through the $A\,^2\Sigma^+$-$X\,^2\Pi$ transition [13]. The rotational state distribution [14] in the OH (v = 0)$^2\Pi_{3/2}^-$ state was measured and the intensities of the individual rotational lines were monitored as a function of total pressure behind the nozzle.

The fluorescence was detected by a photomultiplier (HAMAMATSU, R562) with optical narrow band pass filter (centered at 311.8 nm) positioned perpendicular to the molecular and laser beams. The signal was processed by a boxcar integrator and transferred to a computer. A quadrupole mass spectrometer (UTI) was used for monitoring the total flux of the water in the beam and for obtaining rough estimation of the composition of the beam.

3 Results and Discussion

When monomers of water are illuminated at 212.8 nm, it is known that no dissociation is observed since the far-red side of $H_2O(^1B_1 \leftarrow {}^1A_1)$ absorption spectrum lies at around 190 nm [15]. Vertical transitions at 212.8 nm result in excitation far into the classically "forbidden" regions of the OH stretch potential, where the cross section is already very small and decreases exponentially with the photon energy. In the experiments described here the water partial pressure was kept constant, while the He pressure was varied. The constant flux of water was verified by the mass spectrometer. Figure 1 shows the signal dependence on the backing pressure, for $OH(v = 0, N = 1)$, when the photodissociation was induced by the 212.8 nm laser. No signal could be measured for OH fragments in higher N states. As indicated in the figure, no OH can be detected at low pressures, but as the pressure increases the signal of OH (N = 1) increases and reaches maximum at about 750 mbar. Therefore, the signal shown in Fig. 1 must result from red-shift in the absorption induced by cluster formation as is also indicated by the observation of the LIF signal only at relatively high pressures. The fact that the signal has a maximum as a function of pressure means that the red-shift occurs in one or in a small range of cluster sizes, and that very large clusters do not absorb at this wavelength.

It is known from many reactions that involve clusters that the products emerge with less rotational energy than in the corresponding monomeric process [16]. This effect is manifested in the fact that when the beam containing the water cluster is excited at 212.8 nm, only OH in N = 1 state could be detected. This provides another clue that indeed the red-shift is a result of complex formation.

When the dissociation is initiated at 193 nm it is expected to produce rotationally cold $OH(v = 0)$ and only the first six rotational states of the ground vibronic state were found to be populated. The rotational distribution in the ground $^2\Pi_{3/2}^-$ peaks at N = 2 [11]. Figure 2 shows the LIF spectrum of OH obtained after dissociating the water-containing beam with a 193 nm photon for two different pressures behind the nozzle. The three lines shown correspond to the three lowest rotational states. As the pressure increases the ratio between the population in N = 2 and in N = 1 decreases, namely more OH is produced in the lowest rotational state. Figure 3 presents the pressure dependent LIF signal measured for two rotational states, of the OH, N = 1 (A) and N = 4 (B), produced when water was dissociated at 193 nm. Again, the OH signals were measured under constant partial pressure of water while varying the He partial pressure. All other experimental parameters were kept constant. As can be seen in the figure, two distinct trends are obtained. While the signal from N = 4 state decays monotonically with increasing pressure, the signal from N = 1 state shows a maximum at about 350 mbar. We attribute the monotonic decrease in

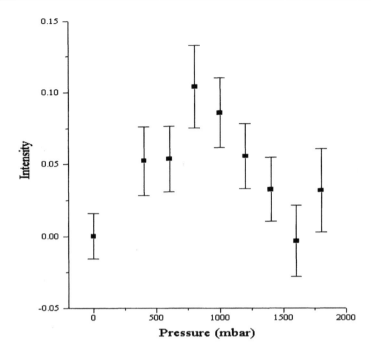

Figure 1: The pressure dependence of the laser induced fluorescence signal from OH(v=0) N=1, following the dissociation at 212.8 nm

the yield of N = 4 product as a function of pressure, to the decrease in the concentration of monomers in the beam. Hence, rotational cooling in cluster dissociation would make the possibility of generating OH in the N = 4 state extremely small. The situation is different for OH in N = 1. This species is probably produced not only when the monomer is dissociating, but also from the dissociation of some fraction of the clusters. Therefore, an increase in the N = 1 population is observed, initially, when the pressure is increased. However, since at higher pressure the signal decreases with increasing of the pressure, one can conclude that large clusters do not dissociate when illuminated at 193 nm. The fact that the signal from N = 1 goes through a maximum indicates the presence of a complex that absorbs 193 nm radiation with a cross section for absorption (and photodissociation) that is larger than that for the monomer. It also means that OH produced from the photodissociation of the complex emerges in its ground rotational state.

Hence, from the results presented so far we have clear evidence that some clusters must absorb to the red of the absorption of the monomers, but larger clusters do not absorb/dissociate when illuminated at 193 nm. In what following we shall try to quantify the observations.

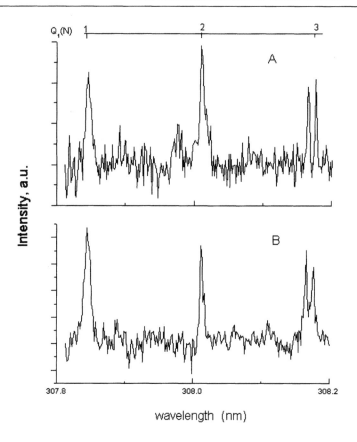

Figure 2: The laser induced fluorescence spectrum obtained after dissociating water containing beam at 193 nm. Spectra are shown when the pressure behind the nozzle was of 18 mbar water vapor and 200 mbar (A) and 700 mbar (B) helium

Assuming that the signal measured for OH(N = 4) at 193 nm is primarily due to the dissociation of the monomers, it is possible to estimate the monomer contribution to the LIF intensity of OH(N = 1) (Fig. 3A) and subtract it from the pressure dependent signal. Figure 4 presents the results obtained by subtracting the normalized signal shown in Fig. 3B from that in Fig. 3A. Hence, Fig. 4 represents the contribution of clusters to the signal shown in Fig. 3A, under the assumptions made above, namely that at low pressure all the signal observed results from monomers and that only the monomer contribute to the population of N = 4. The error bars are substantially higher than in Fig. 3A due to propagation of statistical error. However, one can identify clearly a maximum of the signal as a function of pressure at about 550 mbar.

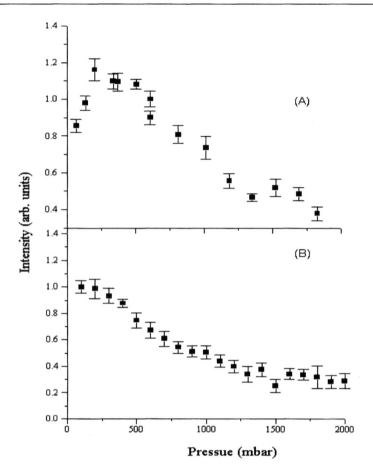

Figure 3: The dependence of the laser induced fluorescence signal on the pressure behind the nozzle. The water was dissociated at 193 nm and the OH(v = 0) product was monitored in two rotational states N = 1 (A) and N = 4 (B) The $Q_1(1)$ and the $Q_1(4)$ transitions were monitored respectively. The pressure behind the nozzle consists of 32 mbar water and the rest He

The results presented above indicate that as clusters are formed, less OH in N = 4 is produced. In addition, there is a cluster, or a range of cluster sizes, for which the absorption at 193 nm is enhanced relative to the absorption of the monomer, and as a result of their excitation the main product is OH(N = 1). The enhancement in the absorption cross-section may result from a red-shift in the spectrum of some of the water clusters. This conclusion is consistent with the observation of production of OH as a result of excitation at 212.8 nm over the same range of pressures. Hence we can

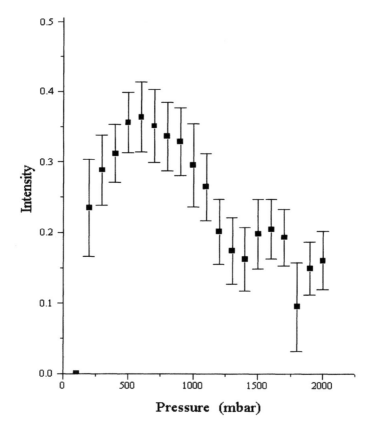

Figure 4: The signal observed from OH(N = 1), after subtracting the contribution from the water monomers. The data shown is obtained by subtracting the signal if Fig. 3B from that shown in Fig. 3A

conclude that there is a specific cluster, or cluster size range, for which the absorption is red-shifted, relative to the absorption of the monomer.

The identification of the cluster size, or size distribution, over which the red shift in the absorption occurred, requires some speculations. The data presented in Fig. 3 help in this respect. If the observed OH is produced only by photodissociation of monomers then the LIF signal, measured for both rotational states (N = 1 and N = 4), is expected to decrease as a function of the total pressure. The decrease in the signal is expected due to the higher dilution and the formation of clusters that reduces the concentration of the monomers in the beam. This trend is indeed observed for OH(N = 4) profile, but the signal from OH(N = 1) goes through a maximum as the pressure is increased. The difference in the pressure dependent of the signal from the two species is a direct evidence that while OH in N = 4 is formed mainly from monomers, the products in N = 1 must also be formed from the dissociation of some clusters. In these clusters the absorption cross section at 193 nm is enhanced relative to that for the monomers. It is also clear that as the clusters reach a certain size, no OH is formed. This statement results from the decrease in the signal from the N = 1 states as the pressure is increased beyond 350 mbar. Next we shall try to estimate the cluster size range responsible for the enhancement in the signal. Imura et al. [17] characterized the molecular beam expansion of pure HCl through the same nozzle used here and found that dimers are the prevalent cluster constituent in the beam in the range 350–550 mbar. Although the binding energy of H_2O dimer [18] is larger than that for HCl dimer [19], 1200 compared to 430 cm^{-1}, the diluted expansion used here ensures that the most abundant cluster in the beam is the dimer. In addition, the relatively low pressure at which the signal that can be attributed to clusters (Figs.1 and 4) reaches maximum suggests that the observed products come from dimers. This conclusion is again consistent with the calculations presented in [10]. Hence we conclude that the enhancement in the photodissociation at 193 and 212.8 nm results mainly from red shift in the absorption of the dimers.

It is possible to estimate the enhancement in the photodissociation cross section of the dimer relative to that of the monomer, when the system is illuminated at 193 nm, by taking the ratio of the signal at the maximum to that at around 100 mbar (first data point) in Fig. 3A. This ratio is approximately 1.4. Based on the mass spectrum we can give a lower limit of 1 % for the concentration of dimers. Therefore, the photodissociation yield of the dimers is enhanced by at least a factor hundred relative to that of the monomer.

The OH(N = 1) formed from clusters that absorbed at 212.8 nm is produced with an efficiency that is $3 \pm 1\%$ of the efficiency for production at 193 nm (Fig. 4), under the same conditions. The observed photodissociation efficiency at 212.8 nm, is much smaller than that predicted by the model calculations [20] in [8]. This discrepancy may result from the difference between the calculated observable as compared to that measured. In the present study we did not measure the absorption cross section directly, but rather the amount of OH produced as a result of the photoabsorption and photodissociation processes. Hence the total dissociation efficiency measured here depends both on the absorption and the dissociation yield.

Beside the cluster-induced shift of $^1B_1 \leftarrow {}^1A_1$ spectrum another mechanism that could contribute to the red-shift in the absorption of water clusters involves the influence of the low lying $^3B_1$ state. The $H_2O(^3B_1 \leftarrow {}^1A_1)$ transition is electric dipole forbidden due to the (S = 0 selection rule, therefore the corresponding transition moment with the ground state is expected to be negligibly small. However, the role of this state in the photodissociation of water was suggested as an explanation for the surprisingly small OD/OH branching ratio in the photodissociation of HOD at 193 nm [21].

When clusters are formed, two effects involving the $^3B_1$ state can take place. First, the increased total mass of the van der Waals complex relative to the isolated molecule could increase the spin-orbit coupling, thereby enhancing the absorption cross section of the $^3B_1 \leftarrow {}^1A_1$ transition in the clusters relative to that in the isolated molecule. Second, the red shift induced by cluster formation on the singlet-triplet transition could be larger than that induced on the singlet-singlet one, resulting in increasing in the dissociation via the triplet surface at this wavelength compared to that through the singlet one.

The experimental data presented here demonstrate that one size, or a narrow distribution of small water cluster sizes, induces a red-shift in the UV absorption spectrum of the constituents. Analysis of the experimental data and their comparison with theoretical calculations in [8] point to the conclusion that dimers of water are responsible for the observed spectral shift. In addition, a decrease in the photodissociation yield is observed for the larger clusters. Based on the calculation this observation may be explained by a blue shift in the absorption of larger water clusters.

4 Conclusions

We report here experimental studies in which we detected OH resulting from photodissociation of water clusters. Water complexes were produced in a molecular beam and were photodissociated either at 193 or at 212.8 nm. Clusters of water were produced by keeping the partial pressure of water constant and varying the pressure of the He, the seeding gas. It was found that while the production of the rotationally excited OH, N = 4, decreases monotonically with the pressure behind the nozzle, the production of the ground rotational state, N = 1, shows a maximum at a pressure of approximately 350 mbar. Our results are consistent with the results from recent calculations that show that while the absorption of most clusters is blue shifted compared to that of the monomer, the dimer absorption is red shifted.

Acknowledgments

We acknowledge fruitful discussions with R.B. Gerber, J.N. Harvey, A.B. McCoy, D.H. Waldeck and V. Vaida. This work was partially supported by the US-Israel Binational Science Foundation. KI thanks the Japan Society for the Promotion of Science for predoctoral fellowship.

References

[1] Y. Hurwitz, R. Naaman: J. Chem. Phys. **102**, 1941 (1995)

[2] L. Brown, V. Vaida: J. Phys. Chem. **100**, 7894 (1996)

[3] P. Pilewskie, F.P.J. Valero: Science **267**, 1626 (1995)

[4] R.D. Cess et al.: Science **267**, 496 (1995)

[5] Z. Slanina: J. Atm. Chem. **6**, 185 (1988)

[6] H.C. Tso, D.J.W. Geldart, P. Chylek: J. Chem. Phys. **108**, 5319 (1998)

[7] G.J. Frost, V. Vaida: J. Geophys. Res. **100**, 803 (1995)

[8] P. Andresen, G.S. Ondrey, B. Titze, E.W. Rothe: J. Chem. Phys. **80**, 2548 (1984); T. Schroeder, R. Schinke, M. Ehara, K. Yamashita: J. Chem. Phys. **109**, 6641 (1998); See also review and references cited therein: P. Andresen, R. Schinke: In Molecular Photodissociation Dynamics, ed. by M.N.R. Ashfold, J.E. Baggott, The Royal Society of Chemistry, London (1987)

[9] D. Hausler, P. Andresen, R. Schinke: J. Chem. Phys. **87**, 3949 (1987)

[10] J.N. Harvey, J.O. Jung, R.B. Gerber: J. Chem. Phys. **109**, 8747 (1998)

[11] D.V. Plusquellic, O. Votava, D.J. Nesbitt: J. Chem. Phys. **101**, 6356 (1994)

[12] Y. Hurwitz, Y. Rudich, R. Naaman: Isr. J. Chem. **34**, 59 (1994)

[13] G.H. Dieke, H.M. Crosswhite: J. Quant. Spectrosc. Radiat. Transfer **2**, 97 (1962)

[14] I.L. Chidsey, D.R. Crosley: J. Quant. Spectrosc. Radiat. Transfer **23**, 187 (1979)

[15] K. Watanabe, M. Zelikoff: J. Chem. Phys. **43**, 753 (1953)

[16] A.B. McCoy, M.W. Lufaso, M. Veneziani, S. Atrill, R. Naaman: J. Chem. Phys. **108**, 9651 (1998)

[17] K. Imura, T. Kasai, H. Ohoyama, H. Takahashi, R, Naaman: Chem. Phys. Lett. **259**, 356 (1996)

[18] S.S. Xantheas, T.H. Dunning, Jr.: J. Chem. Phys. **99**, 8774 (1993)

[19] A.S. Pine, B.J. Howard: J. Chem. Phys. **84**, 590 (1986)

[20] Revised calculations show lower absorption of the dimer on the red side of the absorption peak, compared to the value presented in [8] by J.N. Harvey, J.O. Jung, R.B. Gerber, private communication

[21] T. Schroder, R. Schinke, M. Ehara, K. Yamashita: J. Chem. Phys. **109**, 664 (1998)

Collision Induced Fragmentation of Molecules and Small Na_n^+ Clusters: Competition Between Impulsive and Electronic Mechanisms

J.A. Fayeton, M. Barat, and Y.J. Picard
Laboratoire des Collisions Atomiques et Moléculaires
UMR 8625
Université Paris-Sud, Orsay, France

1 Introduction

The dynamics of interaction between particles and matter involves of two basic mechanisms. Momentum transfer in collisions between atomic cores, a mechanism dominant at low collision velocity (v $\ll$ 1 a.u.), is primarily relevant for pure elastic atomic collisions. This first mechanism, referred to as *the impulsive mechanism* (IM), is primarily responsible for vibro-rotational excitation and reactive processes in molecular collisions. On the other hand, processes induced by excitation of the electron cloud, hereafter referred to as *the electronic mechanism* (EM), usually operated at much larger velocity (v$\approx$1 a.u.). These two mechanisms which are nothing else but the nuclear and electronic contributions of the stopping power of particles in matter [1] are also expected to drive Collision Induced Dissociation[2] (CID) of a simple molecule or a cluster. Let us consider for example the CID of a simple molecule at keV energies. In a first step the molecule is brought into an electronically excited state that can be an unbound or predissociated state. Then in a second step the molecule dissociates far from the collision partner. In contrast at low collision energy, typically below a few eV, dissociation occurs if a large momentum transfer during a close collision to one atomic core of the molecule or cluster is large. As a consequence a stretching of the vibrationnal bond occurs leading to dissociation of the molecule. At intermediate energies (few eV to few 10 keV) both EM and IM mechanisms can be simultaneously active, but up to now no experimental investigation of the relative importance of these two mechanisms was achieved. Actually, the study of CID mechanisms for molecules was widely developed in the last decades when it was realized that velocity measurements of the ionic fragment produced in dissociation of swift molecular ions could provide information on molecular energy levels with meV accuracy[3]. However this accuracy is only obtained when the deflection angles of the molecular fragments are negligible. Therefore this experimental technique excludes dissociation events due to violent collisions strongly reducing the scope of the studie. This limitation does not apply if one measures the velocity vectors of all fragments produced in the CID of fast molecular species in encounters with an atomic target allowing reconstruction of the collision kinematics. Such an experiment was achieved only recently and applied to the study of the fragmentation of small Na_n^+ clusters in collisions with a helium target allowing for the first time the two basic IM and EM mechanisms to be disentangled[4, 5].

2 Experimental Set-up

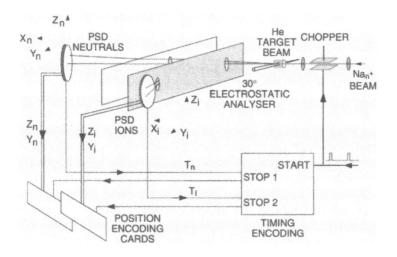

Figure 1: Experimental set-up

Only a brief description of the experimental set-up, shown in Fig. 1, will be given here. A detailed account is reported elsewhere [5]. The cluster beam is produced by adiabatic expansion of a sodium vapor through a sonic nozzle with a diameter of (Φ=0.125mm). The stagnation pressure in the oven is of about 450 torr. Once formed the clusters are ionized by 70 eV electron impact, accelerated at few keV and mass selected by a Wien filter. The cluster ion beam then crosses at 90° a 'cold' He target beam produced by a 'Campargue' supersonic expansion [6]. The He velocity is 1765m.s^{-1}, for a transverse velocity of 45m.s^{-1}. The fragmented clusters then enter a parallel-plate electrostatic analyzer. The neutral fragments fly in straight lines through the analyzer and are received on a position sensitive detector (PSD). The ionic fragments are deflected in the electric field of the analyzer and detected on an additional PSD in coincidence with the neutral fragments. The incident beam is chopped at a 1 Mhz frequency before collision. The chopper clock triggers a multi-stop time-to-digital converter that records the time-of-flight (TOF) of the neutral and ionic fragments. The velocity vectors of the two fragments are determined knowing their locations on the two PSD and their TOF. In order to eliminate fragments produced by collisions with the background gas or by evaporation of hot clusters along the incident trajectory, the collision zone, the analyzing and detection devices are biased with an additional voltage. The electrostatic analyzer can be tuned to select an ionized fragment of a given mass. In contrast the mass of the neutral fragments cannot be directly determined by the present technique. A " fragmentation channel " will be identified then by the mass of the detected ionic fragment. In case of a multi-fragmentation process, the 2 μs time needed for encoding the position information on the MCP prevents detecting the mulitple neutral fragments coming from a given event.; only the first neutral fragment is detected.

3 Na_2^+ Dissociation: Identification of the Basic Dissociation Mechanisms

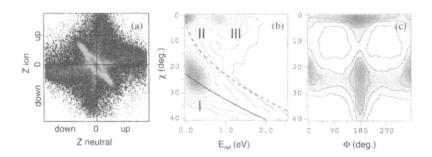

Figure 2: Na_2^+ + He collision-induced dissociation at 80 eV collision energy. **a**: "ZZ" correlation, **b**: $E_{rel}(\chi)$ contour map, *full* and *dashed lines* : prediction of the simple binary model for Na_2^+ in the ground vibrational state and excited at the dissociation limit respectively **c**: $\Phi(\chi)$ contour map. Each contour line shows iso-intensity in a linear scale

A study of the Na_2^+ system has demonstrated the power of the technique. Figure 2a shows the correlation between the locations on the vertical axis Z of the two detected Na and Na$^+$ fragments. Such pattern referred to as the "ZZ correlation" gives a first insight into the fragmentation mechanisms that will be useful to analyze more complex systems (see below). In brief - see[5] for details - points localized inside the bottom-left and top-right squares indicate a significant deflection of the cluster center-of-mass (CCM), a signature of an impulsive character of the fragmentation process. On the other hand, points inside the two other opposite squares correspond to dissociation processes without significant deflection of the CCM, a feature of the electronic mechanism. The two structures seen in the "butterfly" pattern of Figure 2a reflects the importance of these two types of mechanism: the body and the wings of the butterfly reflecting the EM and IM processes respectively. Concerning the impulsive mechanism, the vertical and horizontal wings correspond respectively to the Na$^+$ ion or the Na atom being side scattered.

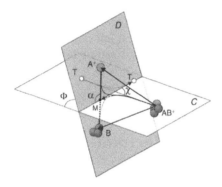

Figure 3: CM velocity diagram for AB$^+$ + T → A$^+$ + B + T collision-induced dissociation. C represents the collision plane, D the dissociation plane

To get more information on the fragmentation mechanisms, a vectorial analysis is needed. Such an analysis is presented as correlation patterns between the relevant parameters of the dissociation, such as χ the deflection angle of the cluster centre of mass (CCM), E_{rel} the relative kinetic energy of the fragments, α and Φ the angles defining the orientation of the dissociating axis. The velocity diagram of Fig.3 shows these parameters in the CM frame. Figure 2b illustrates how the IM and EM are identified in the $E_{rel}(\chi)$ pattern. The two structures (II and III) appearing at very small χ reveal the EM induced by excitation of the active 3sσg electron towards repulsive (III) and weakly bound (II) states. Notice that these two mechanisms could not be distinguished in the "ZZ" correlation. The structure (I) appearing at larger scattering angle χ reveals the IM as shown by the stretching of the contour lines along the $E_{rel}(\chi)$ curve given by a simple kinematics model describing a He Na binary elastic collision[5, 7]. The $\Phi(\chi)$ curve (Fig.2c) shows also that for the impulsive mechanism dissociation occurs in the collision plane (Φ=0 and 180 degrees), the plane inside which the momentum is transferred. This simple analysis has been theoretically confirmed by calculations based on a non-adiabatic quantum-molecular-dynamics approach (NA-QMD) developed by the Dresden group[5, 8]. More recently[9] *ab initio* calculations were used to determine the relevant potential energy surfaces and couplings of Na_2^+-He collisional system . A semiclassical coupled wavepacket method was explored in the treatment of the dynamics. The results nicely reproduce the experimental findings. However, even for such simple molecules, more complex dissociation mechanisms appear that combine both IM and EM as pointed out in CID of the NaK^+ heteronuclear system[10].

4 Cluster Fragmentation

The ZZ correlation patterns for all the Na_n^+-He investigated systems are gathered in Fig.4. The columns correspond to the various fragmentation channels, identified by the size of the ionic fragment. The rows indicate the parent cluster. Such diagrams reveal three typical patterns:

- *(i)* Channels giving Na^+ fragments (first column) show clear indication of a dominant EM (top-left bottom-right lines)

- *(ii)* For all other fragmentation channels, patterns indicate a dominant impulsive character of the processes (horizontal structures)

- *(iii)* However these latter patterns show generally two components: a thin horizontal line corresponding to the ejection of a fast neutral fragment and a more bulky pattern characterizing processes in which the fragments are ejected with a much smaller relative kinetic energy.

The pathways leading to Na_{n-1}^+ fragments located on the main diagonal of in the "ZZ" patterns of Figure 4 correspond to the ejection of a single Na atom as a neutral fragment. These channels, which can then be further analyzed without any ambiguity, will be discussed below.

4.1 The Linear Structure in the "ZZ" Correlation: Direct Impulsive Mechanism

The Na_5^+ and Na_7^+ "ZZ" fragmentation patterns (Fig.4) have a rather straightforward interpretation. They exhibit a horizontal line that corresponds to the ejection of the Na atom after a unique Na^+ - He hit. For example, the $E_{rel}(\chi)$ correlation of Fig.5a shows that the $Na_5^+ \rightarrow Na_4^+ + Na$ fragmentation begins at a threshold angle $\chi = 15°$ and then E_{rel} strongly increases

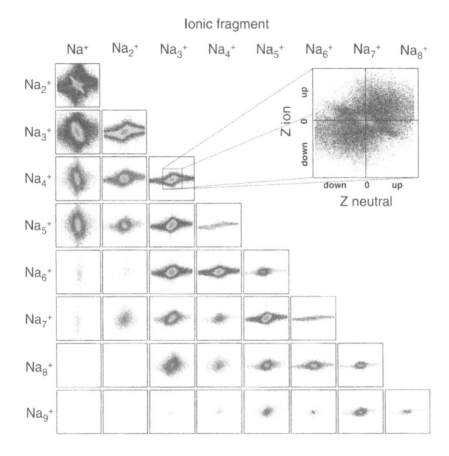

Figure 4: "ZZ" correlation patterns for all investigated clusters. Insert: enlarged central part of the $Na_4^+ \rightarrow Na_3^+$ channel showing the IM and EM components

with increasing χ. This dissociating mechanism is the same as the IM encountered in the Na_2^+ - He system: the Na^+ core hit by the He atom is directly ejected without significant interaction with the other Na cores, just capturing an electron before leaving the cluster. The $E_{rel}(\chi)$ curve based on the same binary He - Na collision model as discussed in Section 3, (full line on Fig.5a) follows well the contour lines of the experimental data showing the cogency of the model. The $\Phi(\chi)$ correlation (Fig.5b) indicates that fragmentation takes place about the collision plane, with the Na atom being side scattered, a feature also in agreement with the binary character of the model. Similar results are obtained for the ejection of a Na atom from a Na_7^+ cluster. It is noteworthy that a such simple "ZZ" correlation only ap-

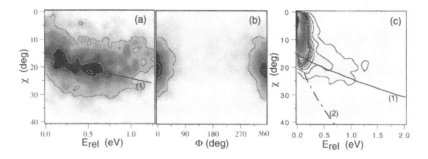

Figure 5: **a** and **b**: $E_{rel}(\chi)$ and $\Phi(\chi)$ contour maps for the $Na_5^+ \to Na_4^+$ fragmentation channel at 80 eV collision energy respectively. **c**: $E_{rel}(\chi)$ contour map for the $Na_6^+ \to Na_5^+$ channel at 67eV collision energy. The *curves (1)* and *(2)* correspond to the IM model, without and with energy redistribution respectively (see text)

pears with Na_5^+ and Na_7^+ clusters for which the release of a single Na atom does not correspond to the less endothermic channel. One also notices that such pathways are only weakly populated. These features will be explained below.

4.2 The Bulky Structure in the "ZZ" Correlation:

large χ contribution
The $E_{rel}(\chi)$ diagram of Fig.5c corresponds to the release of a single Na atom from a Na_6^+ cluster for which the double structure shows up in the "ZZ" correlation. The direct impulsive mechanism discussed above and responsible for the linear structure in the "ZZ" correlation appears in the present $E_{rel}(\chi)$ diagram as contour lines that follow well the curve (1) of the binary impulsive model. However for a given χ angle, the contours extend much below the model curve suggesting that part of the initial momentum given during the Na^+ - He hit is transferred to other Na^+ cores via intra-cluster Na^+ - cores collisions. Such a 2-step mechanism results in the population of the less endothermic fragmentation channels. This momentum redistribution increases the internal energy of the cluster leading to a reduced relative kinetic energy of the fragments. For the largest clusters a statistical redistribution of the transferred momentum into the various degrees of freedom can be assumed[11]. The corresponding curve (2), drawn on the $E_{rel}(\chi)$ diagram of Fig.5c, seems to account rather well for the tail stretching at small E_{rel} and large χ.

Small χ contribution: multiple collisions of the helium atom with several Na^+ cores, electronic transitions and hot clusters

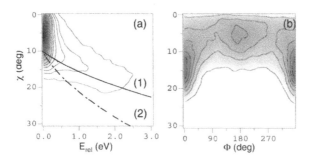

Figure 6: **a** and **b**: $E_{rel}(\chi)$ and $\Phi(\chi)$ contour maps for the $Na_4^+ \rightarrow Na_3^+$ fragmentation channel at 200 eV collision energy respectively. The *curves (1)* and *(2)* correspond to the IM model without and with energy redistribution respectively (see text)

Evidence of the binary impulsive mechanism is the existence of a threshold angle χ_{min} corresponding to the minimum energy transfer needed to overcome the endothermicity of the corresponding fragmentation channel. However χ_{min} is dependent on the temperature of the cluster. Therefore processes appearing at $\chi < \chi_{min}$, as exemplified by the $Na_4^+ \rightarrow Na_3^+$ + Na fragmentation (Fig.6a) is due, at least partly, to fragmentation of hot clusters. Additional information is provided by the $\Phi(\chi)$ pattern (Fig.6b). While for $\chi > \chi_{min} \approx 15$ deg the strong peaking about $\Phi=0$ and 180 deg indicates the IM character, a quasi-isotropic Φ distribution is found for $\chi < \chi_{min}$, with a slight maximum around 180 deg. Such a rather uniform Φ distribution means a loss of the initial Na^+ He collision plane, a feature that can be understood as due to multiple collisions between He and several Na^+ cores. Notice also that the role of peripheral collisions is enhanced for hot collisions. Finally, electronic excitation toward unbound potential energy surfaces, as observed on the "ZZ" pattern (Fig.4), also contributes to this structure and seems to be primarily responsible for the maximum at $\Phi=180$ deg in the $\Phi(\chi)$ contour map. Fig.7 shows the relative contribution of the three types of IM as a function of the size of the cluster. Notice that the direct impulse mechanism decreases with increasing size of the cluster, while the two step mechanisms *(ii)* and *(iii)* become dominant for large clusters as simply expected from the increasing size of the sodium cluster. The overall contribution of the impulsive mechanism to the total fragmentation reaches 70% consistently with the estimation given by the TRIM compilation for stopping power[1]. One must finally stress that the initial temperature of the cluster should play a key role in the relative importance of the various mechanisms and hence in the branching ratios between the pathways.

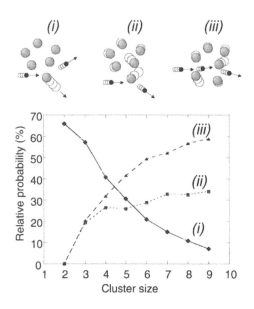

Figure 7: Relative contributions of the three impulsive mechanisms as a function of cluster size

References

[1] J.F.Ziegler and J.P.Biersack,.The Stopping and range of Ions in Solids, Pergamon Press, New York (1985)

[2] J. Los, T.R.Govers , Collision spectroscopy (R.G.Cooks, Plenum press, New York and London, 1978). p. 289.

[3] P. de Bruijn and J. Los, Rev. Sci. Instrum. **53**, 1020 (1981)

[4] J.C. Brenot, M.Barat, H Dunet, J.A. Fayeton and M. Winter, Phys.Rev.Lett. **77**, 1246 (1997)

[5] J.A. Fayeton, M. Barat, J.C. Brenot, H. Dunet, Y.J. Picard, R. Schmidt, U. Saalmann, Phys. Rev. A, **57**, 1058 (1998)

[6] R. Campargue, J. Phys. Chem. **88**, 4466 (1984)

[7] B.H. Mahan, J. Chem. Phys. **52**, 5221 (1970)

[8] U. Saalmann, R. Schmidt, Z. Phys. D. **38**, 153 (1996)

[9] D. Babikov, F. Aguillon, M. Sizun, and V. Sidis, Phys. Rev. A, **59**, 330 (1999)

[10] M. Barat, J.C. Brenot, H. Dunet, J.A. Fayeton and Y.J. Picard, E.P.J.D. **1**, 271 (1998)

[11] P.C. Engelking, J. Chem. Phys. **87**, 936 (1987)

Charge Exchange in Atom–Cluster Collisions

C. Bréchignac, Ph. Cahuzac, B. Concina,
J. Leygnier, I. Tignères

Laboratoire Aimé Cotton CNRS II, Bâtiment 505,
Campus d'Orsay, 91405 - Orsay Cedex, France.

1 Introduction

Charge exchange processes which occur in collisions either between atoms and a surface or between atoms and molecular ions have been extensively studied since they act as precursors of any chemical reactions. The first studies in the field of atom–surface interaction were initiated by Taylor and Langmuir in 1933 for cesium adsorption on tungsten. As the ionization potential of cesium is less than the work function of tungsten, cesium atom transfers its electron to tungsten leading to positive ion which interacts with the metal through image potential [1]. More recently charge transfer in scattering atoms, molecules or ions from surface have been the subject of numerous theoretical and experimental studies for surface analysis and for local probe. The general approach in describing charge transfer in scattering atoms from a metal surface assumes that the metal electrons are free electrons in a potential box and that the atom is a one-electron hydrogen like atom. When the atom is brought close to the metal surface the originally sharp valence level shifts and broadens into a band of finite width making resonant electron tunneling through the potential barrier between the atom and the metal possible [2, 3]. However if the atomic level still stay off resonance from the fermi level of the metal surface, an increase of incident velocity parallel to the surface can shift the fermi sphere into resonance with the atomic state as clearly illustrated in [4]. The influence of the velocity component of the colliding particle parallel to the surface is already known for a long time in atomic collisions, where one studies electron transfer during the collision as a function of relative velocity between the two particles. For particles of different species the energy defect between initial and final state is not zero leading to the so called "non resonant charge transfer" [5]. In this case an increase of the cross section is observed as a function of relative velocity of the two collisional partners up to a maximum value which can reach the same order of magnitude as that for the resonant transfer. An interesting aspect of such colliding systems is that the interacting collision time lies in the range of 10^{-15}s involving changes in electronic state without the involvement of nuclear motion as it is now possible in experiments with femtosecond laser pulses.

Charge exchange in atom–cluster collisions is an intermediate case between atom–atom collisions and atom surface collisions. Since cluster is viewed as a free particle, experimental methods usely done for atomic collision can be transfered to atom–cluster collisions. However metallic clusters may also be viewed as a small piece of metal with delocalized electrons [6]. During the collision, the relative velocity between the two partners brings atomic states into kinematic resonance with the conduction band of the cluster making the electron transfer efficient. Such tunneling process between a cluster and the colliding atom is sensitive for probing cluster electronic structure in 10^{-15}s time scale.

Up to now few experimental studies and even less theoretical ones have been performed on charge exchange between atom and cluster. Most of them concern the neutralization of mass selected cluster ions in order to generate neutral monodisperse cluster beam [7–14]. In this paper we present an extensive study of charge exchange between either homogeneous metallic clusters or heterogenous metal oxide and hydroxide clusters and alkali atoms. The velocity dependence of the charge exchange cross section between a cluster ion and a neutral atom reflects the typical features of non-resonant atom-ion charge transfer cross section profile. The maximum of the cross section as well as the relative velocity at which this maximum occurs are discussed as a function of cluster size and cluster composition for different atomic neutral targets.

2 Experimental

2.1 Experimental setup

The experimental apparatus is shown in Fig. 1. Neutral clusters are formed by gas aggregation technique. Alkali vapor created in a crucible effuses into a carrier gas (He). Downstream from the oven, the nucleation takes place in a copper tube, cooled with flowing liquid nitrogen, leading to a pure neutral alkali cluster distribution. Oxidized, or hydroxided heterogeneous clusters are obtained when adding less than 1% oxygen or water molecules to the carrier gas. The clusters that condensed out of the quenched vapor are carried out by the gas stream through differential pumping chamber. The neutral cluster distribution enters a multiplate ionizing/accelerating region where clusters are ionized. The ionizing photons are delivered by

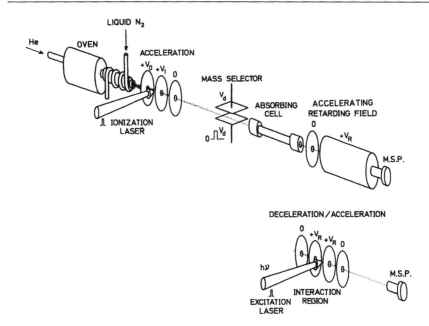

Figure 1: Basic elements of the experimental set-up. Two different configurations are used for product dispersion following UD, CID and CT processes by retarding (accelerating) potential V_R (*upper part*) and for reionization of neutral products (*lower part*).

Nd-YAG laser at an energy $h\nu_i = 3.50\,\text{eV}$. The ionizing fluence is kept high enough to ionize, photoexcite and warm the clusters during the 15 ns laser pulse duration. Rapid sequential evaporation steps occur during the 1 μs residence time in the ionizing region resulting in a cluster distribution shifted down to lower masses. Cluster ion distribution produced in those conditions forms an "evaporative ensemble" of clusters [15] containing internal energy corresponding to a temperature of 600 K for lithium and 350 K for sodium clusters.

The plate voltage determines the ion beam energy and can be set between 1 and 10 keV. After acceleration and focusing, the ions enter a drift region where they spatially resolve into individual mass packets. The ion packets pass between a two electric-field plates mass selector, which deflects all charged particles from the beam path except the packet of interest for which the field momentarily pulses off to allow the selected packet to proceed further. The cluster group then enters a 15 cm long alkali metal heat-pipe kind cell [16] where a stable atomic density can be maintained at a pressure low enough to insure single collision conditions, ranging from 5×10^9 to $5 \times 10^{12}\,\text{at/cm}^3$.

Two diaphragms downstream from the cell limit the observed scattering angle to 10 mrad. in the forward direction, and thus discriminate against large scattering angle collision events. After passage through the angle limiting diaphragm, residual ions and neutral products enter a second electric field region. There we choose either to redisperse and measure the fragmentation of the surviving products (ionic and neutral ones) or to reionize the neutral products with a second UV laser pulse, before subjecting them to a second acceleration field. A final field free region again spatially disperses the product ions and neutral packets which are at last detected by a microsphere plate detector (El Mul Company) of diameter 2.5 cm, wider than the cluster beam diameter.

2.2 Unimolecular Dissociation

The ion clusters are produced with an amount of internal energy. They are metastable and they stabilize via fragmentation. This well known unimolecular dissociation (UD) process, occurs before clusters enter the cell. It consists, for alkali-atom clusters, of an evaporative process, leading to monomer (or dimer) neutral light product [17]:

$$X_n^+ \to X_{n-1}^+ + X \quad \text{or} \quad X_{n-2}^+ + X_2$$

The monomer evaporation predominates at large n. The unimolecular dissociation rate is proportional to the parent size. At small n it is negligeable, whereas at large n ($n \geq 25$) it leads to an ion signal product X_{n-1}^+ comparable to the ion parent X_n^+ signal. We must keep in mind that for large sizes, the measured charge exchange profiles will then concern two successive masses. On the other hand, neutral products X (or X_2) give a contribution superimposed to the neutral packet resulting from the charge exchange process. Due to their low masses, the neutral monomers (or dimers) are detected with a poor efficiency. We checked that their contribution is negligible as compared to the one of the charge transfer itself.

The Fig. 2 (upper traces) illustrates a pure unimolecular dissociation spectrum of Li_9^+ after redispersion but without reionization, when the charge exchange cell is not activated. It shows, from the shorter time of flights toward the larger ones, charged products, residual ion parent and fast neutral signals.

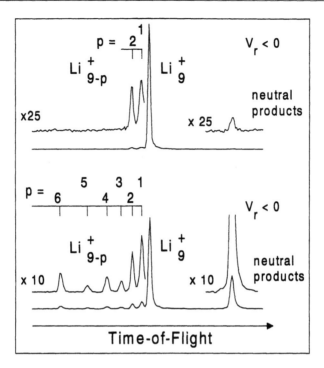

Figure 2: Examples of UD spectrum for Li_9^+ clusters when the cell is at low temperature (non activated) and V_R "on" – *upper traces* – and of CID and CT spectrum again for Li_9^+ clusters propagating in cesium vapor and V_R "on" – *lower traces*.

2.3 Collisional charge transfer and collisional induced dissociation

When the charge exchange cell is activated, two inelastic collisional processes take place and compete. The charge transfer (CT):

$$X_n^+ + Y \to X_n + Y^+$$

and the collisional induced dissociation (CID):

$$X_n^+ + Y \to X_{n-p}^+ \to p_1 X + p_2 X_2 + Y, \quad p_1 + 2p_2 = p.$$

This latter results from an energy transfer into the parent cluster during a collision. As for the unimolecular dissociation process, the warming of the parent promotes dissociation via a sequential evaporation of monomers and dimers [9, 10]. We have checked that the corresponding contribution to the whole neutral signal is small, whatever the atomic target vapor density is. Any change in this density does not affect the relative weights between charge exchange and collisional induced dissociation signals. The example of Li_9^+ propagating in Cs vapor (charge exchange cell activated), is given Fig. 2 (lower traces). A series of ion fragments due to CID is clearly visible, as well as the residual ion parent and a signal due to fast neutral products, mainly CT contribution.

2.4 Charge transfer

In order to analyze the composition of the neutral packet, the incident ions are swept away and neutral products are reionized with a second pulsed laser. The reionization step can be used not only to detect the result of the collisional process but also to probe the internal states of the neutralized parent. Fig. 3 shows an example of the reionization of neutral packet following the interaction between Na_{21}^+ clusters and Cs atoms. After reionization of the neutral packet the size of the main ion peak is Na_{20}^+ and not Na_{21}^+. This shift down has two origines (i) the neutralization of Na_{21}^+, formed during the charge exchange leads to Na_{21} with 21 valence electrons which has a reduced

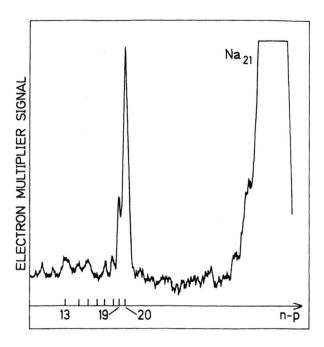

Figure 3: Reionization mass spectrum of the neutral packet following the reaction $Na_{21}^+ + Cs$.

stability against the Na_{21}^+ having a closed shell with 20 valence electrons. The internal energy contained in Na_{21}^+ charged parent is large enough to promote the loss of one monomer after its neutralization before the reionization. This is the typical situation for neutral odd-numbered alkali clusters, which have a lower dissociation energy than the corresponding singly charged species. (ii) The reionization process itself may deposit an excess of energy when the photon energy $h\upsilon$ is higher than the cluster ionization potential I.P. (X_n) as it is the case here ($h\upsilon = 4.67\,\text{eV}$; $IP(Na_{21}) = 3.4\,\text{eV}$ [18, 19]). The warming of the cluster may induce a dissociation. The number of atom lost depends on the excess energy value. From our observations from reionization mass-spectra [9], we conclude that charge exchange process leads to neutral clusters with at most one unit less than the selected ionic parent, depending on the relative stability between charged and neutral species. This is in agreement with the charge transfer picture as a process free of internal energy transfer [7, 8]. Such as result is confirmed through recent studies based on kinetic energy release measurements. In all cases the neutral peak corresponds to a mass close to the mass of the selected ionic parent. Its intensity well represents the charge transfer efficiency.

The charge transfer cross section is obtained from the measured neutralization ratio $r_n(n_y) = S_n/(S_i + S_n)$ where S_i is the intensity of the ion fragments, S_n is the intensity of the neutral signal and $S_i + S_n$ is the peak intensity of the mass selected parent when the potential of the dispersing plates is maintained to zero. The neutralization ratio evolves with target density n_y as the Beer's law.

$$r_n(n_y) = r_n(o)(1 - \exp -\sigma n_y l)$$

where l is the active length of the collision cell and σ the cross section of the charge transfer. In order to insure single collision regime, n_y varies in a range keeping $r_n(n_y)$ lower than 20%. The slope of the curve $r_n(n_y)$ at the origin is proportional to the charge exchange cross section, which is obtained for different ion cluster velocities provided that the product $n_y l$ is well determined. This is not always the case with such an open heat-pipe kind cell. The atomic density profile along the cell axis is not accurately known and the same holds for the extension of active length l. To overcome this problem we measured the cross-section velocity profile $\sigma(v)$ for the atom-atom collisions $X^+ + Y$ and compared it with the values determined in the work of Perel and Daley in the same velocity range [20]. All our $\sigma(v)$ values are then referred to that ones.

3 Charge transfer involving bare alkali-atom clusters

Fig. 4 shows the absolute cross-section profiles for Na_5^+ and Na_9^+ propagating in cesium vapor versus the relative collisional velocity v. Increasing the collisional velocity results in a fast increase of $\sigma(v)$ followed by a relatively flat maximum and a slow decrease at larger velocities. Analyzing experimental data with two parameters universal cross-section profiles developed for atom–atom non resonant collisions [21–23], gives the continuous lines in Fig. 4. The two parameters characterize the profiles by the cross-section value at the maximum σ_m and by the corresponding velocity v_m. Fig. 4 shows that σ_m and v_m are both lower for the $Na_9^+ + Cs$ case than for the $Na_5^+ + Cs$ one. Such a behavior differs from the one expected from atom–

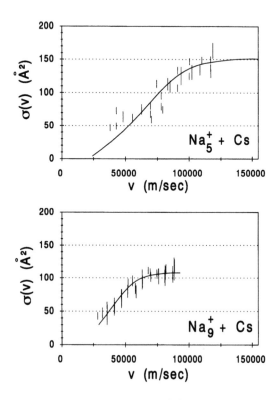

Figure 4: Cross-section velocity profiles $\sigma(v)$ for charge transfer involving $Na_5^+ + Cs$ and $Na_9^+ + Cs$ collisions. Experimental data correspond to two independent series of measurements and illustrate the reproducibility of the these measurements after a calibration procedure (see the text). Continuous lines are the fit of the profiles with classical two-level models.

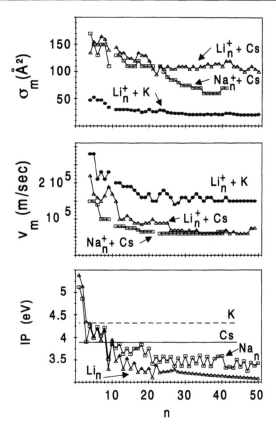

Figure 5: Size evolution of the maximum cross-section σ_m (*upper traces*) and the velocity at the maximum v_m (*middle traces*) for the various studied systems. *Lower traces*: comparison with the ionization potential evolution for Li_n, $n \geq 27$ [19] and Na_n, [18, 19]. For larger sizes, metallic drop model has been used.

atom collision models [23] for which a maximum at small v_m value correlate with a large cross-section σ_m. The Fig. 5 shows the trends observed for σ_m and v_m for the relevant collisional systems $Li_n^+ + Cs$, $Li_n^+ + K$ and $Na_n^+ + Cs$ as the cluster size increases. For a given couple of collision partners, σ_m and v_m simultaneously decrease when n increases and tend to roughly constant values at largest sizes. These asymptotic values depend strongly on the nature of the target.

As for atom–atom collisions, the energy defect ΔE of the process, i.e. the ionization potential difference between the cluster and the target, is of primary importance [23]. Fig. 5 (bottom trace) shows the ionization potentials for Li_n [19] and Na_n [18, 19] clusters together with the ionization potentials of the atomic targets Cs and K. It is clear that for cluster sizes larger than $n = 20$ the larger the energy defect the larger the corresponding maximum velocity, in agreement with the charge transfer models. Moreover the idea that a large velocity is required to compensate a large ΔE value is nothing else but the well known Massey criterion [24]. It tels us that v_m is given by:

$$v_m = \frac{a}{h}\Delta E$$

where a is a typical interaction length. For alkali-atom clusters ionization potentials decrease significantly as n increases. In the studied cases, they are lower than the target-ionization potentials when $n > 8$ (Cs target) and $n > 2$ (K target) so that ΔE increases with n. The decrease observed for v_m led us conclude to a decrease of the interaction length a when n increases. We found for a few tens of angströms at small sizes and a lower limit around 5 Å for the largest values of n, whatever the system is. It is remarkable that these values are comparable to the values found for atomic systems [25] and for atom–surface interactions [3], respectively. This illustrates the intermediate and specific nature of clusters, in between free atoms and condensed phase, as it has been already mentioned for most of their properties. The v_m, ΔE interplay is also illustrated by the fact that, whatever the cluster size one has:

$$v_m(\mathrm{Na}_n^+ + \mathrm{Cs}) \leq v_m(\mathrm{Li}_n^+ + \mathrm{Cs}) \ll v_m(\mathrm{Li}_n^+ + \mathrm{K})$$

while

$$\Delta E(\mathrm{Na}_n^+ + \mathrm{Cs}) < \Delta E(\mathrm{Li}_n^+ + \mathrm{Cs}) < \Delta E(\mathrm{Li}_n^+ + \mathrm{K}).$$

In any case it has to be notice that v_m and σ_m follow the trends of cluster ionization potentials for all sizes without taking into account the ionization potentials of the target (Fig. 5).

Looking at the σ_m values for Li_n^+ charge transfer, their general decrease when n increases support the idea of a ΔE dependence. Especially the large $\sigma_m(\mathrm{Li}_n^+ + \mathrm{Cs})/\sigma_m(\mathrm{Li}_n^+ + \mathrm{K})$ ratio strengthens the above idea at least for large n. However a detailed analysis led us believe in a more complex situation, in which ΔE does not play systematically the predominant role. For example, in the spirit of a correlation between σ_m and ΔE, one should observe $\sigma_m(\mathrm{Li}_n^+ + \mathrm{K})$ in the range $n = 4$–8 larger than $\sigma_m(\mathrm{Li}_n^+ + \mathrm{Cs})$ for any $n > 11$. One finds the opposite, suggesting that the electronic structure of the target plays the major role. Actually, the larger polarizability of atomic cesium against the atomic potassium polarizability (screening effect) is responsible for a larger efficiency of the collisional charge transfer involving Cs atoms.

Another characteristic example is found comparing $\sigma_m(\text{Na}_n^+ + \text{Cs})$ in the range $n = 20\text{--}30$ and $\sigma_m(\text{Li}_n^+ + \text{Cs})$ in the range $n = 11\text{--}15$. Despite very close ΔE values, the charge exchange cross-section is larger with lithium clusters than with sodium clusters. Here the electronic structure of the cluster projectile could explain the difference. During a collisional event, once the target valence electron is captured by the cluster, there is a probability that one of the n equivalent delocalized electrons of the cluster jumps back into the charged target, leading to a lowering of the efficiency. This propensity to give back an electron increases with the polarizability. Recent measurements report on larger static polarisabilities for sodium clusters [26, 27] than for lithium clusters [27]. In that sense, the charge exchange process is a sensitive probe of the cluster "metallicity".

4 Charge transfer involving heterogenous clusters

Oxydation processes are of primary interest in surface physics as they are relevant to many physico-chemical properties (catalysis, photoelectric process ...). By mixing a small amount of oxygen with the carrier gas, we produce metal-rich heterogeneous clusters $X_{n+2p}^+ O_p$ (X = Li, Na). The control of the oxygen quantity determines the p/n ratio. From a fundamental point of view it is interesting to understand how develops itself a non-metallic/metallic character, i.e. the size evolution of the oxygen-metal bonding inside the droplet.

The metallicity of these clusters has been studied through ionization potential measurements [28], shell closing and odd-even alternation in mass spectra ion signals, and dissociation process [29]. Their behavior led to the conclusion that they contain $n - 1$ delocalized electrons and insulating entities $(X_2O)_p$, corresponding to a bonding of two alkali atoms by each oxygen anion. This is imaged by labeling these clusters $X_n^+(X_2O)_p$. Recent calculations give more insight in the X-0 bonding in the droplet and show for example that metal-rich species develop a metal segregation [30].

In the light of our results on bare clusters and following the pioneering work of Arnold et al. [7] and of Abshagen et al. [8] showing an increase of the charge transfer cross-section when the size increases for small non metallic S_n^+ and Pb_n^+ clusters, we have measured the cross-section profiles for $X_n^+(X_2O)_p + \text{Cs}$, under the same experimental conditions as above. The shape of the profiles is similar to that obtained for alkali-atom-clusters. In the Fig. 6 is shown the evolution of σ_m when oxygen atoms are added to a given number of lithium atoms. The main feature is a drastic decrease of the cross-section maximum when the p/n ratio increases. As observed previously, the velocity v_m decreases simultaneously. Similar results have

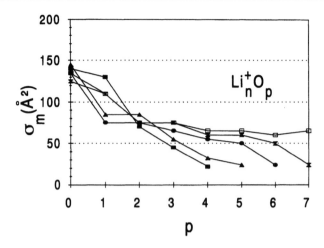

Figure 6: σ_m evolution versus the number p of added oxygen atoms for some heterogeneous lithium based clusters: ■ $n = 9$; ▲ $n = 11$; ● $n = 13$; ✖ $n = 15$; □ $n = 17$.

been obtained for the reactions $Na_n^+(Na_2O)_p + Cs$. At first glance this could correlate with an increase of ΔE. There is no direct measurement of the oxydized lithium or sodium atom cluster ionization potentials (except for $p = 1$, [31]), but such data exist for $Cs_{n+2p}^+O_p$ species [28]. They show a lowering of the ionization potential when increasing oxidation. This is understood in term of confinement of the free electrons in a smaller volume because of the potential barrier describing the inner oxygen atoms and explains the lowering of the work function in bulk cesium suboxides. Applying a scaling low between species based on Li or Na atoms and having the same stoechiometry, it is possible to evaluate the ionization potentials of the heterogeneous clusters studied here. Actually, the IP evolution with p leads to an increase of ΔE, but the change is less than $0.8\,\text{eV}$ from $p = 1$ to $p = 7$ in the $n = 9$ to 13 range. This change is too small for interpreting the drastic decrease by a factor of four for σ_m. Therefore we believe that the σ_m evolution does reflect the change in the nature of bonding inside the droplet and not only the energy defect.

Another hint is given representing the σ_m evolution with the number n of alkali atoms, for various number of insulating blocks X_2O (Fig. 7). Clearly, increasing p lowers σ_m. Beyond $p = 1$ the effect is very large and depends on the ratio n/p. In the whole the behavior is similar for sodium and lithium-based clusters. The σ_m values of heterogeneous species merge with the cross-sections of the corresponding bare clusters X_n^+ when $n/p \gg 1$ showing a non metal–metal transition. In a rough picture of a droplet constituted with an insulating core surrounded by a metallic envelope, this would correspond to

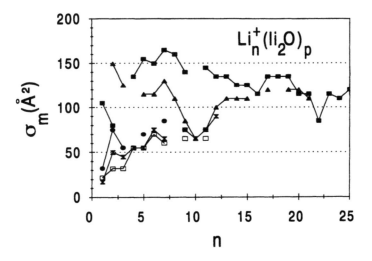

Figure 7: σ_m evolution versus the number of lithium atoms in the "metallic" phase of the heterogeneous clusters for some values of the Li_2O entities number. The corresponding evolution for the particular case of $Li^+(Li_2O)_p$ is shown in the Fig. 9. ■ $p = 0$; ▲ $p = 1$; ● $p = 2$; ✘ $p = 3$; ☐ $p = 4$.

a complete surrounding of the $(X_2O)_p$ core by n "metallic" atoms. Incomplete coverage, i.e. moderate n/p values, gives much smaller charge transfer efficiency as already illustrated in the Fig. 6. A more realistic picture requires ab initio calculations, not yet available. However recent theoretical studies on $Li_{4+p}O_2$ and $Li_{4+p}O_2^+$ clusters by ab initio molecular dynamics [30, 32] show that Li attachment in excess results in metal segregation. A generalization of these results would lead to a topological situation close to the above simple picture.

In order to change more drastically the cluster bound, a small amount of water is mixed with the carrier gas. This produces hydroxyded species with ionic binding. We studied the most abundant ones among the various species present in mass spectra, the $Na^+(NaOD)_p$ clusters. Deuterium is chosen instead of hydrogen for a better discrimination of the species in mass spectra. The Fig. 8 shows the main results. Beyond $p = 6$, suprimposed to an odd even oscillation, σ_m exhibits a quasi linear increase with p. Simultaneously v_m is found constant around 1.8×10^5 m/s whatever p is. This drastically differs from the behavior of $X^+(X_2O)_p$ species, as shown on the Fig. 9 for $Li^+(Li_2O)_p$, $p \le 9$. This cannot be correlated with a ΔE change. Ionization potentials are not known for that species. However measurements performed by Poncharal et al. [33] on $Na(NaF)_{q-1}$ species, having the same kind of bonding inside, indicate a decrease of the ionization potentials when q increases, which would suggest in our case an increase of ΔE when p in-

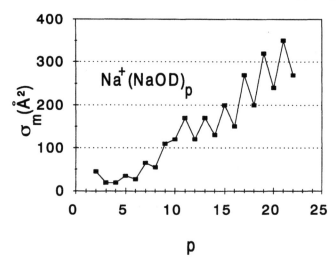

Figure 8: σ_m evolution for the stochiometric hydroxyded species $Na^+(NaOD)_p$.

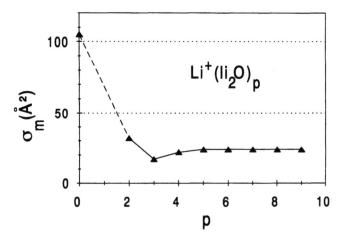

Figure 9: σ_m evolution for the case $Li^+(Li_2O)_p$ for comparison with Fig. 8

creases. We must conclude that this particular behavior characterizes the ionic nature of the bonds inside the droplets $Na^+(NaOD)_p^-$. For such clusters the charge transfer can be done with any of the Na^+ ions constituting the cluster, leading to an increase of the cross-section as p increases. The probability to have an electron jumping back to the target atom is not pertinent for ionic species. On the contrary, in the case of covalent bound as it is for $Na^+(Na_2O)_p$ or $Li^+(Li_2O)_p$, only the extra Na^+ or Li^+ ion participates to the charge transfer, leading to a constant cross-section as p increases, for $p \geq 2$.

5 Conclusion

We have clearly demonstrated the potentiality of charge transfer measurements for probing the electronic structure of clusters. The evolution of the charge transfer efficiency as increasing the size of the droplet as well as their change as modifying the nature of the bond in the cluster, show the sensitivity of the method. It can be used as a new tool for studying the non metal–metal transition in various species. Moreover, collisional aspects are also of primary interest, for a fundamental point of view. It appears necessary to develop ab initio calculations for a more complete understanding of the phenomena.

References

[1] J.B. Taylor and I. Langmuir, Phys. Rev., **44**, 423 (1933)

[2] J.W. Gadzuk, Surf. Science **6**, 133 (1967); **6**, 159 (1967)

[3] J. Los and J.J.C. Geerlings, Physics Reports, **190**, 133 (1990)

[4] H. Winter and R. Zimny in: Coherence in atomic collision physics, eds. H.J. Beyer, K. Blum and R. Hippler (Plenum, New York 1988) p. 283

[5] J. Perel, R.H. Vermon, H.L. Daley, Phys. Rev. **138**, A937 (1965)

[6] M. Brack, Rev. Mod. Physics **65**, 677 (1993) and references therein

[7] J. Arnold, J. Kowalski, G. zu Putlitz, T. Stehlin, F. Träger, Z. Phys. A **322**, 179 (1985)

[8] M. Abshagen, J. Kowalski, M. Meyberg, G. zu Putlitz, J. Slaby, F. Träger, Chem. Phys. Lett. **174**, 455 (1990)

[9] C. Bréchignac, Ph. Cahuzac, J. Leygnier, R. Pflaum, J. Weiner, Phys. Rev. Lett. **61**, 175 (1988)

[10] C. Bréchignac, Ph. Cahuzac, F. Carlier, J. Leygnier, I. Hertel, Z. Phys. D **17**, 61 (1990)

[11] N.D. Bhaskar, R.P. Frueholz, C.M. Climcak, R.A. Cook, Chem. Phys. Lett. **154**, 175 (1989)

[12] S. Pollack, D. Cameron, M. Rokni, W. Hill, J.H. Parks, Chem. Phys. Lett. **256**, 473 (1993

[13] O. Knospe, J. Jellinek, V. Saalmann, R. Schmidt, Eur. Phys. Lett. D **5**, 1 (1999)

[14] M. Guissani, V. Sidis, Z. Phys. D **40**, 221 (1997)

[15] C. Klots, J. Chem. Phys. **83**, 5854 (1985); Z. Phys. D **5**, 83 (1987)

[16] M. Bacal, W. Reichelt, Rev. Sci. Instrum. **45**, 769 (1974)

[17] C. Bréchignac, H. Busch, Ph. Cahuzac, J. Leygnier, J. Chem. Phys. **101**, 6992 (1994)

[18] M.L. Homer, J.L. Persson, E.C. Honea, R.L. Whetten, Z. Phys. D **22**, 441 (1991)

[19] E. Benichou, A.R. Allouche, M. Aubert-Frécon, R. Antoine, M. Broyer, Ph. Dugourd, D. Rayane, Chem. Phys. Lett. **290**, 171 (1998)

[20] J. Perel, H.L. Daley, Electronic and Atomic Collisions, ICPEAC VIII proceedings, B.C. Bobic, M.V. Kurepa Editors, Beograd, Vol II (1973)

[21] R.E. Olson, Phys. Rev. A **6**, 1822 (1972)

[22] R.J. Fortner, B.P. Carry, R.C. Der, T.M. Kavanagh, J.M. Khan, Phys. Rev. **185**, 164 (1969)

[23] D. Rapp, W.E. Francis, J. Chem. Phys. **37**, 2631 (1962)

[24] H.S.W. Massey, Rep. Progr. Phys. **12**, 248 (1949)

[25] J. Perel, H.L. Daley, Phys. Rev. A **4**, 162 (1971)

[26] W. Knight, K. Clemenger, W.A. de Heer, W.A. Saunders, Phys. Rev. B **31**, 2539 (1985)

[27] E. Bénichou, R. Antoine, D. Rayane, B. Vezin, F.W. Dalby, Ph. Dugourd, M. Broyer, C. Ristori, F. Chandezon, B.A. Huber, J.C. Rocco, S.A. Blundell, C. Guet, Phys. Rev A **59,** 1 (1999)

[28] H.G. Limberger, T.P. Martin, J. Chem. Phys. **90**, 2979 (1989)

[29] C. Bréchignac, Ph. Cahuzac, M. de Frutos, P. Garnier, Z. Phys. D **42**, 303 (1997)

[30] F. Finocchi, C. Noguera, Phys. Rev. B **57**, 14464 (1998)

[31] P. Lievens, P. Thoen, S. Bouchaert, W. Bouwen, E. Van der Meert, F. vanhoutte, H. Weidele, R.E. Silverans, Z. Phys. D **42**, 231 (1997)

[32] F. Finocchi, C. Noguera, EPJ. D **9**, 327 (1999)

[33] P. Poncharal, J.M. L'Hermite, P. Labastie, Chem. Phys. Lett. **253**, 463 (1996)

Electron Attachment to Oxygen and Nitric Oxide Clusters

G. Senn, P. Scheier, T.D. Märk
Inst. für Ionenphysik, Universität Innsbruck,
A-6020 Innsbruck, Austria

1 Introduction

Electron attachment studies are of fundamental importance to the understanding of electron-molecule and electron-cluster interaction and the mechanisms of negative ion formation. In addition, these studies are important for the elucidation of atmospheric processes. It is known that NO is involved in the destruction of ozone in the lower stratosphere. Besides this catalytic destruction, ozone is also subject to dissociation by photons and electrons. As electron attachment and the properties of anions are involved in these processes we have recently started to study in detail, and with high accuracy, the attachment cross sections of some of the atmospheric molecules and respective clusters [1-6].

2 Experimental

The measurements were carried out in a crossed beam experiment using a trochoidal electron monochromator (TEM) as the electron beam source [1, 2]. The best energy resolution achieved is about 5 meV (FWHM) and electron energies close to zero are possible [2]. In combination with either a temperature controlled effusive molecular beam source or a supersonic nozzle source and with a quadrupole mass spectrometer for analysis of the anions produced, electron attachment spectra can be measured as a function of electron energy and target beam properties such as gas temperature and target composition. The zero energy position of the energy scale and the energy resolution were calibrated and checked with the known cross section curve for Cl^- from CCl_4.

3 Results and discussion

3.1 Electron attachment to oxygen clusters

Electron attachment to O_2 results in a non-dissociative resonance process [7, 8]

$$O_2(X^3\Sigma_g^-; \nu = 0) + e \rightarrow O_2^-(X^2\Pi_g; \nu\prime \geq 4) \quad (1)$$

Because the bond length of the neutral molecule is much smaller than that of the anion [9] and because O_2 has a positive electron affinity (see Ref. [10]), the incoming electron cannot induce a Franck-Condon transition to the ground state of the anion. Only vertical transitions to the fourth and higher vibrational levels of the anion are possible. The molecular anion formed via reaction (1) is unstable with a predicted [11] lifetime towards autodetachment of about 10^{-10}s. In a high pressure environment this anion can be stabilized collisionally to a vibrational level $\nu\prime < 4$ which lies below the $\nu = 0$ level of the neutral (see Fig. 1), thereby making autodetachment impossible. The "effective" cross section for the "three-body attachment" shows a structure with peak energies coinciding with the positions of the vibrational states of the O_2^- compound state with $\nu \geq 4$ [7, 8, 12]. Similarly, O_2^- has been reported to be produced via electron attachment to $(O_2)_n$ clusters [1, 5]. But no anions have been observed previously in electron/O_2 crossed beam experiments under single collision conditions at low energies. So what is the exact attachment mechanism leading to the production of anions in the case of clusters?

The measured relative attachment cross section function [1, 5] for the production of O_2^-, $(O_2)_2^-$ and $(O_2)_3^-$ in the low energy regime (0-1 eV) produced via reaction

$$(O_2)_n + e \longrightarrow (O_2)_m^- + \text{neutral products} \quad (2)$$

exhibits the same characteristic behavior for all measured $(O_2)_m^-$ ions, i.e., the cross section is largest at about zero energy and then strongly decreases with increasing energy. Moreover, the decreasing cross section is structured by three additional peaks (1,2,3) whose maxima appear to lie (within the error bar of ±10 meV) in all cases at the same energy. This strong initial decrease is compatible with the energy dependence predicted for s-wave scattering. Thus we conclude that the present observation indicates that s-wave electron capture is a likely mechanism in the electron attachment to oxygen clusters. After the initial s-wave capture of the electron by the entire $(O_2)_n$ cluster the energy gain from the positive electron affinity of O_2 (440±8 meV [10]) leads, via monomer evaporation, to the final reaction product.

Peaks at higher electron energies are attributed to the attachment of an incoming electron to a single oxygen molecule within the target cluster via

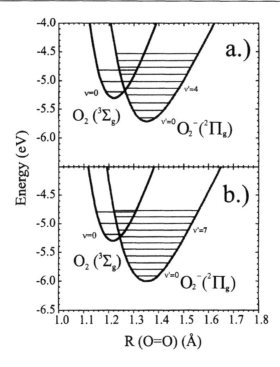

Figure 1: (a) Approximate potential energy curve for O_2 and O_2^- after [7]. The $\nu\prime = 4$ state is located 0.091 eV above the $\nu = 0$ state [13]. (b) Hypothetical potential energy curve for O_2 and O_2^- solvated in an oxygen cluster. The $\nu\prime = 7$ anion state is in this hypothetical drawing 80 meV above the $\nu = 0$ neutral ground state in accordance with the present experimental results.

number of peak	O_2^-	$(O_2)_2^-$	$(O_2)_3^-$	mean value	theoretical value
1-2	111	111	118	113	113
2-3	106	123	99	109	111

Table 1: Spacings in meV between the peaks

direct Franck-Condon transition from the ground vibrationally state to a vibrational excited state of the ensuing anion. Subsequent collisional stabilization of this anion within the cluster environment to a vibrational state below the ground state of O_2 gives a stable O_2^-. Identification of these excited vibrational states populated in the anions can be achieved (for more detail see Ref. [1, 5]) by analyzing the spacings between the different peaks, assuming similar spacings in the cluster and the monomer.

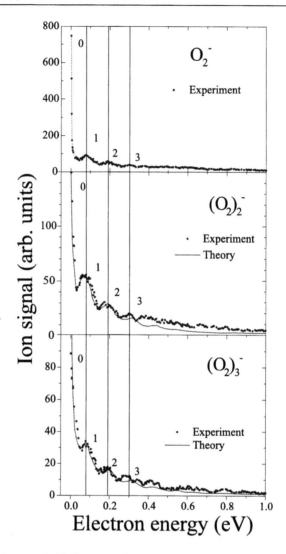

Figure 2: Measured $(O_2)_m^-$ signal produced by electron attachment to a $(O_2)_n$ cluster beam as function of the electron energy. Also shown is the calculated cross section behavior.

The spacings between peak 1,2,3 fit to the vibrational levels 7,8,9 in the O_2^- (see Table 1). A similar result is obtained from a calculation of the downward shift of the anion potential in the cluster environment due to polarization forces. We assume that the excess electron is localized initially on a specific molecule within the cluster. This is a good approximation, because the cluster is only weakly bound by van der Waals forces, and the distance between the molecules is large. The electric field of the negative charge then polarizes the other molecules of the cluster. The corresponding energy V_p has been calculated using the model given in Ref. [14] and strongly depends on the position of the molecular anion in the cluster, and on the size and structure of the cluster. A constant temperature molecular dynamics calculation (T=20 K) yields the distribution f for V_p, which is shown for the three cluster sizes 10, 15 and 20 in Fig. 3 as a function of the adiabatic electron affinity of the cluster (defined as the sum of the polarization energy V_p and the electron affinity of the monomer).

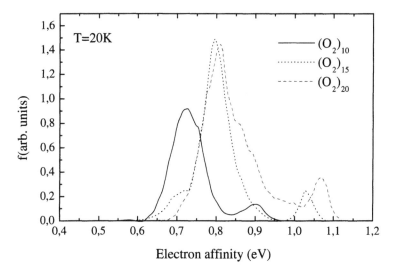

Figure 3: Calculated distribution function of the adiabatic electron affinity for cluster size 10, 15 and 20 at a temperature of 20 K.

As shown in Fig. 3 there exists only a small range of possible polarization energies independent of the precursor sizes probed. Thus the shift induced by this polarization energy is well defined, and the value of 0.8 eV for the total adiabatic electron affinity is close to the value used for constructing the hypothetical potential curves shown in Fig. 1.This leads to a situation where $\nu\prime = 7$ is the first vibrational state accessible via a vertical Franck-Condon transition by electron attachment in accordance with the conclusion drawn above from a comparison between the measured and the known vibrational spacings. Moreover, a further confirmation of this identification comes from a calculation, where we have modeled the energy dependence of the attachment cross section (see theoretical curve in Fig. 2), adding to the s-wave scattering component calculated contributions resulting from the respective Franck-Condon overlap integrals.

3.2 Electron attachment to nitric oxide clusters

The system NO is similar to O_2 in the way that the extra electron occupies an antibonding MO, resulting in a larger equilibrium geometry of the anion. However the adiabatic electron affinity of NO is only 26±5 meV [15] and thus is considerably smaller than that of O_2. In contrast to O_2, NO has a dipole moment (0.16 D) [16], so that an appreciable fraction of direct inelastic scattering may contribute to vibrational excitation of neutral NO. Also vibrationally excited NO^- has a much shorter autodetachment lifetime as compared to the O_2^- which may also affect the electron attachment behavior in clusters. The lifetime for the first autodetachment vibrational level in $O_2^-(\nu\prime = 4)$ is $\approx 10^{-10}$ s [11], while $NO^-(\nu\prime = 1)$ is $3.5 * 10^{-13}$ s [17].

The relative cross section for NO^- formation shows a very sharp peak close to zero eV and a series of further weak peaks (see Fig. 4). While the attachment mechanism is probably similar in both systems, the evolution of the ionized cluster system must be different in the case of NO. The zero energy peak is interpreted as s-wave capture. The positions of the other peaks are located on a basis of a series of experiments at 40 ± 10 meV, 220 ± 10 meV and 410 ± 10 meV, corresponding to a spacing of 180 ± 20 meV and 190 ± 20 meV respectively.

The vibrational frequency of NO^- is given from 166 ± 10 meV [18] to 180 ± 25 meV [19]. The most recent value is obtained from an autodetachment study in NO^- [17] as 159 ± 1 meV. In contrast, for neutral NO the vibrational frequency of 236 meV [20] is considerably larger. From these numbers it is clear that the observed structures cannot directly be correlated to vibrational excitation of neutral NO, although the direct scattering mechanism for vibrational excitation may be operative in the present system. While the vibrational frequency of neutral NO is well established, the numbers given above indicate that this may not be true for NO^-. Moreover the vibrational frequency of a molecule bound in a cluster may be perturbed, in particular in the presence of appreciable coupling as in the case for a molecular ion like NO^-. So we tentatively assign the observed structures as transitions to vibrationally excited solvated NO^-.

Due to the very low adiabatic electron affinity of NO and the strong anion dimer bond,

$$(NO)_n + e \longrightarrow NO^- + \text{neutral products} \quad (3)$$

cannot be a result of an evaporative attachment reaction (i.e., dissociation of monomer units until the single ion is left) which is often used in the

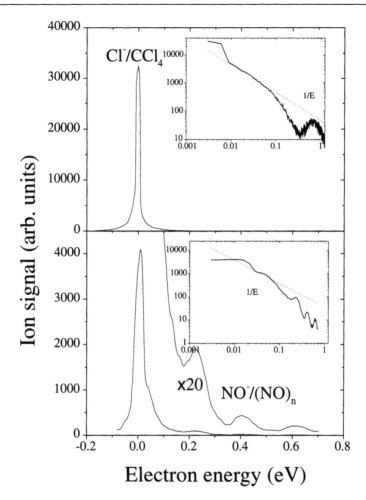

Figure 4: Measured NO$^-$ signal observed from electron attachment to a (NO)$_n$ cluster beam as a function of the electron energy. The signal Cl$^-$ from CCl$_4$ is also shown in the upper panel indicating the energy resolution and reliability of the apparatus.

description of electron attachment to clusters. Evaporative attachment via

$$(\text{NO})_n + e \longrightarrow \text{NO}^- + (n-1)\text{NO} \qquad (4)$$

would require the energy equivalent of (n-1) monomer dissociation energies minus the electron affinity of NO. The binding energy for the NO dimer is known from high resolution photodetachment experiments [21] as D(NO-NO) = 98 meV. Since the experiments with NO are carried out under similar expansion conditions as in the case of oxygen and since the intermolecular

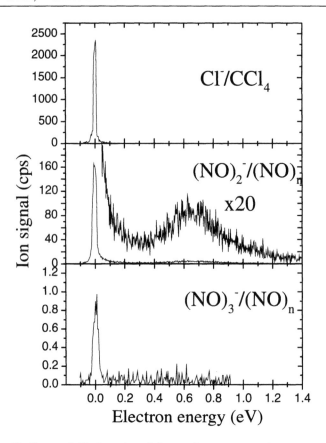

Figure 5: $N_2O_2^-$ and $N_3O_3^-$ signal from electron attachment to a $(NO)_n$ cluster beam as a function of the electron energy. The signal Cl^- from CCl_4 is also shown indicating the energy resolution.

forces are larger in NO, we assume that the neutral clusters on average contain more than 20 molecules. In any case, even for the dimer or under the most unprobable assumption that NO^- is formed leaving a completely undissociated neutral cluster, reaction (4) will remain endothermic since the bond dissociation energy $D[(NO)_m - NO^-]$ is in any case considerably larger than the electron affinity of NO. A possible solution to this problem is the formation of neutrals, which are more stable than their neutral NO counterparts and thereby would provide the necessary energy. Possible reactions are

$$(NO)_n + e \longrightarrow NO^- + N_2 + O_2 + (n-3)NO \qquad (5)$$
$$\longrightarrow NO^- + N_2O + NO_2 + (n-4)NO \qquad (6)$$

which are highly exothermic. It is known that the weakly bound complex $NO^-\cdot NO$ correlates (probably without barrier) to a $[ONNO]^-$ which is about 2 eV below the $NO^-\cdot NO$ system. This $[ONNO]^-$ ion itself is separated by an activation barrier from another isomer $[NNO_2]^-$ of comparable stability. Such ionic intermediates could then be considered to act as catalysts for a reaction generating N_2 and O_2 from the electron/$(NO)_n$ system. This is certainly a highly speculative explanation and one can immediately ask why O_2^- and NO_2^- are not detected. However, from Fig 5 it can be seen that ions of the composition $N_2O_2^-$ and $N_3O_3^-$ are in fact formed in the present experiment (for more detail see Ref [3]).

Work was partially supported by FWF, and BMWV, Wien, Austria. P.S. is presently supported by the ÖAW-APART program.

References

[1] S. Matejcik, A. Kiendler, P. Stampfli, A. Stamatovic, and T. D. Märk, Phys. Rev. Lett. **77** 18 (1996)

[2] S. Matejcik, G. Senn, P. Scheier, A. Kiendler, A. Stamatovic and T.D. Märk, J. Chem. Phys. **107** 8955 (1997)

[3] Y. Chu, G. Senn, S. Matejcik, P. Scheier, P. Stampfli, A. Stamatovic, E. Illenberger and T.D. Märk, Chem. Phys. Lett. **289** 521 (1998)

[4] Y. Chu, G. Senn, P.Scheier, A. Stamatovic and T.D. Märk, Phys. Rev. A **57** R697 (1998)

[5] S. Matejcik, P. Stampfli, A. Stamatovic, P. Scheier and T.D. Märk, J. Chem. Phys. **111** 3548 (1999)

[6] G. Senn, J. D. Skalny, A. Stamtovic, N. J. Mason, P.Scheier and T.D. Märk, Phys. Rev. Lett. **82** 5028 (1999)

[7] D. Spence and G.J. Schulz, Phys. Rev. A **5** 724 (1972)

[8] D.L. McCorkle, L. G. Christophorou and V.A. Anderson, J. Phys. B **5** 1211 (1972)

[9] M.J.W. Boness and G.J. Schulz, Phys. Rev. A **2** 2182 (1970)

[10] R.J. Celotta, R.A. Bennett, J.H. Hall, M.W. Siegel, and J. Levine, Phys. Rev. A **6** 631 (1972)

[11] L.G. Christophorou, D.L. McCorkle and A.A. Christodoulidis, in Electron-Molecule Interaction and their Applications, edited by L.G. Christophorou (Pergamon, New York, 1984) p. 477.

[12] G.J. Schulz, Rev. Mod. Phys. **45** 423 (1973)

[13] J.E. Land and W. Raith, Phys. Rev. Lett. **30** 193 (1973)

[14] D.C. Conway, J. Chem. Phys. **52** 2689 (1970)

[15] M.J. Travers, D.C. Cowles and G.B. Ellison, Chem. Phys. Lett. **164** 449 (1989)

[16] Handbook of Chemistry and Physics, 76th ed. D.R. Lide, Editor-in-Chief, (CRC Press, Boca Raton, Fl, 1995)

[17] M.M. Maricq, N.A. Tanguay, J.C. O'Brien, S.M. Rodday and E. Ridden, J. Chem. Phys. **90** 3136 (1989)

[18] D. Spence and G.J. Schulz, Phys. Rev. A **3** 1968 (1971)

[19] M.W. Siegel, R.J. Celotta, J.L. Hall, J. Levine and R.A. Bennett, Phys. Rev. A **6** 607 (1972)

[20] K.P. Huber and G. Herzberg, Constants of Diatomic Molecules, Van Nostrand Reinhold, New York 1979

[21] I. Fischer, A. Stobel, J. Staecker, G. Niedner-Schatteburg, K. Müller-Dethlefs and V.E. Bondybey, J. Chem. Phys. **96** 7171 (1992)

The Radiative Cooling of C_{60} and C_{60}^+ in a Beam

A.A. Vostrikov[1,2], D.Yu. Dubov[1,2], A.A. Agarkov[1],
S.V. Drozdov[1], V.A. Galichin[1]

[1] Institute of Thermophysics, Novosibirsk, Russia
[2] Novosibirsk State University, Novosibirsk, Russia

1 Introduction

The unique properties of the C_{60} fullerene cluster recommend it for fundamental and applied investigations: (a) The high stability and the low number of atoms in C_{60} facilitate its use in the study of a fundamental phenomenon of the formation of a unified quantum-mechanical system (solid phase) from separate atoms. (b) The high mass, stability, and large cross-sections for C_{60}^- [1] and C_{60}^+ [2, 3] formation by electron impact are favorable for the designing of ion engines. (c) The use of C_{60} holds promise for the improvement of aircraft aerodynamics and for the reduction of heat transfer to their surfaces via either the injection of C_{60} instead of Al_2O_3 particles into the main stream [4] or the ablation of C_{60} from a surface.

Previously [5–7] we studied the radiation of molecular clusters $(CO_2)_n$, $(N_2O)_n$, $(H_2O)_n$, and $(N_2)_n$ formed in supersonic jets and excited by an electron impact. The energy of vibrational [5] and electronic [6, 7] excitation was established to result to rapid cluster heating. The decay rate of excited states increases greatly with the cluster size, and the emission spectrum corresponds to that of the electronically-excited states of molecules which leave the cluster.

The situation is opposite in case of the relaxation of strongly bound particles, namely, clusters of refractory substances or fullerenes. Much higher energy of the dissociation activation (over 5 eV) results in extreme stability of the particle. Coupled with the large density of vibronic states it gives the possibility to reach high internal energy and leads to *thermal* radiation. It has been established that the rate of C_{60} radiative cooling is large enough to affect the kinetics of different relaxation processes, e.g., the thermal dissociation [8]. The optical study of heated $(Nb)_{n\approx 260}$ cluster has revealed a structureless (up to 0.1 nm resolution) spectrum, resembling the

Planck spectrum of black-body radiation with the spectral temperature up to 3200 K [9]. The very similar spectrum has been detected for fullerene clusters heated in laser plasma [10] or by electrons [11]. It should be stressed that the observed radiation was formed by single species, since the concentration of clusters and radiation region dimensions did not exceed 10^{11} cm^{-3} and 1 cm, respectively.

In the present study the radiation of fullerene molecule was induced with the electron impact in crossed C$_{60}$ and electron beams. We measured the radiation intensity and the effective temperature of radiating particles versus the electron energy ranging from 25 to 100 eV. The hot C$_{60}^+$ ions were found to contribute mainly to the radiation. The rate of radiative cooling of the ion was determined to be about $5.5 \cdot 10^5$ eV/s at the temperature of 3150 K. Within the framework of thermal model of radiation it corresponds to the emissivity of $1.1 \cdot 10^{-2}$ at the wave length 550 nm. The results of [9, 10, 11] and this work indicate that the technique of excitation does not influence the emission spectra of the clusters, implying that the black-body-like radiation is a common property of strongly bound clusters.

2 Experimental Technique

The scheme for generation of a C$_{60}$ cluster beam and excitation with an electron beam was described earlier [1, 11]. A schematic diagram of the experimental apparatus supplemented with an optical system is given in Fig.1.

We used a mixture of C$_{60}$/C$_{70}$ fullerenes with the C$_{60}$ content in the gas phase larger than 88 %. The fullerene vapor, effusing from a source 1 heated up to $T_0 \approx 800$ K, was intersected by the electron beam at right angles.

The electron beam was emitted from an oxide-coated cathode 2, shaped by a system of diaphragms 3 – 5 and collimated by a magnetic field (≤ 300 G) of magnets 6. The electron current measured by a Faraday cup 7 did not exceed 60 μA, which avoided multiple electron–cluster collisions. The electron energy E_e was varied from 0 to 100 eV. The currents of the C$_{60}^+$ and C$_{60}^-$ ions were separately measured by a Faraday cup collector 9.

The radiation from the beam-crossing region was detected at right angles to the electron beam; the angle with the direction of the molecular beam was 40^0. After spectral decomposition by a grating monochromator MONO (200–800 nm, inverse dispersion 3.2 nm/mm) the radiation was recorded with a photomultiplier PM. A second optical system with a broadband filter F collected the radiation in a 430 to 580 nm region of wavelength. A calibrated quartz-tungsten lamp was used to obtain the spectral efficiency of the entire optical system. The correction of the raw spectra for this response curve gave us the real intensity of radiation (photon flux) I.

In order to separate the radiation of the hot C$_{60}^{+*}$ ions from that of the neu-

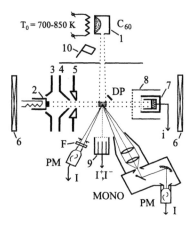

Figure 1: Experimental set-up

tral C_{60}^* clusters, we used an electric field between two deflecting plates (DP) placed along the axes of the molecular and electron beams, there by deflecting the ions. The magnetic collimation ensured that the applied electric field had no influence on the electron beam. To test this fact, we measured in the same scheme the radiation of an N_2^+ ion ($B^2\Sigma_u^+$, $\lambda = 391$ nm). The radiation of this short-lived ($\sim 5 \cdot 10^{-8}$ s) state did not depend on the extracting field.

Particular attention has been given to the elimination of the undesirable radiation from other sources. The electron-induced radiation from the background gas was monitored with the aid of an electromagnetic shutter 10, which can cut off the molecular beam. The IR background from the heated parts of the fullerene source was eliminated by modulating the electron beam and measuring the PM signal in a lock-in mode. As a further precaution we used short-focal-length lenses and collimated the radiation.

3 Results and Discussion

Figure 2 shows typical radiation spectra (curve 1) produced by 66-eV electron impacts on C_{60} cluster in crossed beams. In order to obtained this curve the emission of the background gas (curve 2) was subtracted from the total PM signals and the difference (net signal, curve 3) was normalized to the spectral efficiency of the optical system.

In view of the large number of C_{60} radiating states, the form of the quasi-continuous spectrum is determined not by the structure of the vibrational levels but only by the internal temperature of the molecules. Following [9,

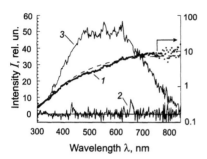 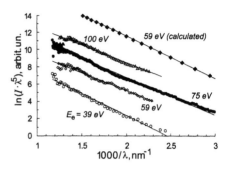

Figure 2: Spectrum of radiation from a fullerene beam

Figure 3: Emission spectra in $\ln(I \cdot \lambda^5) - \lambda^{-1}$ coordinates

10] we described it as a Planck's radiation spectrum of a sphere with diameter $d \ll \lambda$ heated up to some temperature T. In this case the rate of photon emission K_r in the wavelength interval $[\lambda, \lambda + \Delta\lambda]$ is

$$K_r(\lambda, T) = 2\pi c S \Delta\lambda \varepsilon(\lambda, T)/[\lambda^4(\exp(hc/\lambda kT) - 1)], \quad (1)$$

where $S = \pi d^2$ is the surface area, and $\varepsilon(\lambda, T)$ is the emissivity coefficient of the particle. We shall set $\varepsilon(\lambda) = d/\lambda$ [12]. For spectral region in our study $hc/\lambda \gg kT$, thus we derive from (1)

$$K_r(\lambda, T) \sim \lambda^{-5} \cdot \exp(-hc/\lambda kT). \quad (2)$$

The intensity of radiation I measured in experiment is apparently proportional to $K_r \cdot n^*$, where n^* is the density of radiating clusters in the region of observation. The black-body radiation spectrum calculated from Eq. 2 for $T = 3140$ K is shown in Fig. 2 by a dashed curve.

In Fig. 3 the results of the spectral measurements performed with different electron energies E_e are presented in the form of plots of $\ln(I \cdot \lambda^5)$ depending on λ^{-1}. One can see that the results are well approximated by straight lines, whose slope, according to Eq. 2, equals to hc/kT.

At the first glance the good description of measured spectra by black-body-like curves corresponding to uniform temperature is rather surprising due to wide energy deposition function $D(E_v, E_e)$ (determined as the probability per unit energy that during electron impact ionization by electrons with kinetic energy E_e, vibrational energy E_v will be deposited in the C_{60} cluster [13, 14]). Assuming that the linear dependence $D(E_v, E_e)$ on E_v measured in [13] holds over all E_e energies, one can derive that the temperature distribution of radiating clusters has the similar form $D(T) \propto T_{max} - T$. Taking into account data of [14] and considering the energy consumption for ionization we obtain that at $E_e = 59$ eV the value of T_{max} is about 3465 K.

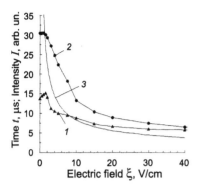
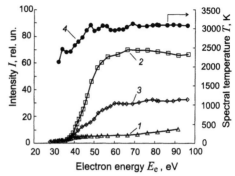

Figure 4: Radiation vs. extracting field

Figure 5: Radiation intensity and temperature vs. electron energy E_e

The result of the calculation of the emission spectrum for the temperature distribution $D(T) \propto 3465 - T$ is shown in Fig. 3 with black diamonds (upper curve). One can see a good linearity of the result. The slope of the line corresponds to the spectral temperature of 2923 K, i.e. much higher than the average temperature of the distribution (1155 K). This is the evident consequence of the very strong dependence of the radiation intensity on the cluster temperature.

Figure 4 shows the intensity of fullerene radiation in a 430 to 580 nm wavelength region depending on the electric field ξ applied between the DP at the energy $E_e = 40$ eV (curve 1) and 65 eV (curve 2). Here is also shown the residence time t of the C_{60} ion in the detection region versus ξ (curve 3). One can see that the ions give the main contribution to the radiation.

The radiation as a function of electron energy for different experimental conditions is shown in Fig. 5. Here curve 1 is the radiation of the neutral clusters C_{60}^* ($\xi = 12$ V/cm), curve 2 is the radiation of C_{60}^* and C_{60}^{+*} ($\xi = 0$ V/cm), curve 3 is the spectral-selected signal ($\lambda = 550$ nm, $\Delta\lambda = 3.2$ nm) of C_{60}^* and C_{60}^{+*} ($\xi = 0$ V/cm).

Describing the radiation with Planck's law and fitting the measured spectra with the formula (2), we have obtained the spectral temperature of radiating clusters depending on the electron energy (curve 4 in Fig. 5). It can be seen that initially $T(E_e)$ increases proportionally with E_e, but at $E_e \approx 47$ eV it reaches the maximum value, $T_m \approx 3100$ K. This value of T_m corresponds to the internal energy of a cluster C_{60} $E_v \approx 36$ eV (the initial internal energy of a cluster at $T_0 = 800$ K is $E_{v,0} \approx 4.6$ eV [13]). The existence of a maximum value T_m may be explained by the competition between the radiative cooling of the electron-heated cluster and other cooling processes, such as the evaporation of fragments and dissociative ionization.

Figure 6 shows the kinetics of the radiation rate at $E_e = 65$ eV for the C_{60}^{+*}

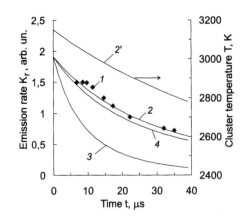

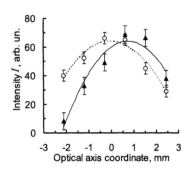

Figure 6: Radiation kinetics

Figure 7: Radiation intensity vs. the optical axis coordinate for fullerene (1) and for $N_2^{+*}(B^2\Sigma^+$, 391 nm) (2)

ion, K_r^+, as a function of the ion flight time to the deflecting plate (curve 1). This dependence was calculated from the data shown in Fig. 4 (curve 2) as follows. First, we subtracted the contribution of the C_{60}^* radiation from the total photon flux of C_{60}^* and C_{60}^{+*}. (The contribution of C_{60}^* was obtained by extrapolation of the $I(t)$ dependence to $t = 0$.) Then for every value of the residence time t in Fig. 4 the intensity of C_{60}^{+*} radiation was divided by the ion density n^+ in the region of observation. The dependence of n^+ on t was determined from the equality of the cluster flux in the beam and those fluxes leaving the observation region due to the electric field ξ and the thermal velocity of C_{60}^{+*} in a beam.

One can see in Fig. 6 that the value of $K_r(t)$ decreases with t. This decrease apparently results from the radiative cooling of a C_{60}^{+*} cluster. Let us obtain the dependence of the radiation intensity on the cooling rate. (The following calculation is based on the analysis given in Ref. [8], but it is refined with taking into account the $\varepsilon(\lambda, T)$ dependence when integrating the radiation energy flux q over λ.) We shall assume for the emissivity coefficient

$$\varepsilon(\lambda, T) = \varepsilon_0 \varepsilon(\lambda) \varepsilon(T) \qquad (3)$$

and take $\varepsilon(T) = (T/T_\varepsilon)^\alpha$,, where ε_0 and T_ε are constants. After integrating $q = (hc/\lambda)K_r$ over λ we obtain the following:

$$q = \varepsilon_0 \cdot \varepsilon(T) \cdot \sigma_c \cdot T^5, \qquad (4)$$

where $\sigma_c = 24.888 \cdot 2\pi hc^2 Sd(k/hc)^5$. Writing $q \cdot dt = -C \cdot dT$, where the heat capacity for our temperature range is $C = dE/dT = 0.0143$ eV/K [13], one can integrate this equation to obtain the time dependence of the cluster temperature due to black-body-like emission

$$T(t, T_i) = T_i \cdot \{1 + t[(4+\alpha)\varepsilon_0 \sigma_c T_i^{4+\alpha}]/(CT_\varepsilon^\alpha)\}^{-1/(4+\alpha)}, \qquad (5)$$

where T_i is the initial temperature of the C_{60}^{+*} cluster. Substituting (5) into (1), we derive (up to a constant) the expression for the $K_r(t)$ dependence

$$K_r(t, T_i) \sim \varepsilon(\lambda, T) \cdot \lambda^{-4} \cdot [\exp(hc/\lambda k T(t, T_i)) - 1]^{-1} \qquad (6)$$

Curves 2–4 in Fig. 6 represent calculated dependencies $K_r(t)$ for $d = 1$ nm (outer diameter of the C_{60} ball), $T_\varepsilon = T_i = 3150$ K, and $\lambda = 550$ nm. The constants were $\alpha = 0$ and $\varepsilon_0 = 6$ (curve 2), $\alpha = 0$ and $\varepsilon_0 = 20$ (curve 3), $\alpha = 2$ and $\varepsilon_0 = 6$ (curve 4).

One can see that the variation of α does not make any qualitative change in the dependence $K_r(t)$ and the value of the ε_0 is the main fitting parameter. Considering that curves 1 and 2 are in agreement, we conclude that $\varepsilon_0 = 6$. Then, assuming $\alpha = 0$ in (3), we obtain for the spectral resolved emissivity of C_{60}^{+*} ion $\varepsilon^+ = 1.1 \cdot 10^{-2}$ at $\lambda = 550$ nm. For this ε^+ the drop of the ion temperature is shown in Fig. 6 with curve 2'.

Figure 7 shows the difference between region of C_{60}^{+*} radiation (1) and emission curve for the short-lived nitrogen states (2), when the optical axis of the detection was shifted at right angles to the electron beam. Measuring the relative intensities of C_{60}^{+*} and N_2^{+*} radiation we were able to estimate the absolute cross-section σ^* of the C_{60}^{+*} formation. At $E_e = 60$ eV the σ^* was found to be about $2 \cdot 10^{-2}$ nm². This value is less than the electron-impact ionization cross-section [1, 2] by a factor of about 20.

Comparing the obtained value of ε^+ with the emissivity reported earlier [13] we should take into account that the original characteristic of the radiating cooling obtainable in experiment was the energy loss rate. After that the emissivity was extracted on the basis of different assumptions (the $q(T)$ dependence, d, T were other than in present study). Therefore we had to restore the original energy loss rates and to re-calculate ε according to our model. By this means we have obtained that in terms of our model the spectral emissivity of neutral C_{60}^* measured in [13] should be about $1.9 \cdot 10^{-4}$.

4 Conclusions

We can draw the following conclusions from our results: (1) up to 44 eV of the electron kinetic energy can be deposited in C_{60} cluster, with a smooth excitation function; (2) the heated C_{60}^* and C_{60}^{+*} clusters emit thermal radiation with a continuous black-body-like spectrum; (3) the radiation rate of C_{60} cluster, heated to $T \approx 3100$ K, is high enough for cluster cooling in tens of microseconds, the emission of an electron enhances the radiation, possibly due to a symmetry breaking.

Acknowledgement

This work was supported by the Russian Foundation for Basic Research (grants 98-02-17804 and 00-03-33028) and a grant from the Education Ministry of the Russia Federation.

References

[1] A. A. Vostrikov, D. Yu. Dubov, A. A. Agarkov, V. A. Galichin, and S. V. Drozdov, Molecular Materials. **20**, 255 (1998)

[2] A. A. Vostrikov, D. Yu. Dubov, and A. A. Agarkov, Tech. Phys. Lett. **21**, 715 (1995)

[3] H. Deutsch, K. Becker, J. Pittner, V. Bonacic-Koutecky, S. Matt, and T. D. Märk, J. Phys. B. **29**, 5175 (1996)

[4] N. P. Gridnev, S. S. Katsnelson, V. M. Fomin, and V. P. Fomichev, Sib. Fiz.–Tekh. Zhurnal. **4**, 36 (1991)

[5] A. A. Vostrikov and S. G. Mironov, Chem. Phys. Lett. **101**, 583 (1983)

[6] A. A. Vostrikov, D. Yu. Dubov, and V. P. Gilyova, Z. Phys. D. **20**, 205 (1991)

[7] A. A. Vostrikov and V. P. Gileva, Tech. Phys. Lett. **20**, 625 (1994)

[8] E. Kolodney, A. Budrevich, and B. Tsipinyuk, Phys. Rev. Lett. **74**, 510 (1995)

[9] V. Frenzel, A. Roggenkamp, and D. Kreisle, Chem.Phys.Lett. **240**, 109 (1995)

[10] R. Mitzner and E. E. B. Campbell, J. Chem. Phys. **103**, 2445 (1995)

[11] A. A. Vostrikov, D. Yu. Dubov, and A. A. Agarkov, JETP Lett. **63**, 963 (1996)

[12] K. Hansen and E. E. B. Campbell, Phys. Rev. E **58**, 54771 (1998)

[13] E. Kolodney, B. Tsipinyuk, and A. Budrevich, J. Chem. Phys. **102**, 9263 (1995)

[14] A. Bekkerman, B. Tsipinyuk, A. Budrevich, and E. Kolodney, J. Chem. Phys. **108**, 5165 (1998)

Internal Energy Distributions in Cluster Ions

James M. Lisy
Department of Chemistry
University of Illinois at Urbana-Champaign
Urbana, Illinois, USA

1 Introduction

Gas phase cluster ions serve as useful models for a number of condensed phase systems. Since they carry a charge, mass spectrometric methods can be used to select by size and/or composition. Thus basic questions of solvation: coordination numbers; size and number of solvent shells; variation with ion size; competition between different solvents, can be probed at the microscopic level. In addition, the timescale over which observations are made can vary from ultrafast measurements in femtoseconds using pump-probe techniques [1] to hundreds of seconds by trapping in an ICR [2].

Most methods of generation such as laser vaporization/ionization, electron impact or ion impact of neutral clusters, produce nascent cluster ions with significant amounts of internal energy. Excess energy can be dissipated by either evaporation or reaction within the cluster [3], with the former the most common method of 'cooling'. Excluding the long-time experiments where the cluster ions are thermally equilibrated by the absorption of blackbody radiation, the amount of internal energy contained by the cluster ions is typically determined by the timescale of observation [4]. Longer times lead to more opportunities for evaporative cooling and less internal energy in the cluster ion. For hydrated ions where competition between ion-water and water-water interactions can lead to a variety of structural isomers, the amount of internal energy is a key factor in determining which isomers are present under a given set of conditions [5].

The internal energy distribution for a given cluster ion becomes an important experimental objective. For small binary cluster ions, high resolution spectroscopy can be used to rotationally resolve vibrational or vibronic transitions. The relative intensities can be used to determine the rotational temperature of the systems. For Ca^+-C_2H_2, band contour analysis of the vibronic band indicated a rotational temperature of 25 K [6]. In the rotationally resolved vibrational spectrum of H_2-HCO^+, the rotational state populations could only be fit by a two temperature model with values of 20 and 160 K [7], an interesting result. This strongly suggests that the amount of internal energy within even a small cluster ion might be a bit more complicated than for neutral clusters. Matters become even murkier when larger clusters are considered. Even for the simplest situation of multiple rare gas atoms in Ar_n-HN_2^+, the rotational structure is lost, and with it any detailed information such as temperatures and structures normally derived from such analyses [8]. Stace has developed an approach which relates the average kinetic energy release to the temperature of the dissociating products of the cluster ion [9]. The application to $SF_6(Ar)_n^+$ examines both the effects of vibrational excitation on the SF_6 moiety on the temperature as well as partial vibrational relaxation [9]. In principle, internal energy information can be obtained by analyzing unimolecular dissociation data from metastable cluster ions. At one limit, blackbody radition can activate this process, a mechanism that has a firm footing [10]. The method presented below can be applied to systems under shorter timescales, less than 1 msec.

A second related issue is the amount of control one has over this distribution of isolated species. It does appear to be possible to vary the amount of internal energy within a cluster ion. Minor changes can be affected by adjusting the conditions of formation [5, 11]. More substantial variation is possible by subtly changing the composition of the ion cluster. While cluster ions of the form $M^+(H_2O)_n$ and $X^-(H_2O)_n$ have considerable binding energies [12], incorporating rare gas atoms into these species can dramatically lower the internal energy. This is reflected in the decrease in the kinetic energy of dissociating fragments [9] and in the infrared absorption profiles of a variety of species. For example, in the case of $X^-(H_2O)_3Ar_m$, X = Cl, Br, I, the cluster ion are 'colder' than their non-argon bearing versions [13], as reflected by sharper and narrower infrared bands and a cyclic $(H_2O)_3$ unit bound to the ion. Given the limited means for *manipulation*, we have worked to establish a procedure for *assessing* the amount of internal energy in these interesting species.

2 Methodology

The successful modeling of the internal energy distribution relies on both experimental data and formulation of unimolecular dissociation rates. The key element is the use of the evaporative ensemble [4] to describe the metastable nature of the cluster ions. In this model, all of the cluster ions have undergone at least one evaporative event, which is treated as a unimolecular dissociation. This is shown schematically in Fig. 1. Each evaporative event reduces the internal energy of the cluster ion through the loss of the binding energy of the solvent and a lesser amount of translational energy.

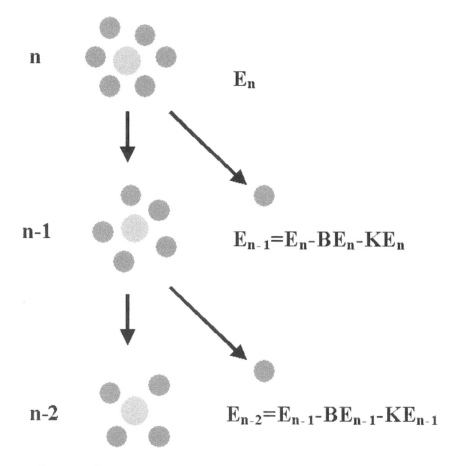

Figure 1: Successive evaporation from a cluster ion with n solvents.

Since the rates depend on the internal energy, each successive evaporation slows the rate of unimolecular loss. Thus the time of observation also plays a role in determining the internal energy distribution. If E_n is too large, $k(E_n)$ will be greater than t_{obs}^{-1} and no clusters of size n will be observed. Conversely if E_{n+1} is too small, $k(E_{n+1})$ will be less than t_{obs}^{-1} and the clusters of size n will not be formed by evaporation from the next largest size. We assume that the initial internal energy distribution of the parent ($n^{th}+1$) cluster is flat, i.e. equal probability at all internal energies. Thus the limits of the internal energy distribution of the nth cluster at a given observation time will be determined by the rates of loss (of the nth cluster) and formation (by dissociation of the $n^{th}+1$). For timescales which are fast compared to blackbody radiative excitation (less than 1 msec), shorter observation times will lead to higher internal energy distributions. As an example, the internal energy distributions for $Na^+(CH_3OH)_{1,3,5,8}$ are shown in Fig. 2 for observation times of 100 - 200 µsec [14].

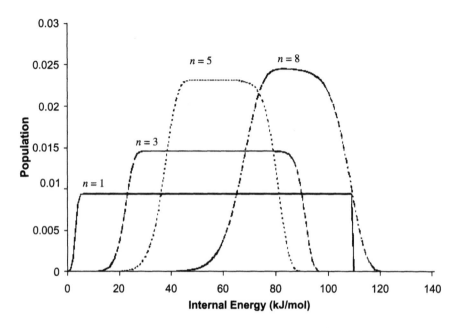

Figure 2: Internal energy distributions for $Na^+(CH_3OH)_n$.

A variety of methods can be used to determine unimolecular dissociation rates. With relatively little knowledge of the internal degrees of freedom, the RRK method can be used with appropriate caveats [15]. More reliable estimates of the internal energy can be made, provided binding energies and vibrational frequencies are known, using the improved RRKM method [14]. For cluster ions, the structure of the transition state can be assumed to be 'loose', which greatly simplifies the calculations [14]. With the relationship between rates and energy well established, it is then possible to estimate the average unimolecular dissociation rate for a given experimental configuration. A tandem mass spectrometer can be used as a convenient example. For a given translational energy, the flight time through the apparatus is known. After selection of the cluster ion of interest by the first mass filter, the flight time to the second mass filter serves as the observation time. The second mass filter is used to determine the fraction of cluster ions that have lost one solvent molecule, thus determining the *experimental* average unimolecular dissociation rate for a given observation time. Using internal energy distributions based on the RRKM methodology given above, the fraction of cluster ions undergoing unimolecular dissociation can be *calculated* by comparing the distributions at two observation times: the time to exit the first mass filter, and the time to exit the second mass filter. Those cluster ions with sufficient internal energy at the upper end of the distribution will undergo dissociation during that time interval. This can be compared with independent experimental results. In Fig. 3, a comparison between observed unimolecular dissociation rates from a tandem quadrupole mass spectrometer and rates calculated using the RRKM method with literature values for the binding energies [16] and vibrational frequencies based on *ab initio* values [14]. The experimental and theoretical rates, while dependent on the timescale of observation, are relatively insensitive to initial conditions in the ion source and energy distributions, respectively. The agreement between the two is quite good for a range of cluster sizes which suggests that the internal energy distributions calculated for this system are reasonable estimates for the experimental system.

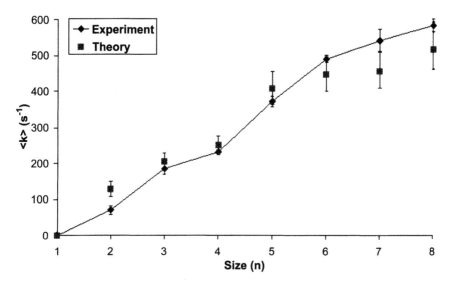

Figure 3: Unimolecular dissociation rates for $Na^+(CH_3OH)_{1-8}$.

3 Applications

The internal energy distribution for the cluster ions can be used to interpret and analyze experimental data from a variety of perspectives. Although cluster ions in molecular beams or ICR spectrometers are not strictly in equilibrium with their surroundings, it is often useful to use a single parameter, temperature, to describe the ensemble. This can be done by first taking the average of the internal energy over the distribution. With knowledge of the internal degrees of freedom, such as the vibrational frequencies

Table 1: $Na^+(CH_3OH)_{1-8}$ averaged properties.

n	$E_{binding}$ (kJ/mol)	E_{int} (kJ/mol)	Temperature (K)
1	110.3	56(6)	890(60)
2	87.3	52(8)	510(40)
3	75.6	56(10)	400(40)
4	68.3	61(11)	340(30)
5	48.5	58(12)	280(30)
6	48.5	68(14)	270(30)
7	48.5	77(16)	260(30)
8	48.5	87(18)	260(30)

of the modes of the cluster ion, this can be related to an average temperature by standard statistical mechanics methods. The temperatures for $Na^+(CH_3OH)_{1-8}$ have been determined by this approach and are shown in Table 1.

It is useful to note that the temperatures steadily decrease with increasing cluster size. This is due to two reasons. First, the binding energy decreases with size approaching the bulk value after the first solvent shell is filled. Second, while the cluster is able to store more energy without undergoing unimolecular dissociation, this energy is spread among more degrees of freedom, so the net energy per mode slightly decreases. Note that the widths of the distributions (see Fig. 2) are closely related to the binding energies of the cluster. This also helps to explain how the presence of a weakly bound species in a cluster ion, such as argon, can 'cool' the system, again for two reasons. First, the distribution is narrowed, since the binding energy has decreased. Second, the distribution is shifted to lower energy, because less energy can be stored in the cluster ion, due to its weakly bound constituent.

The temperatures derived from this analysis can in turn be used in Monte Carlo or Molecular Dynamics simulations [17] to characterize structural properties of the cluster ions, such as the size of the solvent shells and the presence of hydrogen bonds. These structural features can be quite dependent on temperature, as was discussed earlier [5, 11, 13], so an accurate assessment of the temperature is useful and important. When applied to Monte Carlo simulations of $Na^+(CH_3OH)_{1-8}$, the onset of hydrogen bond formation and the types of hydrogen bonded structures [14] were found to be consistent with experimental infrared spectra of these species [18]. A recent theoretical study examined the effect of temperature on the hydrogen bonding network in $Cl^-(H_2O)_2$. The infrared spectral signature of the water-water hydrogen bond disappeared with increasing temperature, a result of the competition between the ion-water and water-water interactions [19].

The experimental infrared spectrum of 'warm' $Cl^-(H_2O)_2$ displayed only features consistent with ion-water interactions [20], which corresponded to a temperature of near 100 K based on the theoretical study [19]. A recent experimental infrared study of 'cold' $Cl^-(H_2O)_2Ar_m$, gave a clear indication of a water-water hydrogen bond [21], consistent with the calculations at a temperature of 5 K [19].

4 Conclusions

We have shown that it is possible to use a variety of experimental and theoretical data to determine internal energy distributions for cluster ions produced under a variety of experimental conditions. Furthermore, dynamics associated with these calculated distributions can be compared with results from experimental unimolecular dissociation as a check for accuracy and consistency. The distributions can then be used to characterize structural aspects of cluster ions for comparison and interpretation of independent experimental observations such as infrared, visible or uv spectra, mass spectral distributions or size-specific reactivities.

Acknowledgments

This work has been supported in part by a grant from the National Science Foundation (CHE 9700722). Dr. Orlando M. Cabarcos and Dr. Corey J. Weinheimer were instrumental in developing and testing this methodology.

References

[1] L.Lehr, M.T.Zanni, C.Frischkorn, R.Weinkauf, and D.M.Neumark. Science, **284**, 635 (1999).

[2] U.Achatz, S.Joos, C.Berg, M.Beyer, G.Niedner-Schatteburg, and V.E.Bondybey, Chem. Phys. Lett., **291**, 459 (1998).

[3] T.J.Selegue, and J.M.Lisy, J. Am. Chem. Soc., **116**, 4874 (1994). X.Zhang, and A.W.Castleman Jr., J. Am. Chem. Soc., **114**, 8607 (1992).

[4] C.E.Klots, Z. Phys. D., **5**, 83 (1987).

[5] J.-C.Jiang, Y.-S.Wang, H.-C.Chang, S.H.Lin, Y.T.Lee, G.Niedner-Schatteburg, and H.-C.Chang, J. Am. Chem. Soc., **122**, 1398 (2000).

[6] M.R.France, S.H.Pullins, and M.A.Duncan, J. Chem. Phys., **109**, 8842 (1998).

[7] R.V.Olkhov, S.A.Nizkorodov, and O.Dopfer, J. Chem. Phys., **107**, 8229 (1997).

[8] O.Dopfer, S.A.Nizkorodov, and J.P.Maier, J. Phys. Chem. A, **103**, 2982 (1999).

[9] S.Atrill, and A.J.Stace, J. Chem. Phys., **108**, 1924 (1998).

[10] R.C.Dunbar, and T.B.McMahon, Science, **279**, 194 (1998).

[11] Y.-S.Wang, H.-C.Chang, J.-C.Jiang, S.H.Lin, Y.T.Lee, and H.-C.Chang, J. Am. Chem. Soc., **120**, 8777 (1998).

[12] A.W.Castleman Jr., and R.G.Keesee, Chem. Rev., **88**, 589 (1986).

[13] P.Ayotte, G.H.Weddle, M.A.Johnson, J. Chem. Phys., **110**, 7129 (1999).

[14] O.M.Cabarcos, C.J.Weinheimer, and J.M.Lisy, J. Phys. Chem. A, **103**, 8777 (1999).

[15] J.A.Draves, Z.Luthey-Schulten, W.-L.Liu, and J.M.Lisy, J. Chem. Phys. **93**, 4589 (1990).

[16] B.C.Guo, and A.W.Castleman Jr., Z. Phys. D, **19**, 397 (1991).

[17] O.M.Cabarcos, and J.M.Lisy, Int. J. Mass Spec., **185/186/187**, 883, (1999).

[18] C.J.Weinheimer, and J.M.Lisy, J. Phys. Chem., **100**, 2938 (1996).

[19] H.E.Dorsett, R.O.Watts, and S.S.Xantheas, J. Phys. Chem. A, **103**, 3351 (1999).

[20] J.-H.Choi, K.T.Kuwata, Y.-B.Cao, and M.Okumura, J. Phys. Chem., **102**, 503 (1998).

[21] P.Ayotte, S.B.Nielsen, G.H.Weddle, M.A.Johnson, and S.S.Xantheas, J. Phys. Chem. A, **103**, 10665 (1999).

Molecular Beams of Silicon Clusters and Nanoparticles Produced by Laser Pyrolysis of Gas Phase Reactants

M. Ehbrecht[1], H. Hofmeister[2], B. Kohn[1], F. Huisken[1]
[1] Max-Planck-Institut für Strömungsforschung, Bunsenstr. 10
D-37073 Göttingen, Germany
[2] Max-Planck-Institut für Mikrostrukturphysik, Weinberg 2
D-06120 Halle, Germany

1 Introduction

In recent years, there has been increasing interest in the synthesis and characterization of nanosized particles [1]. Due to their finite size, they often exhibit physical properties which may significantly differ from those of their bulk counterparts. Research in this direction is strongly motivated by the possibility of designing nanostructured materials that possess novel electronic, optical, magnetic, chemical, or mechanical properties. Since silicon is of great importance for the microelectronics industry a detailed knowledge of the physical and chemical properties of silicon in its mesoscopic state between the atom and bulk matter is particularly desirable.

The observation of photoluminescence from porous nano-structured silicon at the beginning of this decade [2, 3] has triggered an enormous increase in activity. While the early investigations were devoted exclusively to porous silicon produced by electrochemical etching of Si wafers, recent work is concentrating more and more on the production and characterization of silicon nanoclusters constituting zerodimensional quantum dots [4]. Important issues of current research are: what is the smallest cluster size exhibiting photoluminescence, how do the luminescence wavelength and the efficiency vary with the cluster size, and what is the role of surface passivation?

In the past, most experimental studies devoted to small Si_n clusters ($n < 100$) were carried out in molecular beams using laser vaporization sources to produce the silicon clusters [5]. Apart from being restricted to the pulsed mode and having poor long term stability, this source has the disadvantage that the maximum cluster size is restricted to approximately $n = 200$. An alternative technique, capable of producing much larger silicon clusters, is based on the pyrolysis of silicon-containing gas phase precursors. This technique, commonly referred to as chemical vapor deposition (CVD), has been widely used to produce ultrafine powders and thin films. From the many possibilities to dissociate the precursor molecules, CO_2-laser-induced decomposition of SiH_4 in a gas flow reactor has been shown to be particularly useful for the production of ultraclean silicon particles in the nanometer size range [6]. However, with typical diameters well above 10 nm, the particles are too large for many applications, in particular for the study of their photoluminescence.

To close the gap between small silicon clusters ($n < 200$, $d < 2$ nm), produced in laser vaporization sources, and large nanoparticles ($n > 2.5 \times 10^4$, $d > 10$ nm), obtained in laser CVD experiments, and to study the transition from small clusters to nanoparticles, we have recently developed a novel cluster source which combines the laser-driven CVD reactor with a supersonic expansion of the nascent clusters into a high vacuum molecular beam apparatus [7]. Indeed, it could be shown that the source is capable of producing silicon nanoclusters in the desired size regime [8]. In addition, we found that the velocity of the nanoclusters correlates with their mass, enabling us to perform a size selection of the neutral clusters by introducing a slotted chopper wheel [9, 10]. This feature has been successfully exploited to produce thin films of silicon nanoclusters and to study their photoluminescence behavior and Raman spectroscopy as a function of cluster size [10].

2 Experimental

The experiments have been carried out in a molecular beam apparatus which is shown schematically in Fig. 1. The setup consists of three vacuum chambers: a source chamber containing the flow reactor, a differential chamber enhousing the chopper for size selection as well as a movable sample holder for deposition experiments, and an ultra-high-vacuum (UHV) chamber which contains the time-of-flight mass spectrometer (TOF-MS). The silicon clusters are produced in the flow reactor by CO_2 laser pyrolysis of silane (SiH_4) as has been described in detail earlier [6, 7, 9]. Briefly, a flow of silane emanating from a 3-mm-diameter stainless steel tube is subjected to the focused ($f = 200$ mm) radiation of a line-tunable pulsed CO_2 laser. To confine the reactant gas to the flow axis, helium is flushed through a coaxial outer tube. During the course of the experiments, various reactor conditions have been tested. However, the following conditions have been found to be particularly favorable: SiH_4 flow rate: 20 sccm; He flow rate: 1100 sccm; total pressure: 330 mbar; pulse energy of the CO_2 laser: 50 mJ; laser line: $10\mu P(30)$ (934.9 cm^{-1}). The CO_2 laser has a pulse width of 150 ns and is operated with a repetition rate of 20 Hz.

The reaction products are extracted perpendicularly to both the gas flow

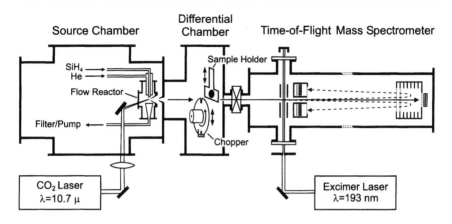

Figure 1: Schematic view of the cluster beam apparatus.

and the CO_2 laser beam through a conical nozzle of 0.3 mm diameter opening, projected into the reaction zone. After skimming, the silicon clusters enter a pressure-reducing differential chamber of 370 mm length. There they pass through the slits of fast-spinning molecular beam chopper which can be vertically moved from the outside. This chopper is used to preselect a small portion of the initially broad cluster pulse, in order to perform (i) a velocity analysis and (ii) to narrow the cluster size distribution. Immediately behind the chopper, a sample holder may be moved into the cluster beam. Thus, silicon particles with preselected size can be deposited on various substrates, suitable to perform transmission electron microscopy (TEM) or Raman and photoluminescence studies. When the chopper is not in place, the silicon particles are deposited without size selection, *i.e.* a broad size distribution is obtained.

If the sample holder is moved out of the beam, the silicon clusters and nanoparticles finally reach the UHV detector chamber where they are ionized by the radiation of an ArF excimer laser ($\lambda = 193$ nm). The mass spectrometer can be operated as a simple Wiley and McLaren TOF-MS (WMTOF-MS), with the ion detector – a pair of channel plates – mounted at the end of the flight tube, or as a linear reflectron-type time-of-flight mass spectrometer (RETOF-MS) if it is operated with reflecting electrostatic field. In the latter case, significantly higher resolution is obtained. The time-of-flight mass spectra are recorded with a multichannel analyzer (EG+G ORTEC, model T914), using a dwell time of typically 200 ns. For every detected ion, the channel corresponding to the ion arrival time is incremented by one. Compared to a previously used digital scope [7], this technique provides considerably higher sensitivity, enabling us to carry out experiments at very low fluence of the ionizing laser and to accumulate the mass spectra over several thousand laser shots [9].

In order to determine the intensity of our Si cluster beam, we have measured the thickness of a film deposited during 45 min without size selection on a silicon wafer, using a surface profiler (Veeco Instruments, DEKTAK^{3}ST). From the measured 200-nm-thickness of the 9-mm-diameter spot, a time-averaged deposition rate of 7 nm/(cm^{2}min) or 4×10^{14} atoms/(cm^{2}s) is calculated. Typical deposition rates for low-pressure inert gas condensation sources and laser vaporization sources are reported by a recent review article on cluster beam deposition [11]. Comparing the intensities, our source is superior to laser vaporization sources [0.1 – 6 nm/(cm^{2}min)] and yields similar output as gas condensation sources [$\approx 10^{13} - 10^{15}$ atoms/(cm^{2}s)] which, however, are operated continuously.

3 Results and Discussion

3.1 Mass Spectrometry

At first we have analyzed the temporal variation of the composition of the silicon cluster pulse extracted from the flow reactor by measuring the TOF mass spectra at different delays $\Delta t = t^- + t^+$ between CO_2 and excimer laser [9]. (t^- refers to the time between the firing of the CO_2 laser and the opening of the chopper while t^+ designates the delay between the chopper opening and the excimer laser pulse.) Three typical TOF mass spectra are shown in Fig. 2. They have been measured with the delays $\Delta t = 280$, 310, and 330 µs while the delay t^- between CO_2 laser and chopper was adjusted in each case so as to obtain maximum transmission. Spectrum (a), which has been taken in the early stage of the silicon cluster pulse, shows a uniform distribution of cluster ions centered at a flight time somewhere between 140 and 160 µs. A closer look to this spectrum, part of which is reproduced with enlarged scale in the inset, reveals an onset of the signal at Si_{22} and a maximum position of the size distribution near $n = 130$. Spectrum (b) has been measured at a somewhat later stage ($\Delta t = 310$ µs). Here the maximum has shifted to ion flight times around 290 µs, corresponding to silicon clusters containing approximately 700 atoms. Furthermore, a secondary distribution, marked by two plus signs (++), is noted at shorter flight times. In spectrum (c) ($\Delta t = 330$ µs), this earlier secondary maximum is even more pronounced.

If the TOF spectra are converted to mass spectra, it is seen that the secondary distributions peak at values which are exactly half of the maximum positions of the primary distributions. Thus, it follows that the secondary distributions must be assigned to doubly ionized silicon clusters which are formed in the intense laser field as a result of the enhanced photoabsorption cross section for larger particles. This conclusion could be confirmed by flu-

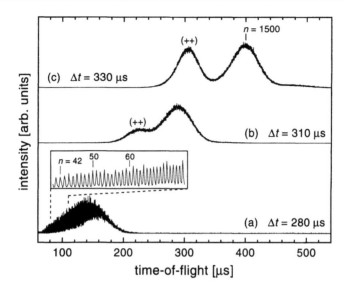

Figure 2: Typical TOF mass spectra revealing the size distributions in different parts of the silicon cluster pulse. The delay between generating CO_2 laser and probing excimer laser was adjusted to $\Delta t = 280$ µs (a), $\Delta t = 310$ µs (b), and $\Delta t = 330$ µs (c). The secondary distributions, appearing at shorter flight times and marked by (++), are assigned to doubly ionized Si clusters.

ence dependence measurements which, in a log-log plot, gave straight lines with slopes $p_1 = 0.8$ for the primary peak and $p_2 = 1.6$ for the secondary peak [9]. The deviation from the ideal fluence dependencies with slopes 1 and 2 (which are expected since the energy of one excimer laser photon is sufficient to ionize all Si_n clusters with $n \geq 22$ [12]) can be explained with saturation effects in the ion detector. Three representative spectra measured during this fluence dependence study are depicted in Fig. 3. To facilitate the discussion, they have now been transformed to mass spectra or cluster size distributions. Going from bottom to top, the spectra were obtained at low (27 µJ/cm^2), medium (2.2 mJ/cm^2), and high (8.6 mJ/cm^2) laser fluence. The delay between CO_2 and excimer laser was adjusted to $\Delta t = 330$ µs. As is seen in spectrum (a), only a narrow distribution of Si_n clusters around $n \approx 1500$ is transmitted by the chopper. At higher laser fluence [spectrum (b)], the secondary distribution, which is assigned to doubly ionized clusters, appears. It should be noted that this spectrum and spectrum (c) of Fig. 2 are extracted from the same measurement ($\Delta t = 330$ µs). The difference in the relative intensities of the two distributions is due to the transformation from TOF spectra to cluster size spectra. The shoulder appearing on the left side of the major peak in spectrum (c) is attributed to triply ionized silicon clusters.

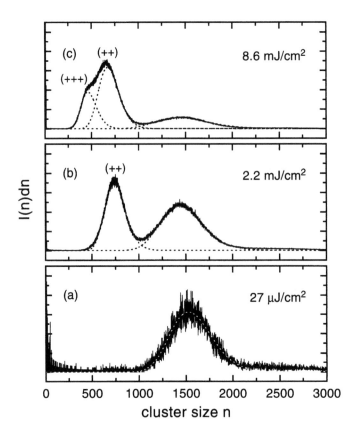

Figure 3: Mass spectra of the same neutral cluster size distribution, measured at different fluences of the ionizing excimer laser.

If the fluence of the excimer laser was further increased, e.g. by focusing, higher than triply charged silicon clusters were not observed. Instead, we found significantly enhanced signal on the mass peaks corresponding to $Si^+ - Si_6^+$. Thus, we obtain the result that, at higher density of positive charges ($z > 3$) on the surface of a large Si_n cluster ($n \approx 1000 - 3000$), the Coulomb repulsion leads to fast evaporation of small silicon cluster cations ($Si^+ - Si_6^+$), with Si_2^+ being the most prominent product [9].

The maximum of the primary peak in spectrum (c) of Fig. 2 (and in Fig. 3) corresponds to Si clusters with an average size of $n = 1500$. Assuming the same crystalline structure and the same lattice parameters as in bulk silicon, a diameter of $d = 3.8$ nm is calculated for this cluster size. When the delay between CO_2 and excimer laser is further increased, even larger silicon clusters, containing several thousand Si atoms, are detected. Working at low laser fluence so as to avoid multiple ionization, the largest cluster size observed in this study was $n = 3000$. However, this value must be considered as a lower limit since the detection probability is greatly reduced for large clusters because of their low velocity which in turn is a consequence of their large mass if the ion energy is constant. When we worked at higher laser fluence the largest (uncorrected) cluster size was around $n \approx 4000$. Assuming that these clusters were triply ionized, a corrected cluster size of $n \approx 12000$, corresponding to $d = 7.7$ nm, is calculated. Thus, we have the result that our cluster source is very well suited to produce a molecular beam of silicon clusters in the size regime between 1 and 8 nm, which are referred to as *nanoclusters*.

Since our silicon clusters are produced by laser pyrolysis of silane and since hydrogen constitutes a major product, the question arises whether the silicon clusters are contaminated with hydrogen. To answer this question, we have carried out a detailed mass spectrometric analysis of the small Si clusters, employing the high resolution mode offered by our reflectron mass spectrometer. In this RETOF-MS mode, it was possible to resolve the isotopic composition of the smaller silicon cluster peaks ($Si_2^+ - Si_{11}^+$) which gave rise to a fine structure consisting of up to six lines. Calculations, based on the natural abundances of silicon [92.23 % for ^{28}Si, 4.67 % for ^{29}Si, and 3.10 % for ^{30}Si], reproduced the experimental results very well. The good agreement indicates that the small silicon cluster ions are not contaminated with hydrogen. Otherwise one would expect higher intensities for the peaks belonging to the masses $m = 28n + 1$ or $m = 28n + 2$ amu. This conclusion is also supported by the close resemblance of our mass spectra (as far as small Si_n clusters with $n \leq 11$ are concerned) to earlier results obtained by other investigators who employed laser vaporization sources containing high-purity silicon [5, 12].

With the help of the chopper, it was possible to determine the velocity of the silicon clusters as a function of their size or mass [9]. The result of this investigation, which is summarized in Fig. 4, shows that the velocity decreases from $v = 2000$ m/s for small clusters ($n < 100$) to $v < 1600$ m/s for silicon nanoclusters with $n > 3000$ ($d > 5$ nm). By operating the chopper at fixed phase t^- with respect to the CO_2 laser, it is therefore possible to select a rather narrow size distribution from the initially much broader cluster size distribution and to perform experiments with *quasi* size-selected silicon clusters. Furthermore, by varying the delay between CO_2 laser and chopper, it is possible to select narrow size distributions peaked at different mean sizes between $d = 1$ and 8 nm. This has already been demonstrated using size-sensitive probing techniques such as transmission electron microscopy (TEM) [8] as well as Raman and photoluminescence spectroscopy [10] to study thin films of silicon clusters produced by cluster

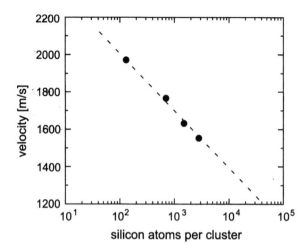

Figure 4: Measured velocities of silicon clusters as a function of their size. The fluence of the excimer laser was sufficiently reduced, and only singly ionized clusters were considered.

beam deposition. The great potential becomes also apparent in Fig. 2: by probing different parts of the cluster pulse, the maximum of the cluster size distribution is shifted from $n = 130$ to $n = 1500$, in this example.

3.2 Electron Microscopy

To characterize the silicon nanoclusters by transmission electron microscopy (TEM), carbon-coated microgrids have been mounted on the sample holder and placed into the cluster beam. In order to obtain a coverage with a broad size distribution, the molecular beam chopper was not used. The TEM studies were carried out at the Max-Planck-Institute for Microstructure Physics in Halle employing a conventional electron microscope (JEM 100C) as well as a high resolution electron microscope (JEM 4000EX) operated at 400 kV. Particle size, oxide shell thickness, and the spacing of the lattice plane fringes were determined from the digitized HREM micrographs by digital image processing.

Figure 5 shows a low resolution TEM micrograph of silicon nanoparticles deposited by cluster beam epitaxy without size selection on the carbon film. Under the conditions mentioned above, the CO_2-laser-induced decomposition of silane yielded spherical particles with diameters between 2 and 30 nm. Remarkably, they did not agglomerate to larger aggregates as it is known from other routes of synthesis. The electron diffraction patterns revealed well established diffraction rings down to lattice plane spacings of less than 0.1 nm, according to the diamond cubic lattice of crystalline silicon [13]. Some diffusive intensity from amorphous material is assigned to the carbon support and to silicon oxide.

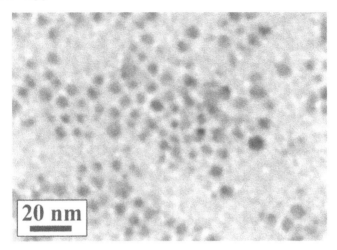

Figure 5: Low resolution TEM micrograph of Si nanoparticles of different sizes, randomly assembled on the supporting carbon film.

Figure 6 shows a high resolution (HREM) image of single crystalline Si particles sticking to the edge of a hole in the carbon film. It is recognized that the particles contain a perfect monocrystalline core and are surrounded by an amorphous shell, which is most probably silicon oxide. Evaluation of a series of images similar to Fig. 6 reveals that the thickness of the oxide shell is a linear function of the particle size. It is found that the shell thickness increases from 0.8 nm (for 7 nm particles) to almost 2.9 nm for particles with 30 nm diameter. The crystalline structure of more than 200 silicon nanoparticles has been carefully analyzed by means of the reciprocal space representation (diffractogram). Aside from being single crystalline and almost completely free of planar lattice defects, the Si nanoparticles with diameters larger than 3 nm were found to exhibit a distinctly reduced lattice spacing as compared to bulk crystalline silicon. For 20 nm particles, for example, a spacing of the {111} lattice plane fringes of $d_{111} = 0.3075$ nm was determined. It follows that for this particle size the spacing is almost 2% smaller than in bulk silicon ($d_{111-\text{bulk}} = 0.3134$ nm). This distinct lattice compression in Si nanoparticles can be explained by the compressive stress exerted by the oxide shell, which may attain several GPa. As the thickness of the oxide shell decreases proportionally with the particle size, the lattice compression is observed to diminish with decreasing size and to turn to lattice dilatation for nanoclusters with a crystalline core smaller than 3 nm [13].

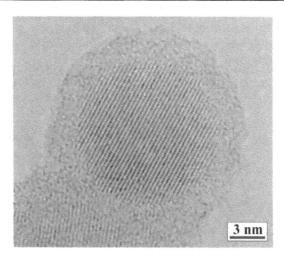

Figure 6: HREM image of single-crystalline Si particles sticking to the edge of a hole in the carbon film. The amorphous layer of silicon oxide may be clearly recognized.

4 Conclusions

It is shown that CO_2-laser-induced decomposition of silane in a flow reactor, combined with early extraction of the reaction products into a high vacuum chamber, is very well suited to produce an intense molecular beam of noninteracting gas phase silicon clusters. The mass spectrometric characterization revealed that this cluster beam contained small Si_n clusters being composed of a few atoms as well as very large units consisting of several thousand silicon atoms. With diameters between 1 and 8 nm, these large clusters are referred to as nanoparticles or nanoclusters. High resolution mass spectrometry with reflecting field (RETOF-MS) further revealed that the silicon clusters were not contaminated with hydrogen. The absence of hydrogen can be explained by the fact that the silicon clusters are quickly transferred from the hot region of the reaction zone into a high vacuum environment. At the rather high temperature of the reaction zone, which is estimated to be 1200 K or higher [6], any chemisorption of atomic hydrogen will result in the associative desorption of hydrogen molecules.

A series of experiments has been conducted to investigate the fragmentation behavior of silicon clusters when they are subjected to the radiation of an ArF laser at medium and high fluences [9]. In agreement with earlier studies carried out with laser vaporization sources [5, 14, 15], it was found that, already at modest laser fluences, medium-sized silicon clusters ($22 \leq n \leq 200$) fragment upon ionization by fission to form $Si_6^+ - Si_{11}^+$ cations. As far as nanoclusters ($n \geq 1000$) are concerned, the formation of doubly and triply ionized clusters constitutes the dominant reaction channel when the cluster size or the laser fluence exceeds a certain value. At very high laser fluence, multiple (higher than triply) ionization may occur; but the strong Coulomb interaction results in the fast evaporation of small positively charged cluster ions ($Si_1^+ - Si_4^+$), with Si_2^+ being produced most frequently [9].

The time-of-flight analysis, carried out with a molecular beam chopper, revealed that the velocity of the neutral silicon clusters ejected from the pulsed source strongly depends on their size. This property enables one to select different portions from the originally broad size distribution and to perform deposition experiments with *quasi* size-selected neutral clusters [8]. Very recently, this particularly attractive feature has been exploited to study the photoluminescence and Raman spectroscopy of silicon quantum dots as a function of their size [10, 16].

The silicon nanoclusters have also been characterized by transmission electron microscopy [13]. The low resolution TEM study showed that the size distribution of the silicon clusters (deposited without size selection) had a maximum at $d = 4.5$ nm and extended up to $d = 30$ nm. Such large silicon clusters could not be observed with the TOF mass spectrometer. This is explained with the greatly reduced detection efficiency for very large and therefore very slow cluster ions. The high resolution study (HREM) revealed that the silicon clusters are composed of a perfect monocrystalline core without any defect and an amorphous shell which can be attributed to silicon oxide. These findings nicely explain the earlier results of the Raman and photoluminescence studies [10] which already indicated that the clusters were crystalline and the dangling bonds of their surface atoms passivated. The computer analysis of the digitized HREM images further revealed that the diamond cubic lattice of the silicon nanoclusters larger than 3 nm exhibited distinctly reduced spacing as compared to bulk c-Si. This phenomenon is ascribed to the surrounding shell of silicon oxide which exerts a pressure of several GPa. With decreasing shell thickness, the lattice contraction is observed to diminish [13].

This work was supported by the Deutsche Forschungsgemeinschaft in the frame of its Schwerpunktprogramm *Fine Solid Particles*.

References

[1] See for example: Nanostructures and Mesoscopic Systems, edited by W. P. Kirk and M. A. Reed (Academic, New York, 1992); Nanophase Materials: Synthesis, Properties, Applications, edited by G. C. Hadijipanyis and R. W. Siegel (Kluwer Academic Publications, London, 1994); Nanomaterials: Synthesis, Properties and Applications, edited by A. S. Edelstein and R. C. Cammarata (Institute of Physics Publishing, Bristol, U. K., 1996).

[2] L. T. Canham, Appl. Phys. Lett. **57**, 1046 (1990).

[3] V. Lehmann and U. Gösele, Appl. Phys. Lett. **58**, 856 (1991).

[4] K. A. Littau, P. J. Szajowski, A. J. Muller, A. R. Kortan, and L. E. Brus, J. Phys. Chem. **97**, 1224 (1993).

[5] J. R. Heath, Y. Liu, S. C. O'Brian, Q.-L. Zhang, R. F. Curl, F. K. Tittel, and R. E. Smalley, J. Chem. Phys. **83**, 5520 (1985).

[6] J. S. Haggerty and W. R. Cannon, in: *Laser-Induced Chemical Processes*, edited by J. I. Steinfeld (Plenum, New York, 1981), pp. 165-241.

[7] M. Ehbrecht, H. Ferkel, V. V. Smirnov, O. M. Stelmakh, W. Zhang, and F. Huisken, Rev. Sci. Instrum. **66**, 3833 (1995).

[8] M. Ehbrecht, H. Ferkel, and F. Huisken, Z. Phys. D **40**, 88 (1997).

[9] M. Ehbrecht and F. Huisken, Phys. Rev. B **59**, 2975 (1999).

[10] M. Ehbrecht, B. Kohn, F. Huisken, M. A. Laguna, and V. Paillard, Phys. Rev. B **56**, 6958 (1997).

[11] P. Melinon, V. Paillard, V. Dupuis, A. Perez, P. Jensen, A. Hoareau, J.P. Perez, J. Tuallion, M. Broyer, J.L. Vialle, M. Pellarin, B. Baguenard, and J. Lerme, Int. J. Mod. Phys. B **9**, 339 (1995).

[12] K. Fuke, K. Tsukamoto, F. Misaizu, and M. Sanekata, J. Chem. Phys. **99**, 7807 (1993).

[13] H. Hofmeister, F. Huisken, and B. Kohn, Europ. Phys. J. D **9** (1999), in press.

[14] Q.-L. Zhang, Y. Liu, R. F. Curl, F. K. Tittel, and R. E. Smalley, J. Chem. Phys. **88**, 1670 (1988).

[15] M. F. Jarrold and E. C. Honea, J. Phys. Chem. **95**, 9181 (1991).

[16] G. Ledoux, M. Ehbrecht, O. Guillois, F. Huisken, B. Kohn, M. A. Laguna, I. Nenner, V. Paillard, R. Papoular, D. Porterat, and C. Reynaud, Astron. Astrophys. **333**, L39 (1998).

Part VI

Spectroscopy and Reaction Dynamics
of Molecules Isolated, Cooled (or Conditioned),
by Techniques
of Molecule, Cluster, Droplets, or Liquid Beams

Spectroscopy in, on, and off a Beam of Superfluid Helium Nanodroplets

J.P. Higgins[1], J. Reho[1], F. Stienkemeier[2],
W. E. Ernst[3], K. K. Lehmann[1], G. Scoles[1]

[1] Department of Chemistry, Princeton University,
Princeton, NJ 08544
[2] Fakultät für Physik, Universität Bielefeld, D-33615 Bielefeld,
Germany
[3] Department of Physics, Pennsylvania State University,
University Park, PA 16802

Abstract

Helium nanodroplet isolation (HENDI) spectroscopy involves the use of a beam of He$_n$ ($10^3 < n < 10^5$) nanodroplets which are doped while passing, largely undeflected, through a pick-up cell containing the low pressure vapor of the substance to be examined. The nanodroplets carry the dopant species downstream, where they can be spectroscopically interrogated. After briefly reviewing the field and providing a few examples of applications, in this paper we show that HENDI is also useful in producing collimated beams of cold, *gas-phase* molecules which, after forming on the cold droplet's surface, spontaneously desorb from it at very low velocities. As this low desorption velocity is added to the relatively large average velocity of the droplets, the desorbed molecules fly forward as part of the main beam and can be used to obtain sub-doppler high resolution spectra. Three examples of applications of this new technique will be reported here showing the spectrum of a well known molecule (ground state Na$_2$), that of a less well known molecule (lowest triplet state Na$_2$) and a third spectrum which is tentatively assigned to a yet unknown molecule: Na$_2$He.

1 Introduction

Despite the absence of external interactions to broaden their transitions, the room temperature spectra of gaseous molecules can be quite complicated due to the presence of many overlapping lines. The introduction of supersonic molecular beam laser spectroscopy has provided a valuable tool to obtain cold spectra of isolated molecules that are vastly simplified compared to their room temperature counterparts [1]. The lowest rotational temperatures that can usually be achieved in the expansion of a supersonic beam source are typically a few degrees Kelvin or less while vibrations are cooled less effectively, especially when the vibrational frequency is high or the number of atoms in the molecule is large.

The introduction of matrix isolation spectroscopy has traditionally permitted the acquisition of low temperature spectra of both stable and unstable radical species [2,3]. The presence of the solid matrix environment has however hindered the usefulness of this technique since matrix effects on the spectra can be important. Large spectral shifts and the inability to obtain rotational resolution (except sometimes in solid hydrogen [4]) severely limit the technique in comparison to molecular beam spectroscopy which does not suffer from these limitations.

Several methods are currently being developed to carry out spectroscopic measurements on gas phase molecules at very low temperatures. These include photoassociative spectroscopy of laser-cooled and trapped atoms [5], helium buffer gas cooling [6], and magnetic trapping of paramagnetic atoms and molecules [7], and the subject of this article: i.e. spectroscopy of atoms and molecules trapped in or on liquid helium nanodroplets. The production of helium nanodroplets and their investigation as finite size superfluid systems dates back to the early sixties [8] but it was only after it was demonstrated that large helium clusters doped with SF_6 molecules could be detected efficiently by laser-induced evaporation [9] that spectroscopy of doped helium nanodroplets became a concrete possibility. A decisive impetus to the field was later provided by the demonstration by Toennies and co-workers [10] that molecules solvated in helium droplets undergo free rotation [11], making it possible to obtain structural information. Two reviews of this new field have recently appeared in the literature[12,13].

The favorable conditions achievable in molecular beams and in low temperature matrices, namely, cooling, isolation, and control over molecular interactions, are all to be found for species attached to or solvated in helium nanodroplets. In fact, one could claim that helium nanodroplets embody the best features of both environments. The cooling achievable in helium surpasses what can be done in a co-expanded molecular beam (due to the small number of collisions available in the latter), while the spectral broadening and shifting induced by the helium droplets are small in comparison with those obtained in standard matrix spectroscopy [14].

HENDI spectroscopy involves the use of a beam of helium nanodroplets (He_n $n \sim 10^3 - 10^5$) which is doped by passage through a pick-up cell where a low vapor pressure ($10^{-5} - 10^{-3}$ torr) of an atom or molecule of interest is maintained. The nanodroplets, largely undeflected by the pick-up process, carry the dopant species downstream where they can be spectroscopically interrogated. In the next section of the present article we will provide a few examples of applications of this new type of matrix spectroscopy and review briefly some of the published results obtained in our laboratory. In the remaining part of the paper we will show that HENDI is also useful in producing gas phase beams of low temperature species because some molecules formed on the nanodroplets spontaneously desorb with very low velocities thus becoming entrained in the molecular beam. For a schematic rendition of the experimental set up see Figure 1.

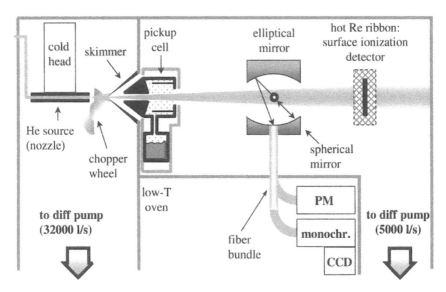

Figure 1: Schematic of the experimental apparatus used in the acquisition of molecular spectra of dopants using helium nanodroplets.

2 Helium Nanodroplet Isolation Spectroscopy: A Short Review

The use of helium nanodroplets as matrices grew out of "gas phase" matrix spectroscopy based on argon clusters [15]. The first Ar cluster-isolation spectra were taken using line-tunable lasers and produced infrared spectra of SF_6 and CH_3F attached to Ar clusters. These spectra showed that oligomer formation could be controlled more easily than using traditional matrix isolation methods. The first spectroscopic studies using helium droplets containing guest species probed the ν_3 vibration of SF_6 and $(SF_6)_2$ using line-tunable lasers and bolometric detection [9]. After demonstrating rotational resolution [10], experiments carried out in the Max Planck Institute in Göttingen [16] in which mass spectrometric detection was employed, made clear that the "free" rotation of molecules in this medium is due to the boson nature of ^{4}He and not merely to the highly quantum nature of liquid helium. Furthermore, the Göttingen experiments provided experimental evidence that the nanodroplets cool to an internal temperature of 0.4 K in a microsecond time scale [17], confirming previous theoretical predictions [18]. Mass spectrometry is not essential to achieve rotational resolution. Indeed, using fiber optic-fed build up cavities or a standard multipass cell in conjunction with color center lasers, rotational resolution is routinely obtained at 1.5 and 3 µm respectively, in our laboratory and in the laboratory of Miller [19] using bolometric detection [11, 20].

Many molecules have since been studied as solutes inside helium droplets, ranging from simple triatomics like OCS to larger molecules such as perylene (see below) and pentacene [21]. As an example, Figure 2 shows the electronic fluorescence excitation spectrum of a medium-sized organic molecule inside helium nanodroplets which was acquired in our laboratory investigating the $S_1 \leftarrow S_0$ electronic transition of perylene. The average droplet size used for perylene was approximately 30 000 atoms/droplet. The 0_0^0 band of the electronic transition in the gas phase occurs at 24 070 cm^{-1} while the spectrum inside the helium nanodroplets displays a shift of the electronic origin of -47 cm^{-1}. The first point to note is the relative simplicity of the spectrum. All transitions that are assigned in Figure 2 originate from the lowest vibrational level of the ground state. This is consistent with the effective vibrational cooling that occurs once the molecule is located inside the helium droplet in the 0.4 K environment. While most of the assigned and unassigned lines in the spectrum are quite narrow due to the weak interaction between the dopant molecule and the liquid helium (FWHM ~ 0.3 cm^{-1}), some phonon broadening is also present (see below).

In addition to vibrational and electronic spectra, microwave [22], microwave-microwave [22] and microwave-infrared [23] double resonance experiments have also been performed on molecules solvated in helium droplets. In addition, the production of He$_2^+$ metastables by ionization of helium droplets has also been studied in detail [24–26]. Recently, progress in modeling the lineshapes of ions [27] and neutral species [28] solvated in the superfluid helium nanodroplets has been made by taking into account the microscopic interactions of such solvated species with the helium droplet. The use of nanodroplets to form unusual conformations of van der Waals complexes was shown by studies conducted by Nauta and Miller in which several HCN molecules were shown to take a linear arrangement in the helium droplet, the maximum length of which is related to the droplet diameter [29].

In contrast with this, when the spectroscopy of HCN oligomers is carried out in a free jet source, the most abundant species in the beam are found to be nonpolar, i.e. to display a cyclic or antiparallel row type of structure.

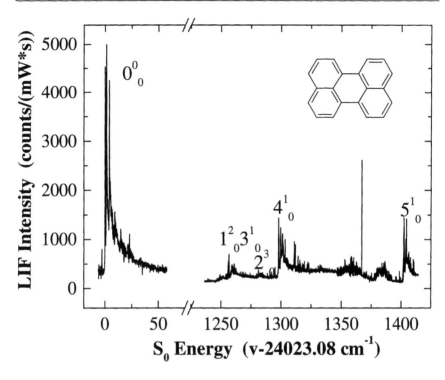

Figure 2: High resolution spectrum of the $S_1 \leftarrow S_0$ electronic transition of perylene inside helium nanodroplets. The vibronic assignments given in the figure are of the form: ν^{upper}_{lower} where ν is the number of the vibrational mode (based on the assignments in Reference [60]) and *upper* and *lower* are the number of vibrational quanta in that mode in the excited and ground electronic state respectively.

Studies of atomic dopants in and on helium nanodroplets have also been made in order to understand their interactions with the liquid helium matrix. Alkali [31, 32] and alkaline earth atoms [33, 34] have been studied both in regard to the shape of their spectral features and their location in or on the nanodroplet. Due to the weak interaction between alkali atoms and the liquid helium nanodroplet, these atoms remain on the droplet surface, while alkaline earth atoms range from magnesium which is definitely solvated, to barium which is likely to be an intermediate case between solvation and expulsion. Other d- and f-block metal atoms and clusters such as silver [35] and europium [36] as well as p-block metal atoms such as aluminum [37] become solvated, in addition to all closed shell molecules studied to date (except the alkali dimers). For spherically symmetric atoms we can state that solvation is correlated with a high value of the ionization potential (I_p).

If the I_p is above 7 eV, solvation is assured, if it is below 6 eV, expulsion is assured. Values of I_p between these two limits correspond to "intermediate" cases.

Building upon work with alkali atoms, experiments carried out in our laboratory have shown that triplet alkali (homonuclear) heteronuclear dimers can be formed using helium nanodroplets through the use of (single) multiple pick-up cells [38]. Both singlet and triplet dimers form on the surface of the helium nanodroplets. The triplet yield is favored by spin statistics and by their larger survival rate which is in turn due to their smaller recombination energy. Figure 3 displays the excitation spectrum of the (9,0) vibronic band of the $(A)1^1\Sigma^+u \leftarrow (X)1^1\Sigma^+g$ electronic transition of Na_2 formed on the surface of helium nanodroplets along with a comparison to the $S_1 \leftarrow S_0$ spectrum of glyoxal (located inside the droplets) obtained by Hartmann and co-workers [39]. The structure observed in both excitation spectra resembles that of a typical matrix-isolated species. At the origin of each band we see a sharp zero phonon line of about $0.8\,cm^{-1}$ width which corresponds to the pure electronic transition, followed by a phonon wing extending to higher frequencies. The glyoxal spectrum was successfully modeled by the Göttingen group on the basis of the dispersion curve for He II [39]. Two peaks superimposed on the phonon side band are observed in both spectra which are blue-shifted by $5.5\,cm^{-1}$ and approximately $9-10\,cm^{-1}$ with respect to the zero-phonon line. These are attributed to the coupling to the roton and maxon excitations which correspond to the extrema in the energy vs. momentum dispersion curve measured for liquid helium by means of neutron scattering [40]. Maxon excitations are those phonons with energy and momenta corresponding to the region of the maxima in the dispersion curve of He II. The presence of roton-like excitations provided the first evidence of superfluidity in helium nanodroplets. The third peak at $1.48\,cm^{-1}$ in the spectrum of singlet Na_2 remains unassigned although it occurs at the same frequency relative to the zero-phonon line as that assigned to rotational excitation of the glyoxal inside the helium droplet [39]. Due to the different rotational structure of the transitions in glyoxal and Na_2, an alternative assignment for the peak in the spectrum of Na_2 must be sought. Since the sodium dimer is located on the surface of the droplet, the peak could be due to a coupling to the surface modes (ripplons) of the helium droplet. However, a series of similar measurements carried out using helium droplets of different average size (see Figure 3B) failed to show the expected shift of this "ripplon" peak towards high frequencies for decreasing droplet size. The question remains unresolved and a definite assignment of the first peak to the right of the zero phonon line will require further theoretical and experimental work.

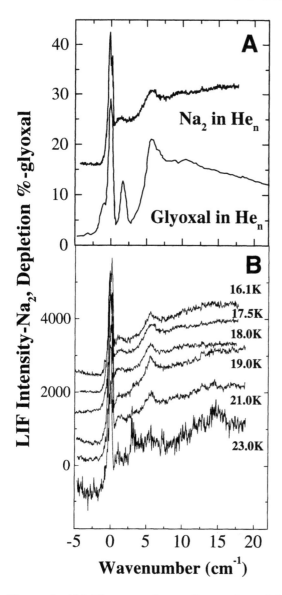

Figure 3: (A) The zero phonon line region of the $(A)1^1\Sigma^+u \leftarrow (X)1^1\Sigma^+g$ electronic transition in Na_2 on the surface of helium nanodroplets and comparison to the $S_1 \leftarrow S_0$ spectrum of glyoxal inside the droplets obtained in Reference [39]. (B) Helium nanodroplet size dependence of the zero phonon region of the $(A)1^1\Sigma^+u \leftarrow (X)1^1\Sigma^+g$ electronic transition in Na_2. The nozzle temperature was varied from 16.1 K to 23.0 K which changes the average cluster size from 15 000 atoms/droplet to 7000 atoms/droplet.

Work with multiple dopants in and on helium nanodroplets has also begun to appear. The incorporation of separate pickup cells along the flight path of the helium nanodroplets provides a convenient way to dope different atoms and molecules into the droplets. This multiple doping scheme was first accomplished by forming NaK dimers on the surface of helium nanodroplets by the separate pick up of single atoms of sodium and potassium onto a nanodroplet. By scanning across a large portion of the visible spectrum with CW dye and Ti:Sapphire lasers and controlling the vapor pressures of sodium and potassium in the pickup cells, three electronic transitions ($2^3\Sigma \leftarrow 1^3\Sigma^+$, $2^3\Pi \leftarrow 1^3\Sigma^+$, $2^1\Pi \leftarrow 1^1\Sigma^+$) can be observed in NaK. In addition to the spectra observed for NaK, the $1^3\Sigma^+g \leftarrow 1^3\Sigma^+u$ transition of Na_2 and the $1^3\Pi g \leftarrow 1^3\Sigma^+u$ transition of K_2 have been observed for the homonuclear triplet alkali dimers. A collection of atomic and dimer excitation and emission spectra that are obtained by doping the helium nanodroplets with both Na and K is displayed in Figure 4. As the surface of the helium nanodroplets can be populated with triplet alkali dimers (see above and later in this section), direct access to the spectroscopy of the triplet manifold is obtained without the usual use of multiple resonance techniques. Spectral shifts were found to be less than 5 cm^{-1} and vibrational constants were nearly unchanged from the gas phase values (in the cases where gas-phase data was available). The spectroscopic data obtained could therefore be used to construct potential energy surfaces of ground and excited states of dimers. Work along these lines is in progress in our laboratory [41].

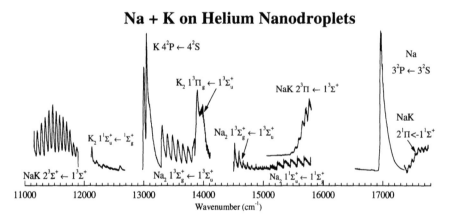

Figure 4: Excitation spectra of sodium and potassium-doped helium nanodroplets in the visible region. Each transition is labeled in the figure. The droplet size ranged from approximately 5000 – 12 000 atoms/droplet which was optimized for each experiment.

For similar reasons to those responsible for the abundant presence of triplet dimers on the nanodroplet surface, the quartet state of Na_3 and K_3 was produced for the very first time [42, 43] by the aggregation of three alkali atoms with parallel electron spins on a single helium nanodroplet. Figure 5A–C displays a total of three excitation spectra of excited quartet states in either Na_3 or K_3. A tentative assignment of the excitation spectra to trimer transitions was accomplished by measuring the pick up pressure dependence of the LIF intensity which diplayed a cubic dependence, consistent with a spectrum arising from three alkali atoms. Positive assignment of these spectra to quartet transitions of Na_3 and K_3 came through the acquisition of their emission spectra. Emission spectroscopy of the $2^4E' \to 1^4A'_2$ transition of Na_3 (see Figure 5A') revealed a very interesting non-adiabatic process where an electron spin-flip is induced upon electronic excitation of the quartet trimer. This spin-flip (which occurs on an nanosecond or subnanosecond time scale) leads to a change in bonding nature from van der Waals to covalent. After spin-flip, predissociation of the doublet trimer leads to formation of both a single sodium atom and a covalently-bound singlet dimer product. Fluorescence into the two product channels coupled with the direct emission back to the lowest quartet state produces the three different fluorescence channels seen in the spectrum (Figure 5A'). In this way, the reaction products of the dissociation could be probed in detail using the tools of spectroscopy. Jahn–Teller analysis of the excited $2^4E'$ state provided accurate information on the potential energy surface of the quartet trimer. Dispersed fluorescence spectra were also acquired for the two quartet transitions observed in K_3 (see Figures 5B' and 5C'). Due to the larger spin-orbit coupling in the potassium trimer, a larger fraction of the total fluorescence is observed in the product channels of the spin-flip process. The quartet trimers therefore serve as a unique and ideal system for the study of a simple unimolecular reaction in the cold helium droplet environment.

3 Alkali Oligomer Formation on the Surface of Liquid Helium Nanodroplets

As it is essential for the second part of this paper, let us now discuss in detail the mechanism of alkali oligomer formation on the surface of a helium nanodroplet which is based on the lack of spin relaxation in open shell atoms after they land on the droplet surface. Consider for instance a helium nanodroplet containing approximately 10^4 atoms after pick-up of two alkali atoms. Alkali dimers can exist in two states, singlet and triplet, resulting in a very simple bonding scheme. If the valence electron spins are oriented in a low-spin (antiparallel) configuration, a covalent chemical bond is formed between the two atoms. A high spin (parallel) configuration of the valence

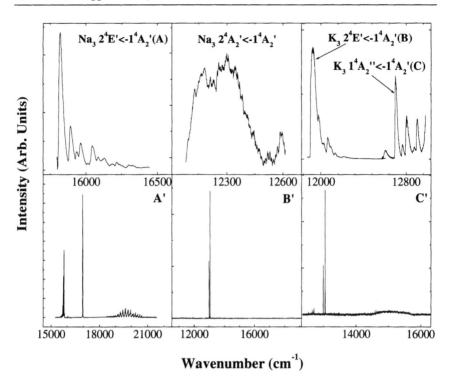

Figure 5: Excitation (top panel) and emission spectra (bottom panels) of four quartet-quartet electronic transitions observed in Na$_3$ and K$_3$ on the surface of helium nanodroplets. Letters (A–C) designate the corresponding emission spectrum for each transition labeled in the excitation spectrum. An average droplet size of 11 000 helium atoms/droplet was used in the experiments.

electrons results instead in weaker van der Waals bonding. In the case of Na$_2$, while the singlet state is bound by 5942 cm^{-1} [44], the triplet state has a dissociation energy of only 161.17 cm^{-1} [45]. In a heat pipe or a molecular beam expansion of alkali vapor, the singlet state of the alkali dimers will form in far greater abundance of the triplet state due to its greater stability. In contrast with this, it is found that when alkali dimers are formed on the surface of helium nanodroplets, triplet states are detected in greater abundance than their singlet counterparts. This can be explained by considering the details of the formation process. In our experiments, after an atom has been captured, its impact energy is dissipated by evaporation of He atoms from the droplet. Under typical conditions approximately 700 cm^{-1} of kinetic energy must be dissipated. Because of the extremely weak alkali-helium interaction the alkali atoms must reside on the droplet surface as theory can predict and experiment has verified. When the helium droplet becomes doped with at least two atoms, they will eventually meet and form a dimer. The binding energy that is released causes heating of the droplet with the subsequent evaporation of helium atoms. Evaporation will cease once the cluster temperature has re-equilibrated to 0.4 K. Assuming that the binding of a helium atom to a large droplet is 5 cm^{-1} [46], approximately 1150 He atoms need to evaporate after the formation of a singlet state, while only 35 will evaporate after a triplet state is formed. The dissipation of the greater binding energy of the singlet states may cause the desorption of the alkali molecule or the spreading of the droplet beam, or even the complete evaporation of the smaller droplets. The helium nanodroplet beam then becomes enriched on axis with droplets doped with triplet dimers over and above the simple enhancement of 3:1 which is expected from the spin statistics.

The dimer formation energy is transferred to the helium nanodroplet and is dissipated through the evaporation of individual helium atoms. A good fraction of the dimers remain adsorbed, but another small fraction recoils and leaves the droplet surface at a very low transverse velocity which is likely to correspond to a temperature a few degrees above the droplet temperature of 0.4 K. Since the forward velocity component of these desorbed dimers remains unchanged from that of the parent helium nanodroplet (typically 480 m/s as measured in our apparatus), most of the free gas-phase dimers will then travel to the LIF detector that is located downstream from the pick-up cell. Their spectra can then be probed at very high resolution since these species are free from any perturbations due to the helium droplet surface and because doppler broadening can be minimized by the collimated nature of the flow.

The discovery of this desorption process of dopants from the helium nanodroplets allows the development of a spectroscopic technique that combines the synthetic and cooling advantages of helium droplet isolation with the high resolution afforded by molecular beam laser spectroscopy. In the following, we will provide three distinct examples of this type of spectroscopy by obtaining high resolution spectra of both singlet and triplet Na$_2$ molecules desorbed from helium nanodroplets and showing what we believe are the spectra of a complex formed by a triplet Na$_2$ molecule bound to one single helium atom.

4 Singlet Na$_2$ Formed on Helium Nanodroplets

The apparatus used to obtain the spectra of alkali dimers formed using the helium droplet isolation technique has been described in detail elsewhere [38]. To generate the largest flux of singlet sodium dimers, helium nanodroplets with a mean size of approximately 11 000 atoms/droplet are produced with a stagnation pressure of 54 bar behind a 10 µm nozzle whose temperature is maintained at 17.5 K and employing a sodium vapor pressure of approximately 10^{-3} torr in the pick-up cell. The average droplet size distribution is obtained from the source conditions by applying the scaling laws found in Reference [47]. The helium droplet beam is then probed downstream of the pick-up cell using laser induced fluorescence employing a continuous wave (CW) Coherent 699-21 ring dye laser (DCM dye) as an excitation source. In the course of obtaining the low resolution (~ 0.16 cm^{-1} laser linewidth) spectrum of the $(A)1^1\Sigma^+u \leftarrow (X)1^1\Sigma^+g$ transition of Na$_2$ on the surface of helium nanodroplets which was first reported in Reference [48], narrow features were observed superimposed on the broad structure of each vibronic band of the dimer. Figure 6 displays an expanded view of the (9,0) vibronic band highlighting the narrow lines. Figure 7 displays a high resolution spectrum (~ 0.0001 cm^{-1} laser linewidth) obtained near the origin of the (9,0) band of the $(A)1^1\Sigma^+u \leftarrow (X)1^1\Sigma^+g$ transition of Na$_2$. The narrow lines turn out to be the rotational structure produced by gas phase Na$_2$ molecules which have desorbed from the helium nanodroplets. These lines are superimposed on a broad fluorescence resulting from the dimers that remain on the droplet surface and are perturbed by it. The use of photon counting techniques coupled with lock-in detection allows the separation of LIF signals from Na$_2$ molecules which have been formed and desorbed from the helium droplets and those that may be formed in the sodium vapor in the pick-up cell and subsequently diffuse into the laser interaction region. Background scans of the laser reveal that the LIF signals originating from dimers formed by gas phase collisions in the pick up cell is below the sensitivity of the detection apparatus and must therefore be very minimal.

Electronic transitions of alkali atoms and molecules possess large transition dipoles that lead to very strong absorptions. For this reason, gas phase lines of electronic transitions can be easily saturated with narrow-band laser radiation. The gas phase lines are saturated at a laser intensity of 0.5 W/cm^2 while the fluorescence of the adsorbed dimers on the helium nanodroplets continues to grow linearly beyond intensities of 7 W/cm^2. Avoidance of excessive power broadening limits the maximum laser intensity that can be used to collect the high resolution spectra to approximately 0.5 W/cm^2. A droplet size dependence of the R(9) rotational line reveals that helium nanodroplets with mean size of 11 000 atoms/droplet (produced by expansion of helium gas at 54 bar pressure through a 10 µm nozzle at 17.5 K) yields the largest quantity of gas phase sodium singlet dimers. As the mean size of the nanodroplets is increased, the probability that a dimer will spontaneously desorb may decrease due to the increased capacity for dissipation of the dimer binding energy which needs to be released upon its formation. The increased capacity for dissipation may in turn be due to the

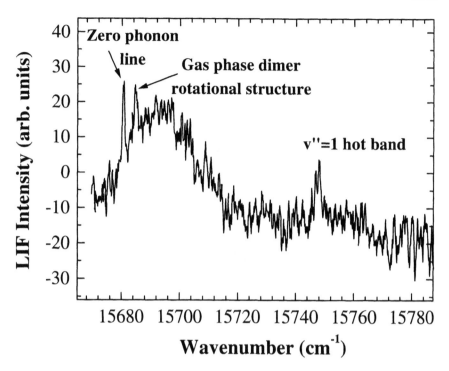

Figure 6: Low resolution spectrum of the (9,0) band of the $(A)1^1\Sigma^+u \leftarrow (X)1^1\Sigma^+g$ transition of Na_2 displaying rotational structure of the gas phase singlet dimer. A weak hot band of the singlet dimer on the helium cluster can also be seen in the spectrum.

increased density of states for the surface excitations, the energy of which decreases as the droplet size increases. While a beam of smaller helium nanodroplets ($< 10^4$ atoms/droplet) will produce a greater percentage of gas dimers compared to those that remain adsorbed on the droplet surface, the expansion conditions that are needed to produce a median size of several hundred helium atoms will not generate a sufficient number of nanodroplets to produce a satisfactory LIF signal arising from the desorbed molecules.

The proportion of dimers that desorb from the surface of the nanodroplets was measured while probing different regions of the helium expansion. This was accomplished by translating the nozzle with respect to the skimmer and monitoring the LIF intensity of the $(A)1^1\Sigma^+u \leftarrow (X)1^1\Sigma^+g$ transition of the dimers in the gas phase relative to those adsorbed on the helium nanodroplet surface. The result is shown in Figure 8. The measurement reveals that larger proportions of desorbed dimers are found in the outer wings of the expansion. From the angular collimation, it can be calculated

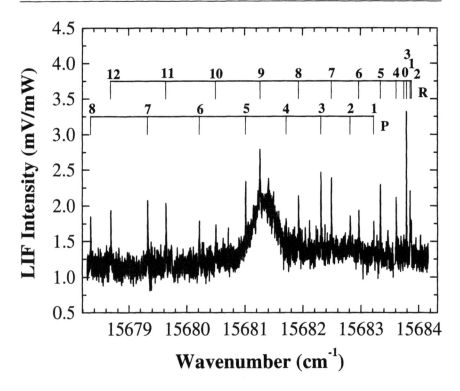

Figure 7: High resolution spectrum of the (9,0) band of the $(A)1^1\Sigma^+u \leftarrow (X)1^1\Sigma^+g$ transition of Na_2 displaying the narrow rotational structure of the gas phase dimers and the wider zero phonon line ($\sim 15\,681 - 15\,682\,cm^{-1}$) of the molecules that remain on the helium nanodroplet surface.

that the dimers are traveling at a perpendicular velocity which is less than approximately 10 m/s. The Maxwell–Boltzmann mean speed of Na_2 at 0.4 K (the temperature of the nanodroplets [17]) is calculated to be 13 m/s which is in good agreement with the number reported above.

The rotational lines of the gas phase singlet dimer can be easily assigned since there is zero net spin or orbital angular momentum in either the ground or excited state. The line positions can be reproduced using the molecular constants for both the $(A)1^1\Sigma^+u$ and $(X)1^1\Sigma^+g$ states of Na_2 given in Reference [49]. The rotational assignments are shown in Figure 7. The prominent band that occurs at $15\,681.3\,cm^{-1}$ in the spectrum of Figure 7 is a high resolution spectrum of the zero phonon line that results from the singlet dimers that remain on the helium nanodroplet surface. It is the pure electronic absorption of the dimer in which no inelastic energy transfer occurs to the internal modes of the droplet. Comparing the position of the zero phonon line with the origin of the gas phase transition (which occurs

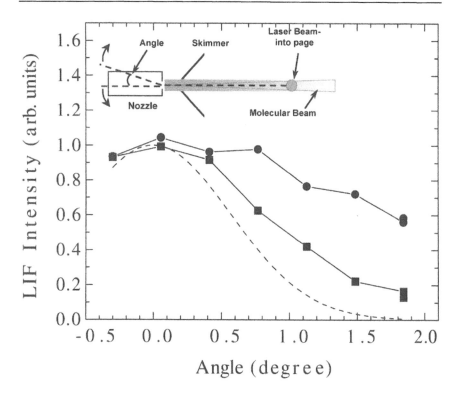

Figure 8: Angular dependence of the molecular beam of alkali-doped helium nanodroplets (filled squares) and gas-phase, desorbed dimers (filled circles) as measured at the laser crossing point in the LIF detector. The approximate profile of the laser beam at the molecular beam crossing point is shown by the dotted line. The geometry of the experiment is shown in the inset.

at $15\,683.5755\,\text{cm}^{-1}$) results in a red-shift of $2.28\,\text{cm}^{-1}$ due to the presence of the helium nanodroplet. Unfortunately, the broadening induced on the zero phonon line by the helium droplet (about $0.3\,\text{cm}^{-1}$) does not allow the resolution of any rotational structure. Only a slight splitting of the peak can be seen at $15\,681.34\,\text{cm}^{-1}$. The rotational constants of the dimer are small and hence the rotational lines are close in frequency near the band head and become masked by the broadening due to the droplet size distribution. The rotational constants are also expected to change due to the influence of the helium nanodroplet (becoming smaller) which contribute to making the spectrum unresolvable. Since the dimers are expected to be in equilibrium with the droplet temperature of $0.4\,\text{K}$, only the lowest rotational levels will be populated and the selection of $\Delta J = \pm 1$ will only allow transitions to the lowest P and R lines.

A degeneracy-corrected plot of the rotational populations could instead, determine the rotational temperature of the gas phase dimers. If the singlet dimers are formed on the surface of the cold helium nanodroplet before desorption into the gas phase, it would be expected that their rotational populations would be consistent with a droplet temperature of 0.4 K. The rotational populations of the gas phase dimers for two different mean droplet sizes (nozzle temperatures 17.5 K, 11 000 atoms/droplet, 22 K, 6000 atoms/droplet) are plotted in Figure 9. The data reveals that the rotational populations do not follow an equilibrium distribution. The slope of the plot for the first several J states is consistent with a temperature of approximately 1 K but the intensities of the higher J's display a temperature of about 20 K. These populations are independent of droplet size (within the experimental error) over the size range probed in the two spectra. Apparently, some dimers are desorbing that have cooled to the low temperature of the droplet (0.4 K) but higher J states also become populated by the transfer of angular momentum to the dimer as it leaves the droplet surface. This situation is similar to the well known phenomenon in supersonic expansions where high rotational states are cooled less effectively than rotational levels with low J. The incomplete cooling of the sodium dimers complicates the interpretation of the data not allowing the evaporated singlet molecules to act as probes of the nanodroplet temperature although they carry valuable information on the dynamics of the desorption process.

5 Triplet Na$_2$ Formed on Helium Nanodroplets

The top panel of Figure 10 displays the low resolution spectrum of the (24,0) to (29,0) bands of the $(c)1^3\Sigma^+g \leftarrow (a)1^3\Sigma^+u$ transition of Na$_2$ on the surface of the helium nanodroplets obtained with a Coherent 699-21 ring dye laser with DCM dye. The low resolution spectrum reveals many narrow gas phase lines superimposed on the broad vibronic bands. A high resolution spectrum (middle panel, Figure 10) was obtained of the structure appearing on the peak of the (29,0) vibronic band while according to our interpretation, the lower panel is the spectrum of a complex which will be described later. The rotational lines are blue degraded due to the increase of the rotational constant resulting from the contraction of the bond distance in the excited electronic state. This is the case of an electronic transition producing a bond length contraction. In covalently bound molecules, excitation to an excited electronic state is usually accompanied by an increase in the spatial extent of the valence electronic wavefunction. The triplet dimer is only bound by van der Waals forces. Viewed from a molecular orbital perspective, the lowest triplet state is non-bonding while electronic excitation places the valence

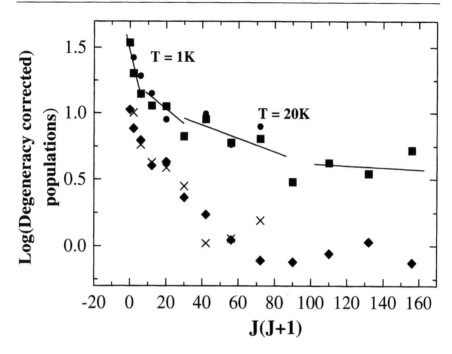

Figure 9: Degeneracy-corrected plot of the rotational populations of the singlet dimers that have desorbed from the helium nanodroplets obtained from the $(A)1^1\Sigma^+_u \leftarrow (X)1^1\Sigma^+_g$ transition of Na_2. Data obtained for nanodroplets produced with a nozzle temperature of 17.5 K are shown as filled circles for the P branch and filled squares for the R branch. The 22 K spectra are shown as filled diamonds for the R branch and crosses for the P branch.

electron in a bonding molecular orbital, resulting in a dramatic increase in binding energy and a contraction of R_e in the excited state.

Following the same procedure as outlined in the previous section, the optimal laser power for the acquisition of the spectrum was determined. The fluorescence from dimers located on the nanodroplet surface does not saturate and continues to grow linearly with laser power, while the gas phase transitions show saturation effects at about 4 mW. The effect of power broadening was measured by scanning across an individual rotational line at different laser powers. By extrapolation to zero laser power, a lifetime-limited linewidth of $\sim 20 \pm 10$ MHz was found. The value of the lifetime can be calculated using the transition dipole moment function of Konowalow [50] which yields 11 ± 5 ns. This result is in agreement with the value of 11.2 ns measured recently in our laboratory using reverse time-correlated photon counting [51].

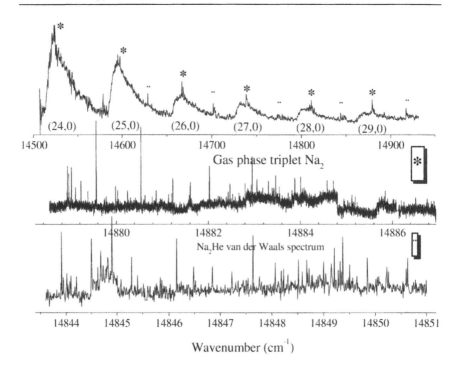

Figure 10: High resolution spectra of the narrow structure observed on the spectrum of the $(c)1^3\Sigma^+g \leftarrow (a)1^3\Sigma^+u$ transition of Na_2 on the surface of helium nanodroplets (gas phase dimer spectra-*, gas-phase Na_2He spectra-••). The low resolution spectrum of the (24,0)–(29,0) vibrational bands of the triplet sodium dimers on the helium nanodroplet is shown in the top panel.

The rotational structure of the (29,0) band was originally assigned using combination differences and the calculated rotational constants derived from the theoretical potential energy curves of the $(c)1^3\Sigma^+g$ and $(a)1^3\Sigma^+u$ states of Na_2 [52,53]. The assignment was confirmed by a contemporary, independent gas phase study of the triplet dimer produced in a molecular beam expansion of sodium vapor [54]. The technique of helium droplet isolation has some advantages over the standard methods of neat or seeded molecular beams for the production of the weakly bound triplet dimers since, as shown in the introduction, the helium nanodroplet beam actually becomes enriched with triplet over the singlet dimers while in a molecular beam, the singlet molecules will be formed in greater abundance due to their larger stability. The measurement of the $(c)1^3\Sigma^+g \leftarrow (a)1^3\Sigma^+u$ transition by Färbert et al. [54]. was only possible through the use of resonant two-photon ionization to select for the weak triplet transition since its fluorescence would be masked by the stronger singlet system present in the same frequency range. An approximate temperature of 28 K can be fit to the rotational populations of the (29,0) band (see Figure 11) which is consistent with the previous observation found for singlet Na$_2$ that desorbed dimers do not show a rotational temperature equal to the internal temperature of the nanodroplets (0.4 K). In addition, the (29,1) hot band of the (c)$1^3\Sigma^+$g ← (a)$1^3\Sigma^+$u transition was observed with a weaker intensity due to the small number of vibrationally excited dimers. The presence of this band also reveals that some vibrationally hot molecules desorb from the nanodroplet surface after formation before complete accommodation to the surface temperature is reached. Since in the lowest triplet state of Na$_2$, $\omega_e = 24\,\text{cm}^{-1}$, it would be expected that only dimers in $v'' = 0$ would be observed if the molecules would accommodate fully to the 0.4 K temperature of the droplet.

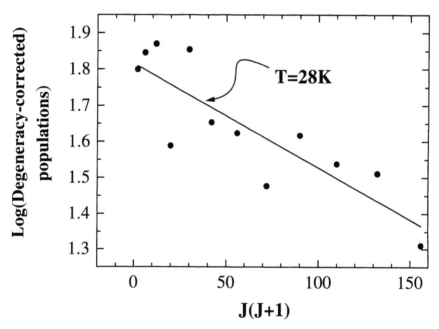

Figure 11: Degeneracy-corrected plot of the populations of the rotational levels of the (29,0) band of the (c)$1^3\Sigma^+$g ← (a)$1^3\Sigma^+$u transition. An approximate temperature of 28 K can be fit the intensity distribution.

Due to the presence of the two unpaired electron spins and the non-zero nuclear spin (I = 3/2), the high resolution rotational spectrum of the gas-phase dimer is complicated by the presence of spin-spin, spin-rotational, and hyperfine interactions that lead to multiple rotational branches and line splittings. The appearance of the fine and hyperfine structure can be detected in the high resolution spectrum of the (29,0) band, but complete assignment of the spectrum is not possible since not all line splittings are easily visible. For this reason it became desirable to record the gas-phase, high resolution spectrum of the triplet dimer of another vibrational band near the peak of the Franck–Condon distribution.

Table 1: Experimental and calculated rotational line positions of the (13,0) level of the $(c)1^3\Sigma^+g \leftarrow (a)1^3\Sigma^+u$ transition of Na_2. Line assignments are given as a function of N. The molecular constants of Färbert et al. [54] were used to calculate the line positions.

Rotational Line (N)	Calculated (cm^{-1})	Experiment (cm^{-1})
P(12)	13 641.040	
P(11)	13 640.330	13 640.22
P(10)	13 639.695	13 639.70
P(9)	13 639.135	13 639.12
P(8)	13 638.649	13 638.68
P(7)	13 638.237	13 638.23
P(6)	13 637.899	13 637.90
P(5)	13 637.636	13 637.62
P(4)	13 637.446	13 637.44
P(3)	13 637.331	13 637.32
P(2)	13 637.290	13 637.26
P(1)	13 637.323	13 637.29
R(0)	13 637.610	13 637.64
R(1)	13 637.866	13 637.86
R(2)	13 638.195	13 638.18
R(3)	13 638.598	13 638.61
R(4)	13 639.075	13 639.14
R(5)	13 639.627	13 639.62
R(6)	13 640.252	
R(7)	13 640.951	
R(8)	13 641.725	
R(9)	13 642.573	

A CW single mode Ti:Al$_2$O$_3$ laser was scanned at a 5 MHz step rate to record the high resolution, rotationally-resolved LIF spectrum of the (13,0) band of the $(c)1^3\Sigma^+g \leftarrow (a)1^3\Sigma^+u$ transition of Na_2 which is displayed in Figure 12. A 20 μm nozzle was used in this case at a temperature of 17.5 K and a stagnation pressure of 20.7 bar. The signal to noise ratio is much larger for the (13,0) band compared to the (29,0) band since the former is near the peak of the Franck–Condon distribution for the $(c)1^3\Sigma^+g \leftarrow (a)1^3\Sigma^+u$

electronic transition. The underlying structure of the spectrum is clearly resolved in this case. The rotational assignments as a function of N are shown in Figure 12. Table 1 lists the experimental line positions along with those calculated from the molecular constants of Färbert et al. [54]. (given in Table 2). The experimental line positions are only given to 0.01 cm^{-1} accuracy since the center frequency of each rotational transition was difficult to determine due to the presence of the molecular fine structure. The band head region (which occurs at the P(2) line for this vibronic band) displays the high complexity of spectrum arising from the non-zero electron and nuclear spins.

Table 2: Molecular constants of the (c)$1^3\Sigma^+$g and (a)$1^3\Sigma^+$u states of Na$_2$ from Reference [55].

State	(c)$1^3\Sigma^+$g	(a)$1^3\Sigma^+$u
D_e	4755.0	174.45
T_e	18240.5	5848.21
ω_e	100.88	24.47
$\omega_e x_e$	0.573	0.6429
B_e	0.10266	0.05352
D_{13}	2×10^{-7}	5×10^{-7}
α_e	0.00090	0.00242
γ	8.672×10^{-5}	1.404×10^{-3}
λ	1.764×10^{-2}	4.316×10^{-2}
b	7.438×10^{-3}	9.773×10^{-3}
c	3.67×10^{-4}	0

Hund's case (b) is a good description of the (c)$1^3\Sigma^+$g ← (a)$1^3\Sigma^+$u transition of the sodium triplet dimer since the electron spins do not interact strongly with the internal field directed along the internuclear axis. Because of the spin multiplicity of the triplet dimer, the spin-rotation coupling of the molecule (N • S) gives rise to a splitting of each rotational level into three components. The spin-rotation and spin-spin coupling therefore give rise to the fine structure splitting of each rotational transition. Applying the selection rules ($\Delta N = \pm 1$, $\Delta J = 0, \pm 1$), 6 rotational branches are found ($R_1, R_2, R_3, P_1, P_2, P_3$). At small values of N, the spin-spin interaction is dominant. The spin-rotation coupling increases as N increases, and becomes significant at high rotational levels. The magnitude of the spin-spin interaction constant is a factor of 200 larger than the spin-rotation parameter in the (c)$1^3\Sigma^+$g state of Na$_2$ [55].

In the (c)$1^3\Sigma^+$g ← (a)$1^3\Sigma^+$u transition of Na$_2$, the hyperfine interaction due to the non-zero nuclear spin of sodium may be of the same order as the

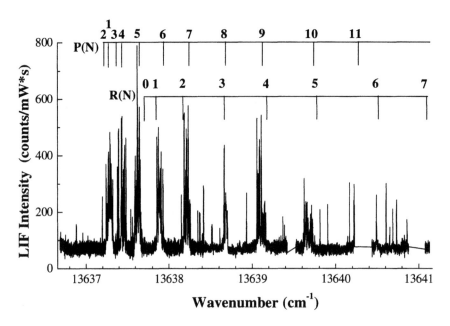

Figure 12: High resolution spectrum of the (13,0) band of the (c)$1^3\Sigma^+$g ← (a)$1^3\Sigma^+$u transition of Na$_2$ with rotational assignments given as a function of N.

fine structure and cannot be neglected when assigning the spectrum. The hyperfine contribution is due to the Fermi contact of the electrons with the nuclei and the electron spin-nuclear spin couplings [55]. Complete resolution of the hyperfine components is not possible due to the large number of possible transitions. The signal to noise level of the high resolution spectrum obtained using the helium droplet technique is at least comparable to that obtained by Färbert et al. [54] while the resolution obtainable is equivalent. An expanded view of the splittings of two rotational lines is shown in Figure 13. Both spectra are complicated by the many transitions which overlap due to lifetime broadening and make the determination of the molecular constants difficult. Figure 14 displays a comparison of the rotational structure of the (13,0) and (29,0) bands showing that the line splittings due to the fine and hyperfine interaction in the spectrum are more clearly seen in the (13,0) vibrational band. A change in the spin-spin, spin-rotation, and hyperfine constants as a function of vibrational level produces a decrease in the splitting observed in each clump of rotational lines in the (29,0) band corresponding to a given value of N.

The measurement of the rotational structure under high resolution of the (c)$1^3\Sigma^+$g ← (a)$1^3\Sigma^+$u transition of Na$_2$ clearly demonstrates the utility of

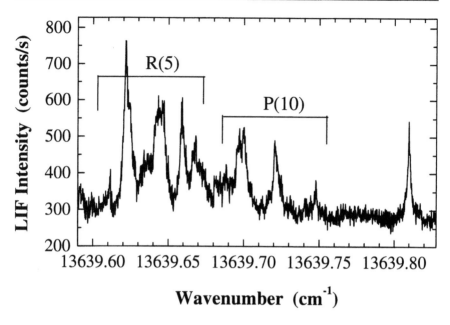

Figure 13: Expanded view of two rotational transitions showing the splittings due to the perturbations in the spectrum.

the helium nanodroplet technique for the production of a weakly bound, van der Waals molecule for which a gas-phase spectrum may be obtained. Detailed analysis of the spin-spin, spin rotation, and hyperfine interactions becomes possible once the effect of the helium nanodroplet surface is removed as the dimer desorbs into the gas phase.

6 Triplet Na$_2$/He Complexes

In the low resolution spectrum of the (c)$1^3\Sigma^+$g $\leftarrow$ (a)$1^3\Sigma^+$u transition of Na$_2$ on the surface of the helium nanodroplets (top panel of Figure 10), the presence of the gas phase rotational structure of the spectrum of the dimers that desorbed from the nanodroplets was detected as a small peak superimposed on the broad phonon side band (marked by $*$). In addition to this structure, other narrow features were seen red-shifted from the gas phase origin of each vibrational band (see bottom panel, Figure 10). These regions were scanned at high resolution and a cluster of narrow lines resulted, the assignment of which was not immediately apparent. The spacing between two successive clusters of lines matches the vibrational level spacing of the Na$_2$ (c)$1^3\Sigma^+$g $\leftarrow$ (a)$1^3\Sigma^+$u transition pointing to the triplet dimer as the species responsible for this spectrum. There are however, high resolution spectra

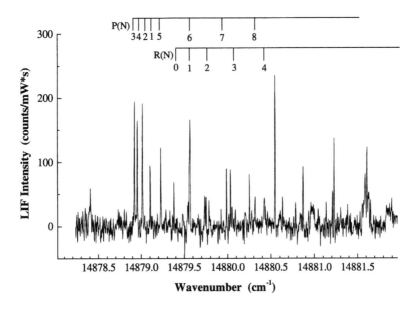

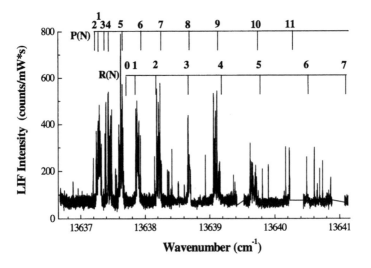

Figure 14: Comparison of the high resolution structure of the $(c)1^3\Sigma^+g \leftarrow (a)1^3\Sigma^+u$ transition of Na_2 observed on the (13,0) and (29,0) bands.

which are inconsistent with this hypothesis (marked with •• in the upper panel of Figure 10). One example is shown in the lower panel of Figure 10 and is believed to be a complex of Na_2 and He. The helium nanodroplets used to prepare the dimers were produced at a nozzle temperature of 17.5 K and a stagnation pressure of 56 bar (corresponding to a mean droplet size of 11 000 atoms/droplet [47]). The laser was scanned a total of 14 cm^{-1} and the resulting spectrum is displayed in Figure 15. A large density of lines is present without a recognizable pattern. The band is blue degraded as in the case of the $(c)1^3\Sigma^+g \leftarrow (a)1^3\Sigma^+u$ transition. A band head occurs at 14 843.8994 cm^{-1} with an identifiable Q branch structure.

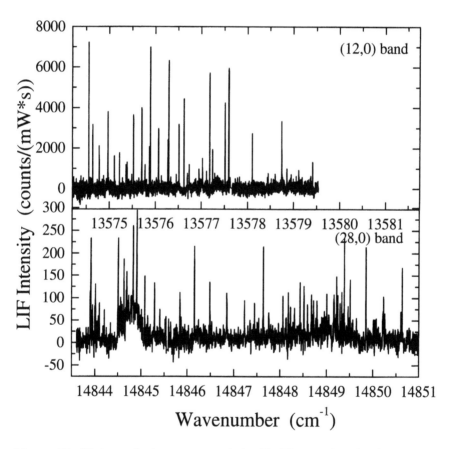

Figure 15: High resolution spectra of the Na_2He van der Waals complex formed and desorbed from the helium nanodroplet surface. Comparison of the spectra obtained on the (12,0) and (28,0) vibrational bands of the $(c)1^3\Sigma^+g \leftarrow (a)1^3\Sigma^+u$ transition of Na_2 of the bare sodium dimer.

A similar spectrum was also recorded at the peak of the (12,0) band. The spectrum is displayed in Figure 15 and shows a similar pattern and line density as that observed on the (28,0) band. The signal to noise ratio of the spectrum on the (12,0) band is improved due the larger Franck–Condon factor of the dimer vibrational transition. The narrow linewidth ($< 0.003 \text{ cm}^{-1}$) reveals that the species responsible for the spectrum is in the gas phase since the linewidth of the alkali dimers on the surface of the helium nanodroplets is approximately 0.6 cm^{-1} and the linewidth observed in the electronic spectra of molecules solvated inside is on the order of 0.03 cm^{-1}. The spectrum is, therefore, likely to be due to a complex of Na_2 and He that has desorbed from the helium droplet surface (see below). The red-shift of this spectrum from the origin of the (13,0) band of the gas phase (c)$1^3\Sigma^+g \leftarrow$ (a)$1^3\Sigma^+u$ transition was determined to be 62.6 cm^{-1} and is displayed in Figure 16. The red shift decreases to 35.0 cm^{-1} for the corresponding spectrum on the (28,0) vibrational band as shown in Figure 10. This must be a consequence of the fact that the anharmonicity of the (c)$1^3\Sigma^+g$ excited state of Na_2 may change when the dimer changes from residing on the droplet's surface to being bound to a single (or few) helium atoms in the gas phase. This is, in turn, due to relatively large changes in the three-body nonadditivity effects in the repulsive part of the potential which has been shown to be large in complexes made of atoms with low ionization potentials [56].

To arrive at a more precise spectral assignment, the vibrational levels of the ground state ($1^3\Sigma^+u$)Na_2He van der Waals complex were calculated using the TRIATOM suite of programs [57]. A potential surface for the lower state of Na_2He was constructed (neglecting three-body effects) by adding the Na-Na triplet (a)$1^3\Sigma^+u$ potential to the Na-He pair potential of Pascale [58]. The potential energy surface for a helium atom interacting with the (X)$1^1\Sigma^+g$ _singlet_ ground state of Na_2 was previoulsy calculated by Schinke et al. and revealed a minimum of approximately 0.8 cm^{-1} in the collinear and perpendicular geometries [59]. The calculation using the potential energy surface of the ($1^3\Sigma^+u$) Na_2He constructed using the pair potentials and a basis set of 900 Morse oscillators gives a D_0 of approximately 1.1 cm^{-1} with respect to dissociation to a _triplet_ dimer ($v'' = 0$) and a single helium atom. The calculation reveals the presence of a single vibrational state as any excited vibrational levels of the Na_2-He bond would result in dissociation. While the binding energy (D_0) of this complex is extremely small, it would still be expected to form on the helium nanodroplet surface due to the low (0.4 K) temperature of the nanodroplets. Hence, kT at the temperature of the helium nanodroplet is less than the binding energy of the Na_2He complex. Since the calculated Na-He two-body potential of Pascale that was used in the TRIATOM calculation may be too deep [41], the actual binding energy of the Na_2He complex may be less than 1.1 cm^{-1}.

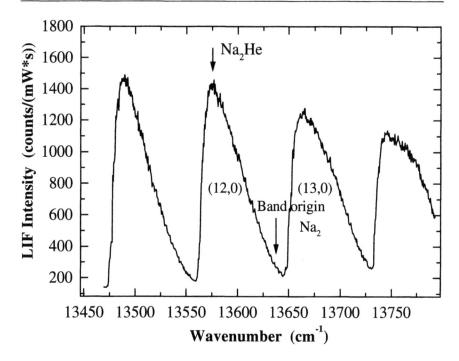

Figure 16: Location of the high resolution structure on the vibrational bands of the (c)$1^3\Sigma^+_g \leftarrow$ (a)$1^3\Sigma^+_u$ transition of Na$_2$ on the surface of helium nanodroplets. The position of the gas-phase rotational lines of the triplet sodium dimer and the red-shifted spectrum of the Na$_2$He molecule are given.

However, the possibility of the existence of a bound trimeric species is clearly demonstrated by these calculations.

The Na$_2$He complex would be extremely floppy and exhibit very large amplitude motion of the He atom with respect to the triplet dimer. For this reason, the spectrum cannot be fit with an asymmetric top model with a rigid structure. However, a calculated spectrum assuming a set of A, B, and C rotational constants in the upper and lower states and a rigid structure all given by the minimum in the ground state potential energy surface calculated above does display some similarities to the experimental spectrum. The calculated spectrum is blue degraded and a Q branch structure is seen near the band head region which are also features of the experimental spectrum. Quantitative agreement between the calculation and experiment would require accurate potential energy surfaces of both the ground and excited state. The red shifts of the spectra with respect to the vibrational band origins of the bare triplet dimer are the result of an increase in the potential energy in the excited state with respect to dissociation to a triplet Na_2 and a He atom. The binding of a helium atom to the excited $(c)1^3\Sigma^+g$ state of Na_2 must account for most of the red shift observed in the spectrum since the depth of the potential between a He atom and the lower $(a)1^3\Sigma^+u$ state triplet dimer is at most $1.1\,\text{cm}^{-1}$. Since the D_e of the $(c)1^3\Sigma^+g$ excited state of Na_2 is approximately $4800\,\text{cm}^{-1}$, a change in the excited state potential energy of only $0.7-1.3\%$ would produce the observed red shifts in the spectra. A blue shift of the $Na_2\text{He}$ spectrum with respect to the vibrational band origin of the triplet Na_2 can be ruled out since a potential leading to a blue shift would be unstable with respect to dissociation into a dimer and a helium atom. The resulting spectrum would therefore display a continuum, not the narrow line structure as observed in the experimental spectrum.

7 Conclusions

The fluorescence excitation spectra of the singlet and triplet sodium dimers on the surface of helium nanodroplets reveal the presence of narrow lines superimposed on the vibrational bands. High resolution spectroscopy shows that this structure is due to dimers that have desorbed from the helium nanodroplets at low lateral velocity and become entrained in the molecular beam. After optimization of this method for dimer preparation, the rotational spectrum of the $(A)1^1\Sigma^+u \leftarrow (X)1^1\Sigma^+g$ transition of singlet Na_2 was recorded and showed that the line positions were identical to those of previous gas-phase studies. The high resolution spectrum provided a very accurate determination of the frequency shift of the adsorbed dimers generated by the presence of the helium nanodroplet. The rotational temperatures of the gas phase dimers that desorb cannot be used as a probe of the droplet temperature since the desorption process is not an equilibrium one as shown by an analysis of the rotational population of the desorbed dimers.

A beam of gas phase triplet Na_2 can also be produced by using the helium nanodroplet as a synthetic tool. A neat or seeded supersonic expansion of sodium vapor will produce an abundance of singlet sodium dimers over the corresponding triplet species. The lack of spin-relaxation on the cold, non-perturbing surface of the helium nanodroplets can instead be exploited for the formation of triplet dimers which can be probed in the gas phase after desorption from the droplet surface. By carefully choosing the expansion conditions for producing the helium nanodroplet beam along with the appropriate vapor pressure of sodium in the pick-up cell, a ratio of $10^4:1$ for the density in the beam of triplet dimers over singlet dimers can be realized. In this way, a collimated beam of gas phase triplet dimers is produced while eliminating the interfering singlet molecules. The $(c)1^3\Sigma^+g \leftarrow (a)1^3\Sigma^+u$ transition of Na_2 yields a complex spectrum which, upon analysis, could provide the magnitude of the spin-spin, spin-rotation, and hyperfine interactions of the electrons in the molecule. The triplet dimer is also found to desorb as a van der Waals complex with one or more helium atoms attached. An analysis of the spectrum of the Na_2He van der Waals complex with a knowledge of the three-body forces present would provide insight into the complex dynamical behavior of the floppy Na_2He trimer.

The discovery of the presence of a small, but useful concentration of molecular dopants that desorb from the helium nanodroplets shows the versatility of the helium droplet isolation technique since it is now possible to use the droplets as a host to synthesize an unstable molecule, cool it to low temperature, and probe it under high resolution once it has re-entered the gas phase. The technique has been shown to work in the case of alkali atoms which reside on the droplet surface. However, if the nanodroplets used as substrates are not too large, the technique may also work with dopant atoms or molecules which reside in the droplet interior. Work to explore this possibility is in progress in our laboratory as the new technique could be become a practical tool in the investigation of the photophysics of many other species such as radicals and metallic clusters.

Acknowledgments

We gratefully acknowledge the help of Matt Radcliff and Carlo Callegari with the experiments described in this chapter.

References

[1] R. E. Smalley, B. L. Ramakrishna, D. H. Levy, and L. Wharton. *J. Chem. Phys.* **61**, 4363 (1974).

[2] M. E. Jaycox. *J. Phys. Chem. Ref. Data*, Monograph 3, Vibrational and Electronic Levels of Polyatomic Molecules. AIP (1994).

[3] E. Whittle, D. A. Dows, and G. C. Pimentel. *J. Chem. Phys.* **22**, 1943 (1954).

[4] T. Oka. *Ann. Rev. Phys. Chem.* **44**, 299 (1993).

[5] P. D. Lett, P. S. Julienne, and W. D. Phillips. *Annu. Rev. Phys. Chem.* **46**, 423 (1995).

[6] M. Mengel, D. C. Flatin, and F. C. De Lucia. *J. Chem. Phys.* **112**, 4069 (2000).

[7] J. Kim, B. Friedrich, D. P. Katz, D. Patterson, J. D. Weinstein, R. DeCarvalho, and J. M. Doyle. *Phys. Rev. Lett.* **78**, 3665 (1997).

[8] W. Becker, R. Klingelhöfer, and H. Mayer. *Z. fur. Natur.* **16a**, 1259 (1961).

[9] S. Goyal, D. L. Schutt, and G. Scoles. *Phys. Rev. Lett.* **69**, 933 (1992).

[10] M. Hartmann, R. E. Miller, J. P. Tonnies, and A. F. Vilesov. *Phys. Rev. Lett.* **75**, 1566 (1995).

[11] C. Callegari, A. Conjusteau, I. Reinhard, K. K. Lehmann, G. Scoles, and F. Dalfovo. *Phys. Rev. Lett.* **84**, 1848 (2000).

[12] J. P. Toennies and A. Vilesov. *Annu. Rev. Phys. Chem.* **49**, 1 (1998).

[13] K. B. Whaley. *Int. Rev. Phys. Chem.* **13**, 41 (1994).

[14] K. K. Lehmann and G. Scoles. *Science.* **279**, 2065 (1998).

[15] T. E. Gough, D. G. Knight, and G. Scoles. *Chem. Phys. Lett.* **97**, 155 (1983) and T. E. Gough, M. Mengel, P. A. Rowntree, and G. Scoles. *J. Chem. Phys.* **83**, 4958 (1985).

[16] S. Grebenev, J. P. Toennies, and A. F. Vilesov. *Science*, **279**, 2083 (1998).

[17] M. Hartmann, R. E. Miller, J. P. Toennies, and A. Vilesov. *Science.* **272**, 1631 (1996).

[18] D. M. Brink and S. Stringari. *Z. Phys. D* **15**, 257 (1990).

[19] K. Nauta and R. E. Miller. *Phys. Rev. Lett.* **82**, 4480 (1999).

[20] K. Nauta and R. E. Miller. *J. Chem. Phys.* **111**, 3426 (1999).

[21] M. Hartmann, A. Lindinger, J. P. Toennies, and A. F. Vilesov. *Chem. Phys.* **239**, 139 (1998)

[22] I. Reinhard, C. Callegari, A. Conjusteau, K. K. Lehmann, and G. Scoles. *Phys. Rev. Lett.* **82**, 5036 (1999).

[23] C. Callegari, I. Reinhard, K. K. Lehmann, G. Scoles, K. Nauta, and R. E. Miller, *submitted for publication.*

[24] B. E. Callicoatt, K. Förde, L. F. Jung, T. Ruchti and K. C. Janda. *J. Chem. Phys.* **109**, 10195 (1998).

[25] J. A. Northby, S. Yurgenson, and C. Kim. *J. Low Temp. Phys.* **101**, 427 (1995).

[26] T. Jiang, C. Kim, and J. A. Northby. *Phys. Rev. Lett.* **71**, 700 (1993).

[27] K. K. Lehmann and J. A. Northby. *Mol. Phys.* **97**, 639 (1999).

[28] K. K. Lehmann. *Mol. Phys.* **97**, 645 (1999).

[29] K. Nauta and R. E. Miller. *Science* **283**, 1895 (1999).

[30] K. W. Jucks and R. E. Miller. *J. Chem. Phys.* **88**, 2196 (1988).

[31] F. Stienkemeier, J. Higgins, C. Callegari, S. I. Kanorsky, W. E. Ernst, M. Gutowski, and G. Scoles. *Z. Phys. D* **38**, 253 (1996).

[32] J. Reho, C. Callegari, J. Higgins, W. E. Ernst, K. K. Lehmann, and G. Scoles. *Discuss. Farad. Soc.* **108**, 161 (1997).

[33] F. Stienkemeier, F. Meier, and H. O. Lutz. *J. Chem. Phys.* **107**, 10816 (1997).

[34] J. Reho, U. Merker, M. R. Radcliff, K. K. Lehmann, and G. Scoles. *J. Chem. Phys. In press.*

[35] A. Bartelt, J. D. Close, F. Federmann, K. Hoffmann, N. Qaas, and J. P. Toennies. *Phys. Rev. Lett.* **77**, 3525 (1996).

[36] A. Bartelt, J. D. Close, F. Federmann, K. Hoffmann, N. Qaas, and J. P. Toennies. *Z. Phys. D.* **39**, 1 (1997).

[37] J. Reho, U. Merker, M. R. Radcliff, K. K. Lehmann, and G. Scoles. *J. Chem. Phys. In press.*

[38] J. Higgins, C. Callegari, J. Reho, F. Stienkemeier, W. E. Ernst, M. Gutowski, and G. Scoles. *J. Phys. Chem. A.* **102**, 4952 (1998).

[39] M. Hartman, F. Mielke, J. P. Toennies, A. F. Vilesov, and G. Benedek. *Phys. Rev. Lett.* **76**, 4560 (1996).

[40] D. Henshaw and A. D. B. Woods. *Phys. Rev.* **121**, 1266 (1961).

[41] J. Reho, A. Ray, J. Higgins, K. K. Lehamnn, and G. Scoles, *manuscript in preparation.*

[42] J. Higgins, C. Callegari, J. Reho, F. Stienkemeier, W. E. Ernst, K. K. Lehmann, M. Gutowski, and G. Scoles. *Science.* **273**, 629 (1996).

[43] J. Higgins, W. E. Ernst, C. Callegari, J. Reho, K. K. Lehmann, and G. Scoles. *Phys. Rev. Lett.* **77**, 4532 (1996).

[44] K. M. Jones, S. Malecki, S. Bize, P. D. Lett, C. J. William, H. Richling, H. Knockel, E. Tiemann, H. Wang, P. L. Gould, and W. C. Stwalley. *Phys. Rev. A: Gen. Phys.* **54**, R1006 (1995).

[45] E.J. Friedman-Hill and R. W. Field. *J. Chem. Phys.* **96**, 2444 (1992).

[46] V. R. Pandharipande, J. G. Zabloitzky, S. C. Pieper, R. B. Wiringa, and U. Helmbrecht. *Phys. Rev. Lett.* **50**, 1676 (1983).

[47] H. Buchenau, E. L. Knuth, J. P. Toennies, and C. Winkler. *J. Chem. Phys.* **92**, 6875 (1990).

[48] F. Stienkemeier, J. Higgins, W. E. Ernst, and G. Scoles. *Z. Phys. B.* **98**, 413 (1995).

[49] M. E. Kaminsky. *J. Chem. Phys.* **66**, 4951 (1977).

[50] D. D. Konowalow, M. E. Rosenkrantz, and D. S. Hochhauser. *J. Mol. Spec.* **99**, 321 (1983).

[51] J. Reho. *Ph.D. thesis*, Princeton University (2000).

[52] S. Magnier, Ph. Milli, O. Dulieu, and F. Masnou-Seeuws. *J. Chem. Phys.* **98**, 7113 (1993).

[53] W. Meyer, *private communication*.

[54] A. Färbert, P. Kowalczyk, H. v. Busch, and W. Demtröder. *Chem. Phys. Lett.* **252**, 243 (1996)

[55] A. Färbert and W. Demtröder. *Chem. Phys. Lett.* **264**, 225 (1997).

[56] J. Higgins, T, Hollebeek, J. Reho, T.-S. Ho, K. K. Lehmann, H. Rabitz, G. Scoles, and M. Gutowski. *J. Chem. Phys. In press.*

[57] J. Tennyson, S. Miller, and C. R. LeSueur. *Comp. Phys. Comm.* **75**, 339 (1993).

[58] J. Pascale. *Technical report, Service de Physique des Atoms et des surfaces (C.E.N. Saclay)*, Gif sur Yvette-Cdex, France (1983). Also in: R. deVivie-Riedle, J.P.J. Driessen, and S.R. Leone. *J. Chem. Phys.* **98**, 2038 (1993).

[59] R. Schinke, W. Muller, W. Meyer, and P. McGuire. *J. Chem. Phys.* **74**, 3916 (1981).

[60] B. Fourmann, C. Jouvet, A. Tramer, J. M. Le Bars, and Ph. Millie. *Chem. Phys.* **92**, 25 (1985).

Spectroscopy of Single Molecules and Clusters Inside Superfluid Helium Droplets

Eugene Lugovoj and J. Peter Toennies
Max-Planck-Institut für Strömungsforschung, 37073 Göttingen, Germany
Slava Grebenev, Nikolas Pörtner, and Andrej F. Vilesov
Ruhr-Universität Bochum, Physikalische Chemie II, 44780 Bochum, Germany
Boris Sartakov
General Physics Institute, Russian Academy of Sciences, 117942 Moscow, Russia

I Introduction

Great progress has been made in recent years in exploring the manifestations of superfluidity of liquid ^{4}He and ^{3}He in confined geometries and low dimensional systems such as in porous substances or on the surfaces of solids [1, 2]. In all these experiments essentially an infinite number of atoms are involved and the experiments are sensitive to collective transport or thermodynamic properties. More local information on superfluidity could be obtained by using a weakly interacting microscopic probe such as an isolated optically active atom or molecule. Since such particles form a point defect they will interact appreciably only with a finite number of He atoms in their immediate vicinity and thus their optical spectra will be sensitive to the local microscopic properties such as the structure and the spectrum of elementary excitations. Because of the small superfluidity coherence length of atomic dimensions effects of superfluidity under these circumstances can also be expected. This is confirmed by extensive theoretical studies of small numbers of about 60 ^{4}He atoms which have been found to be superfluid [3, 4, 5, 6]. Until some years ago the optical spectroscopy of foreign chromophore particles was not feasible due to the extremely low equilibrium solubility of single unassociated atomic species inside liquid helium. Recently new methods have been developed for laser ablation of metals to produce single atoms inside the bulk liquid [7, 8, 9]. However since the metal atoms have open outer electron shells they interact strongly with the helium environment and the spectral features are generally found to be broad. This same technique has so far not been successful when applied to closed shell molecules.
To overcome these problems new molecular beam techniques have recently

been developed to attach single or defined numbers of chromophore particles to He droplets consisting between 10^3 and 10^5 atoms [7]. Whereas "heliophobic" open shell metal atoms (e.g. alkali atoms) remain on the surface [10, 11] closed shell molecules have large negative chemical potentials and reside in the interior near the center of the droplets [12, 6]. As discussed in this review these "heliophilic" molecules inside helium droplets exhibit sharp spectral features. Because of the coupling to at least some of the surrounding He atoms the overall object studied has sizes intermediate between the single free molecules and the macroscopic systems characteristic of low temperature physics thereby providing interesting new perspectives for both of these two mature fields.

The present review is more extensive than one published recently [13]. It is less comprehensive but more up to date than the 1998 review written by two of the present authors [7].

II Apparatus

A schematic diagram of the experimental apparatus used for absorption spectroscopy of molecules immersed in He droplets is shown in Fig. 1 (see Refs. [7, 14, 15]). Both ^{4}He, ^{3}He, and mixed droplet beams are formed by expansion of the corresponding He gas through a small nozzle of 5 μm diameter at source temperatures of $T_0 = 5 - 25$ K and stagnation source pressures of $P_0 = 5 - 45$ bar. The expanding gas in the jet is cooled by the adiabatic isentropic change in state to ambient temperatures and pressures well below the critical point, where extensive condensation to small droplets begins to occur. In the further course of the expansion, at distances of several mm from the source, collisions among the expanding atoms gradually cease and the He droplets now moving in vacuum are rapidly cooled further by evaporation until they reach terminal temperatures of $T = 0.38$ K for pure ^{4}He droplets [14, 16, 17] or $T = 0.15$ K for pure ^{3}He and mixed droplets [16, 18]. By varying the values of T_0 and P_0 aggregates of different average number sizes in the range of $\bar{N} = 2 - 10^8$ atoms can be obtained [19, 20]. Deflection scattering experiments have been instrumental in establishing the average numbers of atoms in the droplets and the distribution in size which for $\bar{N} \leq 10^5$ follows a log-normal distribution, with a half-width ΔN which is approximately given by $\Delta N = 0.8 \cdot \bar{N}$ [19].

By passing the beam of droplets through a scattering chamber filled with the gas to be studied the droplets readily adsorb the foreign molecules with capture cross sections close to the geometrical size of the droplets [19, 20, 21, 22]. Upon increase of the pick-up gas pressure the number of embedded atoms or molecules increases and these have been found to coagulate to form complexes [15, 19, 20, 21, 22, 23] with number sizes which follow a Poisson distribution [15, 20, 22]. The kinetic energy of the impacting particles, their

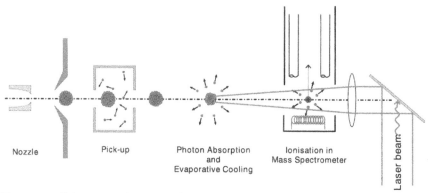

Figure 1: Schematic diagram of a molecular beam apparatus for depletion spectroscopy of molecules embedded in He droplets. Droplets are produced via the nozzle beam expansion of He gas. The pick up of foreign molecules takes place in the pick-up cell, and the droplets are ultimately detected via a quadrupole mass spectrometer with an electron impact ionizer. The output of a tuneable laser is directed antiparallel to the droplet beam.

internal energies, and their binding energy to the droplet are all transferred to the droplet and lead to the subsequent evaporation of several hundreds of atoms [19].

To obtain excitation spectra of the embedded molecules, in the apparatus used in Göttingen shown in Fig. 1, the beam of a tunable diode laser or a dye laser is directed antiparallel and coaxially to the droplet beam. The absorption of a laser photon results in vibrational or electronic excitation of the molecule, which is followed by energy transfer to the host droplet and the subsequent evaporation of about 10^3 atoms. This decrease of the droplet size leads to a "depletion" of the signal of the mass spectrometer, which is placed about one meter from the source. In the visible and UV spectral region laser induced fluorescence detection has been found to be more sensitive especially for very large droplets for which the relative depletion approaches zero. Throughout, the photon energies are given in wavenumber units (cm^{-1}), where 1 cm^{-1}=1.45 K.

III Molecular Spectroscopy in He Droplets

I Electronic Excitation of Molecules in He Droplets

In general on electronic excitation of a molecule the density distribution of its outer electrons will be perturbed. Since liquid helium is strongly repelled by electrons even small changes in the electron density will excite the droplet's volume compressional waves. Related inelastic effects have been observed in the depletion spectra corresponding to the $S_1 \leftarrow S_0$ electronic transition of many molecules. Fig. 2 shows the results for the glyoxal molecule ($C_2H_2O_2$, see insert) in pure ^{4}He (a) and ^{3}He (b) droplets. For both He isotopes the spectrum exhibits a sharp zero phonon line (ZPL) which is shifted to the red by about $\Delta\nu = -31.33$ cm^{-1} and $\Delta\nu = -26.5$ cm^{-1} in ^{4}He and ^{3}He, respectively, compared to the free molecule. The ZPL, which is well known from matrix spectroscopy, corresponds to the pure electronic excitation of the molecule without involvement of the elementary excitations of the droplet. The ZPL in ^{4}He droplets is accompanied by a new feature at higher frequencies separated by a "gap" of about 5.5 cm^{-1} (7.9 K) from the ZPL. Similar features are also known from the spectroscopy of molecules in solid cryomatrices, where they are called a phonon wing (PW). Here they are attributed to the simultaneous excitation of compressional volume vibrations of the entire He droplet, resulting from the perturbations of the medium by small transition-induced changes in the outer electron density of the molecule. Due to the finite size of the droplet the frequency of the lowest compressional mode of a ^{4}He droplet can be estimated to be about $\hbar\omega = 1.2$ K [24] for a droplet with $N_4 = 10^4$ atoms, and thus this internal mode is not appreciably populated at the equilibrium temperature of $T = 0.38$ K. For this reason the interior of the droplets is devoid of excitations thereby greatly simplifying the spectra. A theory [25] based on the Huang-Rhys model for the phonon wings of impurities in solids predicts that under these circumstances the phonon wing intensity is proportional to the density of states of the accessible elementary excitations. The thin line in Fig. 2 (a) shows that the PW could be nicely fitted with the well known dispersion curve of the elementary excitations in bulk superfluid ^{4}He [25]. The gap between the ZPL and the maximum in the PW spectrum is associated with the well known sharp roton energy and the difference between the observed roton energy of 7.9 K as compared to 8.65 K for bulk liquid helium is attributed to the compression of the helium next to the molecule [4, 5, 6, 25]. The observation of a well-defined gap with a maximum at the energy of the roton can only be explained if the roton dispersion curve is also similarly sharp as found in superfluid He II at temperatures well below the temperature of the λ-transition. The observation of the distinct phonon wing provided the first experimental evidence that ^{4}He droplets are superfluid.

This assignment of the PW is nicely confirmed by the spectrum of glyoxal in pure ^{3}He droplets. Since ^{3}He becomes a superfluid at $T_\lambda \leq 3 \cdot 10^{-3}$K it expected to be a normal fluid at the ambient ^{3}He droplet temperatures of $T = 0.15$ K. Thus it is not surprising that the gap in the PW is no longer observed inside ^{3}He droplets (see Fig. 2 (b)). The low energy elementary excitations in ^{3}He are dominated by a broad additional particle-hole excitation branch with features similar to a normal fluid and for this reason the gap seen in ^{4}He is "filled in" by a continuum of low energy excitations.

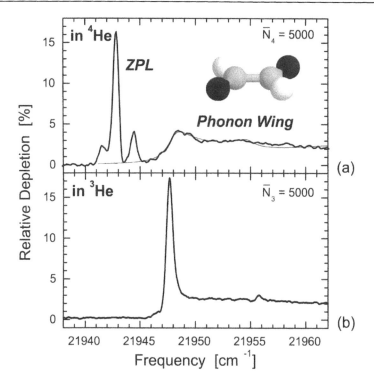

Figure 2: The spectrum of the band origin of the $S_1 \leftarrow S_0$ transition of glyoxal ($C_2H_2O_2$, see insert) embedded in (a) ^{4}He and (b) ^{3}He droplets. The thin line in (a) is the result of the fit from Ref. [25].

Since the PW gap provides a spectroscopic "fingerprint" of superfluidity inside ^{4}He droplets it can be used to study the onset of superfluidity in small ^{4}He clusters by adding well defined numbers N_4 of ^{4}He atoms to pure ^{3}He droplets doped with a single glyoxal at its center. Because of their smaller zero point energy, the ^{4}He atoms attach themselves to the glyoxal molecule thereby displacing the ^{3}He atoms and eventually forming shells of ^{4}He atoms around the probe molecule. The spectra reveal that the distinct gap between the ZPL and PW reappears for $N_4 > 100$ [26] indicating that this number of ^{4}He atoms is required for superfluidity, at least according to this experimental criterium.

Since 1996 when these first spectroscopic studies in the visible were reported the $S_1 \leftarrow S_2$ transitions have been investigated in over 14 different large, mostly organic molecules, listed in Table I, in our institute in Göttingen. In most of the systems such as napthalene, pentacene, tryptophan, tyrosine, porphin, pthalocyanine and C_{60} sharp singular ZPL transitions have been found.

Table I: **Organic molecules in ^{4}He droplets:** Frequency shifts and zero phonon line splitting of origins of $S_1 \leftarrow S_0$ transitions.

Molecule	Frequency in ^{4}He [cm^{-1}]	Frequency shift in ^{4}He [cm^{-1}]	Number of ZPL's	Ref.
Glyoxal	21942.84	-31.33	1	[25]
Naphtalene	32035	$+15$	1	[27]
Tetracene	22293.4	-103.3	2	[28]
Pentacene	18545.0	-104.0	1	[28]
Porphin	16310	-10	1	[29]
Phthalocyanine	15088.9	-46	1	[29]
Indole	35281.5	$+44.7$	3	[30]
3-Methylindole	34943.1	$+57.0$	3	[30]
Tryptophan	34960.8	$+55.0$	1	[31]
NATA	35014	$+59$	4	[30]
Tyrosine	35538.0	$+47.8$	1	[31]
Tryptamine	34959.6	$+44.1$	3	[30]
C$_{60}$	15666	$-14\ldots+14$	1	[32]
BaO (A, v'=0 $\to$ X, v")	16852.7	$+45.3$	1	[33]

A particularly interesting demonstration of the great potential of He droplets as an advantageous cryomatrix is provided by the spectrum measured with the amino acid tryptophan [31]. Tryptophan is an important spectroscopic probe of proteins and peptides and has been studied spectroscopically by cooling the molecule in seeded beams ($T \approx 10$ K) as shown in Fig. 3 (a) [34]. There the spectrum is characterized by a large number of lines, which have been assigned to the many different possible structural conformers of this floppy large molecule [34]. The same molecule deposited in a ^{4}He droplet shows much sharper lines, presumably due to the further reduction in the rotational band contour because of the much lower temperatures [31]. As for the free molecule the spectrum is dominated by a series of ZPL's attributed to different conformers of these floppy molecules. Interestingly the series of lines attributed to the A-conformer in the seeded beam spectrum is not found in the droplet spectrum. This suggests that in He droplets the spectra of large biological molecules may be even simpler than in the gas phase. These spectral studies also demonstrate another advantage, that only low vapour pressures of about 10^{-5} mbar are required of the substances to be studied. For this reason these sensitive biological molecules need only be heated to temperatures well below their decomposition temperatures. The advantage of low vapour pressures has also facilitated the spectroscopic study of metal atoms and small metal clusters [10, 35, 36, 37, 38] as well as of a chemiluminescent chemical reaction inside He droplets [33] (see Section IV).

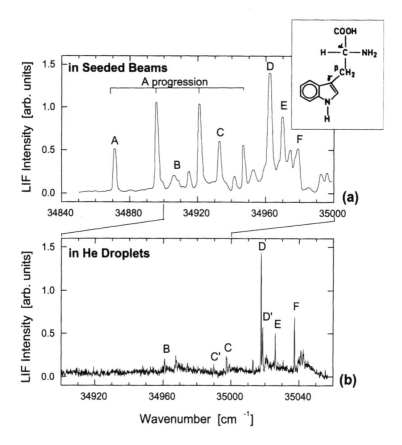

Figure 3: Comparison of (a) the resonant two photon ionization spectrum measured for the free tryptophan molecule in a seeded beam [34] with (b) the laser induced fluorescence spectrum of tryptophan in ^{4}He droplets [31]. The inset in (a) shows the structure of the tryptophan molecule.

In the case of some of the polycyclic aromatic hydrocarbons the electronic spectra also provide evidence for the distortion of the helium structure next to the molecules. Here there are close analogies to the corresponding studies of helium films on the graphite surface. However in the hydrocarbons the interaction potential with He can in principle be obtained with high precision. Indeed peculiarities have been observed in the spectra of planar molecules such as tetracene [28] and indoles [30] in which the ZPL appears as multiplets split by 1-2 cm^{-1}, see Table I. For tetracene and indole, which have no isomers, this can only be attributed to some nonequivalent configurations of the He environment around the molecule. Thus the molecular spectroscopy in helium droplets may also provide a unique opportunity to

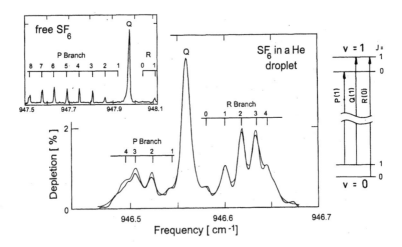

Figure 4: Infrared depletion spectra of SF_6 [16, 17] in ^{4}He droplets. The thin line is the best fit based on the Hamiltonian for a free molecule. The inset in the left hand corner shows for comparison the spectrum of free SF_6 molecules cooled in an Ar seeded beam expansion. The diagram on the right illustrates the transitions involved.

study the local states of a many particle quantum system having several nearly equivalent configurations. Several possible explanations have been proposed [39].

The stronger interaction of He atoms with larger molecules as porphin and phthalocyanine manifests itself in the modified shape of the phonon wing. Porphin shows new sharp features at $\Delta \nu = 1.8$ cm^{-1} and 2.5 cm^{-1}, whereas phthalocyanine has features at $\Delta \nu = 3.7$ cm^{-1}, 4.3 cm^{-1} and 5.1 cm^{-1} on the blue side of the ZPL. These new features have so far not been unambiguously assigned. Most probably they originate from the excitation of the vibrations of He atoms localized near the molecular impurity. Analogous low frequency modes have also been observed in neutron spectroscopy of thin liquid helium films on graphite [40].

II Molecular Vibrations and Rotations

Infra-red vibrational excitation of molecular chromophores has the advantage that because of the very small vibrational amplitudes (< 0.10 Å) the interaction potential with the surrounding helium liquid is not substantially affected. Also, since the characteristic rotational energy is of the order of the energy of the elementary excitations in liquid helium it might be possible to probe directly the elementary excitations spectroscopically.

Up to now vibrational transitions of more than 13 different heliophilic polyatomic molecules have been studied [7]. In many cases the frequencies of the vibrational fundamental are shifted by less than 1 cm^{-1}. This shift is a sensitive function of the He density profile near the molecule which is expected to be considerably distorted [6, 4, 5] and therefore depends on the interaction potential including the coupling to the internal degrees of the molecule in both the ground and excited states. Since these couplings are generally not well known it has not been possible to calculate line shifts with high accuracy up to the present [41].

The first infrared spectroscopy experiments using high resolution continuously tunable laser diodes revealed the clear resolution of rotational structure in the vibrational bands of SF$_6$ (10 μm) [14, 15, 16, 17, 42] and OCS (5 μm) [43, 44]. This indicates that the molecules rotate freely inside the droplets. Figs. 4 and 5 (a) show high resolution ($\delta\nu < 0.002$ cm^{-1}) spectra of SF$_6$ [16] and OCS [43], respectively, in ^{4}He droplets. The spectra exhibit the P, Q and R branches corresponding to $J-1 \leftarrow J$, $J \leftarrow J$ and $J+1 \leftarrow J$ transitions, respectively, where J is the ground state rotational quantum number. The rotational structure was found to be independent of the droplet size for $N_4 > 3 \cdot 10^3$ and therefore the spectra are expected to be the same for even larger He droplets and also in the bulk. The rotational structures of the SF$_6$ spectrum have been accurately fitted by the well known Hamiltonian of the free spherical top molecule, as illustrated by the thin line in Fig. 4. The lack of a splitting of the threefold degenerate ν_3-vibration of SF$_6$ reveals that the liquid helium environment does not influence the symmetry of the vibrations [14].

The relative intensities of the rotational lines in the spectra of Figs. 4 and 5 (a) provide a direct measure of the droplet temperature. Assuming a Boltzmann distribution among the rotational degrees of freedom of the molecule, the corresponding temperature was found to be $T_{\rm rot} = 0.38 \pm 0.01$ K [14, 16, 17] and $T_{\rm rot} = 0.15 \pm 0.01$ K [16, 18] in ^{4}He and ^{3}He droplets, respectively, in good agreement with theoretical estimates of the droplet temperature [24, 45]. The rotational temperature appears to be a characteristic of the droplet itself since the same temperatures, within the experimental errors of about 0.01 K, have been deduced from the spectra of SF$_6$ [16], OCS [43, 44], and propyne and several other molecules [46, 47, 48]. Although the very low temperature of the droplets is advantageous for most spectroscopies, it should be noted that it is difficult to increase the temperature significantly due to the very fast evaporative cooling which essentially fixes the temperatures at the above value [17].

The linear OCS molecule provides an especially advantageous spectroscopic probe. It has a large transition dipole moment and is sufficiently heavy so that several rotational levels are appreciably populated. Moreover being linear, its rotational energy levels are simply given by

$$E_{rot} = B \cdot J(J+1) - D \cdot (J(J+1))^2,$$

Table II: Measured spectroscopic constants (in units of cm^{-1}) of SF$_6$ and OCS in ^{4}He droplets.

Spectr. const.	Free SF$_6$	SF$_6$ in ^{4}He$_N$ [16, 17]	Free OCS	OCS in ^{4}He$_N$ [44]
ν_0	947.9763	946.5633	2062.2012	2061.644
B	0.0909	0.0326	0.202	0.074
D	$2.4 \cdot 10^{-9}$	$0.4 \cdot 10^{-4}$	$0.44 \cdot 10^{-7}$	$0.38 \cdot 10^{-3}$

where B is the effective rotational constant and D is the centrifugal distortion constant. In the case of the SF$_6$ molecule the calculation of the energy levels and line intensities is more complicated due to the additional rotational-vibrational interaction (Coriolis coupling) and the effect of nuclear spin statistics, which are discussed in Refs. [16, 17]. The sharp well resolved series of rotational lines seen in the spectra of OCS shown in Fig. 5 (a) confirms these expectations. As for SF$_6$ the spectral positions and line intensities could be very well fitted by the same Hamiltonian as for a free molecule again implying that their symmetry properties are not effected in any way by the helium environment on closer examination.

From the best fit the values of the vibrational frequency ν_0, B and D and some higher order spectroscopic constants [16, 17] could be determined. The B and D constants of SF$_6$ and OCS in the He droplets are compared with those obtained for the corresponding free molecules in Table II. For both molecules the rotational constants B are decreased by nearly a factor of three relative to their gas phase values. Since the rotational constant B is inversely proportional to the moment of inertia I ($B = h/(8\pi^2 cI)$), the observed decrease of B indicates that the effective moment of inertia of these molecules I_{eff} is about three times larger than for the free molecule. As summarized in Table III the spectral lines for OCS show a line width which increases with J and line shapes with tails which extent to higher frequencies for the P-branch and to lower frequencies for the R-branch [44]. Similar large values have been obtained for other heavy anisotropic molecules [46, 48] whereas for the light, rotationally more energetic and spherical molecules HF, H$_2$O [49, 50] and NH$_3$ [51, 52] the moments of inertia are nearly the same as for the free molecule. Most of the results available at the present time are listed in Table III in order of decreasing rotational energy constant B. For molecules with rotational constants greater than about 9 cm^{-1} which is somewhat greater than the roton energy the decrease in the rotational energy constant is insignificant. This has been attributed to the inability of atoms in the first solvation shell to follow the rapid rotations [53].

At present the large increase of the moment of inertia I_{eff} in the heavy molecules with small B values is not fully understood. Apparently the effect is closely related to the anisotropy of the van der Waals potentials.

Table III: Decrease in the rotational energy constants in He droplets (in order of decreasing of rotational constants for free molecules B_0, i.e. increasing moment of inertia $I = h/(8\pi^2 c\, B)$).

Molecule	B_0 [cm^{-1}]	B_0/B_{He}	Ref.
H$_2$O	27.8, 14.5, 9.3	1.0	[49]
HF	20.9	1.0	[50]
NH$_3$	9.94	1.3	[49]
(CHO)$_2$ (glyoxal)	1.8, 0.16, 0.14	2.7	[26]
HCCCN	1.53	3.0	[48]
HCN	1.47	1.23	[46]
HCCH	1.19	1.13	[48]
CH$_3$CCH (propyne)	0.28	3.9	[48]
OCS	0.20	2.7	[44]
CF$_3$CCH	0.105	2.75	[54]
SF$_6$	0.091	2.8	[14]
(CH$_3$)$_3$CCCH	0.09	3.0	[54]
(CH$_3$)$_3$SiC≡CH (TMSA)	0.0655	4.5	[54]

Recent calculations [55, 53] and measurements in the small ^{4}He clusters inside large ^{3}He droplets [43] indicate that this is a short range effect which is mainly governed by the first He solvation shell around the molecule. When this phenomenon was first observed for SF$_6$ in Ref. [14], the increase was explained by 8 rigidly attached He atoms located at the 8 faces of the octahedron which were determined to be the sites of the global potential minima according to the He-SF$_6$ van der Waals potential determined from molecular beam scattering experiments [56]. In the case of OCS the increase may be modelled by a donut ring of 6 He atoms around the OCS molecule at the waist [44], where the van der Waals potential is strongest [57]. Surprisingly these numbers are considerably fewer than the 22 and 17 He atoms predicted to be in the first solvation shell of SF$_6$ [6] and OCS, respectively. Recently several alternative theories have been proposed to explain the increase in the moments of inertia. One of these is based on a hydrodynamic model in which the backflow of the superfluid induced by the anisotropy of the molecule is shown to be able to explain the effect for linear molecules [48]. The only theory available at the present time which is able to simulate superfluid effects at finite temperatures is based on Feynman's path integral idea. This theory has recently been applied to SF$_6$ inside ^{4}He droplets [55, 58]. The results reveal that the long permutation exchange paths characteristic of the superfluid state penetrate even into the highly compressed layers nearest the impurity molecule despite the rather strong molecular interaction. On the other hand the superfluid fraction near the molecule is somewhat depleted [58, 55] and a substantial nonsuperfluid fraction is expected in the vicinity of the molecule which might well be coupled to the molecular rotations via the anisotropy of the van der Waals potential. Ref. [55] presents the first calculations of the anisotropic nonsuperfluid He density around the SF_6 molecule. Thus the increment of the I_{eff} of the molecule in He provides information on the nonsuperfluid part near the molecule. These results resemble closely the measurements of the build-up of superfluidity in thin He films on a solid substrate, where it has been shown that even for $T \to 0$ K the superfluid fraction in the film is considerably suppressed at the interface [59, 60, 61]. In contrast to solid surfaces molecules have the big advantage that their "surfaces" are well defined and "clean" and that in many cases the interaction potential with He is very well known. The spectroscopic interrogation of molecules makes it possible to study the effect of superfluidity for a system having a small number of He atoms as in the first solvation shell.

Recently it has been possible to gain deeper insight into the structure and dynamics of the first shell ^{4}He atoms by substituting these atoms one-by-one by p-H_2, HD, or o-D_2 molecules and analysing the changes in the highly resolved rotational spectra [62].

III Spectra in Superfluid ^{4}He Droplets versus Nonsuperfluid ^{3}He Droplets

The nearly free rotations of heavy molecules observed in liquid helium droplets has not been observed in other liquid matrices. Initially it was not clear whether the sharp rotational lines might merely be a consequence of the extremely weak van der Waals interactions of helium atoms with molecules. Since the ^{4}He droplets are known to be in a superfluid state the free rotations could also be related to the phenomenon of superfluidity. Unfortunately it is not feasible to heat the droplets to raise the temperature sufficiently to transform the droplets into the normal state since they maintain a constant temperature determined by fast evaporative cooling. Instead a similar effect was achieved by studying the IR spectra of OCS inside nonsuperfluid ^{3}He droplets.

At these low temperatures the viscosity of ^{3}He is known to increase sharply in a manner which is in fact quite similar to what is known about the normal component of liquid ^{4}He. On the other hand the interactions with the molecules is expected to be even weaker than in the case of ^{4}He because of the greater zero point energy and lower temperatures of $T = 0.15$ K (^{3}He) compared to $T = 0.38$ K (^{4}He).

Fig. 5 shows the comparison of the OCS spectra measured in (a) pure ^{4}He and (b) pure ^{3}He droplets. Whereas the OCS spectrum in pure ^{4}He droplets (a) shows a well resolved rotational structure [43] in the pure ^{3}He droplets has only a rather broad ($\delta\nu \approx 0.1$ cm^{-1}) central peak with a hint of a shoulder on the low frequency side is found. This spectrum, which is typi-

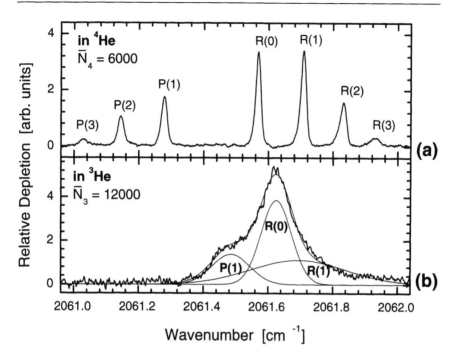

Figure 5: The OCS IR spectrum (a) in pure ^{4}He droplets with $\bar{N}_4 = 6000$ atoms, (b) in pure ^{3}He droplets with $\bar{N}_3 = 12000$ atoms. The thin line in (b) shows a simulation in which the rotational lines were fitted with Gaussians having $\delta\nu = 0.14(2)$ cm^{-1} for the P(1), $\delta\nu = 0.12(2)$ cm^{-1} for the R(0), and $\delta\nu = 0.35(5)$ cm^{-1} for the R(1) lines, respectively [63].

cal for a normal liquid, is interpreted as being due to a partial "collapse" of the rotational structure [63] due to persistent impulsive collisions with the surrounding medium. Thus the appearance of the sharp rotational spectra, although with greatly increased moments of inertia, may be regarded as a microscopic manifestation of superfluidity [43]. At present it is not completely clear how these new superfluid phenomena can be understood in terms of the well-known macroscopic manifestations of superfluidity such as the fountain effect, creep or a very high heat conductivity. For example, it has recently been suggested by Babichenko and Kagan that the free rotations reflect differences in the mean free paths of the elementary excitations in a Boson and a Fermi liquid which determine their inelastic scattering from the microscopic chromophore molecules [64].

IV How Many Atoms are Needed to Make a Superfluid?

The phenomenon of free rotations can also be used to probe for the onset of superfluidity within a small ^{4}He cluster inside a large "inert" ^{3}He droplet using the same approach as described for glyoxal. After embedding a single OCS molecule in a large ^{3}He droplet ^{4}He atoms were added one-by-one [43] replacing the ^{3}He atoms next to the OCS molecule and thereby surrounding and coating the molecule within the large ^{3}He droplet. Fig. 6 shows a series of infra-red spectra measured with increasing numbers of ^{4}He atoms.

With 60 added ^{4}He atoms narrow rotational lines again emerge. But instead of the seven lines seen in pure ^{4}He droplets (Fig. 6 a) only three lines are found (Fig. 6 f-h). This can be explained by the lower temperature of 0.15 K of the outer ^{3}He layer. The appearance of a sharp rotational spectrum is interpreted as indicating the reappearance of superfluidity. These experiments demonstrate that superfluidity sets in gradually, and is almost complete, with the addition of about 60 atoms, which corresponds to about two shells of ^{4}He atoms. This number is smaller than found by observing the PW in glyoxal as described in Section I. Since density functional calculations [65] reveal that the second shell is heavily infiltrated by ^{3}He atoms to about 30%, which may be enough to destroy its superfluidity, only a single shell of ^{4}He atoms (17-20 atoms) may in fact be responsible for the observed effect.

The OCS spectra in the mixed ^{4}He/^{3}He droplets reveal another interesting phenomenon. The moment of inertia of the OCS in pure ^{3}He droplets was found to be a factor of 4.3 larger than for free OCS [63]. Upon the addition of about 300 ^{4}He atoms to the doped ^{3}He droplet the increase in the moment of inertia dropped off gradually to the value of 2.7 found for pure ^{4}He [63]. This gradual change in the moment of inertia is attributed to the presence of a few ^{3}He atoms which penetrate the inner ^{4}He shell and come close to the molecule where they reduce the extent of particle exchange with the surrounding superfluid layers. In many respects these experiments may be treated as a microscopic analogy to Andronikashvili's torsional oscillator measurements in bulk liquid helium which already in 1946 provided clear evidence for the two fluid model of superfluid helium.

V Formation and Study of p-H$_2$ and o-D$_2$ Clusters Inside Helium Droplets

Sequential pick up of several identical or different molecules by He droplets offers a unique opportunity to form and study taylor made molecular clusters spectroscopically at the extremely low temperatures provided by the He droplets. One motivation was to replace one or more of the He atoms in the immediate vicinity of OCS molecule by distinguishable H$_2$ molecules

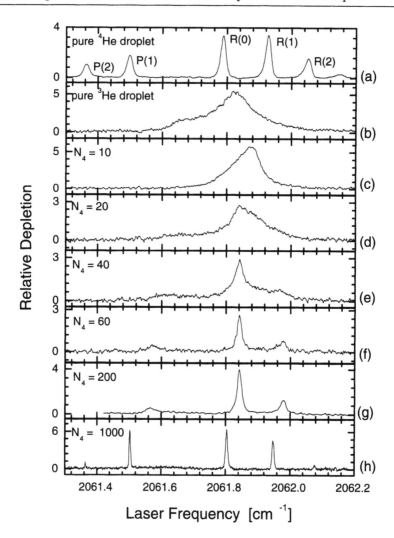

Figure 6: A series of OCS IR spectra in (a) pure ^{4}He droplets, (b) pure ^{3}He droplets, and (c)-(h) mixed ^{4}He/^{3}He droplets with increasing average numbers N_4 of added ^{4}He atoms [43].

and to study their effect on the spectra. Also clusters of para-H$_2$ (p-H$_2$)$_n$ are of special interest since pure clusters having $n < 30$ are predicted to be superfluid [5, 6, 66]. Recently in our laboratory it has been possible to coat a single OCS molecule with a large number of either p-H$_2$ or ortho-D$_2$ (o-D$_2$) molecules inside mixed ^{4}He/^{3}He droplets at temperatures of about 0.15 K and study them spectroscopically [62]. The spectra of OCS-(p-H$_2$/o-D$_2$)$_n$

clusters up to the closure of the first solvation shell of OCS at $n = 18$ show nicely resolved vibrational bands, some of which even exhibit a rotational structure. The most obvious difference is the absence of the Q branch in the OCS-$(p-H_2)_n$ spectra for $n > 11$, whereas at $T = 0.38$ K in pure ^{4}He droplets virtually all of the bands for OCS-$(p-H_2)_n$ are characterized by strong sharp Q branches. The absence of a Q branch is clear evidence that the projection of the angular momentum of the p-H_2 clusters on the OCS molecular axis has vanished in going from $T = 0.38$ K to $T = 0.15$ K. Such a behaviour is consistent with that expected for a transition from the normal to the superfluid state [66].

VI Formation of New Complexes in Helium Droplets

There is now very recent evidence that weakly bound van der Waals and hydrogen bonded complexes may self-assemble inside helium droplets to different structures than those produced in seeded beams. This has been recently vividly demonstrated by Nauta and Miller in studies of clusters formed by HCN molecules [67], and by small numbers of H_2O molecules [68]. In seeded beam expansions of HCN, clusters containing up to four molecules show non-polar, cyclic structures. However, when grown inside He droplets, only linear chains are found. Chains containing more than seven HCN molecules have been detected. Since a droplet diameter of 4 nm is equivalent to a 10-molecule chain, the chain length may be limited by the finite size of the droplets. This preference for the growth of chains is attributed to the very low temperatures, which enable the long-range dipole forces to line up the molecules over large distances. For H_2O, a cyclic hexamer is observed inside the droplets, while in the gas phase a cage structure is seen [68]. These results suggest new strategies for growing nanoscale oligomers with novel structures.

IV Chemical Reactions at T = 0.38 K

Recently it has been demonstrated that He droplets may also be used to cool down both the reactants and products of a bimolecular chemical reaction. The exothermic reaction Ba + N_2O was studied by observing the emission from the highly excited BaO* product molecules [33]. In this experiment the droplets were first doped with Ba and then entered the reaction chamber filled with N_2O and equipped with an optical system to detect the emitted light. Under ordinary conditions the N_2O molecules react with the Ba atoms at the surface since the heliophilic Ba atoms are too weakly bound to penetrate into the interior. This problem was overcome by first embedding about 15 Xe atoms in the interior. The strong van der Waals attraction is then able to pull the Ba atoms into the droplet and the highly resolved vibrational spectra indicated that the reaction occurs exclusively in the interior. Because of the large cross section of the droplets they serve to enhance the rate of the reaction by about a factor 3000 compared to the gas phase. The strong emission observed also indicates that previous determinations of significant activation energies cannot be correct. These initial results point to another area in which He droplets provide new opportunities.

V Conclusions

The technique of helium droplet beams makes it possible to carry out high resolution spectroscopic studies of molecules inside liquid helium. In the relatively short time since the first high resolution spectra were measured in 1994 [42] a wide variety of single atoms and molecules ranging in size from diatomics up to large organic and biological molecules as well as a number of small van der Waals complexes have all been investigated. The sharp rotational structures found in the infrared are well simulated using the effective free molecule Hamiltonian but with different molecular parameters. These provide quantitative information on the interaction of the molecule with the liquid helium environment. In addition the line intensities provide precise information on the very low temperatures of the droplets. Very recent experiments using non-superfluid ^{3}He droplets, in which the sharp rotational structure was not found, suggest that the sharp rotational lines reflect the superfluid nature of the ^{4}He droplets. Evidence that the droplets are superfluid also comes from the observation in electronic spectra of sharp phonon wings separated by a distinct gap from the zero phonon line. This feature, found first in glyoxal, is directly related to the well known sharp phonon-roton dispersion curves characteristic of the superfluid. These experiments demonstrate the great potential for using high resolution molecular spectroscopy as a new probe of microscopic manifestations of superfluidity. At the same time the extensive results now available clearly demonstrate that liquid helium also provides a very gentle almost non-disturbing ultra-cold matrix for molecular spectroscopy. From this point of view liquid helium has the advantage that virtually all the vibrational hot bands and most of the rotational states are frozen out.

In the near future it is hoped that new theoretical work will help to interpret the wealth of information contained in the spectroscopic parameters in terms of microscopic interactions. On the experimental side the big challenge is to find new ways to introduce single molecules into the bulk liquid. This achievement will open up great opportunities to use superfluid helium as a special medium for spectroscopy, and for assembling and manipulating molecules. The large number of unique and unusual phenomena already found suggest that many more exciting and rewarding effects can be expected in the future.

Acknowledgements

The authors are grateful to Brigitta Whaley for many extremely valuable discussions on the interpretation of the experiments.

References

[1] M. Chan, N. Mulders, and J. Reppy. Physics Today, August, p. 30 (1996).

[2] R. B. Hallock. Physics Today, June, p. 30 (1998).

[3] P. Sindzingre, M. L. Klein, and D. M. Ceperley. Phys. Rev. Lett. **63**, 1601 (1989).

[4] F. Dalfovo, A. Lastri, L. Pricaupenko, S. Stringari, and J. Treiner. Phys. Rev. B **52**, 1193 (1995).

[5] K. B. Whaley. Int. Rev. Phys. Chem. **13**, 41 (1994).

[6] K. B. Whaley. in *Adv. in Molecular Vibrations and Collision Dynamics*, edited by J. Bowman (JAI Press, 1998), vol. 3, p. 397.

[7] J. P. Toennies and A. F. Vilesov. Annu. Rev. Phys. Chem. **49**, 1 (1998).

[8] S. I. Kanorsky and A. Weis. in *Advances in Atomic Physics*, edited by B. Bederson and H. Walther (Academic Press, San Diego, 1997), pp. 87–120.

[9] B. Tabbert, H. Günther, and G. zu Putlitz. J. Low Temp. **109**, 653 (1997).

[10] J. Higgins *et al.* Phys. Rev. Lett. **77**, 4532 (1996).

[11] F. Stienkemeier, J. Higgins, W. E. Ernst, and G. Scoles. Phys. Rev. Lett. **74**, 3592 (1995).

[12] F. Dalfovo. Z. Phys. D **29**, 61 (1994).

[13] S. Grebenev *et al.* Physica B **280**, 65 (2000).

[14] M. Hartmann, R. E. Miller, J. P. Toennies, and A. F. Vilesov. Phys. Rev. Lett. **75**, 1566 (1995).

[15] M. Hartmann, R. E. Miller, J. P. Toennies, and A. F. Vilesov. Science **272**, 1631 (1996).

[16] J. Harms *et al.* J. Mol. Spectr. **185**, 204 (1997).

[17] M. Hartmann *et al.* J. Chem. Phys. **110**, 5109 (1999).

[18] J. Harms *et al.* J. Chem. Phys. **110**, 5124 (1999).

[19] M. Lewerenz, B. Schilling, and J. P. Toennies. Chem. Phys. Lett. **206**, 381 (1993).

[20] M. Lewerenz, B. Schilling, and J. P. Toennies. J. Chem. Phys. **102**, 8191 (1995).

[21] A. Scheidemann, J. P. Toennies, and J. A. Northby. Phys. Rev. Lett. **64**, 1899 (1990).

[22] M. Lewerenz, B. Schillings, and J. P. Toennies. J. Chem. Phys. **106**, 5787 (1997).

[23] S. Goyal, D. L. Schutt, and G. Scoles. Phys. Rev. Lett. **69**, 933 (1992).

[24] D. M. Brink and S. Stringari. Z. Phys. D **15**, 257 (1990).

[25] M. Hartmann et al. Phys. Rev. Lett. **76**, 4560 (1996).

[26] N. Pörtner, J. P. Toennies, and A. F. Vilesov. *to be published* (2000).

[27] A. Lindinger. Dissertation, Georg-August-Universität Göttingen (1999).

[28] M. Hartmann, A. Lindinger, J. P. Toennies, and A. F. Vilesov. Chem. Phys. **239**, 139 (1998).

[29] M. Hartmann, A. Lindinger, J. P. Toennies, and A. F. Vilesov. in preparation (2000).

[30] E. Lugovoj, A. Lindinger, J. P. Toennies, and A. F. Vilesov. in preparation (2000).

[31] A. Lindinger, J. P. Toennies, and A. F. Vilesov. J. Chem. Phys. **110**, 1429 (1999).

[32] J. D. Close, F. Federmann, K. Hoffmann, and N. Quaas. Chem. Phys. Lett. **276**, 393 (1997).

[33] E. Lugovoj, J. P. Toennies, and A. F. Vilesov. J. Chem. Phys. **112**, 8217 (2000).

[34] L. A. Philips et al. J. Am. Chem. Soc. **110**, 1352 (1988).

[35] J. Higgins et al. Science **273**, 629 (1996).

[36] F. Federmann, K. Hoffmann, N. Quaas, and J. D. Close. Phys. Rev. Lett. **83**, 2548 (1999).

[37] F. Federmann, K. Hoffmann, N. Quaas, and J. P. Toennies. Eur. Phys. J. D **9**, 11 (1999).

[38] J. Reho et al. J. Chem. Phys. **112**, 8409 (2000).

[39] P. Huang and K. B. Whaley. *private communication* (1999).

[40] B. E. Clements et al. Phys. Rev. B **53**, 12242 (1996).

[41] R. N. Barnett and K. B. Whaley. J. Chem. Phys. **99**, 9730 (1993).

[42] R. Fröchtenicht, J. P. Toennies, and A. F. Vilesov. Chem. Phys. Lett. **229**, 1 (1994).

[43] S. Grebenev, J. P. Toennies, and A. F. Vilesov. Science **279**, 2083 (1998).

[44] S. Grebenev et al. J. Chem. Phys. **112**, 4485 (2000).

[45] A. Guirao, M. Pi, and M. Barranco. Z. Phys. D **21**, 185 (1991).

[46] K. Nauta and R. E. Miller. J. Chem. Phys. **111**, 3426 (1999).

[47] K. Nauta, D. T. Moore, and R. E. Miller. Far. Disc. **113**, 261 (1999).

[48] C. Callegari *et al.* Phys. Rev. Lett. **83**, 5058 (1999).

[49] R. Fröchtenicht, M. Kalloudis, M. Koch, and F. Huisken. J. Chem. Phys. **105**, 6128 (1996).

[50] D. Blume, M. Lewerenz, F. Huisken, and M. Kalloudis. J. Chem. Phys. **105**, 8666 (1996).

[51] M. H. Behrens *et al.* J. Chem. Phys. **107**, 7179 (1997).

[52] M. H. Behrens *et al.* J. Chem. Phys. **109**, 5914 (1998).

[53] E. Lee, D. Farrelly, and K. B. Whaley. Phys. Rev. Lett. **83**, 3812 (1999).

[54] G. Scoles. *private communication* (1999).

[55] Y. Kwon and K. B. Whaley. Phys. Rev. Lett. **83**, 4108 (1999).

[56] R. T. Pack, E. Piper, G. A. Pfeffer, and J. P. Toennies. J. Chem. Phys. **80**, 4940 (1984).

[57] K. Higgins and W. Klemperer. J. Chem. Phys. **110**, 1383 (1999).

[58] Y. Kwon, D. M. Ceperley, and K. B. Whaley. J. Chem. Phys. **104**, 2341 (1996).

[59] P. J. Shirron and J. M. Mochel. Phys. Rev. Lett. **67**, 1118 (1991).

[60] G. Zimmerli, G. Mistura, and M. M. V. Chan. Phys. Rev. Lett. **68**, 60 (1992).

[61] R. A. Crowell, F. W. van Keuls, and J. P. Reppy. Phys. Rev. B **55**, 12620 (1997).

[62] S. Grebenev, B. Sartakov, J. P. Toennies, and A. F. Vilesov. *to be published* (2000).

[63] S. Grebenev, B. Sartakov, J. P. Toennies, and A. F. Vilesov. *to be published* (2000).

[64] V. S. Babichenko and Y. Kagan. Phys. Rev. Lett. **83**, 3458 (1999).

[65] M. Pi, R. Mayol, and M. Barranco. Phys. Rev. Lett. **82**, 3093 (1999).

[66] P. Sindzingre, D. M. Ceperley, and M. L. Klein. Phys. Rev. Lett. **67**, 1871 (1991).

[67] K. Nauta and R. Miller. Science **283**, 1895 (1999).

[68] K. Nauta and R. E. Miller. Science **287**, 293 (2000).

The Spectroscopy of Molecules and Unique Clusters in Superfluid Helium Droplets

Klaas Nauta and Roger E. Miller

Department of Chemistry, University of North Carolina, Chapel Hill, NC 27599

1 Introduction

There has long been interest in the developing methods that allow for the isolation of atoms and molecules in superfluid liquid helium [1]. The low temperatures and weak interactions associated with liquid helium make it an ideal matrix for the isolation of molecules [2]. In addition, the spectroscopy of the molecule can provide important insights into the nature of the solvent. Nevertheless, studies in bulk liquid helium can be problematic, owing to the fact that molecules tend to rapidly aggregate or migrate to the walls of the container [1]. Some progress has been made towards overcoming these difficulties by using laser evaporation within the liquid helium to produce the transient population of the isolated atoms [3]. However, as demonstrated in the following sections and in a number of recent papers [4-10], molecular beam methods provide much better control over the molecular composition of the droplets and yield longer isolation times.

The droplets are produced by free jet expansion from a pinhole source cooled to approximately 20 K. In this way the expanding helium gas liquefies to form a stream of small droplets. Since the droplets have no walls, they are free from the problems discussed above and they cool by evaporation to 0.38 K [7, 11]. Conventional molecular beam methods are used to skim the expansion and produce a beam composed of droplets having a mean size between 1000 and 10,000 atoms [12, 13]. Atoms and molecules are then introduced into these droplets by simply passing them through a pick up cell [14], where the pressure of the gas of interest is maintained between 10^{-6} and 10^{-4} Torr. By adjusting this pressure, one can dope the droplets with different numbers of molecules. This way it is possible to study not only the isolated molecules, but also clusters of molecules grown within the droplets [6, 15-17].

In this paper we discuss the use of infrared laser spectroscopy to study both the dynamics of molecules solvated in liquid helium and the structure of clusters formed following the pick-up of more than one molecule. The homogeneous nature of liquid helium results in highly resolved spectra that provide detailed information about the solvent environment. In many cases rotational fine structure is observed in the spectra, allowing one to use high-resolution spectroscopy to probe the rotational dynamics of the solvated molecules. In the present study we make use of large electric fields to quench the rotational fine structure (pendular state spectroscopy) [18–22] and thus obtain high resolution vibrational spectra that enable us to observe and discriminate between clusters of quite large size.

2 Experimental method

The apparatus used in the present study is shown in Figure 1. It consists of a differentially pumped molecular beam apparatus in combination with a high resolution infrared laser system. In this apparatus, the nozzle, consisting of an electron microscope aperature (5 μm diameter) affixed to the end of a copper tube, is cooled to approximately 20 K by a closed cycle helium refrigerator. Two copper braids provide thermal contact between the closed cycle refrigerator and the nozzle, while allowing for motion of the nozzle for alignment purposes. The helium gas is precooled with liquid nitrogen and filtered several times to avoid nozzle blockages from condensed impurities. At a typical source pressure of 50 atm, helium droplets are formed with an average size of approximately 4,000 atoms [12,13].

The expansion is sampled by an 0.4 mm diameter skimmer to allow the newly formed droplets to pass into a second chamber. The pressure in this chamber must be kept low ($< 10^{-6}$ Torr) to ensure that the droplets are not contaminated by background gas. After traveling approximate 1 cm through this chamber, the droplets enter a 10 cm long pick-up cell, where the pressure is maintained in excess of 10^{-6} Torr. This is done by leaking the gas of interest into the pick-up cell at a rate that is balanced by the pumping speed of a small diffusion pump, at the desired pressure. The number of molecules picked up by the helium droplets is determined by several known parameters, including the cross-sectional area of the droplets, the length of the pick-up cell and the number density of the gas in the cell [23]. It is interesting to note that the energy required to evaporate a single helium atom from the droplet is approximately 7 K, so that the pick-up process results in the evaporation of several hundred helium atoms, due to the dissipation of the molecular kinetic and internal energy. Ultimately this limits the number of molecules that can be captured by a droplet and thus the size of the cluster that can be grown. Nevertheless, the size range can be extended somewhat by operating the source at higher pressures and lower temperatures, in order to make larger droplets. The difficulty with

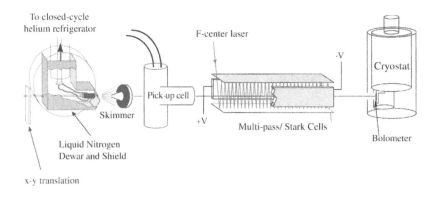

Figure 1: A schematic diagram of the helium droplet apparatus. A closed cyclic helium refrigerator is used to cool the nozzle, enabling the formation of the helium nanodroplets. After pick-up of the molecule(s) of interest, the droplets pass through an F-center laser where vibrational excitation occurs. Subsequent vibrational relaxation results in the evaporation of several hundred helium atoms. This is detected as a depletion of the droplet beam using a bolometer detector.

this approach is that fewer (large) droplets are formed, resulting in lower signal levels.

Upon leaving the pick-up cell, the droplets pass between two parallel multi-pass mirrors, used to obtain multiple laser-molecular beam crossings. In the present study an F-center laser, operated with a RbCl:Li crystal (crystal No. 3) and pumped by approximate 1.4 watts from a krypton ion laser, is used as the infrared excitation source. A detailed discussion of the operation and calibration of this laser can be found elsewhere [24]. All of the molecular beam-laser crossings occur between the two metal electrodes, which can be used to apply a large electric field at the interaction region. As discussed in detail elsewhere [18–22], the large electric field can be used to orient polar molecules and quench their rotational motion. In the case of large clusters grown in liquid helium, this has the advantage of narrowing and intensifying the transitions, enabling us to resolve peaks corresponding to different cluster sizes. This approach also enables us to measure Stark spectra of polar molecules in solution, from which the effective dipole moment can be determined.

Vibrational excitation of the molecules is followed by fast relaxation to the helium droplet. For a 3300 cm^{-1} photon this corresponds to the evaporation of approximately 660 helium atoms. The resulting attenuation of the helium droplet beam is then monitored using a bolometer detector [24]. The spectrum can then be recorded by modulating the laser and using phase sensitive detection of the bolometer signal.

3 Spectroscopy of single molecules in helium nanodroplets

In this section we consider the spectroscopy of a single propyne molecule (H_3CCCH) solvated in a liquid helium nanodroplet. The gas phase spectroscopy of propyne has been studied extensively, most recently in the context of understanding intramolecular vibrational energy redistribution (IVR) [25–28]. In the present study we are interested in exciting the fundamental acetylenic C-H stretch of the molecule, which again has been studied extensively in the gas phase [25, 27, 28]. Propyne is a symmetric top in the gas phase and the C-H vibration in question is well known to be A type, showing P, Q and R branches. The corresponding vibrational origin in the gas phase has been accurately measured to be 3335.059 cm^{-1}[25]. Since the frequency shift associated with solvation in helium is expected to be small, the search for the associated spectrum was carried out in the region of the gas phase origin. Indeed, Callegari et al. [29] have recently reported a study of the C-H stretch overtone spectrum of propyne solvated in helium nanodroplets were they observed a vibrational frequency shift from the gas phase of only -0.26 cm^{-1}.

Figure 2 shows a comparison between the resulting helium droplet spectrum and a reproduction of the gas phase propyne spectrum based on the constants from ref [25]. As expected, the vibrational origin for the helium solvated molecule is only slightly shifted from the gas phase. In addition, the spectrum is clearly rotationally resolved, the most intense transition in the spectrum being easily assigned to the Q branch. As discussed in detail below, the resolution in this spectrum is not laser limited, but instead is indicative of line broadening associated with the liquid helium environment. From the spacing between the P and R branch transitions it is clear that the rotational constant of propyne in solution is considerably smaller than that of the gas phase molecule. Although there is no *a priori* reason to expect that a gas phase Hamiltonian would be appropriate for describing the rotational motion in helium, previous studies have shown that such an analysis often gives an excellent fit to the data. Although the interpretation of the constants resulting from such a fit is still the subject of some debate [30–32], we proceed with such an analysis as a means of efficiently quantifying the data and for the purposes of making comparisons with the corresponding gas phase results.

Figure 3 shows an expanded view of the liquid helium spectrum, along with a fitted spectrum based on the gas phase symmetric top Hamiltonian. Table 1 contains a summary of the constants obtained from this fit, along with

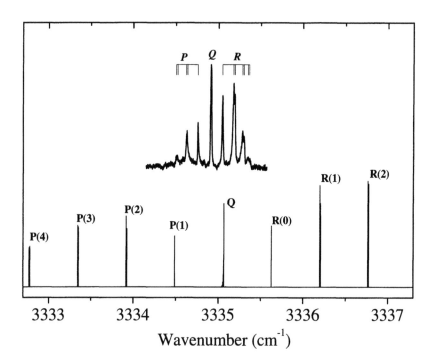

Figure 2: A comparison between the calculated gas phase acetylenic C-H stretch of propyne with that of the helium solvated molecule. The small vibrational frequency shift and substantial change in the rotational constant due to helium solvation are evident.

those for the gas phase molecule. The shift in the vibrational origin due to solvation in helium is thus calculated to be -0.16 cm^{-1}. The rotational constant obtained from this fit is clearly considerably smaller than that of the gas phase molecule, consistent with what has been observed previously for other molecules with similar gas phase rotational constants [7, 8, 16]. This large change in rotational constant has been interpreted as being a result of the fact that the helium atoms participate in the rotational motion of the molecule and thus contribute to the effective moment of inertia of the system [7]. The implication is that the molecule is rotating sufficiently slowly so that the helium atoms can adiabatically follow the motion and thus couple to the rotation of molecule. This is in contrast with HCN, for which we have shown that the rotational constant is only slightly reduced (approximately 15%) by solvation in liquid helium [34]. In the latter case the rotational motion of molecule is so fast that the helium atoms can no

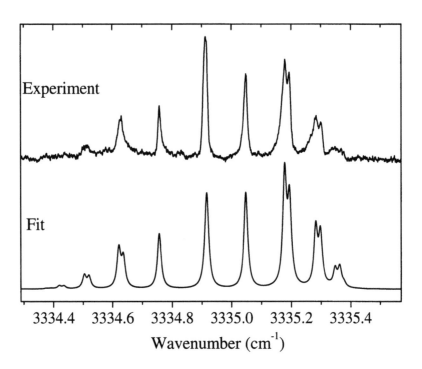

Figure 3: An expanded view of the helium droplet spectrum of propyne, along with a fit to a symmetric top Hamiltonian. The constants obtained from the fit are summarized in Table 1.

longer follow the motion and thus do not contribute to the effective moment of inertia. This behaviour has been noted previously for several other 'light' rotors [9, 10].

Unfortunately, a parallel band of a symmetric top, like the one considered here, does not provide a direct measurement of the A rotational constant. If the temperature is high enough it is possible to estimate A from the relative intensities of the P, Q and R branch transitions. However, the temperature is so low in the present case that the relative populations of the $K = 0$ and 1 states is determined purely by the fact that the latter does not cool into the former, due to nuclear spin statistics. Indeed, if it were not for this fact, all of the molecules would be cooled into $K = 0$. Since the A rotational constant of this molecule is quite large in the gas phase (5.30835 cm^{-1}) it would be interesting to determine the corresponding value in helium, since it is likely that the K rotation will be in the fast, decoupled rotational

Table 1: A comparison of the spectroscopic constants for propyne in the gas phase and in helium droplets.

constant	gas phase [25]	nanodroplet
ν_0 (cm^{-1})	3335.0594	3334.90
B'' (cm^{-1})	0.285060	0.0741
ΔB (cm^{-1})	-0.000675	-0.001
A'' (cm^{-1})	5.30835	0.0015
ΔA (cm^{-1})	0.00879	0.0015
D'' $(\times 10^{-6}$ $cm^{-1})$	0.098[a]	500
D' $(\times 10^{-6}$ $cm^{-1})$	0.098[a]	500

[a] Ground state value from microwave measurement [33].

regime. Indeed, B is small enough for the helium to adiabatically follow the corresponding end-over-end rotational motion of the molecule, while rotation about the molecular axis is much too fast for the helium atoms to follow. Work is presently underway to measure the spectrum of the asymmetric methyl C-H stretch of propyne. This perpendicular band will give a direct measurement of the A rotational constant.

As observed for several other molecules, the centrifugal distortion constant of propyne is orders of magnitude larger in helium than in the gas phase. We interpreted this previously [7] as resulting from the fact that the distortions of this rotating system are primarily associated with the weakly interacting helium atoms that are involved in the motion. Although this centrifugal distortion constant must contain quantitative information about the nature of the molecule-helium interactions, a quantitative theory describing this phenomenon has yet to be developed. This is in contrast with the change in the rotational constants, for which there are now several approaches for estimating the rotational constants of solvated molecules [30, 32], based upon the associated molecule-helium intermolecular potentials.

One of the most remarkable features of helium droplet spectroscopy is its high resolution, which is orders of magnitude higher than that typically obtained in liquid state spectroscopy. Indeed, the fitted spectrum shown in Figure 3 has a Lorentzian linewidth of 0.015 cm^{-1}. For this system, the line width appears to be relatively independent of rotational state, which is certainly not always the case. Rather for many molecules solvated in helium, the line widths and shapes depend strongly upon rotational quantum number [7, 8, 16]. It is fair to say at this point that a comprehensive understanding of the broadening observed in helium droplet spectroscopy is still lacking, although there are cases where the line broadening is almost certainly due to relaxation processes associated with the vibrational excitation [16]. In general, however, there must be several contributions to the lineshape, with the dominant one depending upon the system. It is interesting to note that the lineshapes are often asymmetric, perhaps indicative of inhomogeneous broadening in some cases.

Vibrational relaxation is known to occur in liquid helium, since the methods used for detecting the laser induced signals rely on the fact that the vibrationally excited molecules relax on the time scale of the experiment, resulting in the evaporation of helium atoms from the droplet. Rotational relaxation must also be occurring on the time scale of the experiment, since the molecules end up at the low temperature characteristic of the droplets through some form of rotational relaxation process. It is interesting to note that the line widths obtained from the previous overtone study of propyne in helium are approximately three times broader (0.043 cm^{-1} [29]) than those observed here, suggesting that the vibrational degrees of freedom do play an important role in the line broadening. We must also consider the fact that the liquid helium is not perfectly homogeneous in a finite sized droplet, due to the presence of the droplet surface. Therefore, as the molecule explores the interior of the droplet, its interactions with the helium will change [35], resulting in inhomogeneous broadening, similar to that obtained in normal matrix isolation spectroscopy. At this point it is premature to assign a particular broadening mechanism to the propyne data. Nevertheless, as our understanding of these line widths improve, through future experimental and theoretical studies, they will certainly provide important information about the dynamics occurring in these systems.

4 Pendular state spectroscopy of polar molecules in liquid helium

Although the rotational structure observed in the spectrum discussed above can provide insights into the solvent environment experienced by the molecule, the associated broadening of the vibrational band can present problems when it comes to resolving closely space vibrational bands. Indeed, the overall width of these rotational contours is approximately 1 cm^{-1}, even at the low temperature characteristic of the droplets. In previous gas phase studies [36, 37] we have shown that a large electric field can be used to orient a polar molecule and eliminate the rotational fine structure normally observed in the spectrum. This has the added advantage that, the transition of interest can be pumped more efficiently by adjusting the polarization of the laser to match the transition moment, which is now aligned in the space fixed frame. For example, if the permanent electric dipole moment is parallel to the vibrational transition moment, then the most efficient pumping scheme corresponds to the applied DC electric field being parallel to the laser electric field direction.

Figure 4 shows a comparison between the zero field spectrum of cyanoacetylene (a linear polar molecule) and that obtained in the presence of a large DC electric field. As expected for linear molecule, the zero field spectrum shows no Q branch. Upon the application of the DC field, however, a feature reminiscent of a Q branch appears near the vibrational origin and the entire spectrum collapses into this single transition. This results from the fact that the rotational states of the molecule are strongly mixed by the large electric field, giving rise to a set of hydridized levels that are equally spaced in energy [38]. Although J (corresponding to the total rotational angular moment) is no longer a good quantum number under these conditions, the quantum number defining the projection on the electric field axis (M) is still well defined. When the laser field is parallel to the applied field, such that $\Delta M = 0$, the associated transitions all have approximately the same frequency.

The collapse of the rotational band into a single peak has several important advantages for the studies that follow. First, there is a significant reduction in the overall width of the band, making it easier to distinguish closely spaced vibrational bands associated with different species. Second, the intensity of the Q branch feature is significantly greater than that of the zero field P and R branches for the reasons discussed above. The dependence of the spectrum on the electric field can also provide important structural information, allowing us to differentiate between polar and nonpolar molecules. In fact, studies carried out as a function of electric field strength and polarization direction can provide quantitative information on both the dipole moment and the rotational constant of the molecule.

5 The growth of clusters in liquid helium droplets

As noted previously, the probability of capturing a molecule in a droplet of the given size is related to the cross-sectional area of the droplet and the density of the molecules in the pick-up cell. The subtle details associated with this pick-up cell process, needed to understand it quantitatively, are dealt with in detail elsewhere [23]. More than one molecule can be captured in a single droplet by simply increasing pressure in the pick-up cell or by increasing the size of the droplets. In the latter case the increased heat capacity of the larger droplets facilitates the pick-up, cooling and growth of larger molecular clusters. The capture process obeys Poisson statistics, providing us with a powerful method for determining the cluster sizes [23]. To illustrate several important issues associated with the formation of clusters in liquid helium nanodroplets, we consider the case of the hydrogen cyanide dimer. The gas phase spectroscopy of this complex has been thoroughly studied [39–48] and a number of theoretical calculations have also

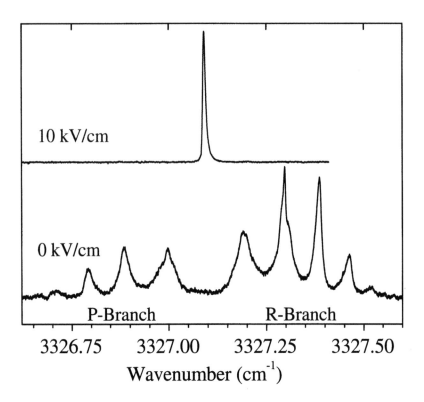

Figure 4: A comparison between the zero field and pendular state spectrum of the (linear, polar) cyanoacetylene molecule solvated in helium. All of the rotational structure in the zero field spectrum is lost as the molecules become oriented in the strong DC electric field, resulting in a substantial increase in both the sensitivity and resolution.

been reported [49–54]. Recently we reported a detailed infrared study of the C-H stretches in helium droplets [16]. The gas phase studies clearly show that this complex is linear and that the two C-H stretches are inequivalent, namely the one associated with the free C-H stretch of the proton acceptor molecule and the other corresponding to the hydrogen bonded C-H stretch of the proton donor. Figure 5 shows a rotationally resolved spectrum of the free C-H stretching vibration, along with a fit to a linear rotor Hamiltonian. Here again we find that the rotational constants of the complex are reduced by a factor of approximately three in comparison with the gas phase.

Although the rotational fine structure observed in the above spectra, combined with the vibrational frequency, give ample information for the as-

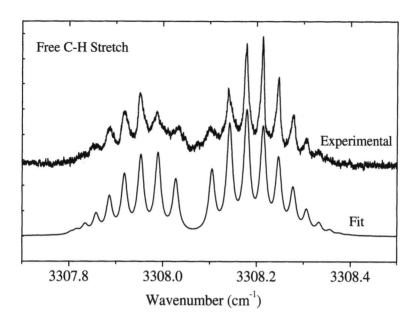

Figure 5: A rotationally resolved spectrum of the linear HCN dimer in helium, corresponding to excitation of the free C-H stretching vibration. The fitted spectrum again shows that the rotational constant in helium is substantially smaller than in the gas phase.

signment of this spectra to the linear dimer of hydrogen cyanide, further evidence supporting this assignment can be obtained from the pressure dependence of the laser induced signals. Figure 6 shows three sets of data, corresponding to the laser induced signals for the HCN monomer and the two vibrational bands of the dimer, recorded as a function of the pick-up cell pressure. The pressures given in the figure were determined directly from an ionization gauge in the pick-up cell and are uncorrected for the ionization efficiency. As expected, the HCN monomer signals peak at lower pick-up cell pressures than those for the two dimer bands. The solid lines through the data correspond to Poisson distributions defined by [23]

$$P(n) = \frac{(\eta \sigma L)^n}{n!} e^{-\eta \sigma L}$$

where $P(n)$ is the probability of picking up n HCN molecules, L is the length of the pick- up cell, η is the gas phase number density in the pick-up cell and σ is the effective cross-section of the helium droplet. Note that there is a single scaling factor in this calculation, which accounts for the absolute calibration of the ionization gauge and includes a velocity averaging correction factor, as discussed in detail elsewhere [23].

The signals associated with the dimer bands peak at approximately twice the pick-up cell pressure that is required for optimization of the monomer. For the monomer and the free C-H stretch of the dimer, the fit to the Poisson distribution is excellent. The deviation between the experimental and fitted curves for the hydrogen bonded mode of the dimer results from the fact that this vibrational band overlaps with a band from a higher order cluster that contributes at the higher pressures. The fact that the dimer signals decrease at the higher pick up cell pressures is indicative of the formation of these larger clusters.

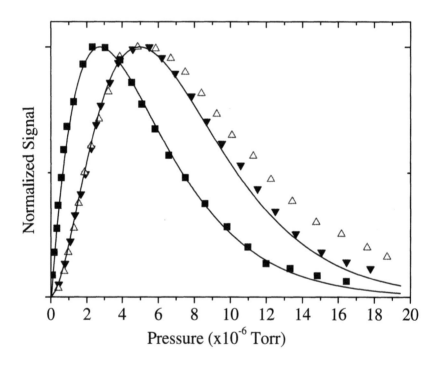

Figure 6: Pick-up cell pressure dependence of the laser induced HCN monomer (solid squares) and dimer ("free" solid triangles, "bonded" open triangles) signals. The solid lines are fits to Poisson distributions (see text), corresponding to $n = 1$ and $n = 2$ for the monomer and dimer, respectively.

6 Non-equilibrium self assembly of unique cluster structures

The process of cluster growth within liquid helium is fundamentally different from that which occurs in a free jet expansion. In the latter case, clusters are formed in the relatively warm region of the expansion and the associated condensation energy is removed from the cluster by subsequent collisions with the carrier gas. This cooling process is relatively slow (compared to the cooling rates in a liquid) so that the warm cluster has plenty of time to find the minimum energy configuration before it is cooled to the point where isomerization can no longer occur. For this reason, free jet expansion studies almost universally result in the formation of the lowest energy isomer. Higher energy isomers are only observed if they lie only slightly above the global minimum.

In liquid helium droplets, the time between pick-up events is quite long and the molecules are cooled to the temperature of the droplet before the next molecule is captured. As a result, clusters grow by sequential monomer addition. At these low temperatures, the long range forces between the molecules tend to orient them prior to formation of the complex [55]. For example, when two polar molecular approach one another at these temperatures, they become mutually oriented in a linear geometry. As the complex is formed, the rapid cooling provided by the liquid helium causes this structural motif to be frozen in. Although more stable isomers may exist, the energy of the system is so low that the system cannot surmount the barriers that lie between the approach geometry and these other structures. Apparently the rapid quenching provided by the helium prevents the system from rearranging in response to the shorter range forces (e.g. dispersion interactions), resulting in the formation of an isomer other than that corresponding to the global minimum on the potential energy surface. We have previously shown that both the linear and cyclic isomers of HCN trimer can be formed in a free jet expansion [56]. The implication is that the energies of these two isomers are approximately the same. Indeed *ab initio* calculations indicate that the linear and cyclic isomers of HCN are essentially isoenergetic, to within the accuracy of the calculations [51, 57]. Although the larger clusters of HCN do not give rotationally resolved spectra that could be used to determine their structure, we have been able to use pendular state spectroscopy to determine whether or not they are polar. Figure 7 shows a gas phase pendular state spectrum obtained under conditions where large HCN clusters are being formed. As expected, linear dimer (LD) and linear trimer (LT) vibrational bands depend strongly on electric field, consistent with the linear structures. The cyclic HCN trimer (CT), on the other hand, is unaffected by the electric field, consistent with the fact that it is non-polar. In fact, all of the remaining peaks in this spectrum, attributed to larger clusters of HCN, shown no dependence on electric field. Clearly the growth process occurring in a free jet expansion results exclusively in the formation of nonlinear structures with zero dipole moment. This is again consistent with *ab initio* calculations [51, 57], which suggest that the cyclic structures are most stable for HCN clusters larger than the trimer, with the trimer being the energetic crossover point between the linear and cyclic motifs.

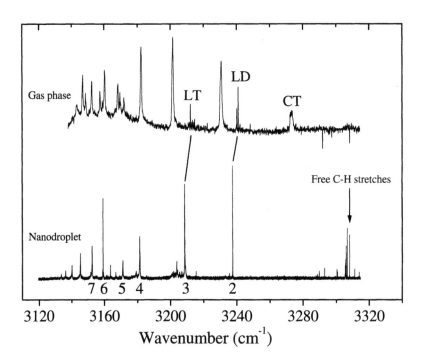

Figure 7: A comparison between the gas phase (upper) and helium droplet (lower) spectra of HCN clusters. In the gas phase, cluster larger than the trimer are all non-polar, while in helium we observe the exclusive formation of linear, polar chains.

Figure 7 also shows a pendular spectrum of HCN clusters grown in liquid helium, covering the same spectral region. Careful examination of the bands in the two spectra quickly reveals that with the exception of the linear dimer and linear trimer bands, the spectrum in helium is completely different from that in the gas phase. Particularly noteworthy is the absence of the cyclic trimer in the helium spectrum. In fact, all of the bands associated with the non-polar complexes are absent in the helium spectrum and instead one observes a regular pattern of bands that are all strongly dependent upon the electric field, indicating that they are all associated with polar complexes. The implication is that the growth in liquid helium gives rise to a completely different structural motif than obtained by gas phase nucleation. In both the region of the free and hydrogen bonded C-H stretches we observe regular progressions that converge to a limiting frequency. Comparisons with *ab initio* calculations confirm the assignment of these bands to linear clusters, at least up to the hexamer [51, 57]. Beyond that, the assignment is based purely on the regularity of the progression. To date we have observed chains up to 12 monomer units in length.

The implications of the above results are quite profound. They suggest that the growth of clusters within liquid helium can provide us access to an entirely different structural motive. Indeed, we have now applied this technique to three different systems, HCN [55], cyanoacetylene [58] and water [59], and in all cases we observe the formation of novel cluster structures that have not been observed in the gas phase. Experiments of this type can clearly provide us with access to regions of the intermolecular potential that are far from the global minimum. Having access to these regions is essential if the global intermolecular potentials are to the obtained from the associated spectroscopy. This is particularly important for the case of water where there is now a great deal of theoretical work available on these higher energy isomeric structures, while the experimental data is relatively sparce.

7 Conclusions

Liquid helium droplet spectroscopy has emerged as an exciting new approach for studying both the properties of liquid helium and the nature of the interactions between the molecule and the solvent, as well as those associated with molecular clusters formed within the liquid. We have shown here that high resolution infrared spectroscopy can be used to study the properties of these solvated systems. In particular, we have emphasized the use of liquid helium droplets as a growth medium for the formation of novel cluster structures. One of the attractive features of this method, is that the vapor pressure of the species of interest required to saturate the helium droplets is only on the order of 10^{-5} Torr. For this reason, it should be possible to use this approach to obtain high resolution infrared spectra for a wide range of systems, including molecules bound to metal atoms and clusters, a wide range of free radical clusters and perhaps even ionic systems. Much of what has been done using traditional matrix isolation spectroscopy should now be possible using helium droplets, with the advantages that the frequency shifts associated with the helium matrix are much smaller and one has greater control over the species that are formed.

Acknowledgments

This work was supported by the National Science Foundation (CHE-97-10026). We also acknowledge the donors of The Petroleum Research Fund, administered by the ACS, for partial support of this research.

References

[1] J. P. Toennies and A. F. Vilesov, Annu.Rev.Phys.Chem. **49**, 1 (1998).

[2] K. K. Lehmann and G. Scoles, Science **279**, 2065 (1998).

[3] J. L. Persson, Q. Hui, M. Nakamura, and M. Takami, Phys.Rev.A. **52**, 2011 (1995).

[4] S. Goyal, D. L. Schutt, and G. Scoles, Phys.Rev.Lett. **69**, 933 (1992).

[5] S. Goyal, D. L. Schutt, and G. Scoles, J.Phys.Chem. **97**, 2236 (1993).

[6] M. Hartmann, R. E. Miller, J. P. Toennies, and A. F. Vilesov, Science **272**, 1631 (1996).

[7] M. Hartmann, R. E. Miller, J. P. Toennies, and A. F. Vilesov, Phys.Rev.Lett. **75**, 1566 (1995).

[8] S. Grebenev, J. P. Toennies, and A. F. Vilesov, Science **279**, 2083 (1998).

[9] R. Fröchtenicht, M. Kaloudis, M. Koch, and F. Huisken, J.Chem.Phys. **105**, 6128 (1996).

[10] M. Behrens et al., J.Chem.Phys. **109**, 5914 (1998).

[11] D. M. Brink and S. Stringari, Z.Phys. D **15**, 257 (1990).

[12] E. L. Knuth, B. Schilling, and J. P. Toennies, in *19th International Symposium on Rarefied Gas Dynamics*, Oxford University Press, 1995.

[13] M. Lewerenz, B. Schilling, and J. P. Toennies, Chem.Phys.Lett. **206**, 381 (1993).

[14] T. E. Gough, M. Mengel, P. Rowntree, and G. Scoles, in *Proc.SPIE-Int.Soc.Opt.Eng.*, volume 669, pages 129–132, 1986.

[15] M. Behrens, U. Buck, R. Fröchtenicht, M. Hartmann, and M. Havenith, J.Chem.Phys. **107**, 7179 (1997).

[16] K. Nauta and R. E. Miller, J.Chem.Phys. **111**, 3426 (1999).

[17] M. Behrens, R. Fröchtenicht, M. Hartmann, J. Siebers, and U. Buck, J.Chem.Phys. **111**, 2436 (1999).

[18] B. Friedrich and D. R. Herschbach, Z.Phys. D **18**, 153 (1991).

[19] B. Friedrich and D. R. Herschbach, At.Mol.Phys. **32**, 47 (1995).

[20] H. J. Loesch and A. Remscheid, J.Chem.Phys. **93**, 4779 (1990).

[21] B. Friedrich, D. P. Pullman, and D. R. Herschbach, J.Phys.Chem. **95**, 8118 (1991).

[22] J. Bulthuis, J. Moller, and H. J. Loesch, J.Phys.Chem. **101**, 7684 (1997).

[23] M. Lewerenz, B. Schilling, and J. P. Toennies, J.Chem.Phys. **102**, 8191 (1995).

[24] Z. S. Huang, K. W. Jucks, and R. E. Miller, J.Chem.Phys. **85**, 3338 (1986).

[25] A. McIlroy and D. J. Nesbitt, J.Chem.Phys. **91**, 104 (1989).

[26] J. S. Go and D. S. Perry, J.Chem.Phys. **97**, 6994 (1993).

[27] A. McIlroy and D. J. Nesbitt, J.Chem.Phys. **100**, 2596 (1994).

[28] E. R. T. Kerstel, K. K. Lehmann, B. H. Pate, and G. Scoles, J.Chem.Phys **100**, 2588 (1994).

[29] C. Callegari, A. Conjusteau, I. Reinhard, K. K. Lehmann, and G. Scoles, in *Rovibrational Bound States in Polyatomic Molecules*, pages 41–49, CCP6, Daresbury, 1999.

[30] E. Lee, D. Farrelly, and K. B. Whaley, Phys.Rev.Lett. **83**, 3812 (1999).

[31] V. S. Babichenko and Y. Kagan, Phys.Rev.Lett. **83**, 3458 (1999).

[32] C. Callegari et al., Phys.Rev.Lett. **83**, 5058 (1999).

[33] V. M. Horneman, G. Graner, H. Fakour, and G. Tarrago, J.Mol.Spectrosc. **137**, 1 (1989).

[34] K. Nauta and R. E. Miller, Phys.Rev.Lett. **82**, 4480 (1999).

[35] K. K. Lehmann, Mol.Phys. **97**, 645 (1999).

[36] P. A. Block, E. J. Bohac, and R. E. Miller, Phys.Rev.Lett. **68**, 1303 (1992).

[37] R. J. Bemish, E. J. Bohac, M. Wu, and R. E. Miller, J.Chem.Phys. **101**, 9457 (1994).

[38] B. Friedrich, A. Slenczka, and D. R. Herschbach, Can.J.Phys. **72**, 897 (1994).

[39] A. C. Legon, D. J. Millen, and P. J. Mjöberg, Chem.Phys.Lett. **47**, 589 (1977).

[40] L. W. Buxton, E. J. Campbell, and W. H. Flygare, Chem.Phys. **56**, 399 (1981).

[41] K. Georgiou, A. C. Legon, D. J. Millen, and P. J. Mjöberg, Proc.R.Soc.London A **399**, 377 (1985).

[42] G. A. Hopkins, M. Maroncelli, J. W. Nibler, and T. R. Dyke, Chem.Phys.Lett. **114**, 97 (1985).

[43] M. Maroncelli, G. A. Hopkins, J. W. Nibler, and T. R. Dyke, J.Chem.Phys. **83**, 2129 (1985).

[44] B. A. Wofford, J. W. Bevan, W. B. Olson, and W. J. Lafferty, J.Chem.Phys. **85**, 105 (1986).

[45] K. W. Jucks and R. E. Miller, Chem.Phys.Lett. **147**, 137 (1988).

[46] K. W. Jucks and R. E. Miller, J.Chem.Phys. **88**, 6059 (1988).

[47] H. Meyer, E. R. T. Kerstel, D. Zhuang, and G. Scoles, J.Chem.Phys. **90**, 4623 (1989).

[48] E. R. T. Kerstel, K. K. Lehmann, J. E. Gambogi, X. Yang, and G. Scoles, J.Chem.Phys. **99**, 8559 (1993).

[49] P. Kollman, J. McKelvey, A. Johansson, and S. Rothenberg, J.Am.Chem.Soc. **97**, 955 (1975).

[50] M. Kofranek, H. Lischka, and A. Karpfen, Mol.Phys. **61**, 1519 (1987).

[51] M. Kofranek, A. Karpfen, and H. Lischka, Chem.Phys. **113**, 53 (1987).

[52] J. G. R. Tostes, C. A. Taft, and M. N. Ramos, J.Phys.Chem. **91**, 3157 (1987).

[53] M. N. Ramos, C. A. Taft, J. G. R. Tostes, and J. W. A. Lester, J.Mol.Struct. **175**, 303 (1988).

[54] R. S. Ruoff, J.Chem.Phys. **94**, 2717 (1991).

[55] K. Nauta and R. E. Miller, Science **283**, 1895 (1999).

[56] K. W. Jucks and R. E. Miller, J.Chem.Phys. **88**, 2196 (1988).

[57] A. Karpfen, J.Phys.Chem. **100**, 13474 (1996).

[58] K. Nauta, D. T. Moore, and R. E. Miller, Faraday Discuss **113**, 261 (1999).

[59] K. Nauta and R. E. Miller, Science **287**, 293 (2000).

Electronic Structure and Dynamics of Solute Molecules on Solution Surfaces by Use of Liquid Beam Multiphoton Ionization Mass Spectrometry

F. Mafuné, N. Horimoto, T. Kondow

Cluster Research Laboratory, Toyota Technological Institute,
717-86 Futamata, Ichikawa, Chiba 272-0001, Japan

1 Introduction

A gas-liquid interface (liquid surface) is an important subject of investigation; chemical and physical properties of molecules on a solution surface are different from those in the liquid and in the gas [1, 2, 3, 4, 5, 6, 7, 8]. The specificity arises mainly from the fact that molecules on the liquid surface are not completely surrounded with the constituent molecules and hence are subject to asymmetric forces. The asymmetry in the force manifests itself in the chemical composition [9, 10, 11, 12, 13], the orientation of surface molecules [14, 15], and dynamical behaviors such as molecular relaxation [16] and chemical reactivity, especially, ion-molecule reactions [17, 18].
We have developed a technique of introducing a continuous liquid flow into vacuum (liquid beam) [17, 18, 19, 20, 21, 22, 23, 24, 25, 26], and combined this technique with laser photoionization and mass spectrometry: Molecules in the liquid beam are ionized by a multiphoton process and ions in the vicinity of the liquid surface are ejected either directly or after chemical reactions. The mass spectrometric findings obtained in the experiments provide information on how the solute molecules form a specific solvation structure [19, 20], and the ions and solvated electrons generated by the multiphoton process react with surrounding solute and solvent molecules [21].
In combination with a pump-probe two-color photoionization, we have performed a decay measurement of an excited aniline molecule (AN) on the surface of a 1-propanol (PrOH) solution by changing a delay time from the photoexcitation to the photoionization [22]. The lifetime of an aniline molecule in S_1 state on the solution surface is comparable to that of an isolated aniline molecule, probably because of a weak interaction between

an aniline molecule and its surrounding propanol molecules on the surface due to its specific orientation. The relaxation dynamics is determined by the short-range interaction in the close vicinity of the excited molecule, while the electronic polarization is expected to operate over a long range around the molecule. In the present work, we measured the threshold ionization potential of an aniline molecule on a 1-propanol solution surface. The ionization potential thus obtained accords with a simple quantum chemical calculation.

2 Experiment

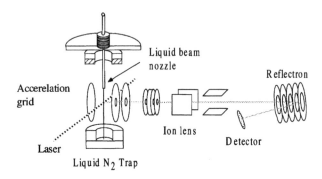

Figure 1: Schematic diagram of the experimental apparatus, which consists of a liquid beam and a time-of-flight mass spectrometer

Figure 1 shows a schematic diagram of the experimental apparatus used in the present study, which consists of a liquid beam and a time-of-flight (TOF) mass spectrometer. A sample liquid was pressurized at ~ 20 atm by a Shimadzu LC-6A pump designed for a liquid chromatograph, and a continuous liquid flow (liquid beam) was formed inside a vacuum chamber through an aperture with $20\,\mu$m in diameter. The liquid beam which is regarded as a stable liquid filament produces an optical Fraunhofer diffraction pattern by laser illumination perpendicular to the liquid beam. The diffraction is generated by the interference of light obstructed by the liquid beam. The diameter of the liquid beam is estimated to be $20\,\mu$m by analyzing the pattern, in agreement with that of the pinhole of the nozzle. The flow velocity of the liquid beam was measured by using an inductive detector located at 5 cm downstream from the nozzle [27]; the flow velocity was estimated to be $10.5\,\mathrm{m\,s^{-1}}$ at the flow rate of $0.2\,\mathrm{mL\,min^{-1}}$ from the arrival time of the ions generated by photoionization, at the detector. The measured flow velocity coincides with that calculated from the flow rate and the diameter of the liquid beam ($20\,\mu$m).

In order to combine the liquid beam with the mass spectrometer, the liquid beam was introduced inside the vacuum chamber from the top, and was captured by a liquid N_2 trap at 10 cm downstream from the nozzle. The liquid N_2 trap works as a cryopump with an effective pumping speed of 5000 L s^{-1}. The chamber was further evacuated down to $10^{-5} - 10^{-6}$ Torr by a 1200 L s^{-1} diffusion pump. In this pressure range, the mean free path of the molecules in the gas phase exceeds 1 m, so that molecules do not stagnate in the close vicinity of the solution surface. This feature is supported by a measurement by Penning electron spectroscopy by Morgner and his coworkers [3, 28]: The Penning electron spectrum contains exclusive information on molecules on the outermost layer of a surface, that is, metastable atoms used for the Penning ionization reach the surface without suffering collisional deactivation by molecules in the gas phase.

Traveling a distance of 1 mm from the aperture, the liquid beam was crossed with pump and probe lasers with a focal length of 400 mm at the first acceleration region of the TOF mass spectrometer. The pump and probe lasers were obtained from the fourth harmonic output (266 nm) of a Quantaray GCR-3 Nd:YAG laser and a MOPO optical parametric oscillator (500–600 nm), respectively. The threshold ionization potential of an aniline molecule on the liquid beam surface was estimated by observing the intensity of the aniline ions as a function of the wavelength of the probe laser.

Ions ejected from the liquid beam were accelerated by a pulsed electric field in the first acceleration region of the TOF mass spectrometer, in the direction perpendicular to both the liquid and the laser beams. A delay time from the ionization to the ion extraction was varied in the range of 0–1.5 μs so as to improve the mass resolution. The ions were then steered and focused by a set of vertical and horizontal deflectors and an einzel lens. After traversing a 1-m field free region, the ions were detected by a Murata EMS-6081B Ceratron electron multiplier. Signals from the multiplier were amplified and processed by a Yokogawa DL 1200E transient digitizer based on an NEC microcomputer. The mass resolution, $m/\Delta m$, was found to be more than 85 at $m = 200$ in the present experimental condition.

3 Result

Figure 2 shows typical TOF mass spectra of ions produced from a 0.5 M aniline solution in propanol. Panel (a) shows the mass spectrum of ions produced by irradiation of only the pump laser at 266 nm. A series of ion peaks ($m/z = 93, 153, 213, 273...$) and ($m/z = 61, 121, 181...$) is assigned to AN$^+$(PrOH)$_n$ and H$^+$(PrOH)$_n$, respectively. Panel (b) shows the mass spectrum of the ions when the pump laser and the probe laser operating at 500 nm irradiate concurrently the liquid beam. Though the mass spectral feature does not change significantly, the intensity of AN$^+$(PrOH)$_n$

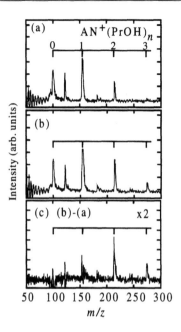

Figure 2: Mass spectrum of ions produced from a 0.5 M aniline solution of 1-propanol under the irradiation of the pump laser at 266 nm (panel (a)), and the pump (266 nm) and the probe (500 nm) laser (panel (b)). A difference spectrum ((b)–(a)) is exhibited in panel (c). The two lasers illuminate the liquid beam concurrently

increases by irradiation of the probe laser along with the pump laser. The difference spectrum shown in panel (c) obtained by subtracting (a) from (b) gives rise to the spectrum of ions produced by absorption of a single photon of the pump laser and a subsequent absorption of a single photon of the probe laser.

The total increment of the $AN^+(PrOH)_n$ intensity with addition of the probe laser, $\Delta I_{\text{total}}^{\text{AN}}$, is defined as

$$\Delta I_{\text{total}}^{\text{AN}} = \sum \Delta I_n^{\text{AN}},$$

where ΔI_n^{AN} stands for the increment of each product ion, $AN^+(PrOH)_n$. The total increment increased linearly with the power of the pump laser and that of the probe laser. Figure 3 shows $\Delta I_{\text{total}}^{\text{AN}}$ as a function of the wavelength of the probe laser with no delay time. The total increment decreases with increase in the wavelength of the probe laser and tends to level off in the vicinity of 650 nm. In the inset, $[\Delta I_{\text{total}}^{\text{AN}}]^{0.4}$ is plotted as a function of the photon energy of the probe laser, where $[\Delta I_{\text{total}}^{\text{AN}}]^{0.4}$ decreases

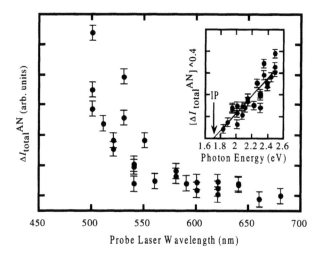

Figure 3: The total increment of the $AN^+(PrOH)_n$ intensity, ΔI_{total}^{AN}, plotted against the wavelength of the probe laser with no delay time. The total increment, ΔI_{total}^{AN}, decreases with the wavelength of the probe laser and tends to almost level off in the vicinity of 650 nm. In the inset, $[\Delta I_{total}^{AN}]^{0.4}$ is plotted as a function of the photon energy of the probe laser, where $[\Delta I_{total}^{AN}]^{0.4}$ decreases linearly with decrease in the photon energy of the probe laser. See text for the definition of ΔI_{total}^{AN}

linearly with decrease in the photon energy. The threshold energy was estimated to be 6.4 eV by extrapolating the linear dependence to $[\Delta I_{total}^{AN}]^{0.4}$.

4 Discussion

4.1 Formation of cluster ions in the gas phase

When an aniline molecule in a propanol solution is irradiated with the focussed laser at 266 nm, the aniline molecule is excited by absorption of one photon to S_1 state and then ionized by absorption of another photon to the continuum (single-color two-photon ionization). A photoelectron generated in the liquid beam has a fate of being (1) captured by a neutral molecule as a negative ion, (2) delocalized in the solution as a solvated electron, (3) neutralized at a cation site via geminate recombination, or (4) liberated as a free electron in the gas phase. As the mean free path of the electron is shorter than ~ 1 nm, only electrons generated in a region as deep as 1 nm from the liquid surface can escape into vacuum. Therefore, only a surface region shallower than 1 nm is positively charged due to the electron depletion. When the repulsive Coulomb energy exceeds the solvation energy of the concerned ion, the ion is ejected from the surface into vacuum (Coulomb ejection model). Accordingly, the mass observed spectrum provide information on the ions generated in the surface region. Even under irradiation of the weakest possible pump laser, ions are inevitably generated via two-photon absorption of the pump laser. However, almost all the excited aniline molecules by the pump laser still remain without suffering the single-color two-photon ionization. In the pump-probe configuration, the remaining excited aniline molecules are ionized by the probe laser; as a result, the abundance of the ions in the liquid beam increases, and hence, an increasing amount of the cluster ions is ejected in the gas phase. It follows that the increment of the $AN^+(PrOH)_n$ intensity is ascribable to two color pump-probe two-photon ionization. In general, the intensity of the ions ejected in the gas phase is not proportional to the abundance of the ions generated inside the liquid beam in the Coulomb ejection scheme. In contrast, in the pump-probe configuration, the increment of the observed ion intensity caused by the probe laser irradiation must be proportional to the increment of the abundance of ions inside the liquid beam, as described elsewhere [22].

4.2 Ionization potential of surface aniline molecule

As shown in Fig. 3, ΔI_{total}^{AN} decreases with increase in the wavelength of the probe laser. The decrease in the increment is ascribable to the decrease in the ionization probability in the vicinity of the ionization potential (IP). An empirical threshold law shows that the ionization probability increases with the photon energy to the power of 2.5 in the threshold energy region because of a finite probability of electron arrival at the solution surface. In fact, $[\Delta I_{total}^{AN}]^{0.4}$ depends linearly on the photon energy. The IP of an aniline molecule on the solution surface is estimated to be 6.3 eV from the plots of $[\Delta I_{total}^{AN}]^{0.4}$ against the photon energy. As the IP of an isolated aniline molecule in the gas phase is 7.7 eV, the IP is lowered by 1.4 eV on the solution surface. The decrease in the IP on the solution surface must arise from the difference in the solvation energy between an aniline molecule and the ion on the solution surface. Note that the IP observed in this measurement is the vertical ionization potential.

4.3 Calculation of threshold ionization potential

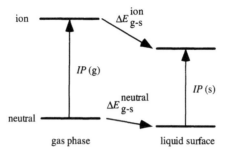

Scheme

An aniline ion on the solution surface is stabilized by the solvent to a more significant extent than a neutral aniline molecule on the same solution surface so that the IP of the molecule on the liquid surface turns out to be lower than that of an isolated aniline molecule. The IP of an aniline molecule on the solution surface, IP(s), is expressed in terms of the IP of an aniline molecule isolated in the gas phase, IP(g), as

$$\mathrm{IP(s) = IP(g) - \Delta E_{g-s}^{neutral} + \Delta E_{g-s}^{ion}},$$

where $\Delta E_{g-s}^{neutral}$ and ΔE_{g-s}^{ion} represent the solvation energies of the neutral molecule and its ion by surface solvent molecules, respectively (see Scheme). The solvation energy, $\Delta E_{g-s}^{neutral}$, is defined as the energy required to remove the molecule from the solution surface to vacuum, and is computed as the energy difference between the molecule in the gas phase and that on the solution surface. Similarly, the solvation energy of the ion, ΔE_{g-s}^{ion} is defined as the energy required to remove the ion from the solution surface, and is computed as the energy difference between the ion in the gas phase and that on the solution surface. According to the Frank–Condon principle, the geometrical structure of the ion right after its birth by the photoionization maintains the structure of the neutral molecule. Therefore, the energy of the ion is calculated by using the optimized geometry of the neutral aniline molecule either in the gas or the liquid phase.

At the first place, the energies of an isolated aniline molecule and its ion were calculated by B3LYP/6-31G(d) (geometrical optimization) and B3LYP/6-311+G(2d,p) (single-point energy calculation): The calculations were performed according to the Becke-style 3-parameter density functional theory using the Lee–Yang–Parr correlation functional (B3LYP), and 6-31G(d) and 6-311+G(2d,p) were used as a basis set, respectively. The zero-point energy was corrected by B3LYP/6-31G(d). At the second place, the energy of the molecule on the solution surface was calculated by using a self-consistent reaction field (SCRF) method [29] with a slight modification (modified SCRF). The modified SCRF treats the solution as a continuum of uniform effective dielectric constant, ϵ_{eff}, instead of the real dielectric constant, ϵ, because the solute aniline molecule on the solution surface is not completely solvated. Note that complete solvation of a given solute molecule is a prerequisite in SCRF. As argued in the Appendix, ϵ_{eff} is found to be approximately 80% of ϵ. In this calculation of the vertical ionization potential, only the electronic polarization is included because the molecular motion is too slow to follow the photoionization; the optical dielectric constants, ϵ_{opt}, are used, instead of the static dielectric constants ϵ_{stat}.

Table 1:

	IP(g)	ϵ_{stat}	$\Delta E_{g-s}^{\text{neutral}}$	ϵ_{opt}	$\Delta E_{g-s}^{\text{ion}}$	IP(g)
homogeneous	7.87	19.6[a]	−0.153	1.5[b]	-1.03	6.82
clustered		6.89[c]	−0.127	2.0[d]	-1.45	6.38
	7.7[e]				6.3[f]	

[a] ϵ_{stat} of the solution is assumed to be proportional to the molar concentration of the component.
[b] 80% of ϵ_{opt} measured by means of an Abbe's refractometer.
[c] ϵ_{stat} of pure aniline liquid.
[d] 80% of pure aniline liquid.
[e] results of photoelectron spectroscopy from [30].
[f] present result.

In the actual calculation, two cases are to be considered as to how the solute aniline molecule is distributed on the surface; (1) the aniline molecule is homogeneously distributed (homogeneous) or (2) the aniline molecule is clustered on the surface (clustered). In case (1), ϵ_{opt} of the aniline solution should be employed. Actually ϵ_{opt} measured by means of an Abbe's refractometer is adopted in the present calculation. In case (2), on the other hand, ϵ_{opt} of pure aniline should be employed.

The calculated IP's together with the other necessary parameters are listed in Table 1. As shown in Table 1, the measured IP agrees well with the calculation on the premise that the solute aniline molecule is clustered on the solution surface [19]. The clustering (inhomogeneous solvation) is supported by our previous result, where aniline polymer ions are found to be generated from an aggregated region of p-bromoaniline [31].

5 Appendix

In order to verify this modified SCRF calculation and to derive the effective dielectric constant, ϵ_{eff}, the regular SCRF method is applied to methyl, ethyl and propyl alcohols. Table 2 lists the comparison of the calculated IP's and the solvation energies with the measured ones. The calculated solvation energies of the neutral molecules, $\Delta E_{g-s}^{neutral}$, agree well with the vaporization energies.

Table 2:

alcohol	IP(g)	ϵ_{stat}	$\Delta E_{g-s}^{neutral}$	ϵ_{opt}	ΔE_{g-s}^{ion}	IP(s)
CH_3OH	10.82	33.62	-0.142	1.76	-1.38	9.70
	10.94[a]					9.99[b]
C_2H_5OH	10.37	25.30	-0.128	1.85	-1.35	9.42
	10.64[a]					9.66[b]
C_3H_7OH	10.22	20.10	-0.130	1.91	-1.32	9.30
	10.49[a]					9.72[b]

[a] results of photoelectron spectroscopy from [30]
[b] results of photoelectron spectroscopy from [32].

The calculated IP(s)'s are systematically smaller to a slight extent than the measured ones. This trend implies that ΔE_{g-s}^{ion} is always overestimated. The systematic overestimation is considered to originate from incomplete solvation of the surface alcohol molecules, so that the dielectric constant, ϵ is replaced by an effective dielectric constant, ϵ_{eff}. Namely a surface alcohol molecule of interest is treated as if it resides in a continuum of effective optical dielectric constant, ϵ_{eff} (see Sect. 4.3). Figure 4 shows the plot of the ΔE_{g-s}^{ion} vs ϵ_{eff}. The measured IP(s) is well reproduced when ϵ_{eff} is set to be 0.8 to 0.9 of ϵ_{opt}.

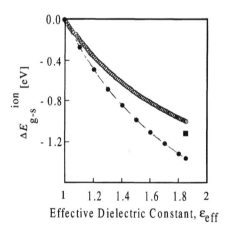

Figure 4: The solvation energies of the ion by the solvent, ΔE_{g-s}^{ion}, calculated by the quantum chemical SCRF method (solid circle), those calculated by the classical Born equation (open circle) at different ϵ_{eff} values, and the experimentally observed energy (solid square)

References

[1] K.B. Eisenthal: Acc. Chem. Res. **26**, 636 (1993)

[2] H. Siegbahn: J. Phys. Chem. **89**, 897 (1985)

[3] H. Morgner, J. Oberbrodhage, K. Richter, K. Roth: J. Electron Spectrosc. Relat. Rhenom. **27**, 61 (1991)

[4] P. Delahay: Acc. Chem. Res. **15**, 40 (1982)

[5] R.E. Ballard, J. Jones, D. Read, A. Inchley, M. Cranmer: Chem. Phys. Lett. **151**, 477 (1988)

[6] M. Faubel, S. Schlemmer, J.P. Toennies: Z. Phys. D. **10**, 269 (1988)

[7] M. Faubel, Th. Kisters: Nature **339**, 527 (1989)

[8] E. King, G.M. Nathanson, M.A. Hanning-Lee, T.K. Minton: Phys. Rev. Lett. **70**, 1026 (1993)

[9] R.E. Ballard, J. Jones, E. Sutherland: Chem. Phys. Lett. **112**, 452 (1984)

[10] R.E. Ballard, J. Jones, D. Read: Chem. Phys. Lett. **121**, 45 (1988)

[11] R.E. Ballard, J. Jones, D. Read, A. Inchley, M. Cranmer: Chem. Phys. Lett. **122**, 161 (1988)

[12] K. Bhattacharyya, A. Castro, K.B. Eisenthal: J. Chem. Phys. **95**, 1310 (1991)

[13] P. Delahay, A. Dziedzic: Chem. Phys. Lett. **108**, 169 (1984)

[14] K. Bhattacharyya, E.V. Sitzmann, K.B. Eisenthal: J. Chem. Phys. **87**, 1442 (1987)

[15] Y.R. Shen: Annu. Rev. Phys. Chem. **40**, 327 (1989)

[16] K.B. Eisenthal: J. Phys. Chem. **100**, 12997 (1996)

[17] J. Kohno, F. Mafuné, T. Kondow, J. Am. Chem. Soc. **116**, 9801 (1994)

[18] N. Horimoto, F. Mafuné, T. Kondow: J. Phys. Chem. B **103**, 1900 (1999)

[19] F. Mafuné, Y. Takeda, T. Nagata, T. Kondow: Chem. Phys. Lett. **218**, 234 (1994)

[20] H. Matsumura, F. Mafuné, T. Kondow: J. Phys. Chem. B **103**, 838 (1999)

[21] H. Matsumura, F. Mafuné, T. Kondow: J. Phys. Chem. **99**, 5861 (1995)

[22] N. Horimoto, F. Mafuné, T. Kondow: J. Phys. Chem. B **103**, 9540 (1999)

[23] N. Nishi: Z.Phys. D **15**, 239 (1990)

[24] N. Nishi, S. Takahashi, M. Matsumoto, A. Tanaka, K. Maruya, T. Takamuku, T. Yamaguchi: J. Phys. Chem. **99**, 462 (1995)

[25] F. Sobott, W. Kleinekofort, B. Brutschy: Anal. Chem. **17**, 3587 (1997)

[26] W. Kleinekofort, J. Avdiev, B. Brutschy: Int. J. Mass Spectrom. Ion Processes **152**, 135 (1996)

[27] J. Kohno, F. Mafuné, T. Kondow: J. Phys. Chem. in press.

[28] H. Morgner, J. Oberbrodhage: J. Phys. : Condens. Matter **7**, 7427 (1995)

[29] J.B. Foresman, A. Frisch: Exploring Chemistry with Electronic Structure Method, Gaussian, Inc., Pittsburgh (1996), p. 237

[30] K. Kimura, S. Katsumata, Y. Achiba, T. Yamazaki, S. Iwata: Handbook of HeI photoelectron spectra of fundamental organic molecules, Japan Scientific Societies Press, Tokyo (1981)

[31] Y. Hashimoto, F. Mafun, T. Kondow: J. Phys. Chem. **102**, 4295 (1998)

[32] M. Faubel, B. Steiner, J.P. Toennies: J. Chem. Phys. **106**, 9013 (1997)

Shedding Light on Heavy Molecules, One by One

Mattanjah S. de Vries
The Hebrew University of Jerusalem, Israel

1 Introduction

One of the major motivations for research employing molecular beams is the capability to work with individual molecules under collision free conditions. However, the production of molecular beams has traditionally been limited to small and stable molecules with sufficiently high vapor pressures to be effectively seeded without extensive heating. While relatively high temperature molecular beam sources have been used, most larger molecules tend to decompose upon heating. New techniques have been developed that produce large molecular ions for analytical purposes, such as laser desorption, matrix assisted laser desorption, and ion spray. Of these, only the first can be used to reliably produce intact neutral molecules, which can be combined with a molecular beam. We capitalize on this capability of laser desorption to bring complex neutral molecules into the gas phase without ionization, for fundamental spectroscopic studies. This becomes particularly powerful when the desorbed material is entrained in a supersonic expansion. The cooling in a supersonic jet (a) stabilizes the molecules by reducing their internal energy, (b) makes it possible to carry out spectroscopy, and (c) allows for the study of clusters. The feasibility of this approach has been demonstrated by previous work and by our own work and on small biomolecules [1-10]. When extending these techniques to larger molecules we need to address the question whether such systems can be photoionized at all [11,12]. In what follows we demonstrate that REMPI spectroscopy of molecules with masses of several thousands of Daltons is possible [13-15], and we present examples of applications to polymers and small biomolecules and complexes containing those.

After describing the technique in section 2, we proceed in the following sections to present REMPI spectra of a number of different types of laser desorbed, jet cooled molecules, which could not otherwise be recorded. In section 3 we discuss REMPI spectroscopy of polymers. In section 4 we show vibronic spectroscopy of dipeptides, while we do the same for DNA bases in sections 5 and 6, with nucleosides and base pairs respectively.

2 Experimental approach

Fig. 1 schematically shows the combination of laser desorption with a pulsed molecular beam and a REMPI setup. Material is laser desorbed from a sample probe in front of a pulsed nozzle. The desorption laser is usually a Nd:YAG laser operated at its fundamental wavelength of 1064 nm. At this wavelength one does not expect photochemical interaction with any of the materials, which we desorb. We have, however, also in many cases successfully used the second and fourth harmonic as well as a 248 nm KrF laser for desorption. Laser desorption involves heating of the substrate, rather than the adsorbate. Therefore it is typically desirable to match the wavelength of the desorbing light with the absorption characteristics of the substrate, while avoiding overlap with the absorption spectrum of the adsorbate. We routinely use graphite as a substrate, although we have also successfully used metal substrates. Typical laser fluences are of the order of 1 mJ/cm^2. or less, which is significantly less than the fluences normally used for ablation. The laser is focused to a spot of the order of 0.5 mm diameter within 2 mm in front of the nozzle. This is important because in a supersonic expansion most of the cooling takes places close to the nozzle by collisions with the drive gas along a distance of about 10 nozzle diameters. The nozzle consists of a pulsed valve with a nozzle diameter of 1 mm. We usually operate with Ar as a drive gas at a backing pressure of about 5 atmospheres.

In earlier work we have optimized the geometry for effective entrainment by mapping entrained perylene with laser induced fluorescence [17,18]. In that work we found that it is possible to entrain a portion of the desorbed material on the axis of the supersonic beam, such that the ionizing laser downstream can interact with a fraction of about 10^{-5} of the desorbed material.

Downstream ionization lasers intersect the beam inside the source region of a reflectron time of flight (TOF) mass spectrometer (Jordan Co.). Excimer lasers and a dye laser, used for two photon ionization, intersect the beam at right angles. A vacuum ultraviolet (VUV) beam, used for single photon ionization, is aligned colinearly with the gas beam.

3 REMPI of polymers

Fig. 2 shows two photon ionization mass spectra of laser desorbed and jet cooled polymers of the form:

P1 = A-O-R$_n$-E (A1)

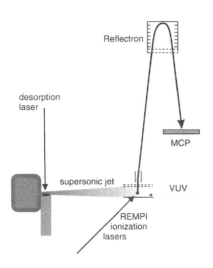

Figure 1: Experimental setup

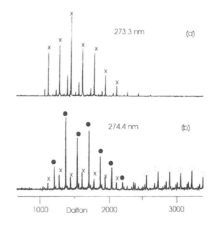

Figure 2: Mass spectra of PFPE polymer sample, two photon ionized at resonant wavelength for (a) monomers and (b) dimers

with a repeat unit R, a non functional end group E and a functional end-group A, consisting of an aromatic ester as follows:

$R = -[CF_2]_3\text{-}O\text{-}$,

$A = -CF_2\text{-}(CO)\text{-}O\text{-}CH_2CH_2\text{-}O\text{-}C_6H_5$,

$E = -O\text{-}CF_2CF_3$.

Bifunctional polymers have the form:

$P2 = A\text{-}O\text{-}R_n\text{-}A.$ (A2)

All major peaks correspond to parent ions and are spaced apart by 166 Da, the mass of a repeat unit. We have obtained parent molecular peaks from polymers with n = 5 (m/z = 1114 Da) to n = 40 (m/z = 6924 Da). The averages of the molecular weight distributions are consistent with NMR as well as with size exclusion chromatography. This is important because of the question of possible fundamental limitations on the size of molecules that can be efficiently photo-ionized [11,12]. We find no evidence of a significant decline in ionization efficiency with molecular size in the current results, which include ionization of molecules up to 7000 Dalton. Nor do we find any evidence for delayed ionization.

By varying the wavelength of the two photon ionization laser, while monitoring polymer parent mass peaks, we obtained excitation spectra. We obtained the S_{0-0} transition of the polymers, which we found to be slightly redshifted with respect to that of the anisole chromophore, $CH_3\text{-}O\text{-}C_6H_5$, and conformationally broadened. We found that the REMPI spectra do not change when the polymer is extended beyond just a few repeat units. This indicates that the excitation is local, in the aromatic chromophore, and completely unaffected by the length of the polymer.

We also obtained spectra of polymer dimers P1-P1, which all show a broadening and a characteristic redshift of about 110 cm^{-1} compared to the monomer origin. Such a shift is characteristic for a van der Waals interaction between two phenyl rings. We find remarkably similar broadening and redshifts in monomer spectra of bifunctional polymers (P2).

The mass spectra in Fig. 2 were obtained from a mixture of monofunctional, P1, and bifunctional, P2, polymers. Crosses indicate monomer peaks, open circles indicate dimer peaks and closed circles P2 type polymers. Panel (a) is obtained at the resonant wavelength for P1 monomers. Panel (b) is obtained at the resonant wavelength for P1-P1 dimers, which is also the resonant wavelength for P2 polymers. The strong similarity in the wavelength spectra of the dimers of the monofunctional polymers on the one hand, and the isolated, doubly functionalized polymers on the other hand, leads to the conjecture that the chromophore environment in both cases are similar. One way of achieving this similarity is to form an *intra*molecular complex in the P2 monomer that resembles the *inter*molecular complex in the P1 van der Waals dimer. This involves similar interactions between pairs of chromophores in each case. In other words, the P2 polymers form *intra*molecular dimers.

For internal complex formation between two endgroup chromophores, the chains must be flexible enough so that during the jet expansion they can efficiently bend to bring the chromophores together. Furthermore, the interaction between chromophores must be strong enough to effectively form the intramolecular complex once the chromophores are brought together. Combinations of polymer chains of variable flexibility with different chromophores can provide an interesting arena for testing models based on barriers to internal rotation and binding energy.

4 REMPI of biomolecules: Peptides

An example of REMPI spectroscopy of laser desorbed, jet cooled dipeptides appears in Fig. 3. We compare the spectra of two dipeptides that contain tyrosine as a chromophore, Tyr-Ala and Ala-Tyr. We have recorded a much larger wavelength range and many more peptide combinations, but here we merely show the origin region of representative examples. The structure in the spectra is due to multiple conformations and possibly to very low energy vibrations. The possible conformations for tyrosine and its derivatives have been discussed by several authors [19-23]. Some tentative assignments have been proposed based on several assumptions and comparisons with derivative spectra. When comparing with the dipeptides it appears that either

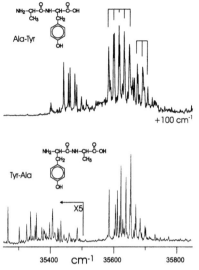

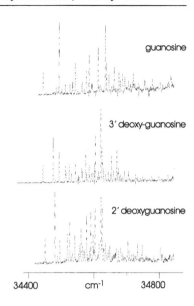

Figure 3: REMPI spectra of dipeptides

Figure 4: REMPI spectra of three nucleosides

the number of conformations increases drastically, or there are sequences of low energy vibrations. The former interpretation seems less likely because these multiple peaks do not occur for dipeptides in which we replaced tyrosine with phenylalanine. The only difference between these molecules is the hydroxyl group which is present in Tyr and absent in Phe. Another remarkable feature is the group of peaks to the red in Ala-Tyr, which may be due to an interaction of the amino terminus with the -OH group in the tyrosine ring, similar to the intramolecular exciplexes observed in tryptophan.

In addition to structural information that can be obtained from these spectra, we notice the tremendous isomeric distinction that can be obtained by wavelength dispersion in addition to mass spectral detection.

5 REMPI of biomolecules: Nucleosides

The study of spectroscopic properties of DNA bases is of fundamental importance. Both vibrational and electronic spectroscopy can help elucidate issues of structure, bonding, and reactivity. UV spectroscopy, especially in the 300 nm region, is also crucial for understanding the photophysics and photochemistry of genetic material. In order to facilitate investigation of complex oligonucleotides, we need detailed information on the basic building blocks, single nucleotides and nucleosides. Most work to date on those molecules has been in solution and in the solid state.

There have been many theoretical studies of vibrational and electronic states of DNA bases [24-26]. These are complicated by multiple lone pair electrons, limited symmetry, and possible tautomerism. Experimentally there have been many reports of absorption spectroscopy [27,28], IR data [29,30], and Raman spectroscopy [31]. Samples have been in the form of vapor, solutions, polycrystalline material, single crystals, and cold matrices. Gas phase studies can offer the advantage of eliminating intermolecular interactions.

Fig. 4 shows the REMPI spectra of three jet cooled guanosines. For comparison the REMPI spectrum of the corresponding base, guanine, appears in Fig. 5. A detailed computation of S1 state vibrations is under way. Absent that, we can make initial assignments by comparison with ground state calculations and with matrix IR data. When studying alkyl substituted guanines, we observed a significant difference between methyl substitution in the 1 and the 9 position. The latter shows a much larger spectral shift, suggesting that the electron pair associated with the 9 nitrogen is more important for the S1 state than that on the 1 nitrogen. The sugar in guanasine is attached in the 9 position and we find that spectral shift in the guanosines is similar to those of the 9 alkyl substituted guanines. The low energy vibrations in these spectra can be understood as vibrations, such as torsion, between the chromophore and the sugar ring.

The spectral differences between the three guanosines, presented here, are small, indicating the electronic excitation is localized in the guanine chromophore. On the other hand the differences are large enough to provide clear spectroscopic distinction between isomers, such as 2' *vs* 3' deoxyguanine.

6 DNA base pairs

In order to form isolated base pairs in the gas phase we laser desorb a mixture of neat guanine (G) and cytosine (C) from a graphite surface, followed by entrainment in the supersonic expansion of argon. Fig. 5 shows the REMPI spectrum of the GC dimer. For comparison we also show the spectra of the GG dimer and of the G monomer, which we recorded simultaneously in the same experiment. The spectra are offset such that the origins (0-0 transitions) are shown at the same position. Actually the GC origin is blue-shifted by 435 cm^{-1} with respect to the G origin, which is at 32878 cm^{-1}. The GG origin is blue-shifted with respect to that of G by 224 cm^{-1}. Neither the G nor the C chromophore of the dimers absorbs in the first 200 wavenumbers above the origin of the complex. The guanine spectrum exhibits its first significant vibronic activity at 235 cm^{-1}. Cytosine absorbs altogether much further to the blue at about 36000 cm^{-1}, similar to uracil and thymine [3]. Therefore, when we excite the guanine chromophore in the GG and GC complexes, we can observe hydrogen bond vibrations built on the 0-0 transition without any interference. Indeed all spectral lines in the first two hundred wavenumbers of these spectra can be understood as hydrogen bond vibrations.

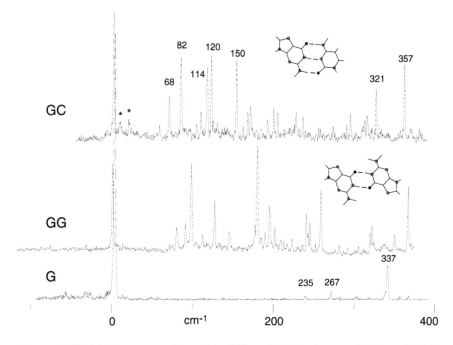

Figure 5: REMPI spectra of guanine (G) and of the base pair formed with cytosine (GC), and with itself (GG)

We find a remarkable agreement between experiment and theory[34-37]. However, we should note that this represents a comparison between excited state measurements and ground state calculations. A final assignment has to be postponed until reliable S1 state calculations, preferably on the CASCF level, are available. We also note the underlying assumption that these bases are predominantly in the keto form, consistent with a single origin in the guanine spectrum [38,39].

Ab initio calculations of the GG structure at different levels of theory agree that the symmetric structure displayed in Fig. 5 is the most stable one [32,33]. When both monomer parts in the dimer are completely identical, only one electronic spectrum is expected. We see no hint of band exciton splittings or a second electronic band system in the investigated spectral range. We therefore conclude that GG has the double hydrogen bonded highly symmetrical structure depicted in Fig. 5.

We desorbed from an equimolar mixture of guanine, adenine, cytosine, thymine, and uracil and found mass peaks for all combinations of G with the other bases at high ionization laser power under non-resonant conditions. We did not, however, observe any vibronic resonances for any of the possible complexes besides those of GG and GC in the investigated spectral range. Ab initio calculations at different levels of theory agree that the GC Watson-Crick base pair with three nearly linear hydrogen bonds and the symmetrically hydrogen bonded GG base pair are by far the most stable base pairs [34].

7 Conclusion

By combining laser desorption and entrainment in a supersonic beam, we can study large and fragile molecules in the gas phase. Such molecules have very low vapor pressures and would normally decompose when heated. By entraining laser desorbed neutral molecules in a supersonic expansion, we can create a beam of large cooled molecules. We demonstrated REMPI spectroscopy of molecules with masses of several thousands of Daltons. We applied this capability to study internal and external dimer formation in polymers. We presented examples of vibronic spectroscopy of jet cooled small biomolecules and their clusters, such as dipeptides, nucleosides and DNA base-pairs.

Acknowledgement This work has been supported by grant no. 96-00217 from the United States-Israel Binational Science Foundation (BSF), Jerusalem, Israel and by the Israel Science Foundation, founded by The Israel Academy of Sciences and Humanities. Valuable contributions are acknowledged of Prof. Karl Kleinermanns, Dr. Louis Grace, Eyal Nir, and Rami Cohen.

References

[1] T.Rizzo, Y.Park, and D.H.Levy, J.Chem.Phys., **85**, 6945 (1986)

[2] J.Cable, M.Tubergen, and D.H.Levy, J.Am.Chem.Soc., **109**, 6198 (1987)

[3] B.B.Brady, L.A.Peteanu, D.H.Levy, Chem.Phys.Lett., **147**, 538 (1988)

[4] R.Tembreull and D.Lubman, Anal.Chem., **59** 1003 (1987)

[5] R.Tembreull and D.M.Lubman., Appl.Spec., **41**, 431 (1987)

[6] L.Li and D.Lubman, Appl.Spec., **43**, 543 (1989)

[7] H.v.Weyssenhoff, H.Selzle, and E.W.Schlag, Naturforsch., **40a**, 674 (1985)

[8] J.Grotemeyer, U.Boesl, K.Walter, and E.W.Schlag., Am.Chem.Soc., **109**, 2842 (1987)

[9] R.Weinkauf, P.Schanen, A.Metsala, E.W.Schlag, M.Buergle, and H.J.Kessler, Phys.Chem., **100**, 18567 (1996)

[10] T.Arikawa and H.Yazawa, Jap.J.App.Phys., Part 1 (Regular Papers & Short Notes)., **35**, 2332 (1996)

[11] E.W.Schlag, J.Grotemeyer, and R.D.Levine, J.Phys.Chem., **96**, 10608, (1992)

[12] E.W.Schlag, J.Grotemeyer, and R.D.Levine, Chem.Phys.Lett., **190**, 521 (1992)

[13] D.S.Anex, M.S.de Vries, A.J.Knebelkamp, J.Bargon, H.R.Wendt, and H.E.Hunziker, Int.J.Mass Spectrom. Ion Processes, **131**, 319 (1994)

[14] H.E.Hunziker, and M.S.de Vries, J.Appl.Surface Science, **106**, 466 (1996)

[15] E.Nir, H.E.Hunziker, and M.S.de Vries, Anal.Chem., **71**, 1674 (1999)

[16] E.Nir, L.Grace, B.Brauer, and M.S.de Vries, J.Am.Chem.Soc. 1999, **121**, 4896 (1999)

[17] P.Arrowsmith, M.S.de Vries, H.E.Hunziker, and H.R.Wendt, Appl.Phys.B, **46**, 165 (1988)

[18] G.Meijer, M.S.de Vries, H.E.Hunziker, and H.R.Wendt, Appl.Phys.B, **51**, 395 (1990)

[19] L.Li, and D.Lubman D., Anal.Chem., **60**, 1409 (1988)

[20] T.Chin-Khuan, and M.Sulkes, J.Chem.Phys., **94**, 5826 (1991)

[21] S.J.Martinez III, J.C.Alfano, and D.H.Levy, J.Molec.Spec., **156**, 421 (1992)

[22] S.J.Martinez III, J.C.Alfano, and D.H.Levy, J.Molec.Spec., **158**, 82 (1993(

[23] G.Alagona, G.M.Ciuffo, and C.Ghio, Theo.Chem., **117**, 255 (1994)

[24] P.R.Callis, Ann.Rev.Phys.Chem., **34**, 329 (1983)

[25] M.P.Fulscher, L.Serrano-Andres, and B.O.Roos, J.Am.Chem.Soc., **119**, 6168 (1997)

[26] J.Leszczynski, J.Phys.Chem.A, **102**, 2357 (1998)

[27] D.Voet, W.B.Gratzer, R.A.Cox, and P.Dotty, Biopolymers, **1**, 193 (1963)

[28] L.B.Clark, G.G.Peschel, and I.Tinoco Jr., J.Phys.Chem., **69**, 3615 (1965)

[29] K.Szczepaniak and M.Szczesniak, J.Molec.Struct., **156**, 29 (1987)

[30] J.Florian, J.Phys.Chem., **97**, 10649 (1993)

[31] J.M.Delabar and M.Majoube, Spectrochim.Acta, **34A**, 129 (1987)

[32] J.Florian and J.Leszczynski, J.Molec.Struct., **349**, 421 (1995)

[33] O.Shishkin, J.Sponer, and P.Hobza, J.Molec.Struct., **477**, 15 (1999)

[34] R.Santamaria, E.Charro, A.Zacarias, and M.Castro, J.Computational Chemistry, **20**, 511 (1999)

[35] P.Hobza, M.Kabelac, J.Sponer, P.Mejzlik, and J.Vondrasek, J.Computational Chemistry, **18**, 1136 (1997)

[36] J.Sponer, J.Leszczynski, and P.Hobza, J.Phys.Chem., **100**, 1965 (1995)

[37] P.Hobza and J.Sponer, Chem.Phys.Lett., **261**, 379 (1996)

[38] T.K.Ha, H.J.Keller, R.Gunde, and H.H.Gunthard, J.Phys.Chem A, **103**, 3312 (1999)

[39] M.J.Nowak, K.Szczepaniak, A.Barski, and D.Shugar, Z.Naturforsch.C., **33**, 876 (1978)

Molecular Beam Studies of DNA Bases

C. Desfrançois and J.P. Schermann
Laboratoire de Physique des Lasers, Institut Galilée Université
Paris 13, Villetaneuse, 93430, France

1 Introduction

The genetic information is coded in a linear polymer, DNA (deoxyribonucleic acid), which is build up of simple units called nucleotides. The native state of DNA is a double helix of two antiparallel strands winding around each other. Each nucleotide possesses the same phosphate-sugar group (pentose) which constitutes the strand backbone and one of the four nucleobases: Thymine (T), Adenine (A), Cytosine, (C) and Guanine (G). Deciphering DNA, which represents the aim of the genome program, is thus reading meaningful base sequences called genes which contain the information. This reading is called transcription and is performed by an enzyme which produces RNA strands, called messenger RNAs, into which T is replaced by a very similar molecule: Uracil (U). In a given strand, bases are stacked one over another while bases on opposite strands are precisely paired: A pairs with T, C with G. Hydrogen bonds between DNA bases ensure this unique pairing and also allow for the separation of the two strands during replication of the genetic content when cells divide [1]. Each strand then tries to replace the nucleotides it has lost by establishing new hydrogen bonds with complementary bases.

The discovery by Watson and Crick of this universal code in living bodies [2] prompted quantum chemistry calculations concerning isolated or paired bases [3, 4]. For example, a possible mechanism for mutations was attributed to the tunneling of one or two protons through H-bonds in an A..T or a C..G pair [5, 6]. Ab initio calculations concerning DNA bases have been recently reviewed by Sponer and Hobza [7]. In addition to their fundamental biological interest, those molecules offer a wide number of possibilities for establishing hydrogen bond or stacking interactions, as compared to simpler systems, such as substituted aromatics which have been widely studied in molecular cluster physics. More and more precise calculations, dubbed "gas-phase calculations", consider isolated or paired molecules, often neglecting the ubiquitous presence of water and of the sugar-phosphate backbone of DNA, at some distance from real biological conditions. Nevertheless, those theoretical results can be directly compared to experimental studies which are the scope of this review and are the starting point of investigation of real nucleotide properties [8].

2 Beam techniques

Gas-phase studies of DNA bases face two major problems, one due the very low vapor pressures of those molecules at room temperature and their ease for decomposition upon heating, the second problem arising from the need of efficient cooling to observe well-resolved spectra. The pioneer experimental work has been performed by the group of L. Sukhodub [9–11] who used glass cell effusion sources to evaporate and mix DNA bases between themselves or to water, leading to homogeneous or hydrated clusters. A tip emitter situated in the mixing zone close to the cell exits allowed for production of cations by means of field-ionization ($10^6 - 10^7$ V/cm). Ions were separated in a mass-spectrometer and ion signal dependencies were monitored as function of cell temperatures, providing association constants and enthalpies for pair formation. Effusive [12] and supersonic expansions have since been used by several groups with different approaches. In a first set of experiments, a carrier gas (few bars of He or Ar) is flown over a heated oven containing crystalline nucleobase powder, typically in between 180 and 230 °C and the seeded beam is produced by means of a continuous or pulsed expansion. Pin holes, conical and slit nozzles [13–20] can be used and must be kept approximately 10 to 20 °C above the oven temperature. When the beam composition is monitored by mass spectrometry, one rapidly observes decomposition products as a consequence of overheating, within tens of °C above the optimum temperature. In laser desorption experiments [21, 22], thin layers of nucleobases deposited on a surface are submitted to 10.6μ CO_2 or 1.06μ YAG laser irradiation (~ 1 mJ/cm^2). The desorbed material is entrained in a pulsed supersonic expansion of argon into which multiple collisions lead to cooling of internal degrees of freedom and production of clusters. Another alternative is offered by the use of very large clusters of rare gases which efficiently pick up molecules when crossing very low density vapors [23–25].

Beams of larger molecular ions containing several nucleobases can be formed by means of several techniques such as electrospray [26], matrix assisted laser desorption [27], laser desorption from liquid beams [28, 29] but will not be here considered.

3 Experimental studies

3.1 Enthalpies of pair formation

Base pairing in the gas-phase may not necessarily be the same as in a double helix since nucleobases are tighten to the DNA backbone by means of covalent bonds between one of their nitrogen atoms (N1 for pyrimidines: T and C, N9 for purines : A and G) and a sugar molecule (ribose). Thus, if one wishes to investigate hydrogen bonding in the gas-phase, any H-bond involving a N1 atom for a pyrimidine or a N9 atom for a purine should be discarded. Even in this case, Watson–Crick (WC) base pairings between A and T or C and G are not the only possibilities. In fact, still 28 base pairs involving at least two cyclic H-bonds between A, T(U), C and G remain possible and their ordering as a function of their enthalpies of formation either theoretically [1, 7] or experimentally is not obvious. A necessary assumption for comparison between "gas-phase" calculations and experimental determinations is that thermal equilibrium is achieved and thus observed configurations correspond to the lowest energy values.

By means of temperature-dependent field-ionization, the group of Sukhodub [10] has measured enthalpies of formation of several nucleobase pairs. The most stable is the C-G pair which involves three hydrogen bonds (Table 1). The number of hydrogen bonds is not the only factor which governs the relative stabilities but the mutual orientation of dipoles is also important [7]. Moreover, the lowest-energy configurations of pairs which are formed in the gas-phase do not necessarily correspond to those which can be formed in a double helix. Those different configurations can be predicted by means of either high-level *ab initio* [7] or semiempirical calculations [40]. For example, the lowest-energy configuration displayed in Fig. 1 of the gas-phase A-T pair has a dipole moment of 3 D, as compared to the biologically relevant Watson–Crick (WC) pair which has dipole moment of 1.8 D. In order to favor formation of WC pairs, Sukhodub's group measured enthalpies of pair formation of bases methylated in N1 and N9 positions, for example, 9-methyladenine (9-MA) and 1-methylthymine (1-MT). Using additional methylation in order to artificially suppress all H-bonding possibilities, stacking pairs were observed into which dispersion interactions became dominant.

Another important mean for modifying H-bonding possibilities is the addition of ribose molecules to nucleobases (nucleosides). Schlag and coworkers [30] have compared formation of nucleotide pairs to that of nucleoside pairs created in a supersonic expansion following IR laser desorption. Neutral isolated bases and pairs were formed in the gas-phase and ionized by single-color multiple-photon absorption. This technique does not allow for the determination of pair configurations and more refined spectroscopic techniques are now available (Sect. 3.5). Only approximate association constants of different pairs were derived from measurements of the relative monomer and dimer ion signals. The observed trends for the ordering of nucleobase pairs are given in Table 1 and compared to predictions of ab initio calculations. This ordering is modified when nucleosides are considered. The most stable pair remains C-G but the A-T pair is strongly destabilized by the addition of sugar molecules. A similar effect appears when a single water molecule is added to the A-T pair which undergoes a transition from the usual planar H-bonded configuration to a stacked configuration.

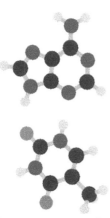
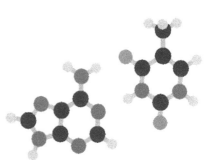

lowest energy configuration of the A-T pair $E = 13.3\,\text{kcal/mol}$ $\mu = 3.1\,\text{D}$

Watson–Crick A-T pair $E = 13.3\,\text{kcal/mol}$ $\mu = 1.8\,\text{D}$

Figure 1: Gas-phase Watson–Crick and lowest energy configurations of the adenine-thymine pair. Structures, binding energies and resultant dipoles are the results of semiempirical calculations [40]

3.2 Ionization potentials

Mutations and carcinogenic effects can be induced by high-energy radiations which are responsible for modifications of DNA [31]. Those radiations can directly ionize nucleobases or produce secondary electrons which can be trapped on bases [32]. Initial damages corresponding to modification of bases occur randomly on DNA but further migrate in the chain of nucleotides and finally end up on specific sites. The ordering of ionization potentials (IP) and electron affinities (EA) of the different nucleobases govern this migration and the localization of damages.

Ionization potentials of adenine and thymine and their hydrated clusters have been experimentally studied by Herschbach and coworkers [19] by means of electron bombardment. The determination of the cytosine ionization potential was hampered by the tautomeric conversion to uracil.

Table 1: Comparison between relative stabilities of H-bonded nucleobase pairs and nucleoside pairs

Nucleobase	C-G	G-G	C-C	A-U
association constant (arb. unit) exp. [30]	17	8.1	1.9	
enthalpy of formation (eV) exp. [10]	0.91		0.76	0.63
enthalpy of formation (eV) theory [7]	1.03 (WC)	0.96	0.76	
Nucleobase	A-T	U-U	T-T	A-A
association constant (arb. unit) exp. [30]	1.1		0.6	0.6
enthalpy of formation (eV) exp. [10]	0.56	0.41	0.39	
enthalpy of formation (eV) theory [7]	0.51 (WC)		0.43	0.48

The measured vertical ionization potentials (energy difference between the neutral base and its cation in the geometry of the neutral) of A and T, respectively equal to 8.45 (0.7) and 9.15 (0.7) eV, are in good agreement with the calculated ab initio values of Sevilla et al. which are respectively equal to 8.28 and 9.97 eV. More precise measurements of vertical and adiabatic ionization potentials by means of two-color resonant ionization at threshold would rely on a detailed interpretation of high-resolution nucleobase UV spectra which have up to now been reported only for guanine (see Sect. 3.5). The ubiquitous presence of water molecules in DNA [33] modifies IP and EA values. If one considers a molecule M solvated by N polar molecules S, the respective ionization potentials IP(M) of the isolated molecule M and IP(N) of the $M..S_N$ cluster satisfy the following relationship:

$$IP(N) = IP(M) + E_{solv}^{neutral}(N) - E_{solv}^{ion}(N).$$

Solvation energies of ions $E_{solv}^{ion}(N)$ being larger than solvation energies $E_{solv}^{neutral}(N)$ of neutrals, a decrease of ionization potentials of A and T is observed when the number of water molecules increases [19]. A similar solvation effect is observed for electron affinities.

3.3 Electron affinities

Nucleobases are strongly polar and it has been observed that they can give birth to different kind of anions. Valence ("conventional") anions correspond to excess electrons localized in valence orbitals while multipole-bound anions correspond to excess electrons in diffuse orbitals. Valence vertical electron affinities which are the energy differences between the neutral parent and the negative ion at the geometry of the neutral have been measured by electron transmission spectroscopy [34] and are negative for the four bases, in agreement with ab initio calculations [8]. Multipole-bound electron affinities of A, T, U and C, which correspond to excess electrons loosely bound by mainly electrostatic interactions (dipole, quadrupole..) at large distances, have been measured by means of photoelectron spectroscopy (PES) [17, 20, 35] and Rydberg electron transfer spectroscopy [36, 37] and the results are in good agreement with calculations [38, 39]. The situation is not as clear for adiabatic valence electron affinities which are those relevant for biology (Table 2). Adenine has a clear negative valence electron affinity. The structures of adenine-water clusters and their valence anions, calculated by means of a semiempirical model, are displayed in Fig. 2. In contrast with the case of multipole-bound anions which retain the geometrical structures of their neutral parents, one can note the drastic geometry modifications which are induced by the presence of extra electrons in valence orbitals. Water molecules tend to reorient their dipoles towards the negative excess charge localized in π orbitals of adenine. Thymine and uracil possess positive valence electron affinities as demonstrated by RET measurements [37, 40] and observation of stable anions in free electron attachment experiments [12]. A DFT calculation provides a value of 63 meV for uracil [37] for the valence electron affinity, smaller than the measured multipole-bound electron affinity of 93 meV [17]. Interestingly, solvation by a single argon atom is sufficient to switch from electron attachment into a diffuse orbital to attachment into a valence orbital. Solvation by a single water molecule leads to valence anions [35, 36]. The valence EA of cytosine is probably negative [41] and the stable C^- anions which have been observed in free electron attachment experiments [12] may be metastable with respect to autodetachment. Calculations of valence electron affinities are difficult and since those of nucleobases are close to zero, even their signs are not always known.

Table 2: Predicted and measured electron affinities of nucleobases

Nucleobase	G	C	A	T
EA_v (theory) [42]	−1.14 eV	−0.37 eV	−0.68 eV	−0.30 eV
EA_v (exp. ETS) [34]	–	−0.32 eV	−0.54 eV	−0.29 eV
EA_{ad} (theory) [42]	−0.67 eV	0.19 eV	−0.19 eV	−0.29 eV
EA_{ad} (exp. RET) [40]	–	−0.55 – 0 eV	−0.45 eV	≈ 0

The ordering predicted by scaled ab initio calculations for nucleobases EA values: A < C < T seems to be confirmed by experiments and is in agreement with the observation that in single stranded DNA at low temperatures, thymine serves as a site for localization of electrons [43].

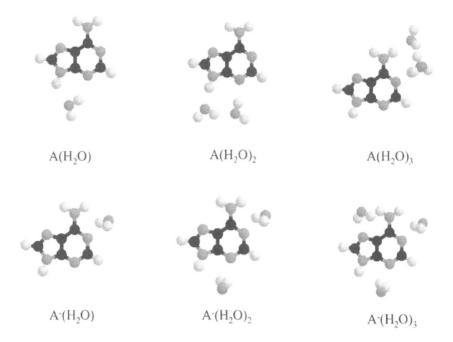

Figure 2: Structures of neutral and negatively-charged hydrated adenine complexes

3.4 Microwave and infrared spectroscopy

Microwave spectroscopic studies of isolated nucleobases have been performed by the group of Brown with supersonic beams crossing a Stark-modulated spectrometer in the region of 60 Ghz [15]. For U and T, only diketo tautomers were observed and their geometries were planar. Thanks to the high sensitivity of the microwave spectrometer, three tautomers of cytosine, among the 6 possibilities, were identified through measurements of rotational constants. From a biological point of view, the existence of such tautomers authorizes a different form of DNA with parallel strands due to formation of G-C base pairs called reverse Watson–Crick (RWC) pairs [44]. A far-infrared ($\approx 1700\,\text{cm}^{-1}$) spectrum of uracil has been recorded at high resolution by using a pulsed slit supersonic jet [18, 45]. Those spectra are generally very rich and their complex assignment strongly benefits from gas [46] or matrix [47, 48] experiments which cover a vast spectral region ($100 - 3700\,\text{cm}^{-1}$) by means of Fourier transform spectroscopy.

3.5 UV spectroscopy

UV spectroscopy of DNA bases is difficult due to possible decomposition and strong overlap of rovibrational bands. Resolved UV fluorescence spectroscopy of isolated uracil was observed in a seeded supersonic beam for by Ito and coworkers [13] but Levy further showed that these observations can easily be plagued by the presence of spurious impurities and that great care must be taken when heating those compounds. The fluorescence yield is also impaired by a very fast non-radiative relaxation channel, most probably ISC. This is supported by the observed efficient phosphorescence of some bases in solution and by the observed increased ionization efficiency of adenine when femtosecond ionization pulses are used as compared to nanosecond ones.

Laser desorption with a YAG laser of guanine and guanine-cytosine mixtures from a graphite surface followed by entrainment in a supersonic expansion of argon gas allows for the observation of sharp features in resonance enhanced multiphoton ionisation (REMPI) spectroscopy [22]. The 0-0 vibrational origin of the $S_1 \leftarrow S_0$ transition of the G monomer has been identified at $32\,878\,\text{cm}^{-1}$. The REMPI spectrum extends up to $35\,000\,\text{cm}^{-1}$ and a group of peaks around $33\,000\,\text{cm}^{-1}$ has been attributed to ring breathing vibrations. The 0-0 origin transitions of the REMPI spectra of the G-G and G-C pairs are respectively shifted by 235 and $446\,\text{cm}^{-1}$ with respect to the G monomer. Since cytosine absorbs much farther in the blue (around $36\,000\,\text{cm}^{-1}$), the only excited chromophore is G. This allows for the first observation of intermolecular vibrational frequencies of the GC pair in the S_1 state. A first attribution of vibrations seems to correspond to those of the most stable configuration of the GC pair in gas-phase which is the Watson–Crick configuration. Complete Active Space SCF (CASSCF) calculations [49] will be necessary to improve this attribution of intermolecular vibrational modes.

Conclusion

A large number of experimental techniques which have been applied to simple model systems can now be applied to complexes of molecules of biological interest, in particular to complexes containing nucleobases. For example, the dynamics of tautomerization has been investigated on a femtosecond timescale in 7-azaindole dimers which contain two hydrogen bonds by Zewail et al. [50, 51]. The same approach could be applied to the study of proton transfer in isolated [52] and hydrated [53] nucleobases or their pairs which has already been theoretically considered. The WC A-T and C-G pairs are not the only possibilities in biological systems and some other associations called wobble base-pairs, such as G-U, are currently encountered in transfer RNA-messenger RNA recognition [1]. Precise measurements of the properties of wobble pairs in the gas-phase, in conjunction with ab initio calculations may provide tools for improving semi-empirical models of RNA.

References

[1] W. Saenger: Principles of Nucleic Acid Structure, Springer-Verlag, 1984

[2] J.D. Watson, F.H. Crick: Nature **171**, 737 (1953)

[3] B. Pullman, A. Pullman: Quantum Biochemistry, Interscience: New York, 1963

[4] J. Langlet, P. Claverie, F. Caron, J.C. Boeuve, Int. J. Quant. Chem. **19**, 299 (1981)

[5] P. Löwdin: Rev. Mod. Phys. **35**, 724 (1963)

[6] J. Florian, V. Hrouda, P. Hobza: J. Am. Chem. Soc. **116**, 1457 (1994)

[7] P. Hobza, J. Sponer: Chem. Rev. **99**, 3247 (1999)

[8] A.O. Colson, M.D. Sevilla: Int. J. Radiat. Biol. **67**, 627 (1995)

[9] L. Sukhodub, I.K. Yanson: Nature **264**, 247 (1976)

[10] L.F. Sukhodub: Chem. Rev. **87**, 589 (1987)

[11] I. Galevitch, S.G. Stepanian, V. Shelkovsky, M. Kosevich, Y.P. Blagoi, L. Adamowicz: J. Phys. Chem. 2000, in press

[12] M.A. Huels, I. Hahndorf, E. Illenberger, L. Sanche: J. Chem. Phys. **108**, 1309 (1998)

[13] Y. Tsuchya, T. Tamura, M. Fuji, M. Ito: J. Phys. Chem. **92**, 17601 (1988)

[14] B.B. Brady, L.A. Peteanu, D.H. Levy: Chem. Phys. Lett. **147**, 538 (1988)

[15] R.D. Brown, P.D. Godfrey, D. McNaughton, A.P. Pierlot: J. Am. Chem. Soc. **111**, 2308 (1989)

[16] C. Desfrançois, H. Abdoul-Carime, C.P. Schulz, J.P. Schermann: Science **269**, 1707 (1995)

[17] J.H. Hendricks, S.A. Lyapustina, H.L. de Clercq, J.T. Snodgrass, K.H. Bowen: J. Chem. Phys. **104**, 7788 (1996)

[18] K. Liu, R.S. Fellers, M.R. Viant, R.P. McLaughlin, M.G. Brown, R.J. Saykally: Rev. Sci. Instr. **67**, 410 (1996)

[19] S.K. Kim, W. Lee, D.R. Herschbach: J. Phys. Chem. **100**, 7933 (1996)

[20] J. Schiedt, R. Weinkauf, D.M. Neumark, E.W. Schlag: Chem. Phys. **239**, 511 (1999)

[21] D.M. Lubman, L. Li: In Lasers and Mass Spectrometry, D.M. Lubman, Ed., Oxford University Press: Oxford, 1990

[22] E. Nir, L. Grace, B. Brauer, M.S. de Vries: J. Am. Chem. Soc. **121**, 4896 (1999)

[23] J.M. Mestdagh, M. Berdah, N. Auby, C. Dedonder-Lardeux, C. Jouvet, S. Martrenchard, D. Solgadi, J.P. Visticot: Euro. J. Phys. D **4**, 291 (1998)

[24] A. Lindinger, J.P. Toennies, A.F. Vilesov: J. Chem. Phys. **110**, 1429 (1999)

[25] F. Huisken, O. Werhahn, A.Y. Ivanov, S.A. Krasnokuski: J. Chem. Phys. **111**, 2978 (1999)

[26] S.A. McLucley, S. Habibi-Goudarzi: J. Am. Chem. Soc. **115**, 12085 (1993)

[27] M. Karas, D. Bachmann, U. Bahr, F. Hillenkamp: Int. J. Mass Spectr. Ion Processes **78**, 53 (1987)

[28] F. Mafuné, J. Kohno, T. Nagata, T. Kondow: Chem. Phys. Lett. **218**, 7 (1993)

[29] W. Kleinekofort, M. Schweitzer, J.W. Engels, B. Brutschy: Int. J. Mass Spectr. Ion Processes **163L**, 1 (1997)

[30] M. Dey, J. Grotemeyer, E.W. Schlag: Zeit. Naturforsch. A **49**, 776 (1994)

[31] S. Steenken: Chem. Rev. **89**, 503 (1989)

[32] A.O. Colson, B. Besler, M.D. Sevilla: J. Phys. Chem. **97**, 8092 (1993)

[33] E. Westhof: Rev. Biophys. Biophys. Chem. **17**, 125 (1988)

[34] K. Aflatooni, G.A. Gallup, P.D. Burrow: J. Phys. Chem. **102**, 6205 (1998)

[35] J.H. Hendricks, S.A. Lyapustina, H. L. Clercqd., K.H. Bowen: J. Chem. Phys. **108**, 8 (1998)

[36] C. Desfrançois, H. Abdoul-Carime, J.P. Schermann: J. Chem. Phys. **104**, 7792 (1996)

[37] C. Desfrançois, V. Périquet, Y. Bouteiller, J.P. Schermann: J. Phys. Chem. **102**, 1274 (1998)

[38] N. Oyler, L. Adamowicz: J. Phys. Chem. **97**, 11122 (1993)

[39] N.A. Oyler, L. Adamowicz: Chem. Phys. Lett. **219**, 223 (1994)

[40] V. Périquet, A. Moreau, S. Carles, J.P. Schermann, C. Desfrançois: J. Electr. Spectr. Rel. Phenom. **106**, 141 (2000)

[41] C. Desfrançois, Y. Bouteiller, R. Weinkauf: to be published

[42] M.D. Sevilla, B. Besler, A.O. Colson: J. Phys. Chem. **99**, 1060 (1995)

[43] M.D. Sevilla, D. Becker, M. Yan, S.R. Summerfield: J. Phys. Chem. **95**, 3409 (1991)

[44] N.U. Zhanpeisov, J. Sponer, J. Leszczynski: J. Phys. Chem. **102**, 10374 (1998)

[45] M.R. Viant, R.S. Fellers, R.P. McLaughlin, R.J. Saykally: J. Chem. Phys. **103**, 9502 (1995)

[46] P. Colarusso, K. Zhang, B. Guo, P.F. Bernath: Chem. Phys. Lett. **269**, 39 (1997)

[47] M. Graindourze, J. Smets, T. Zeegers-Huyskens, G. Maes: J. Mol. Struct. **222**, 345 (1990)

[48] G. Maes, J. Smets, L. Adamowicz, W. McCarthy, M.K.V. Bael, L. Houben, K. Schoone: J. Mol. Struct. **410**, 315 (1997)

[49] J. Lorentzon, M.P. Fülscher, B. Roos: J. Am. Chem. Soc. **117**, 9265 (1995)

[50] A. Douhal, S.K. Kim, A.H. Zewail: Nature **378**, 260 (1995)

[51] R. Lopez-Martens, P. Long, D. Solgadi, B. Soep, J. Syage, P. Millié: J. Chem. Phys. **273**, 219 (1997)

[52] S. Morpugo, M. Bossa, G.O. Morpugo: Chem. Phys. Lett. **280**, 233 (1997)

[53] A.K. Chandra, M.T. Nguyen, T. Uchimaru, T. Zeegers-Huyskens: J. Phys. Chem. **103**, 8853 (1999)

Reaction Between Barium and N₂O on Large Neon Clusters

M.A. Gaveau, M. Briant, V. Vallet, J.M. Mestdagh, J.P. Visticot
CEA/DSM/DRECAM/SPAM
CE Saclay 91191 Gif-sur-Yvette Cedex, France

1 Introduction

Over the past few years, we have developed in our laboratory a new method called CICR (for "cluster isolated chemical reactions"), with the goal of investigating the effect of a reaction medium on reaction dynamics [1, 2, 3]. CICR consists in performing chemical reactions on large van der Waals clusters in a very controlled way. Clusters provide an ideal medium for the study of solvation effects and heterogeneous chemistry at a microscopic level. They are free from any perturbation due to a substrate, they have a well-defined internal energy and they have a finite size (from 10^2 to a few 10^3 atoms or molecules in our experiments). Reactants are deposited on the clusters by sticky collisions between the clusters and a low-pressure buffer gas [4]. This pick-up technique allows us to have strict control over the reactant deposition onto the clusters. On the clusters, reactants will migrate, collide with each other and eventually react. Thus the clusters become true chemical nanoreactors. Reactions of Ba atoms and small Ba aggregates have already been investigated on various clusters, i.e. argon, methane and nitrogen clusters which have similar structures and internal temperatures (between 35 and 43 K) [5, 6, 1]. In particular, it has been shown that the mechanism of the chemiluminescent reaction

$$Ba + N_2O \rightarrow BaO + N_2 \qquad \Delta H = -4.1 \ eV$$

is not substantially affected by the presence of the reaction medium. However, the cluster leads to specific effects. First, there is a dramatic increase of the reaction rate compared to the gas-phase reaction. Second, it leads to the formation of BaO molecules which can either be ejected from the cluster as free molecules or stay solvated on the cluster. As a solvent, it displaces the energy levels of the solute so that the spectrum of the solvated molecule is modified and shifted, and it relaxes the excess energy of the solvated product. The solvated BaO product has an interaction with the solvent, which increases from argon to methane and to nitrogen. In the latter case, the interaction is so strong that chemiluminescence is quenched. Even with argon, the interaction of BaO with the substrate is strong enough to lead to a large shift and broadening of the BaO vibronic emission bands.

This article investigates the same reaction on neon clusters because these clusters are colder (10 K) [7] and are expected to interact less strongly with the BaO product. Two effects are anticipated: the vibronic bands should be narrower and the cooling rate of the hot BaO should be slower. This could allow a better insight into the dynamics of the solvated BaO product.

2 Experiment

The experimental set-up is schematically shown in Fig. 1. Neon clusters are grown by homogeneous nucleation in a continuous supersonic free jet from a molecular beam source of the Campargue type [8]. In this experiment on neon clusters, the nozzle is cooled by liquid nitrogen to decrease the stagnation temperature T_0 of the gas to about 80 K. The gas pressure in the reservoir is $P_0 = 30$ bars and the gas expands through a sonic nozzle of 0.1 mm diameter. After extraction from the supersonic expansion by a 1 mm skimmer, the cluster beam passes through a differentially pumped chamber, which can be used as a pick-up chamber for N_2O. We call it the *early pick-up region*. After passing through a 3 mm collimator, the beam enters the main chamber and flies through a 30-mm-long heated barium

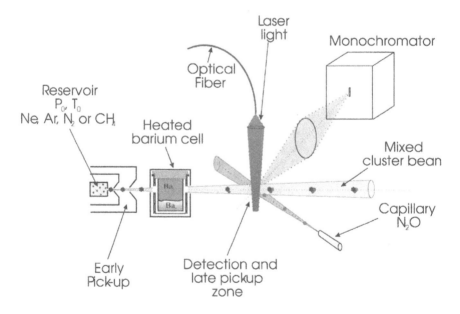

Figure 1: Experimental set-up.

cell, where barium atoms are deposited on the clusters. 17 mm downstream from the exit of the barium cell, it reaches the region of the main chamber (also called *late pick-up region*) where a second pick-up can be performed by crossing the cluster beam with a small effusive beam that outlets from a thin capillary tube at 5 mm from the axis of the cluster beam.

This zone is also the observation region and is imaged onto the entrance slit of a scanning grating monochromator. The collected fluorescence light is thus dispersed by the monochromator and detected by a cooled photomultiplier tube (RCA 31034). The signal is then processed by a photon-counting system, including fast amplifier and discriminator, and accumulated using multichannel cards implemented in a microcomputer. The optical system, the monochromator and PMT cover the visible range (410–900 nm). The recorded spectra can be corrected for the detection efficiency of the optical system. A part of the non-dispersed fluorescence can be directed to a second photomultiplier in order to monitor the total fluorescence signal.

In laser-induced fluorescence measurements, the observation zone can also be illuminated by the light of a tunable cw ring laser (dye or titane-sapphire laser). The laser light is transported into the chamber by means of an optical fiber (0.6 mm core). At the exit of the fiber, the light is refocused in a slightly converging beam which perpendiculary crosses the cluster beam. By flagging the cluster beam in front of the two pick-up regions, the fluorescence spectra can be corrected for undesirable contributions measured with the cluster beam off, i.e. blackbody radiation of the barium cell and gas phase reactions between barium atoms effusing from the cell and the molecular reactant.

Downstream from the main chamber, the beam enters chambers which house a quadrupole mass spectrometer and a time-of-flight mass spectrometer. The latter one allows us to measure directly the cluster size distribution, which is log normal. Its width is about equal to the average size $\langle N \rangle$ of the distribution. Our experimental conditions lead to a mean neon cluster size of 7000. Mass spectrometry is also a way to measure the relative fluxes of the volatile cluster constituents, by scattering the beam in a compression chamber and measuring the resulting increases of partial pressures with a carefully calibrated mass spectrometer [5].

The speed distribution of the clusters can be obtained by measuring the neutral time-of-flight distribution, on a 2275-mm-long path between a mechanical chopper and the quadrupole mass spectrometer [9]. Argon clusters generated from a source at 300 K have an average speed of 500 m/s. In this work, the source temperature is much lower (80 K) and the average speed of neon clusters is about 370 m/s.

We performed induced fluorescence experiments by laser excitation of the barium monomer trapped on neon clusters and reactive experiments when the N_2O molecular reactant is added to the cluster by the pick-up technique. Two types of reactive experiments were performed depending on which pick-up region was used. In the first one, N_2O is deposited onto clusters in the late pick-up region, from which fluorescence signals are observed. Such experiments document the nascent distribution of electronically excited products within a time window of a few microseconds after the reaction, corresponding to the residence time of the clusters in the observation region. In the second type of experiment, the early pick-up region is used and fluorescence signals correspond to emission of excited products 80 microseconds on average after the reaction.

3 Experimental Results

3.1 Laser Induced Fluorescence of Barium Solvated on Neon Clusters

In LIF experiments, barium solvated on Ne clusters is excited in the vicinity of its resonance line ($^1P_1 \to {}^1S_0$ at 553.5 nm) by the light of a cw ring dye laser, whose wavelength can be tuned between 534 and 560 nm by using pyrromethene 556. We recorded two kinds of spectra: excitation spectra and fluorescence spectra. Excitation spectra are obtained by scanning the laser frequency and recording the total fluorescence intensity or the fluorescence intensity at a fixed monochromator wavelength. Fluorescence spectra are recorded by fixing the laser frequency and scanning the monochromator wavelength.

In Fig. 2, the total fluorescence excitation spectra of $BaAr_{2000}$ (top panel) and $BaNe_{7000}$ (bottom panel) are shown. Both of them present some similarities: they are composed of two broad bands distributed about the resonance line. However, the $BaNe_{7000}$ photoexcitation spectrum extends on a smaller range (539–555 nm) than the $BaAr_{2000}$ one (535–562 nm). Moreover, the red band of the former spectrum exhibits a pronounced doublet structure with a thinner component on the red side of the Ba resonance line. This doublet structure is only suggested in the red band of the $BaAr_{2000}$ excitation spectrum, which is nearly totally located on the red side of the Ba line. The structure of the $BaAr_N$ photoexcitation spectrum has been interpreted by the splitting of the barium resonance line into two bands when barium is located at the surface of the argon cluster [10, 11]. In such a case, the blue and red bands on each side of the resonance line can be assigned respectively to the Σ-like and Π-like states of a quasi-diatomic Ba-Ar_N molecule. This interpretation has been confirmed by molecular dynamics simulations which reproduced the experimental excitation spectra for a barium atom diffusing at the surface of an argon cluster [11]. The similar structure of $BaAr_N$ and $BaNe_{7000}$ photoexcitation spectra suggests the same interpretation, i.e. the surface location of the barium atom both on the argon and neon clusters. However, the question of the strong difference between the two red components of the barium excitation spectrum on neon clusters still remains. This observation can be interpreted in two different ways. These two bands can be due to a removal of the twofold degeneracy of the Ba-Ne$_{7000}$ quasi-molecule Π-like state, as seems to be the case for Ba-Ar$_N$ clusters. Or they could arise from a Π-like excitation of barium in two different surface sites of the neon cluster. The respective Σ branches should thus be merged into a single band to account for the blue branch of the experimental excitation spectrum. Although we have no way to choose between these two interpretations without any further experimental or theoretical data, the former one seems to us more likely, particularly because the smaller the argon clusters are, the weaker the barium-cluster binding energy is and the more the Ba-Ar$_N$ excitation spectrum resembles the BaNe$_{7000}$ one [11].

The surface location of barium on neon clusters is also supported by the fluorescence experiment results. On the upper part of Fig. 3, the fluorescence spectrum of Ba on Ne$_{7000}$ when excited at 552 nm (solid line) is compared to the fluorescence spectrum of BaAr$_{2000}$ after excitation at 540.3 nm (dashed line). The difference between the widths of the two spectra is striking: the former one, which ranges between 552.6 and 555 nm, is much narrower than the latter one. The fluorescence spectrum of BaAr$_{2000}$ exhibits a broad band which is almost independent of the excitation wavelength and extends approximately from 550 to 570 nm. It corresponds to the emission from the

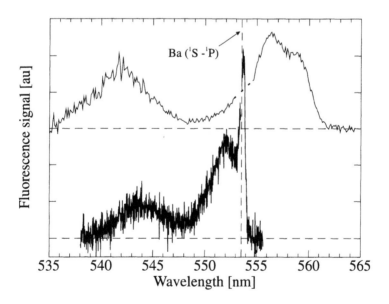

Figure 2: Comparison of the excitation spectra of Ba on Ar$_{2000}$ (*top panel*) and Ne$_{7000}$ (*bottom panel*) clusters.

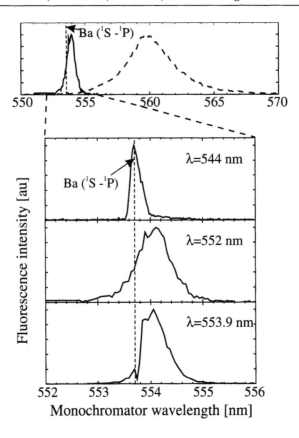

Figure 3: *Upper part*: Comparison of the fluorescence spectra of Ba on Ar$_{2000}$ (dashed line) and Ne$_{7000}$ (solide line) clusters at laser wavelengths of 540.3 nm and 552 nm respectively. *Lower part*: Fluorescence spectra of Ba on Ne$_{7000}$ at laser wavelengths of 544 nm (*top panel*), 552 nm (*middle panel*) and 553.9 nm (*bottom panel*). All spectra have been normalized at their maximum.

bound Π-like state of the system in which the excited barium has relaxed in a time which is short compared to its radiative lifetime (8 ns [12]). The atomic Ba($^1P_1 \rightarrow {}^1S_0$) line is also observed when the Σ-like state of the system is excited [13]. This line is emitted by free excited barium atoms which have been desorbed from the cluster surface just after laser excitation. Photodesorption is enhanced by a higher photon energy and a smaller argon cluster size. Nevertheless, in most of the blue band of the excitation spectra, the photodesorption process always remains a minor channel compared to the broad band fluorescence of barium atoms that has relaxed into the bound Π-like state [14].

The lower part of Fig. 3 zooms in on the BaNe$_{7000}$ fluorescence spectrum (range 552–556 nm) and displays three barium fluorescence spectra at laser wavelengths corresponding to the three maxima of the BaNe$_{7000}$ excitation spectrum: 544 nm (top panel), 552 nm (middle panel), 553.9 nm (bottom panel). At 544 nm, the fluorescence spectrum presents an intense atomic Ba($^1P_1 \to {}^1S_0$) line superimposed on a faint band that extends from 552.6 to 555 nm and corresponds to the emission of excited solvated barium. Just as for BaAr$_N$, there exists a competition between barium desorption and relaxation into a bound excited state when the blue branch of the excitation spectrum is excited. Nevertheless, barium desorption predominates over relaxation on Ne$_{7000}$ clusters. The ratio of the Ba line intensity to the overall emission signal gives a probability of Ba desorption of 83% after excitation at 544 nm. When the excitation is at 552 nm (Fig. 3 middle panel), i.e. on the first maximum of the red branch of the excitation spectrum, barium atom desorption does no longer occur. The emission spectrum is similar to the underlying weak band of the emission spectrum at 544 nm and can therefore be assigned to the emission of excited solvated barium. The emission after excitation at 553.9 nm (Fig. 3 bottom panel) corresponds also to solvated barium fluorescence but the band is a little bit narrower because of the lower laser photon energy.

The similarity between the barium spectroscopy on argon and neon clusters strongly supports the conclusion that barium atoms remain located at the surface of clusters after their capture. So, reaction between a trapped molecular reactant and barium is expected to occur at the surface of the cluster.

3.2 Reaction Between Barium and N$_2$O on Large Argon and Methane Clusters

Previously, we have studied the reaction between Ba and N$_2$O on large argon and methane clusters [5, 6]. The respective chemiluminescence spectra are presented in Fig. 4. Experiments have been performed using the late pick-up region for N$_2$O, which means that reaction occurs in the observation region and that the nascent BaO* product is detected. Let us analyse the chemiluminescence spectrum on argon clusters (bottom panel). It is made of two broad components which are both assigned to BaO* emission. The first one is a broad unstructured band that covers the whole visible range with a maximum at about 21000 cm^{-1}; it resembles the spectrum from the hot gas-phase reaction product. The second one is very structured and extends from 12000 to 18000 cm^{-1}. The spectrum also exhibits the Ba($^1P_1 \to {}^1S_0$) and Ba($^3P_1 \to {}^1S_0$) lines. We have shown that the unstructured component can be assigned to emission of hot BaO molecules which desorbed from the cluster immediately after their formation by the Ba+N$_2$O reaction at the surface of the cluster. The structured component is due to emission of BaO*

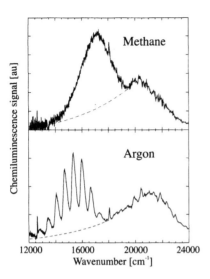
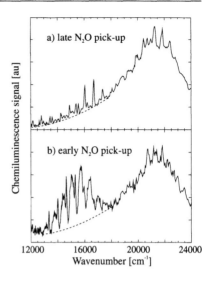

Figure 4: Chemiluminescence spectra of the Ba+N_2O reaction on methane clusters (*top panel*) and argon clusters (*bottom panel*).

Figure 5: Chemiluminescence spectra of the Ba+N_2O reaction on neon clusters : with a late N_2O pick-up (*top panel*) and an early N_2O pick-up (*bottom panel*).

molecules that stay solvated on the argon cluster. They are vibrationally relaxed and the vibrational progression of the spectrum corresponds to the BaO emission from the $A^1\Sigma$ and $a^3\Sigma$ ($v' = 0$) states to the vibrational levels of its ground state [15]. These bands are broadened and blue-shifted by about 750 cm^{-1} from the corresponding transitions of the free BaO molecule, indicating the interaction of the solvated molecule with the argon environment.

On methane clusters (Fig. 4 top panel), the spectral contribution of free ejected hot BaO molecules remains nearly unchanged except for a small red shift of the maximum which peaks now at 20000 cm^{-1}. On the contrary, the solvated BaO* component has dramatically changed: it is now a broad band without any vibrational structure. Its shape coincides exactly with the envelope of the solvated BaO* spectrum from an argon cluster, but it is blue-shifted by 1650 cm^{-1} from the latter one. These changes reflect an interaction between the solvated BaO* molecule and the environment, which is stronger with methane than with argon. It can be shown that the BaO* product associates with at least one CH_4 molecule [6].

3.3 Reaction Between Barium and N_2O on Large Neon Clusters

In this work, the reaction between Ba and N_2O has been performed on neon clusters with two different N_2O pick-up configurations. In the first one, the late pick-up region is used and the reaction region is also the observation region. It produces a chemiluminescence spectrum (Fig. 5 top panel) that documents the nascent distribution of electronically excited products formed on the cluster surface due to the barium location. In the second one, the early pick-up region is used and the reaction occurs in the barium cell. The corresponding chemiluminescence spectrum (Fig. 5 bottom panel) is observed a few tens of microseconds after the reaction. Both spectra are recorded with a 2 nm resolution. Their intensities were adjusted by an arbitrary normalization factor for the sake of visibility. The late pick-up chemiluminescence spectrum exhibits a very intense component covering the whole visible range. It is similar to the free BaO* emission of the Ba+N_2O reaction on argon, methane and nitrogen clusters. It can therefore be attributed to free BaO* ejected from neon clusters just after the reaction. The chemiluminescence spectrum also presents several narrow bands which are regularly spaced between 11500 and 18000 cm^{-1} suggesting vibrational progressions of a solvated BaO* product. The overall intensity of this structured spectrum is weak compared to the emission intensity of the free BaO* emission.

In the early pick-up experiment, the chemiluminescence signal is much less intense. Surprisingly, free BaO* emission can still be observed, though the product is ejected in the reaction region which, in the early pick-up configuration, is located on average inside the barium cell, i.e. far away from the observation region. In our opinion, this emission results from neon clusters carrying N_2O molecules that trap barium atoms effusing from the cell in the observation region. More interesting is the structured spectral contribution ranging from 12000 to 18500 cm^{-1}. Strikingly, it is very different from the structured spectral component of the late pick-up spectrum. It exhibits two vibrational progressions which have the same envelope: however, the first progression is composed of narrow bands, the second one presents broad bands.

In order to analyze the structured part of the chemiluminescence spectra of Fig. 5, we have recorded them at higher resolution (1 nm) in the wavelength range of interest, and subtracted the non-relevant spectral contributions. For the late pick-up case, the free BaO* contribution is subtracted from the spectrum; the resulting structured spectrum is displayed in Fig. 6. It shows BaO molecular bands and two narrower atomic Ba($^1P_1 \rightarrow {}^1S_0$) and Ba($^3P_1 \rightarrow {}^1S_0$) lines. The two most intense molecular progressions correspond to two vibrational progressions of BaO: emission of BaO $A^1\Sigma$ ($v' = 0$) to the different vibrational levels of the ground state, and emission of BaO

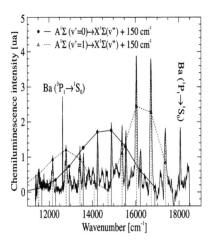
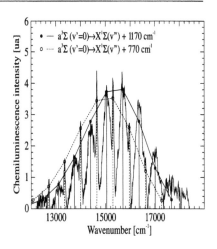

Figure 6: Solvated BaO chemiluminescence spectrum produced by the Ba+N$_2$O reaction on large neon clusters with a late N$_2$O pick-up.

Figure 7: Solvated BaO chemiluminescence spectrum produced by the Ba + N$_2$O reaction on large neon clusters with an early N$_2$O pick-up.

$A^1\Sigma$ ($v' = 1$) [15]. For comparison, the corresponding line intensities calculated for the free BaO molecule are shown, but the emission must be blue shifted by 150 cm^{-1} to account for the solvation. This shift proves that the BaO* product responsible for this structured emission is really solvated on neon clusters. Moreover, weak vibrational progressions are observed; they can be assigned to the emission of BaO in the $a^3\Sigma(v' = 0, 1)$ states, which lie 210 cm^{-1} below the $A^1\Sigma(v' = 0, 1)$ ones [15]. These results are different from what is obtained on argon clusters: the bands are six times narrower (50 cm^{-1} instead of 300 cm^{-1}), the blue shift is smaller (150 cm^{-1} instead of 750 cm^{-1}) and the emission from $v' = 1$ is predominant while only $v' = 0$ is observed on argon clusters.

For the early pick-up case, we assume that the unstructured spectrum is due to residual reactions occurring in the observation region. So, we subtract from the early pick-up spectrum the late pick-up spectrum correctly normalized in the 21000 cm^{-1} region to account for both the free product and the nascent solvated BaO* emission. The remaining spectrum, shown in Fig. 7, is composed of two vibrational progressions of the BaO molecule. It results from the emission of products which have been formed a few tens of microseconds before the observation region. Therefore, the $A^1\Sigma$ state of BaO cannot be involved because its lifetime is too short at about 350 ns [16]. It is composed of two vibrational progressions which can be assigned to the emission of the BaO $a^3\Sigma(v' = 0)$ state and are blue-shifted by 770 and 1170 cm^{-1}, respectively, from the corresponding lines of the free BaO molecule. The calculated relative intensities of these lines are shown for comparison. Surprisingly, one progression exhibits narrow bands (50 cm^{-1}) while the other one presents wide bands (360 cm^{-1}). The narrow bands are on the red side of the corresponding broad transitions. We are therefore not allowed to assign them to zero phonon lines associated to the wide-band progression [17]. It thus appears that they correspond to two different locations of the BaO* product: the latter one to BaO* embedded inside the clusters, the former one to BaO* less solvated but not really on the surface, since the 770 cm^{-1} blue shift is already much larger than for the nascent BaO* in the early pick-up case.

4 Conclusion

In this work, laser induced fluorescence of barium atoms trapped on neon clusters has shown that barium atoms remain at the suface of neon clusters. We have also studied the reaction between Ba and N$_2$O on large neon clusters. Due to the barium location, the BaO reaction product is formed at the surface of the cluster. This reaction was performed either in the observation region, or 80 µs before it. In the former case, we observe the BaO* molecules during the first microseconds after their formation on the cluster surface. The interaction between BaO* and the neon cluster is much weaker than in the case of argon and methane clusters. So, most of the BaO* product is ejected from the neon cluster. A minor part stays solvated in the short-lived A$^1\Sigma$ state. It lies on the surface and keeps some vibrational excitation giving rise to a narrow band spectrum. In the latter case, after several tens of microseconds, only the long-lived a$^3\Sigma$ state remains populated and the vibration is completely relaxed. The spectrum exhibits a broad and a narrow band progression which both have the same Franck–Condon envelope: BaO* is much more solvated at two different locations, one embedded inside the cluster and the other one in an intermediate site. We think that the dramatic change in the solvation of BaO* shown by the spectra between the two reaction configurations reflects the embedding of the BaO* molecule inside the neon cluster on a timescale of a few tens of microseconds.

References

[1] C.Gée, M.A.Gaveau, J.M.Mestdagh, M.Osborne, O.Sublemontier, and J.P.Visticot, J.Phys.Chem. **100**, 13421 (1996).

[2] J.M.Mestdagh, M.A.Gaveau, C.Gée, O.Sublemontier, and J.P.Visticot, Int.Rev.Phys.Chem. **16**, 215 (1997).

[3] M.A.Gaveau, C.Gée, J.M.Mestdagh, and J.P.Visticot, Com.At.Mol.Phys **34**, 241 (1999).

[4] T.E.Gough, M.Mengel, P.A.Rowntree, and G.Scoles, J.Chem.Phys. **83**, 4958 (1985).

[5] A.Lallement, J.M.Mestdagh, P.Meynadier, P.de Pujo, O.Sublemontier, J.P.Visticot, J.Berlande, X.Biquard, J.Cuvellier, and C.G.Hickman, J.Chem.Phys. **99**, 8705 (1993).

[6] M.A.Gaveau, B.Schilling, C.Gée, O.Sublemontier, J.P.Visticot, J.M.Mestdagh, and J.Berlande, Chem.Phys.Lett. **246**, 307 (1995).

[7] J.Farges, M.F.de Feraudy, B.Raoult, and G.Torchet., Surf.Sci. **106**, 95 (1981).

[8] R.Campargue, J.Phys.Chem. **88**, 4466 (1984).

[9] J.Cuvellier, P.Meynadier, P.de Pujo, O.Sublemontier, J.P.Visticot, J.Berlande, A.Lallement, and J.M.Mestdagh, Z.Phys. D **21**, 265 (1991).

[10] J.P.Visticot, J.Berlande, J.Cuvellier, A.Lallement, J.M.Mestdagh, P.Meynadier, P.de Pujo, and O.Sublemontier, Chem.Phys.Lett. **191**, 107 (1992).

[11] J.P.Visticot, P.de Pujo, J.M.Mestdagh, A.Lallement, J.Berlande, O.Sublemontier, P.Meynadier, and J.Cuvellier, J.Chem.Phys. **100**, 158 (1994).

[12] F.M.Kelly and M.S.Mathur, Can.J.Phys. **55**, 83 (1977).

[13] X.Biquard, O.Sublemontier, J.P.Visticot, J.M.Mestdagh, P.Meynadier, M.A.Gaveau, and J.Berlande, Z.Phys. D **30**, 45 (1994).

[14] B.Schilling, M.A.Gaveau, O.Sublemontier, J.M.Mestdagh, J.P.Visticot, X.Biquard, and J.Berlande, J.Chem.Phys. **101**, 5772 (1994).

[15] R.A.Gottscho, J.B.Koffend, and R.W.Field, J.Mol.Phys. **82**, 310 (1980).

[16] S.E.Johnson, J.Chem.Phys. **56**, 149 (1972).

[17] K.K.Rebane, *Impurity Spectra of Solids: Elementary Theory of Vibrational Structure*, (Plenum Press, New-York, London 1970).

Capture and Coagulation of CO Molecules to Small Clusters in Large Supercooled H_2 Droplets

E.L. Knuth[1], S. Schaper[2], J.P. Toennies[2]

[1] Chemical Engineering Department, UCLA, Los Angeles, CA 90095
[2] MPI für Strömungsforschung, 37073 Göttingen, Germany

1 Introduction

The current interest in the creation of superfluid hydrogen stems apparently from the 1972 paper by Ginzburg and Sobyanin [1]. They concluded that p-H_2 can become a superfluid if solidification can be delayed to 6 K. However, since the triple point of p-H_2 is 13.8 K, the production of highly supercooled liquid H_2 would be required. Using the liquid density at the triple point, Maris, Seidel and Huber [2] predicted a slightly higher transition temperature, namely 6.6 K. Then they compared p-H_2 with ^{4}He, noting that the measured transition temperature for ^{4}He is 0.7× the predicted value, and suggested that supercooling to less than 4.6 K might be required in order to achieve superfluid properties. But they note that p-H_2 is a highly quantum system with a quantum-mechanical zero-point energy for the solid greater than that of the liquid, so that the heat of fusion at low temperatures is relatively low. Hence a much larger degree of supercooling is possible than for a classical system. From an examination of homogeneous nucleation of solid formation in low-temperature p-H_2, they concluded that, for temperatures above about 5 K, nucleation is dominated by thermal fluctuations.

Subsequently, Maris et al. [3, 4] conducted experimental studies of supercooled liquid-H_2 droplets, with diameters from about 30 μ to about 1000 μ, levitated in pressurized ^{4}He. They determined nucleation rates for droplet temperatures from 10.6 to 11.2 K at system pressures of the order of 15 bar. Using the thermal fluctuations model developed earlier [2], they then extrapolated their values to lower temperatures and zero pressure, thereby estimating a maximum nucleation rate of 10^{16} cm^{-3}s^{-1} at 7 K and zero pressure. Even this maximum rate would be quite acceptable for experiments on droplets formed in supersonic beams. Since the rate of cooling in typical supersonic beams is exceedingly high, of the order of 10^{10} K/s [5, 6], unusually deep supercooling of the droplets is possible. Since the droplets are relatively small, the characteristic time for homogeneous nucleation at the quoted maximum rate is relatively large – of the order of 10^2 s for a droplet with size of the order of 10^4 molecules. Hence, since the flight time in a typical supersonic cluster beam is of the order of 10^{-3} s, solidification of such clusters via homogeneous nucleation is highly improbable. Furthermore, in a supersonic cluster beam, the probability of heterogeneous nucleation is reduced greatly as a result of the minimal contact with a solid surface during the cooling in the nozzle and the possibility of interactions only with the background gas after leaving the orifice.

Prospects for superfluidity in p-H_2 have been examined by Sindzingre et al. [7] using path-integral Monte Carlo methods. They find superfluid behavior for very small clusters, e.g., for clusters of 13 and 18 molecules, but no superfluid behavior for clusters of 33 molecules. However, their findings do not appear to be applicable to the present studies since their conditions were chosen so that the clusters are in a steady equilibrium state rather than in the transient metastable liquid state which is realized in the present experiments.

Previous experimental studies of the properties of H_2 droplets formed in free-jet expansions are described in earlier papers [8, 9]. The objective of this series of studies is the production of metastable p-H_2 droplets in the liquid state with temperatures below 6 K with the hope that such droplets will exhibit superfluid properties. In the first paper [8], droplet temperatures as low as 6 K were deduced from measured values of terminal droplet velocities; evaporation provides additional cooling. In the second paper [9], dramatically higher terminal droplet velocities were observed for measurements made on p-H_2 which apparently had been inadvertently contaminated in the source with some foreign substance; it was conjectured that this foreign substance provided sites for heterogeneous nucleation of solid p-H_2 during the expansion and that the heat released in solidification was converted into additional terminal velocity. Momentum exchange deduced from time-of-flight measurements made with uncontaminated droplets passing through a scattering gas suggested that these droplets were liquid.

In the measurements described here, a foreign substance, namely CO, was introduced into the H_2 droplets by passing the droplet beam through a scattering chamber containing CO at room temperature and various pressures. Properties of both the H_2 droplets and the CO clusters formed by coagulation within the H_2 droplets were monitored in order to obtain information on whether the droplets are liquid.

2 Apparatus and Measurements

The apparatus used here is essentially the same as used previously [9] except that the length of the chopper chamber has been reduced from 125 mm to 90 mm and that the vacuum was improved in order to reduce residual-gas uptake. Free jets were formed via expansions from supercritical source conditions through a 5-μm-dia orifice. The central core of the free jet passed through a skimmer located 10 mm from the orifice. The resulting droplet beam passed then through seven stages of pumping to a mass spectrometer with resolution adequate for mass spectra up to 360 amu. A catalytic converter raised the p-H_2 concentration to more than 95%.

The beam was characterized by measuring the speed distribution, mean droplet size and mean droplet integral cross section. A chopper located in the first chamber behind the skimmer facilitated determination of the beam speed distribution using the time-of-flight method. The mean droplet size was measured using small-angle scattering deflection by single collisions with an Ar secondary beam crossing the droplet beam at 40° [10]. Values of the H_2-droplet integral cross section were deduced from the attenuation of the droplet beam by the secondary beam [11]. The effective droplet density deduced from the droplet size $\bar{N}$ and cross section s ranged from about 40% of the bulk density for $\bar{N} = 3000$ to about 60% of the bulk density for $\bar{N} = 14000$. Similar values for the effective density were found for ^{4}He droplets, and the deviation from the bulk density was shown to be due to the decrease near the surface [12]. No differences between p-H_2 and n-H_2 were observed.

CO was selected for embedding in the droplets since its ionization potential (14.0 eV) is about 1.4 eV below that (15.4 eV) of H_2 so that it can be ionized indirectly via charge transfer with H_2^+. Experimental arrangements were selected from three source gases (p-H_2, n-H_2 and ^{4}He), four average droplet sizes and two scattering-chamber arrangements. See Table 1. The indicated source conditions all yield expansion paths which cross the binodal line on the gas-phase side of the critical point, so that the measured distribution of droplet sizes was log-normal, in agreement with measurements for ^{4}He [10]. For each experimental arrangement, mass spectra were measured for several values of CO pressure in the scattering chamber. In Fig. 1, CO clusters are observed up to $(CO)_{10}$ for 2.5×10^{-4} mbar CO, up to $(CO)_4$ for 1.2×10^{-4} mbar CO. The peaks are spaced at amu intervals slightly greater than 28, suggesting some hydrogen attachment to the CO clusters.

In order to investigate possible affects of droplet size on the capture and coagulation processes, a series of measurements was made using always the same scattering chamber but three different H_2 droplet sizes. See Table 1, Experiments A, D and E. As a measure of the efficacy of the capture and coagulation process, the maximum number k_{max} of CO molecules observed in the CO clusters is compared with the number of collisions with the CO gas via

$$k'_{max} = k_{max}/n_{CO}\sigma LF$$

where n_{CO} is the CO number density in the scattering region, σ is the integral cross section of the beam clusters, L is the length of the scattering region and F is a factor which takes into account the velocity distribution of the CO molecules in the scattering region [13]. The results are summarized in Fig. 2, where k'_{max} is shown as a function of H_2-droplet size.

Table 1: Experimental arrangements. [a] Scattering Region I: chopper chamber, $L = 90$ mm, entrance at skimmer. [b] Scattering Region II: scattering chamber, $L = 266$ mm, entrance 509 mm from skimmer. Most probable speed of CO molecules in scattering region (300 K) = 422 m/s

Experiment	Beam Species	Droplet Size	Droplet Speed (m/s)	p_0 (bar)	T_0 (K)	Scattering Region
A	p-H_2	5000	891	20	45	I[a]
B	^{4}He	5000	384	40	15	II[b]
C	n-H_2	5000	893	20	45	II[b]
D	n-H_2	1900	1024	20	55	I[a]
E	n-H_2	13700	790	20	40	I[a]
F	n-H_2	5600	887	20	45	I[a]

Possible affects of flight time from the skimmer to the scattering chamber on the capture and coagulation processes were investigated in measurements made using the same H_2-droplet size but two different scattering-chamber locations. See Table 1, Experiments A and C.

Measurements which were intended to facilitate determination of the capture cross section, and hence to facilitate separation of capture and coagulation effects, were made in Experiment F. In this experiment the detector was operated in the stagnation mode, i.e., the fluxes of both the H_2 and the CO into the detector chamber were determined by measuring the increases in the respective partial pressures when the beam was admitted to the chamber but intercepted by a beam flag before reaching the ionizer directly. Details of the experimental procedures were essentially the same as used by Lewerenz, Schilling and Toennies [11]. For the indicated conditions, a capture cross section of 2300 Å^2 was deduced. For the same conditions, the integral collision cross section is about 8700 Å^2.

For comparison, capture and coagulation measurements were made also for ^{4}He droplets, which are definitely liquid. For both the H_2 and the ^{4}He droplets, the source conditions were chosen such as to provide an average droplet size of about 5000 molecules/atoms. For both cases, the CO was admitted to scattering region II. See Table 1, Experiments B and C.

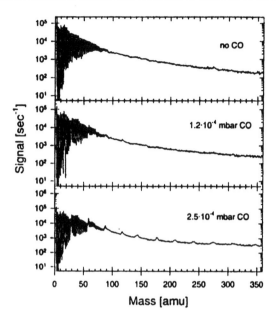

Figure 1: Mass spectra for p-H_2 droplets for several values of CO pressure, p_{CO}, in the scattering chamber

3 Characteristic Times

In order to place the time available for coagulation of the CO molecules within the H_2 droplets in perspective, we compare here several relevant characteristic times. For typical source conditions (20 bar, 45 K) a mean H_2-droplet speed of 891 m/s was measured. Hence, for scattering region I, flight times of 0.05×10^{-3} s and 2×10^{-3} s are realized for the flight from the skimmer to the middle of the scattering chamber and for the flight from the middle of the scattering region to the ionizer, respectively. The mean time between consecutive capture collisions with CO molecules varies inversely as the product of the capture cross section, the mean droplet speed and the number density of CO molecules in the scattering chamber. The integral cross section for H_2 droplets containing 5000 molecules was measured to be about 8000 Å^2. Assume for the present purposes a sticking coefficient of unity. Then, for a CO pressure of 2.5×10^{-4} mbar, the mean time between consecutive capture collisions would be about 2×10^{-6} s.

The characteristic time for coagulation depends upon whether the droplet is a normal fluid or a superfluid. A conservative (i.e., largest) value can be estimated for the normal fluid from the diffusion-limited rate constant k_D, neglecting possible effects of the long-range dipole-dipole potential. Start-

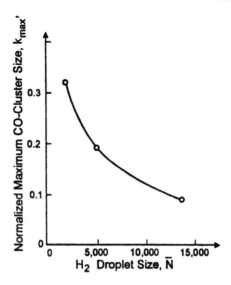

Figure 2: Normalized maximum observed CO cluster size as a function of H_2 droplet size. Data are from Experiments A, D and E. For each droplet size, the indicated value of k'_{max} is the average for several values of CO pressure, p_{CO}, in the scattering region

ing with either Equation (4-26) of Steinfeld, Francisco and Hase [14] or Equation (7-11) of Friedlander [15], one obtains

$$k_D = (2/3)(kT/\mu)(1+n^{1/3})^2/n^{1/3}$$

where k is Boltzmann's constant, T is temperature, μ is viscosity and n is the number of CO molecules in the CO cluster. For one CO molecule moving randomly inside an H_2 droplet of volume V which already contains a CO cluster, the characteristic time is simply V/k_D. Hence, the characteristic time for CO coagulation within an H_2 droplet of 5000 molecules is of the order of or less than 2×10^{-7} s. Thus the CO molecules have time for coagulation between consecutive captures of CO molecules – also during the flight from the scattering chamber to the ionizer – as long as he droplet is liquid.

4 Discussion

The decrease of k'_{max} with increase of $\bar{N}$ shown in Fig. 2 is perhaps an indication of an increase in the probability of forming more than one CO cluster within the H_2 droplet as $\bar{N}$ increases. If the H_2 droplet is small enough, the probability of the CO molecules forming only one cluster is very high, perhaps unity. As the size of the H_2 droplet increases, the mean time between consecutive capture collisions decreases and the characteristic time for coagulation increases, so that the probability of forming only one CO cluster decreases.

Monte Carlo simulations were made [16] in order to determine if scattering of the droplet beam in collisions with the CO background gas contributed significantly to a reduction of the value of k'_{max}. No significant contributions were found.

A continuing interest in the cooling of p-H_2 droplets to temperatures below 6.6 K motivates a comparison of the temperatures realized by the H_2 droplets traversing a scattering region for the experimental conditions used here with the temperature reached by evaporative cooling and no collisions with a scattering gas. For the latter case, the temperature at the end of the flight path is estimated using a model in which (a) the enthalpy decrease of the fluid in the expansion to the terminal velocity is equated to the increase of the kinetic energy as calculated from the measured terminal velocity and (b) the subsequent enthalpy decrease of a droplet is equated to the heat required for evaporation into a vacuum [9]. For an H_2 droplet with $\bar{N} = 5000$ formed by expansion from 20 bar and 45 K, one finds that the droplet is cooled to about 5 K during the adiabatic expansion and by evaporation [17, 18] to 4 K by the end of an additional 1.5-m flight path. Equating the heat loss to the heat of vaporization, one finds that the evaporative cooling results in the loss of about 90 H_2 molecules, i.e., about 2% of the droplet molecules. In the case of multiple collisions of the beam droplets with CO molecules in the scattering region, the droplet temperatures are elevated as a result of energy transfer to the droplets during the collisions. The droplet temperatures approach asymptotically a temperature which can be determined by equating (a) the rate of energy transfer to the droplets due to the sequence of collisions to (b) the rate of energy loss due to the temperature-dependent evaporation process. For the capture and coagulation experiments corresponding to Fig. 2, the droplet temperatures at the exit of the scattering region were calculated to be about 6 K, i.e., about 1 K higher than at the end of the adiabatic expansion and about 2 K higher than the temperature reached ultimately as a result of evaporative cooling.

The CO mobility required to produce CO clusters of the size observed here argues strongly in favor of the H_2 droplets being liquid. This evidence in favor of a liquid state, coupled with the estimated 4 K droplet temperature, suggest that the supercooled H_2 droplets may be superfluid. This opens up the possibility of seeking evidence of superfluidity by spectroscopic interrogation of probe molecules as demonstrated for ^{4}He droplets [19].

References

[1] V.L. Ginzburg, A.A. Sobyanin: JETP Lett. **15**, 242 (1972)

[2] H.J. Maris, G.M. Seidel, T.H. Huber: J. Low Temp. Phys. **51**, 471 (1983)

[3] G.M. Seidel, H.H. Maris, F.I.B. Williams, J.G. Cardon: Phys. Rev. Lett. **56**, 2380 (1986)

[4] H.J. Maris, G.M. Seidel, F.I.B. Williams: Phys. Rev. B **36**, 6799 (1987)

[5] S. Kotake, I.I. Glass: Prog. Aerospace Sci. **19**, 129 (1981)

[6] J.B. Fenn: In: Applied Atomic Collision Physics, edited by H.S.W. Massey, E.W. McDaniel, B. Bederson, Academic Press, New York (1982) Vol. 5, p. 349

[7] P. Sindzingre, D.M. Ceperley, M.L. Klein: Phys. Rev. Lett. **67**, 1871 (1991)

[8] E.L. Knuth, B. Schilling, J.P. Toennies: In: Rarefied Gas Dynamics edited by A.E. Beylich, VCH Verlagsgesellschaft mbH, New York (1991) p. 1035

[9] (a) T.P. Barrera, E.L. Knuth, L.S. Wong, F. Shünemann, J.P. Toennies: In: Rarefied Gas Dynamics edited by B.D. Shizgal, D.P. Weaver, AIAA, Washington, D.C. (1993) p. 267. See also (b) E.L. Knuth, F. Schünemann, J.P. Toennies: J. Chem. Phys. **102**, 6258 (1995)

[10] M. Lewerenz, B. Schilling, J.P. Toennies: Chem. Phys. Letters **206**, 381 (1993)

[11] M. Leverenz, B. Schilling, J.P. Toennies: J. Chem. Phys. **102**, 8191 (1995). See also Comment by M. Macler, Y.K. Bae in J. Chem. Phys. **106**, 5785 (1997)

[12] J. Harms, J.P. Toennies, F. Dalfovo: Phys. Rev. B 58, in print

[13] K. Berkling, R. Helbing, K. Kramer, H. Pauly, Ch. Schlier, P. Toschek: Z. Phys. **166**, 406 (1962)

[14] J.I. Steinfeld, J.S. Francisco, W.L. Hase: In: Chemical Kinetics and Dynamics. Prentice Hall, Englewood Cliffs (1989)

[15] S.K. Friedlander: Smoke, Dust and Haze. Wiley, New York (1977)

[16] N. Quaas: MPI für Strömungsforschung, Göttingen

[17] J. Farges, M.F. de Feraudy, B. Raoult, G. Torchet: Surface Science **106**, 95 (1981)

[18] J. Gspann: In: Physics of Electronic and Atomic Collisions edited by S. Datz, North Holland, Amsterdam (1982) p. 79

[19] M. Hartmann, R.E. Miller, J.P. Toennies, A.F. Vilesov: Science **272**, 1631 (1996)

Part VII

Interactions of Molecular Beams
and Cluster Beams
with Surfaces, and Applications

Essentials of Cluster Impact Chemistry*

T. Raz and R. D. Levine[†]

The Fritz Haber Research Center for Molecular Dynamics,
The Hebrew University, Jerusalem 91904, Israel

1 Introduction

Laboratory techniques for the study of chemical reactivity under extreme conditions have, so far, been limited. The prime source of information is from shock tube studies [1–4] but the upper limit of kinetic temperatures that can be conveniently reached, is about 5000 K. Fast, focused, laser heating and plasma chemistry is another [5–7]. More specialized techniques include the chemistry of translationally hot atoms [8] produced via the nuclear recoil technique [9, 10] and sonochemistry [11]. The technique sketched in this short overview promises to provide the opportunity of reaching a controllable range of energy density and pressure, and thereby allow a systematic study of both chemical and physical processes that can be driven by extreme conditions [4, 12–64]. We also refer to other relevant theoretical and computational studies [22, 52, 65–85]. Computational studies rely on knowledge of the potential and particularly at the high material and energy densities, which are typical of the proposed approach, this is very much an unknown[1]. We therefore note that the phenomena that we shall discuss have, at the time of writing, been explored experimentally [27, 30]. We ask for the help of the molecular beam community for definitive experiments in this promising new direction.

---

*Work supported by the AirForce Office of Scientific Research.
[†]Corresponding author. Fax: 972-2-6513742; e-mail: rafi@fh.huji.ac.il
[1]There are two reasons why so much is unknown. First, at high densities three (and even four) body forces are important. This is particularly so when chemically reactive atoms are present. Then, even for two body forces, the strongly repulsive regime is not well understood and, in addition, close in, as one approaches the united atom limit, there is considerable promotion of molecular orbitals. This is a 'universal' mechanism for electronic excitation which means a breakdown of the Born Oppenheimer approximation for close collisions.

The concept of cluster impact is a simple one: a cold cluster, moving at a supersonic velocity, impacts a hard surface. The impact leads to a rapid thermalization of the initially directed energy of the cluster, see section 4. The medium in which the reactions take place is the impact-heated cluster. It is not obvious that cluster induced chemistry is possible because, as we discuss in section 5, the hot cluster rapidly fragments.

Cluster impact provides a new regime of dynamics, where the activation process is thermal (due to collisions) but occurs on a very short time scale (comparable to vibrational periods). Hitherto this was only possible by photoactivation. Ordinary chemistry occurs in a regime where collisions are fast on the rotational time scale but slow on the vibrational one [86]. The new coupling regime is described in section 2.

Sections 3, 4 and 5 are arranged according to the consecutive stages of the impact of a cluster with a surface. Section 3 notes the new and unique features of reaction dynamics under the special condition within the impact-heated cluster. Section 4 reports that the finite cluster does very rapidly reach thermal equilibrium and suggests a simple mechanism for this facile equilibration. The final stage, that is the fragmentation process of the hot cluster is discussed in section 5. The sudden onset of a shattering regime can be shown using both molecular dynamics simulations and an information theory [87,88] analysis. This theoretical prediction was the first new phenomenon to be experimentally verified [25, 30, 34]. Section 6 presents a comparison between the two theoretical procedures (molecular dynamics simulation and the maximum entropy information theory procedure) that were used in the study of the burning of air [79, 80, 89]. Finally, section 7 raises the question whether the activation during cluster impact can be fast enough to induce electronic excitation.

Cluster Impact

The essential idea of the cluster impact technique is simple. A cold (glass-like) cluster moving at a supersonic velocity (which experimentally can reach tens of $km\,s^{-1}$, $10\,km\,s^{-1} = 0.1\,\text{Å}\,fs^{-1}$) is incident on a hard surface. The cluster itself is either made up of the reactants (e.g., cluster containing only N_2 and O_2 molecules, see sections 3 and 6) or the reactants are solvated in a chemically inert medium (e.g., diatomic molecules embedded inside rare gas clusters, or, say, in clusters of CO_2 [61, 62]). As the leading edge atoms of the cluster reach and then recoil from the surface, the rest of the cluster is still moving forward. Therefore,, immediately after the beginning of the impact, high relative kinetic energies and high local densities (the density can about double), prevail. A short time after the collision with the surface (typically even less than 100 fs), the excess energy is thermalized, see Fig. 1, and a short time later the hot cluster fragments. The final act of rapid

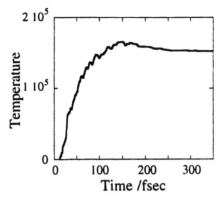

Figure 1: The temperature in °K (computed as the mean kinetic energy relative to the center of mass) vs. time (in fs) during a collision of an initially cold Ar_{125} cluster at an impact velocity of $10\,km\,s^{-1}$. Note the sharp rise of the temperature as a result of the impact with the surface and the cooling of the cluster as it expands.

fragmentation is as important to what one can achieve as the initial fast heating, because fragmentation freezes the composition to be that of the hot and dense cluster.

Before the collision with the surface, the direction of the velocity of all the components of the clusters is the same, normal to the surface, in most of the figures that we show. Although the velocity of the center of mass is high, the relative kinetic energies of the components of the cold cluster, before the collision, are very low. As a result of the impact with the surface, the initially directed velocity is rapidly randomized A rough estimate of the temperature rise within the cluster is provided by the equivalence of temperature to the random part of the kinetic energy. The cluster temperature can thus reach a value that is V_0^2 times room temperature where V_0 is the initial velocity in units of the velocity at room temperature. A cluster impacting the surface at $10\,km\,s^{-1}$ can therefore be heated to well over $10^5\,K$. Our molecular dynamics simulations, for a cluster of interacting but otherwise structureless particles, verify this temperature range is accessible even when energy loss to the surface is allowed, cf. Fig. 1.

True temperature is also a measure of the distribution in the speeds of the molecules. Figure 2 compares the distribution after the impact, computed from molecular dynamics simulations, to the Maxwell–Boltzmann functional form. In section 4 we discuss the reasons why the initially directed energy is so rapidly thermalized.

As a result of the impact with the surface, the cluster is initially quite compressed. At the velocity range of interest (below $10\,km\,s^{-1}$), the density

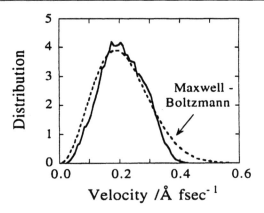

Figure 2: The three dimensional velocity distribution of all the atoms after impact of a cold cluster of 125 Ar atoms at a surface at a velocity of 25 km s^{-1} (= 0.25 Å/fs). Shown for comparison is a Maxwell–Boltzmann functional form for the same mean energy. Even at this high velocity of impact, the velocity distribution after the impact is essentially isotropic.

can almost double. The dynamics under the high densities and high relative velocities, which result from the impact of a cluster with a surface, is the subject of the next section.

2 A New Regime of Dynamics

The important generalization that emerges from our studies is that cluster impact at high velocities provides a new regime of collisional activation, where the intermolecular coupling is comparable or faster than intramolecular motions. At ordinary velocities, only the rotations are in this, 'sudden', regime. Indeed the rates for rotational-translational energy transfer are rather high [86]. The range of relative velocities in the cluster, immediately after the impact, is 10–20 km s^{-1}, which means (for example) for a rare gas atom-halogen molecule (Rg-X$_2$) collision at these velocities distances of less than 2/3 of the range of the Lennard–Jones Rg-X potential can be reached.

The duration of such a close-in collision is not more than a few fs's, and such collisions are sudden-like with respect to vibrational (and, of course with respect to the rotational) motion of the reactants. It follows that vibrotational excitation in a rare-gas atom reactant collision will be impulsive like and hence very efficient [90]. At these high velocities the sequential collisions are quite distinct in time. Despite the high density, the duration of such a close-in collision is shorter than the time between successive collisions and this is also due to the short range of the repulsive force. The constituents of the

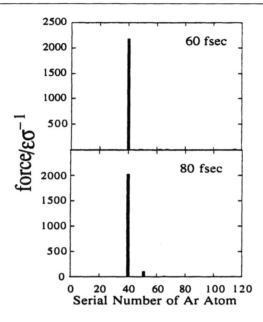

Figure 3: The force (in reduced units) along the bond of a halogen I_2 molecule, applied by the different rare gas atoms after impact at $v = 5\,\mathrm{km\,s^{-1}}$, vs. its (arbitrary) serial number 60 and 80 fsec after the impact with the surface. Note how, at any given time, only one or two atoms act to vibrationally excite the molecule.

cluster are 'smaller' than we are used to. In other words, it is typically one rare gas atom at a time that undergoes a close in collision with a reactant, (Fig. 3).

An additional factor that contributes to the validity of the picture of sequential binary collision is the repulsion between the rare gas atoms. It limits the number of atoms that can effectively couple to the vibration of the diatomic molecule.

Under the binary approximation we can break the whole history of each cluster impact into a series of elementary steps. Each collision can be treated as an isolated one like in gas phase, including the collision between the reactants. In summary, despite the high densities, gas phase dynamics can provide useful insights.

The high material and energy density conditions that can be established very rapidly, lead to heating on a time scale comparable or shorter than typical intramolecular motions. The important conclusion from our (theoretical) research on cluster–surface collisions, is that the short period while the cluster is hot and compressed is long enough for chemical reactions to take place. In both classes of reactions which we examined (the simple bond dissociation of diatomic molecules [82, 91] and various kinds of four center reactions [79, 80, 89, 92, 93], see section 3), molecular dynamics simulation shows high reactivity.

3 Chemistry Within an Impact Heated Cluster

The purpose of our work was to examine whether the compressed hot cluster provides an effective medium for reactions with high barriers. In the simplest case of cluster impact experiments, the reactants are embedded inside a rare gas cluster, and this cold droplet is incident on an inert surface at various velocities. Using standard molecular dynamics simulation we first looked at the dissociation of a halogen molecule[2].

The computed [95] high yield of bond breaking encouraged us to examine a much more complicated chemical process - a four-center reaction. Four center reactions [96–105], involving the switching of covalent bonds, are expected to have a high barrier [94, 106]. Beyond any potential barrier to reaction, four center reactions are also subject to a strong kinematic constraint [107]. However, under the unusual combination of conditions made possible within an impact-heated cluster, the four-center reaction can be driven. A variety of high barrier processes were examined: the bimolecular $N_2 + O_2$ and $H_2 + I_2$ reactions and the unimolecular norbornadiene $\rightarrow$ quadricyclane isomerization. Our primary finding is that the yield of the concerted reaction is quite high because the cluster has a rather essential role to play. So much so that we consider that it is reasonable to speak of 'a cluster catalyzed reaction'. Figure 4 shows how even the mass of the rare gas atoms has an influence.

Both dissociations and four center reactions have an energy threshold. Since the diatom molecule inside the cluster (before the collision) is cold, the nominal energy threshold for the bond dissociation will be the binding energy. For the four-center reaction, the threshold will be the minimum energy required to surmount the reaction barrier. However, although these energetic conditions are necessary for the reaction to occur, there are other considerations that are often kinematic in origin. These result in an effective threshold that is higher than the nominal one. On the other hand, it is characteristic of reactions within the cluster that the reactants are

---

[2]Concerted four center reactions are known to have a high energetic barrier due to an unusual orbital correlation [94]. This is due to curve crossing between the state that originates in the ground state reactants and correlates to the electronically excited products and the state that originates in the electronically excited reactants and correlates to the ground state products.

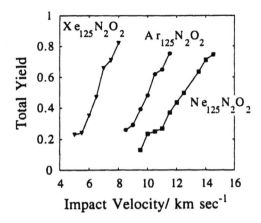

Figure 4: The reactive possible outcomes in the $N_2 + O_2$ rearrangement collisions are:
$$N_2 + O_2 \to \begin{cases} NO + NO \\ NO + N + O \\ N + N + O + O \end{cases}$$
We count the dissociation channels as reactive because all the trajectories we examined yielded such final states only when it was preceded by the formation of NO, however briefly. Shown is the total yield vs. the impact velocity for $N_2 + O_2$ embedded in a cluster of 125 rare gas atoms.

preconditioned by vibrational activation. The actual collision that leads to reaction will therefore have a lower threshold than that for cold reactants.

The sudden-like nature of reactive exchange inside the cluster is illustrated in Fig. 5.

Preceding the reaction is an activating collision that brings the reactants to a relative velocity of 10–30 km/sec. At these energies the entire transversal of the potential energy surface lasts for only 10 fs or so. The reaction is therefore sudden-like with respect to the internal motions of even the lighter reactants (N_2, O_2, H_2). The heavier I_2 molecule is fully standing still during the reaction. Therefore, the bond switching can be described by kinematic considerations [108–115]

The efficiency of cluster impact in driving four center reactions is due to a matching between what the cluster can do to the reactants, or products, and the very selective energy requirements and the specific energy disposal in a concerted reactive collision. The cluster serves to provide both the steric and energetic conditions necessary for reaction to occur. In terms of the impact parameter of the relative motion of the two reactants, their confinement by the cluster keeps it low, so that they do not miss one another.

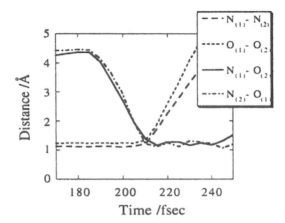

Figure 5: Bond distance vs. time (in fs) plot for the two old bonds and the two new bonds during the $N_2 + O_2 \to 2NO$ reactive collision in a cluster of 216 Ne atoms for an impact at $10.5\,\mathrm{km\,s^{-1}}$. The bond switching occurs in a concerted fashion in less than 10 fs and in physical units it is clearly faster than the vibrational period of the newly formed NO bonds.

That confinement within the cluster favors low impact parameter collisions is a key ingredient in why such processes are so efficient. Furthermore, both the activation of the reactants before the reaction and the stabilization of the hot product after it, are due to the cluster atoms.

After the activation stage, the reactants collide. If reaction occurred, two excited product molecules are rapidly receding from one another. In order to achieve high products yield, the role of the cluster atoms is to cool the internal excitation.

The role of the cluster in making reactants that failed to react on the first try, try again, is enhanced if the rare gas atoms heavier because they are better able to reverse the velocity of the center of mass of the reactive atoms. Here too is the correlation that heavier rare gases favor overall reactivity. On the other hand, a heavier rare gas is more likely to induce dissociation of an internally hot nascent product molecule.

4 The Rapid Thermalization of the Impact Energy

As a result of the impact with the surface, the leading edge atoms of the cluster reverse their direction of motion and collide with the atoms of the cluster which are still moving forward. Further collisions result in an ultra

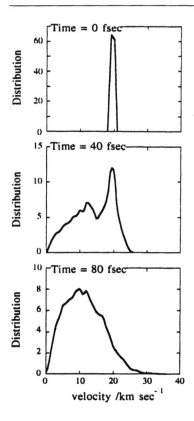

Figure 6: The distribution of the (scalar) velocity of atoms at different times in a molecular dynamics simulation of the impact of a 125 atom Ar cluster at a surface where the surface is simulated by the hard cube model at the temperature of 30 °K. The impact velocity is 20 km s^{-1} or 1 Å per 5 fs where the range parameter of the Ar–Ar potential is 3.41 Å. The mean free path is, very roughly of the same magnitude. Thermalization is essentially complete by $\approx$ 80 fs or, after roughly four collisions.

fast conversion of the initially directed energy to random motion and ultimately to the extensive or complete breaking apart of the cluster. We first note that the finite cluster does reach thermal equilibrium in the very short time that is available before all collisions cease due to the rapid expansion of the cluster. Then we discuss a simple mechanism which suggests an answer to what brings this fast relaxation about.

Figure 6 shows typical output of the simulations. It is seen that by 80 fs the velocity distribution looks like a three dimensional Maxwell–Boltzmann one.

What brings the fast relaxation about? In order to answer this question, we examined a simple model of hard sphere collisions in a plane.

Model

A simple mechanical model can exhibit many of the physical features found in the simulation. It is a set of hard spheres (five suffices for all practical purposes) which can have velocities in a plane, and bounded by two rigid walls. The rigid walls reverse the x component of the velocity of the end atoms that collide with them, and each collision between two hard spheres reverses the components of the two velocities along the line of the center of the two. Each hard sphere has either two adjacent neighbors or, for the two end ones, a neighbor on one side and a wall on the other. In short, the spheres are arranged in a sequence but not along a straight line. Knowing the initial velocities condition of each hard sphere in the chain, one can analytically calculate [116] the velocity components of all hard spheres at any given point of time, meaning after any number of collisions. The central point in this model is that the line joining the centers of any two consecutive spheres in the chain is randomly oriented - the spheres are not arranged exactly along a straight line. Note that in this model when two hard spheres collide, the other spheres are spectators.

The results from the model mimic those obtained by the MD simulations. The main observations are that (i) Thermalization is complete in about five collisions. (ii) Atoms inside the cluster thermalize sooner than the atoms of edges and (iii) Thermalization occurs practically just as fast in all directions.

Our conclusion is that it is the lack of structure (or spatial order) of the cluster that is the cause of the rapid thermalization. In all our molecular dynamic simulations, the clusters are glass-like and therefore no spatial order exist. The implication is that a perfectly ordered array of hard spheres will not relax. If a perfect crystal of hard spheres at 0 K is directed towards a rigid surface in such a manner that all the atoms of the front face hit the surface at exactly the same instant then there is no dissipation. The cluster rebounds from the inert surface as if it is rigid [116].

Not discussed herein are recent experiments [61, 63] on the velocity distribution of molecules and atoms of the cluster after the impact. There are two reasons why these results should at least be mentioned. One is that the experiment used a cluster beam incident at an angle θ with respect to the normal. It measured the component of velocity parallel to the surface. But in the plane containing the incident beam, there are two such components. One at an angle $\pi/2 - \theta$ with the cluster beam and one at an angle $\pi/2 + \theta$. The observed velocity distribution in both directions was identical. This is a computation-free proof of the equilibration. The other point is the magnitude of the measured temperature which was of the order of 15,000 °K, same as that determined for electrons boiling out of the impact [35].

5 The Shattering of Clusters

A short time after the collision with the surface the cluster fragments. An important aspect of the cluster impact induced reactions is that the fragmentation of the cluster prevents the collision-induced dissociation of the energy rich products, by collisions with the atoms of the cluster. Cluster impact heating can induce a tremendous energy deposition in a time much shorter than the rate of expansion of the cluster, as has been shown in the previous section. In this section we discuss the subsequent evolution of this extreme non-equilibrium state, a state that cannot be reached by ordinary heating rates where expansion of the system is faster than the rate of energy deposition. The discussion is not by historical order. The shattering transition that we shall discuss was first seen in a purely theoretical approach, using information theory [117] and was almost immediately verified experimentally [34]. The molecular dynamics simulations [81] and better experiments [25, 26, 30] came later. The experimental signature of the transition to shattering is that up to a certain velocity of impact the hot cluster rebounds intact from the surface. If one waits long enough, this hot cluster may cool by evaporation but there is a clear time interval for which the cluster is intact. More on this below. At a slightly higher impact velocity the parent cluster disappears and many very small fragments appear.

The shattering phenomenon is given much attention because it provides a direct experimental proof of the unusual conditions that are achieved by the impact.

Molecular dynamics simulations show a sharp transition, as a function of the impact velocity, between two behaviors: recoil of the intact parent cluster with the possibility of evaporation (i.e., the departure of a small subcluster ,one or two atoms) and shattering of the cluster to small fragments, mainly monomers. The hyper radius is a convenient measure for the size of an n body system, defined by [118]

$$\rho^2 = \left(\prod_{i=1}^{n} m_i\right)^{-1/n} \sum_{i=1}^{n} m_i(\mathbf{r}_i - \mathbf{r}_{cm})^2. \tag{1}$$

$\mathbf{r}_i$ is the position vector of particle i and $\mathbf{r}_{cm}$ is the position of the cluster center of mass. Figure 7 shows the value of ρ at a large distance from the surface, for a rebounding cluster of 125 Ar atoms as a function of the velocity of impact. The onset of shattering is clear:

The change in ρ at low velocities is due to two processes. Due to the energy deposition the cluster undergoes first a shock wave oscillation, which, upon thermalization, becomes shape changes involving the stretching and contraction of the interatomic distances. This causes a slight increase in the value of ρ with superposed oscillations. After some delay, the warm cluster can cool by the evaporation of one or two atoms (Fig. 8).

At this point there is an experimental bound on the duration of shattering of less than 80 ps [27]. The simulations however suggest that shattering is even faster, cf. Fig. 8.

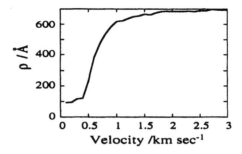

Figure 7: The hyper radius ρ, equation (1), of a cluster of 125 Ar atoms computed at a point where the center of mass has receded to 75 Å from the surface, vs. the velocity of impact. The positions of the atoms are taken from a molecular dynamics simulation. The onset of shattering is at $0.45\,\text{km}\,\text{s}^{-1}$ and the shattering is to individual atoms by about $1.0\,\text{km}\,\text{s}^{-1}$.

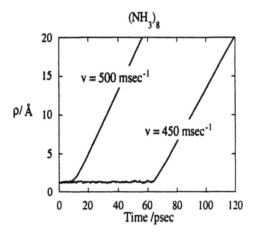

Figure 8: The hyperradius ρ of the cluster, defined in equation (1), in Å, vs. time, in ps, for the same cluster impacting a cold surface at two slightly different low supersonic velocities of 450 and $500\,\text{m}\,\text{s}^{-1}$ respectively. This figure illustrates how a small increment in the energy of impact changes the outcome. The cluster evaporates one monomer at the lower energy but fully shatters at the higher energy.

The point about the technique of cluster impact is that it enables one to 'heat' the cluster on a time scale short compared to that needed for expansion and ipso facto for evaporation. In this way one can prepare 'superheated' clusters with enough energy for breaking most or all intermolecular bonds so that the cluster shatters into its constituents. The sharp transition from evaporation regime to the shattering one, shown in Fig. 9, indicates

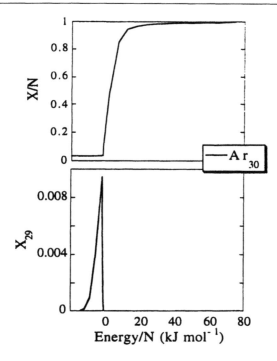

Figure 9: Information theory computations for a cluster of $N = 30$ Ar atoms. *Top panel:* The total number, X, of fragments (plotted scaled by the size of the original cluster since $0 < X/N \leq 1$) vs. the energy. The sudden increase in the number of fragments is the shattering transition. *Bottom panel:* Plotted is the number of clusters of 29 atoms vs. the energy. When a cluster of 30 atoms evaporates a single atom it necessarily leaves an intact cluster of 29 atoms. At a sharp value of the energy, evaporation gives over to shattering where the cluster fragments to many small fragments without going through the intermediate formation of larger clusters. Note that evaporation occurs only over a narrow range in impact velocities. In our definition of the energy, the total energy will have a negative value as long as the total kinetic energy is below the binding energy of the cluster. The threshold for shattering is thus at zero.

a fast translational thermalization of the extreme disequilibrium established immediately after the surface impact.

From the information theory point of view, the switch over from an intact cluster to independent monomers is due to a competition between two entropic effects. On the one hand the large entropy due to the many possible configurations of a large cluster and on the other the entropy due to the translational motion of the many monomers. Each one of these contributions is exponentially large, but they act in opposite direction. The very many possible isomers of a parent cluster of a given size enables it to soak up considerable energy. The translational entropy favors the formation of as many fragments as possible. At the point where the two effects are balanced, a small change in conditions (e.g., in the velocity of impact) results in very large variation in the probabilities of the two outcomes.

6 The Burning of Air: A Theoretical Study Using Two Complementary Procedures

In this section we compare the results from molecular dynamic simulations and an information theory analysis on the study of the impact of clusters containing several N_2 and O_2 molecules and clusters containing only N_2 and O_2 molecules. We still have no definitive results for this experiment but we hope that this challenge will be taken up soon.

The reason we employ two rather distinct methods of inquiry is that neither, by itself, is free of open methodological issues. The method of molecular dynamics has been extensively applied, inter alia, to cluster impact. However, there are two problems. One is that the results are only as reliable as the potential energy function that is used as input. For a problem containing many open shell reactive atoms, one does not have well tested semi empirical approximations for the potential. We used the many body potential [119, 120]. The other limitation of the MD simulation is that it fails to incorporate the possibility of electronic excitation. More on this below. The second method that we used is, in many ways, complementary to MD. It does not require the potential as an input and it can readily allow for electronically excited as well as for charged products. It seeks to compute that distribution of products which is of maximal entropy subject to the constraints on the system (conservation of chemical elements, charge and energy). Both methods consider the surface on which the cluster impacts to be chemically inert. It can exchange energy but not atoms nor charges with the cluster.

The results from both methods at velocities of impact below $10\,\text{km}\,\text{s}^{-1}$ are remarkably consistent: As a result of the ultrafast heating and strong confinement achieved by cluster impact, air burns and forms primarily NO, cf. Fig. 10.

A very interesting result of the maximum entropy distribution, is that at higher velocities of impact there is a steep onset of copious production of electronically excited and of charged species, Fig. 11: This behavior is also seen experimentally [29, 35] and is the reason why an MD simulation involving only ground state species may be realistic only at impact velocities

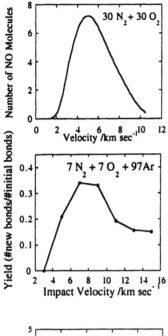

Figure 10: *Top panel:* The yield of NO, computed by the maximum entropy formalism, when a cluster of $30\,N_2$ and $30\,O_2$ molecules is superheated. The velocity scale corresponds to the case in which there is no energy dissipation to the surface. The experimental 'impact velocity' is necessarily equal or higher than this velocity. *Bottom panel:* The yield of new bonds (as a fraction of the number of initial bonds) computed by a molecular dynamics simulation when a cluster of $7\,N_2$ and $7\,O_2$ molecules, embedded in 97 Ar atoms, impacts on a cold (30 °K) surface, vs. the impact velocity.

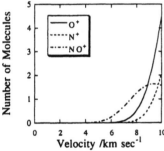

Figure 11: The yield of charged products computed by the maximum entropy formalism for a cluster of $30\,N_2$ and $30\,O_2$ molecules. The velocity scale is the same as in Fig. 10. Note the steep onset for the appearance of charged atoms.

below say, $10\,\mathrm{km\,s^{-1}}$. Since our primary interest is the chemistry under extreme conditions, it appears to put an upper limit on the useful energy range. Even so, it is a quite wide, extending to velocities of impact of about 30 Mach.

7 Electronically Non Adiabatic Tranitions

The information-theoretic prediction, shown in Fig. 11, is that by about an impact velocity of $10\,\mathrm{km\,s^{-1}}$, there is a steep rise in the yield of electronically excited species. Full quantum mechanical computations [121], for four center binary collisions at relative velocities $\geq 10\,\mathrm{km\,s^{-1}}$, also show a high effective

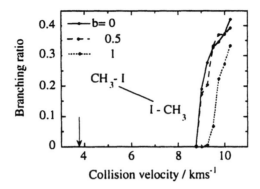

Figure 12: The probability to exit on the reactive side of the upper electronic potential energy surface, as a function of the relative velocity in a near collinear $CH_3I + CH_3I$ collision, (see inset). The results are shown for three impact parameters as indicated. The arrow indicates the nominal energy threshold for accessing the upper electronic surface. (Computed by the quantal FMS method.)

threshold with a steep post threshold rise, (Fig. 12). Below the effective threshold the probability of a non-adiabatic process is not zero but it is exponentially small. As for ground state four center reactions, here too there is a kinematic constraint. The total energy is conserved. If the ground and excited potential energy surfaces do not intersect then there is a change in the potential energy upon electronic excitation. This is necessarily accompanied by a change in the kinetic energy of the nuclei. Such a change is penalized by a low Franck–Condon factor. Past the effective threshold, the energy is high enough that the fractional change in the momenta of the nuclei is small. The, so called, exponential gap behavior [86] is the origin of the steep onset seen in Fig. 12. The computations use the 'Full Multiple Spawning', FMS, methodology.

Concerted four center reactions are expected to have a high energetic barrier due to an unfavorable orbital correlation. Our original suggestions for cluster impact induced chemistry have been that such nominally 'forbidden' reactions could be thermally driven by cluster impact. The high barrier for such reactions is due to curve crossing between the state that originates in the ground state reactants and correlates to the electronically excited products, and the state that originates in the electronically excited reactants and correlates to the ground state products. These reactions offer therefore a good example where there will be a failure of the adiabatic approximation and the dynamics will follow a diabatic correlation.

The quantal computations (Fig. 12) show that for vibrationally cold reactants, there is a steep onset of crossing to the upper surface in the velocity range 9–10 km s^{-1}. The (avoided) crossing region is 'late' and requires extended bonds. Hence, and as also shown by the computations [121], the crossing is favored by vibrational excitation.

Designing a cluster that, upon impact, will favor electronic excitation is of interest because, so far, all the experiments for detecting visible/UV light emission from the hot cluster have, as far as we know, failed. Possibly, the phenomenon of shuttle afterglow [42] can be considered as an exception to this statement. It could be that any electronically excited states that are formed are very effectively quenched by the coated surfaces that are used as targets. It is however of some concern that the lowest energy indication of non-adiabaticity is electron emission. Of course, this emission could be due to the population of highly promoted (nearly united atom [122]) orbitals in close collisions. In that case there will not be a whole set of electronically excited states that are formed enroute to ionization. The only conflicting result is that we have clear experimental evidence that the velocity distribution of the emitted electrons is characterized by a temperature that is similar to the *translational* temperature of the ions formed upon shattering of the cluster. This suggests an equilibration of the electronic and translational degrees of freedom. Yet, so far, experimental evidence for excitation of bound, excited, electronic states is lacking.

High available energy is only one reason for the new chemistry that is made possible in the hot and compressed cluster. The simulations show that equally important is that, in the cluster, collisions necessarily occur with rather low impact parameters. (If two molecules are moving relative to one another with a higher impact parameter then before they come near, one or the other or both will collide with other molecules). The same is true for non-adiabatic transitions. The computed probability of crossing to the upper electronic state decreases rapidly with increasing impact parameter. It is because the cluster favors low impact parameter collisions that the yield is not small.

8 Concluding remarks

Cluster–surface collisions at a supersonic velocity were shown to induce chemical reactions (on the computer). The special conditions made possible within the rapidly heated cluster open a window to new regime of dynamics, where the intermolecular motion are comparable or faster than the intramolecular ones. This 'sudden' coupling is the key for understanding the special dynamics inside the super hot and super dense cluster. Despite the high density inside the cluster a short time after the collision, the dynamics is more like in gas phase than in a condensed one since the duration of each collision is shorter than the time between successive collisions, so that each collision is essentially isolated. That the dynamics is like in gas phase does not mean that the surroundings do not play a role. Quite the contrary. The role of the cluster's atoms is essential in inducing the reactions. Not just by creating the necessary energetic conditions for the reactions to take place, but also by activating the reactants and deactivating the products. When the environment is part of the reactive system, like in the case of the liquid air clusters, new multi-center mechanisms are possible.

A short time after the collision the cluster fragments. Fast translational thermalization precedes this fragmentation process. This ultra fast energy dissipation is shorter than the rate of expansion of the cluster. This phenomenon is demonstrated by the transition from evaporation to shattering as the impact velocity increases. This sharp transition to shattering has the characterization of phase transition known from macroscopic systems. The system is small enough which enables one to follow each atom during the history of each trajectory, but it is large enough to have a macroscopic characterization. Therefore, impact heated clusters can be an optimal tool to bridge between microscopic to macroscopic system.

The present theoretical and computational studies on cluster–surface collisions at various impact velocities, are only the beginning of what can be examined using this system. Despite the relatively simple research tools, our preliminary work has already shown that many unexpected features can arise and opened windows to interesting, unexplored, research areas. We are very gratified that in the short time since our work began there are already published definitive experimental studies that show that cluster impact chemistry is a viable real world possibility. The purpose of this brief overview is to encourage the molecular beam community to follow additional experimental studies of this new regime of dynamics.

Acknowledgments

We thank Uzi Even, Tamotsu Kondow and Karl Kompa for many discussions and encouragement. This work was supported by the Alexander von Humboldt Stiftung.

References

[1] *Shock Wave in Chemistry*, edited by A. Lifshitz (Dekker, New York, 1981).

[2] J. Gardiner, W.C., *Combustion Chemistry* (Sprinber, Berlin, 1984).

[3] W. Tsang and A. Lifshitz, *Ann. Rev. Phys. Chem.* **41**, 559 (1990).

[4] J. Sakata and C. A. Wight, *J. Phys. Chem.* **99**, 6584 (1995).

[5] E. Grunwald, D. F. Dever, and P. M. Keehn, *Megawatt Infrared Laser Chemistry* (Wiley, New York, 1978).

[6] V. N. Kondratiev and E. E. Nikitin, *Gas-Phase reaction* (Springer, Berlin, 1981).

[7] *Nonlinear Laser Chemistry with Multiple Photon Excitation*, Vol., edited by V. S. Letokhov (Springer, Berlin, 1983).

[8] G. W. Flynn and R. E. J. Weston, *Ann. Rev. Phys. Chem.* **37**, 551 (1986).

[9] J. E. Willard, *Ann. Rev. Phys. Chem.* **6**, 141 (1955).

[10] R. Wolfgang, *Ann. Rev. Phys. Chem.* **6**, 15 (1965).

[11] A. Henglein, *Ultrasonic* **25**, 6 (1987).

[12] P. U. Andersson and J. B. C. Pettersson, *Z. Phys. D* **41**, 57 (1997).

[13] R. D. Beck, P. S. John, M. L. Homer, and R. L. Whetten, *Chem. Phys. Lett.* **187**, 122 (1991).

[14] R. D. Beck, P. Weis, G. Brauchle, and J. Rockenberger, *Rev. Sci. Instrum.* **66**, 4188 (1995).

[15] R. D. Beck, J. Rockenberger, P. Weis, and M. M. Kappes, *J. Chem. Phys.* **104**, 3638 (1996).

[16] M. Benslimane, M. Chatelet, A. D. Martino, F. Pradere, and H. Vach, *Chem. Phys. Lett.* **237**, 323 (1995).

[17] E. R. Bernstein, *J. Phys. Chem.* **96**, 10105 (1992).

[18] R. J. Beuhler, *J. Appl. Phys.* **54**, 4118 (1983).

[19] R. J. Beuhler and L. Friedman, *Chem. Rev.* **86**, 521 (1986).

[20] A. W. Castleman, Jr., and K. H. Bowen, Jr., *J. Phys. Chem.* **100**, 12911 (1996).

[21] M. Chatelet, A. D. Martino, J. Pettersson, F. Pradere, and H. Vach, *Chem. Phys. Lett.* **196**, 563 (1992).

[22] M. Chatelet, A. D. Martino, J. Petterson, F. Pradere, and H. Vach, *J. Chem. Phys.* **103**, 1972 (1995).

[23] S. Chen, X. Hong, J. R. Hill, and D. D. Dlott, *J. Phys. Chem.* **99**, 4525 (1995).

[24] W. Christen, K. L. Kompa, H. Schroder, and H. Stulpnagel, *Ber. Bunsenges. Phys. Chem.* **96**, 1197 (1992).

[25] W. Christen, U. Even, T. Raz, and R. D. Levine, *Int'l J. Mass Spectrom. & Ion Proc.* **174**, 35 (1998).

[26] W. Christen, U. Even, T. Raz, and R. D. Levine, *J. Chem. Phys.* **108**, 10202 (1998).

[27] W. Christen and U. Even, *J. Phys. Chem.* **102**, 9420 (1998).

[28] W. Christen, U. Even, T. Raz, and R. D. Levine, *J. Chem. Phys.* **108**, 10202 (1998).

[29] U. Even, P. J. d. Lange, H. T. Jonkman, and J. Kommandeur, *Phys. Rev. Lett.* **56**, 965 (1986).

[30] U. Even, T. Kondow, R. D. Levine, and T. Raz, *Com. At. Mol. Phys.* **1D**, 14 (1999).

[31] Y. Fukuda, *Ph.D. thesis, University of Tokyo*, (1998).

[32] J. Gspann and G. Krieg, *J. Chem. Phys.* **61**, 4037 (1974).

[33] M. Gupta, E. A. Walters, and N. C. Blais, *J. Chem. Phys.* **104**, 100 (1996).

[34] E. Hendell, U. Even, T. Raz, and R. D. Levine, *Phys. Rev. Lett.* **75**, 2670 (1995).

[35] E. Hendell and U. Even, *J. Chem. Phys.* **103**, 9045 (1995).

[36] P. M. S. John and R. L. Whetten, *Chem. Phys. Lett.* **196**, 330 (1992).

[37] H. Kishi and T. Fujii, *J. Phys. Chem.* **99**, 11153 (1995).

[38] C. Lifshitz,,, edited by T. B. C.Y. Ng, I. Powis (Wiley, New York, 1993).

[39] T. Lill, H.-G. Busmann, B. Reif, and I. V. Hertel, *App. Phys.* **A55**, 461 (1992).

[40] T. Lill, F. Lacher, H.-G. Busmann, and I. V. Hertel, *Phys. Rev. Lett.* **71**, 3383 (1993).

[41] T. Lill, H.-G. Busman, B. Reif, and I. V. Hertel, *Surf. Sci.* **312**, 124 (1994).

[42] E. Murad, *Ann. Rev. Phys. Chem.* **49**, 73 (1998).

[43] J. A. Niesse, J. N. Beauregard, and H. R. Mayne, *J. Phys. Chem.* **98**, 8600 (1994).

[44] J. B. C. Pettersson and N. Markovic, *Chem. Phys. Lett.* **201**, 421 (1993).

[45] M. Svanberg and J. B. C. Pettersson, *Chem. Phys. Lett.* **263**, 661 (1996).

[46] M. Svanberg, N. Markovic, and J. B. C. Pettersson, *Chem. Phys.* **220**, 137 (1997).

[47] H. Tanaka, T. Hanmura, S. Nonose, and T. Kondow, in *Similarities and Difference between Atomic Nuclei and Clusters*, edited by A. Abe, Lee, Yabana (The American Inst. of Physics, 1998), p. 193-.

[48] A. Terasaki, T. Tsukuda, H. Yasumatsu, T. Sugai, and T. Kondow, *J. Chem. Phys.* **104**, 1387 (1996).

[49] A. Terasaki, H. Yamaguchi, H. Yasumatsu, and T. Kondow, *Chem. Phys. Lett.* **262**, 269 (1996).

[50] A. Terasaki, H. Yasumatsu, Y. Fukuda, H. Yamaguchi, and T. Kondow, *Riken Rev.* **17**, 1 (1998).

[51] H. Vach, A. D. Martino, M. Benslimane, M. Chatelet, and F. Pradere, *J. Chem. Phys.* **100**, 8526 (1994).

[52] H. Vach, M. Benslimane, M. Chatelet, A. D. Martino, and F. Pradere, *J. Chem. Phys.* **103**, 1 (1995).

[53] V. Vorsa, P. J. Campagnola, S. Nandi, M. Larsson, and W. C. Lineberger, *J. Chem. Phys.* **105**, 2298 (1996).

[54] A. A. Vostrikov, D. Y. Dubovend, and M. R. Pretechenskiy, *Chem. Phys. Lett.* **139**, 124 (1987).

[55] A. A. Vostrikov and D. Y. Dubov, *Z. Phys. D.* **20**, 61 (1991).

[56] C. Walther, G. Dietrich, M. Lindinger, K. Luetzenkirchen, L. Schweikhard, and J. Ziegler, *Chem. Phys. Lett.* **256**, 77 (1996).

[57] H. Weidele, D. Kreisle, E. Recknagel, G. S. Icking-Konert, H. Handschuh, G. Gantefor, and W. Eberhardt, *Chem. Phys. Lett.* **237**, 425 (1995).

[58] R. Wörgötter, B. Dunser, P. Scheier, T. D. Mark, M. Foltin, C. E. Klots, J. Laskin, and C. Lifshitz, *J. Chem. Phys.* **104**, 1225 (1996).

[59] R. Wörgötter, J. Kubitsa, J. Zabka, Z. Dolejsek, T. D. Märk, and Z. Herman, *Int'l J. Mass Spectrom. & Ion Proc.* **174**, 53 (1998).

[60] G. Q. Xu, R. J. Holland, and S. L. Bernasek, *J. Chem. Phys.* **90**, 3831 (1989).

[61] H. Yasumatsu, S. Koizumi, A. Terasaki, and T. Kondow, *J. Chem. Phys.* **105**, 9509 (1996).

[62] H. Yasumatsu, A. Terasaki, and T. Kondow, *J. Chem. Phys.* **106**, 3806 (1997).

[63] H. Yasumatsu, S. Koizumi, A. Terasaki, and T. Kondow, *J. Phys. Chem. A* **102**, 9581 (1998).

[64] C. Yeretzian, R. D. Beck, and R. L. Whetten, *Int. J. Mass Spectr. Ion Proc.* **135**, 79 (1994).

[65] R. Alimi, R. B. Gerber, and V. A. Apkarian, *J. Chem. Phys.* **92**, 3551 (1990).

[66] F. Amar and B. J. Berne, *J. Phys. Chem.* **88**, 6720 (1984).

[67] M. B. Andersson and J. B. C. Pettersson, *J. Chem. Phys.* **102**, 4239 (1995).

[68] R. S. Berry, *J. Phys. Chem.* **98**, 6910 (1994).

[69] E. E. B. Campbell, T. Raz, and R. D. Levine, *Chem. Phys. Lett.* **253**, 261 (1996).

[70] C. L. Cleveland and U. Landman, *Science* **257**, 355 (1992).

[71] U. Even, I. Schek, and J. Jortner, *Chem. Phys. Lett.* **202**, 303 (1993).

[72] F. A. Gianturco, E. Buonomo, G. Delgado-Barrio, S. Miret-Artes, and P. Villarreal, *Z. Phys. D* **35**, 115 (1995).

[73] M. R. Hoare and J. A. McInnes, *Adv. Phys.* **32**, 791 (1983).

[74] C. E. Klots, *Z. Phys. D* **20**, 105 (1991).

[75] L. Liu and H. Guo, *Chem. Phys.* **205**, 179 (1995).

[76] N. Markovic and J. B. C. Pettersson, *J. Chem. Phys.* **100**, 3911 (1994).

[77] L. Ming, N. Markovic, M. Svanberg, and J. B. C. Pettersson, *J. Phys. Chem. A* **101**, 4011 (1997).

[78] L. Qi and S. B. Sinnott, *J. Phys. Chem.* **B1997**, 6883 (1997).

[79] T. Raz and R. D. Levine, *J. Am. Chem. Soc.* **116**, 11167 (1994).

[80] T. Raz and R. D. Levine, *J. Phys. Chem.* **99**, 7495 (1995).

[81] T. Raz and R. D. Levine, *J. Chem. Phys.* **105**, 8097 (1996).

[82] I. Schek, T. Raz, R. D. Levine, and J. Jortner, *J. Chem. Phys.* **101**, 8596 (1994).

[83] I. Schek and J. Jortner, *J. Chem. Phys.* **104**, 4337 (1996).

[84] M. Svanberg, N. Markovic, and J. B. C. Pettersson, *Chem. Phys.* **201**, 473 (1995).

[85] G. Q. Xu, S. L. Bernasek, and J. C. Tully, *J. Chem. Phys.* **88**, 3376 (1988).

[86] R. D. Levine and R. B. Bernstein, *Molecular Reaction Dynamics and Chemical Reactivity* (Oxford University Press, NY, 1987).

[87] R. D. Levine and A. Ben-Shaul, in *Chemical and Biochemical Applications of Lasers, Vol. II*, edited by C. B. Moore (1977).

[88] R. D. Levine, in *Theory of Reactive Collisions*, edited by M. Baer (CRC Press, 1984).

[89] T. Raz and R. D. Levine, *Chem. Phys. Lett.* **246**, 405 (1995).

[90] M. Ben-Nun, T. Raz, and R. D. Levine, *Chem. Phys. Lett.* **220**, 291 (1994).

[91] T. Raz, I. Schek, M. Ben-Nun, U. Even, J. Jortner, and R. D. Levine, *J. Chem. Phys.* **101**, 8606 (1994).

[92] T. Raz and R. D. Levine, *J. Phys. Chem.* **99**, 13713 (1995).

[93] T. Raz and R. D. Levine, in *Heidelberg conference: "gas phase reaction systems: experiments and models - 100 years after Max Bodenstein* (1995).

[94] R. Hoffmann, *J. Chem. Phys.* **49**, 3739 (1968).

[95] T. Raz, I. Schek, M. Ben-Nun, U. Even, J. Jortner, and R. D. Levine, *J. Chem. Phys.* **101** (1994).

[96] W. Altar and H. Eyring, *J. Chem. Phys* **4**, 661 (1936).

[97] H. C. Andersen, D. Chandler, and J. D. Weeks, *Adv. Chem. Phys.* **34**, 105 (1976).

[98] J. B. Anderson, *J. Chem. Phys.* **100**, 4253 (1994).

[99] S. H. Bauer, *Science* **146**, 1045 (1964).

[100] S. H. Bauer, *Ann. Rev. Phys. Chem.* **30**, 271 (1979).

[101] J. M. Bowman and G. C. Schatz, *Ann. Rev. Phys. Chem.* **46**, 169 (1995).

[102] D. C. Clary, *J. Phys. Chem.* **98**, 10678 (1994).

[103] J. Jaffe and S. B. Anderson, *J. Chem. Phys.* **49**, 2859 (1968).

[104] L. M. Raff, L. Stivers, R. N. Porter, D. L. Thompson, and L. B. Sims, *J. Chem. Phys.* **52**, 3449 (1970).

[105] L. M. Raff, D. L. Thompson, L. B. Sims, and R. N. Porter, *J. Chem. Phys.* **56**, 5998 (1972).

[106] R. J. Pearson, *Symmetry Rules for Chemical Reactions* (Wiley, New York, 1984).

[107] T. Raz and R. D. Levine, *Chem. Phys. Lett.* **226**, 47 (1994).

[108] P. J. Kuntz, M. H. Mok, and J. C. Polanyi, *J. Chem. Phys* **50**, 4623 (1969).

[109] R. D. Levine, *Quantum Mechanics of Molecualar Rate Processes* (Clarendon Press, Oxford, 1969).

[110] B. H. Mahan, *J. Chem. Phys.* **52**, 5221 (1970).

[111] M. T. Marron, *J. Chem. Phys.* **58**, 153 (1973).

[112] M. G. Prisant, C. T. Rettner, and R. N. Zare, *J. Chem. Phys.* **81**, 2699 (1984).

[113] S. A. Safron, *J. Phys. Chem.* **89**, 5713 (1985).

[114] I. Schechter, M. G. Prisant, and R. D. Levine, *J. Phys. Chem.* **91**, 5472 (1987).

[115] I. Schechter and R. D. Levine, *J. Chem. Soc. Faraday Trans. 2* **85**, 1059 (1989).

[116] T. Raz and R. D. Levine, *Chem. Phys.* **213**, 263 (1996).

[117] T. Raz, U. Even, and R. D. Levine, *J. Chem. Phys.* **103**, 5394 (1995).

[118] L. M. Delves, *Nucl. Phys.* **8**, 358 (1958).

[119] D. W. Brenner and et al., *Phys. Rev. Lett.* **70**, 2174 (1993).

[120] J. Tersoff, *Phys. Rev. B* **37**, 6991 (1988).

[121] M. Chajia and R. D. Levine, *Chem. Phys. Lett.* **304**, 385 (1999).

[122] W. Lichten, *Phys. Rev.* **164**, 131 (1967).

Probing the Dynamics of Chemisorption Through Scattering and Sticking

A.W. Kleyn

FOM Institute for Atomic and Molecular Physics,
Kruislaan 407, 1098 SJ Amsterdam, The Netherlands
present address: Leiden Institute of Chemistry, Leiden
University, PO Box 9502, 2300 RA Leiden, The Netherlands

1 Introduction

The interaction between O_2 molecules and silver surfaces has been studied for a long period in the history of surface science. These studies are important for a number of reasons. Firstly, the system is an example of precursor mediated dissociative chemisorption, in which more than one precursor can be involved. The O_2-Ag systems are examples of so-called triple well systems. As such, the system can serve as a prototype of oxidation reactions [1]. In addition, the epoxidation reaction between ethene and oxygen is carried out at silver catalysts [2]. Apparently, a unique state of weakly bound oxygen exists exclusively on silver surfaces, which is very hard to characterize.

Several different states of oxygen on Ag(111) have been identified, as recently summarized [1]. Most weakly bound is physisorbed O_2 which desorbs at a surface temperature $T_s \approx 50\,\text{K}$. The corresponding trapping coefficient for thermal O_2 is of the order of unity [3,4]. Molecular chemisorbed O_2 desorbs at $T_s \approx 200\,\text{K}$ [5,6]. Its vibrational frequency is very much lower than that of gas phase O_2 and O_2^-, which has been interpreted as due to significant electron transfer to the molecule [7]. Here we will refer to this state by O_2^-. The sticking coefficient for molecular chemisorption is on the order of 10^{-6} [5,8,9]. Dissociative chemisorption can take place at the surface leading to recombinative desorption at $T_s \approx 600\,\text{K}$ and in subsurface sites, desorbing at $T_s \approx 700\,\text{K}$ [1,10]. The dissociative sticking coefficient is on the order of 10^{-7} [5,8,9,11,12].

A one dimensional representation of the O_2-Ag(111) interaction derived by Campbell [5] from thermal desorption and other data is shown in Fig. 1. The physisorption (O_2), molecular chemisorption (O_2^-), and dissociative ($O + O$) states are clearly indicated. The diagram indicates that molecular

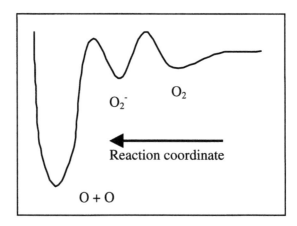

Figure 1: Schematic potential energy diagram for the interaction between O_2 and Ag(111). The three states (see text) into which O_2 can adsorb at the surfaces are depicted as a function of a reaction coordinate. Adapted Campbell [5]

chemisorption is an activated process. The energetic barriers have heights on the order of 0.2 eV or less with respect to the unbound molecule. The precise nature of the reaction coordinate in Fig. 1 is not known.

2 Direct-inelastic scattering

Fast ion beam experiments are insensitive to any subtle barriers to electron transfer and molecular chemisorption because of the large excess translational energy. Thus we studied the formation of negative ions in collisions of fast (100 – 500 eV) ion and neutral beams in collisions with Ag(111). Facile formation of O_2^- was observed [13]. Therefore, we attempted to study adsorption by a scattering experiment of hyperthermal (~ 1 eV) O_2. If the barrier to molecular chemisorption in Fig. 1 can be overcome by translational energy, a dramatic decrease of scattered O_2 should be observed upon an increase of the beam energy above 0.2 eV. However, Spruit et al. could not observe any major change when the beam energy of O_2 was increased substantially above 0.2 eV [14, 15]. In a direct study of dissociative chemisorption Spruit and Kleyn only observed sticking at the 10^{-5} level at energies around 1 eV [16].

Based upon these observations Spruit has suggested connecting the diagram in Fig. 1, containing an undefined reaction coordinate, to a possible potential energy surface. Here only the molecule-surface separation and the internuclear distance in the molecule are parameters. The molecular chemisorption (O_2^-), and dissociative ($O + O$) states are simply accessed from the physisorption (O_2) state by stretching the molecular bond in a vibrationally activated process. Thus, according to this potential surface the barrier to O_2^- formation is lowest along the vibrational degree of freedom. This suggests a dramatic increase of sticking when increasing the vibrational energy content of the beam, and could account for the isotope effect found in dissociative chemisorption [10, 17]. A large effect of vibrational excitation has not been found more recently by Raukema et al. [8], but an isotope effect hinting at the effect was found [10, 17].

Spruit et al. did observe a change in the width of the angular distribution for direct-inelastic scattering with increasing beam energy [14, 15]. Their studies have recently been extended by Raukema et al. [18]. They studied the yield of several atoms and molecules as a function of the final scattering angle Θ_f for several angles of incidence Θ_i and incidence energies E_i. The scattering angles are measured from the surface normal. The dependence of the width $\Delta\Theta_f$ (FWHM) of the angular distribution is plotted as a function of E_i in Fig. 2. A slight decrease of $\Delta\Theta_f$ with E_i is seen for N_2 and Ar, which only interact through a physisorption potential. Very recently, this behavior of Ar has been reproduced in classical molecular dynamics calculations by Lahaye et al. [19–21]. The observed decrease in $\Delta\Theta_f$ with E_i is due to the decreasing influence of the potential well and of the thermal surface motion on the scattering pattern. Without this thermal roughening the surface appears to be effectively very flat, which was also apparent from a computation of the Ar-Ag interaction potential [22]. The dependence of $\Delta\Theta_f$ on E_i closely follows the prediction of cube models.

At low E_i the angular distributions for O_2 scattering are very similar to those of N_2 [18]. However, for $E_i > 0.7$ eV clear differences become apparent. The peak intensity decreases, the angular distribution shows an asymmetric broadening and the final energy exhibits a different dependence on Θ_f. For $E_i > 1.5$ eV there are no further changes; the width is stabilized at a large value such as is also observed for NO scattering from Pt(111), a system dominated by a large chemisorption well [23, 24]. Therefore, Raukema et al. have attributed the very sudden change in $\Delta\Theta_f$ to the fact that part of the flux no longer follows the N_2-like physisorption potential. Lahaye et al. have theoretically studied $\Delta\Theta_f$ for Ar scattering from Ag(111) in the energy range $0.1 - 100$ eV [19, 20]. Such a drastic change in $\Delta\Theta_f$ as a function of E_i is not seen for a single repulsive potential. This observation supports the interpretation that the sudden increase followed by a stabilization of $\Delta\Theta_f$ as a function of E_i is due to a change in the relevant interaction potential: the change in $\Delta\Theta_f$ is attributed to the presence of a second repulsive wall, connected to the chemisorption potential. Whether the second repulsive wall is connected to the O_2^- state or to the $O + O$ state cannot be determined from this scattering experiment.

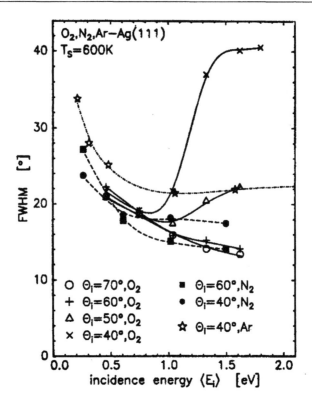

Figure 2: Angular width (Full Width at Half Maximum) of the angular flux distributions of O_2, N_2, and Ar scattering from Ag(111) as a function of the incidence energy. Lines drawn through the data points serve to guide the eye only. From Raukema et al. [18]

A very recent second example of a sudden change in $\Delta\Theta_f$ was found for the system N_2-Ru(0001) by Papageorgopoulos et al. [25]. N_2 can dissociate on this surface, which is relevant for ammonia synthesis, see e.g. [26]. The width $\Delta\Theta_f$ is plotted in Fig. 3. A trend very similar to that found for O_2-Ag(111) is observed. Note the difference with N_2-Ag(111). Clearly the fast N_2 sees a change of potential. In this case this change in potential has been predicted and verified theoretically [26, 27]. The observation of this change for $\Delta\Theta_f$ and the connection to the shape of the interaction potential shows the power of the molecular beams method in the exploration of gas-surface dynamics.

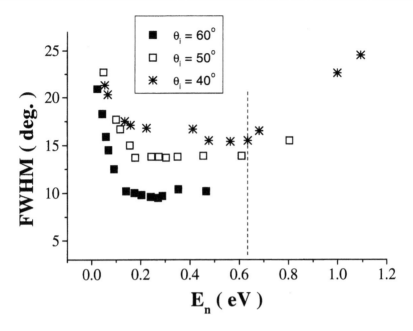

Figure 3: Angular width (Full Width at Half Maximum) of the angular flux distributions of N_2, scattering from Ru(0001) as a function of the incidence energy. From [25]

3 Adsorption-desorption

In the studies of direct-inelastic scattering, discussed above, attention was also paid to the energy exchange between molecule and surface. This can be done using a pulsed molecular beam and time-of-flight (TOF) techniques. Such experiments have been carried out recently by Raukema et al. [3, 28]. These authors focused their attention on experiments at several energies, for $T_s = 150$ K, Θ_i rather large and Θ_f small. For $E_i = 0.1$ eV, the TOF spectrum is dominated by a single broad distribution, which can be fitted very well with a Maxwell–Boltzmann distribution, corresponding to a temperature slightly below the surface temperature as has been observed previously for this system [4]. Clearly, physisorption and complete equilibration followed by desorption is the only important process at these given conditions. When E_i is increased to 0.2 eV, the signal intensity corresponding to physisorption-desorption decreases and is no longer a pure Maxwell–Boltzmann distribution. The TOF distribution can be fitted by two distributions, one of which represents physisorption trapping-desorption. From such experiments the relative intensity of the trapping-desorption signal can be measured as a function of E_i. The result is given in Fig. 4 as a function of the 'normal energy' $E_n = E_i * \cos^2(\Theta_i)$. Scaling of results with E_n indicates that parallel momentum is conserved in the interaction. The characteristic monotone decreasing intensity distributions familiar for physisorption trapping-desorption are seen.

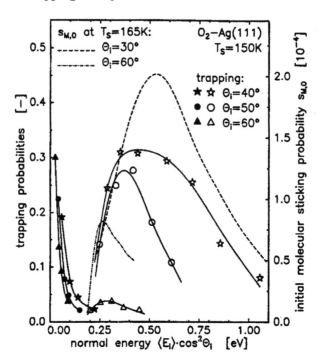

Figure 4: Normal energy dependence of the probabilities for physisorption trapping (*closed symbols*) and transient trapping-desorption (*open symbols*) for three different values of Θ_i. The full lines through the data points serve to guide the eye only. The molecular chemisorption probability S_M is shown by the *dashed* and the *dash-dotted* line. Note the different Y-axes. From Raukema and Kleyn [3]

At higher energies it is observed that other processes arise and the TOF spectra show two narrow peaks, a slow one and a fast one. More detailed analysis suggests that the fast peak corresponds to direct-inelastic scattering of an O_2 molecule with essentially a single Ag atom, and that in the collision considerable rotational excitation occurs [29]. The slow TOF peak corresponds to a final energy $E_f \approx 0.14$ eV, which suggests that these molecules result from activated desorption, that exhibits no measurable surface residence time ($> 10^{-6}$ s). Activated desorption means that the molecules have to overcome a barrier before desorption, and that they are accelerated into the gas phase after crossing the barrier by "rolling down the repulsive potential". Activated desorption is known to result in very peaked angular distributions [30, 31]. As expected, the angular distribution is strongly peaked to the surface normal, and is described very well by $I(\Theta_f) \approx \cos^{15}(\Theta_f)$. The translational energy of the desorbing particle for activated desorption should be independent of initial energy, which was indeed observed. Additional information concerning the activated desorption process can be obtained from the beam energy dependence of its intensity. This is shown in Fig. 4 as a function of E_n. A threshold of 0.2 eV is observed for all Θ_i. This strongly supports the inference that the slow peak is due to an activated adsorption-desorption process. In Fig. 4 it can also be seen that there is a strong Θ_i dependence on the magnitude of the adsorption process. Parallel momentum decreases the adsorption probability. Summarizing: a fast activated adsorption-desorption process is observed. The O_2 molecules are transiently trapped in a metastable state at the surface. We refer to this process by transient trapping-desorption (TTD). More information on the nature of the state follows from the study of molecular chemisorption. The calculations of the potential energy surface for N_2 interaction with Ru(0001) suggest that also in this case TTD should be observable. Using molecular beam methods this could not be verified [25]. It could be verified by looking at recombinative desorption of N-atoms from Ru(0001). The existence of the metastable intermediate is reflected in the translational energy and vibrational temperature of the desorbing N_2 molecules [27].

4 Molecular chemisorption

To get further information about which (metastable) state is responsible for TTD we studied the energy dependence of the sticking coefficient for molecular chemisorption S_M in the O_2^- state, which desorbs at 200 K [8]. It is also reproduced in Fig. 4. It shows a strong similarity to the transient trapping-desorption probability. This suggests that transient trapping-desorption and molecular chemisorption are mediated by the same state. Molecular chemisorption is also an activated process with a threshold energy of about 0.2 eV. It is very hard to molecularly adsorb O_2 at the Ag(111) surface. The Genoa group did not observe any molecular adsorption using EELS [12]. We found a maximum in the molecular sticking coefficient with energy of about 2×10^{-4}, as is shown in Fig. 4. There is no clear explanation for this discrepancy. We feel it might be related to the step density of the crystal, as will be discussed later. The structure of the overlayer formed is unknown. Our study indicates that the layer is dispersed and that adsorption takes place at the terraces. The He reflectivity of the layer is remarkably high. Transient trapping-desorption is about three orders of magnitude more probable than molecular chemisorption.

5 Dissociative chemisorption

In contrast to the (110) face, dissociative chemisorption on the (111) face is very improbable at all experimental conditions. At thermal energies values between 10^{-7} and 10^{-5} are seen [5,10,11,17]. In the energy dependence of the dissociative sticking coefficient at zero coverage S_D for Ag(111) two channels for dissociative chemisorption are seen [8]. At low E_i, S_D is much lower than the molecular chemisorption probability S_M. In addition the T_s dependence of S_D is suggestive of a precursor mediated process. Therefore, it is assumed that for $E_i < 1\,\text{eV}$ molecular chemisorption precedes dissociative chemisorption. Dissociation could very well take place at steps and defects, to be discussed later. For $E_i > 1\,\text{eV}$ the sticking mechanism is different. Increasing T_s leads to enhanced sticking, as does an increase in the energy, and $S_D > S_M$. Here a tunneling or surface motion mediated direct sticking seems to be prominent [32].

6 Interaction dynamics

Perhaps the most remarkable and new observation for the interaction between O_2 and Ag(111) is the occurrence of transient trapping-desorption. Its understanding may shed some light on the question why the chemisorption probabilities on this face are so low. The state involved in transient trapping-desorption seems to be closely related to the molecular chemisorbed state. This inference is based on the very similar energy and angular dependence of the transient trapping probability t_{TTD} and S_M. Clearly transient trapping in this state does in most cases not lead to molecular chemisorption. Several explanations for the nature of this state can be proposed [28,33]. We feel that most likely the O_2^- like state is metastable with respect to desorption. It exists as a local binding minimum in the potential. Recent calculations of the bonding of O_2^- to Ag(110) suggest that the O_2^- binds to Ag-atoms in the same plane [34,35]. Those are abundant on Ag(110) but absent at the Ag(111) face. Therefore the O_2^- state might be a resonance on Ag(111) and decays after transient trapping via desorption. Molecular chemisorption, stabilization of O_2^- at the surface may only occur at special sites, such as at steps or in the presence of ad atoms. The stabilization of the O_2^- by ad-atoms or surface imperfections could explain the difference between the results of the Genoa group for this system and our results. This could be due to a larger step density for the Amsterdam crystal or a better decoration of the steps in the Genoa crystal.

Because the O_2^- well is rather shallow (depth less than 0.5 eV, (transient) trapping in this well will only occur for a limited energy range. This is clearly observed for transient trapping-desorption and for molecular chemisorption in Fig. 4. Above 1 eV trapping in the O_2^- well becomes less probable. Because dissociative chemisorption is also improbable above 1 eV ($S_D < 5 \times 10^{-3}$) most of the incident flux has to be scattered directly. According to Fig. 2 most of the flux is now scattered along the chemisorption potential. However, if dissociation involves defects or restructuring of the lattice, it is very unlikely that dissociation can be described using a two dimensional potential energy surface. Therefore, we believe that the corrugated potential probed in hyperthermal O_2 scattering is related to the O_2^- state at the surface and not by the $O + O$ potential. It should be noted that at even higher energies ($E_n > 5$ eV) direct formation of free O_2^- ions has been observed [13, 36, 37]. This suggests that O_2^- formation (harpooning [38, 39]) at the surface occurs readily at energies above 0.2 eV. Formation of free O^- occurs above $E_n = 10$ eV; according to classical trajectory calculations the dissociation is due to impulsive energy transfer in the collision of an O_2^- and the surface and is not due to spontaneous dissociation at some repulsive $O^- + O^-$ potential energy surface [37, 40].

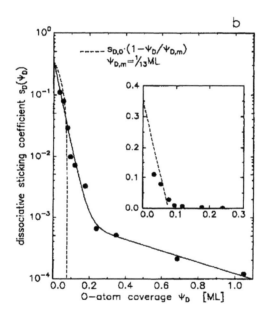

Figure 5: Dissociative sticking coefficient for O_2 on Ag(110) determined as a function of the O-atom coverage. The coverage scale was obtained by differentiation of the O-atom uptake in the experiment. The *solid line* is the sum of two exponential decays. The *dashed line* shows a model for adsorption, which saturated at 1/13 of a monolayer. The *inset* shows the first part of the data on a linear scale. From Butler et al. [41]

7 Fences

Recent experiments on the coverage dependence of the sticking of O_2 to Ag(110) revealed a remarkable drop shown in Fig. 5 [41–48]. This drop has been explained by electrostatic effects [48]. We have attributed the drop to the build-up of added rows at the surface [49]. If the picture of adsorption into a metastable intermediate is valid, it implies that the intermediate should be stabilized at a step edge. Subsequently the molecule can dissociate at the site and form an Ag-O pair, that is inserted into an added row. This involves mass transport at the surface, which has been observed in STM-experiments [50, 51]. Butler et al. have modeled the coverage dependence of the sticking coefficient on Ag(110) in a simple model, which is summarized in Fig. 6. An analytical representation for the dependence of sticking on coverage was derived, which matches the data remarkably well. This analysis shows, that it is not enough to include only energetic barriers in the modeling of dissociative chemisorption. In addition, the dynamics at the surface has to be considered, including the growth of added rows that enhance desorption of molecular intermediates. Although the detailed dynamics of the oxidation of the silver surface in the formation of added rows is not directly visible by molecular beams, is it remarkable that data such as shown in Fig. 5 provide strong evidence concerning the processes occurring at the surface.

8 Conclusions

A variety of processes can occur in the interaction of O_2 molecules and Ag(111). At first scattering from and trapping in the physisorption potential can occur. Secondly, scattering from the chemisorption (O_2^-) potential occurs, together with transient trapping-desorption. The chemisorption potential well is very shallow. From being transiently trapped the molecule can be captured in the molecular chemisorption well; presumably surface imperfections are necessary to stabilize the molecular adsorbate in this case. From the molecular chemisorption well the molecule can proceed to dissociation. In this step ad atoms may be involved on Ag(111). Finally, there is a small probability for direct dissociative chemisorption of O_2 at Ag(111). The formation of added Ag-O-rows (fences) at the surface inhibit further sticking at the surface.

Acknowledgments

This work is part of the research program of FOM and is supported financially by NWO. The author gratefully acknowledges all his co-workers, who have studied the O_2 silver interaction, and whose work forms the basis of this report.

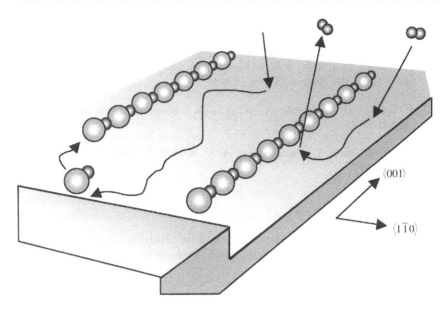

Figure 6: Schematic drawing of the random-walk diffusion model of Butler et al. [49]. It is shown that molecular intermediates that do not hit an added row may dissociative at a step, bind to an Ag-atom and insert in an added row

References

[1] F. Besenbacher, J.K. Nørskov: Prog. Surf. Sci. **44**, 1 (1993)

[2] R.A. van Santen, and H.P.C.E. Kuipers: Adv. Catal. **35**, 265–321 (1987)

[3] A. Raukema, A.W. Kleyn: Phys. Rev. Lett. **74**, 4333–4336 (1995)

[4] M.E.M. Spruit, E.W. Kuipers, F.H. Geuzebroek, A.W. Kleyn: Surf. Sci. **215**, 421–436 (1989)

[5] C.T. Campbell: Surf. Sci. **157**, 43–60 (1985)

[6] K.C. Prince, G. Paolucci, A.M. Bradshaw: Surf. Sci. **175**, 101–122 (1986)

[7] C. Backx, C.P.M. de Groot, P. Biloen: Surf. Sci. **104**, 300–317 (1981)

[8] A. Raukema, D.A. Butler, F.M.A. Box, A.W. Kleyn: Surf. Sci. **347**, 151–168 (1996)

[9] A. Raukema, D.A. Butler, A.W. Kleyn: Surf. Sci. **373**, 127–128 (1997)

[10] P.H.F. Reijnen, A. Raukema, U. van Slooten, A.W. Kleyn: Surf. Sci. **253**, 24–32 (1991)

[11] M. Rocca, F. Cemic, F.B. deMongeot, U. Valbusa, S. Lacombe, K. Jacobi: Surf. Sci. **373**, 125–126 (1997)

[12] F. Buatier de Mongeot, U. Valbusa, M. Rocca: Surf. Sci. **339**, 291–296 (1995)

[13] P.H.F. Reijnen, U. van Slooten, A.W. Kleyn: J. Chem. Phys. **94**, 695–703 (1991)

[14] M.E.M. Spruit, P.J. van den Hoek, E.W. Kuipers, F.H. Geuzebroek, A.W. Kleyn: Surf. Sci. **214**, 591–615 (1989)

[15] M.E.M. Spruit, P.J. van den Hoek, E.W. Kuipers, F.H. Geuzebroek, A.W. Kleyn: Phys. Rev. B **39**, 3915–3918 (1989)

[16] M.E.M. Spruit, A.W. Kleyn: Chem. Phys. Lett. **159**, 342–348 (1989)

[17] P.H.F. Reijnen, A. Raukema, U. van Slooten, A.W. Kleyn: J. Chem. Phys. **94**, 2368–2369 (1991)

[18] A. Raukema, R.J. Dirksen, A.W. Kleyn: J. Chem. Phys. **103**, 6217–6231 (1995)

[19] R.J.W.E. Lahaye, A.W. Kleyn, S. Stolte, S. Holloway: Surf. Sci. **338**, 169–182 (1995)

[20] R.J.W.E. Lahaye, S. Stolte, S. Holloway, A.W. Kleyn: Surf. Sci. **363**, 91–99 (1996)

[21] R.J.W.E. Lahaye, S. Stolte, A.W. Kleyn, R.J. Smith, S. Holloway: Surf. Sci. **309**, 187–192 (1994)

[22] E.J.J. Kirchner, A.W. Kleyn, E.J. Baerends: J. Chem. Phys. **101**, 9155–9163 (1994)

[23] A.E. Wiskerke, A.W. Kleyn: J. Phys. Condens. Matter **7**, 5195–5207 (1995)

[24] R.J.W.E. Lahaye, S. Stolte, S. Holloway, A.W. Kleyn: J. Chem. Phys. **104**, 8301–8311 (1996)

[25] D.C. Papageorgopoulos, B. Berenbak, M. Verwoest, B. Riedmüller, S. Stolte, A.W. Kleyn: Chem. Phys. Lett. **305**, 401–407 (1999)

[26] J.J. Mortensen, Y. Morikawa, B. Hammer, J.K. Norskov: J. Catal. **169**, 85–92 (1997)

[27] M.J. Murphy, J.F. Skelly, A. Hodgson, B. Hammer: J. Chem. Phys. **110**, 6954–6962 (1999)

[28] A. Raukema, D.A. Butler, A.W. Kleyn: J. Chem. Phys. **106**, 2477–2491 (1997)

[29] A.W. Kleyn: Surf. Rev. Lett. **1**, 157–173 (1994)

[30] W. van Willigen: Phys. Lett. **28A**, 80–81 (1968)

[31] G. Comsa, R. David: Surf. Sci. Rep. **5**, 145–198 (1985)

[32] O. Citri, R. Kosloff: Surf. Sci. **351**, 24–42 (1996)

[33] A.W. Kleyn, D.A. Butler, A. Raukema: Surf. Sci. **363**, 29–41 (1996)

[34] P.A. Gravil, D.M. Bird, J.A. White: Phys. Rev. Lett. **77**, 3933–3936 (1996)

[35] P.A. Gravil, J.A. White, D.M. Bird: Surf. Sci. **352**, 248–252 (1996)

[36] P. Haochang, T.C.M. Horn, A.W. Kleyn: Phys. Rev. Lett. **57**, 3035–3038 (1986)

[37] P.H.F. Reijnen, P.J. van den Hoek, A.W. Kleyn, U. Imke, K.J. Snowdon: Surf. Sci. **221**, 427–453 (1989)

[38] A.W. Kleyn: in Invited papers of the XVI International Conference on the Physics of Electronic and Atomic Collisions, New York, 1989,. 1990, Eds: A. Dalgarno, R.S. Freund, P.M. Koch, M.S. Lubell, T.B. Lucatorto, New York: American Institute of Physics Conference Proceedings, Vol. 205, 451–457

[39] J.W. Gadzuk: Comments Atom. Mol. Phys. **16**, 219–240 (1985)

[40] P.J. van den Hoek, A.W. Kleyn: J. Chem. Phys. **91**, 4318–4329 (1989)

[41] D.A. Butler, A. Raukema, A.W. Kleyn: Surf. Sci. **357-358**, 619–623 (1996)

[42] A. Raukema, D.A. Butler, A.W. Kleyn: J. Phys. Condens. Matter **8**, 2247–2263 (1996)

[43] L. Vattuone, C. Boragno, M. Pupo, P. Restelli, M. Rocca, U. Valbusa: Phys. Rev. Lett. **72**, 510–513 (1994)

[44] L. Vattuone, U. Valbusa, M. Rocca: Surf. Sci. **317**, L1120–L1123 (1994)

[45] L. Vattuone, M. Rocca, U. Valbusa: Surf. Sci. **314**, L904–L908 (1994)

[46] L. Vattuone, M. Rocca, P. Restelli, M. Pupo, C. Boragno, U. Valbusa: Phys. Rev. B **49**, 5113–5116 (1994)

[47] L. Vattuone, M. Rocca, C. Boragno, U. Valbusa: J. Chem. Phys. **101**, 713–725 (1994)

[48] L. Vattuone, M. Rocca, C. Boragno, U. Valbusa: J. Chem. Phys. **101**, 726–730 (1994)

[49] D.A. Butler, J.B. Sanders, A. Raukema, A.W. Kleyn, J.W.M. Frenken: Surf. Sci. **375**, 141–149 (1997)

[50] W.W. Pai, N.C. Bartelt, M.R. Peng, J.E. Reutt-Robey: Surf. Sci. **330**, L679–L685 (1995)

[51] W.W. Pai, J.E. Reutt-Robey: Phys. Rev. B **53**, 15997–16005 (1996)

Product State Measurements of Nitrogen Formation at Surfaces

M.J. Murphy, P. Samson, J.F. Skelly, A. Hodgson

Surface Science Research Centre, The University of Liverpool, Liverpool, UK

1 Introduction

Our understanding of the dissociation dynamics of hydrogen at metal surfaces has been transformed over the last decade by the development of measurements which reveal the role of specific molecular degrees of freedom (rotation, vibration and translation) on the dynamical behaviour [1]. This has been achieved by applying state sensitive techniques to probe the role of internal state on adsorption, desorption and scattering of hydrogen at reactive surfaces. Each of these measurements probes a different region of the molecule-surface phase space, providing complementary information about the potential energy surface (PES) and the dynamics of reaction. These measurements have stimulated a large body of theoretical work and the reaction dynamics and the topography of H_2-metal PES are reasonably well understood [2].
Disentangling the adsorption dynamics of heavier molecules, such as N_2, O_2 or small polyatomics, is perhaps more challenging. Unlike hydrogen, where inelastic effects are often ignored and the molecule is assumed to scatter from a rigid metal surface, both energy exchange to the substrate and translational to rotational coupling are efficient. Molecules such as nitrogen and oxygen also show a rich adsorption structure, often having a number of molecular chemisorption states with different adsorption geometries. Dissociation may occur either by direct activated adsorption over a barrier, or by trapping into a molecular well followed by thermally activated dissociation at some constrained transition state. In that case both the topography of the PES and energy transfer cross sections will play a role in determining the overall adsorption behaviour. Whereas previously our ideas about the topography of the PES for heavy molecules have largely been obtained by extrapolating information obtained from structural studies, (which define only the stable adsorption minima), density functional theory calculations are beginning to provide information on the form of the transition states

for dissociation [3, 4]. These ab initio potentials predict the presence of unstable molecular chemisorption states, which have lifetimes too short to be observed directly but can profoundly influence the scattering or dissociation dynamics.

Molecular beam techniques can be used to determine the energy dependence of dissociative chemisorption and to distinguish between activated and trapping-dissociation mechanisms. Scattering measurements provide evidence about changes in the repulsive surfaces with energy [5], but do not provide a particularly detailed test of the shape of the PES. Ideally we would like to obtain (v, J) state resolved information on the reaction channel, but this is complicated by the high density of final states and the efficiency of energy redistribution during scattering of heavy molecules. Here we show how detailed quantum state resolved adsorption data can be predicted from measurements of nitrogen desorption.

Because of its strong bond, nitrogen dissociation is activated on most metal surfaces. However, once dissociated, N atoms are tightly bound to the surface and the combination of a substantial barrier to recombination and a strong molecular bond imply that considerable energy will be released as the bond forms and the molecule scatters from the transition state. This energy will be available to excite internal modes of the product, making these systems good candidates for state selective studies of the desorption dynamics. Two previous studies of nitrogen recombinative desorption, by e^- beam induced fluorescence following N permeation through a hot iron foil [6] and by appearance potential measurements after NH_3 cracking on Pt foil [7], have both indicated that nitrogen can carry considerable internal excitation.

By carrying out simple surface reactions under molecular beam conditions we are able to provide a continuous source of nitrogen molecules from N recombination at well defined metal surfaces. The products are detected by resonance enhanced multiphoton ionisation (REMPI), which gives a complete description of the energy disposal into internal states, while ion time of flight (TOF) provides the translational energy distributions. Here we compare some recent experiments on the recombination of nitrogen from Cu(111), Ru(001) and Pd(110) surfaces and discuss the role of molecular states in determining the differences in energy disposal into the N_2 product.

2 Experimental

Experiments were carried out in a stainless steel ultra-high vacuum chamber with a base pressure of 2×10^{-11} mbar, equipped with conventional surface science diagnostics such as LEED, Auger and a mass spectrometer for residual gas analysis and thermal desorption measurements. A schematic of the system is shown in Fig. 1. A differentially pumped molecular beam doses gas onto the surface of a metal single crystal. The beam source may be a microwave discharge atomic beam source, which was used to provide H/D [8, 9] and N [10] atoms, or may be a conventional source to dose a reactive precursor such as NH_3 or NO onto the surface. The single crystal sample is held on a manipulator which provides heating, cooling and positioning, allowing the surface to be cleaned and prepared such that it reproduces the adsorption-desorption behaviour expected from the literature. During an experiment the sample is usually held at a constant temperature, sufficient for the surface reaction to occur. An equilibrium is established between adsorption of precursor from the beam, reaction and desorption of the products. When dosing reactive atoms, such as H and N, the incident beam must be blocked just before the probe laser fires in order to avoid detecting Eley–Rideal type reaction products where one atom is not equilibrated on the surface. State distributions can also be measured by REMPI during flash desorption from a surface pre-adsorbed with reactant [10].

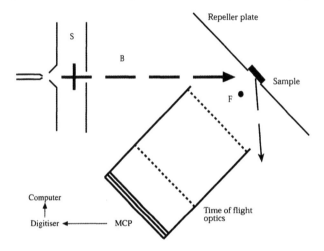

Figure 1: Schematic showing a horizontal section through the desorption system developed in our laboratory to measure quantum state resolved translational energy distributions for the products of surface reactions. The sample is mounted on a manipulator and is cleaned and characterised on an upper level of the chamber. During desorption the crystal sits at 45 to the axis of a molecular beam (B), behind an aperture in the repeller plate of a set of ion TOF optics. The molecular beam can be interrupted by a shutter (S) prior to the laser firing. Ions are created at the laser focus (F) and are detected on a microchannel plate, discriminated and averaged on a computer

The reaction products desorb in a lobular distribution, centred around the surface normal, and are intercepted and ionised by the probe laser. $N_2(v, J)$

was detected using (2 + 1) resonance ionisation on the a″ − X transition, using a frequency tripled Nd:YAG pumped dye laser, operating between 200 and 220 nm. Positive ions, which were emitted from the hot sample, were removed by holding the repeller grid (Fig. 1) at a positive potential to block ions from entering the detector and dropping the potential to that of the sample immediately before the laser fired. The molecular ions retain the same translational energy distribution as the neutral molecules, and were extracted into an ion TOF detector whose dispersion was calibrated using a Knudsen source [11]. Averaging the ion TOF distributions allows us to obtain the translational energy distribution along the detector axis, $P(E)$, for individual quantum states (v, J) of the reaction product. The shape of the TOF distribution also allows the background on an individual quantum state to be determined unambiguously, by fitting this component to the ion TOF distribution of a thermal gas [11].

This experimental scheme allows us to obtain accurate translational energy and population distributions for reaction products from many different surface reactions, whereas previously the technique had largely been limited to H_2, where convenient permeation sources exist.

3 Relationship of Desorption to Adsorption

The energy partitioning into different degrees of freedom of the product will reflect the dynamics of recombination as the molecule forms and scatters from the transition state. For Langmuir–Hinshelwood (LH) reactions, where both reactants are thermally equilibrated to the surface, desorption is driven by thermal activation of the adsorbate from the heat bath of the solid. In this case interpreting the energy disposal requires some care since, unlike an exothermic gas phase reaction for example, the total energy of the products is not well defined. Desorption processes with a low energy requirement but a low probability may compete effectively with those which have a high probability but a large activation barrier. This means that the desorption distributions will depend on temperature, in some cases dramatically so [8]. However, the thermal sampling inherent during a LH surface reaction can be turned to advantage by invoking detailed balance to equate the rates of adsorption and desorption for a surface at equilibrium [12, 13]. This allows us to relate the final state distributions measured for desorption to the detailed sticking probability for each quantum state $N_2(v, J)$ [14]. This approach has been applied very successfully to hydrogen adsorption/desorption [15] but has not been widely exploited for reactions of heavier molecules. For each product quantum state, the translational energy distribution $P(E)$ for desorption (for simplicity assumed here to be along the surface normal) is related to the sticking probability for that state by

$$P(E, v, J) \propto S(E, v, J) \cdot \exp(-Ev, J/kT) \cdot E \exp(-E/kT), \quad (1)$$

where Ev, J is the energy of ro-vibrational level (v, J). This relationship allows us to derive and discuss predicted sticking functions $S(E, v, J)$, rather than interpreting desorption distributions which intrinsically vary with surface temperature, T. This relationship also emphasises that the distribution of final states during desorption is strongly weighted towards low energy channels. The weighting is the same as for adsorption from a Maxwellian gas at the same temperature, $P(E)$ being a direct representation of those molecules which would dissociate under thermal conditions. This is in contrast to molecular beam adsorption experiments where typically only large S(E) can be measured. The principle of detailed balance has been discussed previously and details on its use can be found elsewhere [12, 13, 15, 16, 17].

4 Results

Energy disposal measurements for nitrogen desorption from Cu(111), Ru(001) and Pd(110) will be described briefly and the energy partitioning and dynamics of recombination/dissociation compared.

4.1 Nitrogen Recombination at Cu(111)

Adsorption of nitrogen atoms reconstructs Cu(111) [17], the top layer of Cu forming a distorted Cu(100) overlayer with N atoms adsorbed in the fourfold hollow sites in $ac(2 \times 2)$ arrangement. Dosing N atoms from an atomic beam at 300 K gives rise to a disordered overlayer which will accommodate at least 4 monolayers of N atoms, indicating that N atoms can penetrate into the first few layers of the Cu surface. Above 500 K the excess N desorbs and the surface orders to form a saturated Cu(100)-$c(2 \times 2)$N overlayer with $\Theta_N = 0.42$ ML [17]. For low coverages recombinative desorption occurs from a surface covered by $c(2 \times 2)$N islands as a zero-order peak near 700 K. Desorption shows a rather complex behaviour, the zero-order kinetics being replaced by explosive desorption once the initial surface is completely covered by the $c(2 \times 2)$ overlayer [17]. This behaviour can be understood in terms of a two-phase model, with nitrogen accommodated in Cu(100)-$c(2 \times 2)$N islands and desorbing from a dilute N phase on the bare metal terraces. This implies that detailed balance arguments will relate desorption measurements to nitrogen dissociation on a low coverage Cu(111) surface, rather than at Cu/N islands.

$N_2(v, J)$ state resolved TOF measurements were made by dosing the surface with N atoms from a microwave discharge beam source and detecting the desorbing molecules. The nascent N_2 feels a strong repulsion from the Cu(111) surface and carries a large translational energy release [10], allowing the desorption signal to be distinguished from the background due to undissociated nitrogen in the beam. An identical energy release was observed for thermal desorption of a N overlayer [10]. The nitrogen carries a mean translational energy along the surface normal $\langle E \rangle = 4.2$ eV for $N_2(v=0)$ and 4.0 eV for $(v=1)$, Fig. 2, corresponding to a substantial portion of the N − N bond energy (9.76 eV). The strong repulsion between the molecule and the surface focuses desorbing molecules onto the surface normal, resulting in a sharply peaked angular distribution $P(\theta)$, Fig. 3. However, the distribution is still broader than would be predicted by a simple 1D model [18] if no energy was released into motion parallel to the surface. From the translational and angular distributions we can estimate the energy release into translation parallel to the surface as 0.28 eV, more than an order of magnitude lower than $\langle E \rangle$, but still much greater than the thermal energy available in this co-ordinate. The reaction products also carry internal excitation, with a rotational temperature which is slightly greater than the surface temperature, $T_{\text{rot}} = 910$ K [16], but a large vibrational excitation, $P(v=1)/P(v=0) = 0.52$, corresponding to a vibrational temperature $T_{\text{vib}} = 5100$ K [16]. Predissociation of N_2 $a''(v=2)$ caused interference from fast N^+ in the ion TOF signal and prevented the population of $v = 2$ being measured.

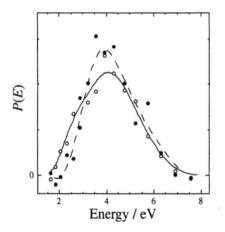

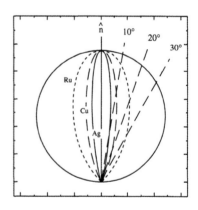

Figure 2: Translational energy distributions $P(E)$ for $N_2(v=0)$ (•) and $(v=1)$ (○) desorbed from Cu(111), taken on the band head, $J = 2$ [10]

Figure 3: Angular distribution for N_2 flash desorbed from Ru(001) [19], Cu(111) [16] and Ag(111) [20]. The circle represents a statistical $P(\theta) = \cos\theta$ distribution

4.2 Nitrogen Recombination at Ru(001)

The Ru(001) surface is not reconstructed by nitrogen but forms (2×2)-N and $(\sqrt{3} \times \sqrt{3})$R30°-N structures, with N bound in the h.c.p. hollow site. Adsorbed nitrogen is deeply embedded in the Ru, with a N – Ru bond length of 1.9 Å and a local relaxation of the top layer of Ru [21]. In comparison with many transition metal surfaces nitrogen is relatively weakly bound, desorbing at around 800 K with density functional calculations giving binding energies of 5.8 and 5.6 eV for the (2×2) and $(\sqrt{3} \times \sqrt{3})$ structures [21]. Dissociation is inefficient, with S reported to be as low as 10^{-12} at thermal energies [22] while Romm and Asscher [23] found that dissociation was activated by both translational and vibrational excitation. Dissociation did not change with surface temperature, suggesting a direct activated adsorption process, rather than activated trapping into a chemisorbed molecular state, as proposed for the dissociation path on iron [24].

Nitrogen was produced by dosing ammonia onto a Ru(001) surface, held at 900 K. On a cold surface ammonia traps into a molecular state and cracks as the surface is heated, forming an NH intermediate which is stable up to 460 K [25]. At temperatures above this the hydrogen atoms recombine and desorb, leaving N atoms adsorbed in the h.c.p. hollow sites, although cracking remains rather inefficient. Exposing the surface to a continuous beam of NH_3 establishes an equilibrium between ammonia trapping and desorption of ammonia, hydrogen and nitrogen. Since the hydrogen and ammonia desorption rates are large, the steady state coverage of H and NH_3 is negligible at 900 K and recombinative desorption occurs from a metal surface with a low coverage of nitrogen.

Nitrogen desorbed from Ru(001) shows an entirely different pattern of energy release to that from Cu(111) [26]. Whereas desorption from Cu(111) gave a large translational energy release (Fig. 2), desorption from Ru(001) creates a translational energy distribution which is peaked at low energy with a tail that extends out towards 2 eV, Fig. 4. However, for the vibrational co-ordinate the situation is reversed, with an inverted vibrational state distribution for N_2 formed at Ru(001), $P(v = 1)/P(v = 0) = 1.4$. The translational energy distributions of the two vibrational states are very similar, both in their shape and the mean energy release, $\langle E \rangle = 0.62$ for $v = 0$ and 0.61 eV for $v = 1$. Again this contrasts with the situation for Cu(111) where the translational energy release for the $v = 0$ and $v = 1$ states was appreciably different. The weaker repulsion leads to a lower energy release into translation away from the surface, and hence to much broader angular distribution for N_2 desorption from Ru(001) compared to that on Cu(111), Fig. 3.

Converting the desorption distributions into relative sticking curves $S(E, v)$ produces the predicted behaviour shown in Fig. 5. Sticking increases exponentially with energy and then starts to saturate near 2.2 eV for $v = 0$.

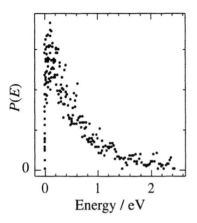
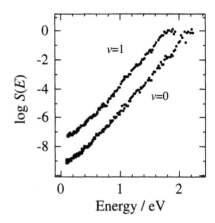

Figure 4: Translational energy distribution $P(E)$ for $N_2(v=0, J=2)$ desorbing from Ru(001) along the surface normal

Figure 5: Calculated $S(E,v)$ for $N_2(v=0,1)$ at Ru(001) from detailed balance. The absolute scale for $S(E)$ is not determined and S has been arbitrarily normalised to 1 at high energy

This behaviour implies that adsorption is highly activated with $S < 10^{-8}$ at low energy, consistent with the very low S observed in thermal measurements [22]. Recent evidence suggests that step sites on the surface have a reduced dissociation barrier [27], perhaps explaining why the dissociation probability is relatively large at low energy in the sticking measurements of Romm and Asscher [23]. The sticking curves for the two vibrational states look very similar in shape, which is why the energy distributions in desorption are so similar, but the curve for ($v=1$) is shifted down in energy by $\sim 0.4\,\mathrm{eV}$ compared to the ground state curve. Unlike the N/Cu(111) [10, 16] or H/Cu(111) [9, 15] systems, the shift in $S(v)$ for $v=0$ and 1 is greater than the vibrational quantum of nitrogen, (0.29 eV), giving rise to a vibrational population inversion during desorption and a predicted vibrational efficacy of 1.3 for sticking [26].

4.3 Nitrogen Recombination at Pd(110)

Palladium is an active catalyst for the NO/CO redox reaction, and finds wide use in car exhaust catalysts. N adsorption forms an ordered Pd(110)-(2×3)N structure when the surface is annealed to 500 K [28]. State distributions were measured for nitrogen formed in the $NO + H_2 \rightarrow 1/2 N_2 + H_2O$ reaction. NO initially traps into a molecular state [29], dissociating only above 500 K. The reaction was run continuously by directing a beam of NO at the hot Pd(110) surface and removing O by reaction with hydrogen, supplied as a background in the chamber. Hydrogen dissociates efficiently on Pd(110) but at temperatures where NO dissociates the steady state concentration of adsorbed H is small, limited by H_2 recombinative desorption. The hydrogen pressure was adjusted until the O removal rate was fast compared to the rate of NO dissociation and N_2 state distributions were not sensitive to the H_2 flux. At high NO coverages N reacts with adsorbed NO to form N_2O, and N_2 which desorbs at 40° to the surface normal, but under low coverage conditions nitrogen atoms recombine and desorb in a lobe centred around the surface normal [30, 31].

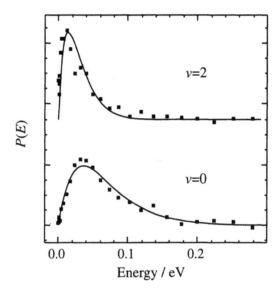

Figure 6: Translational energy released into N_2 desorption for reaction of NO and H_2 at a Pd(110) surface [32]. The solid lines are fits to a thermal energy release for each vibrational state ($T = 425$ K for $v = 0$ and 180 K for $v = 2$), showing how the translational temperature decreases with v

The N_2 state distributions showed a relatively small translational energy release, $N_2(v = 0)$ desorbing from the surface with an energy which is close to thermal, Fig. 6. This is in stark contrast to the N_2 translational energy distribution observed for NO flash desorption [33] and indicates very different reaction dynamics. Instead of being produced at a high coverage surface with considerable adsorbed O, possibly via an N_2O intermediate [33], reaction at high temperatures occurs via the $N + N$ recombination channel [30, 31] on a low coverage N/Pd(110) surface. Just as for desorption from copper and ruthenium, the vibrational co-ordinate is excited, with excess population in the high vibrational states. Unlike the translational energy distributions observed for desorption from Ru(001) [26], the energy release on Pd(110) does depend on the vibrational state, $\langle E \rangle$ decreasing rapidly for excited $N_2(v)$ states [32]. The cold translational energy distributions for vibrationally excited N_2 implies that on Pd(110) translational energy hinders dissociation of these states.

5 Discussion

Nitrogen dissociation is activated on the Ru(001) surface [23], with a very low probability at thermal energies [21], while dissociation has not been reported on either Cu(111) or on Pd(110). This is consistent with the large barrier associated with breaking the N_2 bond, particularly on an inert noble metal such as copper where formation of the nitride is endothermic, but there are no clear measurements of the size of the activation barrier from beam adsorption measurements. The energy disposal into translation of ground state $N_2(v = 0)$ from the surface during desorption provides (via detailed balance) a measure of this activation barrier. The Ru(001) desorption data (Fig. 4) can be inverted to give the sticking curves $S(E)$ (Fig. 5) which show a characteristic exponential increase with energy before saturating at high energy. Since desorption from Ru(001) at 900 K still favours low translational energies the statistics in the region where S flattens out are not particularly good, nevertheless the increase in $S(E)$ is starting to reduce at the same relative S for both vibrational states, providing an estimate of ca. 2.2 eV for the barrier to dissociation of N_2 on Ru(001). On Cu(111) nitrogen desorbs with a very large translational energy $\langle E \rangle = 4.2$ eV. This tells us that $S(E)$ is increasing exponentially in the region near 4 eV, with a much steeper slope than for Ru(001), and that the rate of increase is starting to drop (see (1)). This places the barrier position as 4 eV or greater and also implies that the low energy desorption sites available on Ru are not seen on Cu. The desorption measurements on Pd(110) were carried out in a different TOF detector and the $P(E)$ curves are not sufficiently well defined to determine the shape of $S(E)$ with confidence. However, the temperature of the desorbing gas was less than the surface temperature, implying that sticking is not activated by translational energy, a point which is discussed further below.

Density functional theory calculations performed by Hammer [26] for the dissociation of nitrogen on Cu(111) and Ru(001) show an activated behaviour. Potentials were calculated for nitrogen with its molecular axis parallel to the surface, dissociating in a high symmetry geometry with the N atoms moving from the Cu bridge sites towards the stable threefold hollow sites. This is not necessarily the minimum energy curve, which may involve tilting of the molecular axis [3], but it is a low energy path which has the correct final binding site and allows comparison of the two PES. Both surfaces show a transition state for dissociation which has a greatly extended

N_2 bond, $d = 1.9$ Å for Ru(001) and 2.2 Å for Cu(111), with a barrier of 2.4 and 5.3 eV respectively. These values are consistent with the barriers inferred from desorption measurements.

The shape of the N_2-metal PES also suggests an explanation for the very different vibrational energy disposal in the two systems. In both cases the transition state has an extended N_2 bond, but a vibrational population inversion is only seen for desorption from ruthenium. This implies that vibration is more efficient than translation in aiding dissociation on ruthenium but not on copper, despite the even greater N_2 bond extension on copper. The N_2/Cu(111) PES shows no evidence for a molecular chemisorption state in the flat lying geometry, the surface looking similar to those observed for H_2 on copper [34], but with a larger activation barrier and a very late transition state. However, on Ru(001) the surface shows a weakly bound, metastable molecular state [3], at an energy of 0.5 eV with the nitrogen bond lying parallel to the surface at approximately the gas phase separation [26]. Although there is no evidence that this state is stable, and lying some 1.5 eV below the barrier it can not act as a true precursor to dissociation, its presence nevertheless influences the scattering dynamics. The location of this metastable state allows nitrogen to approach close in to the Ru surface without a large cost in energy, approaching the vertical location of the transition state above the surface before the N_2 bond starts to stretch. During desorption nascent N_2, formed near the transition state at large d_{NN}, will be scattered towards the metastable well, releasing energy into the vibrational co-ordinate. Although the transition state on Cu(111) occurs at even larger d_{NN}, the absence of the metastable well makes the nitrogen-metal surface more repulsive and energy release occurs preferentially into translation.

The sub-thermal translational energy release for nitrogen from Pd(110) suggests that N_2 is not directly scattered from a repulsive surface, but is either trapped or scatters from a molecular state. Two different interpretations of the vibrational excitation are possible, depending on the location of the transition state for desorption from the surface. If the molecular state has a short lifetime then the products may retain some of the original vibrational excitation caused by scattering from a transition state bond at large d_{NN}. Alternatively, if the lifetime of the molecular state is long and N_2 thermalises in the chemisorption well, then the product state distribution will reflect desorption from a molecular chemisorption well which has an extended N_2 bond. The low translational energy release for vibrationally excited $N_2(v = 1)$ formed by desorption indicates that trapping into the molecular well is non-activated for these states. Although a weakly bound molecular state has been seen by electron energy loss spectroscopy (EELS) [35], this state is probably adsorbed atop, perpendicular to the surface and seems unlikely to be the state responsible for trapping during recombinative desorption. The desorption measurements suggest another molecular chemisorption state, with an extended N_2 bond, is responsible for scattering or trapping the desorbing molecules, in a mechanism similar to that discussed by, for example, Gadzuk and Holloway [36].

6 Conclusions

Product state measurements have been extended to study the formation of nitrogen on three contrasting metal surfaces. By carrying out reactions using a molecular beam to supply reactants to a well characterised single crystal surface, we are able to employ detailed balance to discuss the adsorption/desorption behaviour of N_2 and to predict detailed sticking functions $S(E, v, J)$. We find that dissociation is activated by translational energy on Ru(001) and Cu(111) and estimate the barrier to dissociation as 2.2 eV and > 4 eV respectively, consistent with DFT calculations of the barrier [26]. The energy disposal into product vibration is very different on the two surfaces. The presence of a metastable molecular state bound close in to the metal surface with a short bond length on Ru(001) ensures that energy is released preferentially into the vibrational co-ordinate. On Cu(111) the surface is more repulsive in the N_2-metal co-ordinate and energy is released preferentially into repulsion from the surface. This is reflected in the different angular distributions, Fig. 3. Nitrogen recombination from Pd(110) shows an entirely different behaviour with no translational excitation of the surface. In this case desorption appears to proceed via a molecular state, but with considerable vibrational excitation surviving in the product.

References

[1] C.T. Rettner, D.J. Auerbach, J.C. Tully, A.W. Kleyn: J. Phys. Chem., **100**, 13021 (1996)

[2] G.R. Darling, S. Holloway: Rep. Prog. in Phys., **58**, 1595 (1995)

[3] J.J. Mortensen, Y. Morikawa, B. Hammer, J.K. Nørskov: J. Catalysis, **169**, 85 (1997)

[4] J.J. Mortensen, M.V. Ganduglia Pirovano, L.B. Hansen, B. Hammer, P. Stoltze, J.K. Nørskov: Surf. Sci., **422**, 8 (1999)

[5] D.C. Papageorgopoulos, B. Berenbak, M. Verwoest, B. Riedmuller, S. Stolte, A.W. Kleyn: Chem. Phys. Lett., **305**, 401 (1999)

[6] R.P. Thorman, S.L. Bernasek: J. Chem. Phys., **74**, 6498 (1981)

[7] S.N. Foner, R.L. Hudson: J. Chem. Phys., **80**, 518 (1984)

[8] M.J. Murphy, A. Hodgson: Phys. Rev. Lett., **78**, 4458 (1997)

[9] M.J. Murphy, A. Hodgson, J. Chem. Phys., **108**, 4199 (1998)

[10] M.J. Murphy, J.F. Skelly, A. Hodgson: Chem. Phys. Lett., **279**, 112 (1997)

[11] M.J. Murphy, A. Hodgson: Surf. Sci., **390**, 29 (1997)

[12] R.C. Tolman: The principles of statistical mechanics, Oxford University Press, London (1938)

[13] M.J. Cardillo, M. Balooch, R.E. Stickney: Surf. Sci., **50**, 263 (1975)

[14] A. Hodgson: Prog. Surf. Sci., **63**, 1 (2000)

[15] C.T. Rettner, H.A. Michelsen, D.J. Auerbach: J. Chem. Phys., **102**, 4625 (1995)

[16] M.J. Murphy, J.F. Skelly, A. Hodgson: J. Chem. Phys., **109**, 3619 (1998)

[17] J.F. Skelly, A.W. Munz, T. Bertrams, M.J. Murphy, A. Hodgson: Surf. Sci., **415**, 48 (1998)

[18] W. van Willigen: Phys. Letts., 28A, 80 (1968).

[19] T. Matsushima: Surf. Sci., **197**, L287 (1988)

[20] R.N. Carter, M.J. Murphy, A. Hodgson: Surf. Sci., **387**, 102 (1997)

[21] S. Schwegmann, A.P. Seitsonen, H. Dietrich, H. Bludau, H. Over, K. Jacobi, G. Ertl: Chem. Phys. Lett., **264**, 680 (1997)

[22] H. Dietrich, P. Geng, K. Jacobi, G. Ertl: J. Chem. Phys., **104**, 375 (1996)

[23] L. Romm, G. Katz, R. Kosloff, M. Asscher: J. Phys. Chem. B, **101**, 2213 (1997)

[24] C.T. Rettner, H. Stein: Phys. Rev. Lett., **59**, 2768 (1987)

[25] H. Dietrich, K. Jacobi, G. Ertl: J. Chem. Phys., **377**, 308 (1997)

[26] M.J. Murphy, J.F. Skelly, A. Hodgson, B. Hammer: J. Chem. Phys., **110**, 6954 (1999)

[27] S. Dahl, A. Logadottir, R.C. Egeberg, J.H. Larsen, I. Chorkendorff, E. Tornqvist, J.K. Nørskov: Phys. Rev. Lett., **83**, 1814 (1999)

[28] Y. Kuwahara, M. Fujisawa, M. Jo, M. Onchi, M. Nishijima: Surf. Sci., **188**, 490 (1987)

[29] R. Raval, M.A. Harrison, S. Haq, D.A. King: Surf. Sci., **294**, 10 (1993)

[30] M. Ikai, H. He, C.E. Borroni-Bird, H. Hirano, K. Tanaka: Surf. Sci., **315**, L973 (1994)

[31] M. Ikai, K. Tanaka: Surf. Sci., **357**, 781 (1996)

[32] P. Samson, S. Haq, M.J. Murphy, A. Hodgson: submitted

[33] Y. Ohno, K. Kimura, M. Bi, T. Matsushima: J. Chem. Phys., **110**, 8221 (1999)

[34] G. Wiesenekker, G.J. Kroes, E.J. Baerends: J. Chem. Phys., **104**, 7344 (1996)

[35] Y. Kuwahara, M. Jo, H. Tsuda, M. Onchi, M. Nishijima: Surf. Sci., **180**, 421 (1987)

[36] J.W. Gadzuk, S. Holloway: Chem. Phys. Lett., **114**, 314 (1985)

Enrichment of Binary van der Waals Clusters Surviving Surface Collision

E. Fort, A. De Martino, F. Pradère, M. Châtelet, H. Vach

Laboratoire d'Optique Quantique du CNRS,
Ecole Polytechnique, 91128 Palaiseau Cedex, France

1 Introduction

The dynamics of collisions between large van der Waals clusters and solid surfaces has been attracting increasing interest. Recent molecular dynamics simulations have shown that a novel chemistry, involving significant many-body effects, may take place within the cluster itself or with the surface, due to the extreme pressures and temperatures achieved in the impact region for initial cluster velocities of the order of 10 km s^{-1} [1].

The shock wave responsible for the cluster "superheating" and shattering disappears when the incident cluster velocity decreases below a threshold value, typically of the order of 1 km s^{-1}. As a result, when colliding with a surface at low enough kinetic energies, large van der Waals clusters undergo a kind of Leidenfrost process [2, 3] : they "glide" along the surface on a "vapor cushion", with approximate consrvation of tangential velocity, while evaporating essentially monomers, to release "in real time" the collisional heat. When the normal kinetic energy is not sufficient to evaporate the whole cluster, large fragments are eventually "pushed away" from the surface at very low normal velocities, and they can easily by detected at grazing detection angles [3, 4].

In this work, we study this Leidenfrost process with *binary* Ar_nKr_m and Ar_nXe_m clusters incident on a hot graphite surface [5]. The surviving clusters are found to be richer in the "dopant" heavy rare gas, with a nontrivial dependence on dopant molar fraction, size and incidence angle of the incoming binary clusters. The information that can be retrieved from these results is discussed.

2 Experimental

Our setup consists of a Campargue [6] type supersonic beam generator and a UHV chamber containing the surface (a HOPG sample maintained at 500

K) and a Quadrupole Mass Spectrometer (QMS). The beam incidence angle θ_{inc} (from surface normal) can be varied from 0° to 70°. The QMS can be rotated about the sample, to perform angularly and mass resolved flux and velocity measurements, either in the incoming beam or after scattering by the surface. Due its geometry, the QMS ionizer head achieves essentially complete fragmentation of van der Waals clusters into monomers, yielding for each monomer mass a signal proportional to the total flux of the corresponding atoms, independently of whether they are clustered or not prior to detection.

The beam generator is equipped with a 0.25 mm diameter sonic nozzle, operated with neat argon at room temperature, yielding argon clusters with average sizes n adjustable up to 13000 atoms per cluster by varying the argon stagnation pressure P_0, up to 50 bars. The dependence of n on P_0 has been determined directly from the beam angular profile broadening upon scattering by a buffer gas introduced in the third pumping stage of the beam generator [7].

For the present work, a small pipe has been introduced in the beam path, where both argon and dopant (Kr or Xe) can be supplied at independently adjustable pressures. The molar fraction of dopant atoms in the binary clusters after pickup (directly measured by the QMS) is controlled through the dopant pressure, while the cluster velocity is kept constant (at 430 m s^{-1}) through the argon pressure in the pipe. Clearly, the pickup process leads to partial cluster evaporation, with an overall decrease in cluster size which has been estimated with a simple model [5] to be 30% for xenon and 40% for krypton respectively, in spite of the larger binding energy for xenon, as the slowing down of the clusters to a given final velocity requires less many collisions with xenon than with krypton. However, this overall decrease is comparable to the absolute uncertainty ($\pm 50\%$) on the initial argon cluster sizes [7], and we are mainly interested in the changes in cluster *composition*, which can be measured much more accurately than cluster sizes. Therefore in the following our data are reported vs the initial argon cluster size n.

3 Results

The measurements described below have been performed for n ranging from 1000 to 13000, θ_{inc} from 30° to 65°, and Kr or Xe molar fractions x varying from 0.7% to 15%. When these parameters were not varied their values were kept fixed at $n \simeq 4000$, $\theta_{\text{inc}} = 60°$, and $x = 0.7\%$ or 15% respectively.

Figure 1 shows typical angular distributions of scattered particles, obtained at argon and xenon QMS mass settings respectively, with $n \simeq 4000$, $\theta_{\text{inc}} = 40°$, and $x = 15\%$. The angular distributions for both species are clearly sums of two components, a broad one peaked around 60°, due to small fragment evaporation, and a narrow one, at grazing exit angles, due to

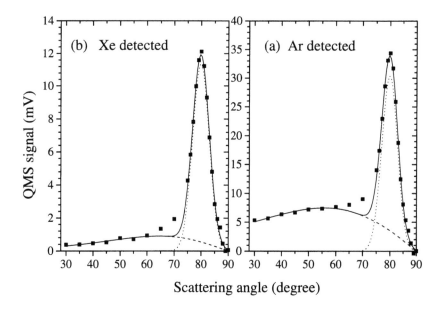

Figure 1: Angular distributions of scattered flux recorded with $n \simeq 4000$, $x = 15\%$ and $\theta_{\text{inc}} = 40°$, for argon (a) and xenon (b) QMS mass settings.

surviving clusters.

Figure 1 clearly shows that argon evaporates more readily than xenon, implying that the surviving clusters are richer in xenon than the incoming ones. We define the enrichment factor $E(X)$ as the ratio of the dopant molar fractions in the surviving clusters and in the incoming ones. This enrichment factor can be measured very accurately, independently of the unknown out-of-plane scattering distributions, simply by comparing the QMS signals at both argon and dopant masses, in the incident beam and in the scattered grazing components. Clearly, if these components are due to surviving binary clusters, their shapes must be identical at the two mass settings, and enrichment factors are evaluated only when it is so.

The broad evaporation component and the narrow grazing one are fitted respectively by the thermokinetic model [8] and by a Gaussian function, shown on Fig. 1 as dashed and dotted lines respectively. The purpose of these fits is twofold. First, they provide a simple way to "extract" the grazing components from the measured distributions, making possible meaningful comparisons of both the shapes and the amplitudes of the grazing components at the two masses. Second, from the reasonable assumptions about out-of-plane scattering distributions involved by these fits we can estimate for each species the total two-dimensional fluxes of both components, and thus the probability for an incoming cluster atom of this species to exit in the surviving fragment. Typically, for the experimental conditions considered, only 10 % of the argon atoms present in the incoming clusters are eventually found in the surviving fragment, in spite of the large grazing peaks observed in the incidence plane.

Comparison of the grazing component shapes at argon and dopant masses shows that for krypton doped clusters binary fragments do survive throughout the investigated range of parameters. Conversely, at high xenon molar fractions binary surviving fragments are found only for incidence angles larger than 45°, indicating that heavy xenon doping seems to render the binary clusters more *fragile* than pure argon clusters or equally heavily krypton doped ones.

Figure 2 shows the variation of the dopant enrichment factor $E(X)$ versus dopant concentration x for $n \simeq 4000$ and $\theta_{inc} = 60°$. Both curves show an enrichment in the solute species larger than one. At the lowest investigated molar fractions, this enrichment is almost the same for Kr and Xe. With increasing dopant molar fraction, $E(Xe)$ decreases much faster than $E(Kr)$. As a result, the enrichment remains higher for krypton than for xenon whatever the dopant concentration.

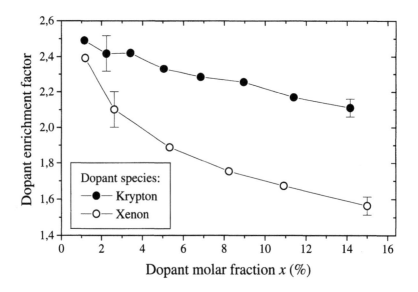

Figure 2: Evolution of enrichment factor with cluster composition. Incoming argon cluster size 4000, incidence angle 60°.

Figures 3 and 4 show the enrichment factor versus incident cluster size and incidence angle, respectively, for both high and low values of x. The essential trends observed in Fig. 2 turn out to be valid throughout the entire investigated range of sizes and incidence angles: at low dopant concentration, the enrichment is the same for krypton and xenon, while at high concentration it is much smaller for xenon than for krypton. Moreover, for both dopants at low molar fraction and for krypton also at high molar fraction, $E(X)$ decreases slowly with increasing incidence angle or initial cluster size. In contrast, for high xenon concentration, it remains essentially constant, close to 1.5, when these parameters are varied throughout the range where binary clusters are likely to survive.

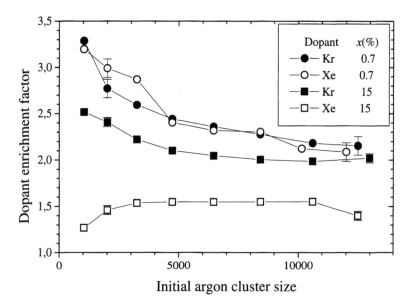

Figure 3: Evolution of enrichment factor with cluster size, for low and high molar fractions of Kr and Xe. Incidence angle 60°.

4 Discussion

As mentioned earlier, for the incident cluster energies used in this work, the collision dynamics belongs to the thermal evaporation regime. Consequently, it could be anticipated that the clusters undergo a *distillation* process leading to selective evaporation of the less tightly bound component. As a result, the dopant enrichment is expected to increase with dopant binding energy, and with the degree of overall cluster evaporation. Indeed, we

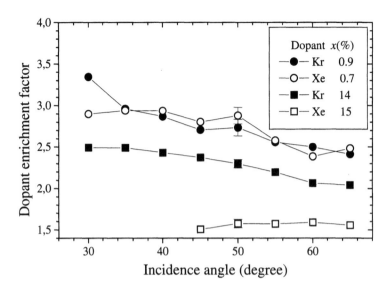

Figure 4: Evolution of enrichment factor with cluster size, for low and high molar fractions of Kr and Xe. Incoming argon cluster size 4000.

always measure an enrichment larger than one for both Kr and Xe. Moreover, the enrichment is found to increase with decreasing incidence angle and/or incident cluster size: in both cases the overall cluster evaporation increases, as shown by earlier experimental and theoretical studies on pure clusters.

However, such simple energetic arguments badly fail to explain an enrichment factor for xenon *lower* than for krypton for all dopant concentrations. This unexpected behavior may be due to differences in incoming binary cluster *structures* and/or surface collision dynamics for the two guest species.

To the best of our knowledge, no detailed computer simulations nor experimental results have been reported to date on the structures of large size ($n = 10^3 - 10^4$) van der Waals clusters having undergone several hundred collisions in a pickup experiment. Quasi-equilibrium structures of clusters picking up single atoms or molecules have been investigated in detail by Perera and Amar [9], in connection with experimental results on the IR spectra of chromophores such as SF_6 and SiF_4 introduced in argon clusters either by coexpansion or by pickup [10]. Basically, the most likely position of a guest atom (or molecule) in a cluster depends on the competition of two opposing forces, one of which being related to the guest size, and the other to the guest-host binding energy. The size effect is due to the energy needed to generate a defect in the cluster lattice to accommodate "large" guest species, which will then experience a kind of "Archimedes force" directed towards the cluster surface. On the other hand, the larger the guest-solvent binding energy, the more the guest will tend to dive into the cluster to optimize the number of bonds with cluster atoms. As a result, with argon clusters the preferred position is inside for SF_6, while SiF_4 remains on the surface, in agreement with IR spectra recorded with the chromophore-argon coexpansions studied by Gu et al. [10].

Clarke et al. [11] extended these results to the case of binary clusters where the molar fractions of both A and B components are significant. Their quasiequilibrium "phase diagram" depends very sensitively on the details of the interaction potentials. Generally speaking, the larger the differences in these potentials, the more the two species behave like nonmiscible liquids. For realistic rare gas clusters, these authors showed that the only systems likely to form stable binary clusters are Kr/Ar and Xe/Kr, while for Xe/Ar the binding energies are probably too different from each other.

As mentioned above, a key issue is the timescale needed to realize binary quasi-equilibrium structures, defined by the diffusion rate of the dopant inside the host cluster. The large argon clusters produced by neat argon expansions are known to be solid, at 32 ± 2 K [12], implying extremely low atom diffusion rates. Indeed, in typical pickup experiments the few guest atoms or molecules deposited on large argon clusters do remain at the surface, as shown for SF_6 by the Scoles group [10] by IR absorption, even though the "stable" positions are inside the cluster.

Therefore, in our experiments the cluster structure at the end of the pickup process is probably defined by diffusion kinetics rather than by quasi-equilibrium thermodynamics. The results of [11] suggest that when the dopant molar fraction x increases, guest diffusion and mixing with argon might be more efficient for krypton than for xenon, implying different binary cluster structures after pickup. If so, for large x values, Xe atoms would be confined to the outer cluster layers more efficiently than Kr atoms, and therefore would evaporate more readily during a cluster-surface collision.

5 Conclusion

We have carried out experiments with large mixed rare gas clusters, obtained by the pickup technique, scattering from a graphite surface. Under our experimental conditions large fragments of the incident clusters survive the collision and are found to be enriched in the dopant species (krypton or xenon) for all the experimental conditions. This enrichment increases when the overall cluster evaporation increases. This result can be understood by simple binding energy considerations. However, these considerations do not explain the lower enrichment in xenon compared with krypton at high concentrations. This effect is to be attributed to the binary cluster structure resulting from the pickup and the dynamics of the cluster surface collision. Surface scattering offers new possibilities to investigate the structure and dynamics of binary clusters, as it provides information on global cluster parameters, in contrast with the more widely used spectroscopic techniques, sensitive essentially to the close environment of the chromophores.

References

[1] I. Schek, J. Jortner, T. Raz, R.D. Levine, Chem. Phys. Lett. **257**, 273 (1996).

[2] J. Gspann, G. Krieg, J. Chem. Phys. **61**, 4037 (1974).

[3] H. Vach, M. Benslimane, M. Châtelet, A. De Martino, F. Pradère, J. Chem. Phys. **103**, 1972 (1995).

[4] M. Svanberg, J.B.C. Pettersson, Chem. Phys. Lett. **263**, 661 (1996).

[5] E. Fort, A. De Martino, F. Pradère, M. Châtelet, H. Vach, J. Chem. Phys. **110**, 2579 (1999)

[6] R. Campargue, J. Phys. Chem. **88**, 4466 (1984)

[7] A. De Martino, M. Benslimane, M. Châtelet, C. Crozes, F. Pradère, H. Vach, Z.Phys. **D 27**, 185 (1993).

[8] H. Vach, A. De Martino, M. Benslimane, M. Châtelet, F. Pradère, J. Chem. Phys. **100**, 3526 (1994).

[9] L. Perera and G. Amar, J. Chem. Phys. **93**, 4884 (1990).

[10] X.J. Gu, D.J. Levandier, B. Zhang, G. Scoles, D. Zhuang, J. Chem. Phys. **93**, 4898 (1990).

[11] A. S. Clarke, R. Kapral, G. N. Patey, J. Chem. Phys. **101**, 2432 (1994).

[12] J. Farges, M.-F. de Feraudy, B. Raoult, G. Torchet, J. Chem. Phys. **84**, 3491 (1986).

Collision Dynamics of Water Clusters on a Solid Surface: Molecular Dynamics and Molecular Beam Studies

A.A. Vostrikov[1,2], I.V. Kazakova[3], S.V. Drozdov[1],
D.Yu. Dubov[1,2], A.M. Zadorozhny[2]

[1] Institute of Thermophysics, Novosibirsk, Russia
[2] Novosibirsk State University, Novosibirsk, Russia
[3] Institute of Computational Technologies, Novosibirsk, Russia

1 Introduction

The electrification of vehicles flying in the atmosphere and their models in supersonic gas flows is considered as an important fundamental and applied problem. It is established that one of the main causes of this phenomenon is the collision of condensed-phase particles with a solid surface. Despite many years of investigations, the mechanisms of electrification have not yet been clarified. This is due to the fact that all laboratory and *in-situ* investigations were carried out using macrodrops of water. There were no laboratory studies of electrification using a supersonic jet as a source of condensed water. It was assumed that only macroparticles of water contribute to the electrification. The effect of electrification during breakdown of water microparticles (($H_2O)_n$ clusters with $n \sim 100$ molecules) was discovered in 1986 in experiments with gas-dynamic (nozzle) molecular beams [1, 2]. It was found that the collision of an $(H_2O)_n$ cluster with a target yields positive and negative ions in the scattered flow. Results obtained in [3 - 5] have drawn attention to the problem of electrification in laboratory gas jets and in the atmosphere, as well as to the problem of generating atmospheric electricity in water cluster collisions.

Considering that water clusters are easily generated even at a low water vapour content in the carrier gas [5], there exists a possibility of uncontrollable electrification in various aerophysical measurements.

This paper presents results on the ion formation and separation during the scattering of neutral water clusters by a solid surface. Firstly, the collision of water clusters $(H_2O)_n$ (where n=32 and 64) with a solid surface was modelled using the methods of molecular dynamics. Data on the kinetics of cluster fragmentation, molecular dissociation, polar ionization of molecules in the cluster, and charge separation were obtained. Secondly, we experimentally studied the formation and separation of charges during the scattering of $(H_2O)_n$ clusters (with n up to 50000) by different targets. Thirdly, we studied the interaction of water clustered jet with a meteorological electric-field sensor.

2 Molecular Dynamics Simulations

Experimentally we can observe only the final result of a collision, i.e., the flows of generated ions and the spatial distribution of neutral and charged fragments of clusters. To study the kinetics and mechanisms of ion formation and separation in detail, we used the classic molecular-dynamic method (MD) of experiment simulation. The experiment was simulated as follows: water clusters were formed on the basis of the polarization model of Stillinger [6]; then a selected velocity of translation towards the target was given to all molecules of the cluster. The collision and scattering patterns were observed with time steps of $2 \cdot 10^{-17}$ s.

According to the Stillinger model, the water molecule consists of two protons and a negative oxygen atom which carries two charge units. An interaction potential includes two parts: the first one is the sum of the Coulomb potentials for each pair of particles, and the second one is the non-additive potential. This non-additive potential represents the polarization energy of the induced dipole moments of the oxygen atoms in the field of the other charges and dipoles. The parameters of Stillinger's potentials are chosen to describe precisely the geometry and binding energy of an isolated water molecule, dimers, and ions.

The cluster was formed in the following manner:

1. The initial configuration of the oxygen atoms in the cluster corresponded to the structure of ordinary ice. The proton orientations were arbitrary.

2. To obtain a stable configuration of the water cluster, relaxation to the given temperature T_0 was performed.

We chose for the interaction potential between the wall and an atom $U(z) = -C/z^9$ [7], where z is the distance between the atom and the surface. The value of C for an oxygen atom was higher than that for the hydrogen atoms. It depended on the ratio of corresponding masses.

We made molecular dynamics simulations MD for $(H_2O)_n$ clusters with $n = 32$ and 64. The collision velocity v was equal to 1 km/s, 3.4 km/s, and 10 km/s. In order to obtain better understanding of the collision kinetics, we visualised the process. We assumed that the formation of ion pairs occurred during the collision of the $(H_2O)_n$ cluster with the surface and was due to the polar dissociation of molecules at the expense of the translational energy

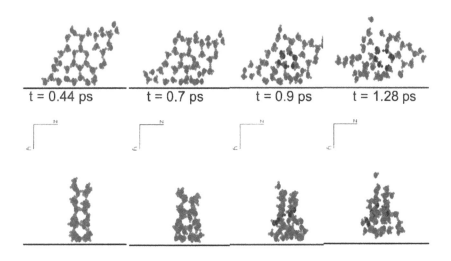

Figure 1: Time evolution of a surface collision

of collision. The energy required for the polar ionization of an H_2O molecule in the gas phase ($H_2O \to H^+ + OH^-$) is $E_a \approx 17$ eV. The energy required for the protolysis in condensed water (($H_2O) \to (H^+)_{aq} + (OH^-)_{aq}$) is equal to $E_a \approx 0.58$ eV. Obviously, the activation energy for the reaction $(H_2O)_n \to H^+(H_2O)_i + OH^-(H_2O)_j$ depends on a size of the solvate shell. The size of the solvate shell depends in turn on the initial size, n, of the cluster. The initial translational energy of the $(H_2O)_n$ in a beam is $E = n \cdot \varepsilon$, where ε is the translational energy per molecule in the cluster. For a velocity $v = 1$ km/s, ε equals to 0.093 eV. Therefore, to produce polar dissociation, the collisional energy E should be distributed nonuniformly among the cluster molecules.

Figure 1 shows four side-view snapshots in the y-z and x-z planes during a standard trajectory for different values of the integration time t. The time t is indicated in each panel. One can see in Fig. 1 the initial structure and the first steps of cluster deformation and fragmentation.

The dependence of the internal temperature, T_i, of an $[r, r + \Delta r]$ layer, $\Delta r = 0.2$ nm, on the distance from the target is shown in Fig. 2 for a cluster with $n = 64$ at $v = 3.4$ km and 10 km. The temperature of the narrow layer is obtained as a measure of the kinetic energy of rotation and vibration of atoms in the cluster, $T_i = (E_k - E_t) \cdot 2/(3n - 3)$.

As is obvious from Fig. 2, the energy distribution among the molecules is non-equilibrium. Mainly, this energy is concentrated in a narrow compressed layer near the wall. This explains the formation and separation of ions in the cluster for relatively low velocities. The fraction of molecules which are

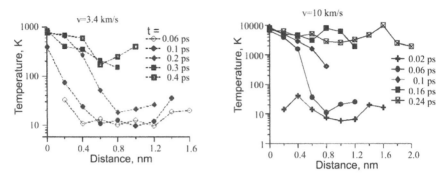

Figure 2: Energy distribution in cluster

dissociated is a stronger function of cluster size than of cluster velocity.

The following pattern was observed. Clusters with a size $n = 32$ dissociated only when the velocity was $v = 10$ km/s. The dissociation did not exceed 15 %, and the generated ions could recombine. Hence the scattered particles were neutral; there were no charged particles among the fragments. This can be explained by the fact that, for $n = 32$, it is impossible to form the complete solvate shell around ions. This leads, firstly, to an increase in the activation energy of polar dissociation in small water clusters, and, secondly, to the increased probability of recombination of generated ions. For clusters with a size $n = 64$, charge separation was observed for all studied velocities.

3 Molecular Beam Measurements

The scheme of the jet and beam experiments is shown in Fig. 3. In order to obtain water cluster with the largest size we used, instead of a sonic nozzle, a supersonic conical nozzle with a full cone angle of $10°$, a throat diameter of 0.2 mm and a supersonic section length of 20 mm. The vapour source operated with the stagnation pressure up to 20 bar. This allowed us to obtain clusters with an average size of up to $5 \cdot 10^4$ molecules without using a carrier gas. A beam of water clusters was formed from the centerline of the jet with a conical skimmer. This beam was directed to the target. Charged fragments reflected from the target were gathered by the collector, which was protected by two screens. The inner screen was a grid with a transparency of 90 %. The collector and screens were coaxial cylinders. The neutral beam of water clusters passed to the target through orifices in electrodes. The incident beam was modulated by a mechanical shutter. The currents to the collector, screen, and target were measured separately in a lock-in mode. The separate measurements were made applying a potential difference U between the collector and inner screen. In the space between

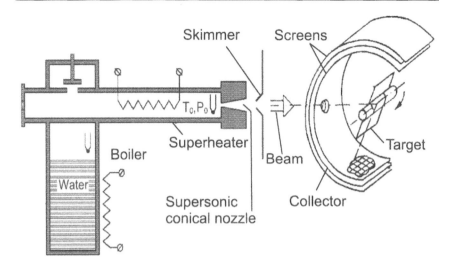

Figure 3: Experimental set-up

the target and the screen there was no electric field, and the ions moved freely. We should note that earlier in [1, 3] we measured only the current of ions reflected from the target, and ions with different signs were separated by the electric field applied between the target and the grid.

The currents I to the collector, screen and target depend on the voltage U and were obtained for different targets, such as plates made from gold, stainless steel covered by colloid graphite and fullerene C_{60}, and plates made from a silicon monocrystal covered by titanium nitride. The dependencies $I(U)$ for clusters of average size $n = 24000$ and incidence angle (relative to the normal to the surface) $\alpha = 75°$ are shown in Fig. 4. It is seen that the sum of currents differs from zero. This can be explained by the fact that some portion of the ions is reflected along the coaxial system of electrodes. The angular distribution of the negative ion current is more diffusive than that for the positive ions [4, 8]. Therefore, losses of negative ions are more significant. A weak dependence of the target current on the voltage U can be explained by the screen scattering of ions.

The data in Fig. 4 allow us to draw conclusions concerning the relative fluxes of ions with different signs scattered by the target. It is obvious that with gold targets, the charged scattered particles are predominantly negative ions. Targets covered by fullerene C_{60} and colloid graphite scatter fewer ions than do metal plates. Currents of scattered ions and currents to metals targets depend also on the target conditions. For instance, it was found out that, after cooling a steel target down to the temperature for the appearance of an ice layer, the current to the target changed its sign.

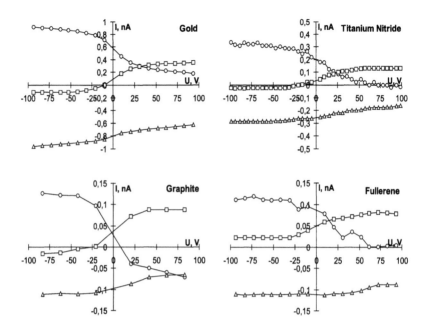

Figure 4: Target ($\triangle$), collector ($\circ$) and screen ($\square$) currents vs. voltage U between collector and screen

4 Charging of Electric Field Mill

The laboratory modelling of the flow of a clustered supersonic jet around a meteorological electric field mill (EFM) sensor [9] was carried out. The EFM used earlier for the rocket measurement of the electric field in the lower mesosphere [10] was installed within the jet axis. The selected distance from the nozzle provided a pressure in the incident flow of about 10^{-1} Pa. The velocity of the incident flow, $v \approx 1.3$ km/s, was of the order of the DECIMALS–B rocket velocity in the mesosphere, ~ 0.8 km/s. The EFM is based on the principle of a rotating condenser. The external electric field induces the discharge on the condenser plates. If this condenser is loaded by a resistor, an alternating current appears in the circuit.

The response of the actual EFM to water cluster bombardment is shown in Fig. 5 together with the raw EFM signal measured in the passage of a rocket through a noctilucent cloud (NLC) layer in the mesosphere during the international rocket/radar campaign "NLC-91" [11]. A strong perturbation of the pure sine-wave signal (expected for the electric field measurement) is clearly seen in both cases.

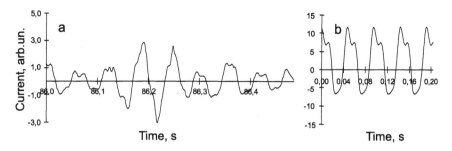

Figure 5: The raw EFM signals, detected during the passage of a rocket through noctilucent clouds on ascent (**a**), and in laboratory experiments in a flow with clusters of about 30000 molecules (**b**)

Acknowledgement

The authors wish to thank Mr. A. A. Tyutin and Mr. O. A. Bragin for the help in the laboratory experiments. This work was supported by the Russian Foundation for Basic Research (grants 98-02-17804 and 98-02-17845) and a grant from the Education Ministry of the Russian Federation.

References

[1] A.A. Vostrikov, D.Yu. Dubov, M.R. Predtechensky, Sov. Phys. Tech. Phys. **32**, 459 (1987)

[2] A.A. Vostrikov, D.Yu. Dubov, M.R. Predtechensky, Chem. Phys. Lett. **139**, 124 (1987)

[3] A.A. Vostrikov, D.Yu. Dubov, M.R. Predtechensky, Sov. Phys. Tech. Phys. **33**, 1153 (1988)

[4] A.A. Vostrikov, D.Yu. Dubov, Sov. Tech. Phys. Lett. **16**, 61 (1990)

[5] A.A. Vostrikov, D.Yu. Dubov, Z.Phys.D. – Atoms, Molecules, and Cluster **20**, 429 (1991)

[6] F.H. Stillinger, C.W. David, J. Chem. Phys. **69**, 1473 (1978)

[7] R.G. Barantsev, in: Interaction of Rarefied Gases with Surfaces (Nauka, Moscow, 1975)

[8] A.A. Vostrikov, D.Yu. Dubov, Z.Phys.D. – Atoms, Molecules, and Cluster **20**, 61 (1991)

[9] A.A. Tyutin, Cosmic Res. **14**, 132 (1976)

[10] A.M. Zadorozhny, A.A. Tyutin, G. Witt, N. Wilhelm, U. Walchli, J.Y.N. Cho, W.E. Swartz, Geophys. Res. Lett. **20**, 2299 (1993)

[11] R.A. Goldberg, E. Kopp, G. Witt, W.E. Swartz, Geophys. Res. Lett. **20**, 2443 (1993)

Hyperthermal Fullerene-Surface Collisions: Energy Transfer and Charge Exchange

A. Bekkerman, B. Tsipinyuk, E. Kolodney

Department of Chemistry, Technion – Israel Institute of Technology Haifa 32000, Israel

1 Introduction

Surface scattering of neutral polyatomic molecules at the hyperthermal impact kinetic energy range of 1−10 eV was studied intensively in the last two decades [1]. The higher energy range of 10 − 100 eV is equally important but was not explored due to technical difficulties related with beam generation and energy analysis of the incident and scattered particles. Recently we reported on the generation and energy analysis of neutral C_{60} seeded molecular beam with kinetic energies up to 60 eV [2, 3]. Kinetic energies were measured using electrostatic energy analyzer (after ionization), and the average vibrational energy content of both incident and scattered C_{60} was measured using mass-spectrometry based vibrational thermometry [4, 5]. During recent years, studies of fullerene collisions and especially the scattering of fullerene ions from surfaces became a highly intensive research field. As large and well-defined covalent clusters, the fullerenes (mainly C_{60}) has turned into a model system for both computational and experimental studies of cluster or large molecule collisions. Interest is also related with the unique behaviour exhibited by the strongly bound carbon cage, e.g. collisional resilience, storing 5−30 eV vibrational energy in a single molecule on the millisecond time-scale, unique decay processes, etc. The scattering dynamics of C_{60}^+ ions colliding with various surfaces (graphite, silicon, diamond and metals) at the impact energies of 80−500 eV, was the subject of a series of recent experiments [6, 7, 8]. The main observations were deep inelasticity (92 − 97% of the impact energy) resulting in nearly constant scattered energy of 5−10 eV and highly resilient behaviour as manifested by intact (non-dissociative) scattering up to 200 eV impact energy. A value of 25−35 % vibrational excitation efficiency (with respect to the impact energy) was assumed, based on molecular dynamics (MD) simulations [9, 10, 11].
In this paper we will present various aspects of the surface scattering dynamics of neutral C_{60} at the hyperthermal impact kinetic energy range of

5 – 50 eV [4, 5, 12, 13]. We have measured scattered beam kinetic energy distributions and angular distributions as a function of impact energy and scattering angle. By combining these with vibrational excitation measurements and simple collisional models we have gained a deeper insight into the various impact-induced energy transfer processes at the surface. Our observations are different from those reported for the high energy C_{60}^+ ion scattering experiments, thus emphasizing the importance of neutral C_{60} molecular beam surface scattering at the 10 – 100 eV range. We have also observed the collisional formation of C_{60}^- following scattering of hyperthermal C_{60}^0 from the graphitized nickel surface (Ni/C) [14]. Collisional charge transfer processes studied in the past were mainly neutralization and negative ion formation during scattering of positive atomic ions from metallic surfaces [15]. Image charge effects were observed and studied [16]. Charge inversion resulting in negative ion formation of both parent and fragments was observed for small molecular ions scattered off metals at impact energies of 200 – 2000 eV [17, 18]. Hyperthermal negative ionization of large neutral polyatomics was studied before at impact energies of 1 – 10 eV [19, 20]. The central role of image charge effects (down shifting the affinity level of the approaching molecule) was assumed but no experimental evidence was reported. We will describe our measurements of image charge effects [14] related with the scattered C_{60}^- and discuss implications of the measured image charge barrier with regard to possible collisional surface-molecule electron transfer mechanism.

2 Experimental Methods

The molecular beam-surface scattering apparatus is shown in Fig. 1. Briefly, C_{60}^0 molecules are accelerated to hyperthermal energies using the seeded beam technique in helium supersonic expansion. The kinetic energy is controlled by varying the nozzle backing pressure while keeping nozzle temperature constant.

The C_{60} molecules are scattered from a fully carbonized polycrystalline nickel surface held in a UHV chamber (5×10^{-10} torr base pressure). All experimental evidence point out that the carbonized phase is a graphite monolayer [13]. Hyperthermal neutral C_{60}^0 scatter from this surface in a typical direct inelastic mode characterised by narrow angular distributions and strongly peaked scattered energy distributions [4, 13]. The surface temperature T_s was 820 – 950 K throughout all the measurements reported here. The scattering configuration is that of a rotating surface with a fixed scattering angle Ψ defined as $\Psi = \pi - (\theta_i + \theta_r)$ where θ_i is the incidence angle and θ_r is the reflection angle, both defined with respect to the surface normal. The most probable values of the corresponding distributions (peak values) will be denoted θ_{im} and θ_{rm}. Both primary

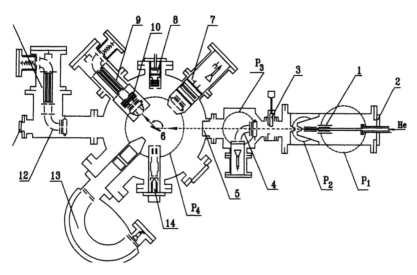

Figure 1: The experimental set-up: (1) Two stage all ceramic nozzle. (2) Nozzle base plate. (3) Beam flag (gate). (4) 90° energy analyzer. (5) Collimator. (6) Rotatable surface. (7) Retarding field energy analyzer. (8) Ion gun. (9,10) Mass filter/Retarding field analyser. (11,12) Mass filter/90° energy analyzer. (13) Hemispherical energy analyzer. (14) Electron gun. (15) Viewport. P_1, P_2, P_3, P_4 are pumping stages

and scattered beam energies were measured using home-made electrostatic energy analyzers mounted between the ionizer and the entrance ion optics of a quadrupole mass spectrometer (QMS). A 90° cylindrical analyzer served for measuring E_o (#12 in Fig. 1) and a retarding field analyzer (RFA) served for scattered kinetic energy E_s measurements of both neutral (post ionized by low energy electron impact) and negatively charged C_{60}^- (#10 in Fig. 1). Energy distributions of scattered neutral C_{60}^0 (using the RFA) were corrected for the ionization efficiency which scales as $E_s^{-1/2}$. Primary beam energy distributions were corrected also for the (cylindrical) analyzer transmission as described before [2]. For each E_o value the corresponding distributions of C_{60}^- and neutral C_{60}^0 (ionized and detected as C_{60}^+) were measured by switching polarities (of both RFA and QMS) and grounding the ionizer as needed. Both types of analyzers were used also in a stand alone configuration (without the mass filter). The 90° cylindrical analyzer (#4 in Fig. 1) served for the on-line measurements of E_o during scattering while the RFA (#7 in Fig. 1) served for E_s measurements at $\Psi = 135°$. Special care was taken to ensure that the ion flight path between surface and RFA was completely field free. In order to check for secondary effects of the analyzer entrance optics on the energy and angle distributions of the negative ion we have repeated the measurements under the same conditions with a different energy analyzer (VG-100 AX hemispherical analyzer, #13 in Fig. 1) and have obtained essentially the same results.

3 Reactive scattering, kinetic energy losses and vibrational excitation

3.1 Reactive scattering of C_{60}^0 from nickel – overlayer growth and C_{60}^- formation

As shown in Fig. 2, upon exposure of a clean nickel surface (sputter-annealed) to the primary C_{60}^0 beam, both C_{60}^- and C_{60}^0 neutral signals were detected and their intensity was gradually rising. The gradual increase of the reflected neutral C_{60}^0 intensity till it saturates at fully carbonized surface (graphite monolayer) was discussed before [13]. All experimental evidence suggests that the carbonized phase is a graphite monolayer [13]. The rise of the C_{60}^- signal intensity was in full correlation with that of the neutral with maximum signal at full coverage. For the purpose of comparison, hyperthermal C_{60}^0 was also scattered from a clean polycrystalline gold surface. Strong scattered C_{60}^0 signal was instanatly observed from the clean gold surface without any surface coverage by carbon species taking place (Auger analysis). However, no scattered C_{60}^- signal could be detected under these conditions. The measured ratio of $\geq 10^4$ of C_{60}^- signal between the graphitized nickel (unknown ϕ) and the clean gold surface ($\phi = 5.0 \pm 0.3$ eV) [21] probably originates from a much lower work function value for the graphitized nickel. The formation of a graphite monolayer on transition metal substrates from gas phase carbon containing molecular precursors is a quiet common and well documented phenomenon [22]. It is usually accompanied by lowering of the work function value by $0.5-1.0$ eV from the bare metal value [23]. Based on this general trend and the reported work function decrease upon formation of a graphite monolayer on rhenium following deep carbonization using C_{60} molecules [23], we roughly estimate $\phi = 3.5-4.0$ eV for our surface.

3.2 Kinetic energy losses

Figure 3 shows the dependence of the peak value of the E_s distributions $E_{sm.}(E_o)$ for $\Psi = 45°$ (near grazing incidence, $\theta_i = 61°$) and $\Psi = 135°$ (near normal incidence, $\theta_i = 15.7°$) measured at the maxima of the corresponding θ_i intensity dependencies ($I(\theta_i)$) of the scattered C_{60}^0. Almost perfect proportionality between $E_{sm.}$ and E_o is demonstrated for both scattering angles. These results are in sharp contrast to the impact energy independence observed for C_{60}^+ scattering at $E_o > 80$ eV, where nearly constant

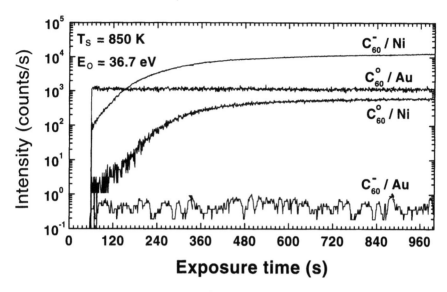

Figure 2: Intensity of scattered C_{60}^0 and C_{60}^- as a function of dosing time (C_{60}^0 beam exposure) starting with initially clean polycrystalline nickel and gold surfaces (sputtered and annealed under Auger analysis) at $T_S = 850$ K. Identical behaviour (slow rise and saturation) is observed for C_{60}^0 and C_{60}^- scattered from nickel. The rise of the scattered C_{60}^0 signal from the gold surface is practically simultaneous with the beam flag "on" time and no C_{60}^- signal could be measured

$E_{sm.} \cong 5$ eV was observed [6, 7, 8]. Our results thus constitute the first evidence for a real "bouncing" scattering of C_{60}^0 from surfaces. For the two different scattering angles we have observed the same energy loss of about 85 – 90% from the normal component and about 25% from the tangential component. Note that a spherical (non-reactive) surface potential barrier (E_b) of 0.9 eV, as measured before was included in the calculation [13]. Full energy losses were about 85% for $\Psi = 135°$ and about 40% for $\Psi = 45°$. This observation of nearly complete decoupling between kinetic energy losses from the normal and tangential components of the incident kinetic energy enable us to discuss separately the C_{60}-surface energy transfer mechanism for the two components [12, 13]. The normal kinetic energy losses were described in terms of a binary model where the collision partners are the C_{60} molecule and some target effective mass unit while the tangential energy losses where modelled by rotational excitation. It was concluded that translational slip is involved during the collision event.

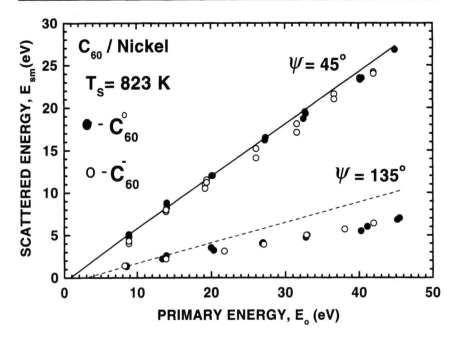

Figure 3: Scattered beam energy E_{sm} (peak values) for C_{60}^0 and C_{60}^- plotted as a function of primary energy E_o for $\psi = 45°$ ($\theta_i = 61°$) and $\psi = 135°$ ($\theta_i = 15.7°$). The dashed line is the calculation for $\psi = 135°$

3.3 Impact Induced vibrational excitation

We have used our capability of independent control over the vibrational and kinetic energy of the incident C_{60} combined with high resolution measurements of both energies before and after collision (± 0.1 eV for kinetic energies and ± 0.25 eV for vibrational energies, most probable values of the corresponding distributions), in order to probe possible coupling effects between the incident C_{60} vibrational energy and impact induced energy transfer processes like kinetic energy loss and collisional vibrational excitation [5]. Average vibrational energies of both incident and scattered C_{60} were measured using vibrational thermometry based on electron impact mass-spectrometry fragmentation pattern and calibrated by effusive C_{60} beam over a wide temperature range [4, 5].

Figure 4 shows the kinetic energy distributions for the primary C_{60} beam and the C_{60} beam scattered off the Ni/C nickel surface at near normal incidence ($\Psi = 135°$), for two different vibrational energies of the incident C_{60}, $E_v = 14.4$ eV and $E_v = 9.8$ eV and the same kinetic energy of $E_o = 33.1$ eV. In this experiment 30.7 eV out of E_o are in the normal component. The total kinetic energy loss is 28.2 eV (85%) out of which 26.9 eV are from

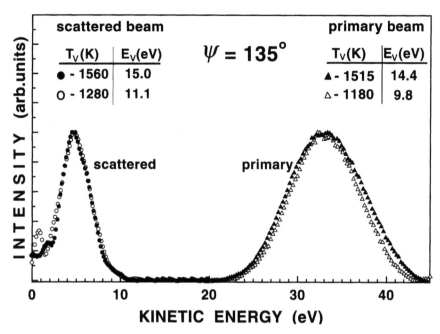

Figure 4: The effect of variable vibrational energy in the primary C_{60} beam on the kinetic energy loss for near-normal scattering ($\Psi = 135°$) from the passivated (graphite covered) nickel surface. Shown are kinetic energy distributions of the primary and scattered C_{60} for two vibrational energies which differ by nearly 5 eV (upper right). The most probable primary and scattered energies are identical to within ± 0.1 eV. Also given are the scattered beam vibrational temperatures (upper left)

the normal component. In spite of the very large energy transfer to the surface and the substantial difference (~ 5 eV) of the incident beam E_v for both experiments, within the experimental accuracy of 0.2 eV we do not observe any coupling between E_v and the kinetic energy loss. The collisional vibrational excitations observed are from 14.4 eV in a primary beam to 15.0 eV in a scattered beam (0.6 ± 0.25 eV) and from 9.8 eV to 11.1 eV (1.3 ± 0.25 eV). The T_v values and associated E_v values are also presented in Fig. 4. We see a clear trend for increasing the relative vibrational excitation with decrease in the incident vibrational energy. For $E_v = 14.4$ eV we measure a vibrational excitation (with respect to E_o) of $2 - 3\%$, while for $E_v = 9.8$ eV an excitation of $4 - 5\%$ was found.

4 Collisional charge exchange – formation of C_{60}^-

4.1 Relative C_{60}^- yield – impact energy dependence

Hyperthermal C_{60}^0 beam was scattered off the Ni/C surface. Due to efficient surface-C_{60}^0 electron transfer the scattered particles included both neutral (C_{60}^0) and negative fullerene ions (C_{60}^-). Negative ions mass spectrum of the scattered particles showed a single C_{60}^- peak with relative abundance higher than 10^4 over the 1–1000 amu mass range. No negative fragments C_n^- ($n = 1 - 59$) were observed in spite of the fact that most of them have higher electron affinity values than C_{60} ($EA(C_{60}) = 2.65$ eV). Energy and angle distributions of both C_{60}^0 and C_{60}^- were measured for the two different scattering angles ($\Psi = 45°$ and $\Psi = 135°$). For near grazing incidence ($\Psi = 45°$) a gradually developing energy dependent shift was observed between the angular distributions of C_{60}^0 and C_{60}^-. A constant small shift between maxima of the corresponding energy distributions was also observed. These shifts were analyzed and explained in terms of image charge effects on the outgoing trajectory (deflection) and exit energy (retardation) of the C_{60}^- [14a] (see Sect. 4.2). Here we will focus on the dependence of the C_{60}^- anion yield on the C_{60}^0 impact energy [14b]. As is shown in Fig. 5, an increase of nearly two orders of magnitudes is observed for the relative C_{60}^- yield starting from an apparent threshold at 3–4 eV.

The conventional picture regarding charge transfer in atom-surface collisions assumes a full memory loss of the initial charge state along the incoming trajectory. The final charge populations are therefore being determined along the exit path. This memory loss is related with a gradual distance dependent evolution of a fast resonant charge exchange between the approaching particle and the surface electrons. At some distance from the surface a complete mixing between the incident particle and the surface electrons sets in and the particle looses track of its initial charge state. Under certain limiting conditions (low normal velocity $V_{S\perp}$ of the scattered ion and low surface temperature) it could be theoretically shown [15] that the ion yield should follow an exponential dependence of the form $\exp[-B/V_{S\perp}]$ where B is a constant (the characteristic velocity). Under well defined collision conditions this expression was verified experimentally before only for medium energy Li$^-$ scattering from cesiated tungsten [24]. In Fig. 6 we show that this relation is obeyed also for a large polyatomic particle at hyperthermal impact energies. A detailed analysis of the slope B can give information about the so called critical distance (Z_C) for ion formation.

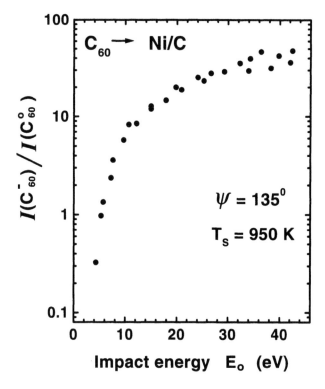

Figure 5: Relative C_{60}^- yield following impact of hyperthermal C_{60} ($E_o = 4-42$ eV) on the Ni/C surface for near normal incidence angle ($\Psi = 135°$, $\theta_i = 16.5°$)

4.2 Image charge effects

In this section, we will focus on the C_{60}^- angular and energy distributions and their analysis in terms of image charge effects. We have measured the energy distributions of scattered neutral C_{60}^0 and C_{60}^- negative ions at the peaks of the corresponding angular distributions. The measurements reported here are for near grazing incidence conditions ($\Psi = 45°$) and impact energy range of $E_o = 5 \div 45$ eV.

All the peaks (most probable values) of the scattered energy distributions (E_{sm}^-) are consistently lower in energy than the corresponding peaks of the neutral C_{60} distributions (E_{sm}^0) by a constant shift of $0.5 \div 0.7$ eV (Fig. 7). The accuracy of the measured E_{sm} values is limited by the unknown contact potentials up to ± 0.5 eV. Although the observed energy shift can be considered only semi-quantitatively it is clearly indicative of a retarding image charge effect. The image potential barrier E_{image} retards the outgoing C_{60}^- such that $E_{sm}^0 = E_{sm}^- + E_{image}$. Assuming now that the image potential

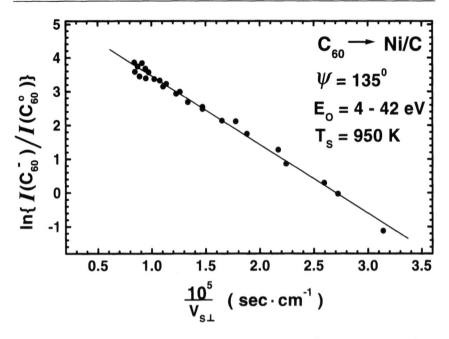

Figure 6: A logarithmic plot of the relative C_{60}^- yield (taken from Fig. 5) as a function of reciprocal normal velocity (peak values) of the scattered C_{60}^-. The straight line (best fit) gives a slope of 2.24×10^5 cm s^{-1}

barrier has a planar geometry one should expect a refractive behaviour manifested by a relative angular shift between outgoing trajectories of C_{60} and C_{60}^-. Indeed, a clear shift is observed between the $I(\theta_i)$ dependencies of neutral C_{60} (solid circles) and those of C_{60}^- (open circles) as presented in Fig. 8 for several E_o values. The shift is also increasing with decrease in E_o (and correspondingly $E_{S\perp}$) as expected. Being formed on the exit path the negative ion suffers an additional deflection (as compared with the scattered neutral) due to the attraction exerted by the image force in the surface normal direction (Z axis) along the outgoing trajectory from the instant of final formation of the ion, at some critical distance Z_C from the surface. For all distances larger than Z_C the formed ion continues on its outgoing trajectory subjected to the attractive image forces while the scattered C_{60} that crossed Z_C as a neutral remains unperturbed by image charge forces. We have analyzed the angular shifts over all the impact energy range studied and have obtained an image charge barrier of $E_{\text{image}}(Z_c) = 0.28 \pm 0.02$ eV corresponding to a critical "ion formation" distance of $Z_c = 13 \pm 1$ Å (assuming that Z_c is measured from the C_{60}^- centre and substracting 5 Å van der Waals radius of C_{60}^0). The tunnelling width of the surface-molecule potential barrier at Z_c is therefore 8 ± 1 Å. The

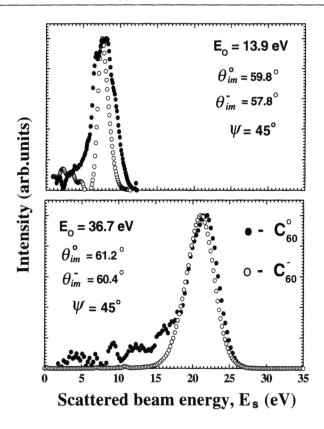

Figure 7: Energy distributions of the scattered neutral and negatively charged C_{60} species at $T_S = 950\,\mathrm{K}$ and two different primary energies. The spectra were measured at the peak (θ_{im}) of the corresponding incidence angle dependencies $I(\theta_i)$. All distributions are normalized to the same maximum intensity

large value of Z_C as measured here is giving us some indication regarding the mechanism for the negative ion formation. Since Z_C is much larger than the C_{60} radius, the incoming C_{60} can cross it at any of the E_o values. Under these conditions, a deeper (E_o dependent), impulsive penetration of the incident molecule into the surface is of no importance in increasing the negative ion yield and the final populations of the different charge states are therefore being determined on the exit path.

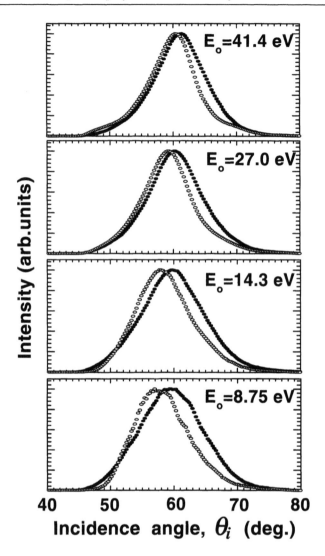

Figure 8: Scattered C_{60}^0 (full circles) and C_{60}^- (empty circles) beam intensity vs. incidence angle $I(\theta_i)$ for $\Psi = 45°$, $T_S = 950$ K and various primary energies E_o. Note the gradual shift between maxima in the $I(\theta_i)$ dependencies with decreasing E_o. All angular dependencies are normalized to the same maximum intensity

5 Summary and conclusions

We have reviewed our recent work on the scattering dynamics of neutral C_{60} molecules from surfaces at the hyperthermal impact energy range of 5–50 eV.

We have measured incidence angle dependencies, kinetic energy losses, vibrational excitation and various features of the charge transfer process. The interaction of hyperthermal C_{60} beams with a clean polycrystalline nickel surface was used to grow graphitic overlayer at high temperatures. This passivated layer on nickel (Ni/C) gave rise to direct inelastic scattering as manifested by sharply peaked angular and energy distributions. The scattered kinetic energy was found to scale linearly with impact energy, and kinetic energy losses varied with scattering angle from $\sim 85\%$ (near normal incidence) to $\sim 40\%$ (near grazing incidence). Analyzing the results, we found nearly complete decoupling between normal and tangential energy losses. These observations describe a real "bouncing" regime for fullerene-surface scattering. Using a newly developed vibrational thermometry method the collisional vibrational excitation of C_{60} was measured as a function of the average vibrational energy of the incident molecule ($\langle E_v \rangle = 7-15$ eV). We have also observed the formation of scattered C_{60}^- anions in collisions with the Ni/C surface. The C_{60}^- ion yield was measured as a function of E_o and was shown to obey an exponential dependence on the inverse of the outgoing normal velocity similar to that observed before for charge exchange in atom-surface collisions. We have compared energy and angle distributions of both C_{60}^- and C_{60}^0 as a function of the impact energy. The shifts observed between the angular and energy distributions maxima of the neutral and negative ion could be analyzed and explained in terms of image charge effects on the outgoing trajectory (deflection) and exit energy (retardation) of the C_{60}^-. The angular deflection analysis yields an image charge barrier of 0.28 ± 0.02 eV corresponding to a rather large "ion formation" critical distance of 13 ± 1 Å. This seems to be the first direct observation of image charge effects in molecule-surface scattering.

We believe that C_{60} can serve as a model system for probing the collisional interaction of very large molecules (or tightly bound clusters) with surfaces. Especially attractive from this point of view is the possibility of controlling the vibrational energy content of the impinging molecule independently of the impact energy. Our future work in this field will be aimed at gaining deeper insight into the charge transfer dynamics and also expanding the impact energy range by using negative fullerene ions as colliders.

Acknowledgments

This research was supported by a grant from the Israel Science Foundation and in part by the James Frank Program.

References

[1] A. Amirav, Comments At. Mol. Phys. **24**, 187 (1990) and references therein

[2] A. Budrevich, B. Tsipinyuk, E. Kolodney: Chem. Phys. Lett., **234**, 253 (1995)

[3] E. Kolodney, A. Budrevich, B. Tsipinyuk: J. Phys. Chem., **100**, 1475 (1996)

[4] B. Tsipinyuk, A. Budrevich, M. Grinberg, E. Kolodney, J.Chem.Phys. **106**, 2449 (1997)

[5] A. Bekkerman, B. Tsipinyuk, A. Budrevich, E. Kolodney,,Int. J. of Mass Spectrom. and Ion Proc., 167/ **168**, 559 (1997)

[6] (a) C. Yeretzian, R.D. Beck, R.L. Whetten,Int. J. of Mass Spectrom. and Ion Proc., **135**, 79 (1994) (b) C. Yeretzian, K. Hansen, R.D. Beck, R.L. Whetten, J. Chem. Phys. **98**, 7480 (1993)

[7] (a) H.G. Busmann, Th. Lill, B. Reif, I.V. Hertel, H.G. Maguire, J. Chem. Phys., **98**, 7574 (1993) (b) Th. Lill, H.G. Busmann, F. Lacher, I.V. Hertel: Chem. Phys., **193**, 199 (1995)

[8] Th. Lill, H.G. Busmann, B. Reif, I.V. Hertel: Surf. Sci., **312**, 124 (1994)

[9] R.C. Mowrey, D.W. Brenner, B.I. Dunlap, J.V. Mintmire, C.T. White: J. Phys. Chem., **95**, 7138 (1991)

[10] G. Galli, F. Mauri: Phys. Rev. Lett., **73**, 3471 (1994)

[11] D. Blaudeck, Th. Frauenheim, H.G. Busmann, T. Lill: Phys. Rev. B., **49**, 11409 (1994)

[12] A. Budrevich, B. Tsipinyuk, A. Bekkerman, E. Kolodney, J. Chem. Phys. **106**, 5771 (1997)

[13] E. Kolodney, B. Tsipinyuk, A. Bekkerman, A. Budrevich Nucl. Instr. Meth. B. 125, 170 (1997)

[14] (a) A. Bekkerman, B. Tsipinyuk, S. Verkhoturov, E. Kolodney, J. Chem. Phys. **109**, 8652 (1998) (b) A. Bekkerman, B. Tsipinyuk, E. Kolodney Phys. Rev. B, in Press.

[15] J. Los, J.J. C Geerlings. Physics reports **190**, 133 (1990)

[16] H. Winter, J. Phys: Condensed Matter **8**, 10149 (1996)

[17] Pan Haochang, T.C.M. Horn, A.W. Kleyne: Phys. Rev. Lett. **57**, 3035 (1986)

[18] W. Heiland in: Low Energy Ion-Surface Interactions. Edited by J.W. Rabalais, (Chichser 1994), p. 313 and references therein.

[19] A. Danon, A. Amirav: J. Phys. Chem. **93**, 5549 (1989)

[20] A. Danon, E. Kolodney, A. Amirav: Surf. Sci. **193**, 132 (1988)

[21] V.S. Fomenko, I.A. Podchernyaeva, " Emission and absorption properties of substrates and materials." (Atomizdat, Moskow, 1975 in Russian).

[22] A.Ya. Tontegode, Prog. Surf. Sci. **38**, 201 (1991)

[23] R.N. Gall, E.V. Rutkov, A.Ya. Tontegode, M.M. Usufov, Tech. Phys. Lett., **23**, 911 (1997)

[24] J.J.C. Geerlings, R. Rodink, J. Los, J.P. Gauyack: Surf. Sci. 186 15 (1987)

Electron Emission Induced by Cluster-Surface Collisions: a Fingerprint of the Neutralization Dynamics?

M. E. Garcia[1,2], O. Speer[2], B. Wrenger[3], K. H. Meiwes-Broer[3],
[1] Departament de Física, Universitat de les Illes Balears,
E-07071 Palma de Mallorca, Spain,
[2] Institut für Theoretische Physik der Freien Universität Berlin,
Arnimallee 14, 14195 Berlin, Germany
[3] Fachbereich Physik, Universität Rostock,
18051 Rostock, Germany

1 Introduction

In the last years the study of cluster-surface collisions has become a subject of intensive research. The collision between a cluster and a solid surface induces many different microscopic processes like photon- and phonon emission, excitation of single- and many-electron states, deformations, bond-breaking and electron emission. During energetic impact there exist nonequilibrium conditions similar to those in shock tubes, but on a very different time scale, a femtosecond scale. New types of chemical reactions, characterized by extremely high density, pressure and kinetic temperature, are expected to occur.
A variety of collision experiments between clusters and solid surfaces were performed, which can be grouped into investigations involving (i) surface modification by cluster bombardment or deposition [1, 2], (ii) scattering of intact species and their fragments [3, 4], and (iii) electron emission [5, 6, 7]. For collisions between ions and surfaces three different mechanisms for electron emission have been proposed: (a) potential emission, involving Auger electrons, (b) kinetic emission, occurring mainly in highly energetic collisions, and (c) thermionic emission, always present and caused by thermalization effects. It is still unclear, if these mechanisms apply to cluster-surface collisions, or if new mechanisms have to be considered.
In addition to the energy-dissipation processes mentioned above, which occur after the impact, quantum effects are important during the collision. For

instance, resonant charge exchange between the projectile and the target may occur. Moreover, if the collision is fast enough, this resonant tunneling may be coherent.

Such quantum effects, which are present whenever two macroscopic objects interact via a tunneling gap are always exciting as they might be important for future nanoelectronic devices. In addition they sometimes allow intriguing insight into the physics of quantum states induced by the charge confinement in a system of reduced dimensionality. However, the dynamics of the electron motion usually remains hidden due to the extremely short time scales involved. Only recently the pump–probe technique with fs laser light pulses started to give some progress in the understanding of such processes.

In this paper we perform an experimental and a theoretical study of low-energy collisions between charged clusters and surfaces. We show that the electron emission yield as a function of the size of the colliding clusters gives, under certain conditions, important information about the neutralization dynamics during the collision.

The paper is organized as follows. In section 2 we describe the experimental setup used to measure the relative electron yield as a function of the size and show the remarkable results obtained. In section 3 we study the collision process theoretically. In particular we determine the neutralization dynamics during the collision. Furthermore, we propose a model for electron emission, based on the assumption that the electron emission yield is sensitive to the neutralization dynamics. We show that using this model, good agreement is obtained between theory and experiment.

2 Experiment

The clusters are produced in a source of the PACIS type [9]. After acceleration to their final collision energy E_{coll}, they are mass separated in a Wien velocity filter (Colutron 600 B) which is followed by a distance of free drift [10]. Neutral clusters are eliminated by deflecting the charged clusters prior to the collision chamber. The selected clusters enter the target region, which is magnetically shielded. Collisions are performed at normal incidence. The emitted electrons are guided from the target to a channeltron. A second detector (channeltron or Daly type) determining the ion intensity is mounted directly behind the movable target. We perform sample preparation by cleaving Highly Oriented Pyrolytic Graphite (HOPG) parallel to the (0001)–plane prior to mounting it into vacuum, or in the case of slightly oxidized aluminum by chemical cleaning. The collision chamber is pumped by a turbo pump and optionally by a LN_2 cryo system. The base pressure in the collision region ranges below 10^{-9} mbar and increases during the experiment due to the He gas from the PACIS.

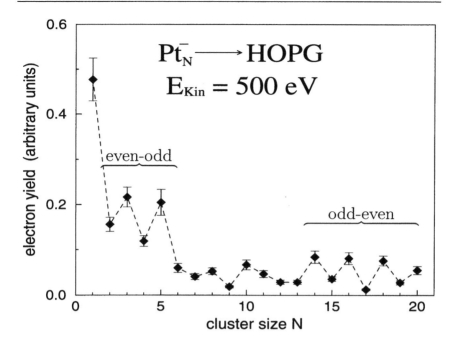

Figure 1: Experimentally determined relative electron yield γ as a function of size N for Pt_N^- clusters colliding with a graphite (HOPG) surface for a collision energy $E_{coll} = 500$ eV

For the measurement of the relative electron yield a Wien filter mass spectrum has to be recorded in order to assign a given filter setting to the corresponding cluster mass. The mass spectra of Pt_N^- are resolved up to $N = 20$. With reduced resolution particles with up to $N = 1000$ can be detected using the present setup. For each selected mass peak, we count cluster ions and emitted electrons under the same conditions. Furthermore, the background noise has to be recorded and subtracted separately. To improve the S/N ratio we count clusters and electrons only within a time window of ≈ 1 ms. The relative electron yield $\gamma(N)$ is given by the number of electrons divided by the number of cluster ions. $\gamma(N)$ may be wrong by a factor which takes into account the ion and electron detection efficiency and which is constant for a defined collision system, i. e. fixed cluster and target type. The collision energy can be varied between 200 eV and 1.1 keV. Fig. 1 shows the measured $\gamma(N)$ of Pt cluster anions colliding with an HOPG surface, at a collision energy of 500 eV in addition to the energy from the supersonic expansion. $\gamma_{HOPG}(N)$ decreases with N, and shows remarkable alternations for $N = 1 \ldots 4$ (maxima for N odd) and $N = 13 \ldots 20$ (maxima for N even). Such non-monotonic behavior has not been observed before.

These odd-even and even-odd alternations in $\gamma_{HOPG}(N)$ for $E_{coll} = 500$ eV would suggest, in principle, that the electronic structure of the colliding clusters plays an important role for electron emission. In a first attempt to understand the remarkable experimental features shown in Fig. 1 one could associate the alternation in $\gamma_{HOPG}(N)$ with the well known "odd-even-effect" observed in the ionization potentials (IP) of some metal clusters. However, the affinity energies of Pt_N clusters $E_A(N)$ (which are also the IP of Pt_N^- clusters) do not exhibit any kind of odd-even alternation [11].

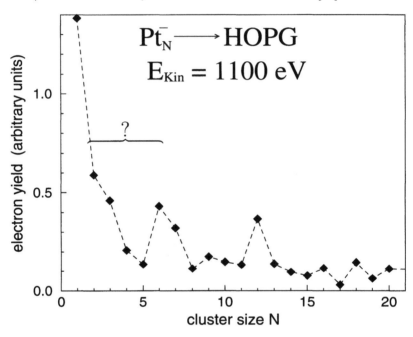

Figure 2: Measured relative electron yield $\gamma(N)$ due to collisions between Pt_N^- clusters and a graphite (HOPG) surface for $E_{coll} = 1100$ eV

Moreover, new remarkable features appear by performing the experiment using the same type of clusters and the same target, but changing E_{coll}. If, instead of $E_{coll} = 500$ eV the impinging clusters have a kinetic energy $E_{coll} = 1100$ eV the dependence of γ_{HOPG} on the cluster size changes dramatically. In particular the even-odd alternations for small clusters disappear, as is shown in Fig. 2. Furthermore, for $E_{coll} < 300$ eV the size oscillations cannot be distinguished any more within experimental resolution.

In contrast to the results of Fig. 1, experiments performed under the same conditions but using Al as target give a smoothly decreasing $\gamma(N)$ ($\gamma \approx 0.03$ already for $N \geq 8$) and further significant size-dependent features are absent [8].

It is clear that the remarkable results presented above cannot be explained if one resorts to the standard theories for electron emission. A rough analysis of the known mechanisms for ejection of electrons upon particle-surface collisions does not help much to understand the results of Figs. 1-2. Potential emission through Auger processes is, in the case of negatively charged clusters, energetically not allowed. Thermionic emission (TE) from the cluster after the collision is, in principle, possible. The corresponding electron yield $\gamma_{TE}(N)$ would reflect the size dependence of the affinity energies $E_A(N)$ of the Pt_N clusters (the less the binding energy of the extra electron, the larger the electron yield). However, as mentioned above $E_A(N)$ of Pt_N clusters is a flat function of N (at least for $N > 3$) and shows no similarities with the electron yields obtained in the present work.

The other possible mechanism to explain the observed ejection of electrons is the kinetic emission (KE). In the last years, KE thresholds for velocities below 10 km/s have been measured [12], which are of the same order of magnitude as the collision energies considered in this work. However, the resulting electron yield $\gamma_{KE}(N)$ would be, in principle, a monotonic function of the cluster size.

Thus, none of the known mechanisms for electron emission, in their simplest form, yields a consistent description of our observed $\gamma(N)$. A further analysis of the collision process is therefore necessary. In the next section we perform such an analysis.

3 Theory

The experimental results presented in the previous section suggest that the electron yield strongly depends on the velocity of the colliding clusters. Now, the velocity of an atomic ion colliding with a surface is the parameter which governs the neutralization dynamics of the projectile [13, 14]. Analogously, during the collision of a cluster the electronic charge may jump from the cluster to the surface and backwards.

Thus, independently of the ultimate mechanism for the ejection of the electrons, we expect the electron yield to be sensitive to the whole neutralization dynamics, and, in particular, to the charge-state of the cluster when it reaches the surface.

Starting from these arguments, and in order to describe the experimental facts shown in Figs. 1-2 we propose a new model based on the following fundamental assumption: the electron yield $\gamma(N)$ for fixed E_{coll} is some functional of the (size-dependent) non-adiabatic survival probability $P_s(N,t)$ of the charged projectiles. The function $P_s(N,t)$ represents the probability that the cluster of size N remains unneutralized until the time t. Thus,

$$\gamma(N) = \mathcal{F}[P_s(N,t)]. \tag{1}$$

Using a linear approximation for the functional $\mathcal{F}$ and assuming that the interactions between the cluster and the surface which lead to electron emission are short ranged [15], we obtain the much simpler expression

$$\gamma(N) \propto P_s(N, t_0), \tag{2}$$

where t_0 is the time at which the cluster reaches the surface. This means that only if a cluster reaches the surface unneutralized, an electron can be emitted.

If this assumption is true we just need to calculate $P_s(N, t_0)$ in order to compare with the experimental values for γ. We now determine $P_s(N)$ by using a microscopic theory. For the description of the dynamics of the extra electron of the Pt_N^- clusters we consider the cluster as a single state (affinity level) which interacts during the collision process with a band of states in the surface. This leads, like in the case of atom-surface interactions [13, 16, 17], to a time-dependent Anderson type Hamiltonian of the form

$$\begin{aligned} H(t) &= \varepsilon_0(t)\, c_0^+ c_0 + \sum_{\mathbf{k}} \varepsilon_{\mathbf{k}}\, c_{\mathbf{k}}^+ c_{\mathbf{k}} \\ &+ \sum_{\mathbf{k}} \left[V_{\mathbf{k}0}(t)\, c_{\mathbf{k}}^+ c_0 + V_{\mathbf{k}0}^*(t)\, c_0^+ c_{\mathbf{k}} \right] \end{aligned} \tag{3}$$

where the subscripts $\mathbf{k}$ and 0 refer to states in the surface and in the cluster, respectively. The time-dependence of H arises from the classical trajectory approximation that the cluster moves with constant velocity v towards the surface. $\varepsilon_0(t)$ corresponds to the affinity level of the cluster. $V_{\mathbf{k}0}(t)$ is the time-dependent matrix element which describes the hopping process between the projectile and the state $\mathbf{k}$ of the target.

The Heisenberg equation of motion leads to the following coupled differential equations

$$i\hbar \frac{\partial c_0}{\partial t} = \varepsilon_0(t)\, c_0(t) + \sum_{\mathbf{k}} V_{\mathbf{k}0}^*(t)\, c_{\mathbf{k}}(t) \tag{4}$$

$$i\hbar \frac{\partial c_{\mathbf{k}}}{\partial t} = \varepsilon_{\mathbf{k}}\, c_{\mathbf{k}}(t) + V_{\mathbf{k}0}(t)\, c_0(t), \tag{5}$$

which have to be integrated to calculate the nonadiabatic cluster-level occupation $n_0(t_0) = \langle c_0^+(t_0)\, c_0(t_0) \rangle$.

For the particular case of negatively charged colliding clusters the nonadiabatic survival probability $P_s(t_0)$ is identical with $n_0(t_0)$, and the subscripts $\mathbf{k}$ refer to the unoccupied surface states. The interaction of the cluster affinity level with surface states below the Fermi level is not taken into account. The neglect of filled bands is justified if the energy difference between the occupied states and the cluster level is sufficiently large. In treatments of atom-metal surface collisions one usually replaces the summation over $\mathbf{k}$

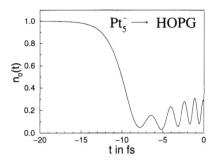 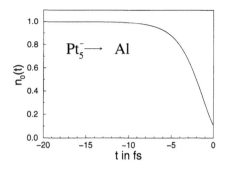

Figure 3: Time-dependence of the cluster-level occupation $n_0(t)$ of Pt_5^- during the collision with a (a) HOPG-target, (b) Al-target

in Eq. (4) by an integral over energies and assumes a constant density of states (DOS) with infinite band-width [13, 17]. This is by no means applicable for materials with DOS containing peaks, like graphite. The relevant matrix elements V_{k0} for $\mathbf{k} \in$ HOPG are those involving the p_z orbitals of graphite (with the z-quantization axis perpendicular to the planes, i.e., parallel to the collision coordinate). These p_z orbitals are mainly responsible for the two peaks (one above and one below the Fermi level) in the DOS of graphite [18, 19]. We have solved numerically the Eqs. (4),(5) by taking into account 20 surface states for HOPG and 200 for Al, distributed in such a way to reproduce the major features of the densities of states (DOS) of the targets. For simplicity, we use rectangular densities of states of different width W, depending on the target material, but in principle, more states and more complicated densities of states can be taken into account. Thus, we represent the density of unoccupied states of HOPG which interact with the cluster by a narrow band of width W_{HOPG}. In contrast, the width of the DOS of Al (W_{Al}) is taken to be large (but not infinite). The quantity W turns out to be essential for description of the experimental features. The time dependence of the hopping matrix elements is given by $V_{k0} = V_s \exp(vt/d)$, which results from the fact that V_{k0} decreases exponentially for increasing distance between the cluster and the surface [16]. The cluster level $\varepsilon_0(t)$ is also a function of the cluster size. Its time dependence is determined by the velocity and the distance dependence of the cluster-surface interactions (image forces, etc) [16]. For fixed cluster size we require $\varepsilon_0(t \to -\infty) = E_A$. The affinity energies $E_A(N)$ are taken from photodetachment experiments [11]. We model the distance dependence of the affinity level when approaching the surface by a constant $E_A(N)$ until the cluster reaches the point z_c and then a decrease with a constant rate $\partial \varepsilon_0 / \partial z = a$.

For the case of HOPG as target we use the following parameters: $W_{HOPG} =$

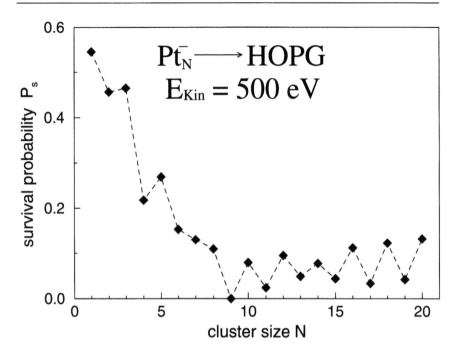

Figure 4: Calculated survival probability $P_s(N)$ of Pt_N^- clusters impinging on a HOPG surface at a collision energy $E_{coll} = 500$ eV.

0.74 eV [20], $V_s = 2.2$ eV, $d = 0.48$, $z_c = 2.2$ and $a = 1.3$ eV/. The position of the narrow p_z-band of graphite is obtained from photoemission experiments [19]. The value for W_{HOPG} is also in agreement with these experimental results [19].

For all cluster sizes $n_0(t)$ exhibits a time-dependence like the one shown in Fig. 3(a). The phase and frequency of these time-oscillations change with the cluster size. In particular, this dynamics results roughly in a periodic dependence of $P_s(t_0)[= n_0(t_0)]$ on the reciprocal velocity of the cluster $1/v$. With increasing bandwidth of the density of states of the target this back and forth tunneling of the charge becomes less important. For very broad densities of states the extra electron of Pt_N^- jumps from the cluster to the surface and delocalizes immediately. Thus, it does not jump back to the cluster. This prevents time oscillations of the cluster occupation, as is shown in Fig. 3(b) for the case of aluminum.

In Fig. 4 we present theoretical results for the survival probability $P_s(N)$ of Pt_N^- clusters impinging on a HOPG surface, with $E_{coll} = 500$ eV. $P_s(N)$ shows oscillations as a function of the cluster size, which reflect the time-oscillations of the neutralization process [Fig. 3(a)]. This effect resembles

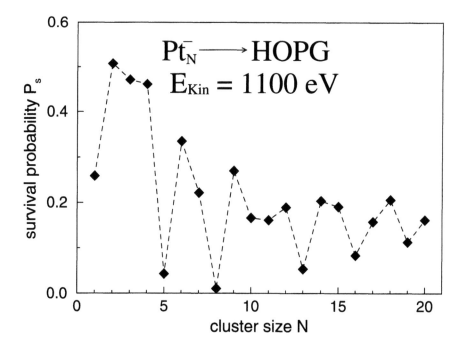

Figure 5: Theoretical results for the survival probability $P_s(N)$ of Pt_N^- clusters impinging on a HOPG surface. The collision energy is $E_{coll} = 1100$ eV.

the well known Stückelberg oscillations, which arise from resonant- or quasi-resonant hopping between two well defined levels corresponding to systems which approach each other with a given velocity. Such oscillations have been experimentally observed in atom-ion collisions [21], and also in collisions between He^+ ions and Pb surfaces [22], where the completely localized d-orbitals of Pb play the role of the well defined level [13]. In the particular case of HOPG, the relevant p_z–band is narrow enough to produce oscillatory behavior in the neutralization dynamics.

As shown in Fig. 4, $P_s(N)$ for HOPG exhibit maxima and minima at the same cluster sizes as $\gamma_{HOPG}(N)$. This results show clearly that the "odd-even" alternation of $P_s(N)$ (and consequently of $\gamma_{HOPG}(N)$ if our assumption is valid) is just a consequence of the oscillating neutralization dynamics for a particular choice of the collision energy. One can understand the results of Fig. 4 qualitatively as follows. Since the resonant charge exchange produces a periodic dependence of P_s on the reciprocal velocity of the cluster $1/v$, and since E_{coll} is the same for all sizes, $P_s(N)$ also oscillates as a function of the size (more precisely of $\sqrt{N}$). Upon a change of E_{coll}, this particular odd-even scenario may change, as we have experimentally observed for $\gamma(N)$. The dramatic dependence of $P_s(N)$ on energy can be nicely observed by comparing Fig. 4 with Fig. 5, in which we show the size dependence of the survival probability for collisions between Pt_N^- clusters and a HOPG surface for $E_{coll} = 1100$ eV.

Of course, the arguments presented below represent only a simplified analysis of our theoretical results. There is an additional dependence of P_s on N due to the affinity energies. Moreover, the size-oscillations of $P_s(N)$ are damped, suggesting an even more complex size dependence. There are two sources of damping. One of them is Landau-Zener-like [$\sim \exp(-const/v)$] involving the crossing of the affinity level of the cluster with the p_z-levels of HOPG. But the most important effect causing damping is the nonzero width of the p_z-band. The larger the width W of the target-DOS, the stronger the damping. If we increase W_{HOPG} to 4eV by keeping the other parameters constant, the oscillations of $P_s(N)$ disappear. On the basis of the previous discussion it is now easy to understand the behavior of $P_s(N)$ for an Al-target. The DOS of Al is characterized by a broad band, which causes a complete damping of the oscillations.

For the characterization of the $Pt_N^- \to$ Al collision we used $W_{Al} = 5$ eV, $V_s = 0.8$ eV and $d = 0.36$. Due to the broad band the results turned out to be independent of the parameters z_c and a. Already a band-width W of 5 eV yields an exponential behavior similar to that obtained in the approximation $W \to \infty$ [8]. For even broader bands we obtain the asymptotic behavior $N_s \sim \exp(-\Delta\, d/\hbar v)$ [13], with $\Delta = \pi \rho V_s^2$ (and $\rho=$ density of states of the target).

4 Conclusions

Summarizing, we have performed measurements of size-dependent electron yields $\gamma(N)$ and calculations of the non-adiabatic survival probability $P_s(N)$ for clusters colliding with surfaces. We have shown that electron emission reflects the neutralization dynamics of the clusters during the collision. It is important to point out that the purpose of our model calculations is to describe the physical picture underlying the experimental results of Fig. 1. The parameters V_s, d, z_c and a have been chosen to give a qualitative demonstration of the effect.

We want to thank Prof. K.H. Bennemann for many fruitful discussions and his continuous support, as well as J. Tiggesbäumker for his help. Financial support by the SFB 198 and SFB 337 is gratefully acknowledged.

References

[1] H. Haberland, M. Karrais, M. Mall and Y. Thurner, J. Vac. Sci. Techn. A **10**, 3266 (1992)

[2] H. R. Siekmann, E. Holub–Krappe, Bu. Wrenger, Ch. Pettenkofer and K. H. Meiwes–Broer, Z. Phys. B **90**, 201 (1993).

[3] T. Tsukuda, H. Yasumatsu, T. Sugai, A. Terasaki, T. Nagata and T. Kondow, J. Phys. Chem. **99**, 6367 (1995).

[4] P. M. St. John and R. Whetten, Chem Phys. Lett. **196**, 330 (1992); P. M. St. John, R. D. Beck and R. L. Whetten, Phys. Rev. Lett. **69**, 1467 (1992).

[5] R. D. Beck, P. St. John, M. M. Alvarez, F. Diederich, and R. L. Whetten, J. Phys. Chem. **95**, 840 (1991); C. Yeretzian and R. L. Whetten, Z. Phys. D **24** 199 (1992).

[6] T. Moriwaki, H. Shiromaru and Y. Achiba, Z. Phys. D **37**, 169 (1996).

[7] P. J. DeLange, P. J. Renkema, and J. Kommandeur, J. Phys. Chem. **92**, 5749 (1988); U. Even, P. de Lange, H. T. Jonkman and J. Kommandeur, Phys. Rev. Lett. **56**, 965 (1986).

[8] B. Wrenger and K. H. Meiwes-Broer, O. Speer and M. E. Garcia, Phys. Rev. Lett. **79**, 2562 (1997).

[9] G. Ganteför, H. R. Siekmann, H. O. Lutz, K. H. Meiwes-Broer, Chem. Phys. Lett. **165**, 293 (1990); H. R. Siekmann, Ch. Lüder, J. Faehrmann, H. O. Lutz, K. H. Meiwes–Broer, Z. Phys. D **20**, 417 (1991); M. Gausa, R. Kaschner, G. Seifert, J. H. Faehrmann, H. O. Lutz and K. H. Meiwes–Broer, J. Chem. Phys. **104**, 9719 (1996).

[10] For details about the Wien velocity filter and its resolving power when connected to a PACIS, see Bu. Wrenger and K. H. Meiwes–Broer, Rev. Sci. Instr. **68**, 2027 (1997).

[11] G. Ganteför and W. Eberhardt, Phys. Rev. Lett. **76**, 4975 (1996).

[12] R. A. Baragiola, Nucl. Instr. and Meth. in Phys. Res. B **88** 35 (1994), and references therein.

[13] W. Bloss and D. Hone, Surface Sci. **72**, 277 (1978).

[14] H. Shao, D. Langreth and P. Nordlander, Phys. Rev. B **49**, 13948 (1994).

[15] O. Speer, M. E. Garcia, Bu. Wrenger and K. H. Meiwes–Broer, to be published.

[16] J.K. Nørskov, D.M. Newns and B.I. Lundqvist, Surface Sci. **80**, 179 (1979).

[17] A. Blandin, A. Nourtier and D. W. Hone, J. Phys. France **37**, 369 (1976).

[18] R. C. Tatar and S. Rabii, Phys. Rev. **B 25**, 4126 (1982).

[19] C. F. Hague, G. Indlekofer, U. M. Gubler, V. Geiser, P. Oelhafen, H.-J. Güntherodt, J. Schmidt-May, R. Nyholm, E. Wuillod and Y. Baer, Synthetic Metals **8**, 131 (1983).

[20] For the simulation of W_{HOPG} 20 equidistant levels turn out to be sufficient. Increasing the number of levels m to 2000 does not result in changes of $P_s(N)$. Notice also, that the parameter V_s must be scaled with $1/\sqrt{m}$.

[21] W. Lichten, Phys. Rev. **139**, A27 (1965).

[22] R. L. Erickson and D. P. Smith, Phys. Rev. Lett. **34**, 297 (1975); N. H. Tolk, J.C. Tully, J. Kraus, C.W. White, S.N. Neff, Phys. Rev. Lett. **36**, 747 (1976).

Supersonic Beam Epitaxy of Wide Bandgap Semiconductors

V.M. Torres, D.C. Jordan, I.S.T. Tsong, R.B. Doak

Department of Physics and Astronomy
Arizona State University, Tempe, AZ, USA 85287-1504

1. Introduction

The advent of wide band gap semiconductors is ushering in an era of high frequency semiconductor devices: blue LED's, high temperature/high power transistors, "visible blind" ultraviolet photodetectors, and blue to ultraviolet solid state lasers [1]. III-N nitrides are the materials of choice for such applications, particularly GaN with its direct band gap of 3.4 eV. Unlike conventional semiconductors, III-N semiconductors cannot be grown from the melt but must be deposited expitaxially onto a suitable substrate under carefully controlled conditions. The established techniques for this are Metal-Organic Chemical Vapor Deposition (MOCVD) and, to a lesser extent, Molecular Beam Epitaxy (MBE) [2]. Alternative techniques based on supersonic beams are now attracting attention, both as a deposition tool and as a probe of the still poorly understood fundamentals of the III-N deposition. Of specific utility are the energy selectivity, state selectivity, enthalpy storage, and high flux attainable in a supersonic free-jet expansion. We discuss two supersonic jets of particular relevance to III-N deposition, seeded supersonic free-jets [3] and discharge supersonic free-jets [4].

2. Selected Energy Epitaxy

By expanding a gas mixture into vacuum through a supersonic nozzle, a heavy "seed" species in a light diluent can be aerodynamically accelerated to suprathermal translational energy. The resulting beam is intense, narrowly distributed in energy, and tunable in energy by adjusting the gas mixture and nozzle temperature. Given their energy specificity and tunability, such beams offer the prospect of selectively regulating gas-surface reaction chemistry. A reaction of considerable interest in III-N semiconductor growth is the decomposition of NH_3 to supply nitrogen to a growing III-N thin film. NH_3 has become the nitrogen species of choice for both MOCVD and MBE growth of III-N compounds. With Ga supplied as elemental metallic vapor or in a metal-organic species, the surface reaction of NH_3 with Ga to deposit GaN initiates heteroepitaxially on an appropriate substrate (sapphire, SiC, or an AlN buffer layer) then continues homoepitaxially on the GaN overlayer. Under standard MBE conditions GaN grows successfully only over a narrow range of surface temperatures, which must exceed ~700 C to provide sufficient surface mobility of reactants yet remain below ~800 C to avoid GaN decomposition. Supplying NH_3 in a seeded beam of tunable kinetic energy provides an additional experimental parameter in adjusting this balance.

The underlying chemistry is easily understood within the Lennard-Jones model of activated dissociative chemisorption [5]. Schematic interaction potentials for physisorption and chemisorption are shown in Fig. 1 for the simple 1-D case of normal incidence onto an uncorrugated surface. Molecular ammonia physisorbs weakly and relatively far from the surface whereas NH_x fragments (including N itself, $x = 0$) chemisorb strongly and correspondingly closer to the surface. Far from the surface the chemisorption potential exceeds the physisorption potential by the dissociation energy. Hence the two curves must cross at some intermediate separation, yielding a local maximum in the overall interaction potential.

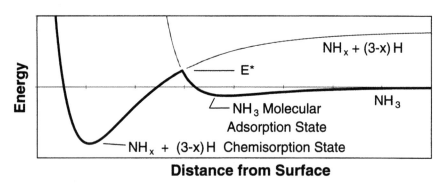

Figure 1: Schematic depiction of activated dissociative chemisorption.

If this curve crossing occurs at positive energy (as depicted) an activation barrier E^* results and a molecule must classically possess at least this energy in order to react. The exact value of the activation energy E^* depends critically on the relative locations, depths, and ranges of the physisorption and chemisorption wells. The 1-D model model is easily extended to scattering at off-normal angles. Still considering a flat surface (1-D barrier), the component of incident velocity normal to the surface, $v_{i\perp} = \cos\theta_i$, is then decisive and $\frac{1}{2}mv_{i\perp}^2 = cos^2\theta_i\, E_i > E^*$ is required to chemisorb. The intent of beam seeding is to provide this needed "normal energy" as beam kinetic energy, choosing the velocity and angle of incidence of the incoming beam appropriately. Within this model, reaction rates measured at various θ_i and E_i should fall onto a common curve when plotted as a function of $cos^2\theta_i E_i$, so-called "normal energy scaling."

The interaction potential of Fig. 1 represents an appropriate average over a family of curves corresponding to different orientations of the molecule, different dissociation states, etc. In effect there is a distribution of E^* values, possibly including even $E^* < 0$ (to yield spontaneous reaction). Since NH_3 pyrolyzes at $T > 650$ C [6] there must exist at least one pathway of very low barrier for NH_3 decomposition. The molecule will follow the minimum energy pathway provided it can rotate and distort rapidly enough on the time scale of the collision. If hindered in its deformation and rotation by conservation laws, quantum laws, or inertial effects, the incoming molecule will sample a higher than minimum barrier.

3. NH_3 Seeded Beams

As it expands adiabatically through a supersonic nozzle of temperature T_0, an ideal gas mixture characterized by a specific heat ratio γ and an average molecular weight $\overline{W}$ attains a terminal velocity [7]

$$V_\infty = \sqrt{\left(\frac{2R}{\overline{W}}\right)\left(\frac{\gamma}{\gamma-1}\right)T_0} \ . \tag{1}$$

The seed species of mass M_i thus accelerates to a kinetic energy which is higher by M_i/M than that of the diluent species (mass M). Translational kinetic energies achievable with typical GaN MOCVD species for a 300 K nozzle in the limit of infinite dilution are provided in Table I. These easily exceed several eV for the heavier MOCVD species. Not all of this energy need channel into cleavage of nitrogen bonds upon surface impact, however, so that a lighter seed species having lower energy but containing fewer internal bonds may dissociate equally well as it strikes the surface.

Seeded supersonic beam epitaxy of the III-N thin films is currently underway in several research groups. Lamb, et al. (NCSU) are growing GaN homoepitaxially using separate seeded beams of triethlygallium and ammonia [8]. Ho and coworkers (Cornell) employ separate seeded beams of triethlyaluminum and ammonia to grow AlN on Si(100) and Si(111) substrates [9]. At ASU we are growing both GaN and AlN on SiC(0001) using ammonia seeded beams. Although not an ideal candidate for seeded beam work due to its relatively low mass, ammonia seeded into helium can nonetheless reach several tenths of an eV in translational energy at quite tolerable T_0.

Species	Symbol	Weight (AMU)	Energy (eV)
Nitrogen	N_2	28.0	0.45
Ammonia	NH_3	17.0	0.27
Trimethylgallium	$(CH_3)_3$ Ga	114.7	1.9
Triethylgallium	$(C_2H_5)_3$ Ga	156.9	2.5
Triethlyaluminum	$(C_2H_5)_3$ Al	114.2	1.8

Table 1: Translational energies for typical MOCVD species at $T_0 = 300$ K seeded into helium carrier gas at near-zero concentration. Kinetic energy increases linearly with T_0.

Deviations from (1) enter through (i) clustering in the supersonic expansion, (ii) variation of C_p due to "freezing out" of internal degrees of freedom as the beam expands and cools, and (iii) real gas enthalpy effects. These contributions are best characterized with experimental time-of-flight (TOF) of the beam as illustrated in Fig. 2. Plotted in each case are the experimental data (gray) and streaming Gaussian fits (solid lines) with the mass spectrometer detector tuned respectively to helium ("Mass 4," light lines) and to ammonia monomer ("Mass 17," heavy lines). At the lower stagnation pressure (top frame) there is relatively little velocity slip between the ammonia and helium in the mixture. Both species in the mixture are slower (longer flight times) than the neat helium beam. This is in accordance with the higher average molecular weight of the mixture, ameliorated somewhat through the mixture's increased heat capacity via energy storage in rotation of the molecule. The relative speeds change dramatically at the higher source pressure (bottom frame). The helium in the mixture becomes significantly faster (shorter flight time) and hotter (broader TOF peak) than in the neat beam. The "Mass 17" ammonia in the mixture shows a bimodal velocity distribution with distinct "fast" and "slow" TOF components. The "fast" component is swifter than the helium in the neat beam yet still "slipping" with respect to the helium in the mixture. The "slow" component is retarded relative to the neat beam yet still faster than the NH_3 of the low pressure beam.

These velocity shifts are largely due to clustering in the beam. Mass spectrometric analysis of the high pressure beam [10] yields the characteristic $(NH_3)_{(n-1)}H^+$ ion signatures known to result from ammonia cluster fragmentation upon electron bombardment ionization in the detector [11]. TOF measurements on the different cluster masses show the velocity slip of clusters to increase with cluster size, allowing the "slow" NH_3 TOF peak in Fig. 2 to be identified with fragmented clusters [10]. The "fast" Mass 17 peak of is then that of true NH_3 monomers. Latent heat is added to the expansion as the clusters form, accelerating and heating both the ammonia and the

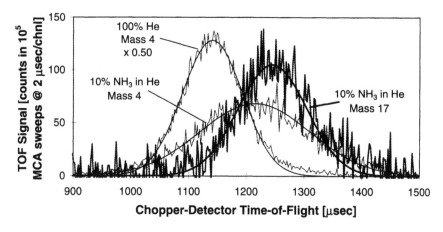

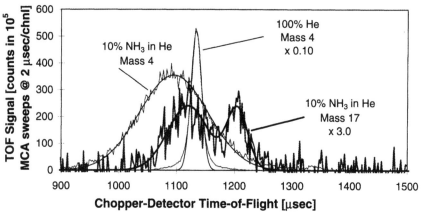

Figure 2: NH$_3$ Seeded beam TOF: $D = 21.5\,\mu$m, $T_0 = 300$ K, $L_{cd} = 1.971$ m. Top: $p_0 = 7$ bar, no clustering. Bottom: $p_0 = 51$ bar, clustering.

helium as evident in TOF shifts and widths of the lower figure.

All of these real gas effects - latent heat of clustering, real gas enthalpy, and release of energy stored in NH$_3$ rotation and vibration - serve to increase the terminal beam energy relative to the simple ideal gas prediction of (1). While the relative TOF shifts are striking, however, their absolute magnitudes remain in the range of $\sim 5\%$ in velocity ($\sim 10\%$ in energy) and, in fact, (1) actually provides a surprisingly good lower estimate. Energy distributions for a 10% NH$_3$ mixture are shown in Fig. 3 for several source temperatures. Clearly it is possible to tune the NH$_3$ translational kinetic energy from 0.25 to 0.65 eV while maintaining energy resolution of ca. 0.1 eV or better [10].

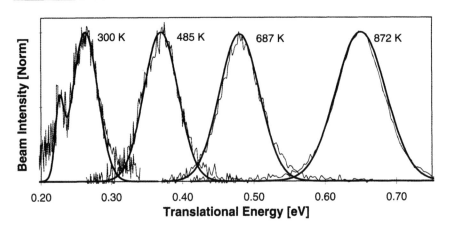

Figure 3: NH$_3$ Seeded beam translational energy distributions; nozzle dia. $D = 20\,\mu$m, pressure $p_0 = 51$ bar, and temperature $T_0 = 300$ K to 872 K.

4. NH$_3$ Seeded Beam Growth of GaN

We have used our seeded beams of 10% NH$_3$ in helium to grow GaN epitaxially on both 6H-SiC(0001) and on AlN buffer layers (grown *in situ* with the seeded beam) on SiC(0001). All deposition was carried out at a substrate temperature of 800 C. The incident NH$_3$ translational energy was varied from 0.034 to 0.44 eV with the beam impinging at $\theta_i = 0°, 30°$, or $75°$ with respect to the surface normal. Ga and Al were supplied in over-abundance from an effusive evaporation source. The thickness and morphology of the resulting films was characterized *ex situ* using standard surface science techniques (RBS, Auger, TEM, and AFM) [3]. *In situ* monitoring with LEEM of the nucleation and growth is now underway. When grown on AlN buffer layers, GaN films of very good crystallinity can be obtained [3].

That the reaction chemistry can indeed be selectively influenced by adjusting the incident NH$_3$ translational energy is seen in Fig. 4. Plotted is the film thickness vs. incident translational kinetic energy E_i for growth under constant normal flux onto the surface at various E_i and θ_i. The shape of this curve is characteristic of activated dissociative chemisorption [12], the interpretation being as follows: At very low incident energy, $E_i \ll E^*$ and the activation barrier cannot be directly surmounted. Yet with only a small loss of perpendicular momentum through either elastic or inelastic interaction with the surface the molecule can become trapped in the physisorption well. This greatly protracts its interaction time with the surface, allowing it to probe configurational reaction space until - possibly activated by energy transfer from the surface - it reacts. This is the "thermal energy" reaction channel. At higher incident energies but still with $E_i < E^*$, the collision time is not only shorter but additional energy/momentum must be transferred to trap the impinging molecule in the physisorption well. This reduces the probability of physisorption, leading to an overall decrease in reaction rate. At still higher incident energies $E_i \geq E^*$, the molecule can directly surmount the activation barrier to chemisorb via the "direct" channel, and the reaction rate rises and plateaus.

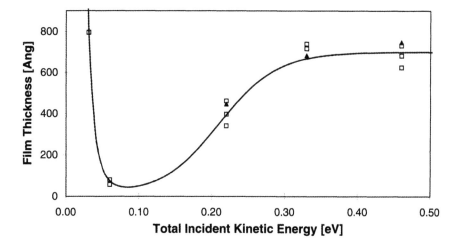

Figure 4: Reaction rate as characterized by film thickness under constant normal flux exposure at $\theta_i = 0°$ (normal incidence, open squares) and $30°$ (solid triangles). The solid line curve is a guide to the eye.

From the data of Fig. 4 the following information is then extracted:

(1) The outer "knee" in the curve may be associated with the activation energy [13] yielding, with ample error bars to reflect the sparsity of the data set, $E^* = 0.25 \pm 0.10$ eV.

(2) A low energy reaction channel does indeed open for incident energy below about 50 meV at this substrate temperature of 800 C. This likely represents physisorption followed by accommodation and reorientation of the molecule to react, indicating that there is at least one trapping site and orientation of very low or negative E^*. It is therefore probably not the NH_3 reaction kinetics which limits the thermal growth kinetics below 750 K, as has been postulated [14] but rather surface mobility of Ga atoms [15].

(3) A reduction factor of $\cos^2(30°) = 0.75$ in the $30°$ (solid triangles) data relative to the $0°$ (open squares) would be readily discernible. Its absence indicates that the NH_3 reaction obeys "total" rather than "normal" energy scaling. Ongoing LEEM measurements at $75°$ support this conclusion. The

total energy scaling presumably results from rotational coupling of the thermal and direct channels, the incident molecule losing momentum perpendicular to the surface by enhanced rotational excitation at grazing angle, thereby trapping and reacting via the thermal channel.

5. Corona Discharge Free Jet

To grow III-N nitrides from molecular nitrogen, the strong N_2 triple bond must be broken or weakened ("activating" the nitrogen in chemistry parlance). Radio frequency (RF) and electron cyclotron resonance (ECR) discharges are the conventional means of doing this, yet both produce a broad spectrum of highly excited molecules, atoms, and ions, the more energetic and reactive of which may damage a growing GaN film [16]. Arc discharge supersonic expansions are therefore under active investigation for GaN growth, both skimmed [17] and unskimmed [18]. We have built and characterized arc-discharge and corona-discharge supersonic free-jet sources, developing special graphite skimmers [19] to skim these discharges.

Theory indicates that the neutral molecular $A^3\Sigma_u^+$ species may be the ideal N_2 activation state for GaN growth [20]. This lowest triplet metastable state de-excites only via the forbidden Vegard-Kaplan bands. It thus has a long intrinsic lifetime (ca. 1 s) making it ideal for beam use. As the molecular $A^3\Sigma_u^+$ reacts, moreover, one N-atom departs the surface, dissipating the strong exothermicity as kinetic energy. This minimizes damage to the film and leaves the second N-atom quiescently attached [20]. A corona discharge supersonic free-jet is the ideal means of producing the $A^3\Sigma_u^+$ state. Temperatures of several thousand degrees may be attained in the discharge, exciting N_2 to high electronic states which relax via the first and second positive series to collect in the $A^3\Sigma_u^+$ state. In the abrupt expansion of a free-jet the $A^3\Sigma_u^+$ molecules can survive collisional deexcitation to appear in the terminal free-molecular flow. To produce exclusively the $A^3\Sigma_u^+$ state (unaccompanied by atom and ions) the initial electronic excitation must be limited by restricting the discharge energy. The corona discharge operates intrinsically in a low energy regime, delivering primarily the $A^3\Sigma_u^+$ state. It is also less expensive and much easier to operate than the arc discharge.

Photographic images of the corona discharge plume are provided in Fig. 5, showing the distinctly different emission which results for negative and positive corona polarity. Note, in particular, the bright streamer attachment points of the positive discharge. We characterized our corona discharge supersonic free jet using TOF analysis and optical spectroscopy. Measurements were made with pure N_2 and with 10% and 20% mixtures of N_2 in Ar. The optical spectrometer viewed the beam transversely through a silica window. Spectra were recorded at the nozzle exit and at various distances

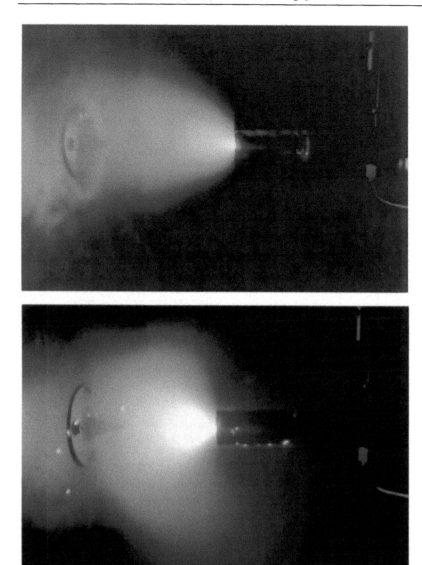

Figure 5: Photographic images of the corona discharge, showing the emission plume resulting when the corona wire is (a) negative and (b) positive with respect to the cylindrical guard anode surrounding the nozzle tube and to the vacuum chamber. The graphite skimmer is visible to the left side of the pictures.

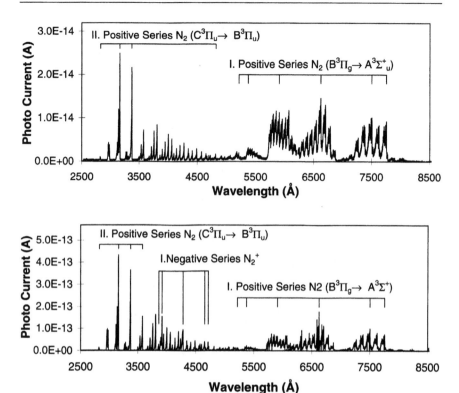

Figure 6: Corona discharge spectra at the nozzle exit. 100% N_2 with corona wire negative (top) and positive (bottom). $D = 100\,\mu\text{m}$, $p_0 = 440$ Torr.

downstream. Typical spectra for a pure N_2 discharge are given in Figs. 6 and 7. The N_2 optical transitions are well known and spontaneous lifetimes have been measured by Ottinger and coworkers [21]. The tip of the corona wire, situated inside the quartz nozzle tube about 15 mm back from the nozzle exit, was operated both negative (top) and positive (bottom) with respect to a ring electrode located just downstream of the nozzle, outside of the free-jet boundary. The discharge current was typically 10 mA at 3 kV. The measured beam intensity was 1.2×10^{18} molecules/sr/s. TOF beam analysis of a pure N_2 discharge with a negative corona wire yielded a velocity of 1940 m/s (0.54 eV kinetic energy) and a velocity spread of 30%, FWHM.

As evidenced by the N_2^+ first negative series, the corona discharge of positive wire polarity produced ions copiously even far downstream of the nozzle (Fig. 7). With the corona wire held negative no ions were produced, as desired for GaN growth. We attribute this partly to electrostatics (posi-

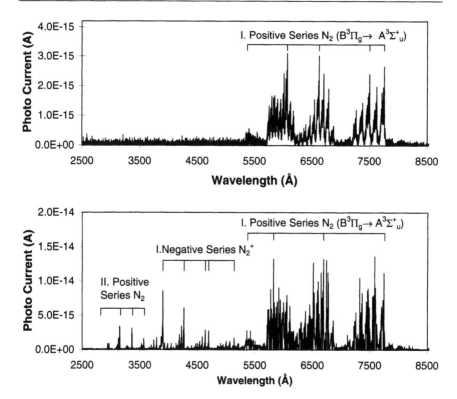

Figure 7: As in Fig. 6 but 1 cm downstream of nozzle

tive ions accelerating downstream, away from the positive corona wire) and partly to the difference in the respective discharges near the wire (uniform glow discharge and electron impact excitation of N_2 with negative polarity; streamer formation and photoexcitation of the N_2 with positive polarity). Collisions within the beam cease ~20 nozzle diameters downstream of the nozzle exit. In the region of spontaneous, collision-free emission farther downstream the spectrum for the negative corona wire shows only the first positive series(Fig. 7, top), revealing the populating of the desired neutral $A^3\Sigma_u^+$ state via relaxation of the $B^3\Pi_u$ and $W^3\Delta_u$ states [21]. In this collision-free region, each photon emitted represents one molecule added to the $A^3\Sigma_u^+$ population. Quantitative information on the excited state content of the corona discharge beam was obtained from appearance potential spectroscopy, distinguishing the various excited species in the beam (including non-emitting states) by virtue of their characteristic ionization thresholds. The negative corona discharge yielded $A^3\Sigma_u^+$ molecules as the sole activated species in a molecular beam containing otherwise only $X^1\Sigma_g^+$ ground state nitrogen molecules and a negligible quantity of $^4S^0$ ground state nitrogen atoms. $A^3\Sigma_u^+$ molecules were present in this discharge at up to 1.74% number fraction of the terminal at a total beam intensity of $6.47 x 10^{18}$ molecules sr^{-1} s^{-1}, providing a flux of up to $2.8 x 10^{13}$ metastables cm^{-2} s^{-1} onto the growth surface [22, 23].

6. Corona Discharge Beam Growth of GaN

With $A^3\Sigma_u^+$ supplied via the corona discharge beam and with Al or Ga provided from simple evaporative sources, films of AlN and GaN were grown on Si-terminated 6H-SiC(0001). GaN was also grown on AlN buffer layers deposited *in situ* on the SiC(0001) substrates [22]. Optimum films were obtained under slightly metal-rich conditions. Over the course of the growth studies, the total beam intensity averaged $4.3 x 10^{18}$ molecules sr^{-1} s^{-1} and the $A^3\Sigma_u^+$ fractional percentage 1.68%, yielding an average $A^3\Sigma_u^+$ flux of $1.7 x 10^{13}$ metastables cm^{-2} s^{-1}. This produced GaN film growth at up to several hundred Ångstrom per hour, more than adequate for studying the reaction chemistry of the N$_2$ $A^3\Sigma_u^+$ state. The deposited films, ranging from 300 to 900 Å in thickness, were investigated *ex situ* by Rutherford backscattering (RBS), scanning and transmission electron microscopy (SEM and TEM), and atomic force microscopy (AFM). Areal densities of the nitride films were extracted from the RBS measurements and combined with absolute measurements of the incident $A^3\Sigma_u^+$ flux to calculate the nitrogen incorporation efficiency, defined as the number of N-atoms attaching to the III-N film per $A^3\Sigma_u^+$ molecule incident. This was found to approach 100% and to be independent of sample temperature over the investigated range of 600 to 800 C, indicating that $A^3\Sigma_u^+$ reaction proceeds by direct molecular chemisorption.

For GaN grown on AlN buffer layers on SiC substrates, AFM measurements of surface roughness showed a distinct minimum at 700 C. The RBS minimum yield also displayed a shallow minimum at this temperature. The films grown at 700 C were composed of a mosaic of islands of approx. 0.5 μm in size, all of the same orientation and with an RMS height variation of approx. 7 nm. Cross-sectional TEM microscopy of the samples revealed sharp, abrupt epitaxial interfaces. Selected area diffraction yielded sharp, distinct diffraction spots, in which individual spots due to the GaN and AlN thin films and the underlying SiC substrate could be distinguished. All evidence indicated that well-ordered epitaxial growth had been achieved. Although further characterization of the films remains to be done (x-ray rocking curves, photoemission spectra) the data would indicate that $A^3\Sigma_u^+$ does indeed provide an ideal precursor species for III-N growth [24].

Acknowledgements

Support for this research by the Office of Naval Research under grants N00014-95-1-0122 and N00014-96-1-0962 is gratefully acknowledged, as is the use of facilities in the ASU Center for High Resolution Microscopy.

References

[1] S.N. Mohammad, A.A. Salvador, and H. Morkoc, Proc. IEEE **83**, 1305 (1995).

[2] S. Strite and H. Morkoc, J. Vac. Sci. Technol., **B 10**, 1237 (1992); R.F. Davis, T.W. Weeks Jr., M.D. Bremser, S. Tanaka, R.S. Kern, Z. Sitar, K.S. Ailey, W.G. Perry, and C. Wang, Solid State Elec. **41** 129 (1997).

[3] V.M. Torres, M. Stevens, J.L. Edwards, D.J. Smith, R.B. Doak, and I.S.T. Tsong, Appl. Phys. Lett., **71** 1365 (1997).

[4] D.C. Jordan, C. Burns, and R.B. Doak, Proc. 45th Int. Symp. Am. Vacuum Soc., Baltimore, MD, 1998 and submited, Rev. Sci. Instrum. (1999).

[5] J.E. Lennard-Jones, Trans. Faraday Soc., **28** 333 (1932).

[6] R. Shekhar and K.F. Jensen, Surf. Sci., **381** L581 (1997).

[7] D.R. Miller in **Atomic and Molecular Beam Methods, Vol. I**, edited by G. Scoles (Oxford Univ. Press, Oxford, 1988).

[8] J. Sumakeris, R.K. Chilukuri, R.F. Davis, and H.H. Lamb, III-V Nitrides Symp., Mater. Res. Soc., Pittsburgh, PA, **Xxi+970** 331 (1996).

[9] L.J. Lauhon, S.A. Ustin, and W. Ho, III-V Nitrides Symp., Mater. Res. Soc., Pittsburgh, PA, **Xxv+2521** 277 (1997).

[10] V.M. Torres, Ph.D. thesis, North Carolina State Univ., Rayleigh, NC, 1998.

[11] U. Buck and H. Meyer, J. Chem. Phys. **88** 3028 (1988).

[12] D.J.D. Sullivan, H.C. Flaum, and A.C. Kummel, J. Phys. Chem. **97** 12051 (1993).

[13] H.A. Michelson, C.T. Rettner, and D.J. Auerbach, in **Surface Reactions** edited by R.J. Madix, Springer-Verlag, Berlin, 1994.

[14] W. Kim, Ö. Aktas, A.E. Botchkarev, A. Salvador, S.N. Mohammad, and H. Morkoc, J. App. Phys. **79** 7657 (1996).

[15] D. Crawford, R. Held, A.M. Johnston, A.M. Dabiran, and P.I. Cohen, Mater. Res. Soc. Internet J., **1** Article 12 (1996).

[16] A. Botchkarev, A. Salvador, B. Sverdlov. J. Myoung, and H. Morkoc, J. Appl. Phys. **77** 4455 (1995).

[17] F.J. Grunthaner, R. Bicknell-Tassius, P. Deelman, P.J. Grunthaner, C. Bryson, E. Snyder, J.L. Guiliani, J.P. Apruzese, and P. Kepple, J. Vac. Sci Technol. **A 13** 1615 (1998).

[18] M.L. Cappelli, A.E. Kull, K. Schwender, H. Lee, S.J. Harris Jr., and J. Mrockowski, Mat. Lett. **31** 161 (1997).

[19] D.C. Jordan, R. Barling, and R.B. Doak, Rev. Sci. Instrum. **70** 1640 (1999).

[20] W.W. Goddard III, Proc. 45th Int. Symp. Am. Vacuum Soc., Baltimore, MD, 1998.

[21] D. Neuschäfer, Ch. Ottinger, and A. Sharma, Chem. Phys. **117** 133 (1987).

[22] D.C. Jordan, Ph.D. thesis, Arizona State University, Tempe, AZ, Dec 2000.

[23] D.C. Jordan, C.T. Burns, and R.B. Doak, in press (2000).

[24] D.C.Jordan, I.S.T. Tsong, David J. Smith, B.J. Wilkens, and R.B. Doak, in press (2000).

Chemical Maps and SEM Images of the Reaction Products on Si Surfaces Irradiated with Cold and Hot C$_2$H$_4$ Beams

I. Kusunoki
Research Institute for Scientific Measurements,
Tohoku University, Sendai, Japan

1 Introduction

For a long time, reactive molecular beams have been detected using chemical reaction with a solid surface. For example, an atomic hydrogen beam is detected using the color change of molybdenum trioxide by reducing it. These techniques are, however, only used to observe the beam position visually, or to measure the beam intensity qualitatively. Recently we have obtained quantitative chemical maps of the products on surfaces reacted with a molecular beam [1,2]. The intensity profiles on the surfaces are measured using scanning X-ray photoelectron spectroscopy (XPS). Especially the carbonization of a Si surface with C$_2$H$_4$ has been studied intensively using this technique [3,4]. We have shown that the technique is useful to study gas–surface reactions.

In addition, molecular beam techniques offer important advantages in studying the mechanisms of surface reactions, since the translational energy and internal states of the reactant molecules can be well controlled independently of the surface conditions. Using these properties, we have already made progress in elucidating the interactions between the Si surfaces and reactant gases such as O$_2$, C$_2$H$_2$, C$_2$H$_4$, NH$_3$, and N$_2{}^+$. Sticking coefficients of the incident molecules on the surface can be easily obtained by measuring the intensity of the scattered beam, which elucidates gas-surface the interaction mode (direct mechanism or trapping-mediated mechanism). The dependence of the initial sticking coefficient on the surface temperature and the sticking coefficient curve as a function of the surface coverage can provide the difference between the activation energies for desorption and chemisorption from a precursor state, $(E_{\text{des}} - E_{\text{ad}})$, and the parameters of surface migration of the adsorbed molecules, respectively [5]. The reaction products evaporated from the surfaces can also be measured using a mass spectrometer. The kinetics of the evaporation of products from etching reactions were studied using a O_2 pulsed molecular beam for active oxidation of Si and Ge [6,7]. In order to study the effect of the translational energy in an extreme case, we have compared the evolutional XPS spectra during the nitridation of a Si surface using a thermal NH_3 beam and a N_2^+ beam in the range of 100 -1000 eV [8]. It was demonstrated that the reaction mechanisms are quite different.

In this work, the beam temperature was varied by heating the nozzle of the beam source up to 900 °C for studying the gas temperature effects for the carbonization of the Si surface. The experimental results obtained in the present work show that the beam temperature and the beam flux, as well as the surface temperature, have a great influence on the surface products.

2 Experimental

The experimental apparatus is shown schematically in Fig. 1. We have attached a homemade molecular beam apparatus to the sample treatment chamber (STC) of a conventional multisurface analysis system (Shimadzu-Kratos XSAM 800) [3]. A molecular beam is produced by three-stage differential pumping. The nozzle aperture for expanding the beam gas is about 50 μm in diameter. The nozzle temperature can be varied from room tem-

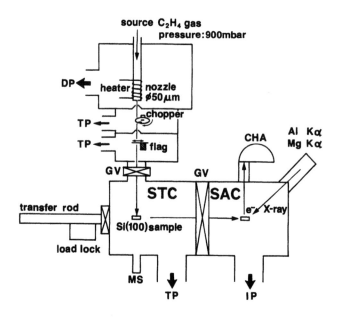

Figure 1: A schematic drawing of the experimental apparatus

perature to 900 °C. A skimmer, with an open tip of 0.5 mm diameter, is placed 5 mm from the nozzle. The gas leaving the nozzle is pumped off by a high speed diffusion pump (10,000 $l\,s^{-1}$) with a liquid nitrogen cold trap, backed by a mechanical booster pump (16,700 $l\,\min^{-1}$ at 5×10^{-2} Torr) and an oil rotary pump (1,080 $l\,\min^{-1}$). The second and third chambers are differentially pumped by turbomolecular pumps having pumping speeds of 500 $l\,s^{-1}$ and 300 $l\,s^{-1}$, respectively. The pressure of the STC is maintained in the range of 10^{-9} Torr during the beam experiment. A quadrupole mass spectrometer in the STC provides measurements of the beam intensity and its compositions. The beam flux was calculated by assuming an ideal gas, at the low pressure of 100 mbar. The mass pattern of C_2H_4 did not change with the nozzle temperature up to 900 °C, which shows that the dissociation of C_2H_4 is negligible under our experimental conditions.

Silicon samples were cut from polished p-type Si(001) wafers, chemically cleaned with an ammonium peroxide solution (H_2:NH_4OH:H_2O_2=5:1:1) at 80 °C for 15 min, and etched with hydrofluoric acid to remove most of the thin oxide layer on the surface. The sample was then transferred into the STC through the load-lock system and placed on the beam axis. The vacuum of the STC was better than 2×10^{-10} Torr using a turbomolecular pump (500 $l\,s^{-1}$) and a titanium sublimation pump. The Si surface was cleaned by heating with electron bombardment in ultrahigh vacuum. Sample cleanliness was checked using XPS to confirm the absence of C 1s and O 1s peaks, and sometimes by low energy electron diffraction (LEED).

The clean silicon surface was irradiated with the C_2H_4 beam at a fixed temperature for a known period of time. The sample temperature in the range of 600 – 1000 °C was measured with an infrared radiation thermometer. After the beam irradiation, the sample was transferred under vacuum into the surface analysis chamber (SAC), which incorporates scanning XPS with an electrostatic lens system and scanning Auger electron spectroscopy (SAM). For the XPS we can use three types of X-ray sources (Mg Kα: 1254 eV, Al Kα: 1487 eV: monochromatized Al Kα). For the scanning XPS the Mg Kα source is usually used, while the monochromatized Al Kα X-ray is used to take spectra of high energy resolution (0.5 eV FWHM). The surface area that is reacted with the molecular beam can be searched using the scanning XPS. The largest scanning area is 10×10 mm^2. It can be divided into 128×128 elements maximum, with spatial resolution of 0.1 mm FWHM. However, we usually used 50×50 elements with a resolution of 0.6 mm to make the measurements feasible in a reasonable time ($<$ 30 min). To observe the surface morphology with high resolution, a scanning electron microscope (SEM) was used. For the SEM observation, however, the samples were exposed to air during transfer into the microscope (LEO 982).

In this paper, we report some experimental results on the carbonization of

Si(100) surfaces by C_2H_4 beams produced from a room temperature nozzle [3] and from a high temperature nozzle [4].

3 Results

A clean Si(100) surface at 675 °C was irradiated with the room temperature C_2H_4 beam (flux: 3×10^{15} molecules cm^{-2} s^{-1}) for 30 min. A chemical map of the surface was drawn by C 1s XPS of the reaction product (SiC). It is shown in Fig. 2a. This circular spot of about 4 mm diameter reflects the beam profile formed through three skimmers placed on the walls which separate the differentially pumped sections. The intensity profile on a line through the center is shown in Fig. 2b. The intensity in the central part is almost constant, which indicates that the direct beam passed through the skimmers. In the half shadow region formed in the arrangement of the three skimmers, intensity gradation occurs. The gradient of the beam flux in this region provides a convenient method for studying the flux effect on surface reactions.

The C 1s and Si 2p XPS spectra taken in the central region of the beam spot are shown in Fig. 3. These spectra were measured with the high resolution of 0.5 eV. The C1s spectrum consists of a single peak at the binding energy 282.9 eV with 1.7 eV FWHM, while the Si 2p spectrum can be divided into two components associated with the Si substrate (99.5 eV) and the SiC products (100.7 eV) on the surface. The chemical shift of the Si 2p level induced by carbonization is about 1.2 eV. The mean thickness of the carbide layer can be conventionally determined from the intensity ratio of the two Si 2p peaks [3]. In the case shown in Fig. 3b, the thickness of the carbide layer is about 40 Å. The growth rate of the carbide has a maximum at around 675 °C as shown in a previous paper [3].

Some surface areas in and around the beam spot were observed using SEM. The SEM images taken from the center, the boundary region, and just outside the beam spot are shown in Figs. 4a-c, respectively. The center is fully covered with epitaxially grown SiC crystals. In the boundary region, the surface is partially covered with the crystal grains. Even in the outside region there are small grains, which are produced by reaction with the background beam gas. The number density of the grown crystals and their sizes in the outside region are, however, much smaller than within the boundary. These images suggest that the SiC film is formed via island growth. Some small islands aggregate at some points and coalesce to form a large island (grain), and finally to produce a SiC thin film. The beam flux affects the number density of the grains and their sizes, which reflects the kinetics of crystal growth as discussed below.

In order to study the effects of the beam temperature on the surface reaction, a hot C_2H_4 beam, produced from a 900 °C nozzle, was directed on a clean

VII.9 Chemical Maps and SEM Images of Carbonization of Si Surfaces

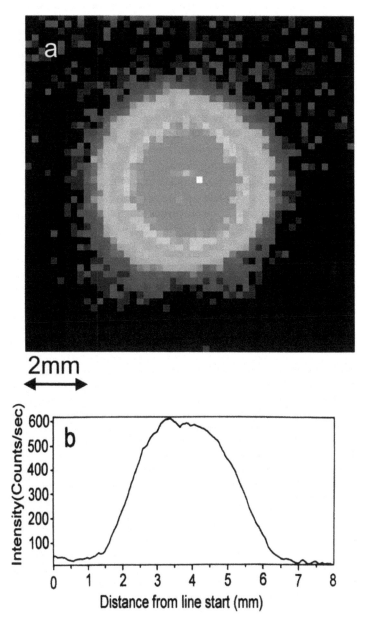

Figure 2: (a) A chemical map of C1s(SiC) XPS of a Si surface reacted with a room temperature C_2H_4 beam, (b) a line profile of the map drawn in (a)

Si(100) surface at 670 °C for 30 min. After the irradiation, two states of C atoms at the surface were observed by XPS in this case (see Fig. 5a). The state at the binding energy of 282.9 eV is from SiC as before, while another one (284.4 eV) is from graphitic (or amorphous) carbon [8]. The line profiles taken through the beam irradiation center are shown in Fig. 5b. The graphitic or amorphous carbon exists only in the central region, while the intensity distribution from the SiC products has maxima at the boundary region. These facts suggest that the beam flux and the beam temperature affect the reaction mechanism and the formation of the different products. The beam flux in the present experiment was smaller than that in the previous experiment using the room temperature nozzle, because the distance from the nozzle to the sample was twice as large due to insertion of an additional differential pumping section. Thus the carbon film formation observed with the hot beam should not result from a beam flux effect. Since the mass pattern measured by the quadrupole mass spectrometer showed no difference between the room temperature and the high temperature beams, the different vibrational excitation of the C_2H_4 molecules may play a role. The vibrational mode is not expected to relax during the nozzle expansion. The excited molecules will dissociate more easily and supply the bare C atoms on the surface. The dense C adatoms produced from the hot beam recombine with each other to form a carbon film. The recombination rate is here assumed to be larger than the reaction rate with the surface Si atoms. The carbon film formed seems to suppress the out-diffusion of the Si adatoms from the bulk, resulting in a very thin C and SiC mixed layer.

High magnification SEM images of the reacted surfaces (Fig. 6a-c) show

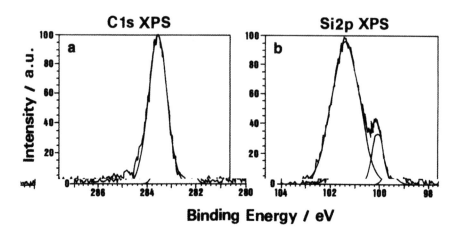

Figure 3: C 1s and Si 2p XPS spectra of the Si surface carbonized at 675 °C for 30 min with a room temperature C_2H_4 beam

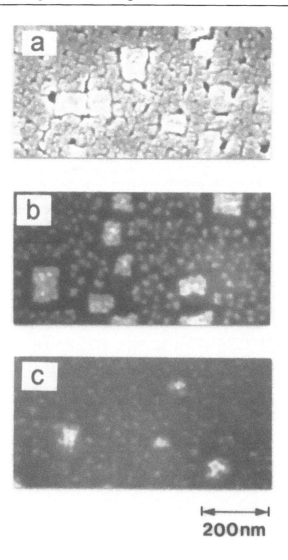

Figure 4: SEM images of the Si surface carbonized at 675 °C for 30 min with a room temperature C_2H_4 beam. (a) the irradiation center, (b) the boundary region, (c) outside of the boundary region

that square SiC crystals grow epitaxially on the Si(100) surfaces. This was also confirmed by the reflection high energy electron diffraction (RHEED) pattern, but no evidence for graphite formation was obtained by this technique. Therefore we have tentatively assigned the carbon state at the binding energy 284.4 eV to amorphous carbon. The number density of the SiC

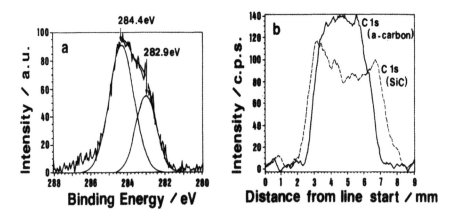

Figure 5: (a) C 1s XPS spectrum of the Si surface irradiated at 670 °C for 30 min with the hot C_2H_4 beam, (b) line profiles of the chemical maps of the Si surface

SEM image in Fig. 6	(a)	(b)	(c)
Mean length in diagonal /nm	20	39	37
Mean area of the grains /nm^2	193	756	679
The number of grains $10^4/\mu m^2$	18.0	22.4	16.9

Table 1: Mean size and the number density of the SiC grains observed in the SEM images in Fig. 6

grains and their sizes in the central region covered with the amorphous carbon film are much smaller than in the boundary region at the maxima of the SiC XPS intensity, as seen in Figs. 6a,b. Their size distributions are plotted in Figs. 6d-f. Their mean values are also given in Table 1.

4 Discussion

It is known that C_2H_4 molecules are chemisorbed on a Si surface, but they do not form SiC below the surface temperature of 550 °C [3]. In the molecular beam scattering experiments which were carried out by us in another apparatus, we have observed that most of the incident molecules were either scattered or desorbed after being trapped for a short time even on high temperature surfaces. Only less than 1% of the molecules reacted with the Si surface to form the SiC layer [3]. The reaction produced discrete SiC islands in the initial stage and their number density depended on the beam flux and the surface temperature as seen in Figs. 4 and 6. This means that the products are not formed by direct reaction with the beam molecules. The reaction must be of the Langmuir-Hinshelwood type. The trapped molecules migrate on the surface and reach the reaction sites. Since the beam flux and the surface temperature affect the mean diffusion length of

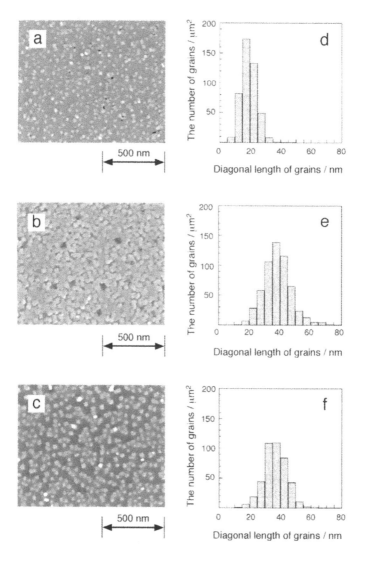

Figure 6: SEM images of the Si surface carbonized at 670 °C for 30 min with a hot C_2H_4 beam (nozzle temperature 900 °C), and the size distributions of the SiC grains. (a) and (d): the irradiation center, (b) and (e): the boundary region, (c) and (f): outside of the boundary region

the adsorbed molecules before desorption or reaction, the formation of the products is sensitive to these factors. The C adatoms produced by decomposition of the C_2H_4 molecules may also be mobile on the high temperature surface.

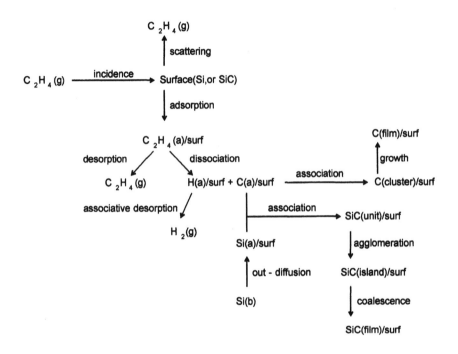

Figure 7: A scheme for SiC and carbon film growth

A hypothetical SiC film growth process consistent with our experimental results is schematically described in Fig. 7. Most of the trapped C_2H_4 molecules are mobile on the surface and desorb after the residence time determined by the surface temperature and the activation energy for the desorption. Some of them are, however, decomposed into atomic hydrogen and other fragments such as CH_x at specific sites, e.g. surface defects. The fragment hydrocarbons decompose further into bare carbon atoms. The hydrogen atoms recombine on the surface after some migration, and desorb as hydrogen molecules. The bare C atoms remaining on the surface may be also mobile on the high temperature surface, which is denoted by C(a)/surf. If there are many mobile Si adatoms on the surface (Si(a)/surf), the C adatoms interact with them to form SiC units. The Si adatoms are supplied from steps of the Si surface in the initial stage when the surface is almost bare. If the number density of Si adatoms is much smaller than that of C(a), the C adatoms recombine with each other to form C-C nets on the surface as observed in the hot C_2H_4 beam experiment. The SiC units may be also mobile and associate to form SiC clusters. Some of the clusters are anchored at sites such as defects. By accumulating the SiC units at the anchored clusters, SiC grains grow epitaxially, changing the lattice constant from that of Si to SiC [10]. In the SiC growing stage, the Si adatoms are supplied from the Si bulk by out-diffusion. Corrosion zones (sometimes voids) formed after Si atoms were pumped out by out-diffusion have usually been observed by TEM beneath the SiC layers [1-3]. The grains coalesce when they come into contact with each other as they grow, resulting in SiC film formation. Using the room temperature C_2H_4 beam, we obtained a maximum growth rate of the SiC film (grains) at around 675 °C [3]. The maximum can be explained by competition between the decomposition rate and the desorption rate of the C_2H_4 molecules on the surface. The supply rate of the C adatoms near the maximum seemed to be balanced by the supply rate of Si adatoms and to form a stoichiometric SiC film.

In the experiment using the 900 °C C_2H_4 beam, however, a carbon film was formed in the central region irradiated with the hot beam. This means that the supply rate of C adatoms was much higher than that of the reactant Si adatoms from the Si bulk. The hot beam must have produced much more bare C adatoms on the surface than the room temperature beam. This is probably due to the high dissociation rate of the hot C_2H_4 molecules. In the high temperature beam, the mean translational speed of the molecules is about twice that of the room temperature beam, but the density in the high temperature nozzle is lower than at room temperature. Therefore a different flux cannot be the reason for the C film formation. The population of the vibrationally excited molecules in the hot beam is much higher than in the room temperature one. The populations at v = 1 in $\nu_s(CH_2)$, $\nu_a(CH_2)$, $\nu(CC)$, $\delta(CH_2)_{scis,twist,rock,wag}$ modes at 900 °C are about 3, 3, 12, 16, 23, 23, and 23 %, respectively. The vibrationally excited molecules, especially in the $\nu(CC)$ mode, should be more easily decomposed on the hot surface than the unexcited molecules. The present results also suggest that the vibrational excitation energy of the molecules is not easily relaxed on the surface and is available for the dissociation at the reaction sites.

Thus the beam flux and the beam temperature, as well as the surface temperature, evidently have a great influence on the surface products.

5 Conclusions

- The C_2H_4 beam profiles were measured from chemical maps of the surface reaction products using scanning XPS.
- This technique is useful to study the effect of beam flux on a surface.
- The reaction of C_2H_4 with a Si surface produced discrete SiC grains.

This means that the products are not formed by direct reaction with the incident molecules. The number density of the grains is, however, related to the beam flux, which make the beam profile measurement possible.

- The chemical state of the reaction product changes with the beam flux and the beam temperature as well as with the surface temperature. In the present experiments, an amorphous carbon film was produced at the central region irradiated with the hot beam, while only SiC grains were formed in the boundary region of the hot beam.

Acknowledgments

We would like to thank Prof. Ch. Ottinger for the careful revision of the manuscript. This work was financially supported in part by grants in aid for scientific research from the Ministry of Education, Science and Culture of Japan and by CREST (Core Research for Evolutional Science and Technology) of the Japan Science and Technology Corporation (JST).

References

[1] I. Kusunoki and Y. Igari, Appl. Surf. Sci. **59**, 95 (1992).

[2] I. Kusunoki, Jpn. J. Appl. Phys. **32**, 2074 (1993).

[3] I. Kusunoki, T. Takagaki, S. Ishidzuka, Y. Igari, and T. Takaoka, Surf. Sci. **380**, 131 (1997).

[4] I. Kusunoki, Y. Igari, S. Ishidzuka, T. Mine, T. Takami, and T. Takaoka, Surf. Sci. **433-435**, 167 (1999).

[5] T. Takaoka and I. Kusunoki, Surf. Sci. **412/413**, 30 (1998).

[6] K. Ohkubo, Y. Igari, S. Tomoda, and I. Kusunoki, Surf. Sci. **260**, 44 (1992).

[7] K. Sugiyama, Y. Igari, and I. Kusunoki, Surf. Sci. **283**, 64 (1993).

[8] I. Kusunoki, S. Ishidzuka, Y. Igari, and T. Takaoka, Surf. Rev. Lett. **5**, 81 (1998).

[9] C. W. Wagner, W. M. Riggs, L. E. Davis, J. F. Moulder, and G. E. Muilenberg, in Handbook of X-ray Photoelectron Spectroscopy, (Perkin Elmer Co., 1979), p.40.

[10] T. Takaoka, H. Saito, Y. Igari, and I. Kusunoki, J. Crystal Growth **183**, 175 (1998).

Looking Ahead

Macro, Micro and Nanobeams

John C. Polanyi
Department of Chemistry, University of Toronto,
Toronto, Ontario, M5S 3H6, Canada

"I believe that the new method of Molecular Rays will, in time, find far wider application."
Otto Stern, July 1931

1 Introduction

The objective of the pioneers of molecular-beam chemistry [1] was, in the first place, to do chemistry under single-collision conditions, where all but the primary reactive events were eliminated. Their second aim was to localise the collision events in space, so that the speeds and angular distributions of the scattered species could be measured with respect to that localised region. Their third objective was to control the reagent attributes such as reagent energies and angles of approach (defined by electromagnetically aligning the molecule under attack).

It was not envisaged at the outset that it would be possible to restrict the impact parameter—that is to say the distance by which the molecules of beam A missed the centre-of-mass of the molecules of beam B. This Chapter can be read as being concerned with this intimate detail of scattering and reaction. The results, though crude, indicate that this quest should succeed.

2 Surface-Aligned Photochemistry, SAP

Before the most primitive molecular beam chemistry could become a reality it was necessary to achieve conditions under which the low-density collision-free processions of molecules that comprised the beams would give rise to a significant number of collisions at their point of intersection. "All that crossed beams do," George Kistiakowsky at Harvard complained famously, following an early failed attempt at crossed-beam chemistry, "...all that crossed beams do, is cross."

Beamists soon learned that in order to fight the inverse-square diminution in beam density away from the beam orifices they needed to cross the beams as close to the source as possible. Success depended on crowding the beam sources into one corner of the molecular beam chamber, along with the differentially-pumped stages that would reduce the pressure of background molecules. This line of thought gave rise to the 'super-machine' that opened up a broad swath of chemistry to examination by the crossed-beam technique [2]. It can also be seen as inspiring the approaches discussed here.

If it is advantageous to bring beam sources close together, might it not be possible to design sources that operate effectively at micron separations? Such novel sources should be pulsed, so as to give maximum beam intensity in brief time intervals. This is advantageous since some of the best methods of product characterisation make use of pulsed lasers for detection [3].

We describe two proposals made to bring the sources microscopic distances apart. The first of these, though simple, has yet to be implemented. It is mentioned here because it provides a conceptual stepping-stone to a method which we discuss in more detail, in which the sources are not microns but nanometres apart.

The concept of micron-separated beam sources is illustrated in Figure 1. The method goes by the name of Crossed Ablated Beams (CAB) [4]. The two faces of each groove can be separately covered with adsorbate by dosing using beams directed normal to a side of the groove. Depending on the case, either the molecular adsorbate itself or photofragments arising from the adsorbate can be released into the gas by laser-ablation or by photodissociation. A profitable application of the method would be to the study of single-collision events, at controllable collision-energy, between free radicals, since these are readily formed by photolysis. An illustrative calculation showed that reactive cross-sections as low as ~ 0.01 Å^2 should be detectable, due to the high density at the crossing point.

This microbeam approach provides a basis for a nanobeam extension in which the two coated surfaces of the CAB method are replaced by one surface, and the distance travelled by a photofragment is to be measured in angstroms rather than microns since it is merely the separation between co-adsorbed molecules.

In this case the angle of approach of reactive species, A, to the axis of the co-adsorbed molecule, BC(ad), is determined by the bond-direction within the adsorbed precursor molecule, P–A, whose photodissociation yields A. In contrast to the examples cited up to this point, the impact-parameter of the encounter is controlled. The reason that it is controlled is that P–A and B–C have preferred locations relative to one another, as well as preferred alignment. Their ordering is determined by adsorbate-adsorbate forces together with adsorbate-substrate forces.

This approach, termed Surface-Aligned Photochemistry, SAP [5], is illustrated schematically for two neighbouring adsorbate molecules, H_2S, in Figure 2 (lower panel) [6]. The method has been applied most assiduously to the case of physisorbed adsorbates on inert (halide or oxide) surfaces, for

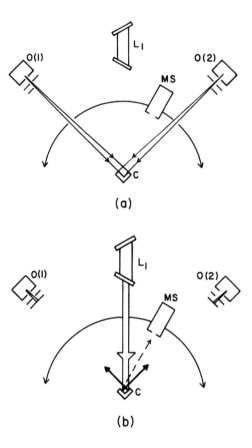

Figure 1: Schematic of proposed crossed ablated beam machine. (a) Two crystal faces (C) at 90° to one another are shown being dosed from oven sources, 0(1) and 0(2). (b) Laser L_1 triggers the pulsed beams by inducing photodissociation, photoejection or thermal desorption of adsorbate from the opposed crystal faces (in the more general case L_1 and L_2 illuminate crystal faces 1 and 2). Unreacted species and also reaction products are identified at the mass-spectrometer (MS) and characterised as to their time-of-flight to the MS. From [4].

which the perturbation of the adsorbate charge-cloud is modest. Under these circumstances a parallel can be drawn to the molecular dynamics as determined by crossed-beam chemistry in the gas-phase.

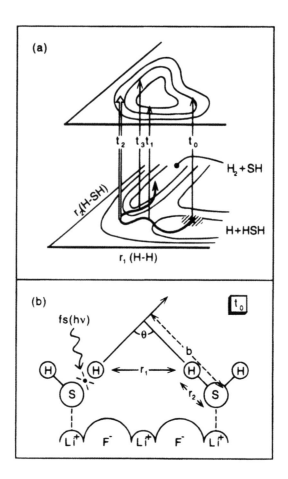

Figure 2: (a) If the trajectories across the lower state originate in a localised region (shaded), then the system will conform to a restricted path, as indicated. This can be interrogated at successive times t_0, t_1, ... , etc. by time-resolved excitation to the excited bound state. Lingering of the trajectory at a turning point is shown leading to enhanced absorption at t_2. (b) The necessary restriction of initial state geometry (r_1(H–H), r_2(H–SH), θ, and b) is shown schematically in the lower panel as being achieved by way of surface-aligned femtosecond photoreaction in $H_2S(ad)/LiF(001)$ starting at time t_0. From [6].

The SAP approach is not restricted to the photoinduced reactions of molecules adsorbed on insulators or semi-conductors. An elegant demonstration of the effect of adsorbate alignment in selecting those reaction-partners within the reactive range of impact parameters, has been obtained on a metal surface [7]. Aligned oxygen and CO were adsorbed on Pt(779). The photorecoiling O formed by SAP gave evidence of reacting only with the sub-set of CO(ad) at which it was aimed. This selectivity in SAP is illustrated in Figure 3.

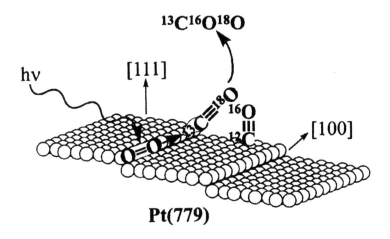

Figure 3: Schematic picture of the selective occupation of sites by O_2 and two CO isotopomers, followed by CO photo-oxidation reaction, on a stepped Pt surface. The step sites have [100] orientation. Adsorption of O_2 and CO molecules occurs on step edges. The experiment shows that some of the nonthermalized oxygen atoms produced by photodissociation of O_2 retain their directionality on the surface and consume CO preferentially on the step sites, exemplifying Surface Aligned Photochemistry, SAP. (Reproduced from E.M. Tripa and J.T. Yates Jr., J. Chem. Phys. 112, 2463 (2000); ref. [7]).

The intention in SAP—still, of course, to be achieved—is to move toward a chemistry in which the reacting species can, so to speak, be taken in either hand and brought together in a known geometry. Additionally, time zero for the reactive encounters will be definable to the same precision as the photolytic light pulse, i.e., to femtoseconds. The ability to trigger reactive events at a known time in a known geometry intermediate between reagent and product makes this approach a candidate for time-resolved Transition State Spectroscopy, TSS [6] (Figure 2; top panel), as well as for experiments in Coherent Control that depends on the lock-step evolution of matter-waves [8].

Surface-aligned photochemistry has its parallel in the photochemistry of clusters [9, 10, 11]. Commonly a precursor constituent, P–A, is weakly bound in the cluster to the target molecule, B–C. Photolysis of P–A··B–C causes A to recoil toward B–C with a preferred angle to the B–C axis, and with a preferred impact parameter.

Since the Van der Waals bond (indicated by dots, in the complex) is only roughly one Angstrom longer than a covalent bond, the initial A··B separation is intermediate between that of separated reagents and fully-formed product A–B. This is, therefore, once more, TSS. The absence of a neighbouring substrate simplifies the dynamics, but the indefinacy in geometry of the loose complex results in a higher degree of averaging. Nonetheless, it is encouraging that so much can be learned from one molecular beam, rather than the classic two.

3 Localised Atomic Scattering, LAS

Molecular-beam scattering at surfaces, as the present volume attests, has provided a valuable general tool for probing surface roughness as a function of collision-energy and approach-angle. This Chapter, by contrast, explores the unique advantages that derive from the use of scatterers originating in molecules tethered to the surface by adsorption forces. The method is termed Localised Atomic Scattering, LAS [12, 13].

This approach to determining localised corrugations at surfaces has the capability of SAP, permitting the study of collisions having a restricted range of impact parameters. At present LAS has only been investigated for hydrogen halides adsorbed on LiF. It is exemplified here for the case of HCl(ad)/LiF(001).

In the chosen example HCl was adsorbed at ~ 0.1 monolayers, coverage on a clean LiF(001) crystal cooled to 50 K, mounted in ultrahigh vacuum [13]. The crystal and its adsorbate layer were then irradiated with 193.3 nm (6.41 eV photon energy) excimer laser radiation. Using a sensitive detection method, namely Rydberg-atom time-of-flight [14], the photoproduct, atomic H, could be characterised as to its translational energy and angular distribution as it scattered from the surface.

The high-energy elastically-scattered H-atoms exhibited peak energies of 1.85 and 1.65 eV, leaving behind Cl and Cl* respectively. The difference in energy between the photon and this scattered H was largely that required to break the H–Cl bond. The elastically-scattered H-atoms, rather than being scattered off the LiF(001) surface at the specular angle of 71° from the surface normal, were found to leave the surface with maximum probability at 40°.

This remarkable discrepancy was attributed (Figure 4) to the fact that the observed scattering, rather than being averaged over the crystal plane, arose from a *localised* encounter at a preferred atomic site on the surface.

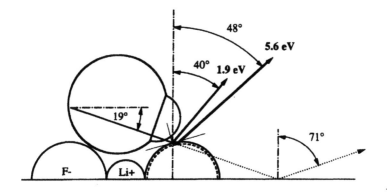

Figure 4: Localised Atomic Scattering, LAS, of H from HCl on LiF(001). Photolysis by 193.3 nm gave an elastic scattering angle of 40° whereas 121.6 nm yielded a scattering angle of 48°; 71° is the specular angle. The drawing is to scale, using van der Waals radii for chlorine $r_{Cl}=1.75$ Å and $r_H \approx 0.7$ Å calculated from the internuclear distances in solid hydrogen halides at 80 K. From [13].

Experiment and theory are in agreement as to the nature of the site. Polarised FTIR (Fourier Transform Infrared Spectroscopy) studies of hydrogen halides adsorbed on LiF (001) [15] showed a downward tilt of the X–H axis. A substantial red-shift in the peak infrared absorption wavelength (by ~ 300 cm^{-1}) gave evidence of hydrogen-bonding of the halide to the LiF surface through X–H··F$^-$.

This indicated that the photorecoiling H-atom was at the downward-tilted end of the X–H and that it was aimed at F$^-$, as illustrated in Figure 4. A theoretical study of the adsorption forces led to the same conclusion [16]. From the figure it is evident that the tangent to the surface at the localised region of H-atom impact upon F$^-$ is tilted toward the normal, providing grounds for the observation of H-atom scattering at an angle deflected toward the normal.

Using a shorter photolysis wavelength (121.6 nm) the H was ejected toward the surface with greater energy and was detected with a larger peak scattering angle of 48°. Figure 4 interprets this in terms of scattering from an inner potential-contour.

A pair of classical trajectory calculations using a potential-energy surface approximately that for HI(ad)/LiF are given in Figure 5. The trajectory at the top is characteristic of the single-collision elastic scattering described above. The trajectory below serves as a reminder that for steep tilt angles (as also for closely-adjacent adsorbate molecules) inelastic events will complicate the picture. These, however, have their own signature, namely a reduced translational energy and broadened angular distribution for the scattered H.

Theoretical studies of this system using a potential-energy surface appropriate to HCl/LiF(001), are lacking. Until these exist the interpretation of these experiments remains speculative. Nonetheless, the power of the LAS approach as a means for the experimental determination of potential-energy contours in atomically-localised regions of surfaces, is evident.

4 Localised Atomic Reaction, LAR

Given the high energy of the H-atom projectiles discussed in section 2, it may be asked whether in addition to the elastic scattering observed there may also have been events in which F atoms were dislodged from the underlying LiF. Evidence is lacking. However, the question is a valid one and opens the way to discussion of reactive encounters between a photorecoiling atom or radical coming from an adsorbed molecule, and the surface beneath. Chemical reaction between impinging gaseous molecular beams as one reagent and surfaces as the other, have been extensively studied. As for elastic scattering, the principal variables are the collision energy at the surface and the angle of approach. The latter can be particularly informative in gas-surface beam-chemistry, since the target species, being part of a crystal, has a fixed alignment. Once again, however, an important detail is lacking, namely the role of the point of impact at the surface—the analogue of the gas-phase 'impact parameter' (or miss-distance). If the impinging species originates in adsorbed molecules attached at preferential atomic sites to a surface, then the photorecoiling reagent, for example atom A coming from some precursor P–A (ad), will approach the surface at characteristic *atomically-localised* sites, with a correspondingly restricted region of impact. This statement does not go beyond that made in the previous section. The interesting question is whether the atomically-localised impact results in what might be termed Localised Atomic Reaction, LAR, at the surface rather than the elastic scattering described previously. Alternatively it is possible that the attacking atom or radical, though it impacts a localised point, moves form site to site across the surface loosing its energy at successive encounters and adjusting its trajectory for reactive attachment at a distance remote from its initial locale. In brief, it may migrate.

There is evidence from thermal reaction at surfaces that migration of highly-reactive fragments of an adsorbed molecule can occur prior to bonding with the surface. Dissociative chemisorption in beam-surface reaction viewed by Scanning Tunneling Microscopy (STM) has been reported in some cases to give rise to reaction at neighbouring or near-neighbouring surface atomic-sites [17, 18, 19] and in others to result in 'ballistic' recoil of the fragments

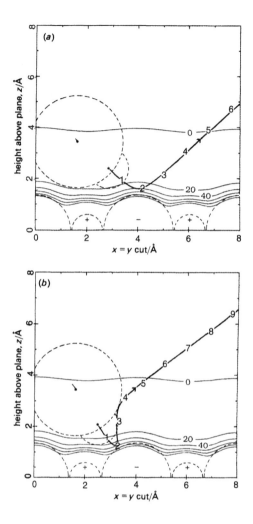

Figure 5: Calculated trajectory of an H-atom after photolysis of HI(ad) to show the effect of increasing tilt angle. The cut along the LiF surface is in the (111) plane. The H–I bond length is fixed at its gas-phase value of 1.609 Å. Dashed lines indicate van der Waals atomic and ionic radii, and the solid lines show the non-electrostatic corrugation contours for the surface at 20 kcalmol^{-1} intervals. Excess energy of photolysis = 3.4 eV. Numbers along the trajectories indicate 5 fs intervals. (a) Initial tilt angle $\theta = 130°$ leads to a direct (single-collision) scattering event; (b) initial tilt angle $\theta = 145°$ leads to an indirect scattering event. From [12].

across the surface [20, 21, 22].

Ballistic recoil would seem much more likely in the case under discussion here in which an adsorbed molecule is dissociated with considerable excess energy by light (in Surface-Aligned Photochemistry, SAP) or by the comparable process of electron-impact. We have described, above, a case in which $O_2(ad)/Pt(779)$ was irradiated with 260–320 nm radiation and the recoil energy of atomic O was sufficient to induce oxidation of a co-adsorbed submonolayer of CO [7]. Similarly in the ultraviolet photofragmentation of chemisorbed O_2 on Pt(111), the recoil of the resulting 'hot' oxygen atoms was found to induce collisional desorption of co-adsorbed noble gases [23]. Induced dissociation of O_2 adsorbed on Pt(111) [24, 25] and Cu(110) [26] has been reported to give 'transient ballistic motion' of the oxygen atoms whose mean displacement of 1–3 lattice constants is limited principally by the rate of energy-loss from the migrating atomic fragment to the surface of the underlying metal [27].

The field of molecular reaction dynamics at surfaces is a sufficiently young one that there exist at present no generalisable models. It is evident, however, that in the case of adsorbed organic halides RX(ad) on silicon, which have been the subject of recent studies, photon- and electron-induced surface reaction occurs by Localised Atomic Reaction, LAR. What this means is that the most likely position for X–Si bond formation is closely adjacent to the adsorption site of RX(ad). This is illustrated below for cases of chlorobenzenes undergoing both electron-induced and photon-induced reaction, and a rationale for this LAR behaviour is given.

The existence of LAR following the electron-impact-induced reaction of chlorobenzene, ClPh(ad), on Si(111)7 × 7 to yield Cl–Si is readily seen in Figure 6 [28]. A Si surface covered with 0.4 ML (ML=monolayers) of ClPh has been induced to react by the application of a series of -4 V electron pulses from an STM tip. These voltage pulses are close to the threshold energy for the reaction to form Cl–Si at the surface. They were delivered to the adsorbate-covered surface at intervals of 60 Å.

This is the only case described here in which the reaction energy was delivered in a localised fashion. The question being addressed was whether the product of reaction, Cl–Si, would be similarly localised. The answer is evident in the post-irradiation STM scans of Figures 6(a) and (b). The dark spots in Figure 6(a) (-1 V on the tip) that 'light up' in 6(b) when the tip voltage is increased to -3 V, are Cl–Si. They consist of groups of (2–3) Cl–Si bonds spaced at the same 60 Å intervals that separated the regions of prior electron irradiation. The electron-induced reaction to form Cl–Si from the parent ClPh is seen to be localised to the site of -4 V pulsed irradiation, as expected for LAR.

Figure 6(c) is included to indicate that this is indeed a chemical reaction to form Cl–Si as a result of the -4 V pulses. When the surface is coated with benzene and subjected to many -4 V pulses a -3 V STM scan fails to show

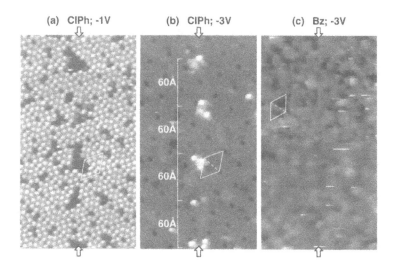

Figure 6: STM images of the Si(111)7 × 7 surface with adsorbed chlorobenzene in (a) and (b), and adsorbed benzene in (c). The images in (a) and (b) were taken after application of a series of Ũ4 V voltage pulses to the STM tip while the tip was moved 60 Å to successive positions along a line in the y-direction, extending between the arrows. The separation between pulses was 30 Å for (c). Dual-bias-voltage images of identical areas of the surface are shown in (a) and (b). The tip bias-voltages were Ũ1 V for (a), and Ũ3 V for (b) and (c). A 7 × 7 unit cell with sides 26.9 Å in length is outlined in each image. From [28].

the bright spots, since there is no Cl present.

Further experiments were performed using *generalised* electron impact (-4 V) to induce reaction. This time the observed pattern of chlorination was determined not by the STM tip but by the self-assembly of the ClPh(ad) on Si(111)7 × 7, which favoured the so-called faulted half (F) of each until cell rather than the unfaulted (U). The reason that the imprinted pattern of Cl–Si favours F rather than U is due to the fact that the distribution of 'daughter' Cl–Si over F and U mirrors the prior distribution of 'parent' ClPh. Localised reaction, LAR, ensures that the 'imprint' is closely related to the distribution of the 'ink.'

Viewed at the atomic level (rather than with respect to areas, F and U) the distribution of daughter Cl–Si differs in its details from the distribution of ClPh. Whereas the parent ClPh self-assembles preferentially on the middle atoms, 'M,' of the unit cell, the daughter Cl–Si formed by electron- or photon-induced reaction attach preferentially to the corner atoms, 'C.'

There is, moreover, a quantitative equivalence between the preference of the parent for 'M,' and that of the daughter for 'C.' This can readily be understood in terms of Figure 7 in which it is proposed that the electron- or photon-induced reaction is a concerted one in which the formation of the new Cl–Si bond occurs concurrently with the breaking of the old C–Cl bond of ClPh [28]. This would explain why the daughter Cl–Si is adjacent to the parent ClPh. Preliminary results from a Density Functional Theory, DFT, calculation support this model.

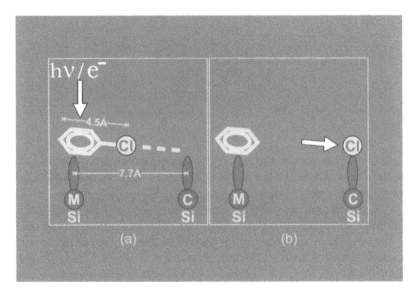

Figure 7: Schematic diagram of the photon or electron-induced reaction of chlorobenzene with Si(111)7 × 7. Interatomic distances are to scale. In (a) the chlorobenzene molecule is adsorbed on a middle adatom (M) with the Cl weakly bound to a neighboring corner adatom (C). Irradiation induces a concurrent breaking of a C–Cl bond, and formation of Si–Cl at the neighboring corner adatom. From [28].

Qualitatively the same type of result has been obtained using illumination by light, in a slightly more sophisticated scenario [29]. In this work light was used as generalised irradiation to break the two C–Cl bonds of a dichlorobenzene (a single-photon process). As before, STM was employed to locate the Cl–Si. Two different molecular 'inks' were compared; 1,2- and 1,4-dichlorobenzene adsorbed alternatively on Si(111)7 × 7. Different chlorine nearest-neighbour separations were observed in the Cl–Si photoformed chemical 'imprints' for these two isotopically-related adsorbates. The most probably separation of neighbouring Cl's in the case of the 1,2 isotope was 8 ± 3 Å, whereas in the case of the 1,4 isotope it was 14 ± 3 Å.

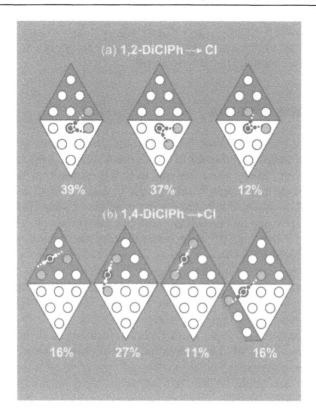

Figure 8: Most probable configurations of the Cl–Si pairs that were observed after laser irradiation of the surfaces (a) for 1,2-diClPh and (b) for 1,4-diClPh. Within each diamond shaped 7 × 7 unit cell the white circles are the bare Si adatoms and the green circles are Cl–Si. The black circles suggest locations for the parent molecule (a) for 1,2-diClPh and (b) for 1,4-diClPh. These parent molecules are drawn to scale to show how each could result in the observed Cl–Si imprint by an approximately linear extension of C–Cl bonds. From [29].

These differing separations can be understood if the pairs of C–Cl bonds in the respective adsorbates extend roughly linearly to chlorinate a nearby Si dangling bond. The postulated molecular reaction dynamics are illustrated schematically in Figure 8 based on the observed statistical distribution of pairs of Cl–Si. In this instance of marking by way of LAR the pattern of Cl–Si bonds derives not from the uneven molecular distribution of the 'ink' (as was the case above), but from the differing distribution of the atoms within each molecule of ink. In the language of molecular beams each molecule, when irradiated with ultraviolet, expels two beams of Cl. In the case that the pair of beams is aimed at more closely adjacent atomic sites, the reactive marks are found to lie closer together. It remains to be seen to what extent this very simple model can be generalised.

5 Cosmic Beams

A child (sadly there will be few leafing through this volume) might question the scientists' unflagging pursuit of smaller and smaller details. It would be a valid question. Much of our delight in living comes from the large; the long view and the night sky. What we hope is that the small will illuminate the large. With this in mind, the closing example of nanobeams relates to pulsed crossed-beams in which each pulse has a diameter of $\sim 10^{26}$ Å. This is 1 parsec or 3.10^{13} km; $10^5 \times$ the distance from the earth to the sun.

The travelling pulses in these pulsed beams are the dense interstellar clouds, which have been known for decades past. Their density is $\sim 10^8 \times$ that of the intercloud region. This is a more favourable density ratio than in most laboratory molecular beam machines, for which the beam is distinguished from background by having $\sim 10^4 \times$ greater density.

Is the crossing point of these giant pulses sufficiently well defined to permit the measurement of the differential cross-sections, central to crossed-molecular-beam studies? The answer would appear to be a tentative 'yes,' since the products might be detected 100 parsec away, with an angular resolution of $\sim 1°$ [30]. Are they forward, backward or sideways scattered? This should be revealing of the dynamics of the encounter. Has the question yet been asked?

We would not pursue science with such devotion if we did not believe it to encompass the very large as well as the very small. Since we are here only briefly, we should make the entire universe our playground.

References

[1] D.R. Herschbach, Reactive Scattering in Molecular Beams, in 'Molecular Beams', Ed. J. Ross (Wiley Interscience, New York, 1966) Advances in Chem. Phys., Vol. X, Ch. 9, pp. 319.

[2] Y.T. Lee, J.D. McDonald, P.R. LeBreton and D.R. Herschbach, Rev. Sci. Instrum. **40**, 1402 (1969).

[3] R.N. Zare and P. Dagdigian, Science **185**, 739 (1974).

[4] I. Harrison and J.C. Polanyi, Z. Phys. D, 'Atoms, Molecules and Clusters' **10**, 383 (1988).

[5] J.C. Polanyi and H. Rieley, Photochemistry in The Adsorbed State, in 'Dynamics of Gas-Surface Interactions' Eds., C. Rettner and M.N.R. Ashfold (Royal Society of Chemistry, London, 1991) Ch. 8, pp. 329; J.C. Polanyi and Y. Zeiri, Dynamics of Adsorbate Chemistry in 'Laser Spectroscopy and Photochemistry on Metal Surfaces', Part II, Eds. H.-L. Dai and W. Ho (World Scientific, Singapore, 1995) Ch. 26, pp. 1241.

[6] J.C. Polanyi and A.H. Zewail, Acc. Chem. Res. **28**, 119 (1995).

[7] E.M. Tripa and J.T. Yates, Jr., J. Chem. Phys. **112**, 2463 (2000).

[8] M. Shapiro and P. Brumer, Chem. Phys. Lett. **126**, 541 (1986); Trans. Faraday Soc. **93**, 1263 (1997).

[9] B. Soep, C.J. Whitham, A. Keller and J.P. Visticot, Faraday Disc. Chem. Soc. **91**, 191 (1991); A. Keller, P. Lawruszczuk, B. Soep and J.P. Visticot, J. Chem. Phys. **105**, 4556 (1996).

[10] S.K. Shin, Y. Chen, S. Nikolaisen, S.W. Sharpe, R.A. Beaudet and C. Wittig, Advances in Photochem., Eds. D. Volman, G. Hammond and D. Neckers (Wiley Interscience, New York, 1991) Vol. 16, pp. 249.

[11] X.Y. Chang, R. Ehlich, A.J. Hudson, P. Piecuch and J.C. Polanyi, Faraday Disc. **108**, 411 (1997), and references therein.

[12] V.J. Barclay, W.-H. Hung, J.C. Polanyi, G. Zhang and Y. Zeiri, Faraday Disc. **96**, 129 (1993).

[13] J.B. Giorgi, R. Kühnemuth and J.C. Polanyi, J. Chem. Phys. **110**, 598 (1999).

[14] L. Schnieder, W. Meier, K.H. Welge, M.N.R. Ashfold and C.M. Western, J. Chem. Phys. **92**, 7027 (1990).

[15] P.M. Blass, R.C. Jackson, J.C. Polanyi and H. Weiss, J. Chem. Phys. **94**, 7003 (1991).

[16] J.C. Polanyi, R.J. Williams and S.F. O'Shea, J. Chem. Phys. **94**, 978 (1991).

[17] J.A. Jensen, C. Yan, and A.C. Kummel, Phys. Rev. Lett. **76**, 1388 (1996).

[18] I. Lyubinetsky, Z. Dohnálek, W. J. Choyke, and J. T. Yates, Jr., Phys. Rev. B **58**, 7950 (1998).

[19] T.W. Fishlock, J.B. Pethica, F.H. Jones, R.G. Egdell, and J.S. Foord, Surf. Sci. **377–379**, 629 (1997).

[20] U. Diebold, W. Hebenstreit, G. Leonardelli, M. Schmid, and P. Varga, Phys. Rev. Lett. **81**, 405 (1998).

[21] H. Xu and I. Harrison, J. Phys. Chem. B **103**, 11233 (1999).

[22] H. Brune, J. Wintterlin, J. Trost, G. Ertl, J. Wiechers, and R.J. Behm, J. Chem. Phys. **99**, 2128 (1993).

[23] A.N. Artsyukhovich and I. Harrison, Surf. Sci. **350**, L199 (1996).

[24] J. Wintterlin, R. Shuster and G. Ertl, Phys. Rev. Lett. **77**, 123, (1996).

[25] B.C. Stipe, M.A. Rezaei and W. Ho, J. Chem. Phys. **107**, 6443 (1997).

[26] R.G. Briner, M. Doering, H.-P. Rust and A.M. Bradshaw, Phys. Rev. Lett. **78**, 1516 (1997).

[27] G. Wahnström, A.B. Lee and J. Strömquist, J. Chem. Phys. **105**, 326 (1996).

[28] P.H. Lu, J.C. Polanyi and D. Rogers, J. Chem. Phys. **111**, 9905 (1999).

[29] P.H. Lu, J.C. Polanyi and D. Rogers, J. Chem. Phys. **112**, 11005 (2000).

[30] J.C. Polanyi, Reaction Dynamics and the Interstellar Environment, in 'Molecules in the Galactic Environment', Eds. M.A. Gordon and L.E. Snyder (Wiley Interscience, New York, 1973) pp. 329-351.

Subject Index

The subjects indexed are those chosen by the authors, using the PACS Index of the American Institute of Physics. The numbers (as IV.6) refer to the Part No. (as IV) and paper No. (as 6) in the volume.

Ab initio calculations (31.15.Ar) IV.6
Alignment of electronic wave functions IV.3
Alkali metals, electronic structure (71.20.D) IV.11
Angular momentum disposal (34.50.Lf) IV.7
Atmospheric chemistry
  (82.40.We, 94.10.Fa) III.3
  (82.40.We) III.1
Atom and molecule optics (03.75.Be, 39.20.+q) V.1
Atom and neutron interferometry (03.75.Dg) I.3
Atom and neutron optics (03.75.Be) I.3, I.12
Atom interferometry techniques (39.20.+q) I.4, I.9, IV.3
Atom optics (03.75.Be) I.11
Atom transfer (82.30.Hk) IV.10
Atom-cluster collisions (34.90.+q) V.5, V.6
Atomic and molecular beam reactions (82.40.Dm) II.6, VII.2
Atomic and molecular beam sources and detectors (07.77.Gx) I.11
Atomic and molecular beam sources and techniques (39.10.+j) Introductory Perspectives, I.9, II.2, II.4, II.5, III.1, V.4, V.7, VI.1, VI.2, VI.3, VI.5, Looking Ahead
Atomic and molecular clusters (36.40.–c) II.8, V.3, V.6, VII.4
Atomic and molecular data, astrophysics (95.30.K) IV.9

Atomic beam sources and techniques (39.10.+j) I.4
Atomic beams-interactions with solids (79.20.R) VII.3
Atomic clusters (36.40.–c) V.2
Atomic ionization (34.50.F) IV.11
Atom-molecule collisions IV.6
Atom-surface interactions (34.50.Dy) V.1
Autoionization, photoionization, and photodetachment (33.80.Eh) III.8

Barrier heights (33.15.Hp) III.7
Black-body radiation V.8
Bond strengths, dissociation energies, hydrogen bonding (35.20.G) V.4, VI.6

Caging processes IV.2
Carbonization VII.9
Charge transfer (34.70.+e) V.6, VII.6
Charged clusters (36.40.Wa) V.5, V.6, V.7, V.9, VII.7
Chemical analysis (82.80.–d) VII.9
Chemical kinetics (82.20.–w) II.6
Chemical reactions (34.50.Lf) III.2
  (82.30, 82.40) VI.7
Chemical reactions, energy disposal, and angular distribution, as studied by atomic and molecular beams (34.50.Lf) Introductory Perspectives, II.6, III.1, IV.1, IV.8, Looking Ahead
Chemiluminescence, chemical reactions (82.40.T) VI.7

Chemisorption (82.65.My) VII.2
Clouds-interstellar (98.38.D) IV.9
Cluster formation
  (36.40.–c, 82.30.Nr, 64.70.Fx) V.1
Cluster impact chemistry VII.1
Cluster impact with surfaces
  (79.20Rf) VII.7
Cluster stability/fragmentation
  (36.40Qv) V.4, V.7, V.9, VI.3
Clusters (61.46.+w) VI.3, VI.5, VII.5
Clusters (atomic and molecular)
  (36.40) Introductory Perspectives,
  V.3, VI.3, VI.7
Clusters, nanoparticles and
  nanocrystalline materials
  (61.46.+w) V.10, VI.1, VI.2
Coherent control (32.80.Qk) I.8
Cold collisions I.6
Collision cross sections (34.50) IV.11
Collision induced molecular alignment
  II.4, IV.4
Control of reactions IV.1
Corona discharges (52.80.Hc) VII.8

Decomposition reactions (82.30Lp) V.9
Density-functional theory
  (31.15.Ew) V.3
Differential alignment IV.4
Differential cross section (82.20.p) IV.7
Differential cross-section IV.4
Diffraction by a microfabricated
  transmission grating (39.10.+j) V.1
Diffuse interstellar bands (98.8.–j) III.6
Diffuse spectra; predissociation,
  photodissociation (33.80.Gj)
  Looking Ahead
Diffusion and dynamics of clusters
  (36.40.Sx) V.5
Diffusion at Surfaces (68.35.F) I 9
Dissociation and dissociative
  attachment by electron impact
  (34.80Ht) V.7
Dissociative chemisorption VII.8
Doppler effect (43.28.p) IV.7
Dynamic polarizability I.7

Electric and magnetic moments (and
  derivatives), polarizability, and
  magnetic susceptibility
  (33.15.Kr) V.3
Electron diffraction (61.14.Lj) II.8
Electron impact ionization
  (34.80.Gs) V.7, V.8
Electronic excitation (34.50.Gb) V.8
Electronic finestructure IV.4
Electronic properties of clusters
  (36.40.Cg) V.3, V.6
Electronic visible spectra
  (33.20.Kf) III.6
Elementary chemical processes IV.8
Emission, absorption and scattering of
  ultraviolet radiation (52.25.Qt) II.1
Energy distribution and transfer;
  relaxation (82.20.Rp) II.5, VI.7
Energy transfer VII.6
  (34.50) IV.4
  (34.50.Ez) IV.2
Evaporation (68.10.Jg) VII.4
Excited electronic states of molecules
  (31.50.+w) III.5, VII.8
Excited harpooning mechanism III.4
Extreme conditions VII.1

Fast and ultrafast reactions
  (82.40.Js) III.8, IV.1
Femtochemistry IV.1
Femtosecond spectroscopy
  (42.65.Re) I.8, IV.1, IV.2
Film growth and epitaxy
  (81.15.–z) VII.8
Fondations, theory of measurement
  (03.65.Bz) I.3
Four center reactions VII.1
Fragmentation of clusters
  (36.40.Qv) V.2
Fullerenes
  (61.48.+c, 78.66.Tr) V.8
  (61.46.+w) VII.6

Gas surface interactions
  (34.50.D, 82.40.Dm) I.11
Gas surface scattering I.10
General theories and models of
  atomic and molecular collisions
  and interactions (including
  statistical theories, transition state,

stochastic and trajectory models, etc.) (34.10.+x) Introductory Perspectives, IV.1, VII.1, Looking Ahead

Helium droplets VI.3
High resolution translational spectroscopy IV.8
Homogeneous nucleation (64.60.Qb, 82.60.Nh) V.1

Impurities and other defects in liquid helium(67.40.Yv) VI.1, VI.2
Inelastic collisions IV.4
Information theory VII.1
Infrared spectra (3220F) VI.6
Insertion (82.30.HK) IV.7
Instrumentation for fluid dynamics (47.80.+v) II.7
Interaction of atoms, molecules and their ions with surfaces (34.50.Dy) I.10, VII.2, VII.9
Interaction of charged clusters with surfaces; (femtosecond) neutralization dynamics; electron emission (34.50.Dy) VII.6, VII.7
Interactions of atoms, molecules and ions with surfaces (34.50.Dy) I.10
Interactions of biosystems with radiations (87.50.B) VI.6
Interatomic and intermolecular potentials and forces, potential energy surfaces for collisions (34.20.-b) I.6
Interatomic potentials and forces (34.20.Cf) IV.3
(34.30) IV.4
Interferometers (07.60.Ly) I.2
Intermediates VII.2
Intermolecular and atom-molecule potentials and forces (34.20.Gj) III.5, III.7
Internal Conversion (33.50.Hv) III.6
Internal energies of clusters V.2
Interstellar clouds (diffuse) (98.38.D) Introductory Perspectives
Intracluster reactions III.4

Intramolecular energy transfer; intramolecular dynamics; dynamics of van der Waals molecules (34.30.+h) Introductory Perspectives, II.6, III.5, IV.1, IV.5
Ion beams (41.75.A,C) IV.11
Ionization by electron impact (34.80.Dp) V.2
Ionization potentials molecules (33.15.R) VI.4
Ionization potentials, electron affinities (35.20.V) V.7, VI.6
Ion-molecule reactions (82.30.Fi) IV.10

Kinetic and transport theory of gases (51.10.+y) II.3
Kinetic theory (05.20.Dd) II.3
Kinetics (68.10.Jy) VII.9

Laser control of reaction pathways Introductory Perspectives
Laser modified scattering (34.50.Rk) IV.2
Laser modified scattering and reactions (34.50.Rk) IV.3
Line and band widths, shapes, and shifts (33.70.Jg) III.7
Line shapes, widths and shifts (32.70.Jz) II.3
Lithography I.7, I.12
Long-range interactions (34.20.Cf) I.5, I.6

Magneto-optical trap (32.80.Pj) I.5
Mass spectrometers (07.75.+h) IV.10, VI.5
Mass spectrometers and related techniques (07.75.+h) IV.5
Mass spectrometry (07.75.+h) I.11
(82.80.K) II.8, VII.4
(82.80.Ms) V.10
Matter waves (03.75.2b) I.1
(03.75.-b) Introductory Perspectives, I.2, I.4, I.10
Mechanical effects of light on atoms, molecules, and ions (32.80.Lg) I.1, I.7

Metallic clusters V.5
Metastable phases (64.60.My) VI.8
Meteors (meteor trails) (96.50.K) Introductory Perspectives
Methods of crystal growth (81.10.–h) VII.9
Methods of deposition of films and coatings; film growth and epitaxy (81.15.–z) VII.8, VII.9
Metrology (06.20.–f) I.2
Microelectronics deposition technology (85.40.Sz) VII.8
Mirror optics I.11
Mixtures (64.75.+g) VII.4
Molecular beam scattering IV.4
Molecular dynamics (31.15.Qg) VII.5
Molecular excitation and ionization by electron impact (34.80.Gs) V.7
Molecular infra-red spectra (33.20Ea) VI.3
Molecular optics (33.80.Wz) I.7
Molecular orientation IV.6
Molecular spectra (33.20.–t) I.6
Multiphoton ionization (33.80) IV.4
Multiphoton ionization and excitation molecular spectra (33.80.R) VI.4
Multiphoton ionization and excitation to highly excited states (e.g., Rydberg States) (33.80.Rv) III.8, IV.5
Multiphoton processes
 (33.80.Wz) III.2, IV.2
 (32.80.Rm, 32.80.Wr) I.8

Nanometer-scale fabrication technology (85.40.Ux) I.12
Nightglow (94.10.R) Introductory Perspectives
Nitric oxide IV.4, V.7
Nitride semiconductors (III–V) VII.8
Nitrogen compound IV.6
Nonadiabatic transitions IV.3
Nonequilibrium gas dynamics
 (47.20.Nd) II.3
 (47.70.Nd) II.2, II.5
Nonrelativistic scattering theory (03.65.Nk) I.10

Nuclear magnetic resonance (molecules) (33.25) Introductory Perspectives
Nucleation (64.60.Qb) II.7, VI.8
Nucleation and condensation (64.60.Qb) II.8

Optical (ultraviolet, visible, infrared) measurements (52.70.Kz) II.2
Optical cooling of atoms; trapping (32.80.Pj) I.1, I.4, I.6, I.12
Optical waveguides and couplers (for matter waves) (42.79.Gn) I.2
Organic molecules in the heavens Introductory Perspectives
Oscillator strengths (33.70.Ca) III.6
Other topics in optical properties, condensed matter spectroscopy and other interactions of particles and radiation with condensed matter (78.90.+t) Looking Ahead
Oxidation VII.2
Ozone layer (82.40.W) III.2

Phase coherent atomic ensembles; quantum condensation phenomena (03.75.Fi) I.1
Phase transitions in clusters (36.40.Ei) VI.8
Photoassociation (33.20.Kf) I.5, I.6
Photochemistry (82.50) IV.8
Photochemistry and radiation chemistry (82.50.–m) Looking Ahead
Photodissociation
 (33.80.G) III.2
 (82.50.Fv, 33.80.Gj) III.3
Photoelectron spectra(33.60.–q) IV.5
Photofragmentation dynamics (34.30.+h) III.4
Photoinduced reactions (33.80.–b) III.4
Photolysis, photodissociation, and photoionization by infrared, visible, and ultraviolet radiation (82.50.Fv) III.1, III.8, V.4, V.10, VI.5

Plasma diagnostic techniques and instrumentation (52.70.–m) II.2
Plasma production and heating by laser (52.50.Jm) II.2
Plasma properties (52.25.–b) II.1
Plasma sources (52.50.Dg) II.1
Polycyclic aromatic hydrocarbons (PAH) cations III.6
Positive-ion beams (41.75.Ak) IV.10
Potential energy surfaces
(34.20.–x) II.4
(34.20Gj) III.3
(82.20.Kh) IV.10
Potential energy surfaces for collisions
(34.20.Gj) III.1
(34.20.Mq) IV.6
Pulse techniques (82.40.Mw) III.8

Quantum mechanics
(03.65.–w) I.10, II.4
Quantum optics (42.50.–p) I.2
Quasiresonant charge transfer VII.7

Radiative cooling V.8
Raman and Rayleigh spectra (33.20.Fb) II.7
Ramsey interferences (39.20.+q) I.8
Reaction rate constants (82.20.Pm) IV.9, V.9
Reactive flows (47.70) IV.9
Reactivity of clusters (36.40.Jn) Introductory Perspectives, IV.1
Relaxation processes (31.70.Hq) IV.2
Rotational analysis
(33.20.Sn) II.5, VI.1, VI.2
Rotational and vibrational energy transfer (34.50.Ez) II.6, IV.5
Rotational excitation IV.4

Scanning electron microscopy (SEM) VII.9
Scattering of atoms, molecules, and ions (34.50.–s) IV.6, Looking Ahead
Seeded supersonic beams VII.8
Semiconductors III-V (81.05.Ea) VII.8
Semiconductors (81.05.D) VII.8
Shattering of clusters VII.1
Shock waves (82.40.Fp) II.7

Silicon surfaces I.11
Size selction of clusters V.2
Solid surfaces and solid solid interfaces (68.35.–p) I.10
Solid-liquid transitions (64.70.Dv) VI.8
Solvent effects (31.70.D) VI.4
Specific chemical reactions; reaction mechanisms (82.30.b) VI.7
Spectroscopic techniques (39.30.+w) II.2, II.5
Spectroscopy and geometrical structure of clusters (36.40.Mr) VI.3
Spin echo (76.60.L) I.9
Spin precession (42.50.Md) I.8
Stability and fragmentation of clusters (36.40.Qv) V.5, V.7, VII.5
Stability of atoms molecules and ions (34.50.Ss) V.5
State-to-state scattering analysis (34.50.Pi) II.4, III.1, IV.8, VII.3
Statistical theory (82.20.Db) IV.7, V.9
Steric effects IV.6
Sticking coefficients I.10
Stueckelberg oscillations VII.7
Supersonic and hypersonic flows (47.40.Ki) II.2, II.7, IV.9
Surface and interface dynamics and vibrations (68.35.Ja) VII.2
Surface chemistry (82.65.–i) VII.9
Surface collisions (34.50.D) II.8, VII.3, VII.4
Surface diffusion VII.2
Surface dynamics (68.35.J) VI.4, VII.3
Surface dynamics, vibrations (68.35.J) I.9
Surface interaction (34.50.Dy) VII.5
Surface reactions (82.65.J) Introductory Perspectives, VII.3, VII.9, Looking Ahead
Surface science Looking Ahead
Surface structure (68.35.B) VI.4

Time and frequency (06.30.Ft) I.1
Trajectory models (34.10) IV.11
Transition state dynamics III.4, IV.1
Transition state spectroscopy III.4
Transmission, reflection, and scanning electron microscopy (61.16.Bg) V.10

Transsonicflows (47.40.Hg) II.1
Transport properties (52.25.Fi) II.1

Ultracold molecules (33.80.Ps) I.5, I.6
Ultra-low temperature scattering I.10
Ultraviolet spectra
 (33.20.Lg) III.5, III.7, IV.2, VI.5, VI.6

van der Waals Clusters
 (36.20.Kd) III.6
van der Waals molecules (36.40.c) III.4
Vibrational dynamics VI.3

Vibrational excitation (82.65.Pa) VII.6
Vibration-rotation analysis
 (33.20.Vq) III.7
Visible spectra (33.20.Kf) VI.1, VI.2

Water clusters VII.5
 (36.40.Qv) V.4
Wavepacket dynamics I.8
Wavepacket revival I.7
X-ray photoelectron spectroscopy
 (XPS) VII.9

Zeeman and Stark effects (32.60+i) I.3

Author Index

Agarkov, A.A. V.8
Aguilar, A. IV.11
Ahern, M.M. II.6
Ahmed, M. III.3
Albertí, M. IV.11
Allison, W. I.11
Allouche, A.R. V.3
Andrés, J. de IV.11
Antoine, R. V.3
Aquilanti, V. II.4
Ascenzi, D. II.4
Aubert-Frecon, M. V.3

Barat, M. V.5
Barbé, R. I.4
Barker, J.R. IV.5
Bassi, D. IV.10
Baudon, J. I.3
Bazalgette, G. III.8
Becucci, M. III.7
Bekkerman, A. VII.6
Belikov, A.E. II.6
Benichou, E. V.3
Blanchet, V. I.8
Bocanegra, J.M. IV.11
Bossennec, J.-L. I.4
Bouchene, M.A. I.8
Boudin, N. III.6
Boustimi, M. I.3
Bréchignac, C. V.6
Bréchignac, Ph. III.6, III.7
Briant, M. VI.7
Brodsky, K. I.3

Broyer, M. V.3
Bruch, L.W. V.1
Buck, U. V.2
Bulthuis, J. IV.6

Cahuzac, Ph. V.6
Campargue, R. II.2
Canosa, A. IV.9
Cappelletti, D. II.4
Castro, M. de II.4
Chandezon, F. V.3
Chandler, D.W. III.1, IV.5
Châtelet, M. VII.4
Chizmeshya, A.V.G. I.10
Cohen-Tannoudji, C. I.1
Concina, B. V.6
Crubellier, A. I.6

Dagdigian, P.J. III.5
DeKieviet, M. I.9
Dalibard, J. I.1
De Martino, A. II.8, VII.4
Desfrançois, C. VI.6
Doak, R.B. I.10, VII.8
Drabbels, M. III.8, IV.6
Drozdov, S.V. V.8, VII.5
Dubbers, D. I.9
Dubov, D.Yu. V.8, VII.5
Dudeck, M. II.2
Dugourd, Ph. V.3

Ehbrecht, M. V.10
Engeln, R. II.1
Engel, V. IV.2

Ernst, W.E.   VI.1

Fayeton, J.A.   V.5
Fernández, J.M.   II.7
Fort, E.   II.8, VII.4

Gadea, F.X.   IV.11
Galichin, V.A.   V.8
Galonska, R.   V.2
Garcia, M.E.   VII.7
Gaveau, M.A.   VI.7
Gerasimov, I.   III.5
Girard, B.   I.8
González Ureña, A.   III.4
Gorceix, O.   I.4
Grebenev, S.   VI.2
Groenenboom, G.C.   III.1
Grosser, J.   IV.3
Guet, C.   V.3
Guibal, S.   I.4

Hafner, S.   I.9
Hancock, G.   III.2
Harich, S.   IV.8
Haubrich, D.   I.12
Heck, A.J.R.   IV.5
Herschbach, D.   Introductory Perspectives
Higgins, J.P.   VI.1
Hitsuda and Y. Matsumi, K.   II.3
Hodgson, A.   VII.3
Hoffmann, O.   IV.3
Hofmeister, H.   V.10
Holst, B.   I.11
Horimoto, N.   VI.4
Huber, B.A.   V.3
Huisken, F.   V.10
Hulsman, H.   II.5
Hwang, D.W.   IV.8

Imura, K.   V.4

Janssen, M.H.M.   III.1, III.8, IV.5
Jie Lei   III.5
Johnson, R.D.   III.2

Jordan, D.C.   VII.8

Kasai, T.   V.4
Kazakova, I.V.   VII.5
Keller, J.-C.   I.4
Kim, H.-J.   V.2
Kleyn, A.E.   VII.2
Knuth, E.L.   VI.8
Kohn, B.   V.10
Kolodney, E.   VII.6
Kondow, T.   VI.4
Kusunoki, I.   VII.9
Kurzyna, J.   II.2

Lago, V.   II.2
Lambrechts, S.   IV.6
Lang, F.   I.9
Lange, M.J.L. de   IV.6
Lauenstein, C.   V.2
Lebéhot, A.   II.2
Lee, S.-H.   IV.7
Lehmann, K.K.   VI.1
Leuken, J.J. van   IV.6
Levine, R.D.   VII.1
Leygnier, J.   V.6
Lin, J.J.   IV.8
Lison, F.   I.12
Lisy, J.M.   V.9
Liu, K.   IV.7
Liu, X.   IV.8
Lohbrandt, P.   V.2
Long, R.   I.4
López-Tocón, I.   III.7
Lorenz, K.T.   IV.5
Lucas, J.M.   IV.11
Lugovoj, E.   VI.2

Mafuné, F.   VI.4
Maréchal, E.   I.4
Märk, T.D.   V.7
Masnou-Seeuws, F.   I.6
Maté, B.   II.7
Mathevet, R.   I.3
Mazouffre, S.   II.1

Meiwes-Broer, K.H. VII.7
Meschede, D. I.12
Mestdagh, J.M. VI.7
Meyer, H. IV.4
Miller, R.E. VI.3
Montero, S. II.7
Murphy, M.J. VII.3

Naaman, R. V.4
Nauta, K. VI.3
Neyer, D.W. III.1, IV.5
Nicole, C. I.8

Perales, F. I.3
Peterka, D.S. III.3
Picard, Y.J. V.5
Pietraperzia, G. III.7
Pillet, P. I.6
Pinot de Moira, J.C. III.2
Pino, Th. III.6
Pirani, F. II.4
Polanyi, J. C. Looking Ahead
Pörtner, N. VI.2
Pradè, F. II.8
Pradère, F. VII.4

Ramos, A. II.7
Rayane, D. V.3
Raz, T. VII.1
Rebentrost, F. IV.3
Rebrion-Rowe, C. IV.9
Reho, J. VI.1
Reinhardt, J. I.3
Rijs, A.M. III.8
Ristori, C. V.3
Ritchie, G.A.D. III.2
Robert, J. I.3
Roeterdink, W. III.8, IV.5
Romero, T. IV.11
Rowe, B.R. IV.9

Samson, P. VII.3
van de Sanden, M.C.M. II.1
Sartakov, B. VI.2
Schaper, S. VI.8

Scheier, P. V.7
Schermann, J.P. VI.6
Schmidt, M. V.2
Schmiedmayer, J. I.2
Schöllkopf, W. V.1
Schram, D.C. II.1
Scoles, G. VI.1
Seideman, T. I.7
Senn, G. V.7
Shizgal, B.D. II.3
Skelly, J.F. VII.3
Skowronek, S. III.4
Smith, M.A. II.6
Snijders, J.G. IV.6
Sogas, J. IV.11
Speer, O. VII.7
Stienkemeier, F. VI.1
Stolte, S. III.8, IV.5, IV.6
Stwalley, W.C. I.5
Suits, A.G. III.3

Tejeda, G. II.7
Teule, J.M. III.1
Tignères, I. V.6
Toennies, J.P. V.1, VI.2, VI.8
Torres, V.M. VII.8
Tossi, P. IV.10
Tsipinyuk, B. VII.6
Tsong, I.S.T. VII.8
Tyley, P.L. III.2

Vach, H. VII.4
Vallet, V. VI.7
Veneziani, M. V.4
Viaris de Lesegno, B. I.3
Vilesov, A.F. VI.2
Visticot, J.P. VI.7
Vostrikov, A.A. V.8, VII.5
Vries, M.S. de VI.5

Wasylczyk, P. III.8
Wiskerke, A. III.8
Wrenger, B. VII.7

Xin Yang III.5

Yang, X.F. IV.8
Yang, X. IV.8
Yoder, L.M. IV.5

Zadorozhny, A.M. VII.5
Zamith, S. I.8
Zewail, A.H. IV.1

Location: http://www.springer.de/phys/

You are one click away from a world of physics information!

Come and visit Springer's

Physics Online Library

Books

- Search the Springer website catalogue
- Subscribe to our free alerting service for new books
- Look through the book series profiles

You want to order?   Email to: orders@springer.de

Journals

- Get abstracts, ToC´s free of charge to everyone
- Use our powerful search engine LINK Search
- Subscribe to our free alerting service LINK *Alert*
- Read full-text articles (available only to subscribers of the paper version of a journal)

You want to subscribe?   Email to: subscriptions@springer.de

Electronic Media

- Get more information on our software and CD-ROMs

You have a question on an electronic product?   Email to: helpdesk-em@springer.de

● Bookmark now:

http://www.springer.de/phys/

Springer

Springer · Customer Service
Haberstr. 7 · D-69126 Heidelberg, Germany
Tel: +49 6221 345 200 · Fax: +49 6221 300186
d&p · 6437a/MNT/SF · Gha.

Lightning Source UK Ltd.
Milton Keynes UK
UKHW020015051019
351033UK00003B/149/P